Geometry

HOLT, RINEHART AND WINSTON

A Harcourt Classroom Education Company

Austin · New York · Orlando · Atlanta · San Francisco · Boston · Dallas · Toronto · London

Requests for permission to make copies of any part of the work should be mailed to the following address: Permissions Department, Holt, Rinehart and Winston, 1120 South Capital of Texas Highway, Austin, Texas 78746-6487.

For permission to reprint copyrighted material, grateful acknowledgment is made to the following sources:

Associated Press: From "Astronomers Dispute NASA Gauge of Universe's Age," June 2, 1999. Copyright © 1999 by Associated Press.

Discover Magazine: From "Egg Over Alberta" by Paul Hoffman, photos by Annette Del Zoppo, from *Discover*, May 1988. Text copyright © 1988 by Paul Hoffman. Photos copyright © 1988 by Annette Del Zoppo.

Dover Publications, Inc.: "6 Accomplishments" and "31 Digits Are Symbols" from *My Best Puzzles in Logic and Reasoning* by Hubert Phillips. Copyright © 1961 by Dover Publications, Inc.

The Griffith Institute, Ashmolean Museum, Oxford: Adapted excerpt from Howard Carter's Notebook.

National Council of Teachers of Mathematics: Adapted from "Activities: Spatial Visualization" by Glenda Lappan, Elizabeth A. Phillips, and Mary Jean Winter from *Mathematics Teacher*, vol. 77, no. 8, November 1984. Copyright © 1984 by the National Council of Teachers of Mathematics.

The New York Times Company: From "Math Problem, Long Baffling, Slowly Yields" by Gina Kolata and map, "The Efficient Traveling Salesman," from *The New York Times*, March 12, 1991. Copyright © 1991 by The New York Times Company.

Oxford University Press, London: Figures 2.5 and 3.7 from *Chinese Mathematics: A Concise History* by Lǐ Yän and Dù Shírán, translated by John N. Crossley and Anthony W.-C. Lun. Translation copyright © 1987 by John N. Crossley and Anthony W.-C. Lun.

Greg Stec: From "Message of the Maya in Modern Translation" by Greg Stec from *The Christian Science Monitor*, June 22, 1989. Copyright © 1989 by Greg Stec.

Printed in the United States of America

ISBN 0-03-052219-6

3 4 5 6 7 048 02 01

AUTHORS

James E. Schultz, *Senior Series Author*
Dr. Schultz has over 30 years of experience teaching at the high school and college levels and is the Robert L. Morton Professor of Mathematics Education at Ohio University. He helped to establish standards for mathematics instruction as a co-author of the *NCTM Curriculum and Evaluation Standards for School Mathematics* and A Core Curriculum: *Making Mathematics Count for Everyone.*

Kathleen A. Hollowell, *Senior Author*
Dr. Hollowell is an experienced high school mathematics and computer science teacher who currently serves as Director of the Mathematics & Science Education Resource Center, University of Delaware. Dr. Hollowell is particularly well versed in the special challenge of motivating students and making the classroom a more dynamic place to learn.

Wade Ellis, Jr.
Professor Ellis has co-authored numerous books and articles on how to integrate technology realistically and meaningfully into the mathematics curriculum. He was a key contributor to the landmark study *Everybody Counts: A Report to the Nation on the Future of Mathematics Education.*

Paul A. Kennedy
Dr. Kennedy is a professor of Mathematics Education, director of the Institute for Mathematics, Science, and Technology Education, and is the holder of the prestigious Adams Chair at Texas Christian University. A leader in mathematics education reform, Dr. Kennedy is often invited to speak and conduct workshops on the teaching of secondary mathematics.

CONTRIBUTING AUTHORS

Martin Engelbrecht
A mathematics teacher at Culver Academies, Culver, Indiana, Mr. Engelbrecht also teaches statistics at Purdue University, North Central. An innovative teacher and writer, he integrates applied mathematics with technology to make mathematics accessible to all students.

Kenneth Rutkowski
A mathematics teacher at St. Stephen's Episcopal School, Austin, Texas, Mr. Rutkowski is an innovative geometry teacher. He sponsors his school's mathematics honor society, serves on various professional committees, and conducts creative teacher workshops.

Table of Contents

1 EXPLORING GEOMETRY 2

MATH CONNECTIONS

Algebra 15, 16, 20, 21, 22, 23, 32, 48, 65, 66, 75

APPLICATIONS

Science
Archaeology 44
Computer Programming 67
Engineering 74
Geology 24

Language Arts
Communicate 13, 21, 29, 39, 45, 54, 63

Business and Economics
Construction 48

Life Skills
Navigation 20, 21, 33, 60
Travel 74

Sports and Leisure
Hobbies 14
Scuba Diving 33

MATH CONNECTIONS

Algebra 81, 83, 84, 85, 87, 107, 108, 110, 112, 113, 122, 124

Patterns in Data 98

APPLICATIONS

Science
Biology 90, 103, 105, 130
Botany 98
Chemistry 103
Computer Programming 130
Geology 103
Weather 94

Language Arts
Communicate 82, 94, 102, 111, 120

Life Skills
Navigation 116

Sports and Leisure
Entertainment 120
Fine Arts 97
Games 86
Hobbies 111
Humor 96
Music 97
Photography 115

MATH CONNECTIONS

Algebra 145, 152, 160, 174, 181, 187, 191, **Technology** 174
192, 195

APPLICATIONS

Science
Agriculture 142, 143
Archaeology 185, 186
Architecture 159
Biology 180, 189
Botany 141
Cartography 176
Civil Engineering 166
Engineering 188, 204
Gemology 182

Language Arts
Communicate 142, 151, 159,
164, 173, 180, 186, 193

Social Studies
Geography 146, 172

Life Skills
Carpentry 161, 167
Construction 195, 196
Drafting 166
Furniture 145
Navigation 161, 167, 173,
175
Traffic Safety 146
Transportation 182

Sports and Leisure
Art 204
Crafts 204
Gardening 153
Music 189
Optical Illusions 151
Painting 188
Quilting 147
Recreation 145
Sports 176, 184, 186

MATH CONNECTIONS

Algebra 215, 234, 238, 240, 246, 278   **Coordinate Geometry** 216, 242

APPLICATIONS

Science
Archaeology 233
Astronomy 229, 230
Engineering 220, 233, 243
Forestry 226, 230
Language Arts
Communicate 213, 220, 230,
   238, 245, 257, 264, 275

Life Skills
Carpentry 215, 222, 255
Construction 223, 238, 239,
   245, 260
Design 279, 288
Fashion 216
Navigation 233, 269
Road Signs 242
Surveying 242, 288

Sports and Leisure
Art 251, 260
Marching Band 279
Quilting 215, 223
Recreation 223
Sports 234, 251

MATH CONNECTIONS

Algebra 297, 299, 309, 317, 319, 326, 327, 329, 331, 332, 333, 343, 344
Coordinate Geometry 309
Maximum/Minimum 310, 352

Technology 311, 359
Number Theory 327
Probability 354, 357

APPLICATIONS

Science
Agriculture 297, 301, 325
Astronomy 319
Automobile Engineering 319
Aviation 298
Civil Engineering 337, 345, 352
Environmental Protection 345
Farming 311
Irrigation 318
Meteorology 347, 358
Solar Energy 300

Language Arts
Communicate 298, 307, 317, 326, 335, 342, 349, 356

Life Skills
Construction 300
Drafting 337
Home Improvement 302
Landscaping 301, 306
Meal Planning 318
Public Safety 329
Safety 323, 366
Travel 329
Transportation 358

Sports and Leisure
Gardening 297
Recreation 298, 359
Skydiving 358
Sports 329, 366
Theater Arts 304

MATH CONNECTIONS

Algebra 377, 393, 400, 401, 403, 405, 406, 408   **Coordinate Geometry** 394, 407
Technology 425

APPLICATIONS

Science
Archaeology 378
Architecture 377
Astronomy 401
Chemistry 383, 390, 391, 394
Geology 424
Optics 394
Physical Science 401

Language Arts
Communicate 375, 383, 391, 399, 405, 413

Business and Economics
Packaging 393
Fund-raising 424

Life Skills
Design 424
Graphics 400
Hobbies 407
House Painting 406
Mechanical Drawing 372
Navigation 385, 407

Sports and Leisure
Recreation 377

MATH CONNECTIONS

Algebra 431, 434, 442, 444, 450, 456, 457, 459, 475

Maximum/Minimum 432, 434, 457, 466, 468, 483

Coordinate Geometry 444

Technology 466

APPLICATIONS

Science
Archaeology 448
Architecture 444
Biology 430, 433, 435
Botany 435
Cartography 471
Chemistry 435
Engineering 453, 455
Geology 460, 463
Marine Biology 458
Optics 484
Physics 492
Physiology 435

Language Arts
Communicate 433, 441, 448, 456, 464, 473, 482

Social Studies
Geography 475

Business and Economics
Coins 454, 456
Manufacturing 433, 434, 443, 458, 467
Printing 492
Product Packaging 458
Small Business 467

Life Skills
Carpentry 459
Construction 449, 451
Cooking 435
Food 475
Metalwork 475

Pottery 483
Public Health 458
Transportation 484

Sports and Leisure
Aquariums 439
Hobbies 434, 484
Hot-Air Ballooning 469, 473
Recreation 443, 444, 467
Scale Models 468
Sports 474, 475, 492

8 SIMILAR SHAPES 496

MATH CONNECTIONS

Algebra 499, 504, 513, 516, 524, 529, 551   **Coordinate Geometry** 506, 530, 532

APPLICATIONS

Science
Architecture 510, 511
Astronomy 501, 549
Biology 531, 546
Earth Science 505
Engineering 505, 532, 533
Optics 505, 541
Paleontology 540
Wildlife Management 514

Language Arts
Communicate 502, 511, 520, 528, 536, 547

Business and Economics
Business 550
Packaging 550
Storage 550

Life Skills
Food 550
Graphic Design 523
Interior Decorating 515
Landscaping 515
Photography 498

Sports and Leisure
Fine Art 515
Music 531
Quilting 504
Sports 550

Other
Indirect Measurement 516

9 CIRCLES 562

MATH CONNECTIONS

Algebra 570, 575, 578, 579, 594, 595, 597, 601, 602, 604, 606, 607, 609, 610, 611, 614, 615, 617

Coordinate Geometry 612
Technology 611, 613
Trigonometry 572

APPLICATIONS

Science
Agriculture 624
Archaeology 597
Astronomy 616
Communications 578, 597, 624
Cartography 587
Civil Engineering 571
Engineering 579, 608
Geology 600, 604
Lunar Exploration 608
Optics 583
Space Flight 578
Wildlife Management 603

Social Studies
Demographics 571

Language Arts
Communicate 569, 576, 585, 592, 605, 613

Business and Economics
Package Design 608

Life Skills
Carpentry 580, 584, 586
Computer Graphics 616
Design 578, 597

Landscaping 571
Navigation 588, 591, 593, 603, 624
Photography 585
Stained Glass 586
Structural Design 617
Surveying 624

Sports and Leisure
Games 567
Sports 571

MATH CONNECTIONS

Algebra 635, 645, 667, 679, 684
Coordinate Geometry 669, 679, 685
Technology 644

Polar Coordinates 662
Trigonometry 669, 685

APPLICATIONS

Science
Agriculture 673
Architecture 661, 684
Astronomy 641, 652, 653, 692
Aviation 678
Engineering 636, 692
Forestry 645, 661
Indirect Measurement 636, 668, 692

Physics 673
Rotary Motion 649
Wildlife Management 660

Language Arts
Communicate 634, 643, 651, 657, 666, 676, 683

Life Skills
Construction 645
Landscaping 668

Navigation 657, 668, 679, 692
Public Safety 685
Surveying 636, 661

Sports and Leisure
Recreation 640, 645, 673
Scuba Diving 678
Sports 668, 675, 676

MATH CONNECTIONS

Algebra 700, 704, 710, 727
Coordinate Geometry 705

Trigonometry 719
Knot Theory 728

APPLICATIONS

Science
Architecture 762
Genetics 721
Telecommunications 741
Wildlife Management 719

Language Arts
Communicate 702, 709, 716,
  724, 733, 742, 752

Social Studies
Geography 739, 762

**Business and
Economics**
Fund-raising 719
Manufacturing 727
Marketing 705

Life Skills
Commuting 710

Food 728
Law Enforcement 718
Photography 705
Public Safety 710

Sports and Leisure
Fine Arts 705
Hobbies 744

12 A CLOSER LOOK AT PROOF AND LOGIC 766

MATH CONNECTIONS

Algebra 796, 804   **Number Theory** 795

APPLICATIONS

Science
Computer Architecture 798
Computer Databases 781
Electricity 803
Environmental Science 804
Physics 788
Telecommunications 796

Language Arts
Communicate 772, 779, 787, 794, 801

Business and Economics
Advertising 780
Criminal Law 784
Law 795

Life Skills
Academics 790
Landscaping 774

Sports and Leisure
Sports 774

INFO BANK 817

Exploring Geometry

GEOMETRY IS BOTH ANCIENT AND MODERN. From its traditional beginnings as a systematic study in the works of Euclid, through its develpoment in the works of the French philosopher and mathematician Rene Descartes, to its present-day study using sophisticated computers and calculators, geometry has an unbroken tradition in the West that spans well over two thousand years.

This first chapter is for you to get acquainted with geometry. You will be introduced to a number of tools for exploring geometry, such as paper folding, compass and straightedge, and geometry drawing software. In this way, you will discover what geometry is by experiencing it.

Lessons

PORTFOLIO ACTIVITIES
CHAPTER PROJECT

As you begin your study of geometry, set up your own portfolio. The first lesson of this book, Lesson 1.0, tells you how to do this. Throughout this book you will find many suggestions of things to include in your portfolio.

About the Chapter Project

Origami is an ancient art form that originated in Japan. Using a few basic folds, masters of the art have created elegant and intriguing forms that have become the classics of the tradition.

One of the most popular and enduring of the origami classics is the paper crane, which you will study in the Chapter Project. As you fashion the bird, notice how paper folding is used to create the symmetry of the finished form.

After completing the Chapter Project, you will be able to do the following:

- Fold the classic origami crane.
- Analyze the patterns formed on the paper by the folds of the crane.

About the Porfolio Activities

Throughout the chapter, you will be given hands-on projects to do that will enhance and extend your understanding of the material in the lessons.

- In the Portfolio Activity on page 42, you will learn how to create a regular hexagon by folding paper.

- In the Portfolio Activity on page 49, the center of mass of a triangle is found by using the midpoints of the sides of the triangle. The triangle can be balanced on this point.

- Two of the basic rigid transformations are used to construct "snowflake" patterns in the Portfolio Activity on page 58.

Building Your Geometry Portfolio

Objective

- Begin to construct a geometry portfolio that will help you to organize your work throughout this course.

Why *Artists and other professionals often keep portfolios of their work. Although you are probably not yet a professional in any area, the work you do in school may help you decide on your future work and career.*

Building a portfolio will help you organize and display your work. Design it to show your work in a way that reflects your interests and your strengths. You should concentrate on the things your enjoy; these will probably be the things you do best. You may want to create geometric constructions on a computer, study the geometry of beehives and spider webs, or explore the geometry found in works of art.

Geometry in Nature

People have long been attracted to geometric figures in nature, such as the spiral shell of the chambered nautilus. The larger the shell grows, the more closely its proportions approach the value of the golden ratio, which is a very important number in mathematics. The underlying geometric principles of natural objects often seem to be the reason for their visual appeal. As you look around yourself, you will find many examples of geometric beauty in nature.

Geometry in Art

Many artists create works of art by using pure geometric forms. The work of art at right, by a famous artist of the Bauhaus School in Germany, uses rectangular solids and flat surfaces. As the work illustrates, pure geometric forms have their own beauty. Make your own collection of works of art that use pure geometric forms.

Herbert Bayer, *Structure with Three Squares*. 1967. Painted aluminum. 62.9 cm h × 62.9 cm w × 5.7 cm d. Gift of Jan van der Marck in loving memory of his wife Ingeborg. Photograph © 1999 The Detroit Institute of Arts.

Geometry in Architecture

The dimensions of the Parthenon reveal the ancient Greeks' fascination with geometry. The ratio of the height of the original structure to its width is very close to the golden ratio. Geometry is still an important element in architecture today. By applying principles from geometry and physics, architects design structures that are both strong and beautiful.

Your Notebook and Journal

Mathematicians keep records of their theories and discoveries. You, too, should keep a notebook of your work, including tests, homework activities, and special activities such as research projects. Your teacher might also want you to keep a journal.

The photos above are details from two pages of Leonardo da Vinci's journals.

Putting Your Portfolio Together

Various containers can be used for your portfolio. File folders, accordion files, and even cereal boxes will work very well. You may have a number of string designs or physical models that will not fit into your portfolio container. These should still be considered a part of your portfolio.

On the following pages are five different things to include in your portfolio. You should begin right away and continue to add to your portfolio throughout this course—don't wait until the last minute.

You Can Begin Now

1. You can create a star with almost any number of points by using the following method: Start by drawing a circle. Draw the desired number of points spaced evenly around the circle. Connect the points, skipping the same number of points each time. Experiment by skipping different numbers of points. You may wish to color your stars or draw two or more stars in the same circle.

   Try to determine when the star can be drawn without picking up your pencil.

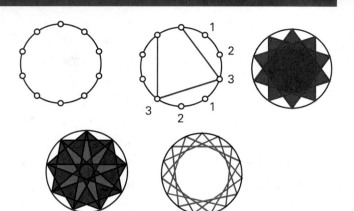

2. Circular designs known as *mandalas* (from the Sanskrit word for "circle" or "center") appear in the Hindu and Buddhist traditions as symbols of the wholeness of the universe. The Aztec calendar, a huge carved stone from ancient Mexico, bears striking resemblances to the mandalas of the East. Write a report on mandalas and their history, and then try creating mandalas of your own.

Aztec calendar (with color added)

3. Interesting designs can be created by using only straight lines. One type of line design is made from string and is known as "string art." Make your own design, using either string or pencil, paper, and a straightedge.

Student project

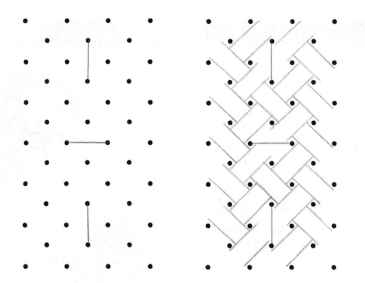

4. A special kind of art is the creation of Celtic "knot" designs such as those shown here. Experiment with your own designs. Study the diagrams above to help you get started.

5. Islamic culture is extremely rich in geometric and calligraphic art. Collect examples of Islamic or other geometric art for your portfolio. You may want to create some designs of your own in the style of Islamic art.

The Building Blocks of Geometry

Objectives

- Understand and identify the undefined terms *point, line,* and *plane.*

- Define *segment, ray, angle, collinear, intersect, intersection,* and *coplanar.*

- Investigate postulates about points, lines, and planes.

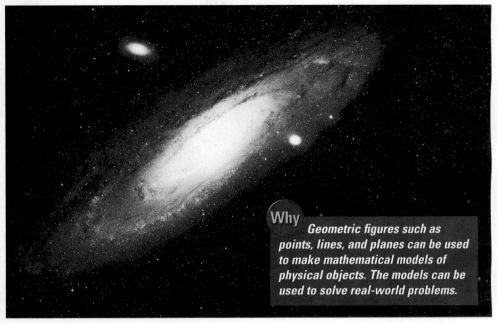

Why *Geometric figures such as points, lines, and planes can be used to make mathematical models of physical objects. The models can be used to solve real-world problems.*

The spiral galaxy M31 in the constellation Andromeda is the companion galaxy of our galaxy, the Milky Way. Just as a galaxy is made up of stars, a geometric figure is composed of points.

Basic Geometric Figures—Undefined Terms

The most basic figures of geometry are *points, lines,* and *planes.* They can be thought of as building blocks for other geometric figures. Because they are so basic, they are not defined in terms of other figures. In fact, they are often referred to as *undefined terms.* But even though points, lines, and planes are undefined, they can be explained.

The first thing to realize about geometric figures is that they are not real-world objects. Lines and planes have no thickness, and points have no size at all. This book contains many illustrations, but the illustrations are not the same as the geometric figures they represent. Theoretically, geometric figures exist "only in the mind."

Points When you look at the night sky and see the stars, the tiny dots of light seem like points. Points are often shown as dots, but unlike physical dots, geometric points have no size. Points are named by capital letters such as *A* or *X*.

Lines A geometric line has no thickness, is perfectly straight, and extends forever. A line can be named by two points on the line, with a double-headed arrow ($\leftrightarrow$) over the two letters, or by a single lowercase letter.

Planes A geometric plane extends infinitely in all directions along a flat surface. You can think of any flat surface, such as the top of your desk or the front of this book, as representing a portion of a plane.

Plane *MNO* or ℛ

In the figure at right, the flat surface represents a portion of a plane. A plane can be named by three points that lie in the plane, such as *M*, *N*, and *O*, and that are not on the same line. A plane can also be renamed by a script capital letter, such as ℛ.

Points are said to be **collinear** if a single line can contain them all. (Any *two* points are collinear.) In the figure at right, *A*, *B*, and *C*, but not *D*, are collinear. Points are said to be **coplanar** if a single plane can contain them all. (Any *three* points are coplanar.)

Points *A*, *B*, and *C* are collinear.

Defining Figures in Terms of the Basics

It is now possible to define three important geometric figures. Notice how each of the following definitions builds on the undefined terms *point*, *line*, and *plane*.

Definition: Segment

A **segment** is a part of a line that begins at one point and ends at another. The points are called the **endpoints** of the segment.

1.1.1

A segment is named by its endpoints. A bar (⁻) is drawn over the two letters representing the endpoints.

Definition: Ray

A **ray** is a part of a line that starts at a point and extends infinitely in one direction. The point is called the **endpoint** of the ray.

1.1.2

The light-gathering capacity of the large mirror of the Mount Palomar telescope makes it possible to see distant stars as points of light.

A ray is named by its endpoint and one other point that lies on the ray. The endpoint is named first. An arrow (→) is drawn over the two letters representing the points.

Definition: Angle

An **angle** is a figure formed by two rays with a common endpoint. The common endpoint is called the **vertex of the angle**, and the rays are the **sides of the angle.**

An angle divides a plane into two regions: the **interior** and the **exterior** of the angle. If two points, one from each side of an angle, are connected by a segment, the segment passes through the interior of the angle.

1.1.3

An angle can be named with the angle symbol ($\angle$) and three letters: one point from each side of the angle plus the vertex, with the letter for the vertex in the middle. If there is only one angle with a given vertex, the angle can be named with the angle symbol and the single letter that represents the vertex. Angles can also be named with a number shown in the angle's interior.

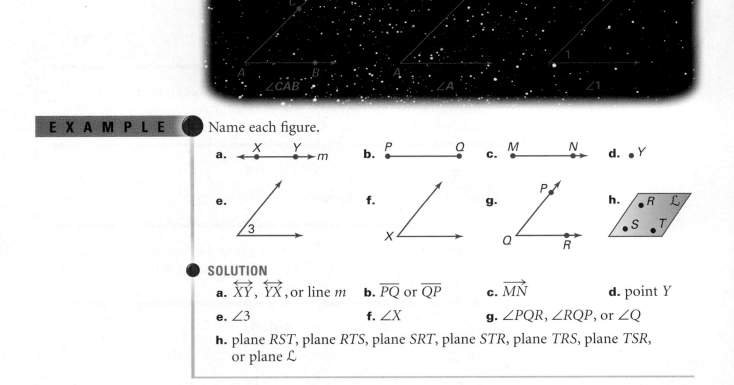

EXAMPLE Name each figure.

a. $\xleftrightarrow{} X \quad Y \; m$
b. $P \quad Q$
c. $M \quad N$
d. $\bullet Y$

e. $\angle 3$
f. X
g. P, Q, R
h. R, S, T (plane $\mathcal{L}$)

SOLUTION

a. $\overleftrightarrow{XY}$, $\overleftrightarrow{YX}$, or line m b. $\overline{PQ}$ or $\overline{QP}$ c. $\overrightarrow{MN}$ d. point Y

e. $\angle 3$ f. $\angle X$ g. $\angle PQR$, $\angle RQP$, or $\angle Q$

h. plane RST, plane RTS, plane SRT, plane STR, plane TRS, plane TSR, or plane $\mathcal{L}$

Intersections of Geometric Figures

When geometric figures have one or more points in common, they are said to **intersect**. The set of points that they have in common is called their **intersection**. In the Activity that follows, you will discover some fundamental geometry ideas, or **postulates**, involving intersections of geometric figures. Postulates are statements that are accepted as true without proof. Postulates, like undefined terms, are building blocks of geometry. (Postulates are also known as *axioms*.)

The illustration at left may be thought of as a model of a real-world object such as a box or a classroom. Complete each postulate below.

1. Examine the illustration. Identify the places where lines intersect each other. What kind of geometric figure is the intersection of two lines?

CHECKPOINT ✔

Postulate

The intersection of two lines is a _____?_____.   **1.1.4**

How many lines intersect at each corner of the figure? Do you think there is a limit to the number of lines that can intersect at a single point?

2. Identify the places in the illustration where planes intersect each other. What kind of geometric figure is the intersection of two planes?

CHECKPOINT ✔

Postulate

The intersection of two planes is a _____?_____.   **1.1.5**

3. Look at points *A* and *B*. How many lines pass through both of these points? Could there be another line, different from the one shown, that passes through both points *A* and *B*?

CHECKPOINT ✔

Postulate

Through any two points there is exactly one _____?_____.   **1.1.6**

4. Look at points *A*, *B*, and *C*. How many planes pass through these three noncollinear points? Could there be a plane, other than the one shown, that passes through all three points?

CHECKPOINT ✔

Postulate

Through any three noncollinear points there is exactly one _____?_____.

1.1.7

5. Pick any plane in the illustration. Then pick two points that are in the plane. Name the line that passes through these two points. Is the line in the plane that you picked?

CHECKPOINT ✔

Postulate

If two points are in a plane, then the line containing them _____?_____.

1.1.8

Exercises

● Communicate

1. Explain how geometric figures are different from real-world objects.

2. Examine the room you are in. Name some objects that could be represented by points, lines, and planes.

3. Explain why one point is not enough to name a line.

4. Explain why two points are not enough to name a plane.

5. Why is the order of the letters important in the name of a ray? You may wish to illustrate your reasoning with a diagram.

Automotive designers often create mathematical models on computers before building actual physical models.

● Guided Skills Practice

6. Refer to the figure at right. Name a point, a line, a segment, and a ray in the figure. **(EXAMPLE)**

7. Give four names for the angle in the figure at right. **(EXAMPLE)**

8. Give three names for the plane in the figure at right. **(EXAMPLE)**

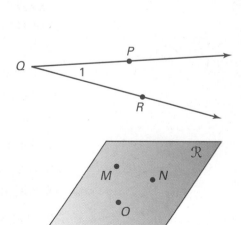

In Exercises 9–12, refer to the triangle below.

9. Name all of the segments in the triangle.

10. Name each of the angles in the triangle in three different ways.

11. Name the rays that form each side of the angles in the triangle.

12. Name the plane that contains the triangle.

HOBBIES In Exercises 13–17, refer to the aquarium shown below. State whether each object in the aquarium is best modeled by a point, a line, or a plane.

13. an edge of the aquarium

14. a grain of sand

15. a side of the aquarium

16. the surface of the water

17. a corner of the aquarium

In Exercises 18–25, classify each statement as true or false, and explain your reasoning.

18. Lines have endpoints.

19. Planes have edges.

20. Three lines that intersect at the same point must all be in the same plane.

21. Two planes may intersect a third plane without intersecting each other.

22. Three planes may all intersect each other at exactly one point.

23. Any two points are contained in exactly one plane.

24. Any three points are contained in exactly one plane.

25. Any four points are contained in exactly one plane.

Refer to the figure below for Exercises 26–30.

26. Name a line in the figure. Give three other names for the line.

27. Name a point on line *m*.

28. Name the intersection of lines *m* and *n*.

29. Name an angle in the figure. Name the vertex of this angle and the two rays that form the sides of the angle.

30. Can an angle in the figure be named ∠*A*? Why or why not?

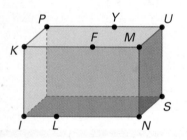

In Exercises 31–40, refer to the figures below.

31. Name the intersection of planes $\mathcal{P}$ and $\mathcal{Z}$.

32. Name a line in plane $\mathcal{P}$.

33. Name a point in plane $\mathcal{P}$.

34. Name a point on $\overline{KM}$.

35. Name the intersection of $\overline{MU}$ and $\overline{MN}$.

36. Name three collinear points in the figure.

37. Name two coplanar segments in the figure.

38. Name the intersection of line *n* and $\overline{AI}$.

39. Name the intersection of planes $\mathcal{Q}$ and MPT.

40. Name three coplanar points in the figure.

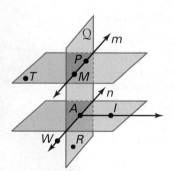

How many different segments can be named in each figure below? Name each segment.

41.

42.

43.

44. Write a general rule or formula for finding the number of segments that can be named by a given number of points on a line. Can you explain why the rule works?

How many different angles can be named in each figure below? Name each angle.

45.

46.

47.

48. Write a general rule or formula for finding the number of angles that can be named by a given number of rays with the same endpoint. Can you explain why the rule works? (Assume that all rays lie on one side of a straight line, as shown.)

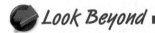

Look Back

Simplify each expression. You may wish to draw a number line.

49. $22 + (-6)$

50. $7 + 15$

51. $11 - (-4)$

52. $-81 - (-30)$

53. $|-14 + (-35)|$

54. $|13 - 10|$

55. $-123 - 41$

56. $|21 + (-35)|$

57. $|-54 + (-20)|$

Look Beyond

PROBLEM SOLVING

58. You can **make a diagram** to find the answer to the following problem. Suppose that 4 people are exchanging cards. Each person exchanges one card with each of the other 3 people. How many exchanges are made? (Hint: Draw 4 noncollinear points. Then draw line segments between the points. How many segments did you draw? How does this relate to the exchanges of cards?)

CHALLENGE

59. If 5 people exchanged cards as described above, how many exchanges would there be? Explain how to determine the number of exchanges when n people exchange cards.

Measuring Length

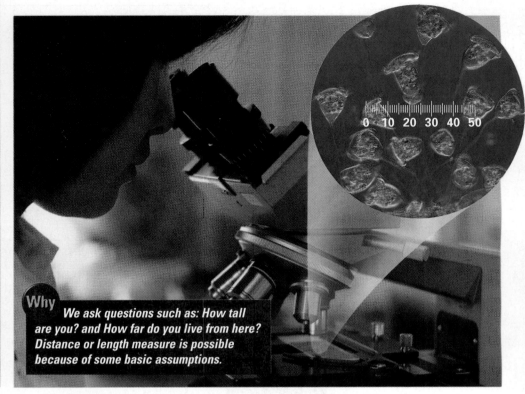

1.2

Objectives

- Construct a geometry ruler.
- Define *length* and *congruent*.
- Identify and use the Segment Addition Postulate.

Why We ask questions such as: *How tall are you?* and *How far do you live from here?* Distance or length measure is possible because of some basic assumptions.

A ruler in the eyepiece of the microscope makes it possible to measure in units of one-millionth of a meter; these units are known as microns.

The Length of a Segment

In defining the length of a segment, we will use a *number line*, which is like a ruler. A **number line** is a line that has been set up to correspond with the real numbers.

A geometry "ruler"

The **coordinate of a point** on the number line is a real number. In the illustration, -3 is the coordinate of point A and 4 is the coordinate of point B.

How would you find the distance between points A and B? Try $|-3 - 4|$. Also try $|4 - (-3)|$. What do you notice about the absolute values? This leads to a definition of the distance between two points.

Definition: Length of $\overline{AB}$

Let A and B be points on a number line, with coordinates a and b. Then the measure of $\overline{AB}$, which is called its **length**, is $|a - b|$ or $|b - a|$.

m$\overline{AB}$, or $AB =$
$|a - b|$ or $|b - a|$ **1.2.1**

The measure, or length, of $\overline{AB}$ is written as m$\overline{AB}$ or, more commonly, as just AB.

① Find the measures (lengths) of $\overline{AB}$, $\overline{AX}$, and $\overline{XB}$ on the number line below.

● SOLUTION

AB (or $\mathrm{m}\overline{AB}$) $= |-4 - 4| = |-8| = 8$ or $AB = |4 - (-4)| = |8| = 8$

AX (or $\mathrm{m}\overline{AX}$) $= |-4 - 1| = -5| = 5$ or $AX = |1 - (-4)| = |5| = 5$

XB (or $\mathrm{m}\overline{XB}$) $= |1 - 4| = |-3| = 3$ or $XB = |4 - 1| = |3| = 3$

TRY THIS   Find ST.

Fair Rulers

The rulers you will use in this book—with a few interesting exceptions (see Exercises 47–49 and the Portfolio Project on page 75)—are "fair" rulers. That is, they have equal intervals from one integer to the next. In the Activity below, you will use a compass and straightedge to construct a simple fair ruler.

Activity
Constructing and Using a Fair Ruler

YOU WILL NEED

compass and straightedge

1. Use your straightedge to draw a line. Choose any point on the line and label it 0. Adjust your compass to an appropriate spacing, and set the point of the compass on the point labeled.

Use the pencil part of the compass to draw a short mark that crosses the number line to the right of 0. Label the point of intersection 1.

2. Set the point of the compass on the new point, and draw another mark to its right. Label the new intersection 2.

3. Repeat the previous step as many times as desired, adding 1 to the label each time.

4. Construct the negative numbers to the left of 0.

5. The distance from 0 to 1 on a ruler is known as the **unit length**. Two common unit lengths are *inches* and *centimeters*. Make up a name for your unit length.

CHECKPOINT ✔ **6.** Use your ruler to measure an object in your classroom. Estimate the fractional part of the measurement. Compare your measurement with those of your classmates. What do you observe?

Congruent Segments

Congruent figures are figures that are the same size and shape. If you move one of them onto the other, they will match exactly, like the segments in the figure at right. The segments on the ruler that you constructed in the Activity were all congruent because the same compass setting was used for each one.

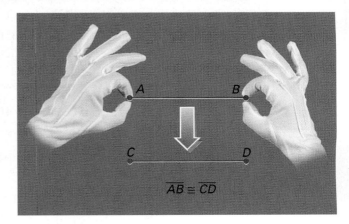

$$\overline{AB} \cong \overline{CD}$$

The symbol for congruence is $\cong$. $\overline{XY} \cong \overline{YZ}$ is read as "Segment $\overline{XY}$ is congruent to segment $\overline{YZ}$."

In geometry, tick marks are used to show which segments are known to be congruent. Within a given illustration, segments that have a single tick mark are congruent. Similarly, segments that have two tick marks are congruent, and so on.

From your exploration of a fair ruler, the following important principle should be obvious:

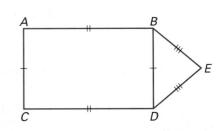

Which segments are congruent?

Segment Congruence Postulate

If two segments have the same length as measured by a fair ruler, then the segments are congruent. Also, if two segments are congruent, then they have the same length as measured by a fair ruler.   **1.2.2**

**E X A M P L E ② ** Use the Segment Congruence Postulate to complete the following:

If $XY = YZ$, then ____?____.   If $\overline{XY} \cong \overline{YZ}$, then ____?____.

● **SOLUTION**

If $XY = YZ$, then $\overline{XY} \cong \overline{YZ}$.   If $\overline{XY} \cong \overline{YZ}$, then $XY = YZ$.

CRITICAL THINKING   What could happen if the intervals on a ruler were not evenly spaced? Could you be sure that segments with the same measure were congruent or that congruent segments had equal measures? Explain your answer.

Segment Addition

Look again at the number line in Example 1. Notice that X is between A and B. The relationships among the lengths AX, XB, and AB depend on an important assumption known as the Segment Addition Postulate.

Segment Addition Postulate

If point R is between points P and Q on a line, then $PR + RQ = PQ$.

1.2.3

CRITICAL THINKING

What is being added in this postulate? In geometry, addition and other arithmetic operations are defined for numbers, not for geometric figures.

Which of these statements make sense?

$$\begin{cases} AB + CD = 5 \\ \overline{AB} + \overline{CD} = 5 \\ m\overline{AB} + m\overline{CD} = 5 \end{cases}$$

E X A M P L E ③

APPLICATION
NAVIGATION

Algebra

The towns of Dyersberg, Newton, and Saint Thomas are located along a straight portion of Ventura Highway. Newton is between Saint Thomas and Dyersberg. The distance from Dyersberg to Saint Thomas is 25 miles. The distance from Dyersberg to Newton is 1 mile more than 3 times the distance from Newton to Saint Thomas. Find the distance from Dyersberg to Newton and from Newton to Saint Thomas.

SOLUTION

PROBLEM SOLVING

Write an equation. First represent each town as a point on a line segment. Let Dyersberg be A, let Saint Thomas be B, and let Newton be C. Let x be the distance in miles from C to B, or CB. Then the distance from A to C, or AC, is $3x + 1$.

Since C is between A and B, $AC + CB = AB$.

$$AC + CB = AB$$
$$(3x + 1) + x = 25$$
$$4x + 1 = 25$$
$$4x = 24$$
$$x = 6 \text{ miles, the distance from Newton to Saint Thomas}$$

The distance from Dyersberg to Newton is found as follows:

$$AC = 3x + 1 = 3(6) + 1 = 19 \text{ miles}$$

As a check, note that the total distance is $6 + 19 = 25$.

Exercises

Communicate

1. The unit length on a ruler can be any desired size. Give some commonly used unit lengths.

2. Explain why it is important for a ruler to have equal intervals.

3. Suppose that the centimeter were the only unit of measure for length. What problems would this create? Discuss why it is useful to have different units for measuring length.

4. Once you have constructed a ruler, why might you want to divide the unit length into smaller intervals?

Computer software often allows you to select the unit of measure.

5. Explain why each of the following statements does or does not make sense:

   a. $\overline{MN} + OP = 30$ **b.** $MN + OP = 30$ **c.** $m\overline{MN} + m\overline{OP} = 30$

Guided Skills Practice

In Exercises 6–8, find the lengths of the segments. *(EXAMPLE 1)*

6. $\overline{MN}$ **7.** $\overline{NP}$ **8.** $\overline{MP}$

9. Complete the statements below. *(EXAMPLE 2)*

   a. If $\overline{AB} \cong \overline{CD}$, then ____?____.   **b.** If $AB = CD$, then ____?____.

10. **NAVIGATION** The cities of Bloomington, Forsyth, and Decatur are located on a straight road from Bloomington to Decatur, with Forsyth between them. The distance from Bloomington to Decatur is 40 miles. The distance from Bloomington to Forsyth is 7 times the distance from Forsyth to Decatur. Find the distance from Bloomington to Forsyth and from Forsyth to Decatur. *(EXAMPLE 3)*

In Exercises 11–16, find the length of $\overline{AB}$.

11.
```
        A           B
 ←──┼──┼──┼──┼──●──┼──┼──┼──●──┼──┼──→
   -4 -3 -2 -1  0  1  2  3  4  5  6
```

12.
```
                      A     B
 ←┼──┼──┼──┼──┼──●──┼──●──┼──┼──┼─→
 -7 -6 -5 -4 -3 -2 -1  0  1  2  3
```

13.
```
                A           B
 ←──┼──┼──┼──●──┼──┼──●──┼──┼──┼──→
   -2 -1  0  1  2  3  4  5  6  7  8
```

14.
```
   A                 B
 ←●──┼──┼──┼──┼──●──┼──┼──┼──┼──┼─→
 -7 -6 -5 -4 -3 -2 -1  0  1  2  3
```

15.
```
        A               B
 ←──┼──●──┼──┼──┼──┼──●──┼──┼──┼──→
   -5 -4 -3 -2 -1  0  1  2  3  4  5
```

16.
```
              A  B
 ←──┼──┼──┼──●──●──┼──┼──┼──┼──┼──→
   -5 -4 -3 -2 -1  0  1  2  3  4  5
```

17. Find the length of each segment determined by points A, B, and C on the number line below. Show that the order of the coordinates does not matter.

In Exercises 18 and 19, name all congruent segments.

18.

19.

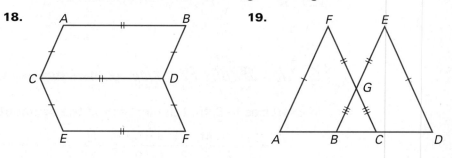

In Exercises 20–22, point A is between points M and B on $\overline{MB}$. Sketch each figure and find the missing lengths.

20. $MA = 30$ $AB = 15$ $MB = \underline{\ ?\ }$

21. $MA = 15$ $AB = \underline{\ ?\ }$ $MB = 100$

22. $MA = \underline{\ ?\ }$ $AB = 13.3$ $MB = 29.6$

 Find the indicated value in Exercises 23–25.

23. $PR = 25$ $x = \underline{\ ?\ }$

24. $PQ = 25$ $PR = \underline{\ ?\ }$

25. $PQ = 25$ $PR = \underline{\ ?\ }$

Towns *A*, *B*, *C*, and *X* are located along a straight highway. Town *B* is between *A* and *C*, and the distance from *A* to *C* is 41 miles. *BC* is 2 miles more than twice *AB*.

26. Write an equation and solve it to find *AB* and *BC*.

27. Town *X* is between *A* and *B*, 6 miles from *A*. Find *XC*.

In Exercises 28–33, explain why each statement does or does not make sense.

28. $XY = 5000$ yd

29. $\overline{PQ} = 32$ in.

30. m$ST = 6$ cm

31. $XY + XZ = 32$ cm

32. m$\overline{PR} = 46$ cm

33. $\overline{XY} - \overline{XZ} = 12$ cm

CULTURAL CONNECTION: AFRICA The Egyptian *royal cubit* (1550 B.C.E.) was subdivided into 28 units known as *digits* or *fingers*. From this basic unit, a number of others were created:

4 digits = 1 palm

12 digits = 1 small span

5 digits = 1 handsbreadth

14 digits = 1 great span

6 digits = 1 fist

16 digits = 1 foot (t'eser)

8 digits = 1 double palm

24 digits = 1 short cubit

34. A digit is approximately 1.9 centimeters. Draw a line and use a centimeter ruler to mark a unit length of 1 digit. Then use a compass to construct a digit ruler.

35. A cubit is the distance from the elbow to the tip of the middle finger. Measure your own cubit with your digit ruler. How does it compare with the Egyptian royal cubit?

You can divide the unit length of a ruler by folding the paper. If you fold the paper so that the points at 0 and 1 line up, the fold marks the point $\frac{1}{2}$ on your ruler.

36. What point would be marked if you folded the paper again so that the points at 0 and $\frac{1}{2}$ line up?

37. Construct a ruler with a fairly large unit length. Divide the interval from 0 to 1 to find the points for $\frac{1}{2}, \frac{1}{4}, \frac{3}{4},$ and $\frac{1}{8}$.

38. What are some points on your ruler that you *cannot* find by using this method?

Choose any two different positive numbers.

39. Add the two numbers together. Is the sum positive or negative?

40. Subtract the smaller number from the larger number. What kind of number do you get? Now take the absolute value of this number.

41. Subtract the larger number from the smaller number. What kind of number do you get? Now take the absolute value of this number.

42. Compare your answers from Exercises 40 and 41. Explain why they are the same or why they are different.

Choose any two different negative numbers.

43. Add the two numbers together. Is the sum positive or negative?

44. Subtract the smaller number from the larger number. What kind of number do you get? Now take the absolute value of this number.

45. Subtract the larger number from the smaller number. What kind of number do you get? Now take the absolute value of this number.

46. Compare your answers from Exercises 44 and 45. Explain why they are the same or why they are different.

Look Beyond

APPLICATION

GEOLOGY Here is one example of a ruler that does not have evenly spaced divisions. The *Richter scale* is used to measure the intensity of earthquakes. An increase of 1 unit on the Richter scale indicates an increase in ground motion by a factor of 10. You can represent this with an "unfair" ruler.

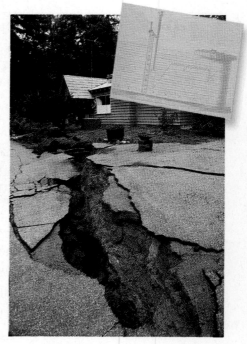

47. Draw part of a ruler to represent values on the Richter scale. Start with the distance from 0 to 1 as 1 mm. The distance from 1 to 2 will be 10 mm, the distance from 2 to 3 will be 100 mm, and so on. (You will quickly run out of room on your paper!) How far would the distance be from 3 to 4? from 4 to 5?

48. How much more ground motion is caused by an earthquake with an intensity of 8 than by one with an intensity of 6?

49. Earthquakes can also be measured by the amount of energy released. An increase of 1 unit on the Richter scale indicates an increase by a factor of 32 in the amount of energy released. How much more energy is released by an earthquake with an intensity of 8 than by one with an intensity of 6?

Two of Earth's continental plates meet at the San Andreas fault. When the plates slip, the resulting shock waves may be felt all over the world.

Measuring Angles

Air speed indicator

Altitude indicator

Altimeter

Automatic direction finder

Vertical speed indicator

Directional gyro

Objectives

- Measure angles with a protractor.

- Identify and use the Angle Addition Postulate.

Why Angle measure is used in many professions. For example, pilots use an angle measure known as the "heading" of an airplane to navigate safely through the skies.

The heading of this airplane, as indicated by the directional gyro in the photo above, is 166, a direction that is a little east of due south. (The readings on the dial must be multiplied by 10.)

Defining Angle Measure

A protractor is used to measure angles. As on a ruler, the intervals on a protractor must be equal. Then you can be sure that if two angles have the same measure, they are congruent, and vice versa.

To understand how a protractor is used, study the following example:

EXAMPLE 1 Use a protractor to find the measure of ∠CAB.

SOLUTION

1. Put the center of the protractor at the vertex.

2. Align the protractor so that $\overrightarrow{AB}$ passes through 0 on the protractor.

3. Read the measure of ∠CAB (in degrees) at the point where $\overrightarrow{AC}$ intersects the scale on the protractor.

The measure of ∠CAB is 121°, or m∠CAB = 121°.

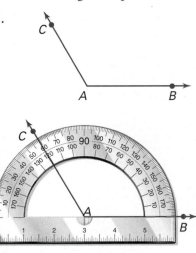

CRITICAL THINKING Why isn't the measure of ∠CAB 59° instead of 121°?

Angles, like segments, are measured in standard units. The most common unit of angle measure is the **degree**. This is the unit of measure that results when a half-circle is divided into 180 equal parts.

A protractor may be thought of as another type of geometry ruler. This "ruler" is a half-circle with coordinates from 0° to 180°. You can use a protractor to define the measure of an angle such as ∠AVB.

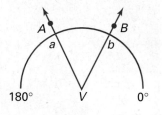

Definition: Measure of an Angle

Suppose that the vertex, V, of ∠AVB is placed on the center point of a half-circle with coordinates from 0° to 180° so that $\overrightarrow{VA}$ and $\overrightarrow{VB}$ intersect the half-circle. Let a and b be the coordinates of the intersections.

Then the **measure of the angle**, written as m∠AVB, is $|a - b|$ or $|b - a|$.

1.3.1

CRITICAL THINKING

In creating a protractor, does the size of the half-circle make a difference? Why or why not?

E X A M P L E 2 Use a protractor to find the measures of ∠BAC, ∠CAD, and ∠BAD. What is the value of m∠BAC + m∠CAD?

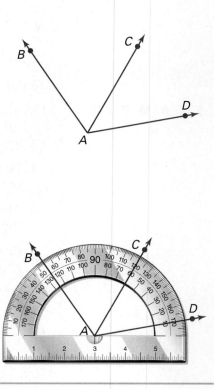

SOLUTION

1. To measure ∠BAC, find the points where $\overrightarrow{AB}$ and $\overrightarrow{AC}$ pass through the scale of the protractor (55° and 120°). Using these coordinates, m∠BAC = $|55 - 120| = |-65| = 65°$.

2. The coordinate of $\overleftrightarrow{AD}$ is 170°, so m∠CAD = $|120 - 170| = |-50| = 50°$.

3. Similarly, m∠BAD = $|55 - 170| = |-115| = 115°$.

4. Using the answers for Steps 1 and 2, m∠BAC + m∠CAD = 65 + 50 = 115°. Notice that this agrees with the answer for Step 3.

Congruent Angles

$\angle A \cong \angle D$ and $\angle B \cong \angle C$

Angles, like segments, are congruent if one can be moved onto the other so that they match exactly. Tick marks are used to show which angles are known to be congruent.

Note: Angles are said to match if their sides match. The length of the sides, which are rays that go on forever, does not matter.

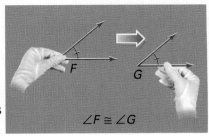

$\angle F \cong \angle G$

From the discussion of a protractor, the following important geometry principle should be obvious:

Angle Congruence Postulate

If two angles have the same measure, then they are congruent. If two angles are congruent, then they have the same measure.

1.3.2

E X A M P L E ❸ Complete the following statements by using the Angle Congruence Postulate:

If $m\angle ABC = m\angle DEF$, then ___?___.

If $\angle ABC \cong \angle DEF$, then ___?___.

● **SOLUTION**

If $m\angle ABC = m\angle DEF$, then $\angle ABC \cong \angle DEF$.

If $\angle ABC \cong \angle DEF$, then $m\angle ABC = m\angle DEF$.

Angle Addition

Look again at Example 2. Step 4 shows that $m\angle BAC + m\angle CAD = m\angle BAD$. This suggests the following postulate:

Angle Addition Postulate

If point S is in the interior of $\angle PQR$, then $m\angle PQS + m\angle SQR = m\angle PQR$.

1.3.3

CRITICAL THINKING

As with the Segment Addition Postulate, you should ask yourself, What is really being added—angles or measures of angles?

Which statement makes sense?

$$\begin{cases} \angle A + \angle B = 180° \\ m\angle A + m\angle B = 180° \end{cases}$$

Pairs of Angles

Sometimes it is important to consider pairs of angles.

Special Angle Pairs

Complementary angles are two angles whose measures have a sum of 90°. Each angle is called the **complement** of the other.

Supplementary angles are two angles whose measures have a sum of 180°. Each angle is called the **supplement** of the other.

1.3.4

EXAMPLE 4 Name all complementary and supplementary angle pairs below.

60°
A

30°
B

120°
C

● **SOLUTION**

$\angle A$ and $\angle B$ are complementary angles.

$\angle A$ is the complement of $\angle B$. $\qquad$ $\angle B$ is the complement of $\angle A$.

$\angle A$ and $\angle C$ are supplementary angles.

$\angle A$ is the supplement of $\angle C$. $\qquad$ $\angle C$ is the supplement of $\angle A$.

If the endpoint of a ray falls on a line so that two angles are formed, then the angles are known as a **linear pair**.

1 / 2

CRITICAL THINKING   If two lines intersect at a point, how many linear pairs are formed? Explain how this situation fits the definition of a linear pair.

The definitions above lead to the following property:

Linear Pair Property

If two angles form a linear pair, then they are supplementary.

1.3.5

You will use the Linear Pair Property to prove an important conjecture in Lesson 2.5.

Classification of Angles

Angles can be classified according to their measure.

Definitions: Three Types of Angles

A **right angle** is an angle whose measure is 90°.
An **acute angle** is an angle whose measure is less than 90°.
An **obtuse angle** is an angle whose measure is greater than 90°.

1.3.6

The symbol for a right angle is a small square placed at the vertex of the angle.

— Right angle symbol

Extending Angle Measure

You may have noticed that so far the angle measures studied have been greater than 0° and less than 180°. One reason for this is that the definition of the interior and exterior of an angle doesn't apply outside this range. (Try it and see!)

However, it is sometimes necessary to talk about angles of 0° and less or of 180° and greater. Compass headings, for example, range from 0° to 360°. You can think of them as being based on a 360°, or circular, protractor. Angles with a measure of 180° are called *straight angles*, and angles with a measure greater than 180° are called *reflex angles*.

In Chapter 10, angles are studied in terms of *rotation*, in which case their measures can be greater than 360° and can even be negative.

Exercises

● Communicate

1. A magnetic compass is used in navigation to determine which direction is north. Once you know which direction is north, how could you use angles to describe the direction in which you are traveling?

2. Think of a memory aid, or *mnemonic*, to help you remember the difference between complementary and supplementary angles.

Compass headings are used to navigate at sea as well as in the air.

3. Discuss the similarities and differences between segment length and angle measure.

4. Use an example to explain why the Angle Addition Postulate fails if point S is not in the interior of $\angle PQR$.

5. Classify each of the following statements as true or false, and explain your reasoning:
   a. All right angles are congruent.
   b. All acute angles are congruent.
   c. All obtuse angles are congruent.

6. Explain why one of these statements makes sense and the other does not:
   a. $m\angle A + m\angle 2 = 190°$
   b. $\angle X + \angle Y = 150°$

7. Explain why the definitions of interior and exterior of an angle do not apply to straight angles and reflex angles.

Guided Skills Practice

8. Use the protractor in the figure at right to find the measure of $\angle WVX$.
   (EXAMPLE 1)

In Exercises 9–11, use the protractor in the figure at right to find the measures below. *(EXAMPLE 2)*

9. $m\angle QPR$

10. $m\angle RPS$

11. $m\angle QPR + m\angle RPS$

12. Complete each statement by applying the Angle Congruence Postulate.
   (EXAMPLE 3)
   a. If $m\angle UVW = m\angle XYZ$, then _____.
   b. If $\angle UVW \cong \angle XYZ$, then _____.

13. Name a complementary angle pair and a supplementary angle pair in the figure below. **(EXAMPLE 4)**

Practice and Apply

Refer to the figure below. Find the measure of each angle.

14. m∠AVB

15. m∠AVC

16. m∠AVD

17. m∠BVC

18. m∠BVD

19. m∠CVD

Use a protractor to find the measures of the indicated angles. You may trace the figures or use a piece of paper to extend the rays if necessary.

20. m∠X

21. m∠Y

22. m∠ABC

23. m∠2

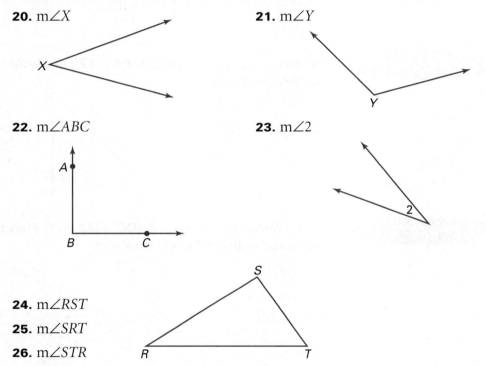

24. m∠RST

25. m∠SRT

26. m∠STR

In the figure below, m∠CED = 25° and ∠AEB and ∠BED form a linear pair. Find the following:

27. m∠BEC

28. m∠AEB

29. m∠AEC

In the figure, m∠EDG = 70° and m∠FDH = 60°. Find the following:

30. m∠2

31. m∠3

32. m∠1

33. m∠4

34. m∠EDJ

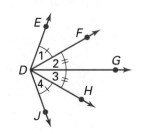

35. In the figure at right, m∠*ALE* = 31°
and m∠*SLE* = 59°. What is the
measure of ∠*SLA*? What is the
relationship between ∠*SLE* and
∠*ALE*?

Find the missing angle measures in Exercises 36–38. Refer to the figure below.

36. m∠*WZX* = 80°   m∠*XZY* = 34°   m∠*WZY* = _?_

37. m∠*WZX* = 21°   m∠*XZY* = _?_   m∠*WZY* = 43°

38. m∠*WZX* = _?_   m∠*XZY* = 34°   m∠*WZY* = 52°

**In the diagram below, m∠*KNM* = 87°, m∠*LNM* = (2*x* − 8)°, and
m∠*KNL* = (*x* + 50)°.**

39. What is the value of *x*?

40. What is m∠*KNL*?

41. What is m∠*LNM*?

**In the diagram below, m∠*ADC* = (43 + *x*)°. Find the value of *x*, and then
find each indicated angle measure.**

42. m∠*ADC*

43. m∠*ADB*

44. m∠*BDC*

45. What is the angle between the minute and hour hands on a clock at 3:00?

46. What is the angle between the minute and hour hands on a clock at 5:00?

47. Another unit of angle measure, used primarily in engineering, is called a
gradian. There are 100 gradians in a right angle. Is 1 gradian smaller or
larger than 1 degree? Why?

CHALLENGE **48.** Name all sets of congruent angles in the figure below.

NAVIGATION Exercises 49–56 involve *compass headings*. Headings are numbered in the clockwise direction from 000 to 360, starting at north. The headings from 000 to 180 are equal to the measure of the angle formed by the compass needle and a ray that points in the direction of travel. The headings from 180 to 360 are found by subtracting the measure of this angle from 360.

Find the heading for each of the following compass directions:

49. N

50. E

51. S

52. W

53. NE

54. SW

55. NNE

56. SSW

57. NAVIGATION A pilot is flying to Chicago on a heading of 355. Refer to the diagram at left. Would the pilot's heading be greater or less than 355 if the plane was traveling to Chicago from point *X*?

58. SCUBA DIVING Scuba divers often navigate in a square pattern at a constant depth. Examine the figure at right. If the diver starts out at a heading of 315, what compass headings are needed to navigate around the rest of the square? Other than using the headings, what else must a diver do in order to navigate the square?

Underwater compasses, like this programmable computer, are used by divers to navigate.

Refer to the figure below for Exercises 59–62. *(LESSON 1.1)*

59. Give a different name for ∠1.

60. Identify a line that is coplanar with ∠1.

61. Name all lines that are formed by the intersection of two planes.

62. Identify three collinear points.

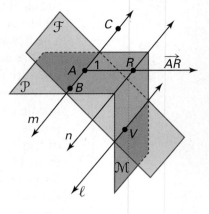

For each figure below, find the length of the indicated segment.
(LESSON 1.2)

63.

64.

65. If $AB = 27$, find AC and BC. *(LESSON 1.2)*

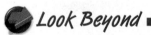

66. CULTURAL CONNECTION: ASIA Our unit measure of degrees comes from the ancient Babylonians. The Babylonians based the measurement of an angle on a circle divided into 360 equal parts.

What are some advantages to using this unit of measurement? (Hint: List all of the factors of 360.)

67. Just as the units on a ruler are subdivided, so are the units on a protractor. Astronomers, architects, and surveyors often need angle measurements that are more precise than degrees. A degree is often divided into 60 units called *minutes*, and a minute is divided further into 60 units called *seconds*.

a. How many seconds are in 1 degree?

b. How many seconds are in 1.5 minutes?

c. How many minutes are in 1.75 degrees?

The ancient Babylonians were sophisticated mathematicians.

Geometry Using Paper Folding

1.4

Objectives

- Use paper folding to construct perpendicular lines, parallel lines, segment bisectors, and angle bisectors.

- Define and make geometry conjectures.

Why Paper folding can be used to create precise geometric figures without the use of special drawing instruments.

Origami, *the ancient Japanese art of paper folding, relies on properties of geometry to produce fascinating and often beautiful shapes.*

Paper Folding: The Basics

When created according to certain rules, a diagram is considered to be mathematically precise. These types of diagrams are called **constructions**. In this book, you will learn to construct geometric figures with paper folding, a compass and straightedge, and geometry software.

The term *folding paper* refers to any paper that is suitable to use for folding geometric figures. You should use paper you can see through, such as wax paper, so that you can match figures precisely when you fold. "Patty paper" is ideal because it makes white creases when folded and is easy to write on.

In the following activities, work with a partner. Be sure both you and your partner understand *why* the constructions work. Learning why things work the way they do is one of the most important objectives of this course.

The terms defined below will be used in the Activities in this lesson.

Definitions: Perpendicular and Parallel Lines

Perpendicular lines are two lines that intersect to form a right angle.
Parallel lines are two coplanar lines that do not intersect. **1.4.1**

Making Conjectures in Geometry

A **conjecture** is a statement that you think is true. It is an "educated guess" based on observations. Mathematical discoveries often start out as conjectures. In the Activity that follows, you will make conjectures about perpendicular and parallel lines.

Activity 1
Perpendicular and Parallel Lines

1. Fold the paper once to make a line. Label the line ℓ.

2. Draw a point on line ℓ and label it A. Fold the paper through A so that line ℓ matches up with itself. Label the new line m.

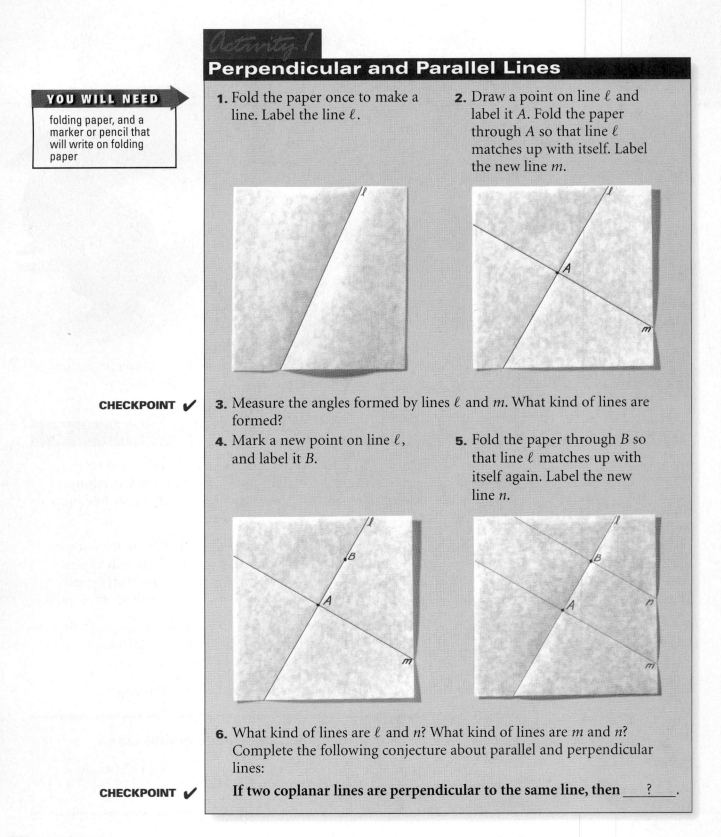

CHECKPOINT ✔

3. Measure the angles formed by lines ℓ and m. What kind of lines are formed?

4. Mark a new point on line ℓ, and label it B.

5. Fold the paper through B so that line ℓ matches up with itself again. Label the new line n.

6. What kind of lines are ℓ and n? What kind of lines are m and n? Complete the following conjecture about parallel and perpendicular lines:

CHECKPOINT ✔

   If two coplanar lines are perpendicular to the same line, then ____?____.

Measuring the Distance From a Point to a Line

How would you measure the distance from the tree to the fence in the photo at right? Would you use the segment $\overline{XA}$, $\overline{XB}$, or $\overline{XC}$?

In geometry, the distance from a point to a line is the length of the perpendicular segment from the point to the line. Thus, the distance from the tree (point X) to the fence ($\overleftrightarrow{AC}$) is the length of $\overline{XB}$.

CRITICAL THINKING

Why do you think the distance from a point to a line is defined along a perpendicular segment?

You can use paper folding to help you measure the distance from a point to a line.

Activity 2

Finding the Distance From a Point to a Line

YOU WILL NEED

folding paper, a marker, and a ruler (for measuring)

1. Fold a line and label it ℓ. Choose a point not on the line and label it P.

2. Fold the paper over so that ℓ lines up with itself, but do not crease the paper yet.

3. Slide the top edge of the paper, keeping ℓ lined up on itself, until P is on the fold. Carefully crease the paper, making sure that P is on the crease.

CHECKPOINT ✔

4. Use a ruler to measure the distance from P to ℓ along the creased fold.

Segment and Angle Bisectors

In Activity 3 below, you will use the following definitions:

Definitions

A **segment bisector** is a line that divides a segment into two congruent parts. The point where a bisector intersects a segment is the **midpoint** of the segment. A bisector that is perpendicular to a segment is called a **perpendicular bisector**. An **angle bisector** is a line or ray that divides an angle into two congruent angles.

1.4.2

Activity 3

Exploring Segment and Angle Bisectors

Part I   Segment Bisectors

1. Fold line ℓ. Choose points A and B on ℓ. Fold the paper so that A matches up with B. Label the resulting line m. What is the relationship between m and $\overline{AB}$?

2. Choose a point on m and label it C. Measure $\overline{AC}$ and $\overline{BC}$. Repeat, choosing several different locations for C. What do you notice? Complete the following conjecture:

CHECKPOINT ✔

The distances from a point on the perpendicular bisector to the endpoints of the segment are _____?_____.

Part II   Angle Bisectors

3. Fold intersecting lines j and k. Label the intersection P. Label a point Q on k and a point R on j. Fold the paper through P so that line j matches up with line k. Label the new line n. What is the relationship between n and $\angle QPR$?

4. Choose a point on n and label it S. Measure the distance from S to j and the distance from S to k. Repeat, choosing several different locations for S. What do you notice? Complete the following conjecture:

CHECKPOINT ✔

The distances from a point on the angle bisector to the sides of the angle are _____?_____.

Exercises

Communicate

1. When you folded line ℓ onto itself in Step 2 of Activity 1, which pairs of angles matched up? Use this to state an alternative definition of perpendicular lines.

2. When you constructed parallel lines m and n in Step 5 of Activity 1, how many right angles were formed? Make a conjecture about how you can determine whether two lines are parallel.

3. When you folded A onto B in Step 1 of Activity 3, how could you tell (without measuring) that the new line divided $\overline{AB}$ into two congruent segments?

4. Explain how you measured the distance from S to j and S to k in Step 4 of Activity 3.

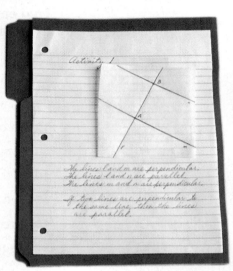

Tape, staple, or glue your folded papers to your work. Label the papers in case they come off.

Guided Skills Practice

For Exercises 5–8, use the figures below and your conjectures from the Activities to complete the following statements:

5. Lines ℓ and m are ___?___ .
   (ACTIVITY 1)

6. The distance from P to ℓ is ___?___ . **(ACTIVITY 2)**

7. *AC* and *BC* are ___?___ .
   (ACTIVITY 3)

8. *BX* and *CX* are ___?___ .
   (ACTIVITY 3)

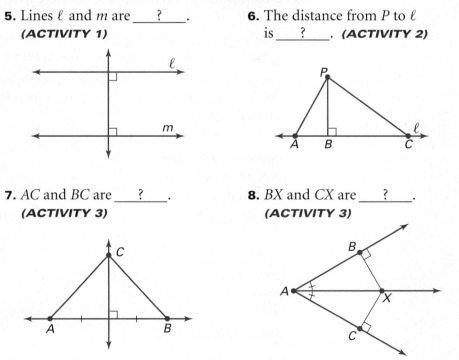

Use a separate piece of folding paper for each exercise. Attach the folded papers to your homework, and be sure to label each piece.

Fold each figure below. Do not use a ruler or protractor. Trace over each figure and label all relevant parts.

9. a 90° angle

10. a 45° angle

11. an acute angle

12. an obtuse angle

13. two complementary angles

14. two supplementary angles

15. two parallel lines

16. a rectangle

17. a square

18. a triangle

19. a right triangle (a triangle with one angle of 90°)

20. a triangle with two equal sides

21. a right triangle with two equal sides

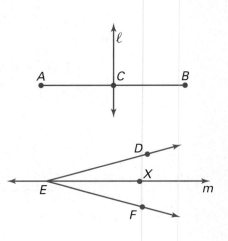

Fold a line and label two points *A* and *B* on the line. Construct the perpendicular bisector of $\overline{AB}$ and label it ℓ. Label a point *C* on ℓ, and fold or draw $\overline{AC}$ and $\overline{BC}$.

22. Using the first conjecture you made in Activity 3, what can you conclude about *AC* and *BC*?

23. Write a conjecture about triangles in which the vertex of one angle is on the perpendicular bisector of the side opposite that angle.

24. How could you use the conjecture from Exercise 23 to construct a triangle with three congruent sides? (Hint: Start by tracing $\overline{AB}$ onto another piece of folding paper.)

25. Suppose that ℓ is the perpendicular bisector of $\overline{AB}$ and that $AB = 10$.

Find *AC* and *BC*.

26. Suppose that *m* is the angle bisector of $\angle DEF$ and that $m\angle DEX = 15°$.

Find m$\angle FEX$ and m$\angle DEF$.

Fold two intersecting lines, and label them as shown in the figure.

27. Fold the bisector of ∠XVY and label it ℓ. What is the relationship between ℓ and ∠WVZ?

28. Fold the bisector of ∠WVX and label it *m*. What is the relationship between ℓ and *m*?

29. Repeat Exercises 27 and 28 with several different pairs of intersecting lines. Make a conjecture about the bisectors of angles formed by intersecting lines.

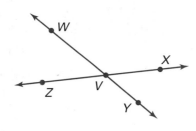

For Exercises 30–32, fold or draw two perpendicular lines on a piece of folding paper and label them ℓ and *m*. Label the intersection of ℓ and *m* point *A*.

30. Fold the angle bisector of one of the angles formed by ℓ and *m*. Choose a point on the angle bisector and label it *B*. Fold two lines through *B*, one perpendicular to ℓ and one perpendicular to *m*.

31. Make a conjecture about the shape you formed in Exercise 30. Measure any sides or angles necessary to confirm your conjecture.

32. $\overline{AB}$ is called a *diagonal* of this shape. Fold the other diagonal of the shape. Make a conjecture about the diagonals of the shape you constructed.

Look Back

Name the geometric figure that each item suggests. *(LESSON 1.1)*

33. the edge of a table

34. the wall of a classroom

35. the place where two walls meet

36. the place where two walls meet the ceiling

Name each figure, using more than one name when possible. *(LESSON 1.1)*

37.

38.

M ●————————● N

39.

40.

P ●————————● S →

41.

42.

A ●

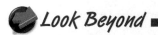
In Chapter 4, you will examine properties of triangles. In Lesson 4.7, you will use a compass and straightedge to construct a triangle with given side lengths. You can also use tracing paper to construct such a triangle, as directed below.

43. Trace each segment at left onto a separate piece of tracing paper. Place the paper with $\overline{AB}$ on top of the paper with $\overline{CD}$, and line up A and C. You may want to poke a pin through A and C to keep them aligned.

Place the paper with $\overline{EF}$ on top of the other two, line up E and B, and then line up F with D.

Trace the complete triangle on the top piece of paper.

44. Can you create a different triangle from these segments? Why or why not?

PORTFOLIO ACTIVITY

A *regular hexagon* is a six-sided figure with equal sides and equal angles. You can create one by following the steps below.

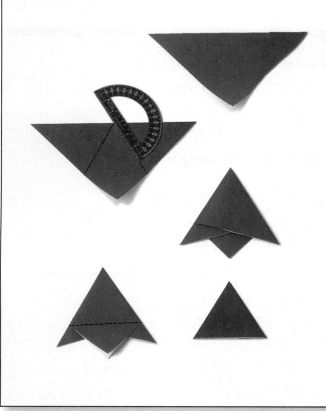

1. Fold a square piece of paper along one diagonal.

2. Place your protractor on the center point of the fold and mark off dashed lines at 60° angles.

3. Fold the right and left portions along the dashed lines.

4. Turn the figure over and mark a new dashed line. Cut along this line.

5. Open the figure. The finished product is a hexagon.

You can also create a snowflake pattern from these folds. Do not open your figure in Step 5. Instead, fold the figure vertically through the center another time. Cut notches into each side of the folded figure. When you unfold the figure, you will have a snowflake.

Special Points in Triangles

Objectives

- Discover points of concurrency in triangles.
- Draw the inscribed and circumscribed circles of triangles.

Why *Special points in triangles have many applications, from art to telecommunications.*

Archaeologists at this site in Belize have uncovered part of a circular structure. In this lesson, you will learn a method for finding the center of a circle when only a part is given.

Triangles and Circles

In this lesson, you will discover some interesting facts about triangles and circles. When you study proofs, you will understand why they are true.

Activity 1

Some Special Points in Triangles

YOU WILL NEED

folding paper
OR
geometry software

In this Activity, you will discover some special points in triangles. Save your triangles for the next Activity. If you are using geometry software, you can "drag" the vertices of the triangles to explore different triangles.

1. Draw or fold a triangle. Then construct the perpendicular bisector of each side of the triangle. What do you notice about the perpendicular bisectors?

2. Draw or fold another triangle. Then construct the angle bisector of each angle of the triangle. What do you notice about the angle bisectors?

3. Share your results with your class. Complete the conjectures below.

CHECKPOINT ✔

The perpendicular bisectors of a triangle ____?____ at a single ____?____ .

The angle bisectors of a triangle ____?____ at a single ____?____ .

Special Circles Related to Triangles

For any triangle, you can draw an *inscribed circle* and a *circumscribed circle*. An **inscribed circle**, as the name suggests, is *inside* the triangle and just touches its three sides. A **circumscribed circle** is outside the triangle (*circum-* means "around") and contains all three vertices.

Inscribed circle

Circumscribed circle

Activity 2
Constructing the Special Circles

YOU WILL NEED

folding paper and compass
OR
geometry software

1. Use the triangles you created in Activity 1. Draw circles with centers at the special points you discovered in Activity 1. Can you draw the special circles defined above?

2. Complete the conjectures below.

Inscribed and Circumscribed Circles

CHECKPOINT ✔

The intersection point of the angle bisectors of the angles of a triangle is the center of the _____?_____ circle of the triangle.

The intersection point of the perpendicular bisectors of the sides of a triangle is the center of the _____?_____ circle of the triangle.

EXAMPLE

APPLICATION
ARCHAEOLOGY

An archaeologist wants to find the original diameter of a broken plate. How can she do this by applying one of the conjectures from Activity 2?

SOLUTION

The archaeologist draws an outline of the broken plate. Then she selects three points on the circumference of the plate and connects them to form a triangle. The intersection of the perpendicular bisectors of the sides of the triangle is the center of the plate. (Only two bisectors are needed, however, the third acts as a check.)

The radius of the circle is the distance from the circle's center, *O*, to the edge of the plate. By measuring this distance and doubling it, the archaeologist can find the diameter of the plate.

The Metropolitan Museum of Art, Rogers Fund, 1.

Exercises

Using geometry graphics software, you can "drag" the points in a figure to examine different cases.

Communicate

1. The lines you constructed in Activity 1 are said to be **concurrent**. The word comes from the Latin words *con* ("together") and *currere* ("to run"). Why is this word appropriate for these lines?

2. The center of an inscribed circle is called the **incenter** of a triangle. Can the incenter be outside the triangle? Explain why or why not.

3. The center of a circumscribed circle is called the **circumcenter** of a triangle. Can the circumcenter be outside the triangle? Explain why or why not.

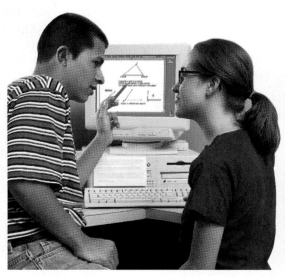

Geometry graphics software has many useful features for creating constructions.

Guided Skills Practice

4. What is the result when you fold or draw the perpendicular bisectors of the three sides of a triangle? *(ACTIVITY 1)*

5. What is the result when you fold or draw the angle bisectors of the three angles of a triangle? *(ACTIVITY 1)*

For Exercises 6 and 7, use your conjectures about inscribed and circumscribed circles to complete the table. *(ACTIVITY 2)*

	Intersecting lines	Type of circle formed	Name of center
6.	perpendicular bisectors	?	?
7.	angle bisectors	?	?

8. Trace the portion of a circle shown at right. Choose three points on the circle and draw a triangle to connect them. Then construct the circumscribed circle around the triangle to complete the figure.

(EXAMPLE)

Draw each of the triangles below onto a separate piece of folding paper, or draw them by using geometry software. Triangles are named by the symbol △ and the names of the vertices, such as △**ABC**.

Find the following:

9. the perpendicular bisectors of each side of △ABC

10. the angle bisectors of each angle of △DEF

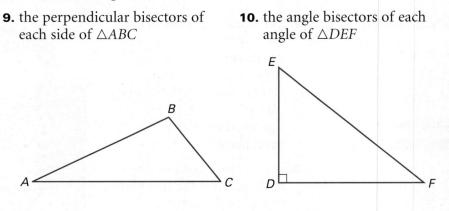

11. the circumscribed circle of △GHI

12. the circumscribed circle of △JKL

13. the inscribed circle of △MNO

14. the inscribed circle of △PQR

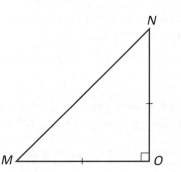

15. the perpendicular bisectors of △STU

16. the angle bisectors of △STU

17. the circumscribed circle of △STU

18. the inscribed circle of △STU

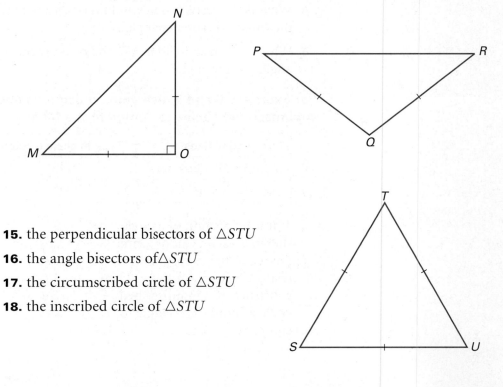

In Exercises 19–23, draw or fold three triangles:
  - **an acute triangle—a triangle with all angles less than 90°**
  - **an obtuse triangle—a triangle with one angle greater than 90°**
  - **a right triangle—a triangle with one angle equal to 90°**

19. Construct the circumcenter of the acute triangle. Is it inside or outside the triangle?

20. Construct the circumcenter of the obtuse triangle. Is it inside or outside the triangle?

21. Where do you think the circumcenter of a right triangle should be? Construct the circumcenter of the right triangle to test your conjecture.

22. Use the circumcenter of the right triangle to draw the circumscribed circle. How does the longest side of the triangle divide the circle?

23. Based on your answer to Exercise 22, make a conjecture about the circumscribed circle of a right triangle. Draw several different right triangles and test your conjecture.

Draw or fold two acute triangles and construct the midpoint of each side. Use one in Exercise 24 and the other in Exercise 25.

24. Connect the midpoints to form another triangle. Your drawing should now contain four small triangles inside the original one. Cut out the four small triangles and compare them. What do you observe?

25. Draw a segment from each midpoint to the opposite vertex. These segments, called the *medians*, are shown in the figure at right. Label point *C* where the medians intersect. Measure the following segments in your triangle and complete the table:

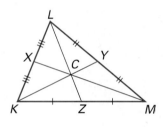

CK = ?	CL = ?	CM = ?
CY = ?	CZ = ?	CX = ?

What is the relationship between the lengths in the first row of the table and the lengths in the second row? Make a conjecture about point *C*, which is known as the *centroid*.

CHALLENGE

26. In the triangle at right, ℓ is the perpendicular bisector of $\overline{AB}$. Suppose that *X* is a point on ℓ. What can you say about *AX* and *BX*?

What can you say about any point on the perpendicular bisector of $\overline{AC}$? of $\overline{BC}$? Explain why the intersection of the perpendicular bisectors is the center of the circumscribed circle.

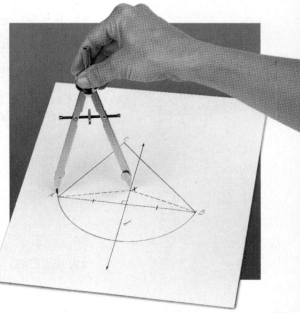

27. In the triangle at right, *m* is the angle bisector of ∠*E*. Suppose that *Y* is a point on *m*. What can you say about the distance from *Y* to sides $\overline{DE}$ and $\overline{EF}$?

What can you say about any point on the angle bisector of ∠*D*? of ∠*F*? Explain why the intersection of the angle bisectors is the center of the inscribed circle.

28. CONSTRUCTION
A contractor is installing a round air-conditioning duct. The duct must pass through a triangular opening, as shown in the figure at right. Trace the figure onto folding paper and construct the largest circle that will fit inside the triangle.

92 in.

68 in.

38 in.

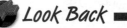 **Look Back**

Complete the statements below. (LESSON 1.1)

29. Points that lie on the same line are said to be ____?____.

30. Points or lines that lie in the same plane are said to be ____?____.

In Exercises 31–33, you may wish to include a diagram to illustrate your answer. (LESSON 1.1)

31. Is it possible for two points to be noncollinear? Why or why not?

32. Is it possible for three points to be noncoplanar? Why or why not?

33. Is it possible for two lines to be noncoplanar? Why or why not?

 Algebra

In Exercises 34–37, refer to the figures below, in which WY = 48 and ∠BAC ≅ ∠CAD. Find the measures listed below. (LESSONS 1.4 AND 1.5)

34. *XY*

35. *WX*

4*x* − 3   3*x* + 2

W   X   Y

36. m∠*BAC*

37. m∠*CAD*

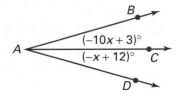

B

$(-10x + 3)°$

A

$(-x + 12)°$ C

D

Look Beyond

Another special point in a triangle is formed by the altitudes of the sides. An *altitude* is a perpendicular line segment from a vertex of a triangle to the line containing the opposite side.

altitude

38. Draw or fold an acute triangle and label the vertices *A*, *B*, and *C*. How would you find the distance from *A* to $\overline{BC}$?

39. Construct the altitude for each vertex of △*ABC*. What do you notice?

40. Draw or fold an obtuse triangle. Construct the altitudes of this triangle. (Hint: Extend the sides of the triangle.) What do you notice? How is this different from the construction you did in Exercise 39?

PORTFOLIO ACTIVITY

A *median* of a triangle is a line segment from a vertex to the midpoint of the opposite side (see Exercise 25). In addition to the property of medians that you already examined, there is another interesting feature of medians.

median

1. Cut out a triangle from stiff cardboard. Draw one median of the triangle. Try to balance the triangle on the line you drew. Make a conjecture about the part of the triangle on each side of the median.

2. Draw the other two medians of your triangle. Try to balance the triangle on the point where the medians intersect, known as the *centroid*. Explain in your own words why the centroid is called the *center of mass*.

3. According to a law of physics, a free-falling body should rotate around its center of mass. Test this theory by tossing your triangle like a Frisbee. What do you observe?

    You will confirm your conjecture from Step 1 when you examine areas of triangles in Chapter 5.

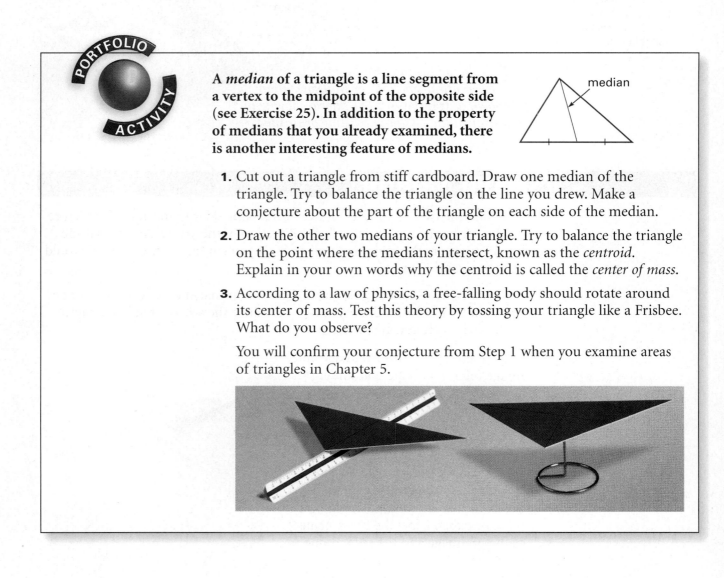

Motion in Geometry

Objective

- Identify and draw the three basic rigid transformations: translation, rotation, and reflection.

This photo illustrates the basic mathematical "motion" or transformation known as translation. *The skier is moving in a straight line, without twisting or turning.*

Rigid Motion

Compare the two "snapshots" above of the same skier in motion. The picture on the left has been *translated* in the direction of the arrow. Notice that the two pictures are the same size and shape. If you cut one of them out, it would fit onto the other exactly.

As with segments and angles, planar figures that match exactly are said to be congruent. Transformations that do not change the size or shape of a figure are known as **rigid transformations**.

In geometry, the terms **preimage** and **image** refer to a shape that undergoes a motion or transformation. Points on the image of an object are usually named by adding a prime symbol (′) to the original name of the point.

You can easily draw a rigid transformation by tracing the same object in two different positions on a piece of paper. Label the first tracing as the preimage and the second one as the image. The only real trick in drawing the basic transformations is to locate the image properly. You will learn how to do this in the Activities that follow.

Translations

In a **translation**, every point of a figure moves in straight line, and all points move the same distance and in the same direction. The paths of the points are parallel.

Activity 1

Drawing a Translation

1. Trace both the object and the line at right onto a piece of tracing paper. Do not move the paper yet.

2. Slide your paper so that your drawing of the line stays on the line in the diagram. Trace the object a second time.

CHECKPOINT ✔ 3. Label your first and second drawings of the object as *preimage* and *image*, respectively.

Rotations

In a **rotation**, every point of a figure moves around a given point known as the **center of rotation**. All points move the same angle measure.

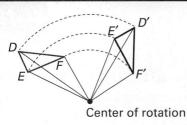

Center of rotation

Activity 2

Drawing a Rotation

1. Trace both the object and the point at right onto a piece of tracing paper. Do not move the paper yet.

2. Place the point of your pencil on the point, and press down to hold the tracing paper in place at that point. Turn the paper around the point as far as you wish. Trace the object a second time.

CHECKPOINT ✔ 3. Label your first and second drawings of the curved object as *preimage* and *image*, respectively.

Reflections

Hold a pencil in front of a mirror and focus on the pencil point, P. The reflection of that point, called the reflection image of the point, P', will appear to be on the opposite side side of the mirror at the same distance from the mirror as P.

In a mathematical **reflection**, a line plays the role of the mirror, and every point in a geometric figure is "flipped" across the line. In Part I of Activity 3, paper folding is used to produce the appropriate motion of a reflected point.

Image

"Mirror" line

Preimage

Drawing a Reflection

Part I   The Reflection of a Point

1. Draw line ℓ and a point, P, not on line ℓ.

$\bullet P$

ℓ

2. Reflect point P across line ℓ as follows: Fold your paper along line ℓ. Use a pointed object such as a sharp pencil or a compass point to punch a hole through the paper at point P. Unfold the paper and label the new point P'.

3. Draw $\overline{PP'}$ and label the intersection of $\overline{PP'}$ and line ℓ as point X.
  a. Measure the length of $\overline{XP}$ and $\overline{XP'}$.
  b. Measure the angles formed by $\overline{PP'}$ and line ℓ.

CHECKPOINT ✔   **4.** What do you notice about the relationship between $\overline{PP'}$ and line ℓ? Complete the statement below, and then use your results from the Activities in Lesson 1.5 to explain why it is true.

The Reflection of a Point Across a Line

If a point is reflected across a line, then the line is the ___?___ of the segment that connects the point with its image.

1. Draw line ℓ with $\triangle ABC$ on one side.

2. Fold your paper along line ℓ and punch holes through points A, B, and C to obtain image points A', B', and C'. Connect the image points to form $\triangle A'B'C'$, the reflection of $\triangle ABC$.

3. Cut out $\triangle ABC$. Place it face down on $\triangle A'B'C'$. Do the two triangles match exactly?

4. Form a conjecture about a triangle and its reflected image. Complete the statement below.

CHECKPOINT ✔

The Reflection of a Triangle

If a triangle is reflected across a line, then the reflected image of the triangle is ____?____ to the original triangle.

TRY THIS   Reflect each figure below across line ℓ. Do your conjectures about the reflection of triangles and segments seem to be true when the figures touch the reflection line?

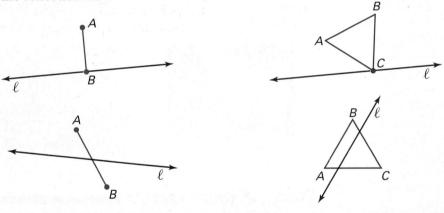

Summary: Three Rigid Transformations

The facts that you discovered by doing hands-on translations, rotations, and reflections in the Activities can be developed into the formal mathematical definitions given below.

A **translation** is a transformation in which every point of the preimage is moved the same distance in the same direction.   **1.6.1**

A **rotation** is a transformation in which every point of the preimage is moved by the same angle through a circle centered at a given fixed point known as the *center of rotation*.   **1.6.2**

A **reflection** is a transformation in which every point of the preimage is moved across a line known as the *mirror line* so that the mirror is the perpendicular bisector of the segment connecting the point and its image.   **1.6.3**

Exercises

● *Communicate*

In Exercises 1–6, determine whether each description represents a translation, a rotation, a reflection, or none of these. Explain why. What is the direction of motion, center of rotation, or line of reflection in each case?

1. a canoe drifting straight ahead
2. a ball rolling down a hill
3. the image of a building in a lake
4. hands moving on a clock
5. a pair of scissors opening and closing
6. a slide projected onto a screen

7. Of the two figures below, which is the best illustration of a reflection across a line? Explain your reasoning.

● *Guided Skills Practice*

In Exercises 8–10, trace the figures onto folding paper.

8. Translate the figure along the given line. *(ACTIVITY 1)*

9. Rotate the figure about the given point. *(ACTIVITY 2)*

10. Reflect the figure across the given line. *(ACTIVITY 3)*

Practice and Apply

In Exercises 11–19, trace the figures onto folding paper.

Translate each figure along the given line.

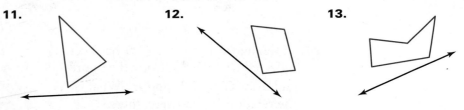

11.   **12.**   **13.**

Rotate each figure about the given point.

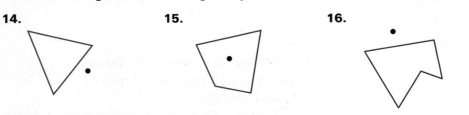

14.   **15.**   **16.**

Reflect each figure across the given line.

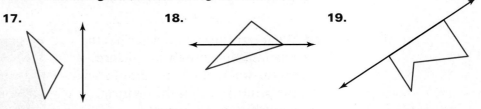

17.   **18.**   **19.**

Copy the figures below onto tracing paper. Fold the paper along the line and trace the reflection of the figure on the back of the paper.

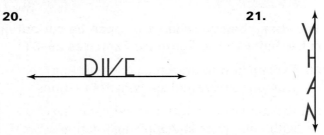

20.   **21.**

22. The word MOM has some unique characteristics when reflected across different lines. Reflect the word across each line.

a.   **b.**   **c.**

23. Write your name in capital letters on a piece of folding paper. Reflect your name across a vertical line. Do any of the letters stay the same? Reflect your name across a horizontal line. Do any of the letters stay the same?

24. In the diagram at right, △*ABC* has been translated to △*A'B'C'*. Copy the figures and draw $\overline{AA'}$, $\overline{BB'}$, and $\overline{CC'}$. What two things can you say about the relationship among these segments?

25. In the figure at right, △DEF has been rotated to △D′E′F′. Copy the figure and draw ∠DCD′ and ∠ECE′. Measure these angles. What do you think is the measure of ∠FCF′?

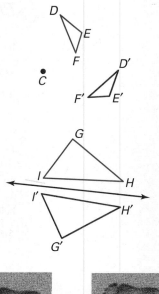

26. In the figure at right, △GHI has been reflected to △G′H′I′. Copy the figure and draw $\overline{GG'}$, $\overline{HH'}$, and $\overline{II'}$. What is the relationship between these segments and the line of reflection?

27. While visiting the beach, Teresa saw the two pairs of footprints shown at right.
   a. Which pair represents a reflection?
   b. Which pair represents a translation?

28. What transformation describes the relationship between the picture on the wood block and the one on the print? Which is the preimage and which is the image?

Wood block   *Print*

The shape below, called a *net*, can be cut out and folded to create a cube. Refer to the figure for Exercises 29–31.

29. Explain how to draw the net by starting with one square and applying translations.

30. Explain how to draw the net by starting with one square and applying rotations.

31. Explain how to draw the net by starting with one square and applying reflections.

A fourth type of transformation is a *glide reflection*. A glide reflection is a combination of a translation and a reflection.

32. The footprints shown below represent a glide reflection. Copy the figure, draw an arrow in the direction of the translation, and draw the line of reflection. What do you notice?

33. The figure below shows a shape that has been transformed by two glide reflections. How could you continue applying glide reflections to form a pattern? Copy the figure and draw two more images of the shape.

Preimage   Second image

First image

![Look Back icon] **Look Back**

For each triangle in Exercises 34–36:
    a. Name all sides and angles of the triangle. (LESSON 1.1)
    b. Measure the lengths of the sides in centimeters. (LESSON 1.2)
    c. Measure the angles of the triangle. (LESSON 1.3)

34. *A*, *B*, *C*   **35.** *D*, *E*, *F*   **36.** *H*, *G*, *I*

37. In the triangle at right, two angle bisectors have been constructed. Trace the figure and draw the third angle bisector. *(LESSON 1.5)*

38. In the triangle at right, two perpendicular bisectors have been constructed. Trace the figure and draw the third perpendicular bisector. *(LESSON 1.5)*

![Look Beyond icon] **Look Beyond**

Sometimes you may want to translate a figure a given distance or rotate a figure by a given degree measure. This can be done with tracing paper, a ruler, and a protractor.

39. Trace the figure and the line below onto a piece of tracing paper. Then translate the figure so that point *T* on your paper matches up with point *S* in the diagram. What distance has the figure been translated?

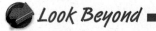

40. Use the method from Exercise 39 and a ruler to translate the figure at right 5 centimeters along the given line.

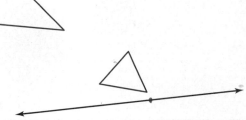

41. Trace the figure and the angle at right onto a piece of tracing paper. Then rotate the figure so that $\overrightarrow{VW}$ on your paper matches up with $\overrightarrow{VX}$ in the diagram. By what angle has the figure been rotated?

42. Use the method from Exercise 41 and a protractor to rotate the triangle at right by 45° about point X.

Snowflakes You may wish to use geometry software to complete this activity.

1. Draw $\overline{AB}$.

2. Draw several more segments attached to $\overline{AB}$, all on the same side of the segment, as shown.

3. Reflect the segments across $\overline{AB}$.

4. Rotate the entire figure 60° about A. Then rotate the image 60° about A. Repeat until you have 6 images, including the preimage.

5. Create other snowflakes with this method. If you are using geometry graphics software, try dragging the points of the snowflake to see how your design changes.

Motion in the Coordinate Plane

The downtown area of Provo, Utah, is laid out as a coordinate grid. Each block is one hundred units on a side.

Objectives

- Review the algebraic concepts of *coordinate plane, origin, x-* and *y-coordinates,* and *ordered pair.*

- Construct translations, reflections across axes, and rotations about the origin on a coordinate plane.

Why A coordinate plane enables you to describe geometry ideas by using algebra. For example, you can use algebra to find the transformation of a figure without using special construction tools.

Operations on the Coordinates of a Point

By applying algebraic operations to the coordinates of a point, you can relocate it on the coordinate plane. For example, if you add 2 to the *x*-coordinate and 4 to the *y*-coordinate of the point $A(2, 3)$, the result is a new point, $A'(4, 7)$.

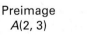

Preimage
$A(2, 3)$

Transformation

Image
$A'(4, 7)$

This operation, which is known as a transformation, can be expressed as a rule by using transformation notation.

$$T(x, y) = (x + 2, y + 4)$$

The letter *T* stands for the transformation, but any other letter could be used. As you may notice, the transformation in the example above is a *function*.

The Three Rigid Transformations

In the Activities that follow, you will discover rules for the three basic rigid transformations of geometry in a coordinate plane.

Activity 1

Translations

YOU WILL NEED

graph paper, a ruler, and a protractor

1. Copy $\triangle ABC$ onto graph paper.

2. Pick a number between −10 and +10 (except 0). Then choose either x or y.

3. Depending on your choice, add the number you picked to either the x- or the y-coordinate of points A, B, and C. The other coordinate remains unchanged in each point.

Plot each of your new points, and then connect them to form a new triangle. Does your new triangle seem to be congruent to the original one? Measure the sides and the angles of each triangle.

PROBLEM SOLVING

4. **Look for a pattern.** The original figure has been translated to a new position. In what direction has it moved and by how much? Draw an arrow to show the translation. Compare your results with those of your classmates.

CHECKPOINT ✔

5. Explain how to translate a figure horizontally or vertically in the coordinate plane. What would you do to the coordinates of each point in the figure? Express your method as a set of rules for moving points in the original figure h units horizontally or v units vertically.

Horizontal and Vertical Coordinate Translations

Horizontal translation of h units: $H(x, y) = (\underline{?}, \underline{?})$
Vertical translation of v units: $V(x, y) = (\underline{?}, \underline{?})$ **1.7.1**

APPLICATION

NAVIGATION

Refer to the map at right. Suppose that you are driving from point A at the intersection of 500 W and 200 S to point B at the intersection of 300 E and 400 N. The rule for the horizontal movement is $H(x, y) = (x + 800, y)$. The rule for the vertical movement is $V(x, y) = (x, y + 600)$. These rules can be combined to show both the horizontal and vertical movements: $T(x, y) = (x + 800, y + 600)$.

Activity 2

Reflections Across the *x*- or *y*-axis

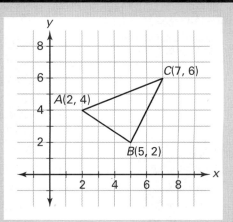

1. Copy △*ABC* onto graph paper.

2. Multiply the *x*-coordinate of each vertex by −1. Leave the *y*-coordinate unchanged.

3. Plot each of the new points, and connect them to form the new triangle. Does the new triangle seem to be congruent to the original one? Measure the sides and the angles of each triangle.

4. Repeat Steps 2 and 3 for the *y*-coordinates, leaving the *x*-coordinates unchanged.

5. The original triangle is reflected in each case. What are the lines of reflection, or the "mirrors" of the reflections?

CHECKPOINT ✔ 6. Explain how to reflect a figure across the *x*- or *y*-axis. What would you do to each point in the figure? Express your method as a set of rules for reflecting points in the original figure across the *x*- or *y*-axis.

Reflection Across the *x*- or *y*-axis

Reflection across the *x*-axis: $M(x, y) = (\underline{?}, \underline{?})$
Reflection across the *y*-axis: $N(x, y) = (\underline{?}, \underline{?})$

1.7.2

TRY THIS   Experiment with figures in other positions, such as those below. Do your rules seem to work for all positions?

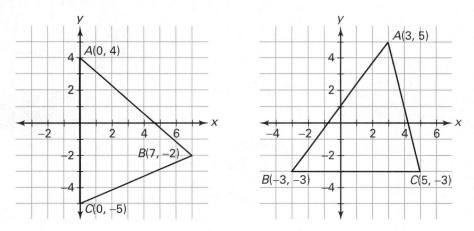

CRITICAL THINKING   How would you write a rule for translating a point and then reflecting it across an axis?

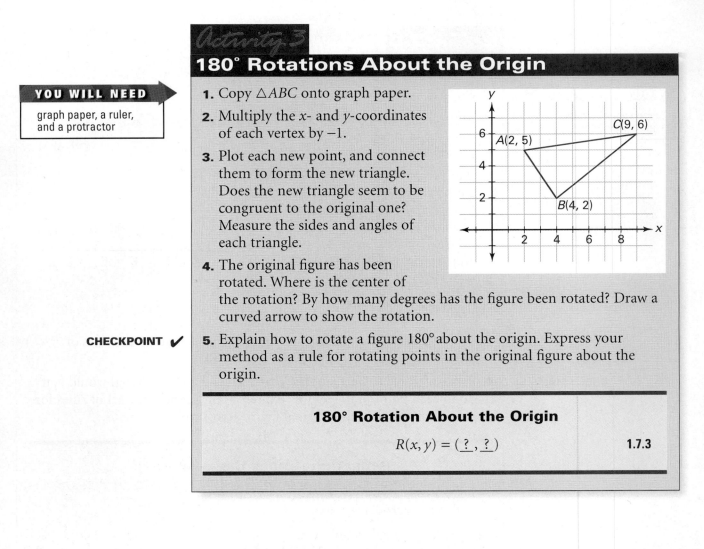

Activity 3

180° Rotations About the Origin

1. Copy △ABC onto graph paper.

2. Multiply the x- and y-coordinates of each vertex by −1.

3. Plot each new point, and connect them to form the new triangle. Does the new triangle seem to be congruent to the original one? Measure the sides and angles of each triangle.

4. The original figure has been rotated. Where is the center of the rotation? By how many degrees has the figure been rotated? Draw a curved arrow to show the rotation.

CHECKPOINT ✔

5. Explain how to rotate a figure 180° about the origin. Express your method as a rule for rotating points in the original figure about the origin.

180° Rotation About the Origin

$$R(x, y) = (\underline{?}, \underline{?})$$

1.7.3

TRY THIS Use your rule from Activity 3 to rotate the triangle with vertices at (1, 5), (−2, 4), and (−3, −2) by 180° about the origin.

CRITICAL THINKING You can think of the figure at right as being formed by two reflections of the upper left panel. First the original panel is reflected across the y-axis. Then the resulting figure is reflected across the x-axis.

Explain why a 180° rotation can be thought of as a combination of these two reflections. Do you think this idea could be extended to describe any rotation as a combination of two reflections?

Exercises

Communicate

1. Explain how you would find the *x*- and *y*-coordinates of points *A* and *B* on the coordinate plane at right.

2. Plot the points (1, 5) and (5, 1). Do you think that the order of the coordinates matters? Why do you think the coordinates of a point are called *ordered pairs*?

3. What are some advantages to using the coordinate plane for transformations?

4. What happens when you reflect a figure across the *x*-axis and then reflect the image across the *y*-axis? Explain the result in terms of a single transformation in a coordinate plane.

Guided Skills Practice

5. Plot the points (2, 4), (−1, 4), and (2, 1), and connect them to form a triangle. What rule would you use to translate the triangle 5 units to the right? Write your rule in the form $H(x, y) = (\underline{?}, \underline{?})$. Draw the translated figure and label its coordinates. **(ACTIVITY 1)**

6. Plot the points (1, 3), (−2, 5), and (0, 0), and connect them to form a triangle. What rule would you use to reflect the figure across the *y*-axis? Write your rule in the form $N(x, y) = (\underline{?}, \underline{?})$. Draw the reflected figure and label its coordinates. **(ACTIVITY 2)**

7. Plot the points (2, 2), (4, 4), and (2, 4), and connect them to form a triangle. What rule would you use to rotate the figure 180° about the origin? Write your rule in the form $R(x, y) = (\underline{?}, \underline{?})$. Draw the rotated figure and label the coordinates. **(ACTIVITY 3)**

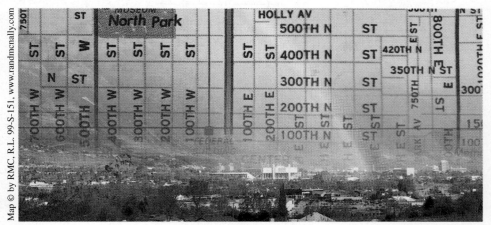

Rainbow over Provo, Utah.

In Exercises 8–17, copy each figure onto graph paper, and then use the given rule to transform the figure. Identify the type of transformation given by each rule.

8. $H(x, y) = (x + 3, y)$

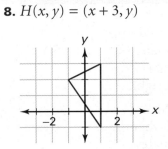

9. $H(x, y) = (x - 2, y)$

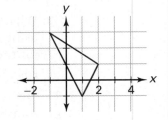

10. $V(x, y) = (x, y + 4)$

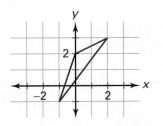

11. $V(x, y) = (x, y - 1)$

12. $M(x, y) = (x, -y)$

13. $M(x, y) = (x, -y)$

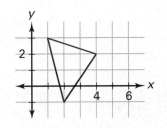

14. $N(x, y) = (-x, y)$

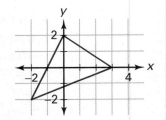

15. $N(x, y) = (-x, y)$

16. $R(x, y) = (-x, -y)$

17. $R(x, y) = (-x, -y)$

In Exercises 18–22, use graph paper to draw the transformation(s) of the figure as indicated. Write the rule(s) you used below each transformed figure.

18. Reflect the figure across the *x*-axis.

19. Reflect the figure across the *y*-axis.

20. Rotate the figure 180° about the origin.

21. Reflect the figure across the *x*-axis, and then reflect the image across the *y*-axis. How does this figure relate to the transformed figure from Exercise 20?

22. Reflect the figure across the *y*-axis, and then reflect the image across the *x*-axis. How does this figure relate to the transformed figures from Exercises 20 and 21?

In Exercises 23–30, describe the result of applying each rule to a figure in a coordinate plane.

23. $F(x, y) = (x + 7, y)$

24. $T(x, y) = (-x, y)$

25. $A(x, y) = (x - 6, y + 7)$

26. $P(x, y) = (x, y - 4)$

27. $C(x, y) = (x, y + 7)$

28. $W(x, y) = (-x, -y)$

29. $Z(x, y) = (x - 7, y)$

30. $K(x, y) = (x, y + 2)$

31. Copy the figure below onto graph paper. Apply Rule 1 to the figure, then Rule 2 to its image, and so on. Draw each image that results.

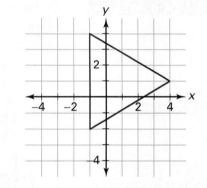

Rule 1: $A(x, y) = (x, -y)$

Rule 2: $B(x, y) = (-x, y)$

Rule 3: $C(x, y) = (x, -y)$

Rule 4: $D(x, y) = (-x, y)$

What is the final result of applying Rules 1–4?

In Exercises 32–35, you will use the idea of the *slope* of a line. Recall from algebra that the slope of a line is the change in the *y*-coordinates divided by the change in the *x*-coordinates for any two points on the line. Graph the triangle with vertices *K*(0, 0), *L*(1, 2), and *M*(0, 3).

32. Apply the transformation $T(x, y) = (x + 4, y + 3)$ to the triangle. Label the new vertices K', L', and M'.

33. Draw a line through K and K'. What is its slope? Draw a line through L and L'. What is its slope? Draw a line through M and M'. What is its slope?

34. How do the slopes of the lines relate to the transformation rule $T(x, y)$?

35. Based on your results from Exercises 32–34, make a conjecture about the slope of the line through K and K' when the transformation $T(x, y) = (x + h, y + k)$ is applied to point K.

In Exercises 36–43, you will discover another rule for transforming an object.

36. On graph paper, graph the line $y = x$. Complete the table of values at right to get started.

x	y
0	?
2	?
−1	?

37. What angle does the line form with the x-axis?

38. Plot each of the following points on the graph, and connect them to form a triangle:

$A(2, 1)$ $\qquad$ $B(5, 2)$ $\qquad$ $C(6, 4)$

39. Reverse the x- and y-coordinates of points A, B, and C to create new points A', B', and C'. Plot these points on the graph with A, B, and C, and connect them to form a new triangle.

40. What is the relationship between the figures you drew?

41. Write a rule in the form $R(x, y) = (\underline{\ ?\ }, \underline{\ ?\ })$ for the transformation. (Hint: The rule should use only the letters x and y, with no signs or numbers.)

42. Draw segments connecting A to A', B to B', and C to C'. What is the relationship between these segments and the line $y = x$?

43. Choose a point (x, y), where $x < y$. Is this point above or below the line $y = x$? Apply the transformation rule you wrote in Exercise 41 to this point. Is the image of the point above or below the line $y = x$? Repeat this exercise for points where $x > y$ and $x = y$.

CHALLENGES

In Exercises 44–47, you will explore glide reflections in a coordinate plane.

44. What two transformations would you apply to $\triangle ABC$ to get $\triangle DEF$? Express these transformations with a single rule of the form $T(x) = (\underline{\ ?\ }, \underline{\ ?\ })$.

45. Apply the transformation rule you found in Exercise 44 to $\triangle DEF$. What is the result? Apply the transformation to the resulting image three or four more times. Describe what happens.

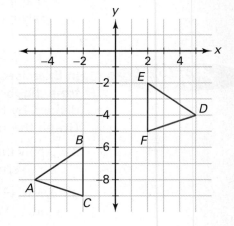

46. Write a general rule for a glide reflection with the y-axis as the line of reflection and with a vertical motion of v.

47. Write a general rule for a glide reflection with the x-axis as the line of reflection and a horizontal motion of h.

COMPUTER PROGRAMING The PostScript computer language is designed to communicate between computers and printers. A page in the PostScript language is laid out as a coordinate plane with the origin at the lower left-hand corner of the page. The following PostScript code draws a segment from a point located at (144, 72) to (144, 432).

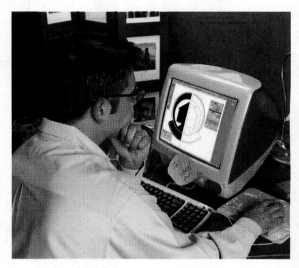

```
newpath
    144 72 moveto
    144 432 lineto
stroke
showpage
```

48. Each unit in the PostScript coordinate system is $\frac{1}{72}$ in. long. What are the coordinates, in inches, of the endpoints of the segment drawn by the code above? Sketch the segment on a graph with axes labeled in inches.

49. Rewrite the code above to draw a segment between points whose coordinates, in inches, are (1, 1) and (3, 4.5).

Look Back

Give the compass heading for each direction below. *(LESSON 1.3)*

50. W

51. NW

52. SW

53. SSW

Tell how you would find each point of a triangle. *(LESSON 1.5)*

54. circumcenter

55. incenter

Describe or draw an example of each type of translation below. If you make a drawing, label the preimage and image. *(LESSON 1.6)*

56. translation

57. rotation

58. reflection

59. glide reflection

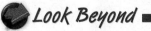

Look Beyond

60. Plot the points (1, 2), (4, 2), and (1, 8) in a coordinate plane, and connect them to form a triangle. Apply the transformation $Q(x, y) = (-y, x)$ to the triangle. What type of transformation is this? Try applying the same transformation to several different triangles to confirm your answer.

61. Based on your results from Exercise 60, write a transformation, $R(x, y)$, for the rotation of a figure 90° about the origin in the clockwise direction.

PROJECT

Origami
Paper Folding

Origami, the ancient Japanese art of paper folding, produces intriguing figures from simple paper folds. In this project, you will use paper folding to create a paper crane. First follow Steps 1–6 to make the crane base. Then continue with the final folds to finish your crane.

Use a 6- to 8-inch square of paper.

1. Fold the paper in half as shown.

2. Fold point *A* forward and down to point *B*. Fold point *C* backward and down to point *B*.

3. Open the paper at point *B* and lay it flat.

4. Crease the paper by folding points *D* and *B* inward to point *F* on the center line.

5. Open the figure and pull *G* upward, reversing the folds you made in Step 4.

6. Turn the paper over and repeat Steps 4 and 5.

You have now completed the crane base. Continue with the steps below to complete the crane.

K ●——— L (underneath)

7. From the crane base, fold points *H* and *I* upward into point *J* on the center line. Turn the paper over and repeat.

8. Pull points *K* and *L* outward and reverse the folds. These will become the tail and neck of the crane.

9. Fold the neck to form the head, and bend the wings down as shown. You can blow into the base of the figure to inflate its body.

Your crane is finished.

10. Unfolding a paper crane will reveal a pattern of creases. Trace over and label the geometric figures formed by the creases. Are any of the figures congruent with one another?

11. The crane base can be used to create many origami figures. Can you figure out how to fold a frog or a fish?

12. Find an origami book at your library and expand your collection of origami figures.

Chapter Review and Assessment

POSTULATES

Lesson	Section	Postulate or Theorem
1.1	1.1.4 Postulate	The intersection of two lines is a point.
	1.1.5 Postulate	The intersection of two planes is a line.
	1.1.6 Postulate	Through any two points there is exactly one and only one line.
	1.1.7 Postulate	Through any three noncollinear points there is exactly one plane.
	1.1.8 Postulate	If two points are in a plane, then the line containing them is in the plane.
1.2	1.2.2 Segment Congruence Postulate	If two segments have the same length as measured by a given fair ruler, then the segments are congruent. Also, if two segments are congruent, then they have the same length as measured by a given ruler.
	1.2.3 Segment Addition Postulate	If point R is between points P and Q on a line, then $PR + RQ = PQ$.
1.3	1.3.2 Angle Addition Postulate	If point S is in the interior of $\angle PQR$, then m$\angle PQS$ + m$\angle SQR$ = m$\angle PQR$.
	1.3.3 Angle Congruence Postulate	If two angles have the measure, then they are congruent. If two angles are congruent, then they have the same measure.
	1.3.5 Linear Pair Property	If two angles form a linear pair, then they are supplementary.

Key Skills & Exercises

LESSON 1.1
Key Skills

Identify and name geometric figures.

In the figure below, A, B, and C are points, $\overleftrightarrow{AC}$ is a line, $\overline{AB}$ and $\overline{AC}$ are segments, $\overrightarrow{AB}$, $\overrightarrow{AC}$, and $\overrightarrow{CA}$ are rays, and $\angle BAC$ is an angle.

Exercises

Refer to the figure below.

1. Name all segments in the figure.

2. Name all angles in the figure.

Refer to the figure below.

3. Name all lines in the figure.

4. Name all rays in the figure.

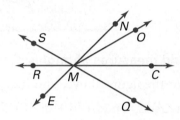

LESSON 1.2
Key Skills

Determine the length of a given segment.

In the figure below, find the lengths of $\overline{CD}$, $\overline{DE}$, and $\overline{EF}$.

$$CD = |-8 - (-2)| = 6$$
$$DE = |-2 - 3| = 5$$
$$EF = |3 - 7| = 4$$

Determine whether segments are congruent.

In the figure above, are $\overline{CD}$, $\overline{DE}$, and $\overline{EF}$ congruent segments?

$\overline{CD} \cong \overline{DE}$, but $\overline{EF}$ is not congruent to the other two segments.

Add the lengths of segments.

In the figure below, $AB = 7$ and $BX = 15$. Find AX.

$$AX = AB + BX = 7 + 15 = 22$$

Exercises

Refer to the figure below.

5. Find the length of every segment in the figure.

6. Name all congruent segments in the figure.

In Exercises 7 and 8, point A is between points R and P on a line. Sketch a figure for each exercise and find the missing measure.

7. $RA = 25$, $AP = 13$, $RP = \underline{?}$

8. $RA = \underline{?}$, $AP = 7$, $RP = 13$

Key Skills

Determine the measure of a given angle.

For the figure below, find the measure of ∠AVB and ∠BVC.

m∠AVB = |40° − 20°| = 20°
m∠BVC = |60° − 40°| = 20°

Add measures of angles.

For the figure above, find m∠AVC.

m∠AVC = m∠AVB + m∠BVC = 20° + 20° = 40°.

Determine whether angles are congruent.

In the figure above, are ∠AVB and ∠BVC congruent angles?

Yes; ∠AVB ≅ ∠BVC because they have the same measure.

Exercises

Refer to the figure below.

9. Find the measure of every angle in the figure.

10. Name all congruent angles in the figure.

In Exercises 11 and 12, find the missing measures.

11. m∠LKM = 12°
    m∠MKN = 26°
    m∠LKN = ?

12. m∠LKM = ?
    m∠MKN = 75°
    m∠LKN = 90°

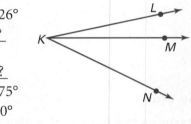

Key Skills

Use paper folding to construct geometric figures.

For the figure below, construct the perpendicular bisector of $\overline{AB}$, the angle bisector of ∠ABC, and a line parallel to $\overleftrightarrow{AB}$ that passes through C.

To construct the perpendicular bisector of $\overline{AB}$, fold the paper so that A matches up with B.

To construct the angle bisector of ∠ABC, fold the paper so that $\overrightarrow{BA}$ matches up with $\overrightarrow{BC}$.

To construct a line parallel to $\overleftrightarrow{AB}$ through C, fold the perpendicular bisector of $\overline{AB}$ onto itself so that C is on the fold.

Exercises

Trace each figure onto folding paper and construct the given figure.

13. a line parallel to $\overleftrightarrow{XY}$

14. a line through P perpendicular to ℓ

15. the perpendicular bisector of $\overline{MN}$

16. the angle bisector of ∠QRS

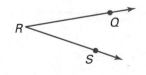

LESSON 1.5

Key Skills

Construct the circumscribed circle of a triangle.

The center of the circumscribed circle of a triangle is the intersection of the perpendicular bisectors of the triangle.

Construct the inscribed circle of a triangle.

The center of the inscribed circle of a triangle is the intersection of the angle bisectors of the triangle.

Exercises

Use geometry software or folding paper and a compass to construct the following:

17. the angle bisectors of an acute triangle

18. the perpendicular bisectors of an obtuse triangle

19. a right triangle circumscribed by a circle

20. a triangle with two congruent sides and its inscribed circle

LESSON 1.6

Key Skills

Identify translations, rotations, and reflections.

Translation

Rotation

Reflection

Translate a figure along a line.

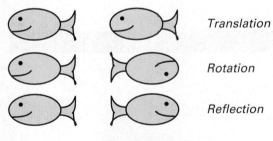

Rotate a figure about a point.

Reflect a figure across a line.

Exercises

21. Identify the following as being best represented by a translation, rotation, or reflection.
  a. geometry ʎɹʇǝɯoǝƃ
  b. a child on a slide
  c. a Ferris wheel

22. Translate the figure along the line.

23. Rotate the figure about the point.

24. Reflect the figure across the line.

Key Skills

Exercises

Use the coordinate plane to transform geometric figures.

Plot the points (2, 3), (−1, 2), and (0, 0), and connect them to form a triangle.

Give the rules for translating the figure 5 units down and for reflecting the figure across the y-axis, and draw the transformed figures.

To translate 5 units down, use the rule $V(x, y) = (x, y - 5)$.

To reflect across the the y-axis, use the rule $N(x, y) = (-x, y)$.

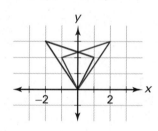

Plot the points (1, 1), (3, 3), and (2, 4), and connect them to form a triangle.

25. Transform the triangle by using the rule $T(x, y) = (x + 2, y - 7)$.

26. What rule would you use to translate the triangle 3 units to the left?

27. Transform the triangle by using the rule $S(x, y) = (x, -y)$.

28. What rule would you use to rotate the triangle 180° about the origin?

Applications

29. TRAVEL Smithville is between Bastrop and LaGrange on a straight highway. The distance from Bastrop to Smithville is 22 kilometers. The distance from Bastrop to LaGrange is 5 kilometers more than 3 times the distance from Bastrop to Smithville. Write an equation for the distances, and then solve it to find the distance from Smithville to LaGrange and from Bastrop to LaGrange.

30. ENGINEERING Jenny has a piece of a broken gear from an antique clock. She needs to find the original size of the gear in order to get a replacement part. Trace the part of the gear shown at right, and construct the complete circle.

Alternative Assessment

Performance Assessment

1. A four-sided shape is inscribed in a circle.

   Copy the figure and then construct the perpendicular bisector of each side by using folding paper or geometry software. What do you notice?

2. Draw some other shapes inscribed in circles. Try a five-sided and a six-sided shape. (Use shapes that have segments for sides, and be sure that all of the vertices are on the circle.) Construct the perpendicular bisector of each side of the shapes. What do you notice?

3. Complete the conjecture below.

   ### Conjecture

   The perpendicular bisectors of the sides of a shape inscribed in a circle _____?_____ .

Portfolio Projects

Slide Rules

1. An example of a useful "unfair ruler" is a *slide rule*. A slide rule is a mechanical device used to make approximate calculations based on the properties of logarithms. You can make your own slide rule from a type of graph paper called logarithm paper.

   Cut out two strips of logarithm paper and label them as shown. The divisions on this paper are not evenly spaced; they get closer together as the numbers get bigger. Put the two strips together so that one slides over the other.

   To multiply the numbers 2 and 3, place strip *D* on strip *C*, as shown below. Align the 2 on strip *D* with the left end, or *index*, of strip *C*. Then find 3 on strip *C* and read the value directly below it on strip *D*.

   Notice that the scale on your slide rule is numbered from 1 to 10. By moving the decimal point, you can approximate any number you wish. For example, the number 3 on the slide rule can represent 0.003, 0.03, 0.3, 3, 30, 300, 3000, etc. If you keep track of your decimal points correctly, you can approximate the product of any two numbers.

2. Choose several pairs of numbers and multiply them by using your slide rule. By moving the decimal point, multiply a one-digit number by a two-digit number and a two-digit number by a two-digit number. Compare your results with the answers you get from a calculator.

3. **ALGEBRA** The slide rule works because of the following property of logarithms:

   $$\log A + \log B = \log (A \times B)$$

   Explain the multiplication of 2×3 on your slide rule in terms of this property.

College Entrance Exam Practice

QUANTITATIVE COMPARISON For Items 1–4, write
A if the quantity in Column A is greater than the quantity in Column B;
B if the quantity in Column B is greater than the quantity in Column A;
C if the two quantities are equal; or
D if the relationship cannot be determined from the information given.

	Column A	Column B	Answers
1.			
	NB	BS	Ⓐ Ⓑ Ⓒ Ⓓ [Lesson 1.2]
2.	$\overline{EF}$ is congruent to $\overline{HG}$.		
	EH	FG	Ⓐ Ⓑ Ⓒ Ⓓ [Lesson 1.2]
3.	$\overline{WZ}$ is the angle bisector of $\angle XWY$.		
	m$\angle XWZ$	m$\angle YWZ$	Ⓐ Ⓑ Ⓒ Ⓓ [Lesson 1.3]

Name each figure. *(LESSON 1.1)*

4.

6.

5.

7.

Trace each diagram onto folding paper and construct the given figure. *(LESSONS 1.4 AND 1.5)*

8. the perpendicular bisector

9. the angle bisector

10. the circumcenter
(Draw the circumscribed circle.)

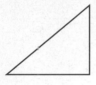

11. the incenter
(Draw the inscribed circle.)

For Items 12–16, identify each transformation as a translation, rotation, or reflection. *(LESSONS 1.6 AND 1.7)*

12.

13.

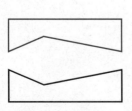

14.

15. $S(x, y) = (x - 1, y)$

16. $T(x, y) = (x, -y)$

FREE-RESPONSE GRID

Items 17–20 may be answered by using a free-response grid such as that commonly used by standardized-test services.

For Items 17 and 18, refer to the figure below. *(LESSON 1.2)*

17. $AB = $ ___?___   **18.** $AC = $ ___?___

19. Refer to the figure below. If $m\angle AVD = 85°$, what is $m\angle AVC$? *(LESSON 1.3)*

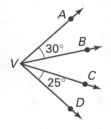

20. Refer to the figure below. What is $m\angle MON$? *(LESSON 1.5)*

Reasoning in Geometry

REGARDLESS OF WHO YOU ARE OR WHAT YOU DO, there are times when you must reason clearly. Whether you are scaling a cliff, or playing a game of nim with a friend, or—like Sherlock Holmes—trying to solve a mystery, the basic principles underlying your reasoning processes are similar.

Geometry presents a unique opportunity for studying the processes of reasoning. Since ancient times, geometry has been for many people the foremost example of a fully reasoned system of knowledge.

About the Chapter Project

In this chapter, you will investigate the principles of logic and apply them to proofs of geometry conjectures. But logic is useful for more than mathematical proofs. As you will see, it is also useful in everyday life and in recreational games and puzzles. In the Chapter Project, *Logic Puzzles and Games*, you will try your hand at the mathematical game known as "sprouts" and two different logic puzzles.

After completing the Chapter Project, you will be able to do the following:

- Analyze certain games and devise strategies for play.

- Solve logic puzzles by using reasoning.

- Solve a type of logic puzzle by setting up a table that displays information in an orderly, structured way.

About the Portfolio Activities

Throughout this chapter, you will be given opportunities to complete Portfolio Activities that are designed to support your work on the Chapter Project.

The rules of the game of nim are given above. The game can be analyzed to develop a strategy for winning. After you have played the game a few times, you will be ready to begin your analysis.

- One of the basic winning patterns for the game of nim is given in the Portfolio Activity on page 87. A second version of the game is also introduced.

- Three different patterns that lead to the winning pattern of the first Portfolio Activity are given in the Portfolio Activity on page 114.

An Introduction to Proofs

Objectives

- Investigate some interesting proofs of mathematical claims.

- Understand the meaning of the term *proof*.

Why *Have you ever needed to prove something you said? If you used a logical argument, you were probably able to make your case. Logical arguments that make a definite point are known as proofs.*

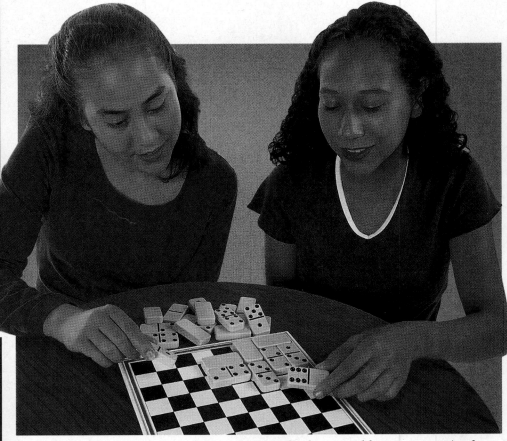

A chessboard has 64 squares, so it can be completely covered by 32 appropriately sized dominoes. This setup is the basis of two famous mathematical questions.

Proving Your Point

PROOF

Suppose that two squares are cut from opposite corners of a chessboard. Can the remaining squares be completely covered by 31 dominoes? If your answer is yes, can you offer a method for showing that it is possible? If your answer is no, can you explain why it cannot be done? In either case, you would be giving a *proof* of your answer.

Here is a proof that the altered chessboard cannot be covered by 31 dominoes:

Because of the pattern of a chessboard, each domino must cover one dark square and one light square. Thus, any arrangement of dominoes must cover the same number of dark squares as light squares.

Notice that the squares which were cut off are the *same color*, leaving more squares of one color than the other. Therefore, it is not possible to cover the altered board with the 31 dominoes.

CHECKPOINT ✔

PROBLEM SOLVING

Three Challenges

Part I

Suppose that you change the chessboard problem so that you cut off one dark square and one light square anywhere on the board.

Use the diagram at right and your own explanation to prove that the altered board can be covered by 31 dominoes.

Part II

How could you find the sum of the first *n* odd counting numbers without actually adding them? **Make a table** like the one below, and see if you can discover the answer.

The pathway through the maze suggests a proof.

n	First *n* odd numbers	Sum of the first *n* odd numbers
1	1	1
2	1, 3	4
3	1, 3, 5	9
4	1, 3, 5, 7	16
5	?	?
6	?	?
n	?	?

CHECKPOINT ✔

The diagram to the right of the table is a "proof without words" of the algebraic result that you may have discovered. Explain in your own words how the diagram proves the result.

Part III

You are given the figure at right, which is built entirely of squares. The area of square *C* is 64, and the area of square *D* is 81.

CHECKPOINT ✔

Use the above information to determine whether the overall figure is a square. (Recall that all four sides of a square must be the same length.)

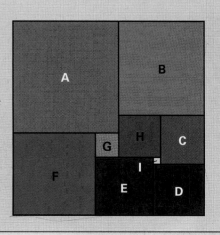

What Is a Proof?

A **proof** is a convincing argument that something is true. But before you allow yourself to be convinced by a supposed proof, you should make sure that it is sound.

In mathematics, a proof starts with things that are agreed on (called *postulates* or *axioms*). Then logic is used to reach a conclusion.

There are many different styles of proofs in mathematics. Some proofs follow a prescribed form and are called *formal proofs*. For example, the calculations below are a formal proof that, for the given equation, $x = 4$.

$$5x + 4 = 24 \qquad \textit{Given}$$
$$5x = 20 \qquad \textit{Subtraction Property of Equality}$$
$$x = 4 \qquad \textit{Division Property of Equality}$$

As you can see, a definite form is followed in this proof. Each statement on the left is given a *justification* in the column on the right. But this is not always the case. The proofs you did in the Activity did not follow any particular form, but they are just as mathematically sound (if they are correct!) as formal proofs.

In addition to the free-form proofs used in this lesson, you will learn the following styles for proofs:

- two-column proofs, Lesson 2.4
- paragraph proofs, Lesson 2.4
- flowchart proofs, Lesson 4.4
- coordinate proofs, Lesson 5.7
- table proofs, Lesson 9.3

Exercises

● Communicate

1. In your own words, describe what a proof is.

2. If the indicated squares were removed from the chessboard at right, explain why you could not cover the remaining squares with 31 dominoes.

3. In Part II of the Activity, a geometric solution is given for an algebraic problem. What do you think might be a advantage for a geometric solution over an algebraic one?

4. Consider the following argument for proving that the figure in Part III of the Activity is a square: I measured the sides and they were the same, so it is a square. Is this a proof? Why or why not?

5. Explain how you proved whether the overall figure in Part III of the Activity was a square. Could anyone find a flaw in your argument?

PROOF

6. Suppose that four squares were removed from a chessboard. What would need to be true about the colors of these squares in order for the remaining squares to be covered with 30 dominoes? *(ACTIVITY)*

7. According to your answer in Part II of the Activity, what is the sum of the first 10 odd numbers? of the first 100 odd numbers? *(ACTIVITY)*

PROOF

8. The diagram at right is built entirely of squares. The overall figure is also a square. If the area of square *A* is 64 and the area of square *I* is 4, what are the areas of the other squares? What is the area of the overall square? *(ACTIVITY)*

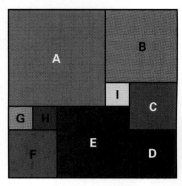

Practice and Apply

For Exercises 9–11, consider a variation of the chessboard problem (pages 80–81) in which the chessboard is covered with square tiles, each the size of 4 squares on the board. The board can then be completely covered with 16 tiles. You may wish to draw diagrams on graph paper.

9. If one square were removed from each corner of the board, could the altered board be covered completely with 15 tiles? Why or why not?

10. If four squares were removed from the top row of the board, could the altered board be covered completely with 15 tiles? Why or why not?

PROOFS

11. State a rule for removing four squares so that the altered board can be covered by 15 tiles. How would you prove your rule?

12. Copy the diagram at right and label each area. Use the diagram and your own explanation to prove the following:

$$(x + a)(x + b) = x^2 + ax + bx + ab$$

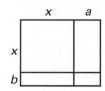

The following procedure is a shortcut for squaring a number that ends in 5: Multiply the number formed by the first digits (all digits except the final 5) by the next consecutive number. Then put 25 at the end of the product. For example, to square 35, multiply 3 × 4 and then put 25 at the end of the product, 12, to get 1225.

13. Use the shortcut to square 25, 75, and 105. Does it work?

14. Explain how the diagram at right illustrates how the shortcut works for squaring 35.

PROOF

15. Use the diagram and your own explanation to prove that this shortcut works for squaring any number ending in 5.

Use the table below to answer Exercises 16–21. The numbers in the table are perfect cubes: $1^3 = 1$, $2^3 = 8$, $3^3 = 27$, $4^3 = 64$, . . .

A	B	C
1	8	27
64	125	216
?	?	?
?	?	?

16. Suppose that you wanted to know which column contains the number 1,000,000,000 (one billion). One method is to fill in the table until you reach 1,000,000,000. What is one disadvantage of this method?

17. Another method is to analyze the pattern of the numbers in the table. One billion (1,000,000,000) is the cube of what number? (Hint: $10^3 = 1000$, $100^3 = \underline{\ ?\ }$, . . .)

18. Look at the columns in the table. Notice that column A contains the numbers 1^3, 4^3, 7^3, . . . What numbers occur in column B? in column C?

19. What is true of every number in column C? Is this true of any number in column A or B?

20. In what column does the number 999^3 occur?

21. Prove that 1,000,000,000 occurs in column A.

Exercises 22–28 involve partial sums of the infinite sequence $\frac{1}{2}$, $\frac{1}{4}$, $\frac{1}{8}$, $\frac{1}{16}$, . . . , in which the denominator of each term is doubled to obtain the next term.

22. What are the next four terms of the sequence?

23. Are the terms getting larger or smaller? Explain your answer.

24. Complete the table below, in which terms of the sequence are added together.

Number of terms	Terms	Sum of terms
1	$\frac{1}{2}$	$\frac{1}{2}$, or 0.5
2	$\frac{1}{2}$, $\frac{1}{4}$	$\frac{3}{4}$, or 0.75
3	$\frac{1}{2}$, $\frac{1}{4}$, $\frac{1}{8}$	$\frac{7}{8}$, or 0.875
4	$\frac{1}{2}$, $\frac{1}{4}$, $\frac{1}{8}$, $\frac{1}{16}$	$\frac{15}{16}$, or 0.9375
5	?	?
6	?	?
7	?	?
8	?	?

25. What number do the sums in the table seem to be approaching?

26. Because the sequence has infinitely many terms, you cannot add them all with your calculator. However, it is still possible to find the sum of the infinite sequence. Start by drawing a square with an area of 1 square unit. Label the lengths of the sides of the square.

27. Divide your square into pieces with areas equal to the terms of the sequence. Use the diagram at right as a guide.

28. Explain how the diagram indicates the sum of the infinite sequence. What is the sum of the sequence?

Suppose that you know the square of a positive integer *n* and you want to find the square of *n* + 1. In Exercises 29–32, you will discover a method.

29. Draw a square array of dots with 5 rows of 5 dots, as shown at right. Explain how the array represents the square of 5, or 5^2.

30. Increase the side lengths of your square to represent the square of the number $(5 + 1)$, or 6^2. Use your diagram to show that $6^2 = 5^2 + 5 + 6$.

PROOF

31. Use your diagram to prove that the square of the number *n* + 1 is found by adding *n* and *n* + 1 to the square of *n*.

32. Use this method to find the square of 21, given that the square of 20 is 400.

Exercises 33–37 refer to the following conjecture:

$$1 + 2 + 3 + 4 + \cdots + n = \frac{n(n+1)}{2}, \text{ where } n \text{ is any positive integer}$$

33. Verify that the conjecture is true for values of *n* from 1 to 5.

34. The triangle at right represents the sum $1 + 2 + 3 + 4 + 5 + 6$. Count the dots to find the sum.

35. Find the sum of the integers 1 through 6 by using the formula in the conjecture. Does the result agree with your answer to Exercise 34?

36. A second triangle is added to the original figure, as shown at right. Find the number of dots in the resulting rectangle.

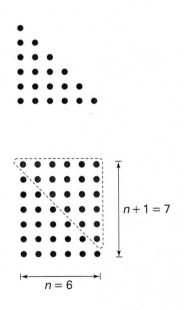

PROOF

37. How does the number of dots in one of the triangles relate to the number of dots in the rectangle? Express this relationship as a formula in terms of *n* and *n* + 1. Explain how this proves the conjecture.

38. The positive integers are written in a triangular array, as shown at right.

Use the conjecture given for Exercises 33–37 to prove that the number 1000 occurs in the 45th row.

```
        1
      2   3
    4   5   6
  7   8   9   10
11 ...
```

39. GAMES In a popular mystery game, players try to solve a murder. The "murder" is represented by three cards: the murderer, the place, and the weapon. The cards are placed in an envelope and kept concealed until the end of the game.

To solve the murder, you must use logic. Suppose you have determined that the crime was committed in the study with the candlestick by either Colonel Mustard or Professor Plum. Then you learn that one of the other players is holding the Professor Plum card, which means that the murderer cannot be Professor Plum. You should now be able to solve the murder. What is your full conclusion?

Your conclusion about the identity of the murderer involves a special kind of argument known as a *disjunctive syllogism*. Explain in your own words when you can use this kind of an argument.

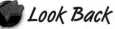 *Look Back*

40. Draw a plane and label it 𝒬. Draw three noncollinear points in plane 𝒬, and label them *A*, *B*, and *C*. **(LESSON 1.1)**

41. In the figure you drew for Exercise 40, draw $\overline{AB}$, $\overline{BC}$, and $\overline{AC}$. What shape is formed? **(LESSON 1.1)**

Use the diagram below for Exercises 42 and 43. (LESSON 1.2)

42. If *LN* = 18, find the value of *x*.

43. If *LM* = 7, find *LN*.

In the diagram at right, m∠*KIT* = 80°. (LESSON 1.3)

44. Find the value of *x*.

45. Find m∠*KIY* and m∠*TIY*.

Algebra

The triangular array of numbers shown below is commonly known as *Pascal's triangle,* named after Blaise Pascal (1623–1662). Pascal's triangle has applications in geometry (see Lessons 11.1 and 11.7), algebra, and probability.

46. How would you find the entries for the next row in the triangle?

47. Write out the first eight rows of Pascal's triangle.

48. For each of the first five rows, add the entries in the row. What is the pattern in the sums?

49. What other interesting patterns can you find in Pascal's triangle?

```
              1
            1   1
          1   2   1
        1   3   3   1
      1   4   6   4   1
    1   5  10  10   5   1
```

CULTURAL CONNECTION: ASIA
The illustration at right is the earliest known version of "Pascal's triangle." It is from a Chinese book printed around 1303 C.E.

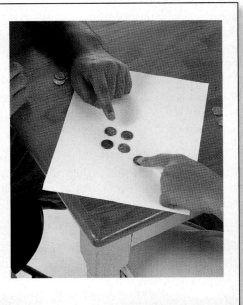

PORTFOLIO ACTIVITY

In the game of nim (see page 79), the object is to take the last remaining counter. In a second version of the game, the object is to force your opponent to take the last counter.

- Play a few games of both versions of nim with a friend. How is your strategy for the second version different from your strategy for the original version of nim?

PROOF

- If you are able to leave two rows with two counters each, you can always win. Show that this is true for both versions of the game.

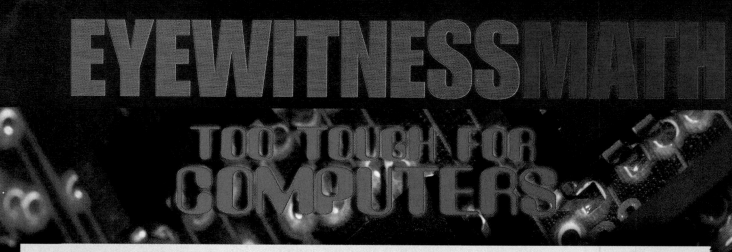
Math Problem, Long Baffling, Slowly Yields

by Gina Kolata, *New York Times*

A century-old math problem of notorious difficulty has started to crumble. Even though an exact solution still defies mathematicians, researchers can now obtain answers that are good enough for most practical applications.

The traveling salesman problem asks for the shortest tour around a group of cities. It sounds simple–just try a few tours out and see which one is shortest. But it turns out to be impossible to try all possible tours around even a small number of cities.

Companies typically struggle with traveling salesmen problems involving tours of tens of thousands or even hundreds of thousands of points.

For example, such problems arise in the fabrication of circuit boards, where lasers must drill tens to hundreds of thousands of holes in a board. What happens is that the boards move and the laser stays still as it drills the holes. Deciding what order to drill these holes is a traveling salesman problem.

Very large integrated circuits can involve more than a million laser-drilled holes, leading to a traveling salesman problem of more than a million "cities."

In the late 1970s, investigators were elated to solve 50-city problems, using clever methods that allow them to forgo enumerating every possible route to find the best one. By 1980, they got so good that they could solve a 318-city problem, an impressive feat but not good enough for many purposes.

Dr. David Johnson and Dr. Jon Bentley of AT&T Bell Laboratories are recognized by computer scientists as the world champions in solving problems involving about 100,000 cities. By running a fast computer for two days, they can get an answer that is guaranteed to be either the best possible tour or less than 1 percent longer than the best one.

In most practical situations, an approximate solution is good enough, Dr. Johnson said. By just getting to within about 2 percent of the perfect solution of a problem involving drilling holes in a circuit board, the time to drill the holes can usually be cut in half, he said.

The researchers break large problems into many smaller ones that can be attacked one by one, and give these fragments to fast computers that can give exact answers.

For example, Dr. Bentley said, "If I ask you to solve a traveling salesman problem for 1,000 cities in the U.S., you would do it as a local problem. You might go from New York toward Trenton and then move to Philadelphia," he explained. Then the researchers would repeat this process from other hubs, like Chicago, and combine the results. "We end up calculating only a few dozen instances per point," Dr. Bentley said. "if you have a million cities, you might do only 30 million calculations."

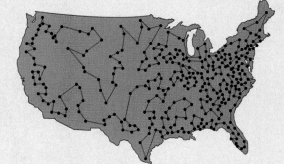

As of 1968, the computer solution for the shortest route connecting 532 cities with AT&T central offices looked like this, It was the biggest such problem calculated up to that time. Now, a route for 2,392 destinations has been computed, and mathematicians are working on a 3,038-city problem. (Source: New York University and Institute for Systems Analysis)

Cooperative Learning

The most obvious way to solve a "traveling-salesman" problem is to try all possible routes and determine which is the shortest. How many routes would you need to consider? One way to find the number of possible routes is to multiply the number of choices for each city along the route. Suppose that a route contains four cities: A, B, C, and D. One of the cities will be the starting and ending point. From that city, there are 3 possible choices for the first city to visit, then from that city there are 2 remaining possible choices for the second city to visit, then from that point there is 1 remaining possible choice for the third city to visit. Thus, the number of possible routes is

One way to visualize the number of routes is by using a tree diagram like the one below.

1	$\times$	3	$\times$	2	$\times$	1	$=$	6
starting city (given)		choices to visit first		choices to visit second		choices to visit third		number of possible routes

In order to reduce the number of computations needed to find the shortest route, notice that a route traveled in reverse order has the same length.

ABCDA has the same length as ADCBA.
ABDCA has the same length as ACDBA.
ACBDA has the same length as ADBCA.

Thus, it is necessary to compute the lengths of only three routes.

1. Copy and complete the table below.

Number of cities	Multiply	Factorial form	Evaluate	Divide by 2
4	$3 \times 2 \times 1$	3!	6	3
5	?	?	?	?
6	?	?	?	?
7	?	?	?	?
n	?	?	?	?

Based on the table above, complete the following formula:
The number of routes, N, that must be calculated to determine the shortest route for visiting n cities is $N =$ ___?___ .

2. Use the formula above to determine the following: How many routes must be calculated to determine the shortest route through 10 cities? through 15 cities? through 20 cities? through 50 cities?

3. Suppose that a computer could calculate the length of 1 billion routes per second. How long would it take to find the shortest route through 20 cities? through 21 cities? through 50 cities? Give your answers in seconds, days, and years.

4. Compare your answers to Exercise 3 with the estimated age of the universe, about 15 billion years. Do you think the traveling-salesman problem is really "too tough for computers"? Why or why not?

An Introduction to Logic

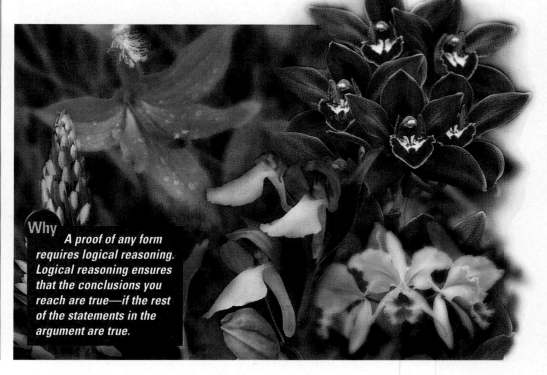

Objectives

- Define *conditionals* and model them with Euler diagrams.

- Use conditionals in logical arguments.

- Form the converses of conditionals.

- Create logical chains from conditionals.

APPLICATION

BIOLOGY

Why *A proof of any form requires logical reasoning. Logical reasoning ensures that the conclusions you reach are true—if the rest of the statements in the argument are true.*

Organisms can be classified according to their structure. For example, all of the flowers shown above belong to the orchid family. Class organization, which also extends to manufactured things, is the basis of logical reasoning.

Drawing Conclusions From Conditionals

The force of logic comes from the way information is structured. For example, all Corvettes are Chevrolets, a fact which can be represented by an Euler (pronounced "oiler") diagram like the one at right. Note: Euler diagrams are often called Venn diagrams.

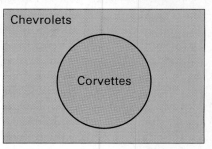

From the Euler diagram, it is easy to see that the following statement is true:

If a car is a Corvette, then it is a Chevrolet.

"If-then" statements like this one are called **conditionals**. In logical notation, conditionals are written as follows:

If *p* then *q*

or

$p \Rightarrow q$ (Read as "*p* implies *q*.")

In a conditional, the part following the word *if* is the **hypothesis**. The part following the word *then* is the **conclusion**.

If a car is a Corvette, then it is a Chevrolet.

Hypothesis   *Conclusion*

Now consider the following statement:

> Susan's car is a Corvette.

By placing Susan's car into the Euler diagram, you can see that it is a Chevrolet. (Anything inside the circle is also inside the rectangle.)

The complete process of drawing the conclusion that Susan's car is a Chevrolet can be written as a *logical argument*. This particular argument, which is known as a *syllogism* (see Lesson 12.1), has three parts.

1. If a car is a Corvette, then it is a Chevrolet.
2. Susan's car is a Corvette.
3. Therefore, Susan's car is a Chevrolet.

The process of drawing *logically certain* conclusions by using an argument is known as **deductive reasoning,** or **deduction.**

E X A M P L E ❶ Recall the following definitions from your earlier studies:

> An *equilateral triangle* is a triangle with three congruent sides.
> An *isosceles triangle* is a triangle with at least two congruent sides.

PROBLEM SOLVING

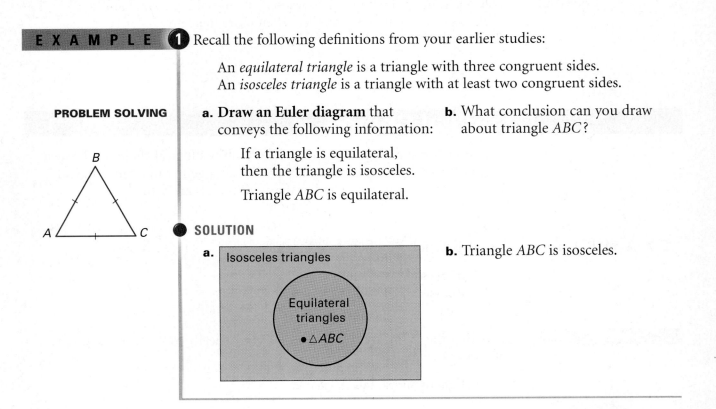

a. Draw an Euler diagram that conveys the following information:

> If a triangle is equilateral, then the triangle is isosceles.
>
> Triangle *ABC* is equilateral.

b. What conclusion can you draw about triangle *ABC*?

● **SOLUTION**

a.

b. Triangle *ABC* is isosceles.

Reversing Conditionals

When you interchange the hypothesis and the conclusion of a conditional, the new conditional is called the **converse** of the original conditional.

Conditional: If a car is a Corvette, then it is a Chevrolet.
Converse: If a car is a Chevrolet, then it is a Corvette.

The original conditional is true. But what about its converse? If there is an example of a Chevrolet that is not a Corvette—and there certainly is—then the converse is false. An example which proves that a statement is false is called a **counterexample**.

EXAMPLE 2 Write a conditional with the hypothesis "a triangle is equilateral" and the conclusion "the triangle is isosceles." Then write the converse of your conditional. Is the conditional true? Is the converse true?

● **SOLUTION**

Conditional: If a triangle is equilateral, then it is isosceles.
Converse: If a triangle is isosceles, then it is equilateral.

The original conditional is true, according to the definition of an equilateral triangle. (If a triangle has three congruent sides, then it has at least two congruent sides.)

The converse is false, however, as the counterexample at right shows. ($\triangle MNO$ is isosceles but not equilateral.)

TRY THIS Write a conditional with the hypothesis "an animal is a snake" and the conclusion "the animal is a reptile." Then write the converse of your conditional. Is the conditional true? Is the converse true?

Logical Chains

Conditionals can be linked together. The result is a logical chain. In the next example, three different conditionals are linked together to form a **logical chain**. (It does not matter whether the conditionals are actually true.)

EXAMPLE 3 Consider the following silly conditionals:

If cats freak, then mice frisk.
If sirens shriek, then dogs howl.
If dogs howl, then cats freak.

PROOF Prove that the following conditional follows logically from the three given conditionals:

If sirens shriek, then mice frisk.

SOLUTION

Identify the hypothesis of the conditional you are trying to prove:

> If sirens shriek, ...

Look for a conditional that begins with "if sirens shriek."

> If sirens shriek, | then dogs howl.

Look for a conditional that begins with "if dogs howl."

> If dogs howl, | then cats freak.

Look for a conditional that begins with "if cats freak out."

> If cats freak, | then mice frisk.

There is a zigzag pattern in the steps of the logical chain.

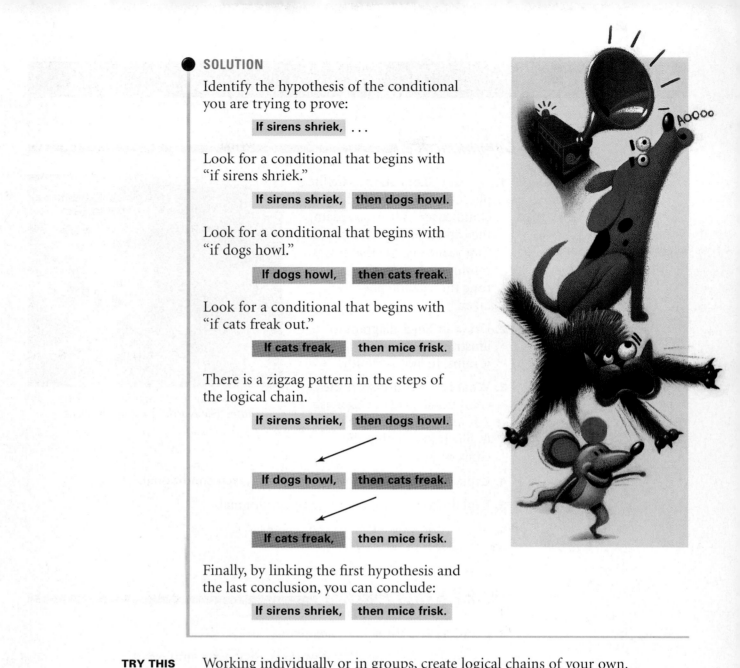

> If sirens shriek, | then dogs howl.
>
> If dogs howl, | then cats freak.
>
> If cats freak, | then mice frisk.

Finally, by linking the first hypothesis and the last conclusion, you can conclude:

> If sirens shriek, | then mice frisk.

TRY THIS   Working individually or in groups, create logical chains of your own.

CRITICAL THINKING   Notice that Example 3 does not prove that mice frisk. What is proven instead? What is lacking in the argument that is necessary to prove that mice actually frisk?

The proof of the conditional in Example 3 relied on the following property:

If-Then Transitive Property

Given:	You can conclude:
If *A* then *B*, and	If *A* then *C*.
if *B* then *C*.	

2.2.1

Notice that the same property is used repeatedly in long chains of conditionals.

Communicate

1. **WEATHER** Look at the satellite photo at right, which shows a cloud cover over Amsterdam, the capital of the Netherlands. Can you conclude that it was raining in Amsterdam at the time the satellite photo was taken?

2. Draw an Euler diagram to illustrate the conditional "If it rains, then it is cloudy."

3. What is the converse of the conditional given in Exercise 2? Use your Euler diagram to illustrate whether the converse is true.

Clouds over the Netherlands, looking south. The North Sea is on the right.

4. Explain how to write the converse of a given conditional.

5. Explain how to disprove a given conditional.

Guided Skills Practice

6. Consider the following statements:

   All United States Postal workers are federal employees.
   John is a United States Postal worker.

   Write the first statement as a conditional and use deduction to show that John is a federal employee. Draw an Euler diagram to illustrate the logic of your deduction. **(EXAMPLE 1)**

7. Write a conditional with the hypothesis "two lines are parallel" and the conclusion "the two lines do not intersect." Then write the converse of your conditional. Is the conditional true? Is the converse true? **(EXAMPLE 2)**

8. Consider the three conditionals below. **(EXAMPLE 3)**

   If a number is divisible by 2, then the number is even.
   If a number is even, then the last digit is 0, 2, 4, 6, or 8.
   If a number is divisible by 4, then the number is divisible by 2.

   Prove that the following conditional follows logically from the three conditionals above:

   If a number is divisible by 4, then the last digit is 0, 2, 4, 6, or 8.

Practice and Apply

For Exercises 9–12, refer to the following statement:

All people who live in Ohio live in the United States.

9. Rewrite the statement as a conditional.

10. Identify the hypothesis and the conclusion of the conditional.

11. Draw an Euler diagram that illustrates the conditional.

12. Write the converse of the conditional.

For Exercises 13–16, use the Euler diagram to write a conditional.

Illustration from The Hunting of the Snark *by Lewis Carroll. A boojum seems to have been a particularly dangerous kind of snark.* (See Exercise 16.)

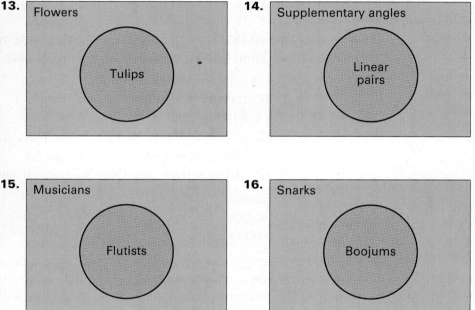

13.

Flowers

Tulips

14.

Supplementary angles

Linear pairs

15.

Musicians

Flutists

16.

Snarks

Boojums

For Exercises 17–20, identify the hypothesis and conclusion of each conditional. Write the converse of each conditional. If the converse is false, give a counterexample to show that it is false.

17. If it is snowing in Chicago, then it is snowing in Illinois.

18. If two angles are complementary, then the sum of their measures is 90°.

19. If the measure of each angle in a triangle is less than 90°, then the triangle is acute.

20. If a figure is rotated, then its size and shape stay the same.

For Exercises 21–23, refer to the diagram below, and write a conditional with the given hypothesis and conclusion.

21. Hypothesis: ∠AXB and ∠BXD form a linear pair.
    Conclusion: ∠AXB and ∠BXD are supplementary.

22. Hypothesis: ∠AXB and ∠BXD are supplementary.
    Conclusion: m∠AXB + m∠BXD = 180°

23. Hypothesis: m∠BXC + m∠CXD = 90°
    Conclusion: m∠AXB = 90°

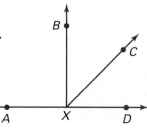

In Exercises 24–29, use the given statements to draw a conclusion. Then draw an Euler diagram to illustrate your conclusion.

24. If an animal is a mouse, then the animal is a rodent. Mikey is a mouse.

25. If someone is human, then he or she is mortal. Socrates is human.

26. If Jennifer goes to the beach, then she will get a sunburn. Jennifer is going to the beach tomorrow.

27. If someone lives in Norway, then he or she lives in Scandinavia. Ingrid lives in Norway.

28. If a figure is a square, then the figure is a rectangle. Figure *ABCD* is a square.

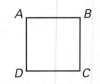

29. If two points are in plane $\mathcal{P}$, then the line containing them is in plane $\mathcal{P}$. Points *S* and *T* are in plane $\mathcal{P}$.

In Exercises 30–33, arrange each set of statements to form a logical chain. Then write the conditional that follows from the logical chain.

30. If it is cold, then birds fly south.
If the days are short, then it is cold.
If it is winter, then the days are short.

31. If the police catch Tim speeding, then Tim gets a ticket.
If Tim drives a car, then Tim drives too fast.
If Tim drives too fast, then the police catch Tim speeding.

32. If quompies plaun, then romples gleer.
If ruskers bleer, then homblers frain.
If homblers frain, then quompies plaun.

33. If you go to a movie, then you will spend all of your money.
If you clean your room, then you will go to a movie.
If you cannot buy gas for the car, then you will be stranded.
If you spend all of your money, then you cannot buy gas for the car.

APPLICATION

34. HUMOR Can logic be used to prove the impossible? Consider the following logical chain:

> The independent farmer is disappearing.
> That man is an independent farmer.
> Therefore, that man is disappearing.

Write the above argument using a conditional statement. How would you criticize the argument?

Is this man disappearing? (See Exercise 34.)

35. The following is an old saying dating back to at least the fifteenth century:

> For want of a nail, the shoe was lost.
> For want of a shoe, the horse was lost.
> For want of a horse, the rider was lost.
> For want of a rider, the battle was lost.
> For want of the battle, the war was lost.

Rewrite the saying as a logical chain of conditionals. Then write the conclusion that follows from the logical chain.

FINE ARTS People have different opinions about what constitutes a work of art. Consider the objects below. Use one of the conditional statements below or one of your own to present a logical proof that each work is or is not a work of art.

If an object displays form, beauty, and unusual perception on the part of its creator, then the object is a work of art.

If an object displays creativity on the part of the person who made it, then the object is a work of art.

36.

Two Open Modular Cubes/Half Off,
Sol Lewitt, 1972

37.

Fond Jaune,
Alexander Calder, 1972

38. MUSIC Write a conditional with the given hypothesis and conclusion.

Hypothesis: A person performs classical music.

Conclusion: The person dislikes jazz.

Write the converse of the conditional.

Yo-Yo Ma is a famous classical cellist who also loves jazz. Is this a counterexample to the original conditional or to the converse? Is either the conditional or the converse true?

39. A floor is best modeled by what geometric figure? *(LESSON 1.1)*

40. With folding paper, construct a segment and its perpendicular bisector. *(LESSON 1.4)*

41. With folding paper, construct an angle and its bisector. *(LESSON 1.4)*

Complete the statements below. *(LESSON 1.5)*

42. The perpendicular bisectors of a triangle meet at the ___?___ .

43. The angle bisectors of a triangle meet at the ___?___ .

44. The center of the circumscribed circle of a triangle is outside the triangle if the triangle is ___?___ , is inside the triangle if the triangle is ___?___ , and is on the triangle if the triangle is ___?___ .

45. Name the three basic types of rigid transformations. *(LESSON 1.6)*

46. Reflect the word **YAM** across a vertical line. *(LESSON 1.6)*

Look Beyond

CONNECTION

PATTERNS IN DATA In this lesson, you learned about deductive reasoning. Another kind of reasoning is called inductive reasoning. Inductive reasoning is based on the recognition of patterns.

For Exercises 47–50, use inductive reasoning to find the next number in each sequence.

47. 5, 8, 11, 14, ?

48. 20, 27, 36, 47, 60, ?

49. 2, 6, 18, 54, ?

50. $3, 1, \frac{1}{3}, \frac{1}{9}, ?$

51. The beginning of the Fibonacci sequence given below. The numbers of this sequence are known as Fibonacci numbers. What are the next five Fibonacci numbers?

$$1, 1, 2, 3, 5, 8, \ldots$$

52. Which terms in the Fibonacci sequence are even? Describe the pattern of even and odd numbers in the sequence, and explain why this pattern occurs.

53. Which terms in the Fibonacci sequence are divisible by 3? by 5? Guess a pattern for the multiples of 3 and of 5, and check your answer by examining the sequence.

APPLICATION

BOTANY *In artichokes and other plants, the number of spirals in each direction are often Fibonacci numbers. This artichoke has 5 clockwise spirals and 8 counterclockwise spirals.*

Definitions

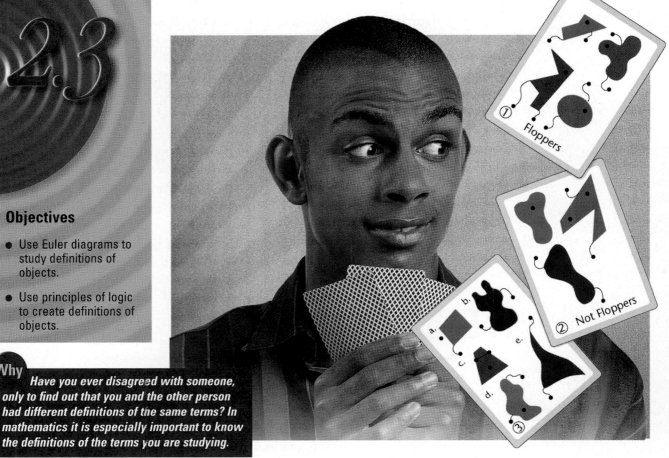

Objectives

- Use Euler diagrams to study definitions of objects.

- Use principles of logic to create definitions of objects.

Why
Have you ever disagreed with someone, only to find out that you and the other person had different definitions of the same terms? In mathematics it is especially important to know the definitions of the terms you are studying.

Definitions and Euler Diagrams

Which of the figures in card 3 above are floppers? Simply by observing the differences between the figures in the card 1 and card 2, it is possible to write a definition of a flopper.

> A flopper is a figure with one "eye" and two "tails."

Using this definition, you can see that figures **d** and **e** are floppers.

Definitions have a special property when they are written as conditional statements. For example:

> If a figure is a flopper, then it has one eye and two tails.

You can also write the **converse** of the conditional by interchanging the hypothesis and conclusion:

> If a figure has one eye and two tails, then it is a flopper.

Notice that both the original conditional and its converse are true. *This special property is true for all definitions.* The two true conditionals can be combined into a compact form by joining the hypothesis and the conclusion with the phrase "if and only if," which is represented by $p \Leftrightarrow q$.

$$p \text{ if and only if } q \qquad or \qquad p \Leftrightarrow q$$

The resulting "if-and-only-if" statement is known as a **biconditional.**

By combining the conditional and its converse, you create the following definition of a flopper, expressed in logical terms:

A figure is a flopper if and only if it has one eye and two tails.

Euler diagrams can be used to represent the two parts of the definition:

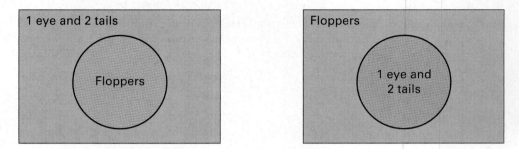

CRITICAL THINKING   Can you create a diagram to represent the fact that both the original statement and its converse are true? How would the parts of the diagram be related?

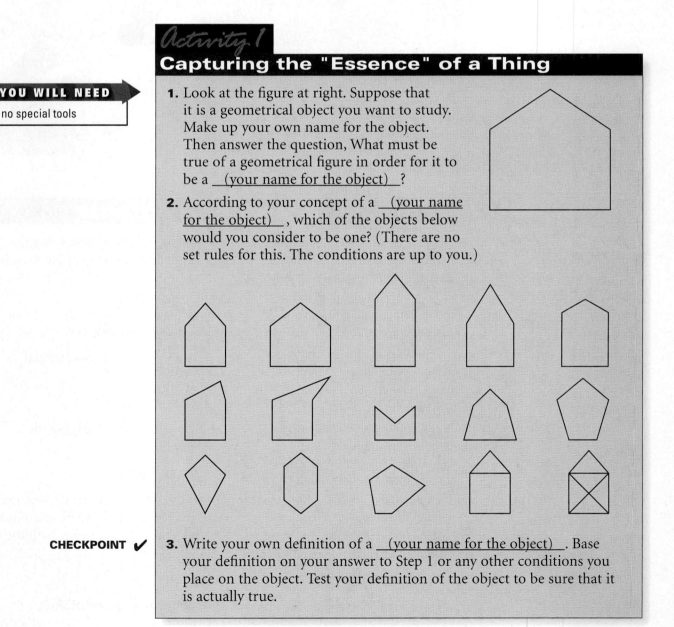

Activity 1

Capturing the "Essence" of a Thing

YOU WILL NEED

no special tools

1. Look at the figure at right. Suppose that it is a geometrical object you want to study. Make up your own name for the object. Then answer the question, What must be true of a geometrical figure in order for it to be a _(your name for the object)_ ?

2. According to your concept of a _(your name for the object)_ , which of the objects below would you consider to be one? (There are no set rules for this. The conditions are up to you.)

CHECKPOINT ✔   3. Write your own definition of a _(your name for the object)_ . Base your definition on your answer to Step 1 or any other conditions you place on the object. Test your definition of the object to be sure that it is actually true.

TRY THIS   Create your own object, and then give it a name and a definition. Your object does not have to be geometrical or even mathematical, but it should be something you can draw. Test your definition to be sure it is valid. Share your definition with others.

Adjacent Angles

An important concept in geometry is that of *adjacent angles*. By examining the figures in the boxes below, you should be able to form an idea about what adjacent angles are—and also what they are not. This information will enable you to write your own definition of adjacent angles.

ADJACENT ANGLES

∠1 and ∠2

∠1 and ∠2
∠2 and ∠3

∠1 and ∠2
∠2 and ∠3
∠4 and ∠5
(There are others also.)

NOT ADJACENT ANGLES

∠1 and ∠2

∠1 and ∠3

∠ADB and ∠ADC

∠1 and ∠2

∠1 and ∠2

YOU WILL NEED

no special tools

1. List the adjacent angles in the figure at right.

2. What do adjacent angles have in common? Can they overlap?

CHECKPOINT ✔

3. Angles that do not overlap have *no interior points in common*. Use this fact and your answers from Steps 1 and 2 to write a definition of adjacent angles.

Definition: Adjacent Angles

Adjacent angles are angles in a plane that have their ___?___ and one ___?___ in common but ___?___.

Exercises

Communicate

1. Explain how a definition is different from a conditional statement.

2. Choose a definition from a dictionary and write it as a biconditional. Is it a valid definition?

3. Explain why the following statement is not a definition:

   A tree is a plant with leaves.

4. The following are blops: The following are not blops:

   Write a definition of a blop and identify which of the following are blops:

   a.   b.   c.   d.

5. Recall from Lesson 1.1 that *points*, *lines*, and *planes* are referred to as undefined terms. Why do you think it is necessary to have undefined terms in geometry?

Guided Skills Practice

6. Use the figure at right to write a definition of a glosh. Write your definition as precisely as possible.

   Examine the shapes below. According to your definition, which (if any) of the figures are gloshes? **(ACTIVITY 1)**

   a.   b.   c.   d.   e.   f.

7. List all pairs of adjacent angles in the figure at right. **(ACTIVITY 2)**

Practice and Apply

In Exercises 8–16, use the following steps to determine whether the given sentence is a definition.

    a. Write the sentence as a conditional statement.
    b. Write the converse of the conditional.
    c. Write a biconditional statement.
    d. Decide whether the sentence is a definition, and explain your reasoning.

  8. A teenager is a person who is 13 years old or older.

  9. A teenager is a person from 13 to 19 years old.

10. Zero is the integer between −1 and 1.

11. An even number is divisible by 2.

12. An angle is formed by two rays.

13. A right angle has a measure of 90°.

14. GEOLOGY Granite is a very hard, crystalline rock.

15. CHEMISTRY Hydrogen is the lightest of all known substances.

16. BIOLOGY An otter is a small furry mammal with webbed feet that are used for swimming.

17. Name all pairs of adjacent angles in the figure at right.

For Exercises 18–22, explain why the indicated angles are *not* adjacent.

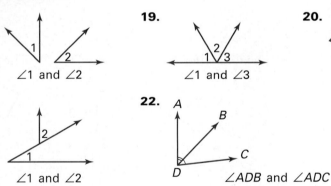

18.

∠1 and ∠2

19.

∠1 and ∠3

20.

∠1 and ∠2

21.

∠1 and ∠2

22.

∠ADB and ∠ADC

23. The following are flishes: The following are not flishes:

Which of the following are flishes?

  a.   **b.**   **c.**   **d.**

24. The following are zobbles:

The following are not zobbles:

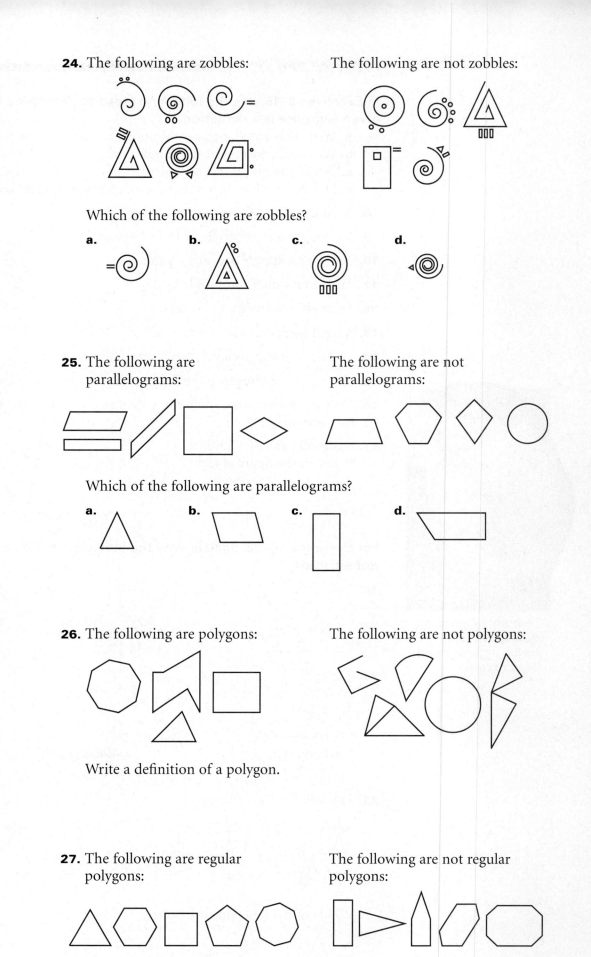

Which of the following are zobbles?

a.   b.   c.   d.

25. The following are parallelograms:

The following are not parallelograms:

Which of the following are parallelograms?

a.   b.   c.   d.

26. The following are polygons:

The following are not polygons:

Write a definition of a polygon.

27. The following are regular polygons:

The following are not regular polygons:

Write a definition of a regular polygon.

28. Examine the figures below. Choose two or more figures with some features in common. Then choose a name and create a definition for the figures you chose. Tell which of the other figures below are examples of the object you defined.

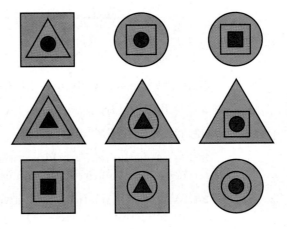

29. BIOLOGY Scientists who study birds are known as ornithologists. They classify birds into 29 different orders. Each order has a number of defining characteristics. Read the descriptions of the three orders below, and classify each of the six birds shown according to their order.

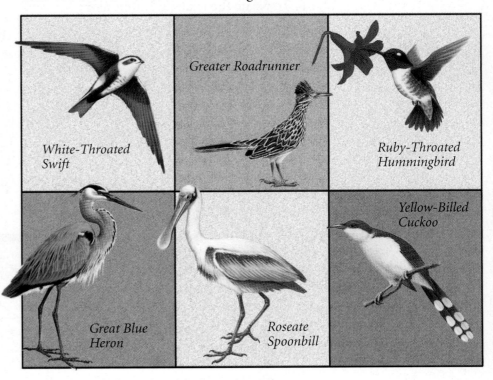

White-Throated Swift

Greater Roadrunner

Ruby-Throated Hummingbird

Great Blue Heron

Roseate Spoonbill

Yellow-Billed Cuckoo

Ciconiiformes: long-legged, long-necked birds that wade in shallow water or, in some cases, feed on open ground

Cuculiformes: small-to medium-sized, slender, usually long-tailed birds with zygodactyl feet (*Zygodactyl* means that the toes are arranged in pairs, with two in front and two in back.)

Apodiformes: small or very small birds with tiny feet, extremely short humeri, and long bones in the outer portion of the wing (*Humeri* are the bones of the upper arm part of the wing extending from the "shoulder" to the "elbow.")

30. Point B is the midpoint of $\overline{AC}$. Find x and AB. **(LESSON 1.2)**

Refer to the figure below for Exercises 31–34. **(LESSON 1.6)**

31. Reflect the triangle across the given line.

32. Draw $\overline{AA'}$, $\overline{BB'}$, and $\overline{CC'}$.

33. What is the relationship among the three segments you drew in Exercise 32?

34. What is the relationship between the line of reflection and each of the segments you drew in Exercise 32?

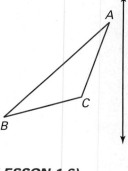

Refer to the figure below for Exercises 35–38. **(LESSON 1.6)**

35. Translate the triangle 5 units along the given line.

36. Draw $\overline{AA'}$, $\overline{BB'}$, and $\overline{CC'}$.

37. What is the relationship among the three segments you drew in Exercise 36?

38. What is the relationship between the given line and each of the segments you drew in Exercise 36?

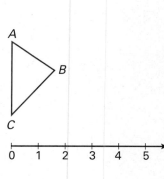

39. Copy the figure at right onto folding paper. Reflect $\triangle ABC$ across line ℓ, and then reflect the image across line m.

40. What is the relationship between $\triangle ABC$ and the final image (after two reflections)? What single transformation would produce the same image?

41. Copy the figure at right onto folding paper. Reflect $\triangle DEF$ across line n, and then reflect the image over line p.

42. What is the relationship between $\triangle DEF$ and the final image (after two reflections)? What single transformation would produce the same image?

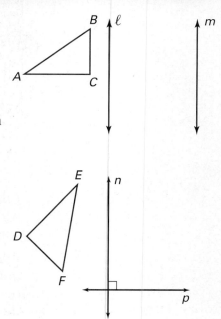

Building a System of Geometry Knowledge

Objectives

- Identify and use the Algebraic Properties of Equality.

- Identify and use the Equivalence Properties of Equality and of Congruence.

- Link the steps of a proof by using properties and postulates.

Why In algebra, you used the Properties of Equality to link the steps of a proof. Each link is a justification which guarantees that each statement you write in a proof is true—assuming that the statements you start with are true.

The first geometry "textbook," The Elements, *was written by Euclid, a Greek mathematician who lived in Alexandria, Egypt, around 300 B.C.E.*

Algebraic Properties of Equality

The importance of Euclid's work lies not so much in what he discovered as in *the way he organized the existing knowledge of geometry* of his time. Starting from simple beginnings, he built up a large system of geometry knowledge.

Euclid began *The Elements* with five basic postulates, or statements that are accepted as true without proof. In addition to the postulates, Euclid included twenty-three definitions and five statements he called "common notions." The second common notion reads as follows:

If equals are added to equals, then the wholes are equal.

You may recognize this from your study of algebra as the Addition Property of Equality. This property is used to solve equations, as in the example below.

$$x - 3 = 5$$
$$x - 3 + 3 = 5 + 3 \qquad \textit{Addition Property of Equality}$$
$$x = 8 \qquad \textit{Simplify.}$$

Notice that *equals* (the 3s) are added to *equals* (the sides of the equation) to give two *wholes* (the sides of the new equation), which are themselves equal.

Algebraic Properties of Equality

Let a, b, and c be real numbers or expressions representing real numbers.

Addition Property	If $a = b$, then $a + c = b + c$.	**2.4.1**
Subtraction Property	If $a = b$, then $a - c = b - c$.	**2.4.2**
Multiplication Property	If $a = b$, then $ac = bc$.	**2.4.3**
Division Property	If $a = b$ and $c \neq 0$, then $\frac{a}{c} = \frac{b}{c}$.	**2.4.4**
Substitution Property	If $a = b$, you may replace a with b in any true equation containing a and the resulting equation will still be true.	**2.4.5**

Algebra

In the figure, the lengths of $\overline{AB}$ and $\overline{CD}$ are equal. What can you conclude about the lengths of the two overlapping segments, $\overline{AC}$ and $\overline{BD}$?

It is easy to see the following:

$$AC = 4 + 5 = 9$$
$$BD = 5 + 4 = 9$$

Thus, $AC = BD$.

CRITICAL THINKING

How does the conclusion illustrate the Addition Property of Equality? In terms of Euclid's second common notion, what *equals* are added to what other *equals*?

Linking Steps to Prove a Theorem

The following example illustrates how each step in a proof can be linked to the information that is given. The proof is a continuation of the example about overlapping segments. The result is called a *theorem*.

EXAMPLE ❶ In the figure, $AB = CD$. Prove that $AC = BD$.

● **SOLUTION**

PROOF

You are interested in the lengths **AC** and **BD**. Because B is between A and C and C is between B and D, the Segment Addition Postulate can be used to write an equation for each length.

$$AB + BC = AC \qquad BC + CD = BD$$

You are given **AB** = **CD**. Notice what happens when you add BC to both sides of the equation.

*The two sides of the new equation express the lengths **AC** and **BD**.*

$AB = CD$	Given
$\underbrace{AB + BC}_{AC} = \underbrace{BC + CD}_{BD}$	Addition Property of Equality
	Segment Addition Postulate

Now you can substitute **AC** and **BD** into the equation to get the conclusion you want: **AC** = **BD**.

The result that was proven in Example 1, along with its converse, can be stated as a theorem. You will be asked to prove the converse in Exercise 21. In mathematics, a **theorem** is a statement that has been proved deductively. In your future work in geometry, you can use a theorem to justify a statement without writing out the whole proof.

Overlapping Segments Theorem

Given a segment with points A, B, C, and D arranged as shown, the following statements are true:

1. If $AB = CD$, then $AC = BD$.
2. If $AC = BD$, then $AB = CD$.

2.4.6

Two-Column Proofs

The proof in Example 1 might be written out as shown below. A **two-column** format has been used. This format is especially convenient for many of the proofs you will do in your study of geometry.

As you may notice, the steps in the final form of a proof may not be in the same order as the steps you followed to discover the ideas.

TWO-COLUMN PROOF

You can use boxes and colors to make the proof easier to understand.

Statements	Reasons
1. $AB = CD$	Given
2. $AB + BC = BC + CD$	Addition Property of Equality
3. $AB + BC = AC$	Segment Addition Postulate
4. $BC + CD = BD$	Segment Addition Postulate
5. $AC = BD$	Substitution Property of Equality

Paragraph Proofs

An alternative to a two-column proof is a **paragraph proof**. An advantage of a paragraph proof is that you have a chance to explain your reasoning in your own words. A paragraph proof of the Overlapping Segments Theorem might read as follows:

PROOF

You are given $AB = CD$. Add BC to both sides of the equation, resulting in $AB + BC = BC + CD$. In the figure, $AB + BC = AC$ and $BC + CD = BD$ by the Segment Addition Postulate. The expressions on the left of these equations match the expressions in the previous equations, so you can substitute the equivalent expressions, AC and BD. The result is $AC = BD$.

The Equivalence Properties of Equality

Algebra

In addition to the Algebraic Properties of Equality, there are three important properties known as the *Equivalence Properties of Equality*. They are so obvious that you probably don't think of them when you use them. In geometry, however, they are often used to justify steps in a proof.

Equivalence Properties of Equality		
Reflexive Property	For any real number a, $a = a$.	**2.4.7**
Symmetric Property	For all real numbers a and b, if $a = b$, then $b = a$.	**2.4.8**
Transitive Property	For all real numbers a, b, and c, if $a = b$ and $b = c$, then $a = c$.	**2.4.9**

There are also equivalence properties for relations other than equality, as you will learn in the next section.

The Equivalence Properties of Congruence

Congruence, like the relation of equality, satisfies the equivalence properties. The congruent shapes below illustrate this fact. Any relation that satisfies these three equivalence properties is called an **equivalence relation**.

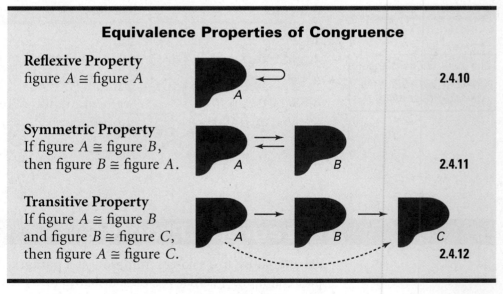

Equivalence Properties of Congruence

Reflexive Property
figure $A \cong$ figure A — **2.4.10**

Symmetric Property
If figure $A \cong$ figure B, then figure $B \cong$ figure A. — **2.4.11**

Transitive Property
If figure $A \cong$ figure B and figure $B \cong$ figure C, then figure $A \cong$ figure C. — **2.4.12**

CRITICAL THINKING

In Lesson 1.4 you used a compass to draw congruent segments on a line. What property of congruence guarantees that the segments are actually congruent? (Hint: Imagine a segment that connects the points of the compass.)

EXAMPLE **2**

APPLICATION
HOBBIES

The first stamp pictured at right measures 3 centimeters by 6 centimeters. The second stamp is congruent to the first. The third stamp is congruent to the second.

From the given information, what can you conclude about the first and the third stamps? State your conclusions by using geometry terms. What property discussed in this lesson justifies your conclusion?

● **SOLUTION**

The first stamp is congruent to the third stamp (and so the measurements of the stamps are equal). The justification for this conclusion is the Transitive Property of Congruence.

Russian stamps commemorating the Russian American Company

TRY THIS Create examples of your own that illustrate the Equivalence Properties of Congruence. Share your answers with your classmates.

Exercises

● *Communicate*

1. Why is it necessary to use postulates in geometry?

2. Explain the difference between a postulate and a theorem.

3. Rephrase Euclid's second common notion in your own words. Give an example that uses this notion.

4. Euclid's first common notion reads as follows:

Things which are equal to the same thing are also equal to one another.

Rephrase this statement in your own words. How is Euclid's first common notion related to the Equivalence Properties of Equality?

● *Guided Skills Practice*

For Exercises 5 and 6, refer to the diagram below. *(EXAMPLE 1)*

W   X   Y   Z

5. $\overline{WX} \cong \overline{YZ}$
$WX = 8$
$WY = 19$
Find XZ.

6. $\overline{WY} \cong \overline{XZ}$
$WX = 30$
$WY = 75$
Find YZ.

For Exercises 7 and 8, refer to the diagram below, in which △CDE ≅ △C′D′E′ and △C′D′E′ ≅ △C″D″E″. *(EXAMPLE 2)*

7. What conclusion can you draw about △*CDE* and △*C″D″E″*?

8. What property justifies your conclusion?

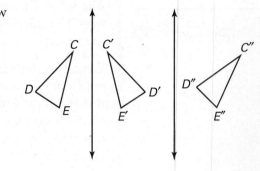

Practice and Apply

PROOFS

In Exercises 9–12, identify the Equivalence Properties of Equality that justify the indicated steps.

9.
$x + 6 = 14$	*Given*
$x + 6 - 6 = 14 - 6$	_?_ *Property*
$x = 8$	

10.
$2x - 3 = 17$	*Given*
$2x - 3 + 3 = 17 + 3$	_?_ *Property*
$2x = 20$	
$2x \div 2 = 20 \div 2$	_?_ *Property*
$x = 10$	

11.
$a + b = c + d$	*Given*
$c + d = e + f$	*Given*
$a + b = e + f$	_?_ *Property*

12.
$AB + CD = XY$	*Given*
$CD + DE = XY$	*Given*
$XY = CD + DE$	_?_ *Property*
$AB + CD = CD + DE$	_?_ *Property*

PROOF

In Exercises 13–19, you will complete and prove the Overlapping Angles Theorem.

Overlapping Angles Theorem

Given ∠*AOD* with points *B* and *C* in its interior as shown, the following statements are true:

1. If m∠*AOB* = m∠*COD*, then _?_ .
2. If m∠*AOC* = m∠*BOD*, then _?_ .

2.4.13

Refer to the figure at right to answer Exercises 13–19.

13. ____?____ + ____?____ = m∠*MLP*

14. ____?____ + ____?____ = m∠*NLQ*

15. What postulate justifies your answers to Exercises 13 and 14?

Part 1: Let m∠MLN = m∠PLQ.

16. Use the Addition Property of Equality to add m∠NLP to both sides of the equation.

17. Use the Substitution Property of Equality, the Angle Addition Postulate, and your conclusions to Exercises 13, 14, and 16 to write the conclusion for the first part of the Overlapping Angles Theorem.

Part 2: Let m∠MLP = m∠NLQ.

18. Use the Subtraction Property of Equality to subtract m∠NLP from both sides of the equation.

19. Use the Substitution Property of Equality, the Angle Addition Postulate, and your conclusions to Exercises 13, 14, and 18 to write the conclusion for the second part of the Overlapping Angles Theorem.

20. Prove the second part of the Overlapping Segments Theorem (see page 109), which is the converse of the first part. Draw and label your own diagram. (Hint: Use the Subtraction Property of Equality.)

Refer to the diagram at right, in which WX = YZ. Use the Overlapping Segments Theorem to complete the following:

21. $WX = a$, $WY = 15$, $XZ = $ _?_

22. $WY = 9$, $XY = 2a + 1$, $YZ = a^2$,

   $a = $ _?_ , $WZ = $ _?_

Refer to the diagram at right, in which m∠EDF = m∠HDG. Use the Overlapping Angles Theorem to complete the following:

23. m∠EDF = 25°, m∠EDG = 85°, m∠HDF = _?_

24. m∠EDG = $(9x - 8)°$, m∠FDG = $(6x + 8)°$, m∠GDH = $(2x - 4)°$,

   $x = $ _?_ , m∠HDF = _?_ , m∠HDG = _?_

Refer to the three triangles below for Exercises 25–28.

25. Complete the conclusion below.

   $AB = DE$

   $DE = GH$

   $AB = ?$

26. Complete the conclusion below.

   m∠ABC = m∠EDF

   m∠EDF = m∠GHI

   m∠ABC = ?

27. Which property justifies your answer to Exercise 25?

28. Which property justifies your answer to Exercise 26?

29. JoAnn wears the same size hat as April. April wears the same size hat as Lara. Will Lara's hat fit JoAnn? What property of congruence justifies your answer?

(See Exercise 29.)

Given: m∠BAC + m∠ACB = 90°
m∠DCE + m∠DEC = 90°
m∠ACB = m∠DCE

Prove: m∠BAC = m∠DEC

Proof:

Statements	Reasons
m∠BAC + m∠ACB = 90°	Given
m∠DCE + m∠DEC = 90°	**30.** ___?___
m∠BAC + m∠ACB = m∠DCE + m∠DEC	**31.** ___?___
m∠ACB = m∠DCE	Given
m∠BAC + m∠ACB = m∠ACB + m∠DEC	**32.** ___?___
33. ___?___	Subtraction Property

Given: m∠1 + m∠4 + m∠5 = 180°
m∠4 = m∠2
m∠5 = m∠3

Prove: m∠1 + m∠2 + m∠3 = 180°

PROOFS

Proof: Since m∠4 = m∠2, you can replace m∠4 with m∠2 in the equation **34.** _____?_____ by **35.** _____?_____. Similarly, you can replace m∠5 with **36.** _____?_____, giving **37.** _____?_____.

CHALLENGE

Given: m∠CBD = m∠CDB
m∠ABD = 90°
m∠EDB = 90°

Prove: m∠ABC = m∠EDC

Proof:

Statements	Reasons
m∠ABD = 90°	Given
m∠ABC + m∠CBD = m∠ABD	**38.** ___?___
39. m∠ABC + m∠CBD = ___?___	Transitive (or Substitution) Property
m∠EDB = 90°	**40.** ___?___
41. ___?___	Angle Addition Postulate
m∠CDE + m∠CDB = 90°	**42.** ___?___
m∠ABC + m∠CBD = m∠CDE + m∠CDB	**43.** ___?___
m∠CBD = m∠CDB	**44.** ___?___
m∠ABC + m∠CBD = m∠CDE + m∠CBD	**45.** ___?___
46. ___?___	Subtraction Property

Examine the relationships below. Determine which of the following properties each relationship satisfies: the Reflexive Property, the Symmetric Property, and the Transitive Property. Then state whether the relationship is an equivalence relation. If a property is not satisfied by a given relationship, give a counterexample.

47. $a < b$, where a and b are real numbers.

48. $a \leq b$, where a and b are real numbers.

49. a is divisible by b, where a and b are real numbers.

50. Figure A is a reflection of figure B.

51. Figure A is a rotation of figure B.

52. A is a sister of B, where A and B are people.

53. A has the same last name as B, where A and B are people.

Look Back

Recall the five postulates about points, lines, and planes that you discovered on page 12 and refer to the figure below to answer Exercises 54–59. *(LESSON 1.1)*

54. Name three points that determine plane $\mathcal{R}$.

55. Which postulate justifies your answer to Exercise 54?

56. Name two points that determine line ℓ.

57. Which postulate justifies your answer to Exercise 56?

58. Name the intersection of plane $\mathcal{R}$ and plane $\mathcal{M}$.

59. Which postulate justifies your answer to Exercise 58?

60. PHOTOGRAPHY A *tripod* is a three-legged stand used by photographers to hold their cameras steady. Why do you think tripods provide more stability than four-legged stands? Which of the five postulates justifies your answer?
(LESSON 1.1)

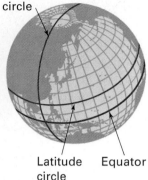

Longitude circle

Latitude circle   Equator

Because the system of logic we use in geometry is based on the postulates of Euclid, it is commonly known as *Euclidean geometry*. It is possible to have geometries in which one or more of Euclid's postulates do not hold. Geometry on a sphere, such as Earth's surface, is known as *spherical geometry*. In spherical geometry, a *line* is defined as a circle that divides the sphere into two equal halves.

61. Using this definition of a line, which of the following are lines?
- lines of latitude
- lines of longitude
- the equator

62. Explain why lines are defined this way in spherical geometry. (Hint: Pick two points on a globe and stretch a string between them to find the shortest distance from one point to the other.)

63. Explain why the following postulate is not true in spherical geometry:

Through any two points, one and only one line passes.

APPLICATION

64. NAVIGATION Pilots navigating long distances often travel along the lines of spherical geometry, called *great circles*. Using a globe and string, determine the shortest route for a pilot traveling from Washington, D. C., to London, England. What do you notice?

PORTFOLIO ACTIVITY

In a game of nim, if you leave two rows of two counters each, then you can always win, in either version of the game. (See pages 79 and 87.)

The following strategies allow you to achieve this winning combination—or an even quicker win, depending on your opponent's moves.

- On your turn, leave two rows with the same number of counters in each row.

    or

- On your turn, leave three rows with one, two, and three counters, in any order.

Explain how these combinations can be converted to eventual wins, no matter what your opponent does, in either version of the game.

In either version of nim, the person who moves first can always win. Try to discover the first move that will allow you to always win. Explain the strategy behind this move.

These are winning combinations.

Conjectures That Lead to Theorems

1 region   2 regions

4 regions   8 regions   16 regions

Objectives

- Develop theorems from conjectures.

- Write two-column and paragraph proofs.

Do you see a pattern in the number of regions in the circles? Do you think the pattern will continue?

Why When you make predictions based on patterns, you are using inductive reasoning.

A Need for Proof

True or false?

From the pattern of circles above, you might make this conjecture:

The number of regions doubles each time a point is added.

This conjecture can be tested by drawing pictures. What do you notice about the figure with 6 points?

6 points
? regions

As you can see, the number of regions for 6 points is not 32 but 31. Thus, the conjecture is false. For a conjecture to be considered true by mathematicians it must first be proven deductively, as in Activity 1.

Activity 1

The Vertical Angles Conjecture

Vertical angles are the opposite angles formed by two intersecting lines. A pair of scissors suggests (approximately) the idea of vertical angles, with the blades making one of the angles and the handles forming the other.

PROOF

1. Draw several pairs of intersecting lines. In the figure at right, $\angle 1$ and $\angle 2$ form one pair of vertical angles, and $\angle 3$ and $\angle 4$ form another.

2. Measure each pair of vertical angles. What do you notice?

3. Make a conjecture about vertical angles.

4. What is the relationship between $\angle 3$ and $\angle 1$? between $\angle 3$ and $\angle 2$?

5. Complete the following:

   $m\angle 1 + m\angle 3 = $ _____?_____ $m\angle 2 + m\angle 3 = $ _____?_____

6. What Property of Equality leads to the following conclusion?

   $m\angle 1 + m\angle 3 = m\angle 2 + m\angle 3$

CHECKPOINT ✔ 7. What Property of Equality leads to the conclusion that $m\angle 1 = m\angle 2$?

Inductive and Deductive Reasoning

Inductive reasoning is the process of forming conjectures that are based on observations. As you have seen, a conjecture can turn out to be false; thus, inductive reasoning is not accepted in mathematical proofs.

Activity 1 began with inductive reasoning and concluded with deductive reasoning. In Steps 1–3, you used inductive reasoning to make a conjecture based on observations. In Steps 4–7, you use deductive reasoning to complete an informal proof. The conjecture you made in Activity 1 is stated and proved formally below.

Vertical Angles Theorem

If two angles form a pair of vertical angles, then they are congruent.

2.5.1

TWO-COLUMN PROOF

Given: $\angle 1$ and $\angle 2$ are vertical angles.
Prove: $\angle 1 \cong 2$

Proof:

Statements	Reasons
1. $\angle 1$ and $\angle 2$ are vertical angles.	Given
2. $m\angle 1 + m\angle 3 = 180°$ $m\angle 2 + m\angle 3 = 180°$	Linear Pair Property
3. $m\angle 1 + m\angle 3 = m\angle 2 + m\angle 3$	Substitution Property of Equality
4. $m\angle 1 = m\angle 2$ $(\angle 1 \cong \angle 2)$	Subtraction Property of Equality

In Activity 2, you will explore another example of an inductive result that is proven as a theorem.

Activity 2

Reflections Across Parallel Lines

1. Draw two parallel lines, ℓ_1 and ℓ_2, and $\triangle ABC$ as shown in the figure at right.

2. Reflect $\triangle ABC$ across ℓ_1. Label its image $\triangle A'B'C'$.

3. Reflect $\triangle A'B'C'$ across ℓ_2. Label its image $\triangle A''B''C''$.

4. Study the relationship between $\triangle ABC$ and $\triangle A''B''C''$. What single transformation do you think would produce $\triangle A''B''C''$ from $\triangle ABC$?

5. Measure the distances AA'', BB'', and CC''. What do you notice?

6. Measure the distance between ℓ_1 and ℓ_2. How does this distance relate to the distances you measured in Step 5?

7. Do all of the points on the triangle seem to have moved in the same direction? Explain your answer.

CHECKPOINT ✔ 8. Complete the following theorem, which is proven below:

Theorem

Reflection across two parallel lines is equivalent to a ___?___ of ___?___ the distance between the lines and in a direction ___?___ to the lines.

2.5.2

CRITICAL THINKING

What must you prove about the reflection of a point across two parallel lines in order to show that the theorem you completed above is true?

PROOF

The diagram at right suggests a proof of the result in Step 6 of Activity 2.

1. Which of the indicated distances are equal? Why?

2. What is the distance between ℓ_1 and ℓ_2?

3. What does the expression $2D_1 + 2D_2$ represent in the diagram? How does it compare with the distance between ℓ_1 and ℓ_2?

CRITICAL THINKING

Do you think the result from Step 6 of Activity 2 is true for any point on the figure? How could you prove that each point moves in the same direction?

The Importance of Theorems

Conjectures, like the prediction about the regions of a circle at the beginning of this lesson, may turn out to be false. This is because they are based on a finite number of observations or measurements. You should always ask yourself whether there is a case that might prove your conjecture to be false.

Theorems are different from conjectures. In the Vertical Angles Theorem, the measures of the angles in the diagram do not matter. They could be any size at all, and the proof would still work. This is why the theorem is true for all possible pairs of vertical angles.

Exercises

● Communicate

1. Consider the following statement: Every integer is less than 1,000,000,000. Does the statement seem true for integers you use in everyday life? Is it really true?

2. Explain the difference between inductive and deductive reasoning.

3. Why is deductive reasoning the only type of reasoning allowed in proofs?

4. Explain the difference between theorems and conjectures.

5. Describe how you can translate a given figure by using only reflections.

APPLICATION

6. **ENTERTAINMENT** If you have enjoyed all of the films by a certain director that you have seen so far, does this mean that you will enjoy the next film of his that you see? Explain your answer.

Films by one director

7. Use the results of Activity 1 to find the measures of all of the angles in the figure at right. *(ACTIVITY 1)*

Copy the figure below onto folding paper and use it in Exercises 8 and 9.

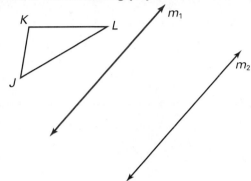

8. Reflect △JKL across m_1, and then reflect the image across m_2. *(ACTIVITY 2)*

9. Reflect △JKL across m_2, and then reflect the image across m_1. How is the final image different from the final image in Exercise 8? *(ACTIVITY 2)*

Practice and Apply

Refer to the diagram below, which consists of three intersecting lines. For Exercises 10–12, tell which angle is congruent to the given angle.

10. ∠LEI

11. ∠SEF

12. ∠PED

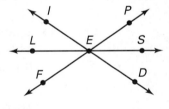

For each pair of intersecting lines, find m∠ABC.

13.

160°

14.

112°

15.

16.

For Exercises 17–20, find the value of *x* and m∠*ABC*.

17.

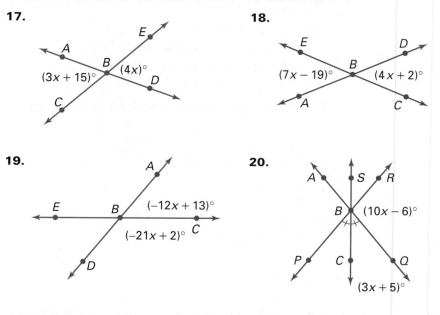

A
E
B (4x)°
(3x + 15)° D
C

18.

E D
B
(7x − 19)° (4x + 2)°
A C

19.

A
E B (−12x + 13)°
(−21x + 2)° C
D

20.

A S R
B (10x − 6)°
P C Q
(3x + 5)°

Tell whether each argument is an example of inductive or deductive reasoning. Is the argument a proof? Why or why not?

21. Every time John eats strawberries, he breaks out in hives. Therefore, John is allergic to strawberries.

22. If Erika did not turn in her homework, then she made a bad grade. Erika did not turn in her homework. Therefore, Erika made a bad grade.

23. Angles that form a linear pair are supplementary. Therefore, ∠1 and ∠2 are supplementary.

1 2

24. The number of black-footed ferrets has increased each year since 1985. Therefore, there will be more black-footed ferrets next year than there are this year.

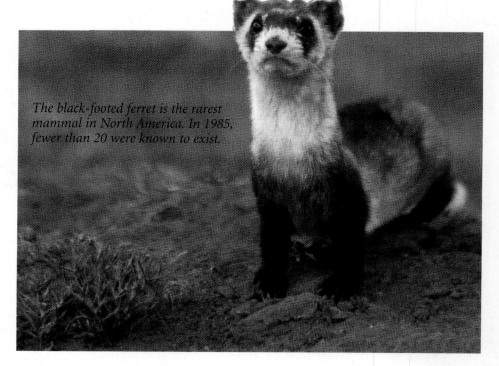

The black-footed ferret is the rarest mammal in North America. In 1985, fewer than 20 were known to exist.

In Exercise 25–27, complete the two-column proof of the following theorem:

Congruent Supplements Theorem

If two angles are supplements of congruent angles, then the two angles are congruent.

2.5.3

Given: ∠1 ≅ ∠3, ∠1 and ∠2 are supplementary, and ∠3 and ∠4 are supplementary.

Prove: ∠2 ≅ ∠4

Proof:

Statements	Reasons
m∠1 + m∠2 = 180° m∠3 + m∠4 = 180°	Definition of supplementary angles
m∠1 + m∠2 = m∠3 + m∠4	**25.** ?
∠1 ≅ ∠3 (m∠1 = m∠3)	**26.** ?
m∠1 + m∠2 = m∠1 + m∠4	Substitution Property
m∠2 = m∠4 (∠2 ≅ ∠4)	**27.** ?

PROOF

28. Write a paragraph proof of the Congruent Supplements Theorem.

In Exercises 29–34, refer to Activity 2 and the proof outlined on page 119. Refer to the figure below. Suppose that △MNO is reflected across line *m*, and then its image is reflected across line *n*. What is the distance from △MNO to △M″N″O″ if the distance from *m* to *n* is

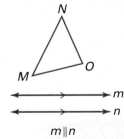

29. 5 cm?

30. 10 cm?

31. *x* cm?

Refer to the figure below. Suppose that you wish to use two reflections to translate the figure in the direction of the arrow.

32. If you wish to translate the figure 10 cm, how far apart should the parallel lines used in the reflections be?

33. How can you determine the location of the parallel lines?

34. Is there more than one pair of parallel lines that will give a translation of 10 cm? Use diagrams to illustrate your answer.

For Exercises 35–38, copy each figure onto folding paper. Reflect each figure across ℓ_1, and then reflect the image across ℓ_2. Draw an arrow that represents the direction and distance that the figure has been translated.

35.

36.

37.

38.

CHALLENGE

Exercises 39–43 refer to the pattern in the number of regions in the circles on page 117.

Draw circles with 6 and with 7 points, and connect the points in each circle with segments. To be sure that you have the maximum number of regions, make sure there are no points where three segments intersect inside the circle. (Hint: Do not space the points regularly on the circle.) How many regions are possible for a circle with

You may want to use geometry graphics software for Exercise 40.

39. 6 points?

40. 7 points?

One way to find a pattern in a sequence is by subtracting each term from the next. This gives a new sequence, called the *first differences*. This method can be continued to find second differences, third differences, etc., until the pattern is clear.

Algebra

41. Complete the chart below. (Work from the bottom up to fill in the chart.)

Number of regions	1	2	4	8	16	?	?	?
First differences		1	2	4	8	?	?	?
Second differences			1	2	4	?	?	?
Third differences				1	2	3	4	?

42. Using the pattern, what conjecture would you make about the number of regions in a circle with 8 points? with 9 points? with 10 points?

43. If you found that the pattern in the chart above was correct for 100, 1000, or even more points on a circle, would you have proven that the pattern is correct for any number of points? Why or why not?

Refer to the figure below for Exercises 44–46.
(LESSON 1.1)

44. Name four points in the figure.

45. Name four segments in the figure.

46. Name four planes in the figure.

Exercises 47 and 48 refer to the statements below. *(LESSON 2.2)*

> If I am well rested in the afternoon, then I am in a good mood.
> If I sleep until 8:00 A.M., then I am well rested in the afternoon.
> If it is Saturday, then I sleep until 8:00 A.M.

47. Arrange the statements to form a logical chain.

48. Write the conditional statement that follows from the argument formed by the logical chain in Exercise 47.

49. Draw an Euler diagram to illustrate the argument below. *(LESSON 2.2)*

> Every member of the Culver High School football team goes to Culver High School. Brady is a member of the Culver High School football team. Therefore, Brady goes to Culver High School.

50. Below is a chain of conditionals. Construct an Euler diagram for these statements, and complete the concluding statement. *(LESSON 2.2)*

> If you are taking geometry, then you are developing good reasoning skills.
>
> If you are developing good reasoning skills, then you will be able to succeed in many different careers.
>
> If you are able to succeed in many different careers, then you will be able to choose your profession.
>
> Therefore, if you are taking geometry, then ____?____.

(See Exercise 49.)

Look Beyond

51. Copy the figure at right. Reflect △ABC across ℓ_1, and then reflect its image across ℓ_2. Label the final image △A″B″C″.

52. What single transformation would produce the same final image as the two reflections?

53. What seems to be special about the intersection point, D?

54. Use a protractor to measure ∠ADA″, ∠BDB″, and ∠CDC″. What seems to be the relationship between the measures of these angles and the measure of ∠1?

55. Write a conjecture about the reflection of a figure across two intersecting lines. Include your results from Exercise 54 in your conjecture.

Logic Puzzles and Games

Activity 1

The game of sprouts was invented by mathematicians John Conway and Michael Paterson. The object is to be the last player to make a legal move.

Rules of the Game
Start by drawing three spots on a piece of paper.
The game is played by two people, who take turns.

To make a move, draw a curve joining two spots or starting and ending at the same spot, and place a new spot on the curve.
• No curve can cross another curve.
• No spot can have more than three curves coming from it.

Play a few games of sprouts with a friend. Try to figure out a strategy that will enable you to always win if you have the first turn.

You can vary the game by starting with four, five, or any other number of spots.

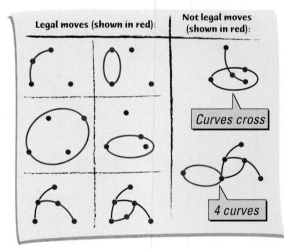

Activity 2

In the following puzzle, each letter represents a different digit. Use the clues given to determine the digits, and then arrange the letters in numerical order to solve the puzzle.

$$O + O = E$$
$$O \times O = E$$

```
  HAT
+ MAT
-----
 LEVE
```

$$T \times V = OL$$

$$M \times M = OM$$

$$H \times I = I$$

Activity 3

The following puzzle is by the British writer known as "Caliban."

Accomplishments

"My four granddaughters are all accomplished girls." Canon Chasuble was speaking with evident self-satisfaction. "Each of them," he went on, "plays a different musical instrument and each speaks one European language as well as—if not better than—a native."

"What does Mary play?" asked someone.

"The cello."

"Who plays the violin?"

"D'you know," said Chasuble, "I've temporarily forgotten. Anno Domini, alas! But I know it's the girl who speaks French."

The remainder of the facts which I elicited were of a somewhat negative character. I learned that the organist is not Valerie; that the girl who speaks German is not Lorna; and that Mary knows no Italian. Anthea doesn't play the violin; nor is she the girl who speaks Spanish. Valerie knows no French, Lorna doesn't play the harp, and the organist can't speak Italian.

What are Valerie's accomplishments?

	Cello	Violin	Organ	Harp	French	German	Italian	Spanish
Mary								
Valerie								
Lorna								
Anthea								
French								
German								
Italian								
Spanish								

Chapter Review and Assessment

SUMMARY OF POSTULATES AND THEOREMS

Lesson	Number	Theorem or Postulate
2.2	2.2.1 If-Then Transitive Property	Given: "If A then B, and if B then C." You can conclude: "If A then C."
2.4	2.4.1 Addition Property	If $a = b$, then $a + c = b + c$.
	2.4.2 Subtraction Property	If $a = b$, then $a - c = b - c$.
	2.4.3 Multiplication Property	If $a = b$, then $ac = bc$.
	2.4.4 Division Property	If $a = b$ and $c \neq 0$, then $\frac{a}{c} = \frac{b}{c}$.
	2.4.5 Substitution Property	If $a = b$, you may replace a with b in any true equation containing a and the resulting equation will still be true.
	2.4.6 Overlapping Segments Theorem	Given a segment with points A, B, C, and D (in order) the following statements are true: 1. If $AB = CD$, then $AC = BD$. 2. If $AC = BD$, then $AB = CD$.
	2.4.7 Reflexive Property of Equality	For any real number a, $a = a$.
	2.4.8 Symmetric Property of Equality	For all real numbers a and b, if $a = b$, then $b = a$.
	2.4.9 Transitive Property of Equality	For all real numbers a, b, and c, if $a = b$ and $b = c$, then $a = c$.
	2.4.10 Reflexive Property of Congruence	figure $A \cong$ figure A
	2.4.11 Symmetric Property of Congruence	If figure $A \cong$ figure B, then figure $B \cong$ figure A.
	2.4.12 Transitive Property of Congruence	If figure $A \cong$ figure B and figure $B \cong$ figure C, then figure $A \cong$ figure C.
	2.4.13 Overlapping Angles Theorem	Given $\angle AOD$ with points B and C in its interior, the following statements are true: 1. If m$\angle AOB =$ mCOD, then m$\angle AOC =$ mBOD. 2. If m$\angle AOC =$ mBOD, then m$\angle AOB =$ mCOD.

2.5	2.5.1 Vertical Angles Theorem	If two angles form a pair of vertical angles, then they are congruent.
	2.5.2 Theorem	Reflection across two parallel lines is equivalent to a translation of twice the distance between the lines and in a direction perpendicular to the lines.
	2.5.3 Theorem	Reflection across two intersecting lines is equivalent to a rotation about the point of intersection through twice the measure of the angle between the lines.

Key Skills & Exercises

LESSON 1.1

Key Skills

Give a proof of a conjecture.

The following table shows the number of points and the corresponding number of segments:

Points	2	3	4	5
Segments	1	3	6	10

Conjecture: For n points on a circle, the number of segments needed to connect all points is $\frac{n(n-1)}{2}$.

Proof: For each point, a segment connects it to every other point, giving $n-1$ segments. Therefore, the total number of segments should be $n(n-1)$. However, $\overline{AB}$ is same segment as $\overline{BA}$, so each segment is counted twice. Thus, the total must be divided by 2, giving $\frac{n(n-1)}{2}$.

Exercises

The horizontal line intersects the curve at three points and divides it into four pieces.

1. Into how many pieces will the curve be divided by two horizontal lines that each intersect the curve at three points each? by three horizontal lines?

2. Complete the table below, subtracting to find the first differences.

Lines	1	2	3	4
Sections	4	?	?	?
First differences		?	?	?

3. Based on the table, make a conjecture about the pattern in the number of sections.

4. Write a proof of your conjecture.

LESSON 2.2

Key Skills

Draw a conclusion from a conditional.

What conclusion follows from these statements?
    If an animal is a cat, then it has four legs.
    Dinah is a cat.
Conclusion: Dinah has four legs.

State the converse of a conditional.

Converse of the conditional above:
    If an animal has four legs, then it is a cat.

Exercises

Refer to the following statements:
    If a "star" doesn't flicker, then it is a planet.
    The evening star doesn't flicker.

5. Give the conclusion that follows from the statements above.

6. Write the converse of the conditional above. Is the converse true?

Arrange statements into a logical chain and draw a conclusion.

  a. If John gets his driver's license, then he will buy a car.

  b. If John buys a car, then he will not have any money.

  c. If John passes his driver's license exam, then he will get his driver's license.

The order of the logical chain is **c, a, b**. The conclusion is "If John passes his driver's license exam, then he will not have any money."

Exercises 7 and 8 refer to the following statements:

  a. If June tells Andrew a secret, then Andrew tells Tina the secret.

  b. If Charles tells June a secret, then June tells Andrew the secret.

  c. If Darren tells Charles a secret, then Charles tells June the secret.

7. Arrange the statements into a logical chain.

8. What conditional statement does the argument formed by this logical chain prove?

LESSON 2.3
Key Skills

Write a definition of an object.

These objects are trapezoids.

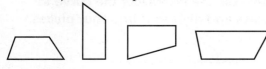

These objects are not trapezoids.

Which of the following objects are trapezoids?

Write a definition of a trapezoid.

Objects **b** and **c** are trapezoids. A trapezoid is a four-sided figure with exactly two parallel sides.

Determine whether a statement is a definition.

Is the following statement a definition?
    A right angle measures 90°.
If a statement is a definition, then the conditional form and its converse are both true.
Conditional: If an angle is a right angle,
           then it measures 90°.
Converse: If an angle measures 90°,
           then it is a right angle.
Since both are true, the statement is a definition.

Exercises

These objects are tori (singular, *torus*).

These objects are not tori.

9. Which of the following objects are tori?

10. Write a definition of *torus*.

Determine whether the following statements are definitions:

11. A square is a figure with four sides.

12. A point that divides a segment into two congruent parts is the midpoint.

LESSON 2.4
Key Skills

Use the Properties of Equality and Congruence to write proofs.

In the figure below, $\angle 1 \cong \angle 3$. Prove that $\angle 2 \cong \angle 4$.

Two-column proof:

Statements	Reasons
$m\angle 1 + m\angle 2 = 180°$	Linear Pair Property
$m\angle 3 + m\angle 4 = 180°$	Linear Pair Property
$m\angle 1 + m\angle 2 = $ $m\angle 3 + m\angle 4$	Substitution Property
$m\angle 1 = m\angle 3$	Given
$m\angle 2 = m\angle 4$ $(\angle 2 \cong \angle 4)$	Subtraction Property

Paragraph proof:

By the Linear Pair Property, $m\angle 1 + m\angle 2 = 180°$ and $m\angle 3 + m\angle 4 = 180°$. By the Substitution Property, $m\angle 1 + m\angle 2 = m\angle 3 + m\angle 4$. You are given $m\angle 1 = m\angle 3$, so by the Subtraction Property, $m\angle 2 = m\angle 4$, or $\angle 2 \cong \angle 4$.

Exercises

In the figure below, $m\angle 1 = 90°$.

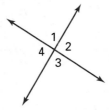

13. What property would you use to prove that $m\angle 2 = 90°$?

14. What property would you use to prove that $m\angle 4 = 90°$?

15. What theorem would you use to prove that $m\angle 3 = 90°$?

16. Use your answers from Exercises 13–15 to write a two-column or paragraph proof of the following conjecture:

> If two lines intersect to form a right angle, then all of the angles formed are right angles.

LESSON 2.5
Key Skills

Use deductive reasoning to prove a conjecture.

Make a conjecture about the angle formed by the angle bisectors of a linear pair, and prove your conjecture.

By measuring the angles, you can make the following conjecture: The angle bisectors of a linear pair form a right angle.

Exercises

Three numbers, *a*, *b*, and *c*, are called a *Pythagorean triple* if $a^2 + b^2 = c^2$.

17. Show that the numbers 3, 4, and 5 are a Pythagorean triple and that the numbers 5, 12, and 13 are another Pythagorean triple.

Given: the figure below, in which $\overrightarrow{OB}$ bisects $\angle AOC$ and $\overrightarrow{OD}$ bisects $\angle BOE$

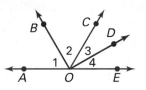

Prove: $m\angle 2 + m\angle 3 = 90°$

Proof:

Statements	Reasons
$m\angle 1 = m\angle 2$	Definition of angle bisector
$m\angle 3 = m\angle 4$	Definition of angle bisector
$m\angle 1 + m\angle 2 +$ $m\angle 3 + m\angle 4 = 180°$	Linear Pair Property
$m\angle 2 + m\angle 2 +$ $m\angle 3 + m\angle 3 = 180°$	Substitution Property
$2(m\angle 2 + m\angle 3) = 180°$	Distributive Property
$m\angle 2 + m\angle 3 = 90°$	Division Property

Four numbers, *a*, *b*, *c*, and *d*, are called a Pythagorean quadruple if $a^2 + b^2 + c^2 = d^2$.

18. Show that the numbers 3, 4, 12, and 13 are a Pythagorean quadruple.

19. If *a*, *b*, and *c* are a Pythagorean triple and *c*, *d*, and *e* are another Pythagorean triple, make a conjecture about *a*, *b*, *d*, and *e*.

20. Use Properties of Equality to prove the conjecture you made in Exercise 19.

Applications

21. COMPUTER PROGRAMMING The following is part of a computer program used to evaluate the quadratic formula:

```
INPUT A
INPUT B
INPUT C
IF B² – 4*A*C < 0
THEN
DISP "NO REAL SOLUTIONS"
END
```

Suppose that $A = 1$, $B = 3$, and $C = 4$. Can you determine what the program would do?

22. BIOLOGY In biological classification, animals are grouped into categories according to physical characteristics. Consider the following definitions:

Mammals are animals that have hair and produce milk. *Monotremes* are mammals that lay eggs.

Suppose that a biologist is examining a platypus and finds that it has hair, produces milk, and lays eggs. Is a platypus a monotreme? Why or why not?

The platypus, about half the size of a house cat, is a mammal that lays eggs. The young are nourished by milk from the mother's mammary glands.

Alternative Assessment

Performance Assessment

1. Consider the number of regions formed by intersecting lines in a plane.

Draw planes with 3, 4, and 5 intersecting lines, and count the number of regions formed. To form the maximum possible number of regions, there should be no parallel lines, and no three lines should intersect at one point. Extend each line as far as necessary to intersect all other lines.

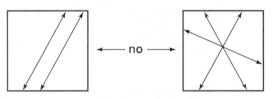

2. Complete the table below.

Lines	1	2	3	4	5
Regions	2	4	?	?	?

First differences   2   ?   ?   ?

3. Describe the pattern in the number of regions.

4. The pattern above can be found in Pascal's triangle. Complete at least the first eight rows of Pascal's triangle (see page 87).

   Subtract 1 from each number in the second row of the table in Exercise 2. Look for these numbers in Pascal's triangle. What do you notice?

Portfolio Projects

1. **BIOLOGY** Sometimes even when you know what something is, it can be difficult to define. For example, try defining the words *dog* and *cat*.

   List some characteristics of dogs and cats. In order to have a true definition, you must be able to write a biconditional statement.

   Compare your list of characteristics with your classmates' lists. Try to come up with definitions of *dog* and *cat* by completing the following conditionals:

   If an animal is a dog, then ____?____.

   If an animal is ____?____, then it is a dog.

   If an animal is a cat, then ____?____.

   If an animal is ____?____, then it is a cat.

© The Granger Collection, New York

2. The following phrase is part of a dictionary definition of *dog*: a domesticated canine. However, the definition of *canine* contains the following phrase: of or like a dog.

   What is the problem with these definitions? How does this relate to the use of undefined terms in geometry?

---

internet**connect**

The HRW Web site contains many resources to reinforce and expand your knowledge of the use of proofs in geometry. The Web site also provides Internet links to other sites where you can find information and real-world data for use in research projects, reports, and activities that involve proofs. Visit the HRW Web site at **go.hrw.com**, and enter the keyword **MG1 CH2**.

QUANTITATIVE COMPARISON For Items 1–4, write

A if the quantity in Column A is greater than the quantity in Column B;

B if the quantity in Column B is greater than the quantity in Column A;

C if the two quantities are equal; or

D if the relationship cannot be determined from the information given.

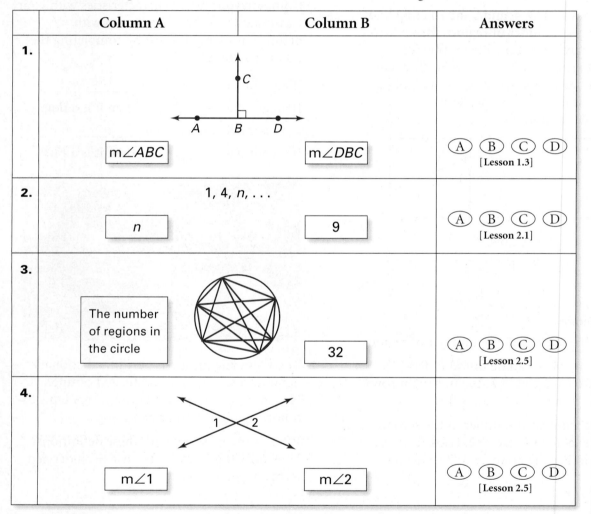

	Column A	Column B	Answers
1.	$m\angle ABC$	$m\angle DBC$	Ⓐ Ⓑ Ⓒ Ⓓ [Lesson 1.3]
2.	1, 4, n, . . .		
	n	9	Ⓐ Ⓑ Ⓒ Ⓓ [Lesson 2.1]
3.	The number of regions in the circle	32	Ⓐ Ⓑ Ⓒ Ⓓ [Lesson 2.5]
4.	$m\angle 1$	$m\angle 2$	Ⓐ Ⓑ Ⓒ Ⓓ [Lesson 2.5]

A square is divided into regions by pairs of perpendicular lines as shown. *(LESSON 2.1)*

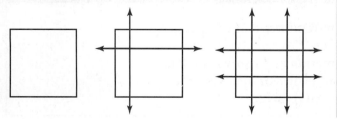

5. How many regions are formed by three pairs of perpendicular lines? by four pairs of perpendicular lines?

6. Write a conjecture about the number of regions formed by n pairs of perpendicular lines.

7. Write a proof of the conjecture you made in Item 6.

A triangle has vertices at (4, 1), (2, 2), and (3, 0). What are the vertices of the image after the following transformations? *(LESSON 1.7)*

8. reflection across the *x*-axis

9. translation 3 units to the left and 1 unit down

10. Write a conditional statement based on the Euler diagram below. *(LESSON 2.2)*

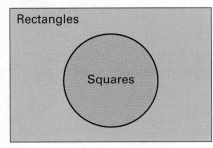

Rectangles

Squares

11. The figures below are splorts.

The figures below are not splorts.

Write a definition of *splort*. *(LESSON 2.3)*

For Items 12–15, refer to the diagram at right. *(LESSON 1.1)*

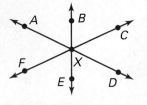

12. Name five points in the figure.

13. Name three lines in the figure.

14. Name four rays in the figure.

15. Name eight angles in the figure.

16. Which of the following is a definition? *(LESSON 2.3)*

　a. The midpoint of a segment is a point that divides the segment into two equal parts.

　b. An acute triangle is a triangle that has an angle of less than 90°.

　c. Two points determine exactly one line.

　d. An equilateral triangle is an isosceles triangle.

FREE-RESPONSE GRID
The following Items may be answered by using a free-response grid such as that commonly used by standardized-test services.

For Items 17 and 18, refer to the figure below. *(LESSONS 1.3 AND 2.5)*

17. m∠1 = _____?_____

18. m∠3 = _____?_____

For Items 19 and 20, refer to the information and the figure below. *(LESSON 1.3)*

m∠QPT = 140°
m∠QPR = (2x − 5)°
m∠RPS = (4x + 10)°
m∠SPT = (3x)°

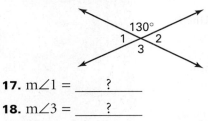

19. x = _____?_____

20. m∠RPS = _____?_____

Parallels and Polygons

RECTANGLES, TRIANGLES, AND HEXAGONS ARE examples of *polygons*. Patterns of polygons are often used for decorative purposes. The tiling pattern at right below is from the Alhambra, a famous Islamic fortress in Spain.

The interplay of parallel beams and polygons in the photograph of the Thorncrown Chapel below suggests the theme of this chapter.

As you will learn, parallel lines and their properties provide a basis for classifying and exploring four-sided polygons known as *quadrilaterals*.

Thorncrown Chapel, Eureka Springs, Arkansas Designed by E. Fay Jones

Piet Mondrian. Composition with Red, Blue and Yellow. *1930. Oil on Canvas. 20″ × 20″*

About the Chapter Project

In this chapter, you will investigate the properties of polygons, The key that unlocks the study of polygons is found in the properties of parallel lines, which you will explore in Lessons 3.3 and 3.4. In the Chapter Project, *String Figures,* you will learn to make one of the world's most popular string figures, which is known in the United States as "Jacob's ladder." The figure consists of polygons between two (roughly) parallel lines.

After completing the Chapter Project, you will be able to do the following:

- Make popular string figures by following a set of instructions.

- Appreciate the skills and ingenuity of the ancient peoples that created such figures.

About the Portfolio Activities

Throughout the chapter, you will be given opportunities to complete Portfolio Activities that are designed to support your work on the Chapter Project.

The theme of each Portfolio Activity and of the Chapter Project is geometric art.

- In the Portfolio Activity on page 147, you will study artful ways of forming the surface of a quilt with patches. The repetition of the individual shapes often results in intriguing overall designs.

- Tessellation, the art of covering a surface with congruent shapes that fit together without gaps or overlapping, is the subject of the Portfolio Activity on page 154. Translation tessellations are studied in depth.

- Rotation tessellations are the subject of the Portfolio Activity on page 197.

Symmetry in Polygons

Objectives

- Define *polygon*.

- Define and use *reflectional symmetry* and *rotational symmetry*.

- Define *regular polygon, center of a regular polygon, central angle of a regular polygon,* and *axis of symmetry.*

Why
Polygons appear all around you in man-made objects. Understanding the mathematical properties of polygons will help you understand how to use them for artistic and practical purposes.

Symmetrical polygons give this Native American blanket design an attractive appearance. They also have interesting mathematical properties.

Defining Polygons

In Lesson 2.3 you learned how to use examples and "nonexamples" to write definitions. Use the following figures to define a *polygon*:

These are polygons.   These are not polygons.

Compare your definition of a polygon with the definition below. Does your definition have all the requirements it needs? Does it have more than it needs?

Polygon

A **polygon** is a plane figure formed from three or more segments such that each segment intersects exactly two other segments, one at each endpoint, and no two segments with a common endpoint are collinear. The segments are called the **sides of the polygon**, and the common endpoints are called the **vertices of the polygon**. **3.1.1**

A polygon is named according to the number of its sides. Familiarize yourself with information in the table below.

Polygons Classified by Number of Sides

Triangle	3	Nonagon	9
Quadrilateral	4	Decagon	10
Pentagon	5	11-gon	11
Hexagon	6	Dodecagon	12
Heptagon	7	13-gon	13
Octagon	8	*n*-gon	*n*

Equiangular

Equilateral

An **equiangular polygon** is one in which all angles are congruent. An **equilateral polygon** is one in which all sides are congruent. A **regular polygon** is one that is both equiangular and equilateral.

The **center of a regular polygon** is the point that is equidistant from all vertices of the polygon. A **central angle** of a regular polygon is an angle whose vertex is the center of the polygon and whose sides pass through two consecutive vertices.

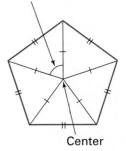

Central angle

Center

Reflectional Symmetry

Is the cat's face symmetrical?

Lay your pencil on the photograph of the blanket on page 138 so that the part of the photo on one side of your pencil is the mirror image of the part on the other side. The line of your pencil is an *axis of symmetry.*

Reflectional Symmetry

A figure has **reflectional symmetry** if and only if its reflected image across a line coincides exactly with the preimage. The line is called an **axis of symmetry**. 3.1.2

Imagine reflecting each of the following figures across the lines shown. As you can see, the reflected image will coincide exactly with the preimage. Therefore, each figure has reflectional symmetry, and each line is an axis of symmetry.

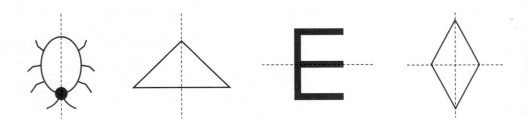

For Activity 1, recall the following classification scheme for triangles:

Triangles Classified by Number of Congruent Sides

Three congruent sides:	equilateral
At least two congruent sides:	isosceles
No congruent sides:	scalene

Equilateral   Isosceles   Scalene

Activity 1
Reflectional Symmetry in Triangles

YOU WILL NEED

folding paper, ruler, and protractor

1. Draw an example of each type of triangle shown above, and find any axes of symmetry that exist. How many axes of symmetry does each type of triangle have?

2. Fold each triangle along its axes of symmetry, if it has any. After folding, which angles seem to coincide?

3. Complete a table like the one below.

Type of triangle	Number of axes of symmetry	Number of congruent angles
equilateral	?	?
isosceles	?	?
scalene	?	?

CHECKPOINT ✔ 4. Study the relationship between each axis of symmetry and the side it intersects. Use your ruler and protractor to make measurements. Complete the following conjecture:

Triangle Symmetry Conjecture

An axis of symmetry in a triangle is the ____?____ ____?____ of the side it intersects, and it passes through the ____?____ of the angle opposite that side of the triangle.

Rotational Symmetry

Rotational Symmetry

A figure has **rotational symmetry** if and only if it has at least one rotation image, not counting rotation images of 0° or multiples of 360°, that coincides with the original image.   **3.1.3**

The flower at right has approximate rotational symmetry. Notice that it will coincide with itself 5 times if it is rotated completely about its center of rotation, which is also known as its center of symmetry. Thus, the figure of the flower is said to have *5-fold rotational symmetry*.

Note: *All* geometric figures have 0° and 360° rotational symmetry. A figure that has only 0° and 360° rotational symmetry is said to have only *trivial* rotational symmetry. For such a figure in a plane, *any* point in the plane of the figure is a center of rotation.

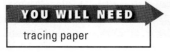

Rotational Symmetry in Regular Polygons

YOU WILL NEED

tracing paper

1. Draw or trace a regular pentagon such as the one in the illustration. Label the center *P* and the vertices *Q*, *R*, *S*, *T*, and *U*. Draw $\overrightarrow{PQ}$. Then copy the figure onto a sheet of tracing paper.

2. Use a pencil point to anchor the traced figure on top of the original at their centers so that the figures coincide.

3. Rotate the top figure counterclockwise until point *Q* on the traced figure coincides with point R on the original. Continue to rotate point *Q*, stopping at each point—*S*, *T*, *U*, and finally *Q*. How many rotations did you make in all?

CHECKPOINT ✔   4. Do the central angles of a regular polygon seem to be congruent? (You will prove that they are congruent in Lesson 4.2.)

Assuming that the central angles of a regular polygon are congruent, complete the rule below for finding the measure of a central angle of a regular polygon.

The Central Angle of a Regular Polygon

The measure, θ, of a central angle of a regular polygon with *n* sides is given by the following formula: $\theta = $ _____?_____ .

A windmill is a device for pumping water by using energy from the wind. If you connected the tips of the blades with segments, what figure would be formed? What is the measure of a central angle of the figure? For what degrees of rotation does the rotational image of the figure coincide with the preimage? What is the rotational symmetry of the figure?

SOLUTION

The windmill has 4 blades, so the segments connecting the tips of the blades will form a regular 4-sided polygon, that is, a square. Use the formula to find the measure of central angle:

$$\theta = 360 \div 4 = 90°$$

With each rotation of the figure through a central angle, the image of the figure coincides with its preimage. This happens for rotations of 90°, 180°, 270°, and 360°. At 360°, the figure is at its original position again. Counting the numbers of rotations in the list, the figure has 4-fold rotational symmetry.

Exercises

Communicate

Taj Mahal, India

1. Describe the position of all the axes of symmetry in the picture at left.

2. Explain how the picture at left illustrates the definition of symmetry. Identify a point and its preimage. What is the relationship between the point and preimage and their axis of symmetry?

3. Why are 0° and 360° rotations not used to define rotational symmetry?

4. What kinds of symmetry does a regular hexagon have? Describe all possible rotations and reflections of a regular hexagon.

5. Identify regions of the Escher woodcut at right that have rotational symmetry. Where are the centers of rotation?

6. Draw the axis of symmetry in the triangle at right. Does this triangle confirm your conjecture from Activity 1? Why or why not? *(ACTIVITY 1)*

7. Find the measure of a central angle for each regular polygon below. *(ACTIVITY 2)*

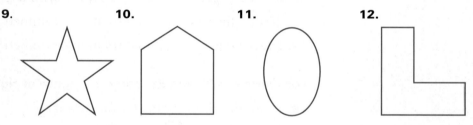

8. AGRICULTURE What is the rotational symmetry of a windmill with 4 blades? with 6 blades? *(EXAMPLE)*

● *Practice and Apply* ▬▬▬▬▬▬▬▬▬▬▬▬▬

For Exercises 9–12, copy each figure and draw all of the axes of symmetry.

9. **10.** **11.** **12.**

13. How many axes of symmetry does a circle have? Explain your answer.

Each figure below shows part of a shape with reflectional symmetry, with its axis of symetry shown as a black line. Copy and complete each shape.

14. **15.** **16.**

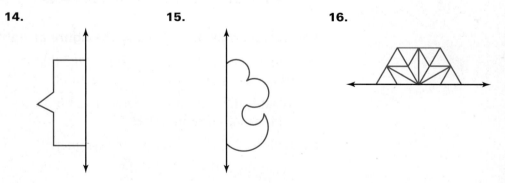

17. Which of the completed shapes from Exercises 14–16 also have rotational symmetry?

If a figure has *n*-fold rotational symmetry, then it will coincide with itself after a rotation of $\left(\frac{360}{n}\right)°$. For example, an equilateral triangle has 3-fold symmetry, so it will coincide with itself after a rotation of $\left(\frac{360}{3}\right)° = 120°$.

Each figure below shows part of a shape with the given rotational symmetry. Copy and complete each shape.

18.

4-fold

19.

2-fold

20.

6-fold

21.

3-fold

22. Which of the completed shapes from Exercises 18–21 also have reflectional symmetry?

23. Draw a figure with exactly 1 axis of symmetry.

24. Draw a figure with exactly 2 axes of symmetry.

25. Draw a figure with exactly 3 axes of symmetry.

26. Draw a figure with 5-fold rotational symmetry.

27. Draw a figure with 8-fold rotational symmetry.

For Exercises 28 and 29, copy the figure at right.

28. Draw an axis of symmetry for $\overline{AB}$ that passes through $\overline{AB}$ at a single point. What is the relationship between $\overline{AB}$ and the axis of symmetry?

$A \rule{2cm}{0.4pt} B$

29. Reflect $\overline{AB}$ over the axis of symmetry.

What point is the image of *A*? What point is the image of *B*? What segment is the image of $\overline{AB}$?

For Exercises 30 and 31, copy the figure at right.

30. Draw an axis of symmetry for $\angle ABC$ that passes through $\angle ABC$ at a single point.

What is the relationship between $\angle ABC$ and the axis of symmetry?

31. Reflect $\angle ABC$ over the axis of symmetry.

What ray is the image of $\overrightarrow{BA}$?
What ray is the image of $\overrightarrow{BC}$?
What angle is the image of $\angle ABC$?

32. Describe how to draw an axis of symmetry for any given angle.

Figure *ABCD* below is equilateral. Determine whether each of the following could be the result of a reflection, a rotation, or either:

33. $\overline{DX}$ is the image of $\overline{BX}$.

34. $\overline{AB}$ is the image of $\overline{AD}$.

35. $\overline{CB}$ is the image of $\overline{AD}$.

36. $\angle BAX$ is the image of $\angle BCX$.

37. $\angle DCX$ is the image of $\angle BCX$.

38. $\angle DAX$ is the image of $\angle BCX$.

39. $\angle CXD$ is the image of $\angle AXD$.

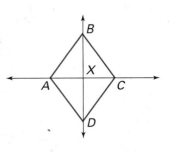

Algebra

Graph each equation below on a graphics calculator or graph paper. Then write an equation for the axis of symmetry of each graph.

40. $y = (x - 1)^2 + 3$

41. $y = 2(x - 4)^2 + 3$

42. $y = -(x + 2)^2 + 3$

43. $y = -2(x + 5)^2 + 3$

44. $y = |x| + 3$

45. $y = |x + 3|$

46. What is the measure of a central angle of a regular chiliagon (1000 sides)? What is the measure of a central angle of a regular myriagon (10,000 sides)?

47. Draw a quadrilateral that is equilateral but not equiangular.

48. Draw a quadrilateral that is equiangular but not equilateral.

APPLICATIONS

49. FURNITURE A table is in the shape of a regular octagon. The table can be extended by putting in a leaf, as shown. Is the extended table regular? equilateral? equiangular?

50. RECREATION The Ferris wheel shown at right has 16 cars. Imagine a polygon formed by connecting the cars. What is the measure of a central angle of this polygon?

CULTURAL CONNECTION: AFRICA The designs below are taken from Egyptian bowls dating back to 3500 B.C.E. Describe the symmetries in each design.

51.   **52.**   **53.**   **54.**

TRAFFIC SAFETY Examine each road sign below. Identify the type of polygon. Is it regular? Describe all of its symmetries (ignore any figures or words on the sign).

55. **56.** **57.** **58.**

Look Back

Refer to the diagram at right for Exercises 59–61.
(LESSON 1.1)

59. Name the intersection of and $\overleftrightarrow{AB}$ and $\overleftrightarrow{MN}$.

60. Name three points that determine plane $\mathcal{T}$.

61. Name the intersection of planes $\mathcal{S}$ and $\mathcal{T}$.

Refer to the diagram at right for Exercises 62–64.
(LESSON 1.1)

62. $m\angle ABD = 80°$, $m\angle ABC = 30°$, $m\angle CBD = \underline{\ ?\ }$

63. $m\angle ABC = 25°$, $m\angle CBD = 60°$, $m\angle ABD = \underline{\ ?\ }$

64. $m\angle ABC = m\angle CBD$, $m\angle ABD = 88°$, $m\angle ABC = \underline{\ ?\ }$

Look Beyond

France   Japan   Syria   United Kingdom   St. Lucia

Qatar   Micronesia   Canada   Macedonia   Israel

United States   Guyana   Russia   Ethiopia   Iceland

GEOGRAPHY Name all of the countries listed above whose flags have the type of symmetry given below.

65. reflectional, with a horizontal axis of symmetry

66. reflectional, with a vertical axis of symmetry

67. rotational symmetry of 180°

68. no symmetry

QUILTING The following designs are called 9-patch quilt blocks because they are based on a grid of 9 congruent squares.

Ohio Star

Shoo Fly

Card Trick

Star Cross

Friendship Star

1. Describe all of the types of symmetry in each quilt block shown above.

2. The quilt design at right consists of Ohio Star blocks. Copy one of the 9-patch blocks shown above or create your own. Design your own quilt by using the block you chose. It may be helpful to use graph paper.

3. Describe the symmetry of your overall quilt design. Did any interesting patterns occur that were not part of the original block?

Quilting is a traditional American social activity.

Properties of Quadrilaterals

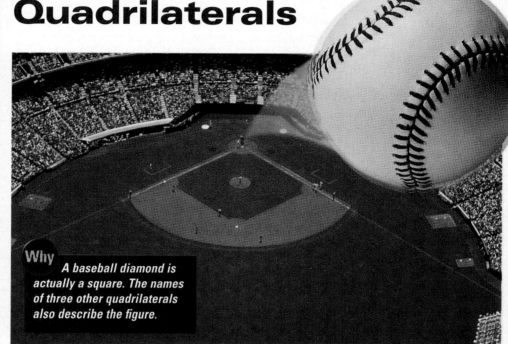

Objectives

● Define *quadrilateral, parallelogram, rhombus, rectangle, square,* and *trapezoid.*

● Identify the properties of quadrilaterals and the relationships among the properties.

Why *A baseball diamond is actually a square. The names of three other quadrilaterals also describe the figure.*

The length of each side of a baseball diamond is 90 feet, and the pitcher's mound lies on the diagonal from home plate to second base at a point about 6 feet closer to home plate than to second base.

Special Quadrilaterals

Any four-sided polygon is a **quadrilateral**. Quadrilaterals that have certain properties are called *special quadrilaterals*. Study the definitions below. (In the figures, arrowheads are used like tick marks to indicate that two lines, segments, or rays are parallel.)

A **trapezoid** is a quadrilateral with one and only one pair of parallel sides.

A **parallelogram** is a quadrilateral with two pairs of parallel sides.

A **rhombus** is a quadrilateral with four congruent sides.

A **rectangle** is a quadrilateral with four right angles.

A **square** is a quadrilateral with four congruent sides and four right angles.

Which of the quadrilaterals seem to have rotational symmetry? Which have reflectional symmetry? What can you learn about quadrilaterals from their symmetries? Keep this in mind as you do the Activities that follow.

Activity 1

Parallelograms

YOU WILL NEED

ruler and protractor
OR
geometry graphics software

1. Draw a parallelogram that is not a rhombus, a rectangle, or a square. Measure the angles and the sides of the figure. Which angles and sides appear to be congruent?

2. Draw diagonals to connect the vertices. Measure the diagonals. Which parts of the figure appear to be congruent? If you are using geometry drawing software, see if the conjecture holds when you vary the shape of your figure by dragging one of the vertices.

3. *Consecutive angles* of a polygon are angles that have a side in common. What do you notice about consecutive angles in a parallelogram?

CHECKPOINT ✔ 4. What conjectures can you make about the sides, angles, or other parts of the figure? Complete the conjectures below.

> **Conjectures: Properties of Parallelograms**
>
> Opposite sides of a parallelogram are _____?_____ .
>
> Opposite angles of a parallelogram are _____?_____ .
>
> Diagonals of a parallelogram _____?_____ .
>
> Consecutive angles of a parallelogram are _____?_____ .

Activity 2

Rhombuses

1. Draw a rhombus that is not a square. Draw diagonals to connect the vertices. Make measurements as in Activity 1.

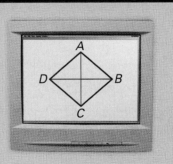

CHECKPOINT ✔ 2. Do the conjectures you made about parallelograms in Activity 1 seem to be true for your rhombus? Discuss why they should or should not be true for rhombuses. Complete the conjecture below.

> **Conjecture: A Property of Rhombuses**
>
> A rhombus is a _____?_____ .

3. What new conjecture can you make about rhombuses that is not true for all parallelograms? Fill in the blank in the conjecture below.

> **Conjecture: A Property of Rhombuses**
>
> The diagonals of a rhombus are _____?_____ .

CHECKPOINT ✔

Activity 3

Rectangles

1. Draw a rectangle that is not a square. Draw diagonals to connect the vertices. Make measurements as you did in Activity 1.

2. Do the conjectures you made about parallelograms in Activity 1 seem to be true for rectangles? Discuss why they should or should not be true for rectangles. Fill in the blank in the conjecture below.

Conjecture: Property of Rectangles

A rectangle is a ____?____.

3. What new conjecture can you make about rectangles that is not true for all parallelograms? Fill in the blanks in the conjecture below.

Conjecture: Property of Rectangles

The diagonals of a rectangle ____?____.

Activity 4

Squares

1. Draw a square with diagonals connecting the vertices. Make measurements as in Activity 1.

CHECKPOINT ✔

2. What conjectures that you made in the preceding Activities seem to be true for squares? Discuss why they should or should not be true for squares. Fill in the blank in the conjecture below.

Conjecture: A Property of Squares

A square is a ____?____, a ____?____, and a ____?____.

3. Fill in the blank in the conjecture below.

Conjecture: A Property of Squares

The diagonals of a square ____?____ each other,
are ____?____, and are ____?____.

The Inheritance of Properties

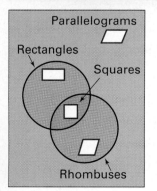

Is a square a rectangle? Is a square a rhombus? In the Activities, you may have made some conjectures about these questions. The Euler diagram at left shows relationships among the classes of quadrilaterals. You will be asked to prove these relationships in Chapter 4.

Certain regions of the Euler diagram "inherit" properties from the larger regions in which they are located. Explain what is meant by the *inheritance of properties*. Illustrate your answer with examples from the conjectures you made in the Activities.

Communicate

1. Explain the relationship between a rhombus and a square.

2. Explain the relationship between a rectangle and a parallelogram.

3. **OPTICAL ILLUSIONS** The drawing at right, composed of three rhombuses, forms an optical illusion. What does the drawing appear to be? Can you see the illusion more than one way? Try drawing your own optical illusions with rhombuses, parallelograms, or trapezoids.

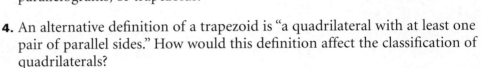

4. An alternative definition of a trapezoid is "a quadrilateral with at least one pair of parallel sides." How would this definition affect the classification of quadrilaterals?

Guided Skills Practice

5. Use your conjectures from Activity 1 and the diagram at right to find the measurements below. **(ACTIVITY 1)**

    a. *AB*   **b.** m∠*ABC*

    c. *AE*   **d.** m∠*BCD*

6. Use your conjectures from Activity 2 and the diagram at right to find the measurements below. **(ACTIVITY 2)**

    a. *FG*   **b.** m∠*FGH*

    c. m∠*GHI*   **d.** m∠*FJG*

7. Use your conjectures from Activity 3 and the diagram at right, in which *KO* is 2 ft, to find *LN*. **(ACTIVITY 3)**

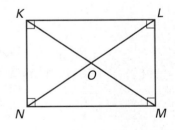

8. Use your conjectures from Activity 4 and the diagram at right to find m∠*PTQ* and m∠*QTR*. **(ACTIVITY 4)**

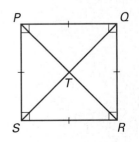

For Exercises 9–32, use your conjectures from Activities 1–4 to find the indicated measurements.

In parallelogram *WXYZ*, *WX* = 10, *WZ* = 4, *WY* = 13, and m∠*WZY* = 130°.

9. *YZ*

10. *XY*

11. *WV*

12. *VY*

13. m∠*WXY*

14. m∠*XWZ*

15. m∠*XYZ*

In rhombus *FGHI*, *FG* = 21, *FH* = 15, and m∠*FGH* = 70°.

16. *GH*

17. *HI*

18. *FJ*

19. *JH*

20. m∠*FIH*

21. m∠*GFI*

22. m∠*FJG*

23. m∠*HJI*

In rectangle *ABCD*, *AB* = 6, *AD* = 8, and *AC* = 10.

24. *CD*

25. *BC*

26. *BD*

27. *AE*

28. *BE*

In square *KLMN*, *KL* = 50 and *KM* ≈ 70.7.

29. *LM*

30. *LN*

31. m∠*KOL*

32. m∠*LOM*

33. In parallelogram *PQRS*, m∠*P* = $(2x)°$ and m∠*Q* = $x°$. Find x and the measure of each angle in *PQRS*.

34. In rectangle *WXYZ*, diagonal *WY* = $x - 2$ and diagonal *XZ* = $\sqrt{x}$. Find x and the length of the diagonals of *WXYZ*.

Use the definitions of quadrilaterals and your conjectures from Activities 1–4 to decide whether each statement is true or false. If the statement is false, give a counterexample.

35. If a figure is a parallelogram, then it cannot be a rectangle.

36. If a figure is not a parallelogram, then it cannot be a square.

37. If a figure is a parallelogram, then it cannot be a trapezoid.

38. If a figure is a trapezoid, then it cannot be a rectangle.

39. If a figure is a square, then it is a rhombus.

40. If a figure is a rectangle, then it cannot be a rhombus.

41. If a figure is a rhombus, then it cannot be a rectangle.

42. In Activity 1, you made the following conjecture: The opposite sides of a parallelogram are congruent. What is the converse of this statement? Do you think that the converse is true?

43. Use your answer from Exercise 42 to explain whether the following statement is true or false: If a figure is a rhombus, then it is a parallelogram.

44. GARDENING Susan is making a for her shelter tomatoes by stretching plastic over a wooden frame. Each wall of the frame is a rectangle, with diagonal braces added for support, as shown. If the brace connecting points *A* and *C* has a length of 73 in., how long is the brace connecting points *B* and *D*?

A

Look Back

Write each statement as a conditional and then write the converse of the conditional. Is the statement a definition? *(LESSON 2.3)*

45. All whales are mammals.

46. All squares are four-sided polygons.

47. All squares are rectangles.

For Exercises 48 and 49, refer to the figure at right. *(LESSON 3.1)*

48. List all nontrivial degrees of the rotational symmetry for the figure.

49. Copy the figure and draw all of its axes of symmetry.

Look Beyond

50. In Activity 2, you made the following conjecture: The diagonals of a rhombus are perpendicular to each other. What is the converse of this statement? Do you think that the converse is true?

51. A figure with two pairs of congruent adjacent sides and opposite sides that are not congruent is called a **kite**. Draw a kite with diagonals connecting the vertices. How does this figure relate to your answer for Exercise 50?

Kite

52. In Activity 3, you made the following conjecture: The diagonals of a rectangle are congruent. What is the converse of this statement? Do you think that the converse is true?

53. A trapezoid in which the nonparallel sides are congruent is called an **isosceles trapezoid**. Draw an isosceles trapezoid and the diagonals connecting the vertices. How does this figure relate to your answer for Exercise 52?

Isosceles trapezoid

The tessellation pattern by M. C. Escher shown below is an example of a **translation tessellation**. Each of the repeating figures is a translation of other figures in the design. You can make your own translation tessellation by following the steps below. You may have adjust your curves to get a pattern that you like.

Draw your figures on graph paper or tracing paper, or use geometry or tessellation software.

1. Start with a square, rectangle, or other parallelogram. Replace one side of the parallelogram with a curve, as shown.

2. Translate the curve to the opposite side of the parallelogram.

3. Repeat Steps 1 and 2 for the other two sides of your parallelogram.

4. Your figure will now fit together with itself on all sides. You can add details to your figure or divide it into two or more parts, as shown below. Translate the entire figure to create an interlocking design.

M. C Escher's "Symmetry Drawing E121" ©1999 Cordon Art B. V. -Baarn-Holand. All rights reserved.

Parallel Lines and Transversals

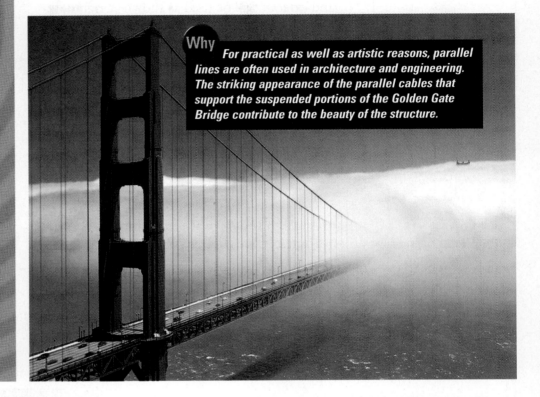

Why *For practical as well as artistic reasons, parallel lines are often used in architecture and engineering. The striking appearance of the parallel cables that support the suspended portions of the Golden Gate Bridge contribute to the beauty of the structure.*

Objectives

● Define *transversal, alternate interior angles, alternate exterior angles, same-side interior angles,* and *corresponding angles.*

● Make conjectures and prove theorems by using postulates and properties of parallel lines and transversals.

Transversals and Special Angles

When parallel lines are taken by themselves, it is hard to imagine how they can be studied. Look, for example, at the parallel lines at right. What conjectures could you be expected to make about them?

Now look at the figure with a third line intersecting the given parallel lines. There are now many discoveries to make. The line that intersects the two parallel lines is known as a *transversal.*

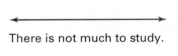

There is not much to study.

The transversal changes the picture.

Definition: Transversal

A transversal is a line, ray, or segment that intersects two or more coplanar lines, rays, or segments, each at a different point.

3.3.1

Notice that, according to the definition, the lines, rays, or segments that are cut by the transversal do not have to be parallel. This will be important in Lesson 3.4.

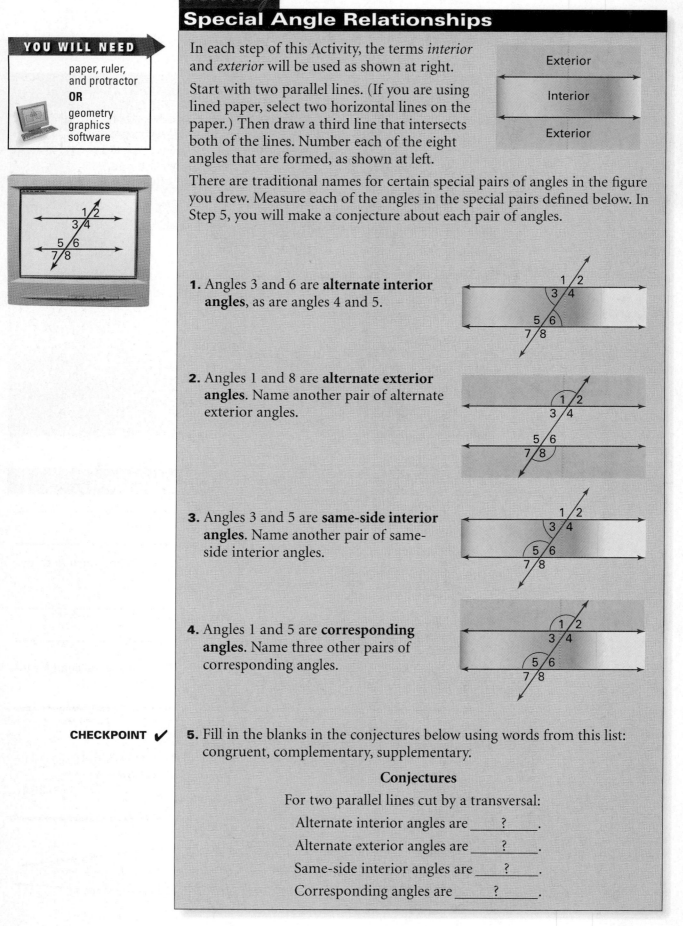

Special Angle Relationships

YOU WILL NEED

paper, ruler, and protractor
OR
geometry graphics software

In each step of this Activity, the terms *interior* and *exterior* will be used as shown at right.

Exterior

Interior

Exterior

Start with two parallel lines. (If you are using lined paper, select two horizontal lines on the paper.) Then draw a third line that intersects both of the lines. Number each of the eight angles that are formed, as shown at left.

There are traditional names for certain special pairs of angles in the figure you drew. Measure each of the angles in the special pairs defined below. In Step 5, you will make a conjecture about each pair of angles.

1. Angles 3 and 6 are **alternate interior angles**, as are angles 4 and 5.

2. Angles 1 and 8 are **alternate exterior angles**. Name another pair of alternate exterior angles.

3. Angles 3 and 5 are **same-side interior angles**. Name another pair of same-side interior angles.

4. Angles 1 and 5 are **corresponding angles**. Name three other pairs of corresponding angles.

CHECKPOINT ✔

5. Fill in the blanks in the conjectures below using words from this list: congruent, complementary, supplementary.

Conjectures

For two parallel lines cut by a transversal:

Alternate interior angles are ___?___.

Alternate exterior angles are ___?___.

Same-side interior angles are ___?___.

Corresponding angles are ___?___.

Indicate whether the pairs below are alternate interior, alternate exterior, same-side interior, or corresponding angles.

a. ∠1 and ∠8
b. ∠7 and ∠3
c. ∠5 and ∠4
d. ∠3 and ∠5

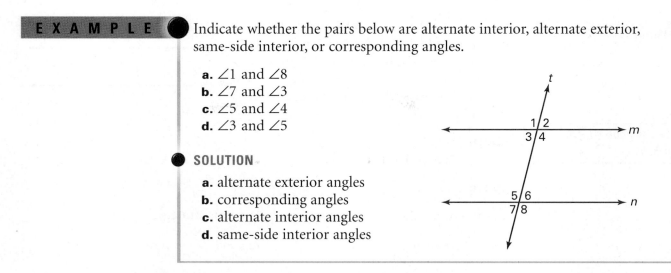

SOLUTION

a. alternate exterior angles
b. corresponding angles
c. alternate interior angles
d. same-side interior angles

TRY THIS List three special pairs of angles not mentioned in the Example above.

One Postulate and Three Theorems

In the Activity on the previous page, one of the four conjectures you may have made is that corresponding angles are congruent. Notice what happens if you slide (translate) one of the parallel lines closer to the other. Eventually, the indicated corresponding angles will overlap. Do you think the corresponding angles will match exactly? Does this diagram seem to support your conjecture?

Move the parallel lines together.

The conjecture about corresponding angles will not be proven in this book, but because it seems obvious, it will be given as a postulate.

Frank Lloyd Wright, *Tree of Life*, 1904. Leaded glass. 41.5 × 26.75 in. Milwaukee Art Museum, Gift of Frederick Layton Art League in memory of Miss Charlotte Partridge and Miss Miriam Frink. © The Frank Lloyd Wright Foundation, Scottsdale, AZ.

Corresponding Angles Postulate

If two lines cut by a transversal are parallel, then corresponding angles are congruent.   **3.3.2**

Because the corresponding angles conjecture has been given as a postulate, it can be used to prove theorems. In particular, you can use it to prove the other three conjectures that you made in the Activity about parallel lines and transversals on the previous page.

To prove your conjecture about alternate interior angles, begin by drawing a figure in which two parallel lines are cut by a transversal. Show that two alternate interior angles, such as ∠1 and ∠2, are congruent.

The symbol ∥ means "is parallel to."

Given: ℓ ∥ m
Line *p* is a transversal.

Prove: ∠1 ≅ ∠2

Plan: Study the figure for ideas, and plan your strategy.

The Corresponding Angles Postulate tells you that ∠1 ≅ ∠3.

∠3 and ∠2 are vertical angles, so ∠3 ≅ ∠2.

From the information in the figure you know the following:

∠1 ≅ ∠3 and ∠3 ≅ ∠2

Thus, ∠1 ≅ ∠2 (the desired result).

What postulate or property allows you draw the final conclusion? (Recall the properties of congruence.)

Now you can write your proof. In proofs of this nature, you may find it convenient to use a two-column format.

Proof:

In writing proofs in geometry, you may wish to develop abbreviations for the reasons. For example, an abbreviation for the reason in Step 2 is "∥s ⇒corr. ∠s ≅."

Statements	Reasons
1. Line ℓ is parallel to line *m*. Line *p* is a transversal.	Given
2. ∠1 ≅ ∠3	If parallel lines are cut by a transversal, then corresponding angles are congruent.
3. ∠3 ≅ ∠2	Vertical angles are congruent.
4. ∠1 ≅ ∠2	Transitive Property of Congruence

The conjecture can now be stated as a theorem.

Alternate Interior Angles Theorem

If two lines cut by a transversal are parallel, then alternate interior angles are congruent.   **3.3.3**

In the exercise set, you will be asked to prove the two remaining conjectures from the Activity.

Exercises

Communicate

The John Hancock Center, Chicago, Illinois, is 1127 feet tall.

1. ARCHITECTURE Describe all of the transversals that you can find in the photo of the John Hancock building at left.

2. In the diagram at right, what type of angle pair are ∠1 and ∠2? What can you say about them? Name three more angle pairs of this type. What type of angle pair are ∠1 and ∠4? What can you say about them? Name three more angle pairs of this type.

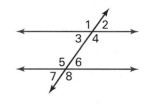

3. In the diagram above, what might you call angles ∠1 and ∠7? What do you think is true of this angle pair? Name another angle pair of this type.

4. For each diagram below, describe all of the transversals or explain why there are no transversals.

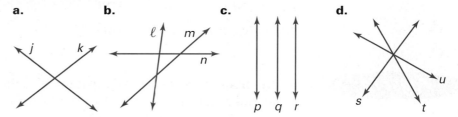

a.   **b.**   **c.**   **d.**

Guided Skills Practice

For Exercises 5–8, refer to the diagram below. List two angle pairs of each indicated type. (ACTIVITY AND EXAMPLE)

5. alternate interior angles

6. alternate exterior angles

7. same-side interior angles

8. corresponding angles

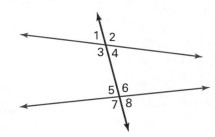

Practice and Apply

In the figure at right, lines *m* and *n* are parallel.

9. List all angles that are congruent to ∠1.

10. List all angles that are congruent to ∠2.

11. Are there any angles in the figure that are not congruent to ∠1 or to ∠2? Explain.

12. If m∠1 = 130°, find the measure of each angle in the figure.

For Exercises 13–17, refer to the diagram below. Lines *p* and *q* are parallel. Name all angles congruent to the given angle, and give the theorems or postulates that justify your answer.

13. ∠1 **14.** ∠2 **15.** ∠3

16. the angle formed by ∠1 and ∠2

17. the angle formed by ∠2 and ∠3

Refer again to the diagram at right. Determine whether the indicated line is a transversal. If so, identify two lines that it intersects.

18. line *p* **19.** line *q*

20. line *r* **21.** line *s*

In △*ABC*, $\overline{DE} \parallel \overline{CB}$ and ∠*ADE* ≅ ∠*AED*. Find the indicated angle measures.

22. m∠*ADE* **23.** m∠*AED*

24. m∠*DEB* **25.** m∠*BDE*

26. m∠*CDB* **27.** m∠*ABD*

In the diagram below, $\ell_1 \parallel \ell_2$, m∠2 = *x*°, and m∠6 = (3*x* − 60)°. Find the indicated angle measures.

28. m∠2 **29.** m∠1

30. m∠3 **31.** m∠5

32. m∠7 **33.** m∠8

Write a two-column or paragraph proof for each theorem below.

34.

Alternate Exterior Angles Theorem

If two lines cut by a transversal are parallel, then alternate exterior angles are congruent.   3.3.4

35.

Same-Side Interior Angles Theorem

If two lines cut by a transversal are parallel, then same-side interior angles are supplementary.   3.3.5

36. In the diagram at right, lines *ℓ* and *m* are parallel. Write a two-column proof that the measure of each angle in the figure is 90°.

CARPENTRY In the diagram of a partial wall frame, the *ceiling joist*, $\overline{PQ}$, and *soleplate*, $\overline{RS}$, are parallel.

37. How is *corner brace* $\overline{PT}$ related to $\overline{PQ}$ and $\overline{RS}$?

38. How are $\angle RTP$ and $\angle QPT$ related?

39. How is $\overline{PT}$ related to the vertical beams that it crosses?

40. How are $\angle 1$ and $\angle 2$ related?

NAVIGATION A *periscope* is an instrument used on submarines to see above the surface of the water. A periscope contains two parallel mirrors that face each other.

41. Identify a transversal in the diagram at left. What does this line represent?

42. Are $\angle 1$ and $\angle 4$ alternate exterior angles? Why or why not?

43. The angle at which light rays are reflected from a mirror is congruent to the angle they form with the mirror. For example, $\angle 1 \cong \angle 2$. Prove that $\angle 1 \cong \angle 4$.

44. Suppose that $m\angle 1 = 45°$. Find the measures of $\angle 2$, $\angle 3$, and $\angle 4$.

Look Back

45. A ____?____ is a part of a line that starts at a point and extends without end in one direction. *(LESSON 1.1)*

46. Two lines that intersect to form a right angle are ____?____. *(LESSON 1.4)*

47. Two coplanar lines that do not intersect are ____?____. *(LESSON 1.4)*

48. ____?____ are the opposite angles formed by two intersecting lines. *(LESSON 2.5)*

49. A ____?____ is a plane figure formed from three or more segments such that each segment intersects exactly two other segments, one at each endpoint. *(LESSON 3.1)*

50. A ____?____ is a quadrilateral with exactly one pair of parallel sides. *(LESSON 3.2)*

51. A ____?____ is a quadrilateral with four congruent sides. *(LESSON 3.2)*

Look Beyond

In the diagram at right, $\overline{AB} \parallel \overline{CD}$.

52. Write a two-column proof that the sum of the interior angles of trapezoid *ABCD* equals 360°.

53. Write a paragraph proof that the sum of the interior angles of any trapezoid, parallelogram, or rectangle is 360°.

Proving That Lines Are Parallel

Objectives

- Identify and use the converse of the Corresponding Angles Postulate.

- Prove that lines are parallel by using theorems and postulates.

Why Suppose that you needed to create a number of parallel lines or that you had to be sure that certain given lines were parallel. The converses of the transversal properties that you learned in Lesson 3.3 will enable you to do these things.

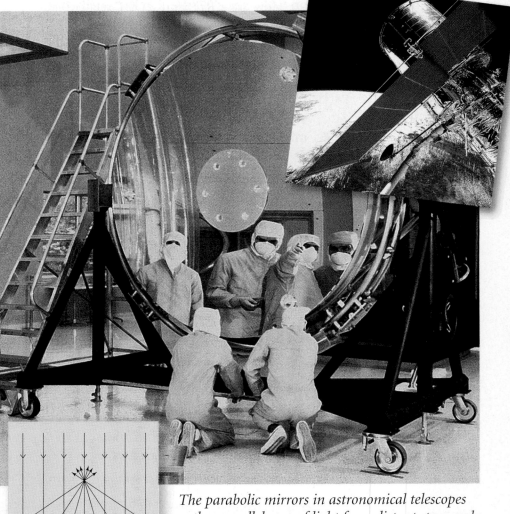

The parabolic mirrors in astronomical telescopes gather parallel rays of light from distant stars and direct them to a central point. In the photograph, technicians are inspecting the mirror of the Hubble Space Telescope.

The Converses of the Transversal Properties

In Lesson 3.3, you studied a postulate and three theorems about parallel lines and transversals. In each of these, the parallel lines and a transversal were given, and conclusions were drawn about certain special angles. In this lesson, the process is reversed in order to write the converses of the postulate and theorems.

E X A M P L E ❶ Write the converse of the Corresponding Angles Postulate.

● **SOLUTION**

Identify the hypothesis and the conclusion of the Corresponding Angles Postulate. Then interchange the hypothesis and the conclusion to form the converse.

Original statement: If two lines cut by a transversal are parallel, then corresponding angles are congruent.

Converse: If corresponding angles are congruent, then two lines cut by a transversal are parallel.

Another way of stating the converse is as follows:

Theorem: Converse of the Corresponding Angles Postulate

If two lines are cut by a transversal in such a way that corresponding angles are congruent, then the two lines are parallel.   **3.4.1**

Notice that the converse is labeled as a theorem. This is because it can be proved with the theorems and postulates you already know. However, the proof will be given later because it involves a special form of reasoning known as indirect proof, which you have not yet studied (see Lesson 12.4).

Using the Converses

In addition to the Converse of the Corresponding Angles Postulate, the converse of each of the other transversal theorems is also true. You will be asked to prove these converses in Exercises 18–25 by using Theorem 3.4.1.

EXAMPLE ❷ Suppose that m∠1 = 64° and m∠2 = 64° in the figure at right. What can you conclude about lines ℓ and m?

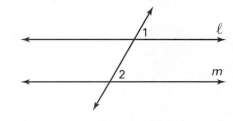

SOLUTION

∠1 and ∠2 are corresponding angles.
The Converse of the Corresponding Angles Postulate states that if corresponding angles are congruent, then the lines are parallel. You can conclude that lines ℓ and m are parallel.

EXAMPLE ❸ Given line ℓ and point P not on the line, draw a line through P that is parallel to ℓ.

SOLUTION

Draw a line through point P and line ℓ.
Label and measure ∠1.

Using a protractor, draw a new line through P such that the new angle corresponds to ∠1 and has the same measure as ∠1. By the Converse of the Corresponding Angles Postulate, the new line, m, is parallel to line ℓ.

Note: You can also do this as a construction with a compass and a straightedge (see Exercises 32 and 33 in Lesson 4.7).

Exercises

Communicate

For Exercises 1 and 2, state the converse of each theorem.

1. If two lines cut by a transversal are parallel, then corresponding angles are congruent.

2. If two lines cut by a transversal are parallel, then same-side interior angles are supplementary.

3. Explain how a theorem that is used to prove that lines are parallel can be used to develop a method for drawing parallel lines, as in the photo at left.

4. Explain why the lines in the diagram below are not parallel.

80°

76°

Guided Skills Practice

5. Write the Converse of the Alternate Interior Angles Theorem. *(EXAMPLE 1)*

6. Are lines m and n in the figure at right parallel? Why or why not? *(EXAMPLE 2)*

7. Copy the figure at right and draw a line through point A that is parallel to $\overleftrightarrow{BC}$. *(EXAMPLE 3)*

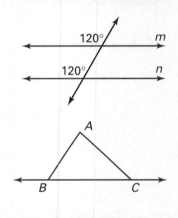

Practice and Apply

For Exercises 8–15, refer to the diagram at left, and fill in the name of the appropriate theorem or postulate.

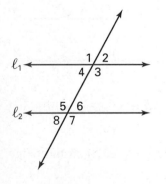

8. If m∠1 = m∠7, then $\ell_1 \| \ell_2$ by the Converse of the _____?_____.

9. If m∠2 = m∠8, then $\ell_1 \| \ell_2$ by the Converse of the _____?_____.

10. If m∠4 = m∠6, then $\ell_1 \| \ell_2$ by the Converse of the _____?_____.

11. If m∠3 = m∠5, then $\ell_1 \| \ell_2$ by the Converse of the _____?_____.

12. If m∠2 = m∠6, then $\ell_1 \| \ell_2$ by the Converse of the _____?_____.

13. If m∠3 = m∠7, then $\ell_1 \| \ell_2$ by the Converse of the _____?_____.

14. If m∠4 + m∠5 = 180°, then $\ell_1 \| \ell_2$ by the Converse of the _____?_____.

15. If m∠3 + m∠6 = 180°, then $\ell_1 \| \ell_2$ by the Converse of the _____?_____.

16. Write a two-column proof that $\overline{DF}\,\|\,\overline{GH}$.

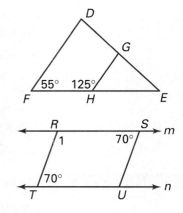

17. Lines m and n in the figure at right are parallel. Write a two-column proof that quadrilateral *RSUT* is a parallelogram.

For Exercises 18–24, complete the two-column proof of the following theorem:

Converse of the Same-Side Interior Angles Theorem

If two lines are cut by a transversal in such a way that same-side interior angles are supplementary, then the two lines are parallel.

3.4.2

Given: $\angle 1$ and $\angle 2$ are supplementary.

Prove: $\ell_1\,\|\,\ell_2$

Proof:

Statements	Reasons
$\angle 1$ and $\angle 2$ are supplementary.	**18.** ___?___
$m\angle 1 + m\angle 2 = 180°$	**19.** ___?___
$m\angle 1 + m\angle 3 = 180°$	**20.** ___?___
$m\angle 1 + m\angle 2 = m\angle 1 + m\angle 3$	**21.** ___?___
$m\angle 2 = m\angle 3\ (\angle 2 \cong \angle 3)$	**22.** ___?___
$\ell_1\,\|\,\ell_2$	**23.** ___?___

Write two-column proofs of the indicated theorems.

24.   ### Converse of the Alternate Interior Angles Theorem

If two lines are cut by a transversal in such a way that alternate interior angles are congruent, then the two lines are parallel.

3.4.3

25.   ### Converse of the Alternate Exterior Angles Theorem

If two lines are cut by a transversal in such a way that alternate exterior angles are congruent, then the two lines are parallel.

3.4.4

For Exercises 26–37, complete the paragraph proofs of the theorems in the boxes below.

Theorem

If two coplanar lines are perpendicular to the same line, then the two lines are parallel to each other.

3.4.5

PARAGRAPH PROOFS

Given: $m \perp \ell$ and $p \perp \ell$

Prove: $p \parallel m$

Proof:

Line ℓ is a **26.** ___?___ of m and p, by definition. $\angle ADC \cong \angle BED$ because **27.** ___?___. Therefore, **28.** ___?___ is parallel to **29.** ___?___ by **30.** ___?___.

Theorem

If two coplanar lines are parallel to the same line, then the two lines are parallel to each other.

3.4.6

Given: $\ell \parallel n$, $m \parallel n$, and p is a transversal of ℓ, m, and n.

Prove: $\ell \parallel m$

Proof:

Because $\ell \parallel n$, $\angle 1 \cong$ **31.** ___?___ by **32.** ___?___. Because $m \parallel n$, then $\angle 3 \cong$ **33.** ___?___ by **34.** ___?___. Thus, $\angle 1 \cong$ **35.** ___?___ by **36.** ___?___, and $\ell \parallel m$ by **37.** ___?___.

APPLICATIONS

38. DRAFTING A T square and a triangle can be used to draw parallel lines. While holding the T square in place, slide the triangle along the T square as shown. How can you prove that the resulting lines, ℓ and m, are parallel?

39. CIVIL ENGINEERING Suppose that you are painting lines for angled parking in a parking lot. How could you make sure that the lines are parallel?

Plumb bob

40. CARPENTRY A *plumb bob* is a weight hung at the end of a string, called a *plumb line*. The weight pulls the string straight down so that the plumb line is perfectly vertical. Suppose that the angle formed by the wall and the roof is 120° and the angle formed by the plumb line and the roof is 120°. Explain why this shows that the wall is vertical.

Roof

120°

120°

Wall

Plumb bob

Look Back

For Exercises 41–45, refer to the following statement: Every rectangle is a parallelogram. *(LESSON 2.2)*

41. Rewrite the statement as a conditional.

42. Identify the hypothesis and conclusion of the statement.

43. Draw an Euler diagram that illustrates the statement.

44. Write the converse of the statement, and construct its Euler diagram.

45. Is the converse true or false? If it is true, write a paragraph proof of it. If it is false, disprove it with a counterexample.

46. NAVIGATION A pilot is flying at a compass heading of 155. What is the heading of a pilot flying in the opposite direction? *(LESSON 1.3)*

47. NAVIGATION A pilot is flying at a compass heading of 235. What are the two possible headings of a pilot flying perpendicular to the first pilot? *(LESSON 1.3)*

Look Beyond

Exercises 48 and 49 refer to the photo of railroad tracks shown below.

48. How could you prove that the railroad tracks are parallel?

49. In the photo, the lines appear to meet at the horizon. Why? Some people use the expression "meet at infinity" when referring to parallel lines. What does this mean? Does it make sense?

50. The drawing at right represents a cube. What is true of the lines that contain the edges shown in red?

51. What would happen if the red edges in the drawing were extended?

Egg Over Alberta

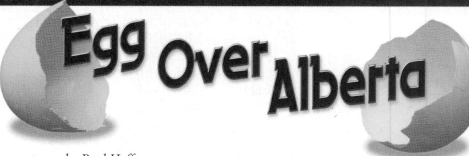

by Paul Hoffman

The town leaders of Vegreville, Alberta, contacted Dale Resch, a computer science professor, for a special project—to build a 31-foot Easter egg.

The problem Resch faced was that no one other than a chicken had ever built a chicken egg. With no formal training in mathematics, Resch relied on his ability to play with geometric abstractions in his mind, then with his hands or a computer, to turn those abstractions into physical reality.

Resch assumed that someone had developed the mathematics of an ideal chicken egg. He soon found, however, that there was no formula for an ideal chicken egg.

After four months of contemplation and simulation, Resch realized that he could tile the egg with 2,208 equilateral triangles and 524 three-pointed stars (equilateral but non-regular hexagons) that varied slightly in width, depending on their position on the egg.

For six weeks, Resch led a team of volunteers in assembling the egg. Residents were afraid it might blow down. Long after the egg was finished, Resch used a computer to analyze the egg's structural integrity and found that it was ten times stronger than it needed to be.

Computer generated drawing of egg

Cutting the egg tiles from aluminum

Partially assembled egg

Many Ukrainians still practice the traditional art of painting pysanki, *eggs decorated with colorful geometric patterns.*

Cooperative Learning

Use folding paper and a compass or geometry graphics software for the following explorations.

1. The tiles on the Vegreville egg have two shapes; equilateral triangles, and three-pointed stars, which are equilateral, but not regular, hexagons. The three-pointed stars may be created as follows:

   Draw or fold an equilateral triangle. Fold the perpendicular bisectors of the sides to find the center of the triangle. Place the point of your compass at the center of the triangle and draw a circle that completely encloses the triangle. The radius of the circle should be at least twice the distance from the center to a vertex of the triangle.

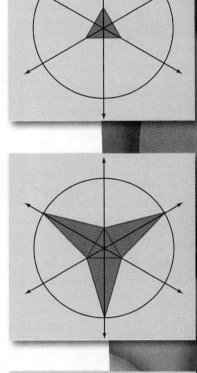

   Draw segments from each vertex to the points where the perpendicular bisectors of the sides intersect the circle to form an isosceles triangle on each side of the equilateral triangle. The figure formed by the equilateral triangle plus the three isosceles triangles is a three-pointed star. Explain why the three-pointed star is equilateral.

2. What are the axes of symmetry of the three-pointed star that you created in Step 1? What type of rotational symmetry does it have?

3. Draw an equilateral triangle on each side of your three-pointed star. Draw three new three-pointed stars in the gaps between the equilateral triangles. Draw an equilateral triangle on each side of these three-pointed stars, and draw three-pointed stars in the gaps between these triangles. Continue drawing three-pointed stars and equilateral triangles in this way until you understand the pattern of shapes in a plane. How is your pattern like the pattern of tiles on the Vegreville egg? (The pattern of tiles is easiest to see near the top and bottom of the egg.) How is it different?

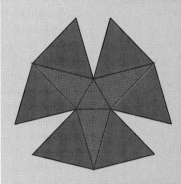

The Triangle Sum Theorem

Objective

- Identify and use the Parallel Postulate and the Triangle Sum Theorem.

Why *Euclidean geometry was founded on five postulates or axioms. One of them, the Parallel Postulate, has been thoroughly investigated by modern mathematicians.*

Three different geometries apply to the triangles drawn on the three different surfaces. The differences in the geometries are based on the Parallel Postulate.

The Parallel Postulate

•*P*

←————————→ *ℓ*

There is one and only one line through point P that is parallel to line ℓ.

The following postulate is a modern equivalent of Euclid's fifth postulate:

The Parallel Postulate

Given a line and a point not on the line, there is one and only one line that contains the given point and is parallel to the given line.

3.5.1

In Lesson 3.4 you drew a line through a given point and parallel to a given line. You probably never questioned whether such a line actually existed, or whether there could be more than one such line. The *assumption* that there is in fact exactly one such parallel line is known as the Parallel Postulate.

The Parallel Postulate is used to prove a theorem about the angles of a triangle. Before you look at the theorem, try the Activity below.

Activity
The Triangle Sum Theorem

1. Cut out a triangle from a piece of paper.
2. Tear two corners off the triangle.
3. Position the two torn-off corners next to the third angle as shown at right.

CHECKPOINT ✔ 4. Make a conjecture about the sum of the measures of the angles of a triangle.

5. In the diagram below, line ℓ has been drawn through a vertex of the triangle so that it is parallel to the opposite side. How does the Parallel Postulate relate to this figure?

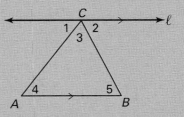

6. Fill in the table below for a figure like the one above. Use geometry theorems, not physical measurements, to find the answers.

m∠1	m∠2	m∠3	m∠4	m∠5	m∠3 + m∠4 + m∠5
40°	30°	?	?	?	?
20°	80°	?	?	?	?
30°	100°	?	?	?	?

7. Does the table support the conjecture you made in Step 4? Explain.

CRITICAL THINKING   Does your work in the Activity prove your conjecture? Why or why not?

The conjecture from the Activity is stated below as a theorem, and it is proven on the following page.

The Triangle Sum Theorem

The sum of the measures of the angles of a triangle is 180°.   3.5.2

Proving the Triangle Sum Theorem

To prove the theorem, begin by drawing line ℓ through a vertex of the triangle so that it is parallel to the side opposite the vertex.

Given: $\triangle ABC$

Prove: $m\angle 1 + m\angle 2 + m\angle 3 = 180°$

Plan: Study the illustration, which is related to the Activity in which you tore off the corners of a triangle. You can use what you discovered in the Activity to write a two-column proof of the Triangle Sum Theorem.

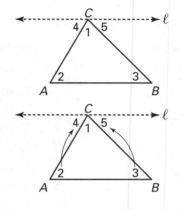

Proof:

Statements	Reasons
1. $\ell \parallel \overline{AB}$	As drawn (justification: the Parallel Postulate)
2. $m\angle 1 + m\angle 4 + m\angle 5 = 180°$	The angles fit together to form a straight line.
3. $\angle 2 \cong \angle 4$ $(m\angle 2 = m\angle 4)$ $\angle 3 \cong \angle 5$ $(m\angle 3 = m\angle 5)$	Alternate Interior Angles Theorem
4. $m\angle 1 + m\angle 2 + m\angle 3 = 180°$	Substitution in Step 2

Another Geometry

It is possible to create geometries in which the Parallel Postulate is not true. On the surface of a sphere, for example, lines are defined differently from the way they are defined on flat surfaces, and there are no parallel lines.

On the surface of a sphere, a line is defined as a **great circle**, which is a circle that lies in a plane that passes through the center of the sphere. (A great circle divides a sphere into two equal parts.) The equator is a great circle on the surface of Earth. Lines of longitude, which run north and south, are also great circles. Notice that any two distinct lines (great circles) intersect at two points. Thus, there are no parallel lines on a sphere.

Discuss the following statement: On the surface of a sphere, the shortest path between two points is not a straight line. What is the shortest path?

Exercises

● *Communicate*

1. Explain how the torn-triangle Activity is similar to a proof of the Triangle Sum Theorem.

2. Explain why the torn-triangle Activity is not a proof of the Triangle Sum Theorem.

3. What role does the Parallel Postulate play in the proof of the Triangle Sum Theorem?

APPLICATION

4. **NAVIGATION** You can use a globe and a piece of string to approximate the distance of the shortest route between two points on Earth. Estimate the shortest route from New York City to Bangkok, Thailand. What might be some difficulties in traveling along this route?

● *Guided Skills Practice*

For Exercises 5–7, refer to the diagram below. (ACTIVITY)

5. Name two pairs of alternate interior angles in the diagram.

6. What is the sum of m∠1, m∠2, and m∠3?

7. If m∠4 = 65° and m∠5 = 50°, what is m∠2?

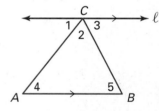

● *Practice and Apply*

For Exercises 8–12, two angle measures of a triangle are given. Find the missing angle measure, or state that the triangle does not exist.

8. m∠1 = 85°, m∠2 = 45°, m∠3 = _____?_____

9. m∠A = 45°, m∠B = _____?_____, m∠C = 90°

10. m∠K = _____?_____, m∠L = 60°, m∠M = 60°

11. m∠X = 90°, m∠Y = _____?_____, m∠Z = 90°

12. m∠F = 105°, m∠G = 80°, m∠H = _____?_____

For Exercises 13–20, refer to the diagram below, in which $\overline{DF} \parallel \overline{BC}$, $\overline{AB} \parallel \overline{FC}$, m∠ADE = 60°, and m∠ACB = 50°. Find the following:

13. m∠B

14. m∠A

15. m∠AED

16. m∠EDB

17. m∠DEC

18. m∠FEC

19. m∠ECF

20. m∠F

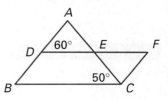

Find the indicated angle measure for each triangle.

21. m∠1   **22.** m∠2   **23.** m∠3

Find *x* and the measure of each angle of the triangle.

24. $x =$ ___?___ , m∠A = ___?___ ,   **25.** $x =$ ___?___ , m∠D = ___?___ ,
m∠B = ___?___ , m∠C = ___?___   m∠E = ___?___ m∠F = ___?___

In Exercises 26–44, you will have an opportunity to discover and prove an important geometry theorem related to the figure at right. Begin by copying and completing the following table:

m∠1	m∠2	m∠1 + m∠2	m∠3	m∠4
30°	70°	**26.** ?	**27.** ?	**28.** ?
30°	80°	**29.** ?	**30.** ?	**31.** ?
40°	80°	**32.** ?	**33.** ?	**34.** ?
40°	90°	**35.** ?	**36.** ?	**37.** ?

38. An angle such as ∠4 that forms a linear pair with an angle of a polygon is called an *exterior angle* of the polygon. How many exterior angles are possible at each vertex of a given polygon? Are the exterior angles at a given vertex congruent? Explain your reasoning.

39. For an exterior angle of a polygon at a given vertex, the angles of the polygon at the other vertices are called **remote interior angles** of the exterior angle. In the diagram above, ∠1 and ∠2 are remote interior angles of ∠4. Use the table above to complete the following theorem:

Exterior Angle Theorem

The measure of an exterior angle of a triangle is equal to ___?___ .

3.5.3

40. TECHNOLOGY Use geometry graphics software to draw the figure shown for Exercises 26–37 so that you can slide point *C* along the ray, displaying m∠1 + m∠2 and m∠4. Describe your results.

Complete the two-column proof of the Exterior Angle Theorem below.

Given: $\triangle ABC$ with exterior angle $\angle 4$ and remote interior angles $\angle 1$ and $\angle 2$

Prove: $m\angle 4 = m\angle 1 + m\angle 2$

Proof:

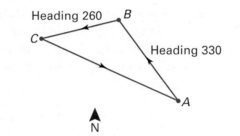

Statements	Reasons
$\triangle ABC$ with exterior angle $\angle 4$	Given
$m\angle 4 + m\angle 3 = 180°$	**41.** ___?___
$m\angle 1 + m\angle 2 + m\angle 3 = 180°$	**42.** ___?___
$m\angle 4 + m\angle 3 = m\angle 1 + m\angle 2 + m\angle 3$	**43.** ___?___
$m\angle 4 = m\angle 1 + m\angle 2$	**44.** ___?___

45. NAVIGATION Towns A, B, and C form a triangle in which $\angle A \cong \angle C$. A pilot flies from town A to town B at a heading of 330, then to town C at a heading of 260. At what heading should the pilot fly to return to town A? (Hint: Draw vertical (north-south) lines through each of the points. Use same-side interior angles.)

Heading 260 · B

C

Heading 330

A

N

Look Back

46. The set of common points of two figures is their ___?___. **(LESSON 1.1)**

47. ___?___ points determine a line. **(LESSON 1.1)**

48. ___?___ points determine a plane. **(LESSON 1.1)**

49. Adjacent supplementary angles form a ___?___. **(LESSON 1.3)**

50. SPORTS The diagram at right shows the layout of a baseball diamond. Suppose that the line from home plate to the third baseman forms a 10° angle with the third-base line and that the line from home plate to the shortstop forms a 30° angle with the third-base line.

If the batter hits the ball so that the path of the ball bisects the angle formed by the shortstop, home plate, and the third baseman, what angle does the path of the ball make with the third-base line? *(LESSON 1.3)*

Look Beyond

CARTOGRAPHY A map that is commonly used in navigation is the Mercator projection, shown below. All straight lines on this map, called *rhumb lines*, have a constant compass heading.

51. Do any rhumb lines correspond to great circles on a globe? If so, which ones?

52. Suppose that you draw a rhumb line that is not a latitude or longitude line. Trace this line (with your finger) on a globe and describe its path.

Gerardus Mercator (1512–1594)

Angles in Polygons

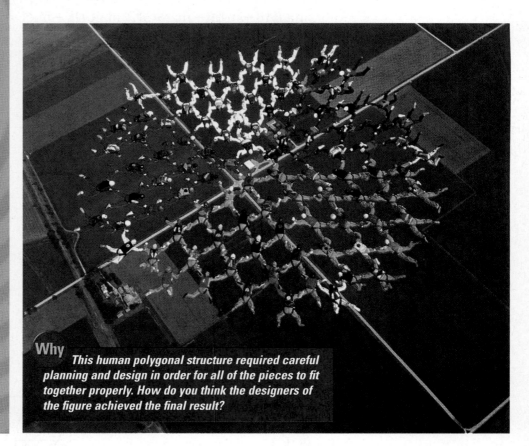

Objective

● Develop and use formulas for the sums of the measures of interior and exterior angles of a polygon.

Why *This human polygonal structure required careful planning and design in order for all of the pieces to fit together properly. How do you think the designers of the figure achieved the final result?*

Angle Sums in Polygons

A **convex polygon** is one in which no part of a line segment connecting any two points on the polygon is outside the polygon. A **concave polygon** does not have this characteristic. In this book, the word *polygon* will mean a convex polygon unless otherwise stated.

Convex polygon   Concave polygon

YOU WILL NEED

calculator (optional)

Activity 1
Sums of Interior Angles

Pentagon *ABCDE* has been divided into three triangular regions by drawing all possible diagonals from one vertex.

1. Find each of the following:

$$m\angle 1 + m\angle 2 + m\angle 3 = \underline{\quad ? \quad}$$
$$m\angle 4 + m\angle 5 + m\angle 6 = \underline{\quad ? \quad}$$
$$m\angle 7 + m\angle 8 + m\angle 9 = \underline{\quad ? \quad}$$

2. Add the three expressions.

$$m\angle 1 + m\angle 2 + m\angle 3 + m\angle 4 + \cdots + m\angle 9 = \underline{\quad ? \quad}$$

3. Use the diagram and the result from Step 2 to determine the sum of the measures of the interior angles of pentagon *ABCDE* (that is, m∠*EAB* + m∠*B* + m∠*BCD* + m∠*CDE* + m∠*E* = _____?_____).

4. You can form triangular regions by drawing all possible diagonals from a given vertex of any polygon. Complete the table below. Use sketches to illustrate your answers.

Polygon	Number of sides	Number of triangular regions	Sum of measures of angles
triangle	?	1	180°
quadrilateral	?	?	?
pentagon	?	3	540°
hexagon	?	?	?
n-gon	?	?	?

CHECKPOINT ✔ 5. Write a formula for the sum of the interior angles of a polygon in terms of the number of sides, *n*. Complete the formula below.

Sum of the Interior Angles of a Polygon

The sum of the measures of the interior angles of a polygon with *n* sides is _____?_____.   3.6.1

Recall that a regular polygon is one in which all the angles are congruent and all the sides are congruent. Equilateral triangles and squares are examples of regular polygons. In an equilateral triangle, each angle has a measure of 60°. In a square, each angle has a measure of 90°.

CHECKPOINT ✔ Complete the chart below. Then complete the formula beneath it.

Regular polygon	Number of sides	Sum of measures of interior angles	Measure of one interior angle
triangle	?	180°	?
quadrilateral	?	?	90°
pentagon	?	?	?
hexagon	?	?	?
n-gon	?	?	?

The Measure of an Interior Angle of a Regular Polygon

The measure of an interior angle of a regular polygon with *n* sides is _____?_____.   3.6.2

Exterior Angle Sums in Polygons

1. Draw a triangle and extend each side in one direction to form an exterior angle at each vertex. Find the sum of the measures of the three exterior angles that you formed. Record your results.

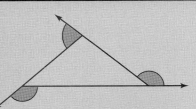

2. Cut out the exterior angles and fit them together. Record your results.

3. Repeat Steps 1 and 2 for a quadrilateral.

4. Repeat Steps 1 and 2 for a pentagon.

5. Make a conjecture about the sum of the measures of the exterior angles of a polygon (one at each vertex). You will prove your conjecture in Steps 6–10.

6. What is the sum of the measures of all interior and exterior angles of the triangle at right?

Interior angles

Exterior angles

7. What is the sum of the measures of all interior and exterior angles of the quadrilateral at right?

8. Using your results from Steps 6 and 7, write a formula for the sum of the measures of all interior and exterior angles of an *n*-gon.

9. Complete the table below.

Polygon	Number of sides	Sum of exterior and interior angles	Sum of interior angles	Sum of exterior angles
triangle	3	540°	180°	360°
quadrilateral	?	?	?	?
pentagon	?	?	?	?
hexagon	?	?	?	?
n-gon	?	?	?	?

CHECKPOINT ✔

10. Use the formula from Activity 1 and the formula from Step 8 above to write an expression for the sum of the measures of the exterior angles of an *n*-gon. Use algebra to simplify the expression. Complete the theorem.

Theorem: Sum of the Exterior Angles of a Polygon

The sum of the measures of the exterior angles of a polygon is ___?___.

3.6.3

Communicate

1. Explain how to tell how many triangular regions can be formed in a polygon by drawing all possible diagonals from one vertex.

2. Is it possible to draw a quadrilateral with three interior angles that measure 60° each? Explain your reasoning.

3. The figures at left can be a "proof without words" of the result you may have discovered in Activity 2. A polygon has its sides extended as rays. Imagine that you are looking at the figure from farther and farther away. Explain why the sum of the exterior angles is 360°.

Guided Skills Practice

4. Find the sum of the measures of the interior angles of an octagon.
   (ACTIVITY 1)

5. Find the sum of the measures of the interior angles of a 13-gon.
   (ACTIVITY 1)

6. Find the sum of the measures of the exterior angles of a heptagon.
   (ACTIVITY 2)

7. Find the sum of the measures of the exterior angles of an 11-gon.
   (ACTIVITY 2)

Practice and Apply

8. Refer to the figure at right to find the indicated measures.

   a. $x =$ _____?_____

   b. $y =$ _____?_____

   c. $z =$ _____?_____

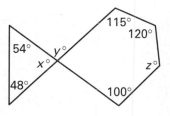

For Exercises 9–14, find the indicated angle measure.

9.

90°   75°

85°   ?

10.

?

132°

120°

11.

?   130°

100°

90°   110°

12.

90°

?

13.

?

110°

100°

14.

105°

75°

?

For each polygon, determine the measure of an interior angle and the measure of an exterior angle.

15. a rectangle

16. an equilateral triangle

17. a regular dodecagon

18. an equiangular pentagon

For Exercises 19–21, an interior angle measure of a regular polygon is given. Find the number of sides of the polygon.

19. 135°

20. 150°

21. 165°

For Exercises 22–24, an exterior angle measure of a regular polygon is given. Find the number of sides of the polygon.

22. 60°

23. 36°

24. 24°

For Exercises 25–37, find the indicated angle measure.

25. m∠A

26. m∠B

27. m∠C

28. m∠D

29. m∠E

30. m∠F

31. m∠G

32. m∠H

33. m∠I

34. m∠J

35. m∠K

36. m∠L

37. m∠M

38. What is the maximum possible number of acute angles in a triangle? Can a triangle have no acute angles? Explain your reasoning.

39. What is the maximum possible number of acute angles in a quadrilateral? Can a quadrilateral have no acute angles? Explain your reasoning.

40. What is the maximum possible number of acute angles in a pentagon? Can a pentagon have no acute angles? Explain your reasoning.

CHALLENGE

41. Find the sum of the measures of the numbered vertex angles of a 5-pointed star polygon. (Hint: First find the measure of the exterior angle indicated by a question mark in each diagram below.)

The brilliant cut was developed by the mathematician Marcel Tolkowsky in 1919.

42. GEMOLOGY Precious stones are often cut in a *brilliant cut* to maximize the amount of light reflected by the stone. The angles of the cut depend on the refractive properties of the type of stone. The optimal angles for a diamond are shown in the cross section below. The cut has reflectional symmetry across the axis shown. Find the measures of the indicated angles in the figure.

Crown angle

Pavillion main angle

Axis of symmetry

43. GEMOLOGY A brilliant cut topaz should have a pavilion main angle of 40° and a crown angle of 37°. Sketch a cross section of such a gem and find the other angles in the cross section.

Look Back

44. How is the distance from a point to a line determined? *(LESSON 1.4)*

45. List all pairs of supplementary angles in the photo below. What are these types of angles called? *(LESSON 1.3)*

46. TRANSPORTATION Due to zoning regulations, the measure of an angle at an intersection cannot be less than 75°. If m∠1 = 75°, what is m∠4? *(LESSON 1.3)*

47. List all pairs of congruent angles in the photo. What are these types of angles called? *(LESSON 2.2)*

Look Beyond

Some regular polygons fit together around a single point with no overlaps or gaps. For example, four squares fit together at a point, as shown at left.

48. What is the measure of each angle at the indicated point? What is the sum of the measures of the angles at this center point?

49. For a regular *n*-gon to form a pattern like the one described above, the measure of its interior angles must be a factor of 360°. Explain why this is true.

50. What other regular *n*-gons will fit together around a point? How can you be sure that you have found all of the possible *n*-gons?

51. BIOLOGY A beehive is constructed from regular hexagons, as shown. What do you think are some advantages of using hexagons?

Midsegments of Triangles and Trapezoids

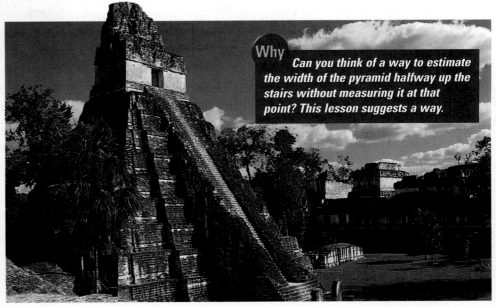

Why Can you think of a way to estimate the width of the pyramid halfway up the stairs without measuring it at that point? This lesson suggests a way.

Objectives

- Define *midsegment of a triangle* and *midsegment of a trapezoid*.

- Develop and use formulas based on the properties of triangle and trapezoid midsegments.

The Mayan Temple of the Giant Jaguar in Tikal, Guatemala, has four approximately trapezoidal faces. A stairway rises 100 feet along one side.

Midsegments of Triangles

Definition: Midsegment of a Triangle

A midsegment of a triangle is a segment whose endpoints are the midpoints of two sides.   **3.7.1**

Activity 1

Triangle Midsegments: A Conjecture

YOU WILL NEED

ruler and protractor
OR
geometry graphics software

CHECKPOINT ✔

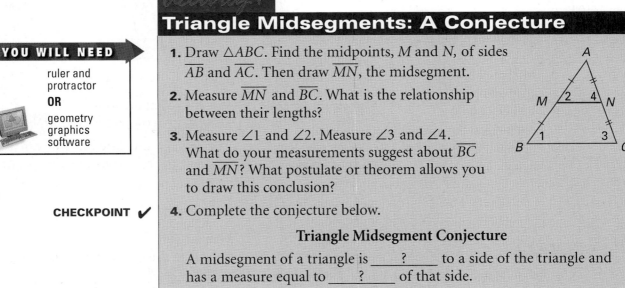

1. Draw △*ABC*. Find the midpoints, *M* and *N*, of sides $\overline{AB}$ and $\overline{AC}$. Then draw $\overline{MN}$, the midsegment.

2. Measure $\overline{MN}$ and $\overline{BC}$. What is the relationship between their lengths?

3. Measure ∠1 and ∠2. Measure ∠3 and ∠4. What do your measurements suggest about $\overline{BC}$ and $\overline{MN}$? What postulate or theorem allows you to draw this conclusion?

4. Complete the conjecture below.

Triangle Midsegment Conjecture

A midsegment of a triangle is ____?____ to a side of the triangle and has a measure equal to ____?____ of that side.

Jose is on his school swim team. During the summer, he enjoys training at a small lake near his house. To evaluate his progress, he needs to know the distance across the lake, XY. How can he use the conjecture from Activity 1 to find this distance?

● **SOLUTION**

Jose can select a point, A, from which he can measure segments $\overline{AX}$ and $\overline{AY}$. Then he can find the midpoints of $\overline{AX}$ and $\overline{AY}$ and measure the distance between them. Since this distance is half the length of $\overline{XY}$, Jose can double the distance to find his answer.

Midsegments of Trapezoids

Definition: Midsegment of a Trapezoid

A midsegment of a trapezoid is a segment whose endpoints are the midpoints of the nonparallel sides.

3.7.2

Activity 2

Trapezoid Midsegments: A Conjecture

1. Draw trapezoid $ABCD$. Find the midpoints, M and N, of the nonparallel sides. Draw $\overline{MN}$, the midsegment.

2. Measure the length of bases $\overline{AB}$ and $\overline{DC}$ and of midsegment $\overline{MN}$.

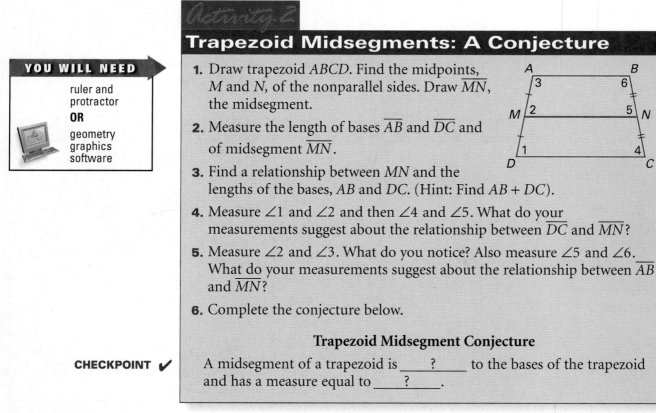

3. Find a relationship between MN and the lengths of the bases, AB and DC. (Hint: Find $AB + DC$).

4. Measure $\angle 1$ and $\angle 2$ and then $\angle 4$ and $\angle 5$. What do your measurements suggest about the relationship between $\overline{DC}$ and $\overline{MN}$?

5. Measure $\angle 2$ and $\angle 3$. What do you notice? Also measure $\angle 5$ and $\angle 6$. What do your measurements suggest about the relationship between $\overline{AB}$ and $\overline{MN}$?

6. Complete the conjecture below.

Trapezoid Midsegment Conjecture

CHECKPOINT ✔

A midsegment of a trapezoid is _____?_____ to the bases of the trapezoid and has a measure equal to _____?_____ .

EXAMPLE **2** The base of the pyramid of the Temple of the Giant Jaguar is a square that is 150 feet on a side. The top is a square that is 40 feet on a side. What is the width of the pyramid at a point midway between the base and the top?

APPLICATION
ARCHAEOLOGY

SOLUTION

Use the trapezoid midsegment conjecture:

$$\text{Length of midsegment} = \tfrac{1}{2}(\text{base } 1 + \text{base } 2)$$
$$= \tfrac{1}{2}(40 + 150)$$
$$= \tfrac{1}{2}(190)$$
$$= 95 \text{ feet}$$

Activity 3

Making the Connection

YOU WILL NEED

ruler and protractor
OR
geometry graphics software

1. Draw trapezoid *ABCD*, which may be any shape or size, with midsegment $\overline{MN}$. Then fill in a table like the one below by gradually reducing the length of $\overline{AB}$. (Choose your own measurements for *AB* and *DC*.)

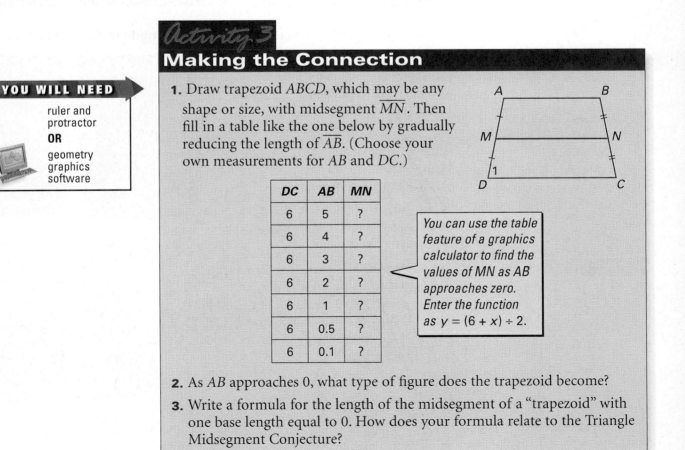

DC	AB	MN
6	5	?
6	4	?
6	3	?
6	2	?
6	1	?
6	0.5	?
6	0.1	?

You can use the table feature of a graphics calculator to find the values of MN as AB approaches zero. Enter the function as y = (6 + x) ÷ 2.

2. As *AB* approaches 0, what type of figure does the trapezoid become?

3. Write a formula for the length of the midsegment of a "trapezoid" with one base length equal to 0. How does your formula relate to the Triangle Midsegment Conjecture?

Exercises

Communicate

1. In the figure at right, the horizontal segments across the large triangle are all midsegments of smaller triangles. What is the ratio of the base of the red triangle to the lower base of the blue trapezoid? Explain your reasoning.

2. If the length of the lower base of the blue trapezoid in the figure at right is 1, what are the lengths of the midsegments in the figure? Describe the pattern in the sequence of lengths you just found.

3. In Activity 3, you saw how a triangle can be considered a *limiting case* of a trapezoid. What do you think *limiting case* means?

4. Consider the following method for finding the length of the midsegment of a trapezoid: First subtract the length of the shorter base from the length of the longer base. Then take half the difference, and add it to the length of the shorter base. Does this method work? Why or why not?

Guided Skills Practice

For Exercises 5–9, refer to the conjectures you made in Activities 1–3.

5. What is the length of the midsegment of a triangle with a base of 12? *(ACTIVITY 1)*

APPLICATION

6. **SPORTS** A swimmer is practicing in the lake shown at right. The red line is the midsegment of the triangle. What is the distance from point *A* to point *B*? *(EXAMPLE 1)*

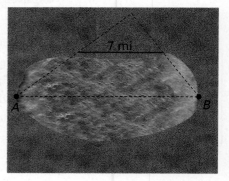

7. What is the length of the midsegment of a trapezoid with bases of 7 and 15? *(ACTIVITY 2)*

APPLICATION

8. **ARCHAEOLOGY** The structure shown at right has four congruent sides that are trapezoids with a height of 96 ft and bases of 180 ft and 48 ft. What is the width of the structure at a point midway between the bottom and the top? *(EXAMPLE 2)*

9. Write the Triangle Midsegment Conjecture and the Trapezoid Midsegment Conjecture as formulas for the length of a midsegment in terms of the length of the base(s). How are the two formulas related? *(ACTIVITY 3)*

● Practice and Apply ▬▬▬▬▬▬▬▬▬▬▬

Use the conjectures you made in Activities 1–3 to find the indicated measures.

10. *AB*

11. *IJ*

12. *PQ*

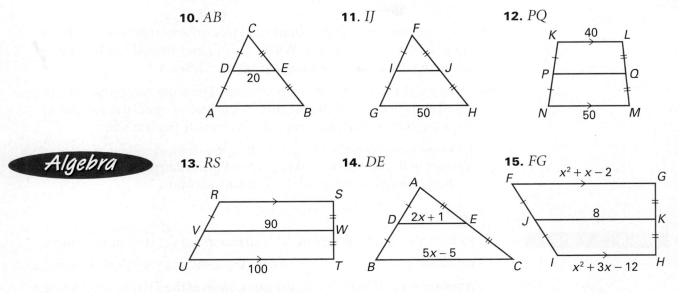

Algebra

13. *RS*

14. *DE*

15. *FG*

In Exercises 16–20, refer to the diagram of △*ABC* at right below.

16. Find *DE*, *FG*, and *HI* in △*ABC* at right.

17. Describe the relationships among the lengths you found in Exercise 16.

18. The midsegment of a triangle divides two sides of the triangle into two congruent segments. Make a conjecture about the lengths of parallel segments that divide two sides of a triangle into four congruent segments.

19. What do you think is true about parallel segments that divide two sides of a triangle into three congruent segments? into eight congruent segments? into *n* congruent segments? Draw several triangles and test your conjecture.

20. Are the conjectures you made in Exercises 18 and 19 also true for trapezoids? Why or why not?

21. Segments $\overline{MN}$ and $\overline{PQ}$ divide two sides of △*JKL* into three congruent segments, as shown. Write a conjecture about $\overline{MN}$ and $\overline{PQ}$. Draw several other triangles and test your conjecture.

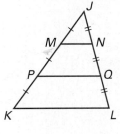

22. In the figure at right, the three red segments are midsegments of the large triangle. Find the length of each midsegment.

Add the side lengths of the outer triangle. Then add the side lengths of the smaller triangle formed by the midsegments. What is the relationship between the two sums?

In Exercises 23–26, write informal arguments based on your conjectures about triangle midsegments and quadrilaterals.

23. Draw a scalene triangle. Draw any two midsegments of the triangle. What type of quadrilateral having these midsegments as adjacent sides is formed? Explain why.

24. Draw an isosceles triangle. Draw the midsegments that connect the two equal sides to the third side. What type of quadrilateral having these midsegments as adjacent sides is formed? Explain why.

25. Draw a right triangle that is not isosceles. Draw the two midsegments that connect the legs to the hypotenuse. What type of quadrilateral having these midsegments as adjacent sides is formed? Explain why.

26. Draw an isosceles right triangle. Draw the two misdegments that connect the legs to the hypotenuse. What type of quadrilateral having these midsegments as adjacent sides is formed? Explain why.

CHALLENGE

27. In Exercises 21–27 of Lesson 2.1, you examined the sum of the infinite sequence $\frac{1}{2} + \frac{1}{4} + \frac{1}{8} + \frac{1}{16} + \cdots$ by using areas of squares and rectangles. Another way of determining this sum involves the Triangle Midsegment Conjecture.

Figure A is a square with a side length of 1, and each horizontal segment is a midsegment of a triangle. Find the length of each midsegment in figure A. Then use figure B to explain why the sum of the infinite sequence is 1.

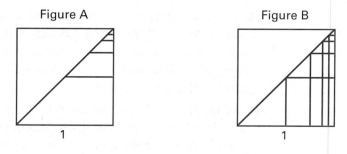

Figure A

Figure B

APPLICATIONS

28. ENGINEERING Is $\overline{FC}$ the midsegment of trapezoid $ABDE$ in the figure at left? Explain your answer.

PAINTING A painter is using a 20-ft ladder whose base is 5 ft from the wall.

29. Determine the distance from the ladder to the wall at a point halfway up the ladder.

30. Determine the distance from the ladder to the wall at a point three-quarters of the way up the ladder.

31. The painter can reach the wall from 2 ft away or less. Estimate the percent of the ladder from which the painter can reach the wall.

32. MUSIC A *hammered dulcimer* is an ancient trapezoidal stringed instrument. The bases of the trapezoid are approximately 17 in. and 38 in. Estimate the length of a string at the center of the dulcimer.

Look Back

33. BIOLOGY Draw an Euler diagram to illustrate the relationships below. *(LESSON 2.3)*

A squirrel is a rodent.
All rodents are mammals.
All mammals are animals.

Use the conjectures you made in Lesson 3.2 to classify each statement as true or false. Explain your reasoning. *(LESSON 3.2)*

34. All rhombuses are squares.

35. All rectangles are parallelograms.

36. All squares are rectangles.

37. All parallelograms are rhombuses.

For Exercises 38–40, find the unknown angle measure in each figure. *(LESSON 3.5)*

38. (triangle with angles $x°$, $60°$, $30°$)

39. (triangle with angles $50°$, $x°$, exterior $120°$)

40. (triangle with angles $86°$, $45°$, exterior $x°$)

Look Beyond

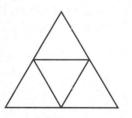

41. Draw a triangle and all of its midsegments, as shown at left. Cut out the four triangles that are formed, and compare them with each other. Write a conjecture about them.

42. Test your conjecture from Exercise 41 by using several other triangles. Based on your conjecture, what is the ratio of the area of each of the four triangles to the area of the original triangle?

43. The figure at left is formed by drawing the midsegments of the three outer triangles in the figure above. The large outer triangle is equilateral and has a side length of 1. Find and add the side lengths of all of the shaded triangles.

44. Is the total area of the shaded triangles in the figure more or less than the area of the outer triangle?

45. If the midsegments of each shaded triangle were drawn and the resulting center triangle of each was "unshaded," would the shaded areas of the new figure be more or less than the shaded areas above? Would the sum of all the side lengths be more or less? Explain your reasoning.

Analyzing Polygons With Coordinates

Objectives

- Develop and use theorems about equal slopes and slopes of perpendicular lines.

- Solve problems involving perpendicular and parallel lines in the coordinate plane by using appropriate theorems.

Why *Geometers use a mathematical method to indicate steepness. This method, while not as colorful as the names of roller coasters, is more precise.*

Amusement parks often have descriptive names for their roller coasters. Names such as "Shocker" or "Wild Thing" give riders an idea about the steepness of the falls they will experience.

Slope: A Measure of Steepness

The *slope* of a line or surface tells you how steeply it rises or falls in terms of a ratio. This ratio is found by using a right triangle. Recall that the sides of a right triangle that are adjacent to the right angle are called the legs of the triangle and that the remaining side is the hypotenuse of the triangle.

Consider a right triangle in a coordinate plane with horizontal and vertical legs. The slope of the hypotenuse is the ratio of the length of the vertical leg, the *rise*, to length of the horizontal leg, the *run*. If the hypotenuse rises from left to right, the slope is positive; if it falls from left to right, the slope is negative.

E X A M P L E **1** Find the slope of the segment with endpoints at (2, 3) and (8, 6).

SOLUTION

Draw a right triangle as shown. By counting squares, you can see that the rise is 3 and the run is 6. Thus, the slope is $\frac{3}{6} = \frac{1}{2}$, or 0.5.

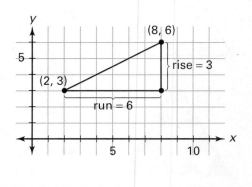

Note: In Example 1, you found that the slope of the segment with the given endpoints is 0.5. To use this method to find the slope of a line, choose any two points on the line and find the slope of the segment with those endpoints.

To find the slope of a line with two given points without drawing a picture, use the definition below.

Definition of Slope

The slope of a nonvertical line that contains the points (x_1, y_1) and (x_2, y_2) is equal to the ratio $\frac{y_2 - y_1}{x_2 - x_1}$.

3.8.1

CRITICAL THINKING

In the case of a vertical line or segment, the slope is undefined. Explain why the slope of a vertical line or segment is undefined.

E X A M P L E ❷ Find the slope of $\overline{AB}$ with the endpoints $A(5, -3)$ and $B(2, 3)$.

SOLUTION

$$\text{slope} = \frac{y_2 - y_1}{x_2 - x_1} = \frac{3 - (-3)}{2 - 5} = \frac{6}{-3} = -2$$

Parallel and Perpendicular Lines

Recall from algebra the slope-intercept form of a line:

$$y = mx + b,$$

where m represents the slope and b represents the y-intercept

The two lines in the graph at left have the same slope, 1.5. As you can see, they seem to be parallel. The theorem below, which will not be proved formally in this book, follows from the algebraic concept of slope.

$y = 1.5x + 2$ $(0, 2)$

$(0, -3)$ $y = 1.5x - 3$

Parallel Lines Theorem

Two nonvertical lines are parallel if and only if they have the same slope. Any two vertical lines are parallel.

3.8.2

The slopes of perpendicular lines also have a special relationship to each other. This relationship is stated in the following theorem, which can be proved using the coordinate proof methods described in Lesson 5.7.

Perpendicular Lines Theorem

Two nonvertical lines are perpendicular if and only if the product of their slopes is -1. Any vertical line is perpendicular to any horizontal line.

3.8.3

The fact that the product of the slopes of perpendicular lines is −1 can be stated another way. In order for the product of two numbers to equal −1, one number must be the negative reciprocal of the other. Thus, if the slope of a line is $\frac{a}{b}$, then the slope of any line perpendicular to that line must be $\frac{-b}{a}$. You can use this relationship to test whether lines are perpendicular without multiplying their slopes.

E X A M P L E 3 Draw quadrilateral *QUAD* with vertices at *Q*(1, 4), *U*(7, 8), *A*(9, 5), and *D*(3, 1). What type of quadrilateral is *QUAD*?

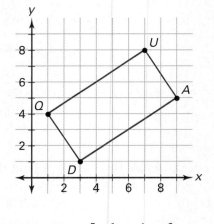

SOLUTION

Based on the figure, it appears that *QUAD* is a parallelogram and perhaps a rectangle. You can test these conjectures by finding the slope of each segment.

slope of $\overline{QU} = \frac{8-4}{7-1} = \frac{4}{6} = \frac{2}{3}$ slope of $\overline{DA} = \frac{5-1}{9-3} = \frac{4}{6} = \frac{2}{3}$

slope of $\overline{QD} = \frac{4-1}{1-3} = \frac{-3}{2}$ slope of $\overline{UA} = \frac{5-8}{9-7} = \frac{-3}{2}$

Because the opposite sides of the quadrilateral have the same slope, the figure is a parallelogram.

Because the slopes of the adjacent sides of the quadrilateral are negative reciprocals, the angles in the figure are right angles. Thus, the figure is a rectangle.

Midsegments in the Coordinate Plane

Coordinates may be used to verify the conjectures about triangle and trapezoid midsegments that you made in the previous lesson.

Algebra

Recall the following formula from algebra for the midpoint of a segment in a coordinate plane:

Midpoint Formula

The midpoint of a segment with endpoints (x_1, y_1) and (x_2, y_2) has the following coordinates:

$$\left(\frac{x_1 + x_2}{2}, \frac{y_1 + y_2}{2}\right)$$

3.8.4

Notice that the coordinates of the midpoint are the averages of the coordinates of the endpoints.

E X A M P L E **4** Draw △ABC with vertices A(2, 6), B(0, 0), and C(4, 0), and use this triangle to test the triangle midsegment conjecture.

SOLUTION

The midpoint of $\overline{AB}$ is $\left(\frac{2+0}{2}, \frac{6+0}{2}\right)$ = (1, 3), and

the midpoint of $\overline{AC}$ is $\left(\frac{2+4}{2}, \frac{6+0}{2}\right)$ = (3, 3).

Thus, the slope of the midsegment is
$\frac{3-3}{3-1} = \frac{0}{2} = 0$.

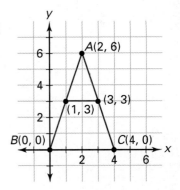

The slope of $\overline{BC}$ is $\frac{0-0}{4-0} = \frac{0}{4} = 0$, so the midsegment is parallel to a side of the triangle.

Because $\overline{BC}$ and the midsegment of the triangle are horizontal, you can determine their lengths by counting squares on the grid. The length of $\overline{BC}$ is 4, and the length of the midsegment is 2, so the length of the midsegment is half of the length of $\overline{BC}$. This confirms the Triangle Midsegment Conjecture.

TRY THIS Draw trapezoid DEFG with vertices D(4, 2), E(7, 2), F(9, 0), and G(0, 0), and use this trapezoid to test the Trapezoid Midsegment Conjecture.

CRITICAL THINKING Are the cases above proofs of the Triangle and Trapezoid Midsegment Conjectures? Why or why not?

Exercises

● *Communicate* ━━━━━━━━━━━━━━━━━━━

1. Describe lines with the following:
   a. a positive slope   **b.** a negative slope
   c. a zero slope   **d.** an undefined slope

2. Explain the meanings of the terms *rise* and *run*.

3. Suppose that nonvertical lines ℓ_1 and ℓ_2 are perpendicular and that the slope of ℓ_1 is m. What is the slope of ℓ_2? Explain your answer.

4. In the diagram at right, the two lines have slopes of 1 and −1, but they do not appear to be perpendicular. Explain why.

5. In Example 4 and the Try This that follows it, one side of the triangle and trapezoid is on the *x*-axis. What is a possible advantage to placing the figures in this position?

6. A segment has endpoints at (1, 2) and (3, 8). Plot the points and draw a right triangle with this segment as the hypotenuse. Use the right triangle to find the slope of the segment. **(EXAMPLE 1)**

In Exercises 7–9, use the definition of slope to find the slope of the segment with the given endpoints. (EXAMPLE 2)

7. (0, 0) and (4, 4)   **8.** (−1, 3) and (4, 5)   **9.** (2, 1) and (4, −6)

10. Draw a quadrilateral with vertices at (0, 2), (1, −1), (5, 1), and (4, 4). What type of quadrilateral is this? Explain your answer. **(EXAMPLE 3)**

11. Draw a trapezoid with vertices at (0, 0), (0, 4), (3, 4), and (5, 0), and use this trapezoid to test the Trapezoid Midsegment Conjecture. **(EXAMPLE 4)**

● *Practice and Apply* ▬▬▬

In Exercises 12–15, the endpoints of a segment are given. Determine the slope and midpoint of the segment.

12. (0, 0) and (4, 2)   **13.** (−1, 1) and (1, −1)

14. (−3, −1) and (3, 3)   **15.** (−5, 2) and (1, −3)

In Exercises 16–19, the endpoints of two segments are given. Determine whether the segments are parallel, perpendicular, or neither.

16. (−1, 1) and (2, 3); (2, 2) and (5, 4)

17. (−2, 1) and (1, −2); (−1, −1) and (3, 3)

18. (−2, 2) and (3, 2); (2, −1) and (2, 4)

19. (−1, 2) and (1, −2); (1, −2) and (2, −1)

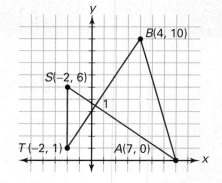

Refer to the diagram at right for Exercises 20–22.

20. Find the slope of $\overline{SA}$.

21. Find the slope of $\overline{TB}$.

22. Is ∠1 a right angle? Explain your answer.

For Exercises 23–28, the vertices of a triangle are given. Use slopes to determine whether each triangle is a right triangle.

23. (−1, 4), (4, 3), (1, 1)   **24.** (1, 3), (2, 0), (−3, 2)

25. (−2, 3), (3, −1), (−2, −1)   **26.** (1, 0), (0, 1), (−1, 0)

27. (1, 2), (3, 3), (4, 0)   **28.** (9, 5), (2, −6), (−1, −1)

PROOFS

Draw the quadrilateral with the given vertices on a graph. Identify the type of quadrilateral and prove your answer.

29. (0, −2), (5, −2), (5, 6), (0, 6)   **30.** (0, 0), (2, 3), (5, 3), (7, 0)

31. (0, 3), (−1, 1), (2, 0), (3, 2)   **32.** (1, 3), (−2, 1), (−1, −3), (2, −1)

33. (0, 6), (3, 9), (9, 3), (6, 0)   **34.** (1, 0), (5, 1), (−1, 4), (−1, 1)

PROOFS

For Exercises 35–37, the vertices of a rhombus are given. Draw the rhombus and use slopes to prove that the diagonals of each rhombus are perpendicular.

35. $(-2, 1), (1, 5), (5, 2), (2, -2)$

36. $(1, 3), (6, 3), (3, -1), (-2, -1)$

37. $(0, 0), (0, a), (a, a), (a, 0)$

For Exercises 38–41, the slope and one endpoint of a segment are given. Give the coordinates of the other endpoint. More than one answer is possible. (Hint: Start by drawing a right triangle with the given point as one of its vertices.)

38. slope $= \frac{3}{4}$; $(0, 0)$ **39.** slope $= 2$; $(6, -1)$

40. slope $= -\frac{5}{3}$; $(-1, 4)$ **41.** slope $= -1$; $(2, 5)$

42. The vertices of a triangle are $A(0, 8)$, $B(2, 0)$, and $C(3, 4)$. Find the midpoints of the sides and prove that each midsegment is parallel to a side of the triangle.

43. The vertices of a trapezoid are $K(0, 0)$, $L(0, 7)$, $M(4, 0)$, and $N(4, 9)$. Find the endpoints of the midsegment and prove that it is parallel to the bases.

44. Parallelogram $ABCD$ has vertices at $A(-1, x-1)$, $B(x, x+1)$, $C(3, 1)$, and $D(x-2, -1)$. Use the slopes of $\overline{AB}$ and $\overline{CD}$ to find x.

For Exercises 45–49, draw a quadrilateral that fits the given conditions. Label the vertices, and give the slope of each side. More than one answer may be possible.

45. trapezoid $ABCD$ with vertices $A(3, 5)$ and $B(8, 5)$

46. parallelogram $EFGH$ with vertices $E(3, 2)$ and $F(-1, 5)$

47. rectangle $JKLM$ with vertices $J(1, 6)$ and $K(3, 2)$

48. parallelogram $NPQR$ with vertices $N(-2, 1)$ and $Q(4, -2)$

CHALLENGE

49. rectangle $STUV$, in which the midpoint of $\overline{ST}$ is $(5, 5)$, the midpoint of $\overline{TU}$ is $(10, 5)$, the midpoint of $\overline{UV}$ is $(7, 1)$, and the midpoint of $\overline{VS}$ is $(2, 1)$

APPLICATION

50. CONSTRUCTION According to the Americans With Disabilities Act, a *ramp* is a route with a slope greater than $\frac{1}{20}$. The maximum allowable slope of a ramp is $\frac{1}{12}$, and the maximum rise is 30 in. What are the minimum and maximum runs for a ramp with a rise of 30 in.?

51. CONSTRUCTION A house is 25 ft wide, and has a peaked roof, as shown at right. City building codes require the pitch (slope) of the roof be at least 0.3 and no greater than 0.7. Use the given coordinates to show that the roof of this house does not meet the building codes. How could the height of the roof be adjusted so that it does meet the codes?

Look Back

52. Explain the difference between a postulate and a theorem. *(LESSON 2.2)*

Refer to the figure at right, in which $\ell_1 \parallel \ell_2$, **and find the indicated angle measures.** *(LESSON 3.5)*

53. m∠3   **54.** m∠2

55. m∠4   **56.** m∠1

57. CONSTRUCTION The diagram at right shows a house whose roof has a pitch of $\frac{1}{3}$. Find the angle that the roof forms with the walls. *(LESSON 3.6)*

Look Beyond

CULTURAL CONNECTION: AFRICA An ancient Egyptian drawing from 2650 B.C.E. shows a rounded vault. The numbers, which are the marks on the diagram, give the height, *y*, of the vault at horizontal intervals, *x*, of 1 cubit.

58. Given that 1 palm is equal to 4 fingers and 1 cubit is equal to 7 palms, copy and complete the table below.

x (in fingers)	Height (from drawing)	y (in fingers)
0 cubit = 0 fingers	3 cubits, 3 palms, and 2 fingers	98 fingers
1 cubit = ?	3 cubits, 2 palms, and 3 fingers	?
2 cubits = ?	3 cubits	?
3 cubits = ?	2 cubits and 3 palms	?
4 cubits = ?	1 cubit, 3 palms, and 1 finger	?
5 cubits = ?	0	?

59. Use the *x*- and *y*-coordinates from the table to plot the points on a graph. Then draw a smooth curve through the points.

In addition to translation tessellations (see page 154), another pattern used by M. C. Escher is known as a **rotation tessellation**. You can make your own rotation tessellation by following the steps below. As in the Portfolio Activity for Lesson 3.2, you may have to make adjustments to your curves in order to get a pattern you like.

Draw your figures on graph paper or tracing paper, or use geometry or tessellation software.

1. Start with a regular hexagon. Replace one side of the hexagon, with a curve, as shown below. Rotate the curve about point *B* so that point *A* lies on point *C*.

2. Replace side $\overline{CD}$ with a new curve, and rotate it around point *D* to replace side $\overline{DE}$.

M. C Escher's "Symmetry Drawing E25" ©1999 Cordon Art B. V. -Baarn-Holand. All rights reserved.

3. Replace side $\overline{EF}$ with a new curve, and rotate it around point *F* to replace side $\overline{FA}$.

4. Your figure will now fit together with itself on all sides. You can add details to your figure, if desired. Rotate the figure to create an interlocking design.

String Figures

The string activity below and the resulting net have appeared in many parts of the world under different names: Osage diamonds among the Osage Indians of North America, the Calabash net in Africa, and the Quebec bridge in Canada. In the United States, it is commonly known as Jacob's ladder.

1. Start with a piece of string 4 to 5 feet in length. Tie the ends together. Loop the string around your thumbs and little fingers as shown.

2. Use your right index finger to pick up the left palm string from below. In a similar way, use your left index finger to pick up the right palm string.

3. Let your thumbs drop their loop. Turn your hands so that the fingers face out. With your thumbs, reach under all the strings and pull the farthest string back toward you.

4. With your thumbs, go over the near index-finger string, and then reach under and pull back the far index-finger string.

5. Drop the loops from your little fingers. Pass your little fingers over the index-finger string and get the thumb string closest to your little fingers from below.

6. Drop the thumb loops. Pass your thumbs over the index-finger strings, get the near little-finger string from below, and return.

7. Loosen the left index-finger loop with your right hand, and place the loop over your thumb. Do the same with the right index-finger loop.

8. Each thumb now has two loops. Using your right hand, lift the lower loop of the left thumb up and over the thumb. Do the same with the lower loop of the right thumb.

9. Bend your index fingers and insert the tips into the triangles that are near the thumbs.

10. Gently take your little fingers out of their loops. Turn your hands so that your palms face away from you. The index-finger loops will slip off your knuckles. Straighten your index fingers. The finished net will appear.

Chapter Review and Assessment

POSTULATES AND THEOREMS

Lesson	Number	Postulate or Theorem
3.3	3.3.2 Corresponding Angles Postulate	If two lines cut by a transversal are parallel, then corresponding angles are congruent.
	3.3.3 Alternate Interior Angles Theorem	If two lines cut by a transversal are parallel, then alternate interior angles are congruent.
	3.3.4 Alternate Exterior Angles Theorem	If two lines cut by a transversal are parallel, then alternate exterior angles are congruent.
	3.3.5 Same-Side Interior Angles Theorem	If two lines cut by a transversal are parallel, then same-side interior angles are supplementary.
3.4	3.4.1 Theorem: Converse of the Corresponding Angles Postulate	If two lines are cut by a transversal in such a way that corresponding angles are congruent, then the two lines are parallel.
	3.4.2 Converse of the Same-Side Interior Angles Theorem	If two lines are cut by a transversal in such a way that same-side interior angles are supplementary, then the two lines are parallel.
	3.4.3 Converse of the Alternate Interior Angles Theorem	If two lines are cut by a transversal in such a way that alternate interior angles are congruent, then the two lines are parallel.
	3.4.4 Converse of the Alternate Exterior Angles Theorem	If two lines are cut by a transversal in such a way that alternate exterior angles are congruent, then the two lines are parallel.
	3.4.5 Theorem	If two coplanar lines are perpendicular to the same line, then the two lines are parallel.
	3.4.6 Theorem	If two lines are parallel to the same line, then the two lines are parallel.

Lesson	Number	Postulate or Theorem
3.5	3.5.1 The Parallel Postulate	Given a line and a point not on the line, there is one and only one line that contains the given point and is parallel to the given line.
	3.5.2 Triangle Sum Theorem	The sum of the measures of the angles of a triangle is 180°.
	3.5.3 Exterior Angle Theorem	The measure of an exterior angle of a triangle is equal to the sum of the measures of the remote interior angles.
3.6	3.6.1 Sum of the Interior Angles of a Polygon	The sum, s, of the measures of the interior angles of a polygon with n sides is given by $s = (n-2)180°$. .
	3.6.2 The Measure of an Interior Angle of a Regular Polygon	The measure, m, of an interior angle of a regular polygon with n sides is $m = 180° - \frac{360°}{n}$.
	3.6.3 Sum of the Exterior Angles of a Polygon	The sum of the measures of the exterior angles of a polygon is 360°.
3.8	3.8.2 Parallel Lines Theorem	Two nonvertical lines are parallel if and only if they have the same slope.
	3.8.3 Perpendicular Lines Theorem	Two nonvertical lines are perpendicular if and only if the product of their slopes is –1.

Key Skills & Exercises

LESSON 3.1
Key Skills

Identify reflectional symmetry of figures.

Draw all of the axes of symmetry of a regular hexagon.

Identify rotational symmetry of figures.

Describe the rotational symmetry of a regular hexagon.

A regular hexagon has 6-fold rotational symmetry. The image will coincide with the original figure after rotations of 60°, 120°, 180°, 240°, 300°, and 360°. After a rotation of 360°, the figure is returned to its original position.

Exercises

Copy each figure below.

1. Draw all axes of symmetry of the figure above.

2. Describe the rotational symmetry of the figure above.

3. Draw all axes of symmetry of the figure above.

4. Describe the rotational symmetry of the figure above.

LESSON 3.2

Key Skills

Make conjectures about the properties of quadrilaterals.

In a parallelogram, the opposite sides and angles are congruent, and the diagonals bisect each other. In a rhombus, the diagonals are perpendicular to each other. In a rectangle, the diagonals are congruent.

Exercises

ABCD is a parallelogram, EFGH is a rhombus, and JKLM is a rectangle. Find the indicated measures.

5. m∠ABC = 48°
   m∠ADC = ___?___

6. AX = 7
   AC = ___?___

7. m∠EYF = ___?___

8. JL = 13
   KM = ___?___

LESSON 3.3

Key Skills

Identify special angle pairs.

In the diagram below, ∠4 and ∠5 are alternate interior angles, ∠2 and ∠7 are alternate exterior angles, ∠4 and ∠6 are same-side interior angles, and ∠2 and ∠6 are corresponding angles.

Find angle measures formed by parallel lines and transversals.

For the diagram above,
m∠4 = m∠5 = m∠8 = 100° and
m∠2 = m∠3 = m∠6 = m∠7 = 80°.

Exercises

Refer to the diagram below.

9. Name a pair of corresponding angles.

10. Name a pair of alternate interior angles.

11. Name all angles that are congruent to ∠1.

12. Suppose that m∠1 = 130°. Find the measure of each angle in the figure.

LESSON 3.4

Key Skills

Use the converses of transversal properties to prove that lines are parallel.

Are lines ℓ₁ and ℓ₂ parallel?

The angles ∠WXU and ∠XUV are congruent alternate interior angles. By the converse of the Interior Angles Theorem, the lines are parallel.

Exercises

Refer to the diagram below.

13. Is ℓ parallel to m? Explain your answer.

14. Is m parallel to n? Explain your answer.

15. Is p parallel to q? Explain your answer.

16. Prove that the opposite sides of a rectangle are parallel.

Key Skills

Use the Triangle Sum Theorem to find angle measures.

Find the measure of ∠F.

∠DEF and ∠CEF form a linear pair,
so m∠F = 180° − 110° = 70.
By the Triangle Sum Theorem,
m∠D + m∠F + m∠DEF = 180°,
so m∠DEF = 180° − 30° − 70° = 80°.

Exercises

Refer to the diagram below. m∠PRQ = 90°, m∠PTQ = 125°, m∠PQR = 57°, m∠PSQ = 83°, and m∠RSQ = 30°.

17. Find all angle measures in △RST.

18. Find all angle measures in △PST.

19. Find all angle measures in △PQR.

20. Find all angle measures in △PQT.

Key Skills

Find interior and exterior angle measures of polygons.

In a regular octagon, what is the measure of an interior angle? What is the measure of an exterior angle?

The sum of the interior angles of an octagon is 1080°. The measures of all the interior angles are equal, so the measure of each interior angle is 1080° ÷ 8 = 135°.

The sum of exterior angles of any polygon is 360°, so the measure of each exterior angle is 360° ÷ 8 = 45°.

Exercises

21. Find the missing angle measure in the diagram at right.

22. What is the measure of an interior angle of a regular pentagon?

23. What is the measure of an interior angle of a regular 17-gon?

24. What is the measure of an exterior angle of a regular dodecagon?

Key Skills

Solve problems by using triangle and trapezoid midsegments.

A trapezoid with one base of 10 has a midsegment of 17. What is the length of the other base?

The midsegment is the average of the bases. Let x represent the unknown length.

$$\frac{10 + x}{2} = 17$$
$$10 + x = 34$$
$$x = 24$$

Exercises

25. What is the length of the midsegment of a triangle with a base of 12?

26. What is the length of the base of a triangle with a midsegment of 45?

Find the indicated lengths in trapezoid *KLMN*.

27. *RS*

28. *TU*

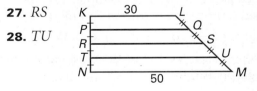

Use slope to determine whether lines and segments are parallel or perpendicular.

A triangle has vertices at $A(0, 0)$, $B(4, 1)$, and $C(3, 5)$. Is $\triangle ABC$ a right triangle?

$\overline{AB}$ has a slope of $\frac{1}{4}$, $\overline{BC}$ has a slope of -4, and $\overline{AC}$ has a slope of $\frac{5}{3}$. Because $\frac{1}{4} \cdot (-4) = -1$, $\overline{AB}$ is perpendicular to $\overline{BC}$. Therefore, $\triangle ABC$ is a right triangle.

Exercises

29. $\triangle FGH$ has vertices at $F(1, 1)$, $G(2, 0)$, and $H(-1, -1)$. Is $\triangle FGH$ a right triangle?

Match each type of special quadrilateral with the correct set of vertices. Explain your reasoning.

   a. trapezoid
   b. parallelogram
   c. rectangle

30. $(0, 2)$, $(2, 3)$, $(5, 2)$, $(3, 1)$

31. $(2, 1)$, $(1, 3)$, $(5, 5)$, $(6, 3)$

32. $(0, 1)$, $(3, 3)$, $(7, 3)$, $(1, -1)$

Applications

33. CRAFTS Ruth is making a quilt with the block design called Bow Ties, shown at right. Describe all lines of symmetry and all rotational symmetries for the block.

34. ART An artist is making a wooden frame for stretching a canvas. To make sure that the frame is rectangular, he measures the diagonals. If the frame is a rectangle, what will be true of its diagonals?

35. ENGINEERING The *grade* of a road is its slope expressed as a percent. For example, a road that rises 6 ft over a horizontal run of 100 ft has a slope of $\frac{6}{100}$, or a grade of 6%. What is the grade of each section of the road represented by the graph at right?

Alternative Assessment

Performance Assessment

1. Explain why a triangle cannot have more than one obtuse angle.

2. Can a quadrilateral have more than one obtuse angle? Can a pentagon have five acute angles? Complete the table below, assuming that all angles are less than 180°. Then write a conjecture about each type of angle.

Polygon	Maximum # of obtuse angles	Maximum # of acute angles
triangle	1	3
quadrilateral	?	?
pentagon	?	?
hexagon	?	?

3. What is the measure of an interior angle of a regular polygon with 10 sides? with 100 sides? with 1000 sides? What number do the angle measures seem to approach as the number of sides increases to infinity?

4. Repeat Exercise 3 for exterior angles.

Portfolio Projects

PENROSE TILES In 1975, a mathematical physicist named Roger Penrose discovered a pair of quadrilaterals that fit together to create interesting patterns.

Dart

36°

72°

Kite

72°

72°

1. What is the measure of the obtuse angle of the kite? One interior angle of the dart has a measure greater than 180°. This angle fits together exactly with the obtuse angle of the kite.

2. Trace the kite and dart patterns given above, or create your own by starting with a rhombus, as shown at right. The acute angles of the rhombus should be 72°.

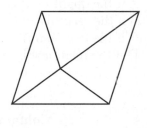

Mark the vertices with two different colors, as shown in the two figures above.

3. When these shapes are put together according to certain rules, the resulting pattern is known as a Penrose tiling. One feature of this type of design is that it is *aperiodic*; that is, it does not have translational symmetry.

Penrose Tiling Rules

- Adjacent vertices must be the same color.
- Adjacent sides must be the same length.

The pattern below was created from Penrose tiles. Use your own tiles or geometry software to create your own design.

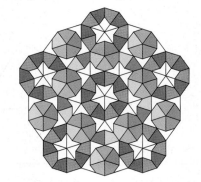

QUANTITATIVE COMPARISON For Items 1–3, write
A if the quantity in Column A is greater than the quantity in Column B;
B if the quantity in Column B is greater than the quantity in Column A;
C if the two quantities are equal; or
D if the relationship cannot be determined from the information given.

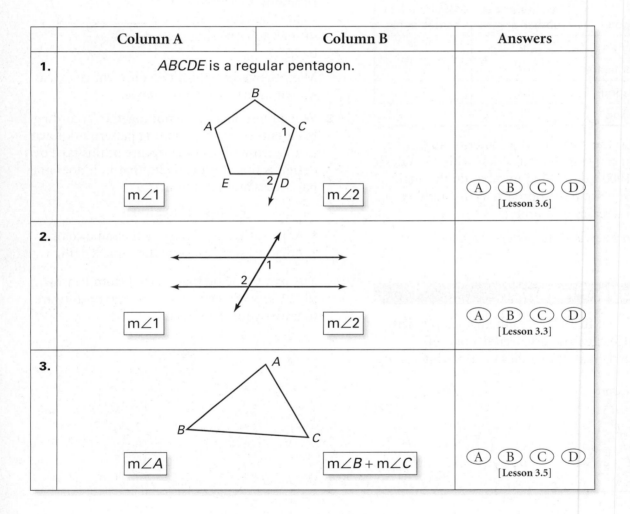

Column A	Column B	Answers
1. *ABCDE* is a regular pentagon. $m\angle 1$	$m\angle 2$	Ⓐ Ⓑ Ⓒ Ⓓ [Lesson 3.6]
2. $m\angle 1$	$m\angle 2$	Ⓐ Ⓑ Ⓒ Ⓓ [Lesson 3.3]
3. $m\angle A$	$m\angle B + m\angle C$	Ⓐ Ⓑ Ⓒ Ⓓ [Lesson 3.5]

Complete the statements in Items 4–7.
(LESSONS 1.1, 1.4, AND 1.5)

4. The intersection of two lines is a ____?____.
The intersection of two planes is a ____?____.

5. If two points are in a plane, then the line
containing them ____?____.

6. The distance from a point to a line is the
length of the ____?____ from the point to
the line.

7. The intersection of the perpendicular
bisectors of a triangle is the center of the
____?____ of the triangle.

For Items 8–11, write the rule for each transformation. *(LESSON 1.7)*

8.

9.

10.

11.

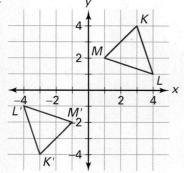

For Items 12–16, refer to the diagram below, in which $\overline{DE} \parallel \overline{FG}$, DE = 3, FG = 8, m∠EGF = 34°, m∠EGD = 30°, and m∠EJH = 42°. *(LESSONS 3.3, 3.5, 3.6, AND 3.7)*

12. Name three angles that are congruent to ∠EGF.

13. What is the ratio of *IJ* to *FG*?

14. *HJ* = _____?_____ , *HI* = _____?_____

15. m∠GFE = _____?_____ , m∠HJF = _____?_____ , m∠GIJ = _____?_____ , m∠GEJ = _____?_____

16. m∠HIE = _____?_____ , m∠GHJ = _____?_____ , m∠GDE = _____?_____ , m∠DHI = _____?_____

FREE-RESPONSE GRID Items 17–20 may be answered by using a free-response grid such as that commonly used by standardized-test services.

17. The measure of an angle is 25°. What is the measure of its complement? *(LESSON 1.3)*

18. The measure of an angle is 75°. What is the measure of its supplement? *(LESSON 1.3)*

19. An angle is its own complement. What is its measure? *(LESSON 1.3)*

20. An angle is its own supplement. What is its measure? *(LESSON 1.3)*

Triangle Congruence

ALL AROUND YOU—IN NATURE, ART, AND HUMAN technology—you find things that are the same shape and size. Such things are said to be congruent. In the photos on these pages, notice that there are many congruent triangles.

Triangles have the property of being rigid, which makes them useful in building bridges and other structures. Also, since any polygon can be divided into a number of triangles, the properties of triangles can be used to study polygons in general.

Buckminster Fuller
(1895–1983)

Lessons

Manhattan Bridge, New York

About the Chapter Project

In 1939 an American graduate student at Princeton made an interesting discovery. He folded strips of paper to form an object called a flexagon.

Flexagons have an interesting mathematical property. The hexaflexagon, for example, has three faces, but only two are visible at any given time. All flexagons have at least one hidden face. To find the hidden face, the flexagon must be folded, or "flexed," a certain way.

After completing the Chapter Project, you will be able to do the following:

• Create a hexaflexagon and a hexahexaflexagon.

• Describe the patterns in the order of faces of a flexagon.

About the Portfolio Activities

Throughout the chapter, you will be given opportunities to complete Portfolio Activities that are designed to support your work on the Chapter Project.

The theme of each Portfolio Activity and of the Chapter Project is congruent polygons.

• In the Portfolio Activity on page 225, you will use a strip of paper to fold a series of triangles. The more triangles you fold, the closer they become to a set of congruent equilateral triangles.

• Tessellations involve covering a surface with congruent shapes that fit together without gaps or overlapping. In the Portfolio Activity on page 252, you will explore tessellations with congruent, nonregular quadrilaterals.

• In the Portfolio Activity on page 281, you will explore tessellations with congruent hexagons that are not regular but that have one pair of parallel and congruent opposite sides.

Congruent Polygons

Objectives

● Define *congruent polygons.*

● Solve problems by using congruent polygons.

Why In earlier lessons you learned about congruent segments and angles. In this lesson you will develop a definition of congruent polygons.

Repeating patterns of congruent polygons are frequently employed in Islamic art.

Polygon Congruence

Polygons 1 and 2 at right are congruent. If you slide one on top of the other, you will see that they match exactly. Can you think of a way to determine whether two polygons are congruent without actually moving them? What measurements would you need to make?

Polygon 1   Polygon 2

If two polygons are congruent, then their respective angles and sides are congruent. The converse is also true: If the respective angles and sides of a polygon match, then the polygons are congruent. These facts will be stated later as a postulate (see page 212), but first you will need to learn some terminology and notation.

Naming Polygons

When naming polygons, the rule is to go around the figure, either clockwise or counterclockwise, and list the vertices in order. It does not matter which vertex you list first.

EXAMPLE **1** **What are all of the possible names for the hexagon at right?**

● **SOLUTION**

PROBLEM SOLVING

Make an organized list. You can approach the question systematically, as follows:

Pick a letter from the figure, such as *A*. Then write the letters of each vertex of the figure, going first in one direction and then in the other.

$$ABCDEF \qquad AFEDCB$$

Then use each of the other letters in the figure as starting points.

BCDEFA	*BAFEDC*
CDEFAB	*CBAFED*
DEFABC	*DCBAFE*
EFABCD	*EDCBAF*
FABCDE	*FEDCBA*

In all, there are 12 possible names for the hexagon.

TRY THIS   What are all of the possible names for the pentagon at right?

Corresponding Sides and Angles

If two polygons have the same number of sides, it is possible to set up a correspondence between them by pairing their parts. In quadrilaterals *ABCD* and *EFGH*, for example, you can pair angles *A* and *E*, *B* and *F*, *C* and *G*, and *D* and *H*. Notice that you must go in the same order around each of the polygons.

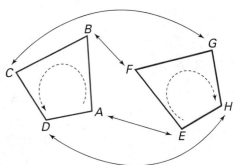

The correspondence of the sides follows from the correspondence of the angles. In this case, side $\overline{AB}$ corresponds to side $\overline{EF}$, and so on.

How many different ways are there of setting up a correspondence between the vertices of the two quadrilaterals on the previous page if they must go in order?

When you write a congruence statement about two polygons, you must write the letters of the vertices in the proper order so that they correspond.

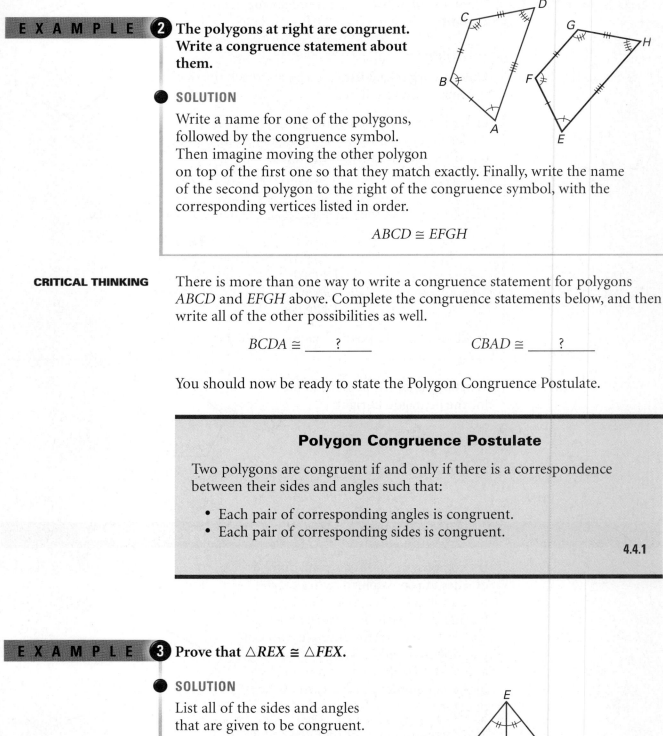

E X A M P L E **2** **The polygons at right are congruent. Write a congruence statement about them.**

● **SOLUTION**

Write a name for one of the polygons, followed by the congruence symbol. Then imagine moving the other polygon on top of the first one so that they match exactly. Finally, write the name of the second polygon to the right of the congruence symbol, with the corresponding vertices listed in order.

$$ABCD \cong EFGH$$

CRITICAL THINKING There is more than one way to write a congruence statement for polygons *ABCD* and *EFGH* above. Complete the congruence statements below, and then write all of the other possibilities as well.

$$BCDA \cong \underline{\quad ? \quad} \qquad\qquad CBAD \cong \underline{\quad ? \quad}$$

You should now be ready to state the Polygon Congruence Postulate.

Polygon Congruence Postulate

Two polygons are congruent if and only if there is a correspondence between their sides and angles such that:

- Each pair of corresponding angles is congruent.
- Each pair of corresponding sides is congruent.

4.4.1

E X A M P L E **3** **Prove that △REX ≅ △FEX.**

● **SOLUTION**

List all of the sides and angles that are given to be congruent.

$\angle R \cong \angle F$ $\overline{RE} \cong \overline{FE}$
$\angle REX \cong \angle FEX$ $\overline{RX} \cong \overline{FX}$
$\angle EXR \cong \angle EXF$

Six congruences are required for triangles to be congruent—three pairs of angles and three pairs of sides. Thus, one more pair of congruent sides is needed for these triangles.

Notice that $\overline{EX}$ is shared by the two triangles. Use the Reflexive Property of Congruence to justify the statement that $EX \cong EX$. This gives the sixth congruence, so you can conclude that $\triangle REX \cong \triangle FEX$.

Exercises

Communicate

1. Sketch a triangle and label the vertices A, B, and C. List all of the possible names for this triangle. Does the order of the vertices matter in naming a triangle? Explain your reasoning.

2. Suppose that quadrilaterals $MNOP$ and $QRST$ are congruent. List all of the pairs of corresponding sides and angles in the two polygons.

3. When are two segments congruent? Explain.

4. When are two angles congruent? Explain.

5. Explain the difference between the statements $\overline{AB} \cong \overline{CD}$ and $AB = CD$.

6. Does $AB \cong CD$ make sense? Explain.

Guided Skills Practice

7. Give all possible names for the pentagon at right. **(EXAMPLE 1)**

8. Use notation to write a congruence statement about both pentagons at right. **(EXAMPLE 2)**

9. Prove that $\triangle ABC \cong \triangle DBC$. **(EXAMPLE 3)**

Practice and Apply

For Exercises 10 and 11, consider a pentagon, *QRSTU*.

10. Give three other names for pentagon *QRSTU*.

11. Name the interior angles of pentagon *QRSTU*, using three letters for each.

Suppose that *QRSTU* ≅ *VWXYZ*.

12. List all pairs of corresponding angles.

13. Name the segment that is congruent to each segment below.

    a. $\overline{ST}$ **b.** $\overline{WX}$ **c.** $\overline{QU}$

Determine whether the pairs of figures below are congruent. Explain your reasoning.

14.

5 cm

6 cm

15.

7 cm   7 cm

16.

17. △*ABC* and △*DBC*

18.

4

4   4

4

4   4

4   4

19.

50°

50°

For Exercises 20–25, refer to the figures at right and complete the statements.

△*ABC* ≅ △*EDF*

20. a. ∠*A* ≅ __?__   **b.** ∠*B* ≅ __?__

    c. ∠*C* ≅ __?__

21. a. $\overline{AB}$ ≅ __?__   **b.** $\overline{BC}$ ≅ __?__

    c. $\overline{AC}$ ≅ __?__

22. a. ∠*MLQ* ≅ __?__   **b.** ∠*LMP* ≅ __?__

    c. ∠*MPO* ≅ __?__   **d.** ∠*NOP* ≅ __?__

23. a. $\overline{PQ}$ ≅ __?__   **b.** $\overline{LQ}$ ≅ __?__

    c. $\overline{MN}$ ≅ __?__   **d.** $\overline{NO}$ ≅ __?__

LMPQ ≅ *MNOP*

214 CHAPTER 4

24. a. $\angle A \cong$ ___?___ **b.** $\angle E \cong$ ___?___

c. $\angle C \cong$ ___?___ **d.** $\angle G \cong$ ___?___

e. $\angle I \cong$ ___?___

25. a. $\overline{AB} \cong$ ___?___ **b.** $\overline{IJ} \cong$ ___?___

c. $\overline{HI} \cong$ ___?___ **d.** $\overline{BC} \cong$ ___?___

e. $\overline{GH} \cong$ ___?___

$ABCDE \cong FGHIJ$

Given: $\angle L \cong \angle P$, $\angle M \cong \angle O$, $\angle MQL \cong \angle ORP$, $\overline{LM} \cong \overline{PO}$, $\overline{MQ} \cong \overline{OR}$, $LR = 5$, $QR = 3$, and $QP = 5$

26. $LQ =$ ___?___

27. $RP =$ ___?___

28. Is $\triangle LMQ \cong \triangle POR$ true? Why or why not?

Use the figure of rhombus ABCD at right for Exercises 29 and 30.

29. If $m\angle BDA = x°$, what is $m\angle DAB$? What is $m\angle DCB$?

30. Write a two-column proof that $\triangle ABD \cong \triangle CBD$.

Given: $\triangle XYZ \cong \triangle FGH$ and $\triangle FGH \cong \triangle JKL$

31. What must be true about $\triangle XYZ$ and $\triangle JKL$?

32. Name a property that justifies your conclusion from Exercise 31.

CHALLENGES

33. Given $\triangle ABC \cong \triangle ACB$, what can you say about $\triangle ABC$? Explain your reasoning.

34. Given $\triangle ABC \cong \triangle BCA$, what can you say about $\triangle ABC$? Explain your reasoning.

APPLICATIONS

35. CARPENTRY A shop makes house-shaped mailboxes like the one shown at right. The walls opposite each other are congruent, as are the two slanted portions of the roof.

  a. Name 6 pairs of congruent polygons on this mailbox.

  b. Name 5 pairs of congruent angles on this mailbox.

  c. Name 5 pairs of congruent sides on this mailbox.

36. QUILTING The quilt shown at right was assembled from 12 identical "Balkan Puzzle" blocks. Make a sketch of one of these blocks. How many different congruent shapes can you find in the quilt? Describe them.

37. FASHION Janet is making a pair of earrings out of polymer clay. She wants the left and right earrings to be congruent. Draw a diagram of both earrings and label the vertices. Write a congruence statement for the two triangles and find the measure of each angle in both earrings.

38. FASHION Janet cuts a triangle out of paper and traces it onto the clay twice to make the two earrings. Will the earrings be congruent? What property of congruence justifies your answer?

Look Back

Given: *ABCD* is a parallelogram. Justify each statement below with a theorem, property, or definition. *(LESSONS 3.2, 3.3, AND 3.4)*

39. $\overline{AB} \parallel \overline{CD}$
40. $\overline{AD} \parallel \overline{BC}$
41. $\angle BAC \cong \angle DCA$
42. $\angle DAC \cong \angle BCA$
43. $\overline{AC} \cong \overline{CA}$

Determine whether each of the following statements is true or false. Explain your reasoning. *(LESSON 3.2)*

44. A square is a rhombus.

45. A parallelogram is a trapezoid.

46. A parallelogram is a rhombus.

47. A rectangle is a parallelogram.

Look Beyond

COORDINATE GEOMETRY Plot the following points on a graph and connect them to form a rectangle: *A*(1, 2), *B*(1, 4), *C*(4, 4), and *D*(4, 2).

48. Use the rule $F(x, y) = (x + 5, y - 3)$ to transform the figure. Is the image congruent to the preimage? Why or why not?

49. Use the rule $F(x, y) = (3x, -2y)$ to transform the figure. Is the image congruent to the preimage? Why or why not?

Triangle Congruence

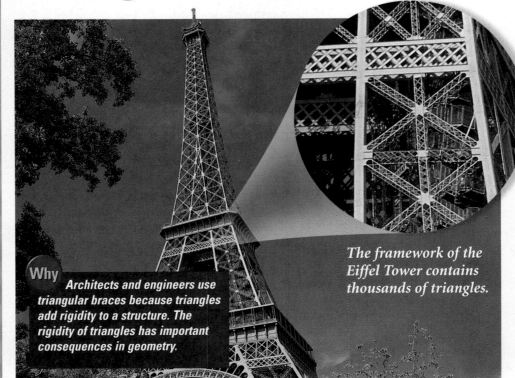

Objectives

- Explore triangle rigidity.

- Develop three congruence postulates for triangles—SSS, SAS, and ASA.

Why Architects and engineers use triangular braces because triangles add rigidity to a structure. The rigidity of triangles has important consequences in geometry.

The framework of the Eiffel Tower contains thousands of triangles.

Triangles in Physical Structures

A physical triangle is rigid. As long as the sides do not change or break loose, the shape of the triangle will not change. The rigidity of triangles contributes a postulate to geometry, which you will investigate in Activity 1 below.

Activity 1
Triangle Rigidity

YOU WILL NEED

drinking straws, string, a ruler, and scissors

1. Construct each triangle described below from drinking straws and string. When you pull the strings tight, your triangles will be rigid.

   $\triangle ABC$:
   $AB = 8$ in., $BC = 9$ in., $CA = 10$ in.

   $\triangle XYZ$:
   $XY = 6$ in., $YZ = 8$ in., $ZX = 10$ in.

2. Compare your triangles with those made by other members of your class. Are your triangles congruent to theirs?

3. Given the lengths of the sides of a triangle, can the triangle have more than one shape? How does your answer to this question relate to the fact that triangles are rigid?

4. Do you need to know the angle measures of a triangle to make a copy of it? Explain.

5. Your work with the triangle in Steps 1–4 should suggest the important geometry postulate that is partially given below. Complete it by filling in the blanks.

SSS (Side-Side-Side) Postulate

If the ___?___ of one triangle are congruent to the ___?___ of another triangle, then the two triangles are ___?___.   **4.2.1**

CRITICAL THINKING

Are any polygons other than triangles rigid? If you repeat Activity 1 with quadrilaterals instead of triangles, would your results be similar? If the corresponding sides of two quadrilaterals are congruent, must the quadrilaterals be congruent?

Useful Geometry Tools

If you use the Polygon Congruence Postulate on page 212 to show that two triangles are congruent, you must show that three pairs of sides are congruent and three pairs of angles are congruent. Postulate 4.2.1, which you discovered in Activity 1, provides a shortcut for showing that two triangles are congruent. You need to show only that three pairs of sides are congruent. In Activity 2, you will discover two more shortcuts for proving triangle congruence.

Activity 2

Two More Congruence Postulates

YOU WILL NEED

6 pieces of tracing paper and a straightedge

Part I

1. Trace each figure at right onto a separate piece of tracing paper near the center of the paper.

2. Arrange the pieces of paper on top of each other so that point *A* of each segment is over the vertex of ∠*A* and the segments fall along the two rays of the angle.

3. On the top piece of paper, draw a segment connecting points *B* and *C*. Trace the rest of △*ABC*.

4. Compare your triangle with those of other members of your class. Are the triangles congruent?

5. In a triangle, an angle formed by two sides of the triangle is called the *included angle* of the two sides. (Informally, it is the angle "between" the two sides.) If two sides and their included angle are fixed in a triangle, is the size and shape of the triangle fixed?

CHECKPOINT ✔ **6.** Your work with the triangle in Steps 1–5 should suggest the important geometry postulate that is partially given below. Complete the postulate by filling in the blanks.

SAS (Side-Angle-Side) Postulate

If two _____?_____ and their _____?_____ in one triangle are congruent to two _____?_____ and their _____?_____ in another triangle, then the two triangles are _____?_____.   **4.2.2**

Part II

1. Trace each figure at right onto a separate piece of tracing paper near the center of the paper.

2. Arrange the pieces of tracing paper on top of each other so that the vertices of the angles are over the endpoints of the segments, one side of each angle lies along the segment, and the other sides of the angles are on the same side of the segment.

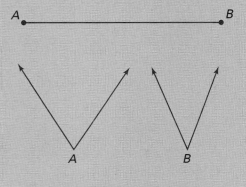

3. On the top piece of paper, extend the sides of the angles so that they meet to form a triangle. Trace the rest of △*ABC*.

4. Compare your triangle with those of other members of your class. Are the triangles congruent?

5. In a triangle, a side that is part of two angles of the triangle is called the *included side* of the two angles. (Informally, it is the side "between" the two angles.) If two angles and their included side are fixed in a triangle, is the size and shape of the triangle fixed?

CHECKPOINT ✔ **6.** Your work with the triangle in Steps 1–5 should suggest the important geometry postulate that is partially given below. Complete the postulate by filling in the blanks.

ASA (Angle-Side-Angle) Postulate

If two _____?_____ and the _____?_____ in one triangle are congruent to two _____?_____ and the _____?_____ in another triangle, then the two triangles are _____?_____.   **4.2.3**

Using the New Postulates

The triangle postulates you discovered in the activities of this lesson allow you to save steps in proofs (compare with the proof on pages 212–213). But much more important is the fact that they allow you to determine triangle congruence from limited information.

E X A M P L E In each pair below, the triangles are congruent. Tell which triangle congruence postulate allows you to conclude that they are congruent, based on the markings in the figures.

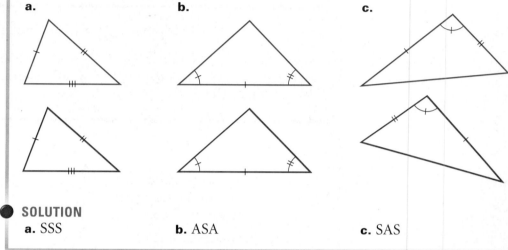

a.

b.

c.

SOLUTION

a. SSS   **b.** ASA   **c.** SAS

Exercises

Communicate

1. What is the advantage of using the SSS, SAS, and ASA Triangle Congruence Postulates instead of the Polygon Congruence Postulate given in Lesson 4.1?

2. Given $\angle A$ at right, describe how you would create $\triangle ABC$ in which $AB = 2$ inches and $AC = 3$ inches.

3. When using the ASA Triangle Congruence Postulate, does it matter which two angles are given? Why or why not?

APPLICATION

4. **ENGINEERING** What is meant by triangle rigidity? How does this property of triangles make them especially useful for building structures such as bridges and towers?

For each pair below, tell which triangle congruence postulate allows you to conclude that the triangles are congruent. *(EXAMPLE)*

5. **6.** **7.**

● *Practice and Apply* ▬▬▬▬▬▬▬▬▬

Determine whether each pair of triangles can be proven congruent by using the SSS, SAS, or ASA Congruence Postulate. If so, write a congruence statement and identify which postulate is used.

8. **9.**

10. **11.**

12. **13.**

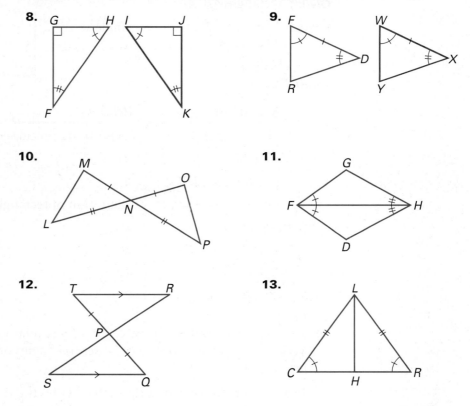

For Exercises 14–19, some measurements of a triangle are given. Is there exactly one triangle that can be constructed with the given measurements? If so, identify the postulate that applies.

14. $\triangle ABC$: $AB = 5$, $AC = 7$, $BC = 10$

15. $\triangle DEF$: $DE = 14$, $EF = 12$, $m\angle E = 75°$

16. $\triangle GHI$: $m\angle G = 60°$, $m\angle H = 60°$, $m\angle I = 60°$

17. $\triangle JKL$: $JK = 3$, $m\angle J = 45°$, $m\angle K = 90°$

18. $\triangle MNO$: $MN = 8$, $m\angle M = 30°$, $m\angle N = 110°$

19. $\triangle UVW$: $m\angle U = 40°$, $UV = 10$, $VW = 7$

In Exercises 20–22, you will create a triangle from three given sides.

20. Draw a segment the same length as one of the segments shown at right. Then set your compass to the length of one of the other segments. Place the compass at an endpoint of the first segment you drew, and construct a circle. What is true of any segment that connects this endpoint of the first segment to a point on the circle? Why?

21. Set your compass to the length of the third segment. Place the compass at the other endpoint of the first segment you drew, and construct a circle. What is true of any segment that connects the endpoint of the first segment to a point on the circle? Why?

22. Connect both endpoints of the first segment to one of the points where the circles intersect. What is true of the figure you created? Why?

TWO-COLUMN PROOF

For Exercises 23–29, complete the two-column proof below.

Given: rectangle $ABCD$

Prove: Diagonal $\overline{DB}$ divides rectangle $ABCD$ into two congruent triangles.

Proof:

Statements	Reasons
$m\angle A = m\angle D = 90°$	Definition of rectangle
$\overline{AB} \parallel \overline{CD}$	**23.** ?
$\angle ABD \cong \angle CDB$	**24.** ?
$m\angle A = m\angle B = 90°$	Definition of rectangle
$\overline{AD} \parallel \overline{BC}$	**25.** ?
$\angle ADB \cong \angle CBD$	**26.** ?
$\overline{DB} \cong \overline{BD}$	**27.** ?
$\triangle ADB \cong$ **28.** ?	**29.** ?

30. Refer to the definition of the center of a regular polygon on page 139. Write a paragraph proof that the central angles of a regular polygon are congruent.

31. Prove that a diagonal of a rhombus divides the rhombus into two congruent triangles.

CHALLENGE

32. Suppose that two congruent triangles share a common side. Is the figure formed by the two triangles sometimes, always, or never a parallelogram? Explain your reasoning.

APPLICATION

33. CARPENTRY A carpenter is building a rectangular bookshelf and finds that it wobbles from side to side. To stabilize the bookshelf, he nails on a board that connects the top left corner to the bottom right corner. Why does this diagonal board stabilize the bookshelf?

34. CONSTRUCTION In framing a house, certain triangles in the roof structure must be congruent. How can you be sure that the triangles are congruent without measuring any angles? Which triangle congruence postulate would you use?

35. QUILTING A quilter is making a quilt out of regular hexagons that are composed of triangles, as shown at right.

    a. Is $\triangle FOA \cong \triangle COD$? Why or why not?

    b. Is $\triangle BOA \cong \triangle COD$? Why or why not?

    c. Explain two ways to prove that $\triangle FOE \cong \triangle COB$.

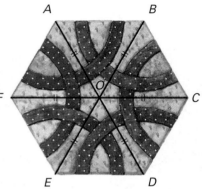

RECREATION In Exercises 36–39, pairs of triangular boat sails are shown. Based on the congruences shown, are the sails in each pair necessarily congruent?

36.

37.

38.

39.

Look Back

Complete the two-column proof below. *(LESSON 3.4)*

Given: $\ell_1 \parallel \ell_2$ and $\angle 2 \cong \angle 3$

Prove: $\ell_1 \parallel \ell_3$

Proof:

Statements	Reasons
$\ell_1 \parallel \ell_2$	**40.** __?__
$\angle 2 \cong \angle 3$	**41.** __?__
$\angle 1 \cong \angle 2$	**42.** __?__
$\angle 1 \cong \angle 3$	**43.** __?__
$\ell_1 \parallel \ell_3$	**44.** __?__

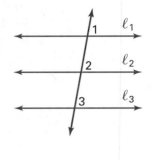

Given: quadrilateral *JKLM* $\cong$ **quadrilateral** *WXYZ* *(LESSON 4.1)*

45. Identify an angle that is congruent to $\angle K$.

46. Identify a segment that is congruent to $\overline{KL}$.

47. Is it impossible, possible, or definite that $\angle W \cong \angle L$?

Look Beyond

Use straws or geostrips to create four segments—two that are 3 inches long and two that are 5 inches long.

48. Form a quadrilateral from these segments. Is your quadrilateral rigid?

49. Make a model. What type of quadrilateral do you seem to get when the congruent segments are opposite each other? What type of quadrilateral do you seem to get when the congruent segments are adjacent to each other?

CREATING EQUILATERAL TRIANGLES

Use folding paper, geometry graphics software, or a ruler and protractor to draw each figure.

1. Draw two parallel lines, ℓ and m, and a transversal, p.

2. Bisect the obtuse angle formed by lines m and p. Label the angle bisector q.

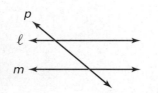

3. Bisect the obtuse angle formed by ℓ and q. Continue bisecting angles as shown, alternating between lines ℓ and m. What seems to be true about the triangles that are created?

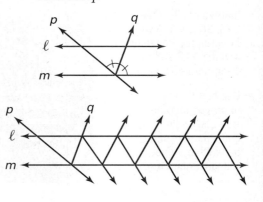

4. Use a long strip of paper with parallel edges. Fold the paper to create a transversal, and then unfold the paper.

To bisect the obtuse angle formed, fold the paper so that one edge matches exactly with the crease from the first fold. Alternate folding the strip up and down to create a long row of triangles.

5. The triangles should become more regular as you go. Keep making triangles until you get a row of nine equilateral triangles. Cut off the irregular triangles from the beginning. Using the row of nine equilateral triangles (without further cutting), how can you make a regular hexagon?

WORKING ON THE CHAPTER PROJECT

You should now be able to complete Activity 1 of the Chapter Project.

Analyzing Triangle Congruence

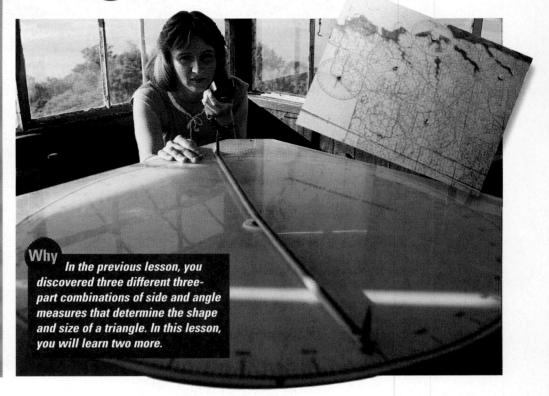

Objectives

- Identify and use the SSS, SAS, and ASA Congruence Postulates and the AAS and HL Congruence Theorems.

- Use counterexamples to prove that other side and angle combinations cannot be used to prove triangle congruence.

Why *In the previous lesson, you discovered three different three-part combinations of side and angle measures that determine the shape and size of a triangle. In this lesson, you will learn two more.*

APPLICATION
FORESTRY

In the method of *triangulation*, observers at two different places each take note of their line of sight to an object. The location of the object is the point where their lines of sight cross. By plotting the lines of sight on a map, the location of the object can be found. Why do you think this method is called *triangulation*? Think about this question as you continue your investigation of triangle congruence postulates and theorems.

Observers in towers often work in pairs.

Three New Possibilities

You may have realized that the three-part combinations of sides and angles you studied in the previous lesson are not the only combinations that are possible. In fact, there are three others. Which of the following combinations do you think can be used to establish triangle congruence?

Valid or not?

1. **AAA combination**—three angles

2. **AAS combination**—two angles and a side that is not between them

3. **SSA combination**—two sides and an angle that is not between them (that is, an angle opposite one of the two sides)

You can rule out the AAA combination by finding a counterexample, and a simple proof will show that the AAS combination is a valid test for triangle congruence. A counterexample can also be found for the SSA combination.

Show that the AAA combination is not a valid test for triangle congruence.

● **SOLUTION**

In order to get a clear idea of what you need to *disprove*, state the combination in the form of a conjecture, as given below.

> Conjecture: If the three angles of one triangle are congruent to the three angles of another triangle, then the triangles are congruent.

Counterexamples to this statement are easy to find. In the triangles at right, there are three pairs of congruent angles, but the two triangles are not congruent. Therefore, the conjecture is *false*.

AAA counterexample

Show that the AAS combination is a valid test for triangle congruence.

● **SOLUTION**

The AAA combination can be converted to the ASA combination.

The triangles at right are an AAS combination. To convert this to an ASA combination, find the measures of the third angle in each triangle. Remember that the sum of the measures of the angles of a triangle is 180°.

$$m\angle O = 180 - (60 + 75) = 45°$$
$$m\angle R = 180 - (60 + 75) = 45°$$

The measure of the third angle is the same in each triangle, so $\angle O \cong \angle R$. The given side, measuring 3 units, is the included side of the 45° and 60° angles in each triangle. Therefore, the two triangles are congruent by the ASA Congruence Postulate.

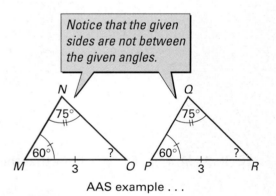

Notice that the given sides are not between the given angles.

AAS example . . .

. . . can be converted to ASA.

CRITICAL THINKING

The two triangles at right represent a version of AAS, but the triangles are not congruent. There is an important difference between the two triangles. What is it?

The missing angle measure is 75°.

Is this an example of AAS?

For an AAS combination to be used, *the congruent parts must correspond.* Notice carefully the wording of the following theorem:

AAS (Angle-Angle-Side) Congruence Theorem

If two angles and a nonincluded side of one triangle are congruent to the corresponding angles and nonincluded side of another triangle, then the triangles are congruent.   **4.3.1**

The three combinations you studied in Lesson 4.2 are postulates, but the AAS combination is a theorem. You will be asked to prove this theorem in the exercise set.

TRY THIS   Which pairs of triangles below can be proven to be congruent by the AAS Congruence Theorem?

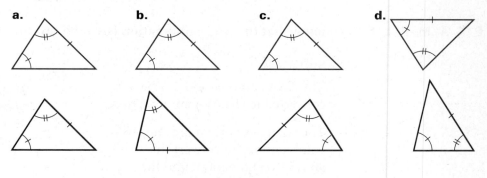

a.   **b.**   **c.**   **d.**

E X A M P L E   ❸   **Show that the SSA combination is not a valid test for triangle congruence.**

● **SOLUTION**

It is a little more difficult to find a counterexample for this conjecture, but the figures at right provide one. The sides and angles are congruent, but the triangles are obviously not congruent. Therefore, the conjecture is false.

SSA counterexample

When you try to draw a triangle for an SSA combination, the side opposite the given angle can sometimes pivot like a swinging door between two possible positions. This "swinging door" effect show that two triangles are possible for certain SSA information.

"Swinging door"

The "swinging door" effect, which invalidates certain SSA information as a test for triangle congruence, does not always pose a problem—as you will see in the next section. (See also Exercises 38–39.)

A Special Case of SSA

If the given angle in an SSA combination is a right angle, then the "swinging door" side cannot pivot to touch the ray in two different places as in the illustration in Example 3.

Thus, if the given angle is a right angle, SSA can be used to prove congruence. In this case, it is called the Hypotenuse-Leg Congruence Theorem. You will be asked to prove this theorem in the exercise set for Lesson 4.4.

> *The perpendicular segment touches the line at just one point because it is the shortest segment that can be drawn from the point to the line.*

HL (Hypotenuse-Leg) Congruence Theorem

If the hypotenuse and a leg of a right triangle are congruent to the hypotenuse and a leg of another right triangle, then the two triangles are congruent.   **4.3.2**

EXAMPLE 4

When Venus rises before the Sun it is known as a morning star. When it sets after the Sun, it is known as an evening star.

> *Venus rising ahead of the Sun as a morning star.*

The diagram shows why Venus can be both a morning star and an evening star. Note: The daily rotation of the Earth and the orbits of the planets are both counterclockwise, viewed from above (from the north).

> *Venus is a morning star when it is to the right of the dashed line in the diagram. It is an evening star when it is on the left.*

The orbits of Earth and Venus are nearly circular. The radius of Earth's orbit is about 1.5×10^8 kilometers, while the radius of Venus's orbit is about 1.1×10^8 kilometers. Make a scale drawing of the positions of the Sun, Earth, and Venus when Venus rises as a morning star 30° ahead of the Sun.

SOLUTION

PROBLEM SOLVING

Make a diagram. There are two possible positions for Venus for a given position of Earth. This is an SSA combination, with the Sun, Earth, and Venus at the vertices of a triangle. The side connecting Venus and the Sun is the "swinging door."

Exercises

● Communicate

1. Think of a memory aid, or *mnemonic* device, to help you remember that SSS, SAS, ASA, and AAS are all valid tests for triangle congruence and that AAA and SSA are not valid tests.

For Exercises 2 and 3, is the given information sufficient to determine whether the triangles are congruent? Why or why not?

2. Any two corresponding angles and one corresponding side are congruent.

3. Any two corresponding sides and one corresponding angle are congruent.

4. Explain why any two corresponding sides are sufficient to determine congruence for right triangles.

APPLICATION

5. **FORESTRY** In the method of triangulation, a triangle has one vertex at the object being observed and one at each observer. Which angles and sides are known in this triangle? Which postulate or theorem guarantees that the triangle is "uniquely determined"—that it is the only possible triangle with the given measurements?

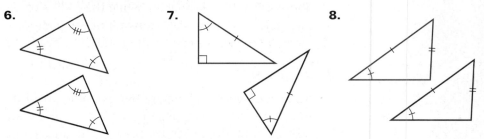

● Guided Skills Practice

For Exercises 6–8, is it possible to prove that the triangles are congruent? Explain your reasoning. (EXAMPLES 1, 2, AND 3)

6. 7. 8.

APPLICATION

9. **ASTRONOMY** Copy the diagram of the positions of the Sun, Earth, and Venus on page 229. Use the diagram to estimate the distance from Earth to Venus for both positions of Venus. *(EXAMPLE 4)*

Practice and Apply

In Exercises 10–19, determine whether each pair of triangles can be proven congruent. If so, write a congruence statement and name the postulate or theorem used.

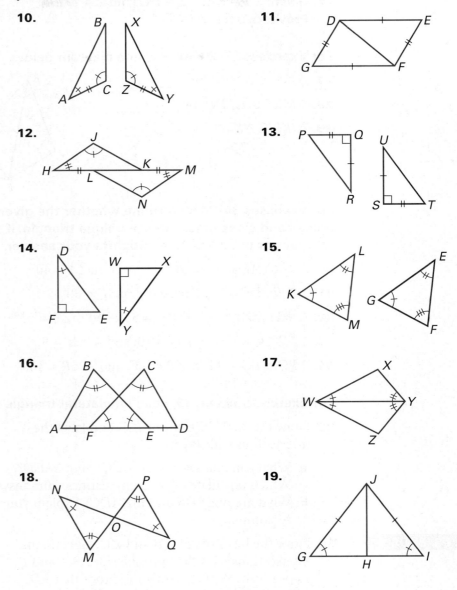

10.

11.

12.

13.

14.

15.

16.

17.

18.

19.

For Exercises 20–24, refer to △*XYZ* below, in which $\overline{XY} \cong \overline{XZ}$. Copy the figure and mark the congruent sides.

20. Draw a segment perpendicular to $\overline{YZ}$ with one endpoint at *X*. Label the other endpoint *W*.

21. Identify the two right triangles formed by $\overline{WX}$ and name the hypotenuse of each.

22. What can you say about the hypotenuses of the two right triangles? Explain your reasoning.

23. Which property proves that $\overline{WX} \cong \overline{WX}$?

24. What can you say about the two right triangles in the figure? Which theorem or postulate justifies this conclusion?

For Exercises 25 and 26, refer to the diagram below.

25. Given: $\angle A \cong \angle D$, $\overline{AF} \cong \overline{DC}$, and
$\angle BFA \cong \angle ECD$
Prove: $\triangle AFB \cong \triangle DCE$

26. Given: $\angle 1 \cong \angle 4$, $\overline{AF} \cong \overline{DC}$, and $\angle A \cong \angle D$
Prove: $\triangle AFB \cong \triangle DCE$

For Exercises 27–29, refer to the diagram below.

27. $\triangle JKL \cong \triangle MKL$ by ____?____

28. $\triangle MKL \cong \triangle MNL$ by ____?____

29. $\triangle JKL \cong \triangle MNL$ by ____?____

For Exercises 30–34, determine whether the given combination of angles and sides determines a unique triangle. If so, identify the theorem or postulate that supports your answer.

30. $\triangle ABC$: $AB = 6$, $m\angle B = 70°$, and $m\angle A = 40°$

31. $\triangle DEF$: $DE = 5$, $EF = 7$, and $m\angle F = 30°$

32. $\triangle JKL$: $m\angle J = 50°$, $m\angle K = 75°$, and $m\angle L = 55°$

33. $\triangle MNO$: $MN = 8$, $MO = 10$, and $m\angle N = 90°$

34. $\triangle PQR$: $PQ = 12$, $m\angle P = 45°$, and $m\angle R = 100°$

For Exercises 35 and 36, copy equilateral triangle *ABC* below.

35. Draw the angle bisector of $\angle A$, and label the intersection with $\overline{BC}$ as D.

   a. What can you say about $\triangle ABD$ and $\triangle ACD$? Which postulate or theorem justifies your answer?

   b. What are $m\angle ADB$ and $m\angle ADC$? Explain your reasoning.

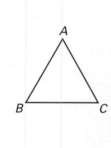

36. Draw the bisectors of $\angle B$ and $\angle C$, labeling the intersections with the opposite sides as E and F, respectively. Write a paragraph proof that $\overline{AD}$, $\overline{BE}$, and $\overline{CF}$ divide $\triangle ABC$ into six congruent triangles. Hint: Use your results from Exercise 35.

37. Write a two-column proof of the AAS Congruence Theorem.

In Exercises 38 and 39 you will further explore SSA combinations. If two sides and a nonincluded angle of two triangles are congruent, then the triangles are not necessarily congruent. However, if certain restrictions are placed on the side lengths or angle measure, it is possible to show that the triangles are congruent (see Lesson 10.4).

38. Try to draw $\triangle ABC$ with each set of side and angle measures below. How many different triangles can you draw for each set of measurements?

   a. $AB = 4$, $BC = 1$, and $m\angle A = 75°$

   b. $AB = 3$, $BC = 2$, and $m\angle A = 60°$

   c. $AB = 5$, $BC = 4$, and $m\angle A = 100°$

39. Try to draw $\triangle ABC$ with each set of side and angle measures below. How many different triangles can you draw for each set of measurements?
  a. $AB = 1$, $BC = 4$, and $m\angle A = 75°$
  b. $AB = 2$, $BC = 3$, and $m\angle A = 60°$
  c. $AB = 4$, $BC = 5$, and $m\angle A = 100°$
  d. Complete the following conjecture:

SSA Conjecture
In an SSA combination, if the given side opposite the given angle is
_____?_____ the other given side, then the triangle is uniquely determined.
Under these conditions, SSA can be used to establish congruence.

Test your conjecture by making up some triangles of your own.

APPLICATIONS

40. ARCHAEOLOGY A student is estimating the height of a pyramid. From a certain distance, the angle of elevation of a point on the highest part of the structure is 25°. From a distance of 190 feet closer, the angle of elevation of the point is 30°. Draw a triangle with the point at the top of the structure as one vertex, and the points where the measurements were made as the other vertices. Which postulate or theorem can be used to show that this triangle is uniquely determined?

The Temple of Kukulkan at Chichen Itza. The height of the structure can be estimated by a method that involves creating a triangle as described in Exercise 40. The actual computations involve trigonometry, which you will study in Chapter 10.

41. ARCHAEOLOGY Make a drawing of the triangle for Exercise 40, using the scale 100 feet = 1 centimeter. Use your drawing to estimate the height of the pyramid.

42. ENGINEERING An embankment rises at an angle of 20° from horizontal. On the embankment, two 10-foot vertical poles are each anchored by a guy wire to the ground directly uphill from each pole. If the guy wire of each pole makes an angle of 60° with the embankment, must the wires be the same length? Which postulate or theorem justifies your answer?

43. NAVIGATION Towns A, B, and C are connected by straight roads. For each set of measurements given below, draw a possible map, and determine whether $\triangle ABC$ on your map is uniquely determined. If so, state which theorem or postulate supports your answer.
  a. $AB = 12$ miles, $BC = 7$ miles, and $AC = 8$ miles
  b. $AB = 6$ miles, $m\angle BAC = 40°$, and $m\angle ABC = 60°$
  c. $m\angle BAC = 75°$, $m\angle ABC = 50°$, and $m\angle ACB = 55°$

44. Draw three lines that intersect at a point. *(LESSON 1.1)*

45. Draw three lines that do not intersect. *(LESSON 1.1)*

46. Draw two lines that intersect a third line without intersecting each other. *(LESSON 1.1)*

In the diagram at right, $\ell \parallel m$.

47. Identify all congruent angles in the diagram at right. *(LESSON 3.3)*

48. Identify a pair of supplementary angles that are not a linear pair in the diagram at right. What is this type of angle pair called? *(LESSON 3.3)*

49. The angles of a triangle measure $(2x)°$, $(3x)°$, and $(4x)°$. Find each angle measure. *(LESSON 3.5)*

50. What can you say about the interior angles of a triangle that has two congruent exterior angles? *(LESSON 3.6)*

Look Beyond

51. Consider two regular pentagons. Are they necessarily congruent? What information would you need to have in order to show that two regular polygons are congruent?

APPLICATION

52. SPORTS A soccer ball is composed of regular pentagons and regular hexagons. Are all of the regular pentagons congruent? Are all of the regular hexagons congruent? Explain your reasoning.

Using Triangle Congruence

M. C. Escher's "Square Limit" ©1999 Cordon Art
B. V. -Baarn-Holand. All rights reserved.

Why *The design elements of this woodcut by M. C. Escher are mathematically related. The size of one part determines the size of another. By using chains of mathematical reasoning, you can deduce things about one part of a figure from information about another part.*

CPCTC in Flowchart Proofs

It follows from the Polygon Congruence Postulate (given in Lesson 4.1) that if two triangles are congruent, then their corresponding parts are congruent. Therefore, if $\triangle ABC \cong \triangle DEF$, you can conclude that $\overline{AB} \cong \overline{DE}$. What other pairs of sides and angles must be congruent?

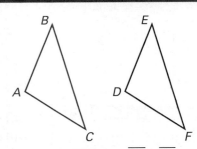

If $\triangle ABC \cong \triangle DEF$, then $\overline{AB} \cong \overline{DE}, \ldots$

This idea is often stated in the following form: *Corresponding parts of congruent triangles are congruent*, abbreviated as **CPCTC**. In each proof in this lesson, you will use a triangle congruence postulate or theorem to establish that two triangles are congruent. Then you will use CPCTC.

In the following examples, the **flowchart proof** is introduced. This type of proof will be especially useful for understanding more complicated proofs that occur later in the book, but it is best to begin with simple examples.

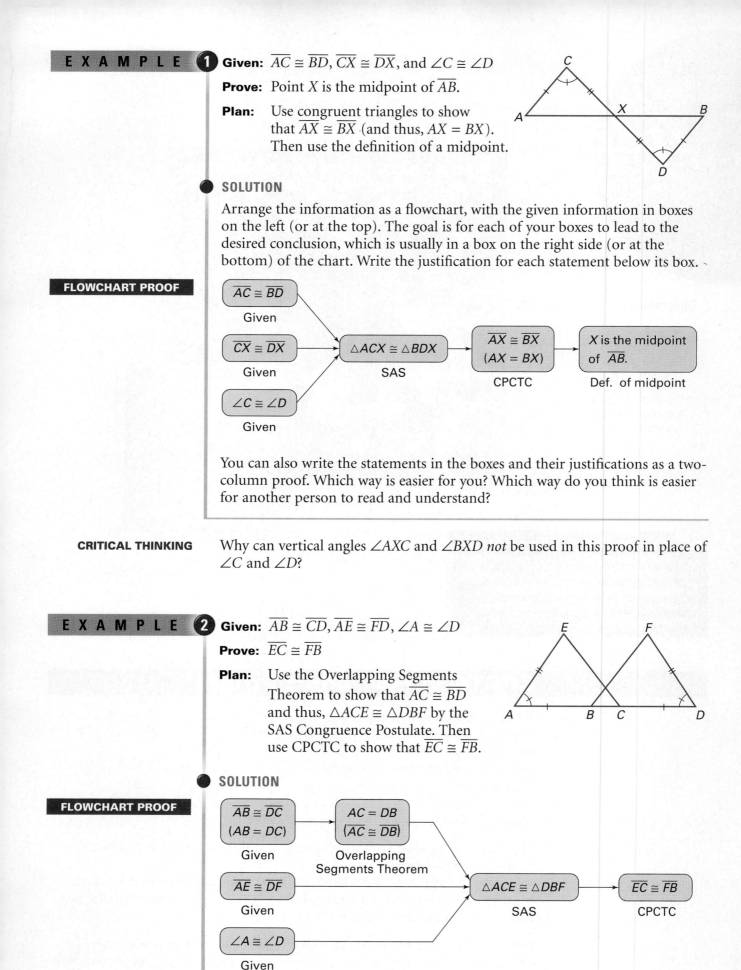

EXAMPLE ❶ **Given:** $\overline{AC} \cong \overline{BD}$, $\overline{CX} \cong \overline{DX}$, and $\angle C \cong \angle D$

Prove: Point X is the midpoint of $\overline{AB}$.

Plan: Use congruent triangles to show that $\overline{AX} \cong \overline{BX}$ (and thus, $AX = BX$). Then use the definition of a midpoint.

● **SOLUTION**

Arrange the information as a flowchart, with the given information in boxes on the left (or at the top). The goal is for each of your boxes to lead to the desired conclusion, which is usually in a box on the right side (or at the bottom) of the chart. Write the justification for each statement below its box.

FLOWCHART PROOF

You can also write the statements in the boxes and their justifications as a two-column proof. Which way is easier for you? Which way do you think is easier for another person to read and understand?

CRITICAL THINKING   Why can vertical angles $\angle AXC$ and $\angle BXD$ *not* be used in this proof in place of $\angle C$ and $\angle D$?

EXAMPLE ❷ **Given:** $\overline{AB} \cong \overline{CD}$, $\overline{AE} \cong \overline{FD}$, $\angle A \cong \angle D$

Prove: $\overline{EC} \cong \overline{FB}$

Plan: Use the Overlapping Segments Theorem to show that $\overline{AC} \cong \overline{BD}$ and thus, $\triangle ACE \cong \triangle DBF$ by the SAS Congruence Postulate. Then use CPCTC to show that $\overline{EC} \cong \overline{FB}$.

● **SOLUTION**

FLOWCHART PROOF

The Isosceles Triangle Theorem

An **isosceles triangle** is a triangle with at least two congruent sides. The two congruent sides are known as the **legs** of the triangle, and the remaining side is known as the **base**. The angles whose vertices are the endpoints of the base are known as the **base angles**, and the angle opposite the base is known as the **vertex angle**.

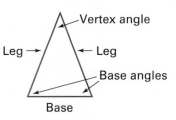

CRITICAL THINKING   Is an equilateral triangle an isosceles triangle? Explain.

The following theorems, which you may have already conjectured, are among the great classics of geometry:

Isosceles Triangle Theorem

If two sides of a triangle are congruent, then the angles opposite those sides are congruent.   **4.4.1**

Converse of the Isosceles Triangle Theorem

If two angles of a triangle are congruent, then the sides opposite those angles are congruent.   **4.4.2**

TRY THIS   Using the plan provided below, write a flowchart proof of the Isosceles Triangle Theorem.

FLOWCHART PROOF

Given: $\overline{AC} \cong \overline{BC}$

Prove: $\angle A \cong \angle B$

Plan: Draw the bisector of the vertex angle and extend it to the base as shown at right. Show that the two triangles that result are congruent by SAS. Then use CPCTC.

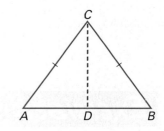

Two Corollaries

A **corollary** of a theorem is an additional theorem that can easily be derived from the original theorem. Once the theorem is known, the corollary should seem obvious. A corollary can be used as a reason in a proof, just like a theorem or a postulate.

There are a number of corollaries to the Isosceles Triangle Theorem that you will be asked to prove. The corollaries on the next page are two of the most important ones. You will be asked to prove them in the exercise set.

Corollary

The measure of each angle of an equilateral triangle is 60°.

4.4.3

Corollary

The bisector of the vertex angle of an isosceles triangle is the perpendicular bisector of the base.

4.4.4

E X A M P L E **3** A resort owner plans to install a gondola ride across a small canyon on her property. Study the diagram below. What is the distance across the canyon?

A P P L I C A T I O N

CONSTRUCTION

Algebra

SOLUTION

The 80° angle is an exterior angle.

By the Exterior Angle Theorem:

$$m\angle X + 40° = 80°$$
$$m\angle X = 40°$$

Because two angles of the triangle are congruent, the sides opposite them are congruent, by the Converse of the Isosceles Triangle Theorem. Therefore, the distance across the canyon is 350 feet.

Exercises

Communicate

1. What is CPCTC? How would you use it in a proof?

2. What is a corollary? How is a corollary related to a theorem?

3. If a triangle has three congruent angles, is it necessarily equilateral? Why or why not?

4. Can a right triangle be isosceles? Can a right triangle be equilateral? Explain your reasoning.

FLOWCHART PROOFS

Complete the flowchart proofs below. *(EXAMPLES 1 AND 2)*

Given: $ABCDE$ is a regular pentagon.

Prove: $\triangle ACD$ is isosceles ($AC = AD$).

Proof:

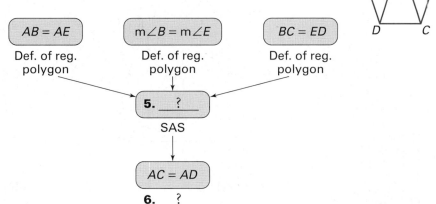

$AB = AE$	$m\angle B = m\angle E$	$BC = ED$
Def. of reg. polygon	Def. of reg. polygon	Def. of reg. polygon

5. _____

SAS

$AC = AD$

6. ___?___

Given: $QRST$ is a rectangle, $TU = VS$, and $QV = RU$.

Prove: $\triangle QTV \cong \triangle RSU$

Proof:

$TU = SV$ → $TV = SU$

Given   **7.** ___?___

$m\angle QTV = 90°$
Def. of rect.

$m\angle QTV = m\angle RSU$
($\angle QTV \cong \angle RSU$)

$m\angle RSU = 90°$
Def. of rect.

Transitivity
or
Substitution

$\triangle QTV \cong \triangle RSU$

8. ___?___

$QV = RU$
Given

APPLICATION

9. CONSTRUCTION Jerome wants to estimate the distance across a river from point A to point B. Starting at point A and facing directly opposite point B, he turns 50° to his right and walks in a straight line to point C, where the angle beween the lines of sight to points A and B is 25°. How does he know that the triangle formed by points A, B, and C is isosceles? If Jerome walked 100 feet from point A to point C, what is the distance across the river? *(EXAMPLE 3)*

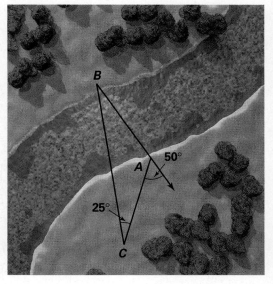

● *Practice and Apply*

Find each indicated measure.

10. m∠Z

11. *KL*

12. *QR*

13. m∠*F*

14. m∠*ABD*

15. *GH*

Algebra

16. *PR*

17. m∠*L*

FLOWCHART PROOFS

For Exercises 18 and 19, write flowchart proofs.

18. Given: ∠*A* ≅ ∠*D*, *AB* = *DE*, and *AF* = *DC*

Prove: ∠*B* ≅ ∠*E*

19. Given: *VW* = *WX* and *VY* = *XZ*

Prove: △*WYZ* is isosceles.

PROOFS

For Exercises 20–22, refer to the diagram below and write a flowchart, paragraph, or two-column proof.

20. Given: $\overline{AB} \parallel \overline{DE}$, and *C* is the midpoint of $\overline{BE}$.

Prove: $\overline{AB} \cong \overline{DE}$

21. Given: △*ABC* ≅ △*DEC*

Prove: $\overline{AB} \parallel \overline{DE}$

22. Given: $\overline{AB} \cong \overline{DE}$ and $\overline{AB} \parallel \overline{DE}$

Prove: *C* is the midpoint of $\overline{BE}$.

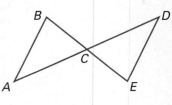

Complete the flowchart proof below of the Converse of the Isosceles Triangle Theorem.

Given: $\angle X \cong \angle Y$ and $\overline{WZ} \perp \overline{XY}$

Prove: $\overline{XZ} \cong \overline{YZ}$

Proof:

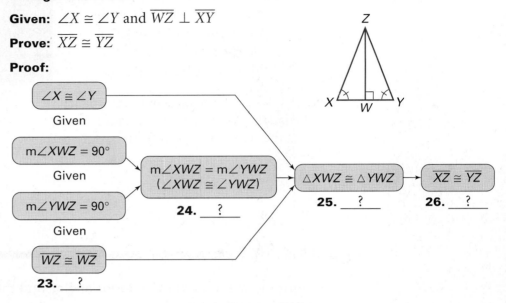

Complete the two-column proof below.

Given: $\overline{AB} \cong \overline{BC}$ and $\angle ABX \cong \angle CBX$

Prove: $\overline{AX} \cong \overline{CX}$

Proof:

Statements	Reasons
$\overline{AB} \cong \overline{BC}$	Given
$\angle ABX \cong \angle CBX$	Given
$\overline{BX} \cong \overline{BX}$	**27.** ?
$\triangle ABX \cong \triangle CBX$	**28.** ?
$\overline{AX} \cong \overline{CX}$	**29.** ?

Based on the two-column proof above, complete the following statement, which you can use in your proof of Corollary 4.4.4 in Exercise 33.

The **30.** _____?_____ of the vertex angle of an isosceles triangle **31.** _____?_____ .

32. Write a paragraph proof of Corollary 4.4.3 on page 238.

33. Write a paragraph proof of Corollary 4.4.4 on page 238.

34. Use the diagram below to prove that $\overline{AK} \cong \overline{GH}$.

APPLICATIONS

35. SURVEYING A surveyor needs to measure the distance across a pond from point *T* to point *B*. Describe how the measurements shown in the diagram at right enable him to determine the distance.

36. ROAD SIGNS A yield sign is an equilateral triangle. What can you say about the angles of the triangle? Explain your reasoning.

 Look Back

Determine the sum of the interior angles and the sum of the exterior angles for each polygon. *(LESSON 3.6)*

37. square   **38.** hexagon   **39.** dodecagon

APPLICATION

40. ROAD SIGNS A stop sign is a regular (equilateral and equiangular) octagon. Are all equilateral octagons equiangular? Are all equiangular octagons equilateral? Explain your reasoning. *(LESSON 3.1)*

CONNECTION

COORDINATE GEOMETRY Points *A*(0, 0) and *B*(3, 6) are endpoints of $\overline{AB}$. Point *C*(3, 1) is an endpoint of $\overline{CD}$. For each point *D* given below, determine whether $\overline{AB}$ and $\overline{CD}$ are parallel, perpendicular, or neither. *(LESSON 3.8)*

41. $D(1, 2)$ **42.** $D(4, 4)$ **43.** $D(2, -1)$

 Look Beyond

PROOFS

Another proof of the Isosceles Triangle Theorem was "discovered" by a computer. This proof had been done by the Greek geometer Pappus of Alexandria (320 C.E.) but had not been known to mathematicians for centuries. It is much simpler than Euclid's proof of the same theorem (which is quite complicated) and even simpler than the one given in this chapter. Discover it for yourself below.

44. Given: $\triangle ABC$ is isosceles, and $\overline{AC} \cong \overline{BC}$.

   Prove: $\angle A \cong \angle B$

   Plan: Show that $\triangle ABC \cong \triangle BAC$.

45. Use the plan given for Exercise 44 to prove the Converse of the Isosceles Triangle Theorem.

Proving Quadrilateral Properties

Objective

- Prove quadrilateral conjectures by using triangle congruence postulates and theorems.

Why *In Chapter 3 you made conjectures about some properties of quadrilaterals. Now, with the help of the triangle congruence postulates and theorems, you are in a position to prove your conjectures—if they are true.*

APPLICATION
ENGINEERING

A property of rhombuses explains why this lamp assembly stays perpendicular to the wall as it moves. You learned this property as a conjecture in Lesson 3.2. Do you know which property it is?

An Important Conjecture and Proof

In the Activity below, you will make a conjecture about an important property of parallelograms. The proof of the conjecture is given on the following page.

Activity

Rotational Symmetry in Parallelograms

1. Trace the parallelogram at right onto two sheets of tracing paper.

2. Place one figure over the other so that they match. Place the point of your pencil at point *C* and hold it firmly in place. Rotate the top piece of paper 180°. Describe the result.

3. Does △*PLM* seem be congruent to △*GML*? Fill in the blank to complete the conjecture below.

Conjecture:

CHECKPOINT ✔ A diagonal of a parallelogram divides the parallelogram into ___?___.

In the diagram below, the diagonal $\overleftrightarrow{ML}$ is a transversal to two different pairs of parallel lines. Name those parallel lines. List the alternate interior angles for each pair of lines.

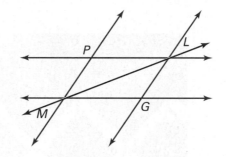

E X A M P L E

PROOFS

Given: parallelogram *PLGM* with diagonal $\overline{LM}$

Prove: $\triangle LGM \cong \triangle MPL$

SOLUTION

Proof A (paragraph format):

$\overline{PL}$ and $\overline{GM}$ are parallel, according to the definition of a parallelogram. Therefore, $\angle 3$ and $\angle 2$ are congruent alternate interior angles. Similarly, $\angle 1$ and $\angle 4$ are congruent because $\overline{PM}$ and $\overline{GL}$ are also parallel. Finally, diagonal $\overline{LM}$ is congruent to itself. Thus, two angles and the included side are congruent in $\triangle LGM$ and $\triangle MPL$. Therefore, the triangles are congruent by the ASA Congruence Postulate.

Proof B (two-column format):

Statements	Reasons
1. Parallelogram *PLGM* has diagonal $\overline{LM}$.	Given
2. $\overline{PL} \parallel \overline{GM}$	Def. of a parallelogram
3. $\angle 3 \cong \angle 2$	$\parallel$ lines $\Rightarrow$ alt. int. angles $\cong$
4. $\overline{PM} \parallel \overline{GL}$	Def. of a parallelogram
5. $\angle 1 \cong \angle 4$	$\parallel$ lines $\Rightarrow$ alt. int. angles $\cong$
6. $\overline{LM} \cong \overline{LM}$	Reflexive Prop. of Congruence
7. $\triangle LGM \cong \triangle MPL$	ASA Congruence Postulate

This result is stated as a theorem on page 247. In Exercises 24-30, you will be asked to complete a flowchart proof of this theorem.

CRITICAL THINKING

How can you use the result from the Example to prove that opposite angles of a parallelogram are congruent? You will be asked to write this proof in the exercise set.

A System of Geometry Knowledge

With the theorems, postulates, and definitions you now know, you can prove all of the conjectures you have made about the properties of parallelograms and the other special quadrilaterals. It is best to start with the simplest ones first. As you progress, you will find that it is often possible to use a previously proven result as part of a proof of a more complicated theorem. In this way, you are building a system of knowledge.

The exercises in this lesson and the next will guide you through a series of proofs of the conjectures about quadrilateral properties that you made in the Activities in Lessons 3.3 and 3.4. The earlier exercises will give you the most guidance, but in the later exercises you will be on your own.

Exercises

● Communicate

1. As you proved in this lesson, a parallelogram has 180° rotational symmetry. Describe all of the types of symmetry of rectangles, rhombuses, and squares.

2. For parallelogram *PQRS*, state all pairs of congruent triangles that are formed by diagonals $\overline{PR}$ and $\overline{QS}$ intersecting at point *X*.

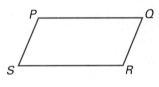

3. Is Theorem 4.5.2 on page 247 true for rectangles? Why or why not?

4. Is Theorem 4.5.3 on page 248 true for rhombuses? Why or why not?

● Guided Skills Practice

Find the indicated measures for parallelogram *WXYZ*. (EXAMPLE)

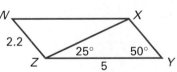

5. m∠*WXZ*   6. m∠*W*   7. *XY*

APPLICATION

8. **CONSTRUCTION** If a ramp has a rise of more than 6 inches, handrails that are parallel to the ramp are required on both sides. If the upright post at the bottom of the ramp is 36 inches tall, how do you know that the post at the top of the ramp is also 36 inches tall?

Practice and Apply

In Exercise 9–16, find the indicated measures for each parallelogram.

9. *CD*   **10.** *DA*

11. m∠*C*   **12.** m∠*D*

13. m∠*Q*   **14.** m∠*RPQ*

15. m∠*SPR*   **16.** m∠*PRQ*

Algebra

In Exercise 17–22, find the indicated measure for each parallelogram.

17. *SV*

18. m∠*A*

19. *QR*

20. *CD*

21. m∠*N*

22. m∠*G*

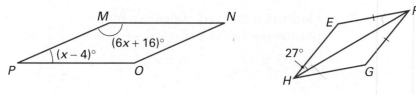

23. State whether each pair of triangles could fit together to form a parallelogram (without reflecting). Justify your answer.

a.   **b.**   **c.**

In Exercises 24–71, you will be asked to prove theorems about quadrilaterals.

Fill in the blanks below to complete a flowchart proof of Theorem 4.5.1 (refer back to the Example on page 244).

Theorem

A diagonal of a parallelogram divides the parallelogram into two congruent triangles.

4.5.1

FLOWCHART PROOF

Given: parallelogram $ABCD$ with diagonal $\overline{BD}$

Prove: $\triangle ABD \cong \triangle CDB$

Proof:

Reflexive Property
of Congruence

Complete the proof below of Theorem 4.5.2.

Theorem

Opposite sides of a parallelogram are congruent.

4.5.2

TWO-COLUMN PROOF

Given: parallelogram $ABCD$ with diagonal $\overline{BD}$

Prove: $\overline{AB} \cong \overline{CD}$ and $\overline{AD} \cong \overline{CB}$

Proof:

Statements	Reasons
$ABCD$ is a parallelogram.	**32.** ?
33. ?	A diagonal of a parallelogram divides the parallelogram into two congruent triangles.
$\overline{AB} \cong \overline{CD}$ and $\overline{AD} \cong \overline{CB}$	**34.** ?

Complete the proof below of Theorem 4.5.3.

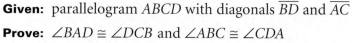

Theorem

Opposite angles of a parallelogram are congruent.   **4.5.3**

Given: parallelogram $ABCD$ with diagonals $\overline{BD}$ and $\overline{AC}$

Prove: $\angle BAD \cong \angle DCB$ and $\angle ABC \cong \angle CDA$

Proof:

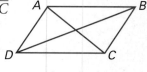

Statements	Reasons
$ABCD$ is a parallelogram.	Given
$\triangle ABD \cong$ **35.** ___?___	**36.** ___?___
$\triangle BAC \cong$ **37.** ___?___	**38.** ___?___
39. ___?___ and **40.** ___?___	CPCTC

41. Write a paragraph proof of Theorem 4.5.4. Begin by drawing an appropriate diagram and writing out what is given and what is to be proved. (Your proof should be very short.)

Theorem

Consecutive angles of a parallelogram are supplementary.   **4.5.4**

Complete the paragraph proof below of Theorem 4.5.5.

Theorem

The diagonals of a parallelogram bisect each other.   **4.5.5**

Given: parallelogram $ABCD$ with diagonals $\overline{AC}$ and $\overline{BD}$ intersecting at point E

Prove: Point E is the midpoint of $\overline{AC}$ and $\overline{BD}$.

Proof:

$\overline{AB}$ and $\overline{CD}$ are parallel by **42.** ___?___, so $\angle BDC$ and $\angle DBA$ are congruent **43.** ___?___ angles. Also, $\angle ACD$ and $\angle CAB$ are congruent alternate interior angles, and $\overline{AB} \cong \overline{CD}$ because **44.** ___?___ sides of a parallelogram are **45.** ___?___. $\triangle ABE \cong \triangle CDE$ by **46.** ___?___. $\overline{BE} \cong \overline{DE}$ because **47.** ___?___, so point E is the midpoint of $\overline{BD}$ by definition. Also, $\overline{AE} \cong \overline{CE}$ because **48.** ___?___, so point E is the midpoint of $\overline{AC}$ by the definition of a midpoint.

In Exercises 49–59, you will complete proofs of the following theorems:

Theorems	
A rhombus is a parallelogram.	**4.5.6**
A rectangle is a parallelogram.	**4.5.7**

Given: rhombus *EFGH*

Prove: $\overline{EF} \parallel \overline{GH}$ and $\overline{FG} \parallel \overline{EH}$

Proof:

Statements	Reasons
$\overline{EF} \cong \overline{GH}$	**49.** ?
$\overline{EH} \cong \overline{GF}$	Def. of a rhombus
$\overline{FH} \cong \overline{HF}$	**50.** ?
$\triangle EFH \cong \triangle GHF$	**51.** ?
$\angle EFH \cong \angle GHF$	CPCTC
$\overline{EF} \parallel \overline{GH}$	**52.** ?
$\angle GFH \cong \angle EHF$	**53.** ?
$\overline{FG} \parallel \overline{EH}$	**54.** ?

Given: rectangle *PQRS*

Prove: $\overline{PQ} \parallel \overline{RS}$ and $\overline{PS} \parallel \overline{QR}$

Proof:

Statements	Reasons
$m\angle P = 90°$	Def. of a rectangle
$m\angle S = 90°$	**55.** ?
$m\angle P + m\angle S = 180°$	**56.** ?
$\overline{PQ} \parallel \overline{RS}$	**57.** ?
$m\angle Q = 90°$	**58.** ?
$m\angle P + m\angle Q = 180°$	Add. Prop. of Equality
$\overline{PS} \parallel \overline{QR}$	**59.** ?

Complete the paragraph proof below of Theorem 4.5.8.

Theorem	
The diagonals of a rhombus are perpendicular.	**4.5.8**

Given: rhombus *DEFG* with diagonals $\overline{DF}$ and $\overline{GE}$ intersecting at point *H*

Prove: $\overline{DF}$ and $\overline{GE}$ intersect to form a right angle.

Proof:

$\overline{EF} \cong \overline{GF}$ by **60.** ? . *DEFG* is a parallelogram (Theorem 4.5.6), and so $\overline{EH} \cong \overline{GH}$ because **61.** ? . $\overline{FH} \cong \overline{FH}$ by **62.** ? . $\triangle EHF \cong \triangle GHF$ by **63.** ? . $\angle FHG \cong \angle FHE$ because **64.** ? . Because $\angle FHG$ and $\angle FHE$ are a linear pair, the sum of their measures is **65.** ? . Because $\angle FHG \cong \angle FHE$, their measures are equal, and thus, $m\angle FHG = m\angle FHE =$ **66.** ? .

67. Write a two-column, flowchart, or paragraph proof of Theorem 4.5.9. (Hint: A rectangle is a parallelogram; therefore, it has the properties of a parallelogram.)

Theorem

The diagonals of a rectangle are congruent.   4.5.9

Given: rectangle *RSTU* with diagonals $\overline{RT}$ and $\overline{US}$ intersecting at point *V*

Prove: $\overline{RT} \cong \overline{US}$

68. Write a two-column, flowchart, or paragraph proof of Theorem 4.5.10. Recall that a kite has two pairs of congruent adjacent sides and that opposite sides are not congruent.

Theorem

The diagonals of a kite are perpendicular.   4.5.10

Given: kite *WXYZ* with diagonals $\overline{WY}$ and $\overline{XZ}$ intersecting at point *A*

Prove: $\overline{WY} \perp \overline{XZ}$

Write a paragraph proof of each theorem below.

Theorem

69. A square is a rectangle.   4.5.11

Theorem

70. A square is a rhombus.   4.5.12

Theorem

71. The diagonals of a square are congruent and are the perpendicular bisectors of each other.   4.5.13

72. Draw a parallelogram and find the midpoint of each side. Connect the midpoints to form a quadrilateral within the parallelogram. Prove that two pairs of congruent triangles are formed. Use this to prove that the quadrilateral formed by connecting the midpoints is a parallelogram. (Hint: Connect one pair of opposite midpoints in the original figure.)

73. ART An artist cuts four congruent right triangles of stained glass. Show how the four triangles can be put together to form each quadrilateral listed below.

   a. a parallelogram that is not a rhombus or a rectangle
   b. a rhombus
   c. a rectangle
   d. a trapezoid
   e. a kite

74. SPORTS Because a baseball diamond is a square, it is also a rectangle, a rhombus, and a parallelogram. Which of the following must be true, based on the properties of special quadrilaterals? Explain your reasoning.

   a. The distance from first base to third base equals the distance from second base to home plate.
   b. The center of the pitcher's mound is on the diagonal from first to third base. (Note: The mound is on the diagonal from home plate to second base at a point 3 feet from the midpoint of that diagonal.)
   c. The path from first base to second base is parallel to the path from third base to home plate.

Look Back

75. Suppose that the angles in one triangle are congruent to the angles in another triangle. Are the triangles necessarily congruent? Why or why not? *(LESSON 4.3)*

Write a congruence statement for each pair of triangles below, and name the postulate or theorem to justify it. *(LESSONS 4.2 AND 4.3)*

76.   **77.**   **78.**

79. Given △KLM ≅ △RST, find the values of x and y. *(LESSONS 3.8 AND 4.4)*

Look Beyond

80. Recall that an isosceles trapezoid has congruent legs. In the diagram at right, two segments perpendicular to the bases have been added to form a pair of right triangles.

Use the information given in the diagram to prove that the base angles of an isosceles trapezoid are congruent. (Hint: First prove that $ABFE$ is a rectangle and therefore a parallelogram.)

81. Use your result from Exercise 80 to prove that the diagonals of an isosceles trapezoid are congruent.

Tessellations With Quadrilaterals

In Chapter 3, you explored several different types of repeating patterns, or tessellations. The only regular polygons that can be used to form tessellations are equilateral triangles, squares, and regular hexagons. (Why?) However, if you use nonregular polygons, there are many more possibilities.

1. Draw a nonregular quadrilateral, such as $ABCD$. Find the midpoint of one of the sides and label it E.

2. Using tracing paper or geometry graphics software, rotate $ABCD$ about point E.

3. Repeat Step 2 for the midpoints of each side of $ABCD$. Then continue the pattern by rotating the images about the midpoints of their sides.

4. Try creating tessellations with different quadrilaterals, including special quadrilaterals and concave quadrilaterals. Do all of the quadrilaterals that you chose work? Can any of them be used to form tessellations in more than one way?

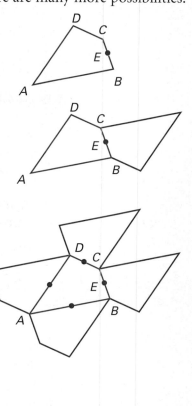

Conditions for Special Quadrilaterals

Objective

● Develop conjectures about special quadrilaterals— parallelograms, rectangles, and rhombuses.

Why *How can you tell whether a given figure is a parallelogram, a rectangle, a rhombus, or a square? In this lesson, you will discover some theorems that provide answers to this question.*

A perfectly rectangular foundation is required for the proper construction of many modern buildings. Here workers are applying an important geometry principle to ensure that a portion of the foundation is a rectangle.

Visitors to non-western cultures as far away as Mozambique, in southern Africa, have reported observing local workers employing the technique shown above to ensure that a foundation of a house is a rectangle. According to their reports, measurements are made by using long poles (see page 255).

The Conditions That Determine a Figure

If you are given a quadrilateral, how can you tell which of the special quadrilaterals, if any, it is? One way to try is to check the given information against the definitions of the special figures. However, the information you are given may not match the information referred to in the definitions. This does not mean that the figure is not one of the special quadrilaterals.

uncooked spaghetti or
another material (such
as soda straws or
toothpicks) to serve as
the sides of model
quadrilaterals, a ruler,
and a protractor

Activity 1

What Does It Take to Make...

In each part of this Activity, decide whether the conjectures listed are true or false. If you believe that a conjecture is false, prove that it is false by giving a counterexample. Make sketches of your counterexamples. If you believe that a conjecture is true, consider how you would prove it.

CHECKPOINT ✔

Part I: What does it take to make a parallelogram?

State whether the following conjectures about parallelograms are true or false:

1. If one pair of opposite sides of a quadrilateral are congruent, then the quadrilateral is a parallelogram.

2. If two pairs of opposite sides of a quadrilateral are congruent, then the quadrilateral is a parallelogram.

3. If one pair of opposite sides of a quadrilateral are parallel and congruent, then the quadrilateral is a parallelogram.

4. If two pairs of sides of a quadrilateral are congruent, then the quadrilateral is a parallelogram.

5. If the diagonals of a quadrilateral bisect each other, then the quadrilateral is a parallelogram.

CHECKPOINT ✔

Part II: What does it take to make a rectangle?

State whether the following conjectures about rectangles are true or false:

1. If one angle of a quadrilateral is a right angle, then the quadrilateral is a rectangle.

2. If one angle of a parallelogram is a right angle, then the parallelogram is a rectangle.

3. If the diagonals of a quadrilateral are congruent, then the quadrilateral is a rectangle.

4. If the diagonals of a parallelogram are congruent, then the parallelogram is a rectangle.

5. If the diagonals of a parallelogram are perpendicular, then the parallelogram is a rectangle.

CHECKPOINT ✔

Part III: What does it take to make a rhombus?

State whether the following conjectures about rhombuses are true or false:

1. If one pair of adjacent sides of a quadrilateral are congruent, then the quadrilateral is a rhombus.

2. If one pair of adjacent sides of a parallelogram are congruent, then the parallelogram is a rhombus.

3. If the diagonals of a parallelogram are congruent, then the parallelogram is a rhombus.

4. If the diagonals of parallelogram bisect the angles of the parallelogram, then the quadrilateral is a rhombus.

5. If the diagonals of a parallelogram are perpendicular, then the parallelogram is a rhombus.

The "Housebuilder" Theorem

Carpenters have long used diagonal measurements to determine whether the sides of a house's foundation formed a rectangle. The building method has been used throughout the world, including the African countries Mozambique and Liberia, where people use long poles instead of measuring tapes to compare lengths. In Activity 2 below, you will apply an important theorem, which you may have conjectured in Activity 1.

Activity 2

Building a Rectangular Structure

YOU WILL NEED

uncooked spaghetti or another material (such as soda straws or toothpicks) to serve as the sides of model quadrilaterals

1. Break off two pairs of sides of two different lengths from your modeling material. Let these be the pairs of opposite sides for your model foundation.

2. Model the diagonals of the foundation with two more pieces of your modeling material. Break off the pieces for the diagonals at the correct length.

CHECKPOINT ✔

3. Compare the lengths of the diagonals. Are they equal? If they are, then your foundation is rectangular, according to one of the conjectures you made in Activity 1. State that conjecture. (This conjecture, when proven, will be called the Housebuilder Theorem.)

4. If the diagonals are not the same length, adjust the sides of your figure until they are. Note: It may be easy to make a reasonably good first guess at a rectangle when you are working with a small model, but it is much more difficult when you are trying to create a large rectangle such as a foundation.

CRITICAL THINKING

The head of the lamp in the picture remains vertical as it moves up and down. Which of the conjectures you made in Activity 1 explains why this is true?

The following theorems are based on your work in Activity 1. You will have the opportunity to prove these theorems in the exercise set.

Theorem

If two pairs of opposite sides of a quadrilateral are congruent, then the quadrilateral is a parallelogram. **4.6.1**

Theorem

If one pair of opposite sides of a quadrilateral are parallel and congruent, then the quadrilateral is a parallelogram. **4.6.2**

Theorem

If the diagonals of a quadrilateral bisect each other, then the quadrilateral is a parallelogram. **4.6.3**

Theorem

If one angle of a parallelogram is a right angle, then the parallelogram is a rectangle. **4.6.4**

The Housebuilder Theorem

If the diagonals of a parallelogram are congruent, then the parallelogram is a rectangle. **4.6.5**

Theorem

If one pair of adjacent sides of a parallelogram are congruent, then the parallelogram is a rhombus. **4.6.6**

Theorem

If the diagonals of a parallelogram bisect the angles of the parallelogram, then the parallelogram is a rhombus. **4.6.7**

Theorem

If the diagonals of a parallelogram are perpendicular, then the parallelogram is a rhombus. **4.6.8**

Exercises

Communicate

1. The definition of a parallelogram is a quadrilateral with two pairs of parallel sides. Consider the following alternative definition:

   A parallelogram is a quadrilateral whose opposite sides are congruent.

   Does this definition work? How many other definitions for a parallelogram can you think of?

2. Choose a conjecture from Part I of Activity 1 that turned out to be false, and explain why it is false.

3. Choose a conjecture from Part II of Activity 1 that turned out to be false, and explain why it is false.

4. Choose a conjecture from Part III of Activity 1 that turned out to be false, and explain why it is false.

Guided Skills Practice

For Exercises 5–8, determine whether each quadrilateral is necessarily a parallelogram, a rectangle, a rhombus, or none of these. Give all of the names that apply to each quadrilateral. (ACTIVITY 1)

5.

6.

7.

8.

9. The figure at right is a parallelogram. What must be true about the figure in order for it to be a rectangle also?
   (ACTIVITY 2)

Practice and Apply

Exercises 10–13 refer to quadrilateral *ABCD* with diagonals $\overline{AC}$ and $\overline{BD}$ intersecting at point *E*. For each set of conditions given below, determine whether the quadrilateral is necessarily a parallelogram. If so, give the theorem that justifies your answer.

10. $\overline{AB} \cong \overline{CD}$, $\overline{AC} \cong \overline{BD}$

11. $\overline{AE} \cong \overline{CE}$, $\overline{BE} \cong \overline{DE}$

12. $\overline{AB} \cong \overline{AC}$, $\overline{AB} \parallel \overline{CD}$

13. $\overline{AB} \cong \overline{CD}$, $\overline{AB} \parallel \overline{CD}$

Exercises 14–19 refer to a parallelogram *FGHI* with diagonals $\overline{FH}$ and $\overline{GI}$ intersecting at point *J*. For each condition given below, determine whether the parallelogram is a rhombus, a rectangle, or neither. Give the theorem that justifies your answer.

14. $\overline{FH} \cong \overline{GI}$ **15.** $\overline{FG} \cong \overline{GH}$ **16.** m$\angle FGH = 90°$

17. $\overline{FG} \cong \overline{FH}$ **18.** m$\angle FJG = 90°$ **19.** m$\angle FHG = 90°$

For Exercises 20–25, refer to the diagram of parallelogram *KLMN* at left. State whether each set of conditions below is sufficient to prove that *KLMN* is a square. Explain your reasoning.

20. m$\angle KLM = 90°$, m$\angle KPN = 90°$ **21.** $\overline{KL} \cong \overline{LM}$, $\overline{KL} \cong \overline{MN}$

22. m$\angle KLN = 45°$, m$\angle LNK = 45°$ **23.** $\overline{KN} \cong \overline{MN}$, $\overline{KM} \cong \overline{LN}$

24. $\angle LKM \cong \angle LMK$, $\angle LKM \cong \angle MKN$ **25.** $\triangle NKL \cong \triangle KLM$

TWO-COLUMN PROOF

Use the diagram at right to complete the following two-column proof of Theorem 4.6.1:

Given: $\overline{AB} \cong \overline{CD}$ and $\overline{AD} \cong \overline{CB}$

Prove: *ABCD* is a parallelogram.

Proof:

Statements	Reasons
$\overline{AB} \cong \overline{CD}$ and $\overline{AD} \cong \overline{CB}$	**26.** ?
$\overline{BD} \cong \overline{BD}$	**27.** ?
$\triangle ABD \cong \triangle CDB$	**28.** ?
$\angle 1 \cong \angle 2$ and $\angle 3 \cong \angle 4$	**29.** ?
$\overline{AD} \parallel \overline{BC}$ and $\overline{AB} \parallel \overline{CD}$	**30.** ?
ABCD is a parallelogram.	**31.** ?

PARAGRAPH PROOF

Use the diagram below to complete the paragraph proof of the Homebuilder Theorem below.

Given: parallelogram *EFGH* with $\overline{EG} \cong \overline{FH}$

Prove: *EFGH* is a rectangle.

Proof:

Opposite sides of a parallelogram are congruent, so $\overline{EH} \cong$ **32.** ? and $\overline{HG} \cong \overline{GH}$ (Reflexive Property). Because $\overline{EG} \cong \overline{FH}$ (given), $\triangle EHG \cong \triangle FGH$ by **33.** ? . $\angle EHG \cong \angle FGH$ because **34.** ? , and $\angle EHG$ and $\angle FGH$ are supplementary because they are **35.** ? angles. $\angle EHG$ and $\angle FGH$ are right angles because **36.** ? . Therefore, $\angle FEH$ and $\angle EFG$ are also right angles because **37.** ? . Thus, *EFGH* is a rectangle by **38.** ? .

PROOFS

Write a two-column, paragraph, or flowchart proof for each of the following theorems:

39. Theorem 4.6.2   **40.** Theorem 4.6.3   **41.** Theorem 4.6.4

42. Theorem 4.6.6   **43.** Theorem 4.6.7   **44.** Theorem 4.6.8

CHALLENGE

Recall the following conjecture from Lesson 3.7, which you will now prove as a theorem:

The Triangle Midsegment Theorem

A midsegment of a triangle is parallel to a side of the triangle and has a measure equal to half of the measure of that side.   **4.6.9**

FLOWCHART PROOF

Complete the flowchart proof below.

Given: In $\triangle ABC$, D is the midpoint of $\overline{AC}$, and E is the midpoint of $\overline{BC}$.

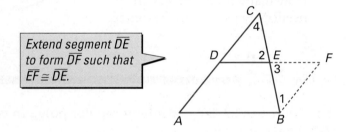

Prove: $\overline{DE} \parallel \overline{AB}$ and $DE = \frac{1}{2}AB$

Proof:

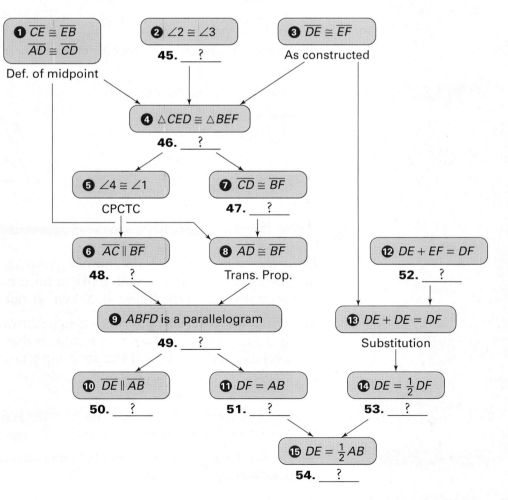

55. ART An artist is making a frame for stretching a canvas. She cuts two boards that are 2 feet long and two boards that are 3 feet long. How can she put the boards together to make sure that the frame is rectangular?

56. CONSTRUCTION In a ladder, all of the rungs are congruent and the two side pieces are congruent. What else is necessary in a ladder in order to ensure that the rungs are parallel?

Corner braces may be used to stabilize the ladder. Explain how corner braces can be placed to ensure that the rungs are perpendicular to the side pieces. How many corner braces are needed? Explain your reasoning.

Look Back

For Exercises 57–59, consider a regular polygon with a central angle of 72°. Find each value below. *(LESSONS 3.1 AND 3.6)*

57. the number of sides of the polygon

58. the measure of an interior angle

59. the measure of an exterior angle

60. Explain the difference between a postulate, a conjecture, and a theorem. *(LESSONS 1.1, 1.4, AND 2.4)*

PROOF

61. Given: $\overline{BA} \cong \overline{BC}$ and $\overline{BD} \cong \overline{BE}$

Prove: $\triangle ADC \cong \triangle CEA$
(Hint: $\triangle ABC$ is isosceles.)
(LESSON 4.4)

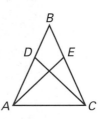

Look Beyond

Draw a general quadrilateral and find the midpoint of each side. Connect the midpoints to form another quadrilateral. Label the figure as shown at right.

62. What type of quadrilateral appears to be formed by the midpoints? Test your conjecture by drawing several quadrilaterals or, if you are using geometry graphics software, by dragging the vertices to different positions.

63. Draw diagonal $\overline{WY}$ to form triangles $\triangle WXY$ and $\triangle WZY$. What part of these triangles are segments $\overline{KL}$ and $\overline{MN}$, respectively?

PROOF

64. Use the Triangle Midsegment Theorem to prove that *KLMN* is a parallelogram.

Compass and Straightedge Constructions

Objectives

- Construct congruent copies of segments, angles, and triangles.

- Construct an angle bisector.

Why *The use of a compass and straightedge to construct figures involves the principles of triangle congruence. As you will learn, compasses are useful for much more than drawing circles.*

A Classical Mathematical Game

For centuries, mathematicians have enjoyed a sort of game. The "game" requires a compass and a straightedge. By using just these tools, you can create precise geometric figures.

Activity 1

CONSTRUCTION
COMPASS and **STRAIGHTEDGE**

Note: A straightedge is not a ruler. It has no marks on it and cannot be used to measure distances.

A Segment Congruent to a Given Segment

Given $\overline{AB}$ at right, follow Steps 1–4 to construct $\overline{CD} \cong \overline{AB}$.

A •————————• B

1. Using a straightedge, draw line ℓ. ℓ ⟷	**2.** Select a point on ℓ and label it C. ℓ ——•——⟶ C
3. Set your compass to the distance AB in the given figure. A B	**4.** Place the point of your compass on C and draw an arc that intersects line ℓ. Label the intersection of the arc and the line as point D. ℓ ⟷ C D

Three Important Assumptions

Three important assumptions were involved in the construction in Activity 1. Keep these assumptions in mind as you do Activity 2.

1. You can place the point of your compass or the edge of your straightedge precisely on a given point or line (or circle).

2. Given two points, you can set your compass so that the distance between the point and pencil of the compass is equal to the distance between the two given points; you can also set your straightedge precisely on both given points.

3. The distance between the point and pencil of the compass does not change once it has been set—until you reset it. (This is why a compass can be used to draw a circle.)

Copying a Triangle

CONSTRUCTION
COMPASS and **STRAIGHTEDGE**

Activity 2

A Triangle Congruent to a Given Triangle

Given △*ABC* at right, follow Steps 1– 5 to construct △*MNO* ≅ △*ABC*.

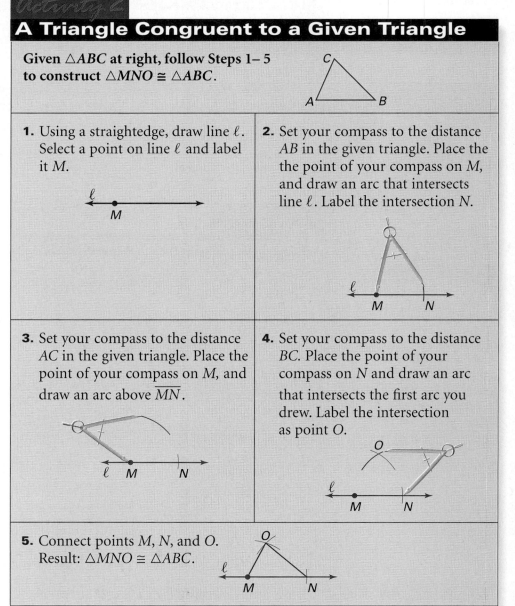

1. Using a straightedge, draw line ℓ. Select a point on line ℓ and label it *M*.

2. Set your compass to the distance *AB* in the given triangle. Place the the point of your compass on *M*, and draw an arc that intersects line ℓ. Label the intersection *N*.

3. Set your compass to the distance *AC* in the given triangle. Place the point of your compass on *M*, and draw an arc above $\overline{MN}$.

4. Set your compass to the distance *BC*. Place the point of your compass on *N* and draw an arc that intersects the first arc you drew. Label the intersection as point *O*.

5. Connect points *M*, *N*, and *O*. Result: △*MNO* ≅ △*ABC*.

Justifying a Construction

The steps in the construction of a figure can be justified by combining the assumptions of compass and straightedge constructions with geometry theorems you already know. The proof below justifies the construction in Activity 1.

TWO-COLUMN PROOF

Given: $\triangle ABC$, with $\triangle MNO$ constructed as in Activity 2.

Prove: $\triangle MNO \cong \triangle ABC$

Proof (refer back to Activity 2):

Statements	Reasons
1. $\overline{MN} \cong \overline{AB}$	Compass set at distance AB was used to construct $\overline{MN}$.
2. $\overline{MO} \cong \overline{AC}$	Compass set at distance AC was used to construct $\overline{MO}$.
3. $\overline{NO} \cong \overline{BC}$	Compass set at distance BC was used to construct $\overline{NO}$.
4. $\triangle MNO \cong \triangle ABC$	SSS

Bisecting an Angle

Activity 3

Angle Bisector

CONSTRUCTION
COMPASS and **STRAIGHTEDGE**

Given $\angle A$ at right, follow the steps below to construct its bisector.

1. Place your compass point at A and draw an arc through the rays of the angle. Label the intersection points as B and C.

2. Place your compass point first at B and then at C. Using one compass setting, draw arcs that intersect in the interior of $\angle A$. Label the intersection as F. Draw a ray from A through F. $\overrightarrow{AF}$ is the bisector of $\angle BAC$.

TRY THIS   Write a proof to justify the construction of the angle bisector in Activity 3.

Exercises

● **Communicate**

1. Explain the difference between a straightedge and a ruler. Why is a ruler not used in constructions?

2. In Step 1 of the construction in Activity 3, which of the construction assumptions were used? Explain.

3. In Step 2 of the construction in Activity 3, which of the construction assumptions were used? Explain.

4. Describe how you would do the constructions from the activities with folding paper or on a computer. Name some advantages or disadvantages of one of these methods.

● **Guided Skills Practice**

Trace the figures below onto paper, and construct a congruent copy of each. *(ACTIVITIES 1 AND 2)*

5.

6.

For Exercises 7–9, refer to the diagram at right. Trace the figure and construct the angle bisector of the indicated angle. *(ACTIVITY 3)*

7. ∠A   **8.** ∠B   **9.** ∠C

● **Practice and Apply**

Construct a congruent copy of each figure below.

10.

11.

12.

13.

14.

15.

Trace each angle below onto your paper and construct the angle bisector of each.

16.

17.

For Exercises 18–21, trace each angle onto paper and follow the steps below to construct a congruent copy.

An angle congruent to a given angle

Given ∠R at right, follow these steps to construct ∠B ≅ ∠R.

R

a. Using a straightedge, draw a ray with endpoint *B*.

B •————→

b. Place your compass point on *R* and draw an arc. Label the intersection points *Q* and *S*.

c. Without adjusting your compass, place the compass point at *B*. Draw an arc that crosses the ray, and label the intersection point *C*.

d. Set your compass equal to the distance *QS* in ∠*R*.

e. Without adjusting your compass, place the compass point at *C*. Draw an arc that crosses the first arc and label the intersection *A*.

f. Draw $\overrightarrow{BA}$ to form ∠*B*. ∠*B* is congruent to ∠*R*.

18.

19.

20.

21.

For Exercises 22 and 23, trace each segment onto paper and follow the steps below to construct the perpendicular bisector and midpoint.

You will prove this construction in Exercise 48.

The perpendicular bisector of a given segment and the midpoint of a given segment

Given $\overline{AC}$ at right, follow these steps to construct line the perpendicular bisector of $\overline{AC}$.

a. Set your compass equal to a distance greater than half of *AC*.

b. Place your compass point on *A* and draw an arc as shown.

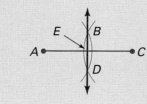

c. Without adjusting your compass setting, place the compass point at *C*. Draw a new arc as shown and label the intersection points of the two arcs *B* and *D*.

d. Use a straightedge to draw $\overleftrightarrow{BD}$. Label the intersection with $\overline{AC}$ as point *E*. $\overleftrightarrow{BD}$ is the perpendicular bisector of $\overline{AC}$, and *E* is the midpoint.

22.   **23.**

Trace the triangles below onto paper and construct the perpendicular bisector of each side. Using the intersection of the perpendicular bisectors, construct the circumscribed circle of each triangle.

24.   **25.**

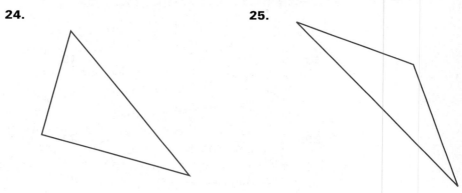

For Exercises 26 and 27, trace each figure onto paper and follow the steps below to construct a line through the given point and perpendicular to each given line.

CONSTRUCTION

You will prove this construction in Exercise 49.

A line through a point perpendicular to a given line

Given point *A* and line *ℓ* at right, follow these steps to construct $\overleftrightarrow{AC} \perp$ line *ℓ*.

a. Place your compass point on *A* and draw an arc as shown. Label the intersection points *D* and *B*.

b. Place your compass point on *D* and draw an arc below the line, as shown. The compass does not need to be at the same setting as in the previous step.

c. Without adjusting your compass setting, place the compass point at *B*. Draw a new arc as shown, and label the intersection point *C*.

d. Use a straightedge to draw $\overleftrightarrow{AC}$. $\overleftrightarrow{AC}$ is perpendicular to line *ℓ*.

26.

27.

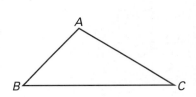

28. A segment from a vertex of a triangle perpendicular to the opposite side is called an *altitude* of the triangle. Trace the triangle at right onto your paper.

a. Construct the altitude from each vertex.
b. What seems to be true about the three altitudes of a triangle?

29. Draw a large scalene triangle, △*ABC*, on your paper. Follow the steps below to construct the inscribed circle.

a. Construct the angle bisectors to find the incenter, *I*.
b. Construct a line through *I* perpendicular to any side. Label the point of intersection *E*.
c. Draw a circle centered at *I* with radius *IE*. This is the inscribed circle of △*ABC*

For Exercises 30 and 31, trace each figure onto paper and follow the steps below to construct a line through the given point and parallel to the given line.

You will prove this construction in Exercise 46.

CONSTRUCTION

A line through a point parallel to a given line

Given point M and line ℓ at right, follow these steps to construct $\overleftrightarrow{AC} \parallel$ line ℓ.

$M_\bullet$

$\ell \longleftrightarrow$

a. Use a straightedge to draw a line through M that intersects ℓ. Label the intersection point P.

b. Place your compass point on P and draw an arc as shown. Label the intersection points R and T.

c. Without adjusting your compass setting, place the compass point at M. Draw a new arc as shown and label the intersection point N.

d. Set your compass to the distance RT. Place your compass point on N and draw an arc as shown. Label the point of intersection O.

e. Use a straightedge to draw $\overleftrightarrow{MO}$. Line $\overleftrightarrow{MO}$ is parallel to line ℓ.

30.

31.

Construct each of the following special quadrilaterals with a compass and straightedge:

32. trapezoid

33. parallelogram

34. rectangle

35. rhombus

36. square

37. kite

Refer to the construction, shown at right, of an angle congruent to a given angle. Complete the two-column proof that ∠B ≅ ∠R.

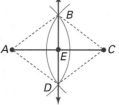

Given: ∠B, as constructed

Prove: ∠B ≅ ∠R

Proof:

Statements	Reasons
$\overline{AB} \cong \overline{QR}$	Same compass setting used
$\overline{BC} \cong \overline{RS}$	**38.** ___?___
$\overline{AC} \cong \overline{QS}$	**39.** ___?___
$\triangle ABC \cong \triangle QRS$	**40.** ___?___
$\angle B \cong \angle R$	**41.** ___?___

PROOFS

42. Refer to the construction, shown at right, of the perpendicular bisector of a given segment. Prove that $\overleftrightarrow{BD} \perp \overline{AC}$ and that $\overleftrightarrow{BD}$ bisects $\overline{AC}$.

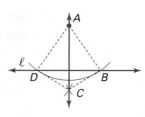

43. Refer to the construction, shown at right, of a line through a point perpendicular to a given line. Prove that $\overleftrightarrow{AC} \perp$ line ℓ.

44. Refer to the construction, shown at right, of a line through a given point parallel to a given line. Prove that $\overleftrightarrow{MO} \parallel$ line ℓ.

APPLICATION

Croton Peak

45. MAP READING A geologist is at point *X* near Croton Peak in Big Bend National Park. He wishes to go back to the road by the shortest path. Make your own sketch of the map and use a compass and straightedge to construct his path. (You may assume that the road is perfectly straight.)

Detail of topographical map

46. Draw a large scalene triangle, and construct at least two of each of the following:

- Perpendicular bisectors: these meet at a point called the *circumcenter*.
- Altitudes: these meet at a point called the *orthocenter*.
- Medians (a segment joining a vertex to the midpoint of the opposite side): these meet at a point called the *centroid*.

If your constructions have been done carefully, these three special points should be collinear. Draw the line through the three points.

Look Back

Identify the type of transformation shown in each figure below.
(LESSON 1.6)

47.

48.

49.

Classify each quadrilateral below as a parallelogram, rectangle, square, rhombus, kite, or trapezoid. List all terms that apply to each figure.
(LESSONS 3.2 AND 3.3)

50.   **51.**   **52.**   **53.**

Look Beyond

54. Trace segment $\overline{AB}$ onto paper and construct equilateral triangle $\triangle ABC$ as follows:

Place your compass point at A, and adjust the compass so that its pencil point is at B. Draw an arc as shown.

Without adjusting the compass setting, place the compass point at B and draw another arc. Label the intersection point C. Draw segments $\overline{AC}$ and $\overline{BC}$.

55. CULTURAL CONNECTION: EUROPE The French general Napoleon (1769-1821) was an amateur mathematician. Napoleon discovered the following construction:

Draw any triangle. Construct an equilateral triangle along each side, as shown. Construct the perpendicular bisectors of each equilateral triangle to find the circumcenters, and join the circumcenters to form a triangle. What seems to be true about this triangle?

Napoleon (1769–1821)

Constructing Transformations

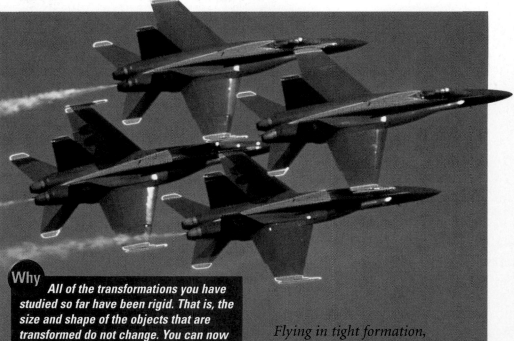

Objectives

- Translate, rotate, and reflect figures by using a compass and straightedge.

- Prove that translations, rotations, and reflections preserve congruence and other properties.

- Use the Betweenness Postulate to establish the Triangle Inequality Theorem.

Why *All of the transformations you have studied so far have been rigid. That is, the size and shape of the objects that are transformed do not change. You can now prove that these transformations preserve congruence and other properties as well.*

Flying in tight formation, these jet planes move as a rigid geometric unit.

Translating Segments and Polygons

Recall from Lesson 1.6 that a translation is a transformation that moves every point of an object the same distance in the same direction.

Activity 1

Translating a Segment

YOU WILL NEED

compass, straightedge, and ruler

Make your own drawing like the one at right. The arrow, known as a translation vector, shows the direction and distance of the translation you are to construct. The distance is the length of the vector.

1. Construct a line, ℓ_1, through point A and parallel to the translation vector. Construct another line, ℓ_2, through point B and parallel to the translation vector. Are lines ℓ_1 and ℓ_2 parallel? Explain.

2. Set your compass to the length, x, of the translation vector. On the right side of $\overline{AB}$, construct points A' and B' that are the same distance, x, from points A and B on lines ℓ_1 and ℓ_2, respectively.

CHECKPOINT ✔ 3. Connect points A' and B'. Measure $\overline{AB}$ and $\overline{A'B'}$. Are the two segments congruent?

The following Example gives a proof that $\overline{AB}$ and $\overline{A'B'}$, as constructed in Activity 1, are congruent:

EXAMPLE

PARAGRAPH PROOF

Given: $\overline{AB}$ and $\overline{A'B'}$, as constructed in Activity 1

Prove: $\overline{AB} \cong \overline{A'B'}$

Proof (refer to Activity 1, page 271):

$\overline{AA'} \parallel \overline{BB'}$ because they lie on lines ℓ_1 and ℓ_2, which were constructed to be parallel. (For the method of constructing a line through a given point and parallel to a given line, see page 268.) $\overline{AA'} \cong \overline{BB'}$ because the same compass setting (the distance x) was use to construct them. Quadrilateral $AA'B'B$ is a parallelogram because a pair of opposite sides are congruent and parallel. Therefore, $\overline{AB} \cong \overline{A'B'}$ because opposite sides of a parallelogram are congruent.

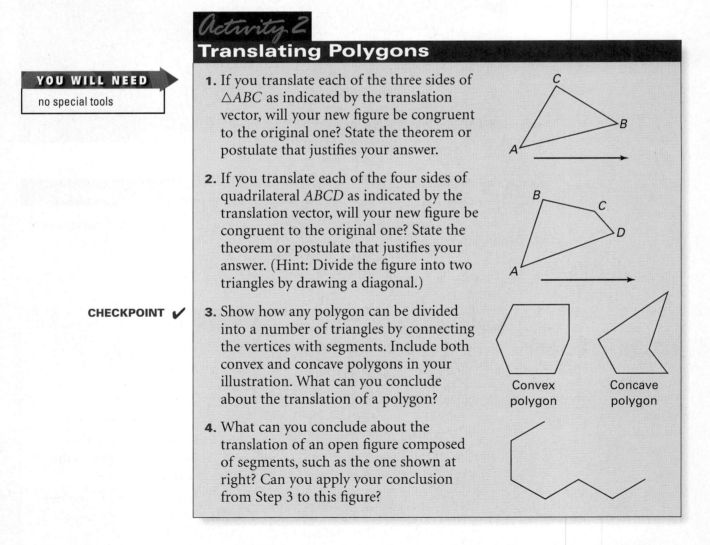

Activity 2
Translating Polygons

YOU WILL NEED

no special tools

1. If you translate each of the three sides of $\triangle ABC$ as indicated by the translation vector, will your new figure be congruent to the original one? State the theorem or postulate that justifies your answer.

2. If you translate each of the four sides of quadrilateral $ABCD$ as indicated by the translation vector, will your new figure be congruent to the original one? State the theorem or postulate that justifies your answer. (Hint: Divide the figure into two triangles by drawing a diagonal.)

CHECKPOINT ✔

3. Show how any polygon can be divided into a number of triangles by connecting the vertices with segments. Include both convex and concave polygons in your illustration. What can you conclude about the translation of a polygon?

Convex polygon   Concave polygon

4. What can you conclude about the translation of an open figure composed of segments, such as the one shown at right? Can you apply your conclusion from Step 3 to this figure?

CRITICAL THINKING   Do you think your conclusion about translations of segments and polygons can be applied to translations of curves? (Hint: How can you use a number of segments to approximate a curve?)

Preservation of "ABCD"

The rigid transformations, which are also known as *isometries*, preserve *Angles*, *Betweenness*, *Collinearity*, and *Distance* ("ABCD").

CRITICAL THINKING

How do Activities 1 and 2 show that translations preserve angles and distance?

If *A*, *B*, and *X* are collinear points, how can you be sure that their image points, *A'*, *B'*, and *X'*, will be collinear in a translated image of the figure? If *X* is between *A* and *B*, how can you be sure that *X'* will be between *A'* and *B'*?

To answer the above questions, you will need a mathematical way to determine which of three given points is between the other two (if any) from information about the distances between them. The postulate below gives a method. It is the converse of a postulate you have already studied.

Betweenness Postulate
(Converse of the Segment Addition Postulate)

Given three points *P*, *Q*, and *R*, if $PQ + QR = PR$, then *P*, *Q*, and *R* are collinear and *Q* is between *P* and *R*.   **4.8.1**

TRY THIS

You are given the points *R*, *S*, and *T*, where $RS = 9.8$, $TR = 9.6$, and $TS = 19.4$. Assuming that the given distances are exact, are the points collinear? If so, which point is between the other two?

Collinearity and Betweenness

You are given segment *AB* with point *X* between *A* and *B*. Points *A*, *B*, and *X* have been translated as shown at right by the translation vector and connected by segments. Are *A'*, *B'*, and *X'* in the translated image collinear? Does *X'* lie between *A'* and *B'*? How can you prove your answers mathematically? Follow the steps below.

1. What do you know about distances *AB* and *A'B'*? about distances *AX* and *A'X'*? about distances *BX* and *B'X'*? Explain your reasoning.

2. From the Segment Addition Postulate you know that $AX + XB = AB$. What can you conclude about $A'X' + X'B'$? Explain your reasoning.

CHECKPOINT ✔
3. Is *X'* collinear with *A'* and *B'*? Is *X'* between *A'* and *B'*? Do translations preserve collinearity and betweenness? Explain your reasoning.

The Triangle Inequality Theorem

What seems to be true about $AB + BC$ as compared with AC? Suppose that point B is relocated so that $AB + BC = AC$. What happens to $\triangle ABC$?

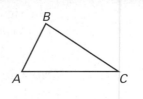

The answers to the questions above suggest the following theorem:

Triangle Inequality Theorem

The sum of the lengths of any two sides of a triangle is greater than the length of the third side. **4.9.2**

This theorem can be argued informally by considering the two other possible cases.

Informal Argument

Case 1
Given three segments, if the sum of the lengths of any two of them is **less than** the length of the third segment, then no triangle can be formed by connecting their endpoints. The endpoints of two of the segments cannot be connected.

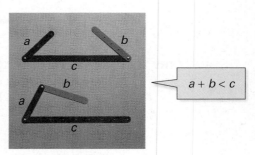

$a + b < c$

The endpoints cannot be connected.

Case 2
Given three segments, if the sum of the lengths of any two of them is **equal to** the length of the third segment, then no triangle can be formed by connecting their endpoints. The endpoints, and hence the segments themselves, are collinear (justification: the Betweenness Postulate).

$a + b = c$

The endpoints are collinear.

Thus, if three segments form a triangle, the sum of the lengths of any two segments must be greater than the length of the third. This is the only remaining possibility.

TRY THIS Which of the following are possible side lengths of a triangle?

a. 14, 8, 25 **b.** 16, 7, 23 **c.** 18, 8, 24

Exercises

Communicate

1. How does the Betweenness Postulate justify the translation of the segment in Activity 1?

2. What does "ABCD" refer to? Explain in terms of the translation of a triangle.

3. In the word *isometry*, the root *-metry* means "measure." What do you think the prefix *iso-* means? Check your dictionary. How does this help explain the term *isometry*?

4. In $\triangle ABC$, $AB = 9$ and $AC = 6$. What can you say about BC? Does it have a minimum possible value? Does it have a maximum possible value? Explain your reasoning.

Guided Skills Practice

Trace the figures below. Translate each figure as indicated by the given translation vector. (ACTIVITIES 1 AND 2 AND EXAMPLE)

5.

6.

7. Trace the figure at right. Translate the angle as indicated by the given translation vector, and measure both angles. What do you notice?
(ACTIVITY 3)

Practice and Apply

Trace each figure below and translate it as indicated by the given translation vector.

8.

9.

10.

11.

A rotation about a point by a given angle

Given $\overline{AB}$, point P, and $\angle R$ shown at right, follow the steps below to construct the rotation of $\overline{AB}$ about P by m$\angle R$.

You will prove this construction in Exercises 26-29.

a. Place your compass point on P. Draw an arc through A and an arc through B.	**b.** Draw $\overline{PA}$. Construct an angle congruent to $\angle R$ with $\overline{PA}$ as one side, as shown. Label the intersection with the arc through A as A'.
c. Draw $\overline{PB}$. Construct an angle congruent to $\angle R$ with $\overline{PB}$ as one side, as shown. Label the intersection with the arc through B as B'.	**d.** Draw $\overline{A'B'}$. $\overline{A'B'}$ is the rotation of $\overline{AB}$ about P by m$\angle R$.

Trace each figure below and follow the steps above to rotate the segment about the given point by the angle below it.

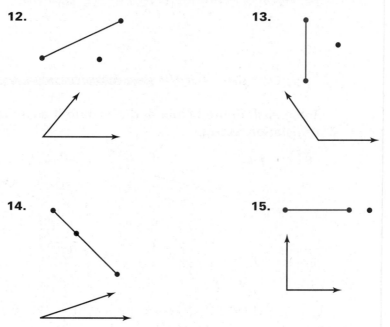

12.

13.

14.

15.

A reflection across a given line

> You will prove this construction in Exercises 30–37.

Given $\overline{AB}$ and line ℓ shown at right, follow the steps below to construct the reflection of $\overline{AB}$ across ℓ.

a. Construct a line, *m*, through *A* and perpendicular to ℓ, and another line, *n*, through *B* and perpendicular to ℓ.

b. Set the compass point on the intersection of ℓ and *m* and the pencil point at *A*. Draw an arc that intersects *m* as shown. Label the intersection point *A′*.

c. Set the compass point on the intersection of ℓ and *n* and the pencil point at *B*. Draw an arc that crosses *n* as shown. Label the intersection point *B′*.

d. Draw $\overline{A'B'}$. $\overline{A'B'}$ is the reflection of $\overline{AB}$ across line ℓ.

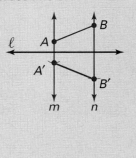

Trace each figure below and follow the steps above to reflect the figure across the given line.

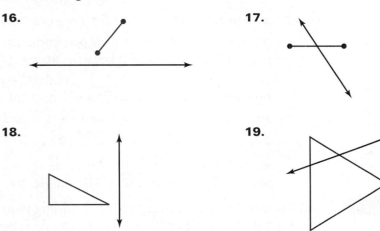

16.

17.

18.

19.

Which of the following triangles are possible?

20. $AB = 7$, $BC = 10$, $AC = 12$

21. $DE = 6$, $EF = 5$, $DF = 14$

22. $GH = 17$, $HI = 9$, $GI = 8$

23. $JK = 10$, $KL = 10$, $JL = 10$

Refer to the diagram below for Exercises 24 and 25.

24. Complete the following inequalities:

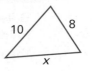

a. _?_ + _?_ > x

b. x + _?_ > 10

c. x + _?_ > 8

25. Solve the inequalities in Exercise 24 and complete the following statement:
_?_ < x < _?_

TWO-COLUMN PROOF

Refer to the rotation of a segment about a point by a given angle (page 276). Complete the two-column proof below.

Given: $\overline{A'B'}$ is a rotation of $\overline{AB}$ about point *P*.

Prove: $\overline{A'B'} \cong \overline{AB}$

Proof:

Statements	Reasons
$\overline{PA} \cong \overline{PA'}$	Same compass setting used
$\overline{PB} \cong \overline{PB'}$	**26.** _?_
$\angle APA' \cong \angle BPB'$	By construction
$\angle APB \cong \angle A'PB'$	**27.** _?_
$\triangle APB \cong \triangle A'PB'$	**28.** _?_
$\overline{A'B'} \cong \overline{AB}$	**29.** _?_

PARAGRAPH PROOF

Refer to the reflection of a segment across a line (page 277). Two lines are added to the figure: one through *A* and parallel to ℓ and one through *A'* and parallel to ℓ, as shown below. Complete the following proof:

Given: $\overline{A'B'}$ is a reflection of $\overline{AB}$ across line ℓ.

Prove: $\overline{A'B'} \cong \overline{AB}$

Proof: By construction, $\overline{AA'}$ and $\overline{BB'}$ are perpendicular to ℓ, $\overline{AX} \cong \overline{A'X}$, and $\overline{BY} \cong \overline{B'Y}$ ($BY = B'Y$). $\overline{AA'}$ is perpendicular to *p* and *q*, and $\overline{BB'}$ is perpendicular to *p* and *q* because **30.** _?_ . Therefore, *ACDA'* is a rectangle by **31.** _?_ , and $\overline{AC} \cong$ **32.** _?_ because opposite sides of a rectangle are congruent. By definition, *ACYX* and *A'DYX* are rectangles, so $\overline{AX} \cong \overline{CY}$ and $\overline{A'X} \cong \overline{DY}$ because **33.** _?_ . Thus, $\overline{CY} \cong \overline{DY}$ ($CY = DY$) by **34.** _?_ , and $\overline{BC} \cong \overline{B'D}$ ($BC = B'D$) by **35.** _?_ . Then $\triangle ACB \cong \triangle A'DB'$ by **36.** _?_ , and $\overline{A'B'} \cong \overline{AB}$ because **37.** _?_ .

CHALLENGE

38. Use a compass and straightedge or folding paper for the following:
- Draw a triangle and label the vertices *K*, *L*, and *M*.
- Rotate the triangle 180° about point *K*. Label *K'*, *L'*, and *M'*.
- Reflect the rotated image, $\triangle K'L'M'$, across $\overline{L'M'}$. Label *K"*, *L"*, and *M"*.
- Construct the perpendicular bisector of $\overline{L''M''}$, and reflect $\triangle K''L''M''$ across the perpendicular bisector. Label *K'''*, *L'''*, and *M'''*.
- Draw a translation vector from *M'''* to *M*, and use it to translate $\triangle K'''L'''M'''$. What happens?

39. MARCHING BAND During a marching-band show, a group of band members moves in a triangular formation, as shown at right. Band members X, Y, and Z are the section leaders. What must the section leaders do to ensure that $\triangle XYZ \cong \triangle X'Y'Z'$? What must the other band members in this formation do to ensure that $\triangle XYZ \cong \triangle X'Y'Z'$?

40. DESIGN Trace the design at right and translate it by using $\overline{AB}$ as a translation vector. Translate the image several times, using the same translation vector, in order to create a border pattern. If you were to use this pattern as a border on a page, describe what translation to use to go around a corner.

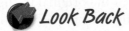

Look Back

Describe all of the types of symmetry in each letter and symbol below. (LESSON 3.1)

41. A

42. N

43. ∞

44. ⌘

45. Which of the following are true for every equilateral triangle? Choose all answers that apply. *(LESSON 4.4)*

    a. A segment joining a vertex to the midpoint of the opposite side divides the triangle into two congruent right triangles.

    b. The angle bisectors are the same as the perpendicular bisectors.

    c. The inscribed circle is the same as the circumscribed circle.

    d. All of the angles measure 60°.

PROOFS

Write a two-column, flowchart, or paragraph proof for Exercises 46 and 47. *(LESSONS 4.5 AND 4.6)*

46. Given: $ABCD$ is an isosceles trapezoid; $\overline{BC} \parallel \overline{AD}$; $\overline{AB} \cong \overline{DC}$; $\overline{BX} \perp \overline{AD}$, $\overline{CY} \perp \overline{AD}$

Prove: $\angle A \cong \angle D$

47. Given: $PQRS$ is a parallelogram; $\overline{QS}$, $\overline{PR}$, and $\overline{XY}$ intersect at Z.

Prove: $\overline{XZ} \cong \overline{YZ}$

APPLICATION

48. NAVIGATION The edges of a parallel ruler, shown at right, always remain parallel as the ruler is opened and closed. Use a geometry theorem to explain why this is true. *(LESSON 4.6)*

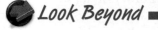 *Look Beyond*

For Exercises 49–52, refer to the diagram at right.

49. Draw a large equilateral triangle. Connect the midpoints of the sides to form a second triangle inside the first. Next, draw a third triangle by connecting the midpoints of the sides of the second triangle. Continue this pattern, drawing as many triangles as you can.

50. Are all of the triangles formed equilateral? Why or why not?

51. Suppose that the length of a side of the original triangle is 1 unit. What is the perimeter of the original triangle? of the second triangle? of the third triangle?

PROBLEM SOLVING

52. Look for a pattern. Add the perimeters of the second, third, fourth, and fifth triangles. How does this compare with the perimeter of the first triangle? What happens as the perimeters of additional triangles are added to this sum?

TESSELLATIONS WITH HEXAGONS

Any hexagon with one pair of opposite sides that are parallel and congruent can be used to create a tessellation. Use a compass and straightedge or geometry graphics software to create the following constructions:

1. Construct a hexagon with one pair of opposite sides that are parallel and congruent, and label the vertices A, B, C, D, E, and F, where $\overline{AB}$ and $\overline{ED}$ are the sides that are parallel and congruent.

2. Translate the hexagon so that $\overline{A'B'}$ coincides with $\overline{ED}$. (Hint: Draw a translation vector from A to E.)

3. Locate the midpoint, M, of $\overline{BC}$. Rotate hexagon $ABCDE$ 180° about point M.

4. By continuing the translations and rotations, you should be able to continue this pattern indefinitely. Try experimenting with different hexagons. If you used geometry graphics software, try dragging the vertices to different locations.

This design was created with concave hexagons.

In 1939, a university mathematics student named Arthur H. Stone was playing with a strip of paper trimmed from a notebook. He discovered something interesting—flexagons.

Flexagons are polygons made from folded paper that show different faces when "flexed." The instructions below will give you a chance to play with two of these unique figures yourself.

For your first flexagons, it may be helpful to use paper with one color on the front and another on the back, as shown. Later, you can get creative by drawing designs on the flexagons, which have a kaleidoscopic effect when flexed.

Activity 1

HEXAFLEXAGON A hexaflexagon has three faces, but only two are visible at a time. Be sure to make your cuts and folds as precisely as possible to ensure that your flexagon will flex smoothly.

1. First cut a strip of paper divided into 10 equilateral triangles. Label the front and back of the strip as shown.

2. Fold the strip so that the triangles labeled 1 face each other.

Front

Back

3. Fold the strip so that the triangles labeled 2 face each other.

4. Fold the strip so that the triangles marked 3 face each other. Carefully glue together the triangles labeled 4, and let the glue dry.

5. Now you are ready to flex your hexaflexagon. Pinch together two triangles and push in the opposite side so that the flexagon looks like a **Y** shape when viewed from above. Open the flexagon from the center. Repeat. How does the arrangement of the faces change as you flex the flexagon?

HEXAHEXAFLEXAGON A hexahexaflexagon has twice as many faces as the hexaflexagon. It is also more complicated to construct, so follow the directions below carefully.

1. First cut a strip of paper divided into 19 equilateral triangles. Label the front and back of the strip as shown.

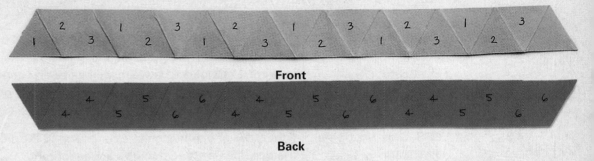

Front

Back

2. Fold the strip so that each pair of adjacent 4s, 5s, and 6s face each other. The strip should coil around itself and look like the one shown below.

3. Fold the strip so that each pair of adjacent 3s face each other.

4. Tuck one end of the strip under the other so that the remaining pair of 3s face each other. Fold down the flap, and carefully glue the unlabeled triangles together.

5. Flex your hexahexaflexagon in the same way as you did the first flexagon. Can you get all six faces of this flexagon to show?

Extension

1. What are the front-back face combinations?

2. Are any combinations of faces not possible?

3. Is there a pattern to the order in which the faces are revealed?

Chapter Review and Assessment

POSTULATES AND THEOREMS

Lesson	Number	Postulate or Theorem
4.1	4.1.1 Polygon Congruence Postulate	Two polygons are congruent if and only if there is a way of setting up a correspondence between their sides and angles, in order, such that (1) all pairs of corresponding angles are congruent, and (2) all pairs of corresponding sides are congruent.
4.2	4.2.1 SSS (Side-Side-Side) Postulate	If the sides of one triangle are congruent to the sides of another triangle, then the two triangles are congruent.
	4.2.2 SAS (Side-Angle-Side) Postulate	If two sides and their included angle in one triangle are congruent to two sides and their included angle in another triangle, then the two triangles are congruent.
	4.2.3 ASA (Angle-Side-Angle) Postulate	If two angles and their included side in one triangle are congruent to two angles and their included side in another triangle, then the two triangles are congruent.
4.3	4.3.1 AAS (Angle-Angle-Side) Congruence Theorem	If two angles and a nonincluded side of one triangle are congruent to the corresponding angles and nonincluded side of another triangle, then the triangles are congruent.
	4.3.2 HL (Hypotenuse-Leg) Congruence Theorem	If the hypotenuse and a leg of a right triangle are congruent to the hypotenuse and corresponding leg of another right triangle, then the two triangles are congruent.
4.4	4.4.1 Isosceles Triangle Theorem	If two sides of a triangle are congruent, then the angles opposite those sides are congruent.
	4.4.2 Converse of the Isosceles Triangle Theorem	If two angles of a triangle are congruent, then the sides opposite those angles are congruent.
	4.4.3 Corollary	The measure of each angle of an equilateral triangle is 60°.
	4.4.4 Corollary	The bisector of the vertex angle of an isosceles triangle is the perpendicular bisector of the base.
4.5	4.5.1 Theorem	A diagonal of a parallelogram divides the parallelogram into two congruent triangles.
	4.5.2 Theorem	The opposite sides of a parallelogram are congruent.
	4.5.3 Theorem	The opposite angles of a parallelogram are congruent.

Lesson	Number	Postulate or Theorem
	4.5.4 Theorem	The diagonals of a parallelogram bisect each other.
	4.5.5 Theorem	A rhombus is a parallelogram.
	4.5.6 Theorem	A rectangle is a parallelogram.
	4.5.7 Theorem	The diagonals and sides of a rhombus form four congruent triangles.
	4.5.8 Theorem	The diagonals of a rhombus are perpendicular.
	4.5.9 Theorem	The diagonals of a rectangle are congruent.
	4.5.10 Theorem	The diagonals of a kite are perpendicular.
	4.5.11 Theorem	A square is a rectangle.
	4.5.12 Theorem	A square is a rhombus.
	4.5.13 Theorem	The diagonals of a square are congruent and are the perpendicular bisectors of each other.
4.6	4.6.1 Theorem	If two pairs of opposite sides of a quadrilateral are congruent, then the quadrilateral is a parallelogram.
	4.6.2 Theorem	If one pair of opposite sides of a quadrilateral are parallel and congruent, then the quadrilateral is a parallelogram.
	4.6.3 Theorem	If the diagonals of a quadrilateral bisect each other, then the quadrilateral is a parallelogram.
	4.6.4 Theorem	If one angle of a parallelogram is a right angle, then the parallelogram is a rectangle.
	4.6.5 Housebuilder Theorem	If the diagonals of a parallelogram are congruent, then the parallelogram is a rectangle.
	4.6.6 Theorem	If one pair of adjacent sides of a parallelogram are congruent, then the quadrilateral is a rhombus.
	4.6.7 Theorem	If the diagonals of a parallelogram bisect the angles of the parallelogram, then the parallelogram is a rhombus.
	4.6.8 Theorem	If the diagonals of a parallelogram are perpendicular, then the parallelogram is a rhombus.
	4.6.9 Triangle Midsegment Theorem	A midsegment of a triangle is parallel to a side of the triangle, and its length is equal to half the length of that side.
4.8	4.8.1 Betweenness Postulate	Given three points P, Q, and R, if $PQ + QR = PR$, then Q is between P and R on a line.
	4.8.2 Triangle Inequality Theorem	The sum of the lengths of any two sides of a triangle is greater than the length of the third side.

Key Skills & Exercises

LESSON 4.1
Key Skills

Identify corresponding parts of congruent polygons.

Given: $ABCDE \cong FGHIJ$

Identify all pairs of congruent sides and angles.

$\overline{AB} \cong \overline{FG}, \overline{BC} \cong \overline{GH}, \overline{CD} \cong \overline{HI}, \overline{DE} \cong \overline{IJ}, \overline{EA} \cong \overline{JF}$
$\angle A \cong \angle F, \angle B \cong \angle G, \angle C \cong \angle H, \angle D \cong \angle I, \angle E \cong \angle J$

Exercises

In the diagram below, $\triangle KLM \cong \triangle PQR$. Complete the following statements about congruence:

1. $\overline{LM} \cong$ ____?____

2. $\overline{RP} \cong$ ____?____

3. $\angle P \cong$ ____?____

4. $\triangle QPR \cong$ ____?____

LESSON 4.2
Key Skills

Use SSS, SAS, and ASA postulates to determine if triangles are congruent.

Which triangles below are congruent to $\triangle ABC$?

$\triangle DEF \cong \triangle ABC$ by ASA. $\triangle GHI \cong \triangle ABC$ by SSS. $\triangle JKL$ cannot be proven congruent to $\triangle ABC$ from the given information. $\triangle MNO \cong \triangle ABC$ by SAS.

Exercises

Are the triangles in each pair below congruent? State the postulate or theorem that supports your answer.

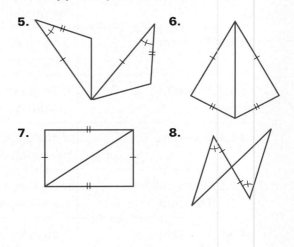

5.

6.

7.

8.

LESSON 4.3
Key Skills

Use AAS and HL Theorems to determine whether triangles are congruent.

In the figure at right, is $\triangle UVW \cong \triangle UST$?

$\triangle UVW \cong \triangle UVT$ by HL and $\triangle UVT \cong \triangle UST$ by AAS, so $\triangle UVW \cong \triangle UST$ by the Transitive Property of Congruence.

Exercises

Are the triangles in each pair below congruent? State the postulate or theorem that supports your answer.

9.

10.

11.

LESSON 4.4

Key Skills

Use triangle congruence in proofs.

In the figure below, $\overline{AB} \cong \overline{DC}$ and $\overline{BD} \cong \overline{CA}$. Prove that $\angle A \cong \angle D$.

$\triangle ABC \cong \triangle DCB$ by SSS, so $\angle A \cong \angle D$ because CPCTC.

Use properties of isosceles triangles in proofs.

$QRST$ is a kite. Prove that $\overline{QS}$ bisects $\overline{TR}$.

Because $\triangle QTR$ is isosceles, $\angle QTR \cong \angle QRT$. Also, $\triangle QTS \cong \triangle QRS$ by SSS, so $\angle TQS \cong \angle RQS$ because CPCTC. Then $\triangle QUT \cong \triangle QUR$ by ASA, so $\overline{TU} \cong \overline{UR}$ because CPCTC.

Exercises

Complete the proof below.

Given: quadrilateral $MNOP$ with $\angle MPN$, $\angle MNP$, $\angle OPN$, and $\angle ONP$ all congruent.

Prove: $MNOP$ is a rhombus.

Proof:

$\triangle MNP$ and $\triangle ONP$ are isosceles by **12.** ___?___, so $\overline{MP} \cong \overline{MN}$ and $\overline{OP} \cong \overline{ON}$ by **13.** ___?___. Also, $\triangle MPN \cong \triangle OPN$ by **14.** ___?___, so $\overline{MP} \cong \overline{OP}$ and $\overline{MP} \cong \overline{ON}$ because **15.** ___?___. Because all of the sides are congruent, $MNOP$ is a rhombus.

LESSON 4.5

Key Skills

Prove properties of quadrilaterals.

Quadrilateral $EFGH$ is a rhombus. Prove that $\overline{ED} \cong \overline{GD}$.

Because $EFGH$ is a rhombus, it is also a parallelogram, so the diagonals bisect each other. Therefore, $\overline{ED} \cong \overline{GD}$.

Exercises

16. In parallelogram $RSTU$, prove that $\angle R$ and $\angle S$ are complementary.

17. In rectangle $CDEF$ with diagonals intersecting at G, prove that $\triangle CDG \cong \triangle EFG$.

18. In rhombus $NOPQ$, prove that $\angle NOQ \cong \angle POQ$.

19. In square $VWXY$, prove that $\triangle VWX \cong \triangle WXY$.

LESSON 4.6

Key Skills

Classify quadrilaterals from given information.

If quadrilateral $ABCD$ has two parallel sides and $\overline{AB} \cong \overline{CD}$, is $ABCD$ a parallelogram?

The information given is not enough to determine whether $ABCD$ is a parallelogram. For example, if the parallel sides are $\overline{AD}$ and $\overline{BC}$, $ABCD$ could be an isosceles trapezoid, as shown below.

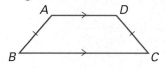

Exercises

Classify each quadrilateral according to the given information. List all terms that apply to each quadrilateral. (You may wish to draw a diagram of each quadrilateral.)

20. In $PQRS$, $\overline{PQ} \parallel \overline{RS}$ and $\overline{PQ} \cong \overline{RS}$.

21. In $KLMN$, $\overline{KL} \parallel \overline{MN}$, $\overline{KN} \parallel \overline{LM}$, and $\overline{KL} \cong \overline{LM}$.

22. In $WXYZ$, $\overline{WY}$ and $\overline{XZ}$ bisect each other, $\overline{WY} \perp \overline{XZ}$, and $\overline{WX} \perp \overline{XY}$.

23. In $EFGH$, $\overline{EF} \parallel \overline{GH}$, $\overline{EG} \cong \overline{FH}$, and $m\angle E = 90°$.

Key Skills

Construct figures by using a compass and straightedge.

Construct an angle with twice the measure of ∠A.

Draw $\overrightarrow{BC}$, and construct ∠CBX ≅ ∠A.

Using $\overrightarrow{BX}$, construct ∠XBY ≅ ∠A. By the Angle Addition Postulate and the Substitution Property, m∠CBY = 2 × m∠A.

Exercises

Use a compass and straightedge to complete the following constructions:

24. Construct an isosceles triangle.

25. Construct the perpendicular bisector of the base of the isosceles triangle.

26. Construct a rectangle.

27. Construct the angle bisector of any angle of a rectangle.

Key Skills

Translate, rotate, and reflect a figure by using a compass and straightedge.

Rotate △PQR about point P.

Construct arcs centered at P through points Q and R. Choose a point on the arc through Q, and label it Q′. Construct ∠Q′PR′ ≅ ∠QPR, and connect Q′ and R′ to form △PQ′R′.

Exercises

Refer to the diagram at right. Trace the figure onto paper for each exercise.

28. Translate △JKL as indicated by the translation vector.

29. Rotate △JKL about point K.

30. Reflect △JKL across the translation vector.

31. Reflect △JKL across $\overline{KL}$.

Applications

32. SURVEYING Refer to the figure at right. What measurement(s) would you need to make in order to determine the distance across the lake? (Assume that the distance cannot be measured directly.)

33. DESIGN Trace the figure shown below. Construct a congruent copy of the figure by using a compass and straightedge.

Alternative Assessment

Performance Assessment

An ancient problem relating to compass-and-straightedge constructions is the *trisection* of an angle, or the construction of an angle with a measure equal to one-third of the measure of a given angle. It has been proved that it is impossible to trisect an angle by using only a compass and an *unmarked* straightedge. However, if you bend the rules to allow a marked straightedge, the trisection of an angle is possible.

1. Given ∠ABC, first construct line ℓ through point A and parallel to $\overrightarrow{BC}$ that passes through point A. Set the compass point at A and draw an arc across line ℓ and through point B.

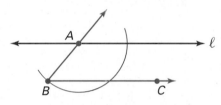

2. Now mark the distance from A to B on your straightedge. Move your straightedge so that it passes through B, and then adjust it until one mark is on the arc and the other mark is on line ℓ. Draw a line along the staightedge in this position. Label the point of intersection with line ℓ as D and with the arc as E.

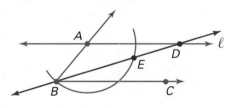

3. Prove that m∠EBC is one-third of m∠ABC. (Hint: Draw $\overline{AE}$. Let m∠ABE = x. $\overline{AB}$, $\overline{AE}$, and $\overline{DE}$ are all congruent, so △ABE and △ADE are isosceles.)

Portfolio Projects

The property of triangle rigidity makes triangles very useful in building structures because they are very stable.

1. Use uncooked spaghetti and miniature marshmallows or other materials to make several different triangles. Test the strength of your triangles by pushing and pulling on them. Which triangles seem to be the strongest? Link several triangles together in different ways. Which links seem to be the strongest?

2. Try to build a bridge that is constructed of only spaghetti and marshmallows and that spans at least 2 feet. Compare your bridge with those of others in your class. Loop a string around your bridge and hang weights from it with a paper clip to test its strength. What type of bridge seems to be the strongest?

---

internet**connect**

(HRW) The HRW Web site contains many resources to reinforce and expand your knowledge of quadrilaterals and constructions. The Web site also provides Internet links to other sites where you can find information and real-world data for use in research projects, reports, and activities that involve quadrilaterals and constructions. Visit the HRW Web site at **go.hrw.com**, and enter the keyword **MG1 CH4**.

College Entrance Exam Practice

QUANTITATIVE COMPARISON For Items 1–3, write
A if the quantity in Column A is greater than the quantity in Column B;
B if the quantity in Column B is greater than the quantity in Column A;
C if the two quantities are equal; or
D if the relationship cannot be determined from the information given.

	Column A	Column B	Answers
1.	slope of $\overline{DE}$	slope of $\overline{AB}$	Ⓐ Ⓑ Ⓒ Ⓓ [Lesson 3.8]
2.	x	y	Ⓐ Ⓑ Ⓒ Ⓓ [Lesson 4.1]
3.	m∠XYZ	m∠TUV	Ⓐ Ⓑ Ⓒ Ⓓ [Lesson 4.3]

4. Which pair of points defines a line perpendicular to $\overleftrightarrow{MN}$? **(LESSON 3.8)**

a. $(0, 7), (8, -4)$
b. $(4, -7), (-4, 4)$
c. $(-7, 0), (4, 8)$
d. $(7, -4), (-4, 4)$

5. What is the measure of ∠ACB? **(LESSON 4.4)**
a. 42°
b. 126°
c. 63°
d. cannot be determined

6. What is the unknown angle measure? **(LESSON 3.6)**

a. 120°
b. 180°
c. 130°
d. 100°

7. Which postulate or theorem justifies the congruence statement $\triangle QDA \cong \triangle UAD$? **(LESSONS 4.2 AND 4.3)**

a. SSA
b. SSS
c. HL
d. SAS

8. Which is not a feature of every rhombus? **(LESSON 4.6)**

a. parallel opposite sides
b. congruent diagonals
c. four congruent sides
d. perpendicular diagonals

9. Write the converse of this statement: If trees bear cones, then the trees are conifers. **(LESSON 2.2)**

10. Use the rule $T(x, y) = (x + 2, y - 2)$ to transform the figure below. What type of transformation results? **(LESSON 1.7)**

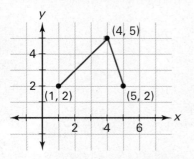

11. Lines ℓ and m are parallel. Find m∠2. **(LESSON 3.3)**

12. Are the triangles below congruent? Which postulate or theorem justifies your answer? **(LESSONS 4.2 AND 4.3)**

13. Which of the following measurements do not define a possible triangle? Choose all that apply. **(LESSONS 3.8 AND 4.8)**

a. m∠A = 50°, m∠B = 85°, m∠C = 45°
b. AB = 12, BC = 7, AC = 9
c. m∠A = 90°, m∠B = 65°, m∠C = 15°
d. AB = 18, BC = 6, AC = 10

FREE-RESPONSE GRID
Items 14–16 may be answered by using a free-response grid such as that commonly used by standardized-test services.

14. $QUAD$ is a rectangle. Find x. **(LESSON 4.5)**

$QC = x$
$DC = 3x - 8$

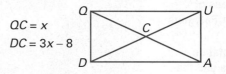

15. Find the measure of an interior angle of the regular polygon below. **(LESSON 3.6)**

16. Point A is the midpoint of $\overline{BF}$. Point D is the midpoint of $\overline{CE}$. Find BC. **(LESSON 3.7)**

Perimeter and Area

THE IMPORTANCE OF THE IDEAS OF PERIMETER AND AREA is suggested by the images on these pages. How long is the oval track around the playing field? How long is the Great Wall of China? How much grain can be grown on a given field?

Sometimes, people require information about area for strange reasons. When the conceptual artist Christo devised his plan of surrounding islands by pink plastic, he needed to know how much material would be required for the project.

You may already be familiar with formulas for perimeter and area. In this chapter, you will deepen your knowledge of them by developing and proving them.

Lessons

Christo's Surrounded Islands, *Biscayne Bay, Miami, Florida, 1980-1983*

PORTFOLIO ACTIVITIES
CHAPTER PROJECT

About the Chapter Project

In this chapter, you will study formulas for the area of common shapes such as rectangles, triangles, and circles. But there are many other shapes that cannot be included in these categories: stars, crescents, and irregular shapes that do not even have names, much less formulas for their areas.

In the Chapter Project, you will study polygons drawn on square grid paper in such a way that every vertex is at a grid point. By determining the area of the figures and searching for a pattern, you will derive the formula for the area of such figures discovered by George Pick in 1899.

After completing the Chapter Project you will be able to do the following:

● Find the areas of certain complex figures without using traditional area formulas.

About the Portfolio Activities

Throughout the chapter, you will be given opportunities to complete Portfolio Activities that are designed to support your work on the Chapter Project.

The theme of each Portfolio Activity and of the Chapter Project is the area of figures in a plane.

● In the Portfolio Activity on page 320, you will find areas of irregular polygons by dividing them into rectangles and triangles.

● In the Portfolio Activity on page 346, you will find the area of a crescent-shaped figure called a *lune*.

● In the Portfolio Activity on page 359, you will use a calculator to generate random points on a grid and use them to test the theoretical probability that a point chosen at random lies within a given area on the grid.

Perimeter and Area

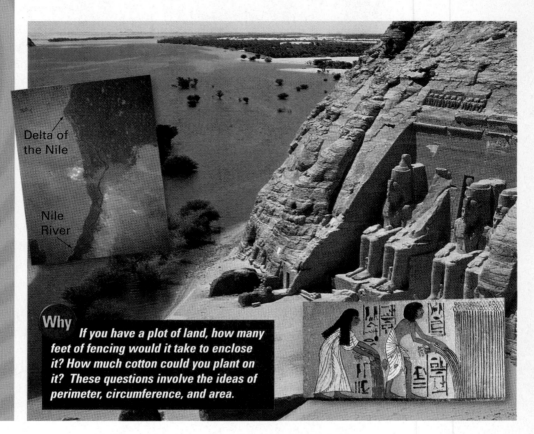

Objectives

- Identify and use the Area of a Rectangle and the Sum of Areas Postulates.

- Solve problems involving fixed perimeters and fixed areas.

Why *If you have a plot of land, how many feet of fencing would it take to enclose it? How much cotton could you plant on it? These questions involve the ideas of perimeter, circumference, and area.*

Delta of the Nile

Nile River

APPLICATION

SURVEYING

In ancient Egypt, the yearly flooding of the Nile River was beneficial for crops, but it damaged and even destroyed property boundaries. This forced the Egyptians to devise methods for redetermining boundaries after each flood. The later development of such methods by Greek mathematicians led to the science of "Earth measure" for which geometry is named.

Perimeter

Perimeter

The **perimeter** of a closed plane figure is the distance around the figure.

5.1.1

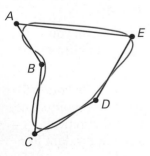

For a polygon, the perimeter is the sum of the lengths of its sides.

The perimeter of a polygon can be used to approximate the perimeter of a closed figure in a plane. For example, the perimeter of the irregular figure at left is approximately equal to the perimeter of polygon *ABCDE*.

CRITICAL THINKING

How could you redraw the polygon in the figure at left above to get a better estimate of the perimeter of the irregular figure?

EXAMPLE ❶ Find the perimeter of each figure.

a. pentagon $ABCDE$

b. parallelogram $WXYZ$

● **SOLUTION**

a. The perimeter is the sum of the sides of the pentagon.

perimeter = $3 + 7 + 2 + 5 + 6 = 23$ units

b. The figure is a parallelogram, so opposite sides are of equal length.

$WX = YZ = 5$ $ZW = XY = 3$

The perimeter is the sum of the sides.

perimeter = $WX + XY + YZ + ZW$
$= 5 + 3 + 5 + 3 = 16$ units

Area

> *Figures are **non-overlapping** if they have no points in common (except for boundary points).*

Area

The **area** of a closed plane figure is the number of non-overlapping squares of a given size that will exactly cover the interior of the figure.

5.1.2

In the floor tile illustration at right, there are 3 rows of tiles with 4 tiles in each row. You can find the total number of tiles by multiplying.

$3 \times 4 = 12$

Number of rows   *Number of tiles in each row*

Twelve 1×1 tiles will exactly cover a 3×4 rectangle.

The following postulate is used to find the areas of nonrectangular shapes. (See Example 2 on page 296.)

Postulate: The Sum of Areas

If a figure is composed of non-overlapping regions A and B, then the area of the figure is the sum of the areas of regions A and B.

5.1.3

Area and Perimeter of Rectangles

The perimeter and area of a rectangle can be found easily by using the definitions on the previous pages. However, it is often more convenient to use the following formulas:

The Perimeter of a Rectangle

The perimeter of a rectangle with base b and height h is given by:
$P = 2b + 2h$.

5.1.4

Postulate: Area of a Rectangle

The area of a rectangle with base b and height h is given by:
$A = bh$.

5.1.5

EXAMPLE 2 Find the area of each figure below. Explain your method in each case.

a.

b.

c.

● **SOLUTION**

a. Multiply the base and the height. The area is 15 square units.

b. Divide the figure into separate rectangles. One way of doing this is shown at right.

$(2.5 \times 6) + (3 \times 8) = 15 + 24$
$= 39$ square units

c. Select an appropriate grid and place it over the figure. Estimate the area by counting the number of squares inside the figure. One method is as follows: Take half the number of squares that are only partially inside the figure (these are identified with dots). Add this to the number of squares that are entirely inside the figure. There are 18 partial squares and 16 complete squares. The area is approximately $\frac{1}{2}(18) + 16 = 25$ square units.

CRITICAL THINKING How is Postulate 5.1.3 (on page 295) used in Example 2?

Algebra

Activity 1

Fixed Perimeter/Maximum Area

A gardener has 24 feet of fencing for a garden. What are the dimensions of the rectangle that will enclose the greatest area?

1. Sketch three different rectangles with a perimeter of 24 units. What is the area of each?

2. Show that $h = 12 - b$ for a rectangle with a perimeter of 24.

3. Fill in a table like the one at right. What do you observe about the area values?

4. Plot the values for b and A from the table on a graph, with b on the horizontal axis and A on the vertical axis.

b	$h = 12 - b$	$A = bh$
1	11	11
2	?	?
3	?	?
⋮	⋮	⋮

5. What are the dimensions of the rectangle that has the maximum area for the given perimeter? If you have a graphics calculator, graph $A = b(12 - b)$ by entering **Y = X(12 – X)**. Trace the graph to find the maximum.

CHECKPOINT ✔ 6. What kind of rectangle is your result from Step 6? Do you think this would be the result for any fixed perimeter?

Activity 2

Fixed Area/Minimum Perimeter

A farmer wants to enclose a rectangular area of 3600 square feet with the minimum amount of fencing. What should be the dimensions of the rectangle?

1. Show that $h = \frac{3600}{b}$ for a rectangle with an area of 3600.

2. Fill in a table like the one at right. Let b range from 10 to 100.

3. Plot the values for b and P from the table on a graph, with b on the horizontal axis and P on the vertical axis.

b	$h = \frac{3600}{b}$	$P = 2b + 2h$
10	360	740
20	?	?
30	?	?
⋮	⋮	⋮

Algebra

4. What are the dimensions of the rectangle that has the minimum perimeter for the given area? If you have a graphics calculator, graph $P = 2b + \frac{7200}{b}$ by entering **Y = 2X + $\frac{7200}{X}$**. Trace the graph to find the minimum.

CHECKPOINT ✔ 5. What kind of rectangle is your result from Step 5? Do you think this would be the result for any fixed area?

Communicate

1. RECREATION Explain a way to estimate the perimeter of the swimming pool shown below.

2. AVIATION Explain a way to estimate the area of a wing of the airplane shown below.

3. Use the figures below to explain the following statement:

area of figure A + area of figure $B \neq$ area of figure C

4. Many applications in mathematics involve finding the area under a curve. How could you estimate the shaded area in the figure at right?

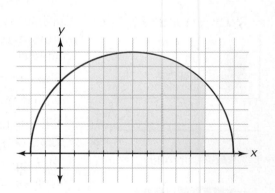

Guided Skills Practice

5. Find the perimeter of the figure at right.
(EXAMPLE 1)

6. Find the area of the figure at right.
(EXAMPLE 2)

7. What is the maximum area of a rectangular garden with a perimeter of 48 feet? **(ACTIVITY 1)**

8. What is the minimum perimeter of a rectangular region with an area of 900 square square feet? **(ACTIVITY 2)**

● *Practice and Apply*

For Exercises 9–18, use the figure and measurements below to find the indicated perimeters and areas:

$AD = 15$ in.	$AC = 13$ in.	$BD = 10$ in.
$DL = 11$ in.	$EI = 3$ in.	$CH = 4$ in.

9. the perimeter of rectangle *ADLI*

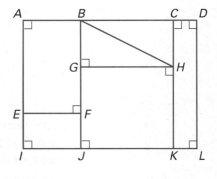

10. the area of rectangle *ADLI*

11. the perimeter of rectangle *CDLK*

12. the area of rectangle *CDLK*

13. the perimeter of hexagon *GHCDLJ*

14. the area of hexagon *GHCDLJ*

15. the perimeter of rectangle *BCHG*

16. the area of rectangle *BCHG*

17. the area of △*BHG* (Hint: Use your answer to Exercise 16.)

18. If points *I* and *D* were connected by a segment, what would be the area of △*ADI*?

19. Write a paragraph proof of the formula for the perimeter of a rectangle.

In Exercises 20–23, find the area of the rectangle with vertices at the given points. You may find it helpful to sketch a graph.

20. $(0, 0), (0, 2), (5, 0), (5, 2)$

21. $(3, 1), (3, 7), (9, 1), (9, 7)$

22. $(-2, -5), (-2, 3), (4, -5), (4, 3)$

23. $(0, 0), (3, 3), (6, 0), (3, -3)$

24. The perimeter of a rectangle is 72 centimeters. The base is 3 times the height. What are the dimensions of the rectangle? What is the area?

25. The area of a rectangle is 27 square feet. The base is 3 more than twice the height. What are the dimensions of the rectangle? What is the perimeter?

26. The perimeter of a rectangle is 80*x*. The base is 7 times the height. In terms of *x*, what are the dimensions of the rectangle? What is the area?

27. The perimeter of a rectangle is equal to its area. The height is 3 more than the base. What are the dimensions of the rectangle? What is the area?

28. For a rectangle with a fixed perimeter of 100, $2b + 2h = 100$. Solve this equation for *b* or *h* and graph the resulting function. What type of function represents this relationship? What values of *b* and *h* do not make sense in the equation?

29. For a rectangle with a fixed area of 100, $bh = 100$. Solve this equation for b or h and graph the resulting function. What type of function represents this relationship? What values of b and h do not make sense in the equation?

CHALLENGES

30. The squares in the grid below measure 0.5 centimeters on each side. Estimate the area of the shaded figure.

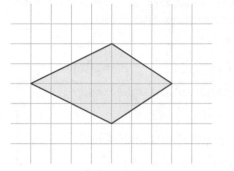

31. The squares in the grid below measure 0.5 centimeters on each side. Estimate the area of the shaded figure.

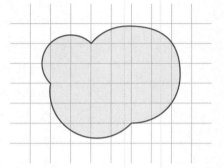

APPLICATIONS

CONSTRUCTION A house has a roof with the dimensions shown at right. Both rectangular halves of the roof will be covered with plywood.

42 ft   25 ft

32. If plywood comes in pieces that measure 8 feet by 4 feet, how many pieces of plywood are needed to cover the roof? Assume that both halves of the roof are the same.

33. Show a possible arrangement of the plywood pieces on one half of the roof. Try to make as few cuts as possible. How much scrap plywood is left over?

34. SOLAR ENERGY Eric uses solar power to heat his home. He wants to provide 15,000 BTUs of heat from 2 solar panels. Eric wants the panels to be equal in size and 10 feet long. If 6 square feet of panels provide 1000 BTUs, what should be the dimensions of the panels that he uses?

The amount of energy produced by a solar panel depends on its area.

35. AGRICULTURE You have 200 feet of fencing material to make a pen for livestock. If you make a rectangular pen, what is the maximum area you can fence in? Extend the table at right to determine the answer.

Base	Height	Perimeter	Area
1	?	200	?
2	?	200	?
5	?	200	?
20	?	200	?

36. AGRICULTURE Suppose that you need an area of 5625 square feet for grazing livestock. What is the minimum amount of fencing needed for a rectangular pen with this area? Extend the table at right to determine the answer.

Base	Height	Perimeter	Area
1	?	?	5625
2	?	?	5625
5	?	?	5625
20	?	?	5625

37. AGRICULTURE Repeat Exercise 35, but suppose that one side of the pen can be left out by placing the pen against the wall of the barn, as shown at left.

Base	Height	Perimeter	Area
1	?	200	?
2	?	200	?
5	?	200	?
20	?	200	?

Barn

Pen

38. AGRICULTURE Repeat Exercise 36, but suppose that one side of the pen can be left out by placing the pen against the wall of the barn, as shown at left.

Base	Height	Perimeter	Area
1	?	?	5625
2	?	?	5625
5	?	?	5625
20	?	?	5625

39. LANDSCAPING Rudy is buying sod for the lawn illustrated at right. Given the dimensions, estimate the number of square feet of sod that he needs. Assume that all angles shown are 90°.

40 ft

21 ft

15 ft

30 ft

40. LANDSCAPING Dian wants to plant a rectangular vegetable garden that measures 10 feet by 16 feet. She will plant the vegetables in rows that are 10 feet long. Dian wants an equal amount of space for lettuce, tomatoes, carrots, zucchini, and peppers. In addition, she wants one 10-foot row of marigolds that will be 1 foot wide. How much space will Dian have for each type of vegetable?

Marigolds are often used by gardeners to repel insects from other plants.

41. HOME IMPROVEMENT Brenda wants to paint her room, which measures 14 feet × 16 feet × 10 feet, as shown. The room has a 6 foot × 4 foot window and a 3 foot × 7 foot door that will not be painted. She will give the walls and ceiling two coats of paint: a base coat and a final coat.

One gallon of the paint for the base coat costs $10 and covers 500 square feet. One gallon of the paint for the final coat costs $20 and covers 250 square feet. The paint is sold only in 1-gallon cans. If the sales tax is 7%, how much will it cost to paint the room?

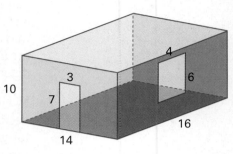

Look Back

42. Draw an Euler diagram to illustrate the relationships among parallelograms, rectangles, rhombuses, and squares. **(LESSON 3.2)**

43. Find the measure of an exterior angle of an equilateral triangle. **(LESSON 3.6)**

44. If the sum of the measures of 3 angles of a quadrilateral equals 300°, find the measure of the fourth angle. **(LESSON 3.6)**

45. Find the sum of the measures of the angles of a polygon with n sides. **(LESSON 3.6)**

46. Find the measure of an interior angle of a regular hexagon. **(LESSON 3.6)**

47. Find the slope of the segment connecting the points $(2, 3)$ and $(4, -1)$. **(LESSON 3.8)**

48. Find the slope of a line perpendicular to the segment given in Exercise 47. **(LESSON 3.8)**

49. What is the slope of a horizontal line? **(LESSON 3.8)**

50. What is the slope of a vertical line? **(LESSON 3.8)**

51. Find the midpoint of the segment connecting the points $(-4, 6)$ and $(6, 4)$. **(LESSON 3.8)**

Look Beyond

52. Which has a greater area, a square with a side length of 4 inches or a circle with a diameter of 4 inches? Explain your answer.

53. The square at right has 1-inch sides. How could you estimate the area of the circle inside the square?

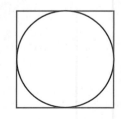

Areas of Triangles, Parallelograms, and Trapezoids

Objectives

- Develop formulas for the areas of triangles, parallelograms, and trapezoids.

- Solve problems by using the formulas for the areas of triangles, parallelograms, and trapezoids.

Why

Designs on graph paper, such as this knitting pattern, suggest a method for estimating areas of geometric figures, but it is often more convenient to use exact formulas.

Areas of Triangles

For each possible choice of the base of a triangle, there is a corresponding altitude and height.

Parts of a Triangle

Any side of a triangle can be called the **base of the triangle**. The **altitude of the triangle** is a perpendicular segment from a vertex to a line containing the base of the triangle. The **height of the triangle** is the length of the altitude.

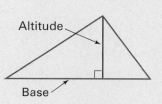

5.2.1

Activity 1

The Area Formula for Triangles

Part I: Deriving a Formula for the Area of a Right Triangle

1. Draw a rectangle on graph paper. Calculate its area.

2. Draw a diagonal of your rectangle to form two right triangles. What do you know about these triangles from your study of special quadrilaterals? Based on this, what is the area of each triangle?

3. If you are given a right triangle, can you always form a rectangle by fitting it together with a copy of itself? Illustrate your answer with examples.

CHECKPOINT ✔ **4.** Write a formula for the area, A, of a right triangle in terms of its base, b, and its height, h.

Part II: Deriving a Formula for the Area of Any Triangle

1. Copy the drawing at right onto graph paper.

2. Draw an altitude of the triangle from the top vertex to the base of the triangle. Is the altitude parallel to the sides of the rectangle? What theorem justifies your answer?

3. The altitude divides the rectangle into two smaller rectangles. Each rectangle is divided into two congruent triangles. What theorem justifies these statements?

4. What is the relationship between the area of the rectangle and the area of the shaded triangle? Explain your answer.

CHECKPOINT ✔ **5.** Write a formula for the area, A, of a triangle in terms of its base, b, and its height, h.

CRITICAL THINKING

How can you use the method from Activity 1 to derive the formula for the area of an obtuse triangle with base b and height h, such as the one at right? (See Exercise 52.)

In Activity 1, you derived the following formula:

Area of a Triangle

For a triangle with base b and height h, the area, A, is given by:
$$A = \tfrac{1}{2}bh$$

5.2.2

E X A M P L E ❶

APPLICATION
THEATER ARTS

You are building a triangular flat, as shown at right, for a dance performance. The flat needs to be covered with cloth. What is the area you need to cover?

35 in.   29 in.
21 in.
48 in.

● **SOLUTION**
$A = \tfrac{1}{2}bh = \tfrac{1}{2}(48)(21) = 504$ in.^{2}

Areas of Parallelograms

Parts of a Parallelogram

Any side of a parallelogram can be called the **base of the parallelogram.** An **altitude of a parallelogram** is a perpendicular segment from a line containing the base to a line containing the side opposite the base. The **height of the parallelogram** is the length of the altitude.

5.2.3

Activity 2

The Area Formula for Parallelograms

1. Copy the drawing at right onto graph paper. Draw an altitude in the parallelogram from point *A* to the base so that a right triangle is formed.

2. Cut out the parallelogram. Cut off the right triangle, translate it to the opposite side, and fit it to the figure.

   What kind of quadrilateral is formed? What is the area of the figure in terms of *b* and *h* of the original parallelogram? How does the area of the parallelogram relate to the area of the figure formed by the translation?

CHECKPOINT ✔

3. Write a formula for the area of a parallelogram in terms of its base, *b*, and its height, *h*.

4. How do you know that the triangle will always fit, as in Step 2? To answer this question, first prove that $\triangle AEB \cong \triangle DFC$. Then prove that *ADFE* is a rectangle.

In Activity 2, you derived the following formula:

Area of a Parallelogram

For a parallelogram with base *b* and height *h*, the area, *A*, is given by:
$$A = bh$$

5.2.4

**E X A M P L E ② ** Find the area of parallelogram *ABCD.*

● **SOLUTION**

$A = bh = 5(13) = 65$ square units

Areas of Trapezoids

Parts of a Trapezoid

The two parallel sides of a trapezoid are known as the **bases of the trapezoid**. The two nonparallel sides are called the **legs of the trapezoid**. An **altitude of a trapezoid** is a perpendicular segment from a line containing one base to a line containing the other base. The **height of a trapezoid** is the length of an altitude.

5.2.5

Activity 3

The Area Formula for Trapezoids

1. Make two copies of the trapezoid at right on graph paper and cut them out. The bases of the trapezoid are b_1 and b_2, and the height is h.

2. Find a way to fit the two copies of the trapezoid together to form a parallelogram. Sketch the parallelogram.

CHECKPOINT ✔

3. Write a formula for the area of the parallelogram involving the expression $(b_1 + b_2)$. Use the formula you wrote for the parallelogram to write a formula for the area of the original trapezoid.

In Activity 3, you derived the following formula:

Area of a Trapezoid

For a trapezoid with bases b_1 and b_2 and height h, the area, A, is given by:
$$A = \frac{1}{2}(b_1 + b_2)h$$

5.2.6

E X A M P L E ❸ A homeowner needs to buy sod for a bare trapezoidal region of the lawn in front of his house. The dimensions of the region are shown at right. What is the area of the region?

APPLICATION

LANDSCAPING

● **SOLUTION**

$A = \frac{1}{2}(b_1 + b_2)h$

$A = \frac{1}{2}(30 + 50)(23)$

$A = 920 \text{ ft}^2$

EXAMPLE **4** Use the diagram and measurements given below to find the areas of the indicated figures.

a. $\triangle VWX$

b. $\triangle WXY$

c. parallelogram VWXY

d. trapezoid WXYZ

SOLUTION

a. area of $\triangle VWZ = \frac{1}{2}bh = \frac{1}{2}(6)(8) = 24$

b. area of $\triangle WXY = \frac{1}{2}bh = \frac{1}{2}(9)(8) = 36$

c. area of parallelogram $VWXY = bh = (9)(8) = 72$

d. area of trapezoid $WXYZ = \frac{1}{2}(b_1 + b_2)h = \frac{1}{2}(9 + 15)(8) = 96$

Exercises

● **Communicate**

1. Draw a parallelogram and one of its diagonals. What can you say about the two triangles formed by the diagonal? Draw the other diagonal to form four triangles. Are any of the four triangles congruent? Which postulates and theorems can you use to prove your answer?

2. Can two triangles have the same base and height and not be congruent? Can two trapezoids have the same base and height and not be congruent? Explain your reasoning.

3. In the parallelogram shown at right, which is longer, AB or h? Use your answer to explain why the parallelogram with a given base and height that has the smallest perimeter is a rectangle.

4. In a trapezoid, what is the average of the lengths of the bases? How is this quantity related to the area of the trapezoid?

5. Suppose that you are given a rectangle, a triangle, a parallelogram, and a trapezoid, each with a base of 16 and a height of 11. Which figures must have the same area? Which area cannot be determined? Explain your reasoning.

6. Find the area of the shaded triangle.
(ACTIVITY 1 AND EXAMPLE 1)

15

40

7. Find the area of the shaded parallelogram. **(ACTIVITY 2 AND EXAMPLE 2)**

5

7

8. The two congruent trapezoids shown fit together to form a parallelogram. Find the area of each trapezoid, and find the area of the parallelogram.
(ACTIVITY 3 AND EXAMPLE 3)

7

6

10

9. Use the diagram below to find the areas of the indicated figures.
(EXAMPLE 4)

  a. △*ADE*

  b. parallelogram *ABCE*

  c. trapezoid *ABCD*

A 20 *B*

26   24   25

D 17 *E*   *C*

● *Practice and Apply* ▬▬▬▬▬▬▬▬▬▬▬

For Exercises 10–12, find the area of each triangle.

10.   **11.**   **12.**

7

4

3

8

5

6

20

14

For Exercises 13–15, find the area of each parallelogram.

13.   **14.**   **15.**

16   13

24

11   13

2

7

6

8

For Exercises 16–18, find the area of each trapezoid.

16.   **17.**   **18.**

7

4   3

9

5

2

4

2

25

40

25

20

For Exercises 19–30 use the diagram and measurements below to find the area of each figure.

$\overline{AC} \parallel \overline{IL}$ $\overline{DH} \parallel \overline{IL}$

$\overline{AI} \parallel \overline{JB}$ $\overline{KC} \parallel \overline{JB}$

$BK = 14$ $KL = 14$

$IL = 32$ $DH = 25$

$FK = 7$ $EH = 17$

$JI = 8$ $DG = 18$

19. $\triangle KCL$
20. $\triangle BJK$
21. $\triangle BCK$
22. $\triangle DIJ$
23. $BCKJ$
24. $EGKJ$
25. $ABED$
26. $ACKI$
27. $EHLJ$
28. $BCHE$
29. $BCLJ$
30. $ACLI$

CONNECTIONS

COORDINATE GEOMETRY For Exercises 31–34, find the area of a triangle with vertices at the given points. You may find it helpful to sketch a graph.

31. $(0, 0), (0, 3), (4, 0)$
32. $(1, 2), (5, 2), (3, 7)$
33. $(1, 0), (3, 0), (0, 6)$
34. $(-2, 1), (4, 1), (1, -3)$

COORDINATE GEOMETRY For Exercises 35–38, find the area of a parallelogram with vertices at the given points. You may find it helpful to sketch a graph.

35. $(0, 0), (4, 0), (6, 2), (2, 2)$
36. $(0, 1), (0, 3), (3, 5), (3, 3)$
37. $(2, 3), (3, -1), (-1, -1), (-2, 3)$
38. $(4, 1), (5, 1), (2, -2), (1, -2)$

COORDINATE GEOMETRY For Exercises 39–42, find the area of a trapezoid with vertices at the given points. You may find it helpful to sketch a graph.

39. $(0, 2), (3, 2), (5, 0), (-1, 0)$
40. $(2, 1), (4, 1), (6, 4), (-1, 4)$
41. $(1, 0), (5, 0), (5, 5), (1, 3)$
42. $(3, 1), (6, -1), (-1, -1), (-2, 1)$

43. Find the base of a triangle with a height of 10 cm and an area of 100 cm².

44. Find the height of a parallelogram with a base of 15 cm and an area of 123 cm².

45. Use the diagram at right of a trapezoid and its midsegment to determine a formula for the area of a trapezoid that uses only the height of the trapezoid and the length, m, of its midsegment.

46. MAXIMUM/MINIMUM The triangles below have the same perimeter. Which has the largest area? Make a conjecture about the triangle with the largest possible area for a given perimeter. Explain your reasoning.

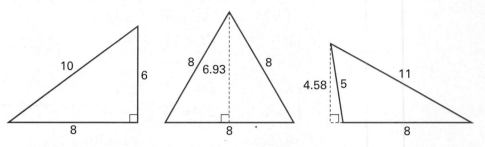

47. MAXIMUM/MINIMUM The parallelograms below have the same perimeter. Which has the largest area? Make a conjecture about the parallelogram with the largest possible area for a given perimeter. Explain your reasoning.

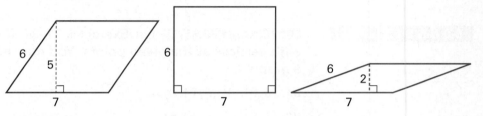

48. Write out a two-column or paragraph proof of the formula you derived in Activity 1.

49. Write out a two-column or paragraph proof of the formula you derived in Activity 2.

50. Write out a two-column or paragraph proof of the formula you derived in Activity 3.

51. A **kite** is a quadrilateral with exactly two pairs of adjacent congruent sides.

Given: The diagonals of a kite are perpendicular.
Prove: The area is equal to one-half of the product of the lengths of the diagonals.

(Hint: Use the formula for the area of a triangle.)

52. Use the diagram at right to prove that the formula for the area of a triangle works for an obtuse triangle.

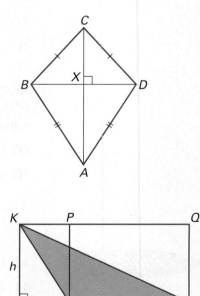

The longest side of the shaded triangle $\triangle KLM$ divides the rectangle into two congruent right triangles. What is the area of each of these triangles?

Right triangle $\triangle KLN$ is divided into two triangles, $\triangle KLM$ and $\triangle KMN$. What is the area of $\triangle KMN$?

Prove that the area of $\triangle KLM$ is $\frac{1}{2}bh$.

53. FARMING In order to fertilize a field, a farmer needs to estimate its area. Estimate the area of the field outlined in the photo at right. If an acre equals 43,560 square feet, how many acres is the field? If it takes 435 pounds of fertilizer to cover 1 acre, how much fertilizer is required to cover the field?

3800 ft

2800 ft

Look Back

54. Construct an Euler diagram to illustrate the relationship between scalene, isosceles, and equilateral triangles. **(LESSON 2.3)**

55. Refer to the diagram below. Prove that $\triangle ABC \cong \triangle DEF$. **(LESSON 4.4)**

56. Find the area of a square with a side length of $x + y$. **(LESSON 5.1)**

57. Find the area of a rectangle with a base of $x + 2$ and a height of $x - 3$. **(LESSON 5.1)**

Look Beyond

CULTURAL CONNECTION: AFRICA Heron's formula, named after a mathematician who lived in Alexandria in around 100 C.E., can be used to find the area of a triangle from the lengths of the sides. The formula is
$$A = \sqrt{s(s - a)(s - b)(s - c)},$$
where s is the *semiperimeter*—that is, half of the perimeter—and a, b, and c are the lengths of the sides.

58. Find the semiperimeter of a triangle with side lengths of 7, 8, and 9.

59. Find the area of the triangle described in Exercise 58.

60. TECHNOLOGY Set up a spreadsheet to find the area of a triangle using Heron's formula when the lengths of the sides are given. Find the areas of the triangles with the sides given below.

=.5*(A2 + B2 + C2)

	A	B	C	D	E
1	a	b	c	s	A
2	7	8	9	?	?
3	3	4	5	?	?
4	2	10	10	?	?
5	x	x	x	?	?

=SQRT(D2*(D2–A2)*(D2–B2)*(D2–C2))

AN ANCIENT WONDER

The ancient Chinese book *The Nine Chapters on the Mathematical Art* is the collective effort of mathematicians over hundreds of years that may have begun long before Euclid compiled his famous work, *The Elements*. The version that exists today has been significantly edited and annotated by many people.

The Nine Chapters on the Mathematical Art has only recently received attention in the West. This ancient text contains algebraic and geometric proofs that are commonly credited to other mathematicians who didn't discover them until hundreds of years later.

On the following page, you will explore how to simplify complex problems by using a "patchwork" method of cutting up figures and rearranging the pieces.

又有田廣十二步從十四步問爲田幾何　　答曰一畝　　百六十八步　圖從十四　廣十二

今有田廣十五步從十六步問爲田幾何

以御田疇界域

方田

九章算經卷第一

魏劉徽注

唐朝議大夫行太史令上輕車都尉臣李淳風等奉　敕注釋

Cooperative Learning

The following problems and methods are the work of Liú Hui , who wrote a commentary for the book around 236 C.E. The commentary, which is now considered to be a part of the book, contains explanations of solutions given in the earlier version.

1. You can begin your exploration of *The Nine Chapters on the Mathematical Art* with the following problem:

Find the side length of a square that is inscribed in a right triangle.

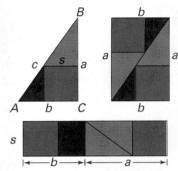

a. Draw a square inscribed in a right triangle as shown at right. Make two congruent copies of the square and triangle and fit them together to form a rectangle.

b. Cut each triangle into three pieces and reassemble them to form a long rectangle as shown at right.

c. What can you say about the area of the rectangle that you formed in part **a** compared with the area of the one that you found in part **b**?

d. Write an expression in terms of a and b for the area of the rectangle in part **a**. Write an expression in terms of a, b, and s for the area of the rectangle in part **b**.

e. Set the two expressions from part **d** equal to each other to form an equation in terms of a, b, and s. Solve the equation for s to find the side length of the square.

2. Now try a harder problem from *The Nine Chapters on the Mathematical Art*.

Find the radius of a circle that is inscribed in a right triangle.

a. Refer to the diagrams at right. Two congruent copies of the triangle fit together to form a rectangle. The rectangle is divided into triangles and squares, which are reassembled to form a long rectangle. Explain why EH in the long rectangle is equal to r, the radius of the inscribed circle.

b. In rectangles $ADBC$ and $EFGH$, explain why the lengths labeled a, b, and c are the same as the side lengths a, b, and c of $\triangle ABC$.

c. Write an expression in terms of a and b for the area of rectangle $ADBC$. Write an expression in terms of a, b, c, and r for the area of rectangle $EFGH$.

d. Set the two expressions from part **c** equal to each other to form an equation in terms of a, b, c, and r. Solve the equation for r to find the radius of the circle.

Circumferences and Areas of Circles

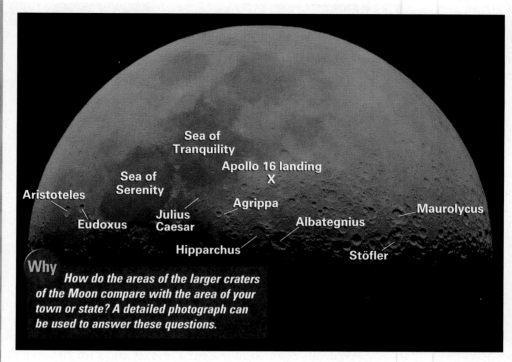

Sea of Tranquility

Apollo 16 landing
X

Sea of Serenity

Aristoteles

Eudoxus

Julius Caesar

Agrippa

Albategnius

Maurolycus

Hipparchus

Stöfler

Why

How do the areas of the larger craters of the Moon compare with the area of your town or state? A detailed photograph can be used to answer these questions.

Objectives

● Identify and apply formulas for the circumference and area of a circle.

● Solve problems using the formulas for the circumference and area of a circle.

The diameter of the Moon is about 1077 miles. You can use this information to determine the scale of this photograph. Then you can estimate the areas of the craters that are visible.

The Definition of a Circle

When you draw a circle with a compass, the distance from the point of the compass to the pencil or pen does not change. Therefore, every part of the circle you draw is the same distance from the *center* of the circle, the point where the compass is fixed. This leads to the following definition:

Definition: Circle

A **circle** is the set of all points in a plane that are the same distance, r, from a given point in the plane known as the **center** of the circle. The distance r is known as the **radius** of the circle. The distance $d = 2r$ is known as the **diameter** of the circle.

5.3.1

Circumferences and Areas of Circles

You are probably already familiar with the formulas for the circumferences and areas of circles from your earlier mathematics studies. The following activities will lead you to an understanding of these formulas.

Activity 1

The Circumference of a Circle

YOU WILL NEED

string and ruler or tape measure and calculator

1. The distance around a circle is called its **circumference**. Measure the circumferences and diameters of several circular objects. Record the results in a table like the one at right.

Object	C	d	Ratio: $\frac{C}{d}$
1. can	31.4	10	?
2. ?	?	?	?
3. ?	?	?	?

2. The ratio $\frac{C}{d}$ is known as π, pronounced "pie." What do you notice about your values for this ratio? Find the average of the values.

3. Compare your result with the results of your classmates. How close are the results to 3.14, an approximate value of π? Press the π key on your calculator, followed by ENTER. What value does the calculator show?

CHECKPOINT ✔

4. Write a formula for the circumference, C, of a circle. Begin by expressing π as a ratio. Then solve for C and write the formula in terms of the radius, r.

In Activity 1, you discovered the following formula:

Circumference of a Circle

The circumference, C, of a circle with diameter d and radius r is given by:
$$C = \pi d \text{ or } C = 2\pi r$$

5.3.2

EXAMPLE ① Find the circumference of each circle below. Give your answers exactly, in terms of π, and rounded to two decimal places.

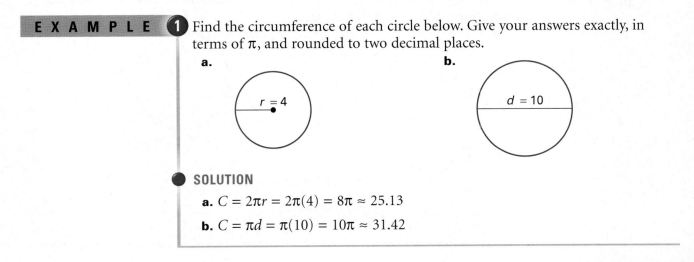

a.

$r = 4$

b.

$d = 10$

● **SOLUTION**

a. $C = 2\pi r = 2\pi(4) = 8\pi \approx 25.13$

b. $C = \pi d = \pi(10) = 10\pi \approx 31.42$

The Area of a Circle

1. Draw a circle. Label its radius r. Using any method you like, such as paper folding, divide the circle into eight congruent pie-shaped parts, or **sectors**.

2. Cut out the sectors and reassemble them into a single figure, as shown at right. If the curved parts of your figure were segments instead of curves, what kind of figure would you have?

3. Divide the circle into 16 congruent sectors by cutting each sector from Step 2 in half. Then reassemble the sectors as in Step 2. As the number of sectors increases, do the curved parts seem straighter?

4. What geometric figure do your sector assemblies in Steps 2 and 3 resemble? The height of your figures is approximately equal to r, the radius of the circle. What happens to this approximation as the number of sectors increases infinitely?

5. The base of your assembled figure is approximately equal to half of the circumference of your original figure. (Why?) Write an expression for the base of the figure in terms of π and the radius, r, of the circle.

CHECKPOINT ✔ 6. Write an expression for the area of the figure in terms of π and r.

Area of a Circle

The area, A, of a circle with radius r is given by:
$$A = \pi r^2$$

5.3.3

CRITICAL THINKING Why does the method you used in Activity 2 become more realistic as you increase the number of sectors?

E X A M P L E ② Find the area of each circle below. Give your answers exactly, in terms of π, and rounded to two decimal places.

a.

$r = 3$

b.

$d = 7$

SOLUTION

a. $A = \pi r^2 = \pi(3^2) = 9\pi \approx 28.27 \text{ units}^2$

b. $r = d \div 2 = 7 \div 2 = 3.5 \text{ units}$
$A = \pi r^2 = \pi(3.5^2) = 12.25\pi \approx 38.48 \text{ units}^2$

Exercises

Communicate

1. Suppose that you have 100 feet of fence to make a play area for your dog. Does a square or a circle provide more area? What other factors might you take into consideration in designing the play area?

2. There are many different approximations for π. Two commonly used values are 3.14 and $\frac{22}{7}$. Compare these values with the value your calculator gives for π. Give a reason you might choose either 3.14 or $\frac{22}{7}$. Why is it necessary to estimate π when calculating the area and circumference of a circle?

3. When the cassette in the photo is rewinding, which moves faster, point A or point B? Explain your reasoning.

Guided Skills Practice

In Exercises 4–7, give your answers exactly, in terms of π, and rounded to two decimal places.

4. Find the circumference of a circle with a radius of 3. *(EXAMPLE 1)*

5. Find the circumference of a circle with a diameter of 25. *(EXAMPLE 1)*

6. Find the area of a circle with a radius of 5. *(EXAMPLE 2)*

7. Find the area of a circle with a diameter of 28. *(EXAMPLE 2)*

Practice and Apply

In Exercises 8–13, find the circumference and area of each circle.

Use 3.14 for π. Round your answers to the nearest tenth.

8. $r = 6$ 9. $r = 10$ 10. $d = 18$

Use $\frac{22}{7}$ for π. Leave your answers in fraction form.

11. $r = 6$ 12. $d = 21$ 13. $d = \frac{35}{8}$

Algebra

Find the radius of the circle with the given measurement. Give your answers exactly, in terms of π, and rounded to the nearest tenth.

14. $C = 12$ 15. $C = 62.8$

16. $C = 50\pi$ 17. $A = 314$

18. $A = 50$ 19. $A = 100\pi$

In Exercises 20–27, find the area of the shaded region. Give your answers exactly, in terms of π, and rounded to the nearest hundredth.

20.

8

21.

9
4

22.

5
3

23.

2
2.8

24.

12
20

25.

30
24

26.

60

27.

8
4
20

28. What happens to the circumference of a circle when the radius is doubled?

29. What happens to the area of a circle when the radius is doubled?

30. **MEAL PLANNING** If a 10-inch pizza is enough to feed 2 people, will an 18-inch pizza be enough to feed 6 people? Why or why not?

31. **MEAL PLANNING** If a 10-inch pizza costs $5 and an 18-inch pizza costs $15, which is the better deal? Explain your reasoning.

32. **IRRIGATION** *Center pivot irrigation* is a method of agricultural irrigation using a long, wheeled arm with many nozzles that pivots about the center of a circle.

   If the area inside the square is one square mile, what is the area, in square feet, of the irrigated circle? (Note: 1 mile = 5280 feet)

Differential gear

AUTOMOBILE ENGINEERING Tires are tested for traction by driving around a small circular track. Refer to the diagram below for Exercises 33–37.

33. What is the circumference of the circle formed by the inside tire tracks?

34. What is the circumference of the circle formed by the outside tire tracks?

35. Based on your answers to Exercises 33 and 34, what can you say about the speed of the inside tires compared with the speed of the outside tires?

Suppose that the tires have a radius of 15 inches.

36. How many revolutions will the inside tires make in one lap around the circle?

37. How many revolutions will the outside tires make in one lap around the circle?

7 ft

$12\frac{1}{2}$ ft

The tires of a car need to turn at different speeds on a curve. To make this possible, a device called a differential gear *is used.*

Detail of the Moon's surface, showing the Hipparchus Crater (circled)

38. **ASTRONOMY** At left is a detail of the photo of the Moon from page 314. The diameter of the Moon in the photo is 12.2 cm. Use your own measurements and the fact that the actual diameter of the Moon is 1077 miles to estimate the actual diameter of the indicated crater.

Use your estimate of the diameter to find the approximate area of the crater. Is it closest in size to New York City (309 miles^2), Los Angeles (469 miles^2), Delaware (1955 miles^2), Lake Superior (31,700 miles^2), or Texas (261,914 miles^2)?

Look Back

Algebra

Recall from algebra that a radical can be simplified by taking the square root of any factors that are perfect squares. For example:

$$2\sqrt{75} = 2\sqrt{25 \times 3} = 2 \times 5\sqrt{3} = 10\sqrt{3}$$

Simplify each expression below.

39. $3\sqrt{8}$ 40. $16\sqrt{32}$ 41. $3\sqrt{500}$

Find the positive solution for x.

42. $x^2 + 16 = 25$ 43. $x^2 + 144 = 169$ 44. $x^2 + 12.25 = 13.69$

45. Given parallelogram $ABCD$ with diagonal $\overline{AC}$, prove that $\triangle ABC \cong \triangle CDA$. **(LESSON 4.5)**

46. Find the area of a triangle with a base of 9 in. and a height of 7 in. **(LESSON 5.2)**

47. Find the area of a parallelogram with a base of 5 cm and a height of 3.5 cm. **(LESSON 5.2)**

48. Find the area of a trapezoid with a height of 3 cm and bases of 6 cm and 5 cm. **(LESSON 5.2)**

Look Beyond

PROOFS

49. In the diagram at left, the centers of all the circles are collinear. Prove that the sum of the circumferences of the small circles is equal to the circumference of the large circle.

50. Show that the result you proved in Exercise 49 is true for any number of smaller circles.

51. CULTURAL CONNECTION: ASIA A *yin-yang* symbol is composed of circles and semicircles, as shown below. Yin and yang represent the two complementary forces, or principles, that make up all aspects and phenomena of life.

Which of the three indicated paths from point *A* to point *B* in the diagram is the longest? Explain your reasoning.

You can find the areas of some irregular polygons by dividing them into rectangles and triangles and adding the areas of these figures.

1. Find the areas of the figures below. The first one has been divided into two rectangles and a triangle to help you.

2. Sometimes it may be easier to subtract areas than to add them. Find the areas of the figures below by subtracting areas from the area of a larger rectangle. The first rectangle has been drawn to help you.

WORKING ON THE CHAPTER PROJECT

You should now be able to complete Activity 1 of the Chapter Project.

The Pythagorean Theorem

5.4

Objectives

- Identify and apply the Pythagorean Theorem and its converse.

- Solve problems by using the Pythagorean Theorem.

Why *A 4000-year-old clay tablet from ancient Babylon—in what is now Iraq—revolutionized our knowledge of ancient mathematics.*

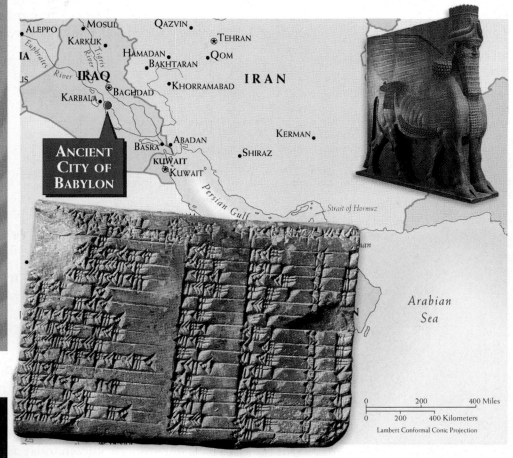

ANCIENT CITY OF BABYLON

When the tablet known as Plimpton 322 was first found, no one understood the significance of the strange columns of numbers—until a mathematician who looked at it made an exciting discovery.

Plimpton 322

CULTURAL CONNECTION: ASIA At the height of its power, according to Greek historian Herodotus (485-427 B.C.E.), Babylon was the world's most splendid city. It was surrounded by walls almost 85 feet thick with eight bronze gates. The main gate and its walls were decorated with figures composed of glazed colored brick that depicted dragons, lions, and bulls.

The tablet in the illustration above, known as Plimpton 322, comes from a much earlier period known as the Old Babylonian Empire, which included the reign of King Hammurabi, who ruled from 1792 to 1750 B.C.E. and became famous for his wise and fair code of laws.

Plimpton 322 is a piece of a larger tablet. Part of the larger tablet, including at least one column of numbers, has broken off and is lost. On the part that remains, there are columns of numbers, including the two columns shown on the next page, with four apparent errors corrected.

Activity

Solving the Puzzle

1. Work in pairs. Each person should pick two numbers at random from 50 to 5000. Use a calculator to square each number. Subtract the smaller square from the larger square. Take the square root of the difference. Is the result an integer?

2. Repeat Step 1 at least five times. How often, if ever, did you obtain an integer as the result?

3. Square the numbers in each row of the tablet at right. Subtract the smaller square from the larger square and take the square root of the difference. How often is the result an integer?

4. Do you think the Babylonians knew which sets of integers were related in this way? Explain.

119	169
3367	4825
4601	6649
12709	18541
65	97
319	481
2291	3541
799	1249
481	769
4961	8161
45	75
1679	2929
161	289
1771	3229
56	106

Columns II and III of Plimpton 322

In the Activity, you discovered that Babylonians knew about the sets of numbers now called **Pythagorean triples**—that is, sets of positive integers a, b, and c such that $a^2 + b^2 = c^2$.

It is also clear, from other evidence on the tablet, that the Babylonians knew that right triangles have the property known today as the *Pythagorean relationship*. But surprisingly, Plimpton 322 was created over a thousand years before the teacher Pythagoras (569–500 B.C.E.), the person to whom the relationship is traditionally attributed.

At present there is no direct evidence that the Babylonians could prove the Pythagorean relationship.

Proving the Relationship for Right Triangles

An ancient Chinese diagram.

CULTURAL CONNECTION: EUROPE The Pythagoreans were members of a secret society of followers of Pythagoras in ancient Greece. According to tradition, Pythagoras said, "Number rules the universe." We do not know how Pythagoras actually proved the theorem that now bears his name, or whether his particular proof—there are many others—originated with him.

CULTURAL CONNECTION: ASIA One very early illustration of the Pythagorean Theorem is found in an ancient Chinese source known as *Chou pei suan ching*. The source does not give details of a proof, but it does include the diagram shown at left, which suggests a proof of the theorem (see page 323). According to some scholars, the date of the diagram is at least as early as Pythagoras.

In outer part of the *Chou pei suan ching* diagram, four congruent right triangles form a large square with a smaller square in the center. The area of the larger square can be found by squaring the length of its sides, which are equal to $a + b$, or by adding the individual pieces that make up the figure.

By setting these two expressions equal to each other and simplifying, you obtain the famous result.

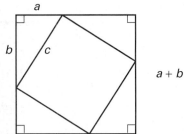

$$(a + b)^2 = 4\left(\tfrac{1}{2}ab\right) + c^2$$
$$a^2 + 2ab + b^2 = 2ab + c^2$$
$$a^2 + b^2 = c^2$$

CRITICAL THINKING Does the shape of the right triangles matter in the proof above? How do you know that the central figure is in fact a square? (See Exercise 39 for another proof of the Pythagorean Theorem based on the *Chou pei suang ching* diagram.)

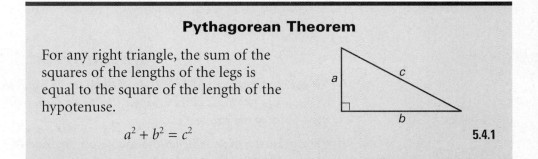

Pythagorean Theorem

For any right triangle, the sum of the squares of the lengths of the legs is equal to the square of the length of the hypotenuse.

$$a^2 + b^2 = c^2$$

5.4.1

E X A M P L E ❶

The following is a rule of thumb for safely positioning a ladder: The distance from bottom of the ladder to the wall should be one-fourth of the length of the ladder. Thus, the bottom of a 16-foot ladder should be 4 feet from the wall. How far up the wall will the ladder reach?

● **SOLUTION**
$$a^2 + b^2 = c^2$$
$$4^2 + b^2 = 16^2$$
$$16 + b^2 = 256$$
$$b^2 = 256 - 16$$
$$b^2 = 240$$
$$b = \sqrt{240} \approx 15.5 \text{ ft}$$

16 ft

4 ft

The Converse of the Theorem

The converse of the Pythagorean Theorem is also a true theorem. It is useful for proving that two segments or lines are perpendicular.

Converse of the Pythagorean Theorem

If the square of the length of one side of a triangle equals the sum of the squares of the lengths of the other two sides, then the triangle is a right triangle.

5.4.2

CULTURAL CONNECTION: AFRICA The following proof of the Converse of the Pythagorean Theorem is taken directly from Euclid's *The Elements*, Book 1, which was translated from the original Greek by Thomas Heath. You will notice that some of Euclid's notation and use of terms are different from our own. Exercises 52–64 will help you follow the proof and put it into a modern form.

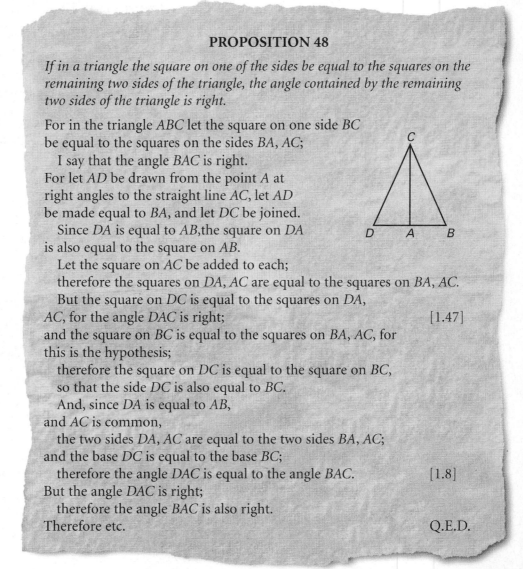

PROPOSITION 48

If in a triangle the square on one of the sides be equal to the squares on the remaining two sides of the triangle, the angle contained by the remaining two sides of the triangle is right.

For in the triangle *ABC* let the square on one side *BC* be equal to the squares on the sides *BA, AC*;
   I say that the angle *BAC* is right.
For let *AD* be drawn from the point *A* at right angles to the straight line *AC*, let *AD* be made equal to *BA*, and let *DC* be joined.
   Since *DA* is equal to *AB*, the square on *DA* is also equal to the square on *AB*.
   Let the square on *AC* be added to each;
   therefore the squares on *DA, AC* are equal to the squares on *BA, AC*.
   But the square on *DC* is equal to the squares on *DA, AC*, for the angle *DAC* is right;   [1.47]
and the square on *BC* is equal to the squares on *BA, AC*, for this is the hypothesis;
   therefore the square on *DC* is equal to the square on *BC*,
   so that the side *DC* is also equal to *BC*.
   And, since *DA* is equal to *AB*,
and *AC* is common,
   the two sides *DA, AC* are equal to the two sides *BA, AC*;
and the base *DC* is equal to the base *BC*;
   therefore the angle *DAC* is equal to the angle *BAC*.   [1.8]
But the angle *DAC* is right;
   therefore the angle *BAC* is also right.
Therefore etc.   Q.E.D.

> The bracketed numbers in the right margin refer to propositions proven earlier in The Elements.

EXAMPLE **2** A plowed field is in the shape of a triangle. If the sides have the lengths shown in the figure, is the field a right triangle?

3.7 mi

● **SOLUTION**

If the field is a right triangle, then $a^2 + b^2 = c^2$, where a, b, and c are the lengths of the sides. The longest side of a right triangle is the hypotenuse, so if the field is a right triangle, the hypotenuse, c, must be 3.7.

$$a^2 + b^2 = (1.2)^2 + (3.5)^2$$
$$= 13.69$$

Since $c = \sqrt{13.69} = 3.7$ miles, the field is a right triangle.

The following helpful inequalities can be derived from the Pythagorean relationship:

Pythagorean Inequalities

For $\triangle ABC$, with c as the length of the longest side:

If $c^2 = a^2 + b^2$, then $\triangle ABC$ is a right triangle.
If $c^2 > a^2 + b^2$, then $\triangle ABC$ is an obtuse triangle.
If $c^2 < a^2 + b^2$, then $\triangle ABC$ is an acute triangle.

5.4.3

EXAMPLE **3** A triangle has side lengths of 7 inches, 8 inches, and 12 inches. Is the triangle right, obtuse, or acute?

● **SOLUTION**

$12^2 \ \underline{?} \ 7^2 + 8^2$
$144 > 113$

Therefore, the triangle is obtuse.

TRY THIS A triangle has side lengths of 8 inches, 8 inches, and 8 inches. Is the triangle right, obtuse, or acute?

Exercises

Communicate

1. State the Pythagorean Theorem in your own words.

2. What are some practical uses of the Pythagorean Theorem and its converse?

3. Explain how something called the "3-4-5 rule" could help carpenters create square corners.

4. Explain how the Greek postage stamp shown at right illustrates the Pythagorean Theorem.

Guided Skills Practice

5. A right triangle has one leg with a length of 48 and a hypotenuse with a length of 80. What is the length of the other leg? *(EXAMPLE 1)*

6. A triangle has side lengths of 7, 10, and 12. Is the triangle a right triangle? *(EXAMPLE 2)*

7. A triangle has side lengths of 8, 15, and 18. Is the triangle right, acute, or obtuse? *(EXAMPLE 3)*

Practice and Apply

For Exercises 8–13, two side lengths of a right triangle are given. Find the missing side length. Leave your answers in radical form.

8. $a = 3, b = 4, c = \underline{?}$ 9. $a = 10, b = 15, c = \underline{?}$

10. $a = 46, b = 73, c = \underline{?}$ 11. $a = \underline{?}, b = 6, c = 8$

12. $a = 27, b = \underline{?}, c = 53$ 13. $a = 1, b = 1, c = \underline{?}$

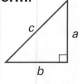

Find the perimeter of each triangle. Round your answers to the nearest tenth.

14.

15.

16.

17.

Each of the following triples represents the side lengths of a triangle. Determine whether the triangle is right, acute, or obtuse.

18. 5, 9, 12 **19.** 13, 15, 17 **20.** 7, 24, 25

21. 7, 24, 26 **22.** 3, 4, 5 **23.** 25, 25, 30

Find the area of each figure. Leave your answers in radical form.

24.

25.

26. What is the length of a diagonal of a square with a side length of 5?

27. What is the side length of a square with a diagonal of 16?

Find the area of the shaded region in each figure. Round your answers to the nearest tenth.

28. **29.** **30.**

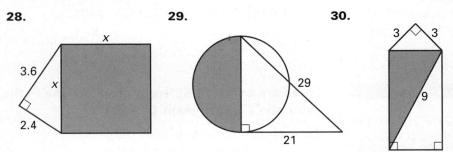

NUMBER THEORY Mathematicians have long been fascinated with techniques for generating Pythagorean triples. For each method below, generate five sets of triples. Then use algebra to show that the method will always work.

31. Method of Pythagoreans
Let m be any odd number greater than 1.

32. Method of Plato
Let n be any integer greater than 1.

33. NUMBER THEORY Test the following conjecture for 10 different Pythagorean triples: In each Pythagorean triple, at least one of the numbers is divisible by 3, and at least one is divisible by 5.

34. NUMBER THEORY If you multiply each number of a Pythagorean triple by the same constant, you get another Pythagorean triple. For example, (3, 4, 5) is a Pythagorean triple. Multiplying each number by 2 results in (6, 8, 10), another Pythagorean triple.

Use algebra to show that if (x, y, z) is a Pythagorean triple, then (ax, ay, az) is also a Pythagorean triple for any positive integer a.

An altar dedicated to the god Vishnu, whose worship is derived from the early Vedic scriptures

35. CULTURAL CONNECTION: ASIA

The *Sulbasutras* were ancient Indian mathematical manuals for the design and construction of Vedic altars.

The diagram at right, from the *Sulbasutras*, demonstrates a method for constructing a square with an area equal to the sum of the areas of two given squares. Use the Pythagorean Theorem to prove that this construction works.

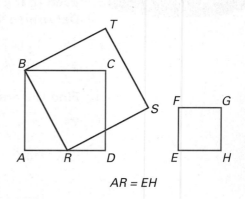

AR = EH

Given: *ABCD*, *EFGH*, and *BRST* are squares.
AR = EH

Prove: area of *BRST* = area of *ABCD* + area of *EFGH*

36. Explain how to use the method from the *Sulbasutras* to construct a square with twice the area of a given square.

James Garfield (1831–1881)

There are over 100 different proofs of the Pythagorean Theorem. Exercises 37–39 present three of them.

37. President James Garfield (1831–1881), a president of the United States, devised a proof of the Pythagorean Theorem that uses a trapezoid.

Using the figure at right, write two different expressions for the area of the trapezoid. Set the two expressions equal to each other to discover Garfield's proof.

38. Explain how the diagram below is a "proof without words" of the Pythagorean Theorem. You may wish to draw the figures on a separate piece of paper and cut them into pieces to help explain the proof.

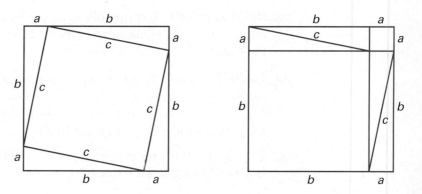

39. The diagram from the *Chou pei suan* suggests at least two different proofs of the Pythagorean Theorem.

a. Write an expression for c^2, the area of square *EFGH*, by adding the areas of the four inner right triangles and square *IJKL*.

$$c^2 = \underline{\ ?\ }$$

b. Write an expression for c^2, the area of square *EFGH*, by subtracting the areas of the four outer right triangles from the area of square *ABCD*.

$$c^2 = \underline{\ ?\ }$$

Simplify each equation to prove the Pythagorean Theorem.

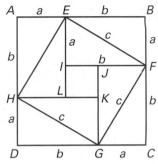

40. PUBLIC SAFETY If the base of the ladder in the photo at right is 8 feet off the ground, how far up the wall will the ladder reach?

41. TRAVEL Starting from his house, Jesse drives north 6 miles and then turns east and drives 2 miles. He turns north again and drives 4 miles and then turns east and drives 7 miles. How far is he from his house? (Hint: You may wish to draw a map.)

42. SPORTS A baseball diamond is a square with 90-foot sides. What is the approximate distance of a catcher's throw from home plate to second base?

2nd base

Catcher (home plate)

Look Back

Algebra

Recall from algebra that a fraction with a radical in the denominator can be simplified by multiplying by a fraction equal to 1. This technique is known as *rationalizing the denominator*. For example:

$$\frac{5}{\sqrt{2}} = \frac{5}{\sqrt{2}} \times \frac{\sqrt{2}}{\sqrt{2}} = \frac{5\sqrt{2}}{\sqrt{2} \times \sqrt{2}} = \frac{5\sqrt{2}}{2}$$

Rationalize the denominator in each expression below.

43. $\dfrac{4}{\sqrt{2}}$ **44.** $\dfrac{13}{\sqrt{3}}$ **45.** $\dfrac{\sqrt{7}}{\sqrt{5}}$

For Exercises 47–48, classify each statement as true or false. If true, explain why. If false, give a counterexample. *(LESSON 4.6)*

46. Every rhombus is a rectangle.

47. Every rhombus is a parallelogram.

48. If a diagonal divides a quadrilateral into two congruent triangles, then the quadrilateral is a parallelogram.

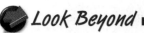
In Exercises 49–64, you will examine Euclid's proof of the Converse of the Pythagorean Theorem. Read through Euclid's proof on page 324.

49. What is the hypothesis of Proposition 48? What is the conclusion?

50. Draw a triangle and label the vertices A, B, and C. Use Euclid's directions to construct the triangle $\triangle DAC$ as in the proof.

51. In your diagram for Exercise 50, can you be sure that the figure connecting D to B is really a single segment? (Hint: Don't assume what you are trying to prove.)

FLOWCHART PROOF

Complete the flowchart proof based on Euclid's proof of the Converse of the Pythagorean Theorem.

Given: $(BA)^2 + (AC)^2 = (BC)^2$

Prove: $\triangle ABC$ is right.

Proof:

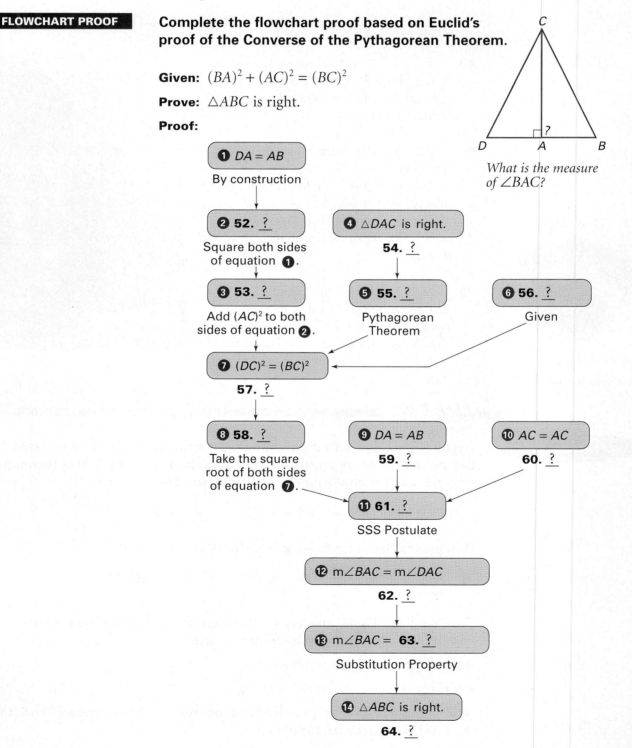

What is the measure of $\angle BAC$?

1 $DA = AB$
By construction

2 52. ?
Square both sides of equation **1**.

3 53. ?
Add $(AC)^2$ to both sides of equation **2**.

4 $\triangle DAC$ is right.
54. ?

5 55. ?
Pythagorean Theorem

6 56. ?
Given

7 $(DC)^2 = (BC)^2$
57. ?

8 58. ?
Take the square root of both sides of equation **7**.

9 $DA = AB$
59. ?

10 $AC = AC$
60. ?

11 61. ?
SSS Postulate

12 $m\angle BAC = m\angle DAC$
62. ?

13 $m\angle BAC =$ **63.** ?
Substitution Property

14 $\triangle ABC$ is right.
64. ?

Special Triangles and Areas of Regular Polygons

Objectives

● Identify and use the 45-45-90 Triangle Theorem and the 30-60-90 Triangle Theorem.

● Identify and use the formula for the area of a regular polygon.

Why *The traditional tools of mechanical drawing include a T square and two special triangles. One of the triangles has angles measuring 30°, 60°, and 90°. The other has angles measuring 45°, 45°, and 90°. The properties of these triangles make them especially useful in geometry as well as in drawing.*

This draftsman uses her T square to establish an imaginary line parallel to the base of her page. Then, using a special triangle, she draws a segment at a 30° angle to this line.

45-45-90 Triangles

If you draw a diagonal of a square, two congruent isosceles triangles are formed. Because the diagonal is the hypotenuse of a right triangle, its length can be found by using the Pythagorean Theorem.

E X A M P L E ❶ Use the Pythagorean Theorem to find the length of the hypotenuse of $\triangle ABC$. What is the ratio of the hypotenuse to a leg?

Algebra

SOLUTION

$$h^2 = 10^2 + 10^2 = 200$$
$$h = \sqrt{200} = \sqrt{100} \times \sqrt{2} = 10\sqrt{2}$$

The ratio of the hypotenuse to a leg is $\frac{10\sqrt{2}}{10}$, or $\sqrt{2}$.

CRITICAL THINKING   What is the length of a diagonal of a square with a side of length s? What is the ratio of the diagonal to the side?

Notice that a diagonal of a square forms a right triangle with two 45° base angles. This triangle is known as a **45-45-90 triangle**. Because this is a right triangle, the hypotenuse can be found by applying the Pythagorean Theorem.

45-45-90 Triangle Theorem

In any 45-45-90 triangle, the length of the hypotenuse is $\sqrt{2}$ times the length of a leg.

5.5.1

30-60-90 Triangles

If you draw an altitude of an equilateral triangle, two congruent right triangles are formed. The measures of the acute angles of each right triangle are 30° and 60°. This triangle is known as a **30-60-90 triangle**. The length of the hypotenuse of a 30-60-90 triangle is twice the length of the shorter leg.

EXAMPLE **2** Find the unknown lengths for the 30-60-90 triangle shown at right.

● **SOLUTION**

In a 30-60-90 triangle, the length of the hypotenuse is twice the length of the shorter leg. Thus, the length of the hypotenuse is 10.

Algebra

Use the Pythagorean Theorem to find the length of the other leg.

$$5^2 + x^2 = 10^2$$
$$25 + x^2 = 100$$
$$x^2 = 100 - 25$$
$$x^2 = 75$$
$$x = \sqrt{75} = 5\sqrt{3} \approx 8.66$$

30-60-90 Triangles

YOU WILL NEED

no special tools

1. Use the Pythagorean Theorem to fill in a table like the one below for several 30-60-90 triangles. Write your answers in simplest radical form.

Shorter leg	Hypotenuse	Longer leg
1	2	?
2	?	?
3	?	?

2. Look for a pattern in the lengths of the longer leg and make a generalization.

3. Let x be the length of the shorter leg.

   a. What is the length of the hypotenuse?

   b. What is the length of the longer leg?

CHECKPOINT ✔   **4.** Use your results to label a general 30-60-90 triangle like the one at right. Then complete the theorem below.

30-60-90 Triangle Theorem

In any 30-60-90 triangle, the length of the hypotenuse is ___?___ times the length of the shorter leg, and the length of the longer leg is ___?___ times the length of the shorter leg.

5.5.2

E X A M P L E ❸ Jake is measuring the height of a tree that his grandfather planted as a boy. Jake uses a special instrument to find a spot where a 30° angle is formed by his line of sight and a ray parallel to the ground. Jake's eye level is 5 feet above the ground. How tall is the tree if Jake is standing 80 feet from the base?

SOLUTION

The line of sight to the tree is the hypotenuse of a right triangle. The length of the longer leg of the triangle is 80 feet. The length, x, of the shorter leg of the triangle is equal to the height of the tree above Jake's eye level.

$$80 = x\sqrt{3}$$

$$x = \frac{80}{\sqrt{3}} = \frac{80}{\sqrt{3}} \times \frac{\sqrt{3}}{\sqrt{3}} = \frac{80\sqrt{3}}{3}, \text{ or about } 46.2 \text{ feet}$$

Now find the height, h, of the tree.

$$h \approx 46.2 + 5 = 51.2, \text{ or about } 51 \text{ feet}$$

Areas of Regular Polygons

To find the area of a regular hexagon, divide the hexagon into 6 congruent, non-overlapping equilateral triangles. Find the area of one triangle and multiply by 6 to find the area of the hexagon. Note that the altitude of the equilateral triangle is the longer leg of a 30-60-90 triangle. The altitude is $\frac{1}{2}$ of the length of the side of the hexagon multiplied by $\sqrt{3}$.

Altitude

EXAMPLE **4** Find the area of a regular hexagon with sides of 20 centimeters.

● **SOLUTION**

Divide the hexagon into 6 equilateral triangles. Because the altitude of one of the triangles forms the longer leg of a 30-60-90 triangle and half of the side of the hexagon forms the shorter leg, the length of the altitude is $10\sqrt{3}$ centimeters.

20 cm

Use the area formula to find the area of one of the triangles.

$$A = \frac{1}{2}(20)(10\sqrt{3})$$
$$A = 100\sqrt{3}$$

10 cm — Altitude = $10\sqrt{3}$ cm

20 cm

Because the hexagon is composed of 6 congruent triangles, the area of the hexagon is found as follows:

$$A = 6(100\sqrt{3}) = 600\sqrt{3} \approx 1039 \text{ centimeters}$$

The method of dividing a regular hexagon into congruent triangles can be applied to find the area of any regular polygon.

An n-sided regular polygon can be divided into n non-overlapping congruent triangles. Each altitude of a triangle from the center of the polygon to a side of the polygon is called an **apothem** of the polygon.

Apothem

The Canadian dollar coin, or "loonie," is a regular 11-gon.

Area of a Regular Polygon

The area, A, of a regular polygon with apothem a and perimeter p is given by:

$$A = \frac{1}{2}ap$$

5.5.3

EXAMPLE **5** Find the area of the regular pentagon shown at right.

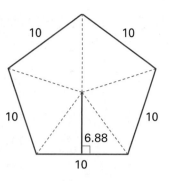

SOLUTION

The perimeter, p, is 5×10, or 50 units.

$$A = \frac{1}{2}ap$$

$$A = \frac{1}{2}(6.88)(50) = 172 \text{ square units}$$

Exercises

● Communicate

1. Describe how two 45-45-90 triangles fit together to form a square. Describe how four 45-45-90 triangles fit together to form a square.

2. An equilateral triangle is a regular 3-sided polygon. Describe its apothem.

3. A square is a regular 4-sided polygon. Describe its apothem.

4. Can the lengths of the sides of a 45-45-90 triangle or a 30-60-90 triangle ever be a Pythagorean triple? Why or why not?

● Guided Skills Practice

5. Find the length of the hypotenuse of △RST. **(EXAMPLE 1)**

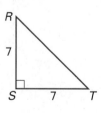

6. Find the missing lengths for △WXY. **(EXAMPLE 2)**

7. How tall is the tree in the diagram at left? **(EXAMPLE 3)**

8. Find the area of a regular hexagon with sides of 12 inches. **(EXAMPLE 4)**

9. Find the area of the regular decagon at right. **(EXAMPLE 5)**

LESSON 5.5 SPECIAL TRIANGLES AND AREAS OF REGULAR POLYGONS **335**

Practice and Apply

For each length given below, find the
remaining two lengths. Give your
answers in simplest radical form.

10. $x = 6$

11. $y = 6$

12. $z = 14$

13. $y = 4\sqrt{3}$

14. $p = 6$

15. $r = 6$

16. $q = 4\sqrt{2}$

17. $r = 10$

18. $k = 3.4$

19. $k = 6\sqrt{3}$

20. $g = 17$

21. $h = 2\sqrt{3}$

Find the area of each figure. Round your answers to the nearest tenth.

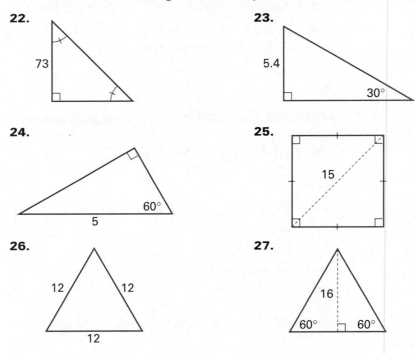

22.

73

23.

5.4

30°

24.

60°

5

25.

15

26.

12   12

12

27.

16

60°   60°

Each of the following triples represents the side lengths of a triangle.
Determine whether the triangle is a 45-45-90 triangle, a 30-60-90
triangle, or neither. Explain your reasoning.

28. $4, 4, 4\sqrt{2}$

29. $6, 3\sqrt{2}, 3\sqrt{2}$

30. $3, \sqrt{3}, 2\sqrt{3}$

31. $2, 1, 2\sqrt{3}$

32. $5, 5\sqrt{2}, 5\sqrt{2}$

33. $\sqrt{2}, 2\sqrt{2}, \sqrt{6}$

For Exercises 34–39, find the perimeter and area of each figure. Give your answers in simplest radical form.

34. a 30-60-90 triangle with a hypotenuse of 18

35. a 45-45-90 triangle with a hypotenuse of 24

36. an equilateral triangle with sides of length 8

37. a regular hexagon with sides of length 13

38. a square with a diagonal of 14

39. a regular hexagon with an apothem of 5

CHALLENGES

For Exercises 40–43, refer to the figure at left of a regular octagon inside a square.

40. From the diagram at left, what can you say about the triangles formed by the corners of the square, outside of the octagon? Find the value of x.

41. Find the length of the apothem of the octagon.

42. Find the area of the octagon.

43. If the sides of a regular octagon have length s, what is the area of the octagon?

APPLICATION

CIVIL ENGINEERING An engineer is in charge of attaching guy wires to a tower. One wire attaches to a point on the ground 60 feet from the tower.

44. If the angle that the wire forms with the tower is 45°, what is the length of the wire?

45. If the angle that the wire forms with the tower is 60°, what is the length of the wire?

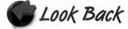 *Look Back*

APPLICATION

DRAFTING In drafting, a T square is used to align drawings. The T square is held against the side of the drawing board, and the straightedge stays perpendicular to the side of the drawing board. *(LESSON 3.4)*

46. An artist makes a series of horizontal lines with the straightedge of the T square by moving it up 1 inch for each new line. What is the relationship between these lines? Why?

47. With the 30-60-90 triangle positioned as shown at right, the artist uses the hypotenuse to draw a line that forms an angle of 60° with the straightedge. Without moving the T square, she slides the triangle 1 inch to the right and draws another line the same way. What is the relationship between the two lines? Why?

48. Find the smallest possible area for a rectangle with a perimeter of 100 and side lengths that are whole numbers. Then find the greatest possible area. State the dimensions of each rectangle. *(LESSON 5.1)*

49. Find the area of the triangles in the figure at right. *(LESSON 5.2)*

50. Find the height of a trapezoid with an area of 103.5 ft² and base lengths of 17.5 ft and 5.5 ft. *(LESSON 5.2)*

51. The base of an isosceles triangle is 8 m and the legs are 6 m. Find the area and perimeter. *(LESSONS 5.2 AND 5.4)*

52. In △PQR, find PR. *(LESSON 5.4)*

53. Find the area and perimeter of a right triangle with legs of lengths *a* and *b*. *(LESSONS 5.2 AND 5.4)*

Look Beyond

> *Be sure your calculator is in degree mode.*

Refer to the diagram at right. The ratio $\frac{y}{x}$ of the lengths of the legs of a right triangle is called the *tangent* of ∠A. The [TAN] button on your calculator is used to calculate tangent ratios.

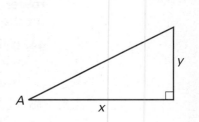

54. Use your calculator to find the tangent of 45°. Does your answer agree with the ratio of the corresponding sides of a 45-45-90 triangle? Why or why not?

55. Use your calculator to find the tangent of 30°. Does your answer agree with the ratio of the corresponding sides of a 30-60-90 triangle? Why or why not?

56. In the regular pentagon at right, with its center at *A,* what is the measure of ∠A?

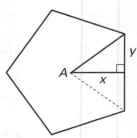

57. Use your calculator to find the tangent of ∠A.

58. If the side length of the regular pentagon is 12, what is *y*?

59. Use your results from Exercises 57 and 58, and the fact that the tangent of ∠A is $\frac{y}{x}$ to find *x*.

60. Use your results from Exercises 56–59 to find the area of a regular pentagon with a side length of 12. (Hint: What is the apothem?)

61. Use the method above to find the area of a regular 9-gon with a side length of 4.

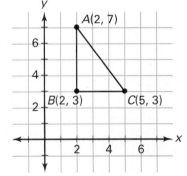

The Distance Formula and the Method of Quadrature

5.6

Objectives

- Develop and apply the distance formula.

- Use the distance formula to develop techiques for estimating the area under a curve.

Why *The ability to compute the distance between two points is important in many situations. In some cases, an estimate based on measurement is sufficient. In other cases, greater precision is necessary.*

To reach his destination as quickly as possible, a helicopter pilot travels the shortest distance between two points.

The Distance Formula

The distance between two points on the same horizontal or vertical line can be found by taking the difference of the *x*- or *y*-coordinates.

The vertical distance between points A and B is $AB = |7-3| = |3-7| = 4$.

The horizontal distance between points B and C is $BC = |5-2| = |2-5| = 3$.

Because AC is the hypotenuse of right triangle $\triangle ABC$, its length can be found by using the Pythagorean Theorem.

$$(AC)^2 = (AB)^2 + (BC)^2$$
$$= 4^2 + 3^2$$
$$= 16 + 9$$
$$= 25$$
$$AC = \sqrt{25} = 5$$

The Pythagorean Theorem can be used to find the distance between any two points in the coordinate plane.

Distance Formula

In a coordinate plane, the distance, d, between two points (x_1, y_1) and (x_2, y_2) is given by the following formula:

$$d = \sqrt{(x_2 - x_1)^2 + (y_2 - y_1)^2}$$

5.6.1

A proof of the distance formula is given below.

PROOF

Given: points $A(x_1, y_1)$ and $B(x_2, y_2)$

Prove: The distance between the two points is $\sqrt{(x_2 - x_1)^2 + (y_2 - y_1)^2}$.

Proof: Draw a right triangle with hypotenuse $\overline{AB}$ and a right angle at point $C(x_2, y_1)$. Let d be the distance between points A and B.

$$AC = |x_2 - x_1|$$

$$BC = |y_2 - y_1|$$

Use the Pythagorean Theorem.

$$d^2 = |x_2 - x_1|^2 + |y_2 - y_1|^2$$
$$d^2 = (x_2 - x_1)^2 + (y_2 - y_1)^2$$

You can drop the absolute-value symbols because the quantities are being squared.

Solve for d.

$$d = \sqrt{(x_2 - x_1)^2 + (y_2 - y_1)^2}$$

EXAMPLE

A helicopter pilot located 1 mile east and 3 miles north of the command center must respond to an emergency located 7 miles east and 11 miles north of the the center. How far must the helicopter travel to get to the emergency site?

SOLUTION

Let $(0, 0)$ be the coordinates of the command center, with the positive part of the y-axis representing north. Then the helicopter's current coordinates are $(1, 3)$. The coordinates of the emergency site are $(7, 11)$. Use the distance formula.

$$d = \sqrt{(7 - 1)^2 + (11 - 3)^2}$$
$$d = \sqrt{6^2 + 8^2} = \sqrt{36 + 64} = \sqrt{100} = 10$$

The helicopter must travel 10 miles to get to the emergency site.

The Method of Quadrature

The area of an enclosed region on a plane can be approximated by the sum of the areas of a number of rectangles. This technique, called **quadrature**, is particularly important for finding the area under a curve.

Activity

Estimating the Area of a Circle

YOU WILL NEED

graph paper and calculator

Part I: Method A (Left-Hand Rule)

1. Draw a quarter of a circle with a radius of 5 units centered at the origin, $(0, 0)$. Draw rectangles as shown, with the upper left vertex of each rectangle touching the curve. This method is called the *left-hand rule*.

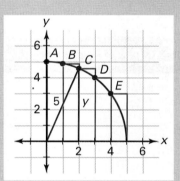

2. Find the y-coordinate of each point, A, B, C, D, and E. The segment connecting each point to the origin is the hypotenuse of a right triangle. Because the radius of the circle is 5 units, this is the length of the hypotenuse of each triangle. For point C, the y-coordinate is found as shown below.

$$y = \sqrt{5^2 - 2^2} = \sqrt{21}$$

The y-coordinate of each point is the height of a rectangle.

3. Find the area of each rectangle. (This is simplified by the fact that the base of each rectangle is 1.)

4. Find the sum of the areas of the rectangles by completing the pattern below.

$$\sqrt{5^2 - 0^2} + \sqrt{5^2 - 1^2} + \sqrt{5^2 - 2^2} + \cdots =$$
$$\sqrt{25} + \sqrt{24} + \sqrt{21} + \cdots \approx \underline{?}$$

CHECKPOINT ✔

5. Multiply your sum by 4. The result is an estimate of the area of a complete circle with a radius of 5 units. Does this method overestimate or underestimate the area of the circle? Explain why.

6. Calculate the true value of the area of a complete circle to four decimal places by using $A = \pi r^2$. Find the relative error of your estimate by using the following formula:

$$E = \frac{|V_e - V_t|}{V_t} \times 100$$

where V_e = estimated value, V_t = true value, and E = percent of error

Part II: Method B (Right-Hand Rule)

CHECKPOINT ✔ Repeat Part I with the rectangles arranged as shown at right. Here, the upper right vertex of each rectangle is touching the curve. This method is called the *right-hand rule*. Does this new method overestimate or underestimate the area of a complete circle? Explain why.

Part III: Combining Methods

CHECKPOINT ✔ Average your estimates for the area of the circle from Parts I and II. What is the relative error of your new estimate?

CRITICAL THINKING Do you think that the average of the results from methods A and B will always give more accurate results than either method by itself? Explain your answer.

Exercises

Communicate

1. Explain the relationship between the distance formula and the Pythagorean Theorem.

2. When using the distance formula, does it matter which point is (x_1, y_1) and which point is (x_2, y_2)? Why or why not?

3. Describe two methods of estimating the shaded area under the curve at right.

4. For each method you described in Exercise 3, how might you make your estimate of the area more accurate?

Guided Skills Practice

Find the distance between the indicated points. *(EXAMPLE)*

5.

6.

For Exercises 7 and 8, estimate the area of the quarter-circle with a radius of 4 by finding the areas of the given rectangles. (ACTIVITY)

7.

8.

Practice and Apply

Find the distance between each pair of points. Round your answers to the nearest hundredth.

9. $(0, 0)$ and $(5, 8)$

10. $(1, 2)$ and $(4, 6)$

11. $(1, 4)$ and $(3, 9)$

12. $(-3, -3)$ and $(6, 12)$

13. $(-1, 4)$ and $(-6, 16)$

14. $(-2, -3)$ and $(-6, -12)$

Refer to the diagram below for Exercises 15–23.

15. $AB = $?

16. $BC = $?

17. $GJ = $?

18. Find the perimeter of $\triangle EJH$.

19. Find the perimeter of $\triangle ABF$.

20. Find the perimeter of quadrilateral $FGHI$.

21. Is $\triangle BCD$ isosceles? Why or why not?

22. Is $\triangle AJE$ isosceles? Why or why not?

23. Is $\triangle BEH$ equilateral? Why or why not?

For Exercises 24–26, use the converse of the Pythagorean Theorem to determine whether the triangle with the given vertices is a right triangle. You may wish to plot the points and connect them to form the triangle.

24. $(2, 1)$, $(6, 4)$, and $(-4, 9)$

25. $(1, 5)$, $(6, 0)$, and $(-2, 2)$

26. $(1, 4)$, $(4, 2)$, and $(6, 6)$

For Exercises 27–30, give your answers in simplest radical form.

27. Suppose that the endpoints of the hypotenuse of a 45-45-90 triangle are $(3, 3)$ and $(9, 2)$. What is the length of the legs?

28. Suppose that the endpoints of one leg of a 45-45-90 triangle are $(0, 3)$ and $(4, -1)$. What is the length of the hypotenuse?

29. Suppose that the endpoints of the hypotenuse of a 30-60-90 triangle are $(4, -2)$ and $(7, 2)$. What is the length of the shorter leg? of the longer leg?

30. Suppose that the endpoints of the longer leg of a 30-60-90 triangle are $(-3, 5)$ and $(2, 1)$. What is the length of the shorter leg? of the hypotenuse?

In Exercises 31–34, you will use coordinate geometry to explore a property of right triangles.

31. Find the midpoint of the hypotenuse of the right triangle with vertices at $(0, 0)$, $(6, 0)$, and $(0, 8)$. What is the distance from this midpoint to each vertex? What do you notice?

32. Repeat Exercise 31, using a right triangle with vertices at $(0, 0)$, $(4, 0)$, and $(0, 7)$.

33. Based on your results from Exercises 31 and 32, make a conjecture about the distance from the midpoint of the hypotenuse of a right triangle to each vertex of the triangle.

PROOF

34. Prove your conjecture from Exercise 33 by using a right triangle with vertices at $(0, 0)$, $(x, 0)$ and $(0, y)$.

35. Use the method of quadrature to estimate the shaded area under the curve shown below. Use both the left-hand and right-hand rules, and then average your results.

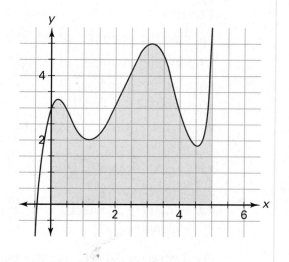

Algebra

36. Use the method of quadrature to estimate the area between $y = x^2$ and the x-axis for $0 \le x \le 4$.

37. Use the method of quadrature to estimate the area between $y = x^2 + 2$ and the x-axis for $0 \le x \le 2$.

38. Use the method of quadrature to estimate the area between $y = -x^2 + 4$ and the x-axis for $-2 \le x \le 2$.

39. ENVIRONMENTAL PROTECTION
An oil pipeline runs through part of a national forest. Because of the danger to the forest, an enviromental group wants to know the length of the pipeline that lies inside the boundaries of the forest. Use the map at right to find the desired length. Each square of the grid represents 1 square mile.

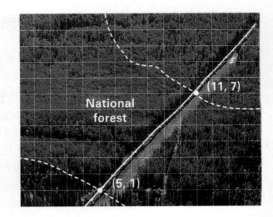

40. CIVIL ENGINEERING A dam has been built on a river, and the river has begun to back up to form a lake as shown in the "map" at right. Each square on the map represents 10,000 sq ft. Use the method of quadrature to estimate the area of the lake.

A lake formed by a dam, viewed from above, will usually have a flat side. Explain why. This makes it convenient to use the method of quadrature. What else must be true about the shape of the lake in order to use the method of quadrature as explained in this lesson?

Look Back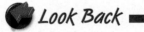

41. Prove that the quadrilateral with vertices at $A(-1, 1)$, $B(7, -5)$, $C(10, -1)$, and $D(2, 5)$ is a rectangle. *(LESSON 3.8)*

42. Suppose that a circle and a square both have a perimeter of 100 meters. Which has the larger area? *(LESSON 5.3)*

43. Suppose that a circle and a square both have an area of 225 square centimeters. Which has the larger perimeter? *(LESSON 5.3)*

44. The legs of a right triangle are 4.5 centimeters and 8 centimeters. What is the length of the hypotenuse? *(LESSON 5.4)*

45. The hypotenuse of a right triangle is 5 centimeters and one of the legs is 4 centimeters. What is the length of the other leg? *(LESSON 5.4)*

46. Each side of an equilateral triangle measures 8 inches. Find the area and perimeter. *(LESSON 5.5)*

47. Each side of an equilateral triangle measures s units. Find the area and perimeter. *(LESSON 5.5)*

The area of an enclosed region can be approximated by using shapes other than rectangles. In the figure below, the area of the quarter-circle is approximated by the areas of four trapezoids and one triangle.

48. The quarter-circle in the figure has a radius of 5 units. Use the *y*-values you found in the Activity on page 341 as the bases of the trapezoids and use 1 as the height to estimate the area under the curve.

49. Is your answer an overestimate or underestimate of the actual area under the curve? Why?

50. Compare your answer to Exercise 48 to your answer to Part III of the Activity. What do you notice?

51. The area found by using trapezoids is equal to the average of the estimates found by using the left-hand and right-hand rules.

Use the diagram at right and the formulas for the area of a rectangle and the area of a trapezoid to prove this fact algebraically.

The figure below contains a 45-45-90 triangle. Three semicircles have been constructed, with their centers at the midpoint of the hypotenuse and the midpoint of each leg. The shaded crescents in the figure are called *lunes*.

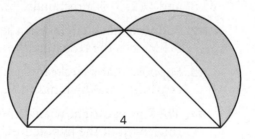

1. Find the area of the 45-90-45 triangle.
2. Find the total area of the small semicircles.
3. Find the area of the large semicircle.
4. Find the total area of the lunes.
5. What other area in the figure is equal to the total area of the lunes?

Proofs Using Coordinate Geometry

Why
The essential attributes of geometric figures can be captured in a coordinate drawing. For this reason, coordinate geometry can be used to prove geometry theorems.

Objectives

- Develop coordinate proofs for the Triangle Midsegment Theorem, the diagonals of a parallelogram, and the reflection of a point across the line $y = x$.

- Use the concepts of coordinate proofs to solve problems on the coordinate plane.

APPLICATION
METEOROLOGY

Meteorologists use latitude and longitude readings to track hurricanes on a map of Earth. Near the equator, these coordinates are very similar to xy-coordinates.

Three Coordinate Proofs

YOU WILL NEED

graph paper

Activity 1

The Triangle Midsegment Theorem

CHECKPOINT ✔

1. Use the coordinates of the triangles given in the table to test the following theorem: The midsegment of a triangle is parallel to a side of the triangle and has a measure equal to half the measure of that side. (Theorem 4.6.9) Draw the first two triangles in the coordinate plane, and complete the table below.

Vertices of triangle	Coordinates of midpoint M and S		Slope		Length	
	$\overline{AB}$	$\overline{BC}$	$\overline{MS}$	$\overline{AC}$	$\overline{MS}$	$\overline{AC}$
$A(0, 0)$, $B(2, 6)$, $C(8, 0)$	$M(?, ?)$	$S(?, ?)$	?	?	?	?
$A(0, 0)$, $B(6, -8)$, $C(10, 0)$	$M(?, ?)$	$S(?, ?)$	?	?	?	?
General case (see Steps 3 & 4)	$M(?, ?)$	$S(?, ?)$	?	?	?	?

2. In each triangle you drew in Step 1, one vertex was at the origin and another was on the *x*-axis. These vertices were chosen to keep the calculations simple. Do you think it is possible to create a triangle of any shape and size that fits these same conditions? Explain your answer.

3. The figure at right represents the general case of the Triangle Midsegment Theorem. The triangle could be of any shape and size, and it could be in any quadrant of the coordinate plane—so don't take the drawing too literally.

The number 2 has been used as a multiplier to keep the calculations simple. Notice that the numbers 2*p*, 2*q*, and 2*r* can represent any desired real number. Explain why this is true.

CHECKPOINT ✔ **4.** Fill in the third row of the table on the previous page. Based on this information, what have you proven about the relationship of $\overline{MS}$ and $\overline{AC}$? Explain why this proves the Triangle Midsegment Theorem.

PROOF

CRITICAL THINKING   Compare this proof of the Triangle Midsegment Theorem with the one that appears in Lesson 4.6. Which proof seems easier to you? Explain why.

Activity 2
The Diagonals of a Parallelogram

YOU WILL NEED

graph paper

1. Use the coordinates of the parallelograms given in the table below to test the following theorem: The diagonals of a parallelogram bisect each other. (Theorem 4.5.4) Draw the first figure in the coordinate plane. Three vertices of a parallelogram are given. Find the fourth vertex and fill in the blanks of the table.

Three vertices of a parallelogram	Fourth vertex	Midpoint of $\overline{BD}$	Midpoint of $\overline{AC}$
A(0, 0), *B*(2, 6), *D*(10, 0)	*C*(?, ?)	(?, ?)	(?, ?)
General case (see Step 2)	*C*(?, ?)	(?, ?)	(?, ?)

CHECKPOINT ✔ **2.** The figure at right represents the general case of the theorem. Use the coordinates of the figure to fill in the blanks of the table. Based on this information, what have you proven about the diagonals of a parallelogram? Explain your answer.

PROOF

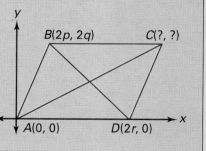

CRITICAL THINKING   How can you prove this theorem without using coordinate geometry? Which proof seems easier to you? Explain why.

Activity 3

Reflection Across the Line $y = x$

You may recall that the effect of reversing the x- and y-coordinates of a point is to reflect the line across the line $y = x$. You can use coordinate geometry to prove this result.

1. If you know the x-coordinate of a point on the line $y = x$, what can you conclude about the y-coordinate? (Filling in a table like the one at right will reveal the pattern.)

2. Find the midpoint between the points $P_1(a, b)$ and $P_2(b, a)$ by using the midpoint formula. Does this point lie on the line $y = x$? Explain your reasoning.

3. Pick two points on the line $y = x$ and use them to find the slope of this line. Record your result.

x	y
1	?
2	?
0	?
−1	?
n	?

4. Find the slope of the line that passes through the points $P_1(a, b)$ and $P_2(b, a)$. Record your result. Hint: You can rewrite $b - a$ as $-(a - b)$.

5. Compare your results in Steps 3 and 4. What can you conclude about the relationship between the two lines?

CHECKPOINT ✔

6. Recall the definition of a reflection from Lesson 1.6. Explain how your results prove that the result of reversing the coordinates of a point is a reflection of the point across the line $y = x$.

Exercises

● *Communicate*

1. In coordinate geometry proofs, why is it possible to place one vertex of the figure at the origin and one side along the x-axis?

2. In a coordinate geometry proof about triangles, could you place one vertex at the origin, one on the x-axis, and one on the y-axis? Why or why not?

3. In a coordinate geometry proof, what do you need to show to prove that two lines are parallel? that two lines are perpendicular?

4. In a coordinate geometry proof, what do you need to show to prove that two segments bisect each other?

Guided Skills Practice

Find the length of the segment with the given endpoints. *(ACTIVITY 1)*

5. $(0, 0)$ and (p, q)

6. (p, q) and (r, s)

Find the midpoint of the segment with the given endpoints.
(ACTIVITIES 1 AND 2)

7. $(0, 0)$ and $(2p, 2q)$

8. $(2p, 2q)$ and $(2r, 2s)$

Find the slope of the segment with the given endpoints. *(ACTIVITY 3)*

9. $(0, 0)$ and (p, q)

10. (p, q) and (r, s)

Practice and Apply

Determine the coordinates of the unknown vertex or vertices of each figure below. Use variables to represent any coordinates that are not completely determined by the given vertices.

11. rectangle $ABCD$
$A(0, 0)$, $B(0, p)$,
$C(\underline{\ ?\ }, \underline{\ ?\ })$, $D(q, 0)$

12. isosceles triangle DEF
$D(0, 0)$, $E(p, q)$, $F(\underline{\ ?\ }, \underline{\ ?\ })$

13. parallelogram $GHIJ$
$G(0, 0)$, $H(p, q)$,
$I(\underline{\ ?\ }, \underline{\ ?\ })$, $J(r, 0)$

14. square $KLMN$
$K(0, 0)$, $L(\underline{\ ?\ }, \underline{\ ?\ })$,
$M(\underline{\ ?\ }, \underline{\ ?\ })$, $N(p, 0)$

15. trapezoid $PQRS$
$P(0, 0)$, $Q(p, q)$,
$R(\underline{\ ?\ }, \underline{\ ?\ })$, $S(\underline{\ ?\ }, \underline{\ ?\ })$

16. rhombus $TUVW$
$T(0, 0)$, $U(p, q)$,
$V(\underline{\ ?\ }, \underline{\ ?\ })$, $W(\underline{\ ?\ }, \underline{\ ?\ })$

In Exercises 17–20, you will prove the Trapezoid Midsegment Theorem by using coordinate geometry.

17. Trapezoid *ABCD* at right has vertices at *A*(0, 0), *B*(2*p*, 2*q*), *C*(2*r*, __?__), and *D*(2*s*, __?__).

18. Find the coordinates of *M*, the midpoint of $\overline{AB}$, and *N*, the midpoint of $\overline{CD}$.

19. Find the lengths *AD*, *BC*, and *MN*. Prove that the length of the midsegment is the average of the lengths of the bases.

20. Find the slopes of $\overline{AD}$, $\overline{BC}$, and $\overline{MN}$. Prove that the midsegment is parallel to the bases.

In Exercises 21–23, you will prove that the center of the circumscribed circle for a right triangle is the midpoint of the hypotenuse.

21. Find the midpoint, *M*, of the hypotenuse of △*JKL* at right.

22. Show that the midpoint of the hypotenuse is equidistant from the three vertices.

23. Explain why point *M* must be the center of the circumscribed circle for △*JKL*.

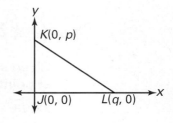

In Exercises 24–26, you will prove that the midpoints of any quadrilateral are the vertices of a parallelogram.

24. In quadrilateral *EFGH* at right, find the coordinates of the midpoints of the sides.

25. Copy the diagram and connect the midpoints of the sides.

26. Prove that the quadrilateral formed by connecting the sides is a parallelogram.

Use coordinate geometry to prove each of the following theorems:

27. The opposite sides of a parallelogram are congruent.

28. The diagonals of a square are perpendicular to each other.

29. The diagonals of a rectangle are congruent.

30. If the diagonals of a parallelogram are perpendicular, then the parallelogram is a rhombus.

31. The medians of a triangle intersect at a single point. (Hint: Find the equations of the lines containing the medians.)

32. CIVIL ENGINEERING Points *A* and *B* are on opposite sides of a straight portion of a river 30 meters wide. Point *B* is 100 meters downstream from point *A*. An engineer is designing a road from point *A* to point *B*. To find the lowest cost road between the points, the engineer positions points *A* and *B* on a coordinate plane as shown. The road costs more to build over the river, so part of the road will be built on land, leaving the river bank at point *X*, as shown.

a. Use the given coordinates to find d_L, the length of the road that lies on land, and d_W, the length of the road that lies over the water.

b. **MAXIMUM/MINIMUM** Suppose that the cost of building the road is $1000 per meter on land and $2000 per meter over the water. Write an equation for the cost of the road in terms of *x*. Use a graphics calculator to graph the cost function, and trace to find the coordinates of *X* for the lowest cost road.

Look Back

Given square *ABCD* with *AB* = 5, and diagonals $\overline{AC}$ and $\overline{BD}$ intersecting at point *E*, find the indicated measures. (LESSON 3.2, 4.5, 5.4)

33. *BC*　　　　　　　**34.** *CD*　　　　　　　**35.** *AD*

36. *AC*　　　　　　　**37.** *BD*　　　　　　　**38.** *AE*

39. *EC*　　　　　　　**40.** *BE*　　　　　　　**41.** *ED*

42. m∠*BEA*　　　　　**43.** m∠*BEC*

You are given trapeziod *ABCD* with bases $\overline{AB}$ and $\overline{CD}$ and midsegment $\overline{MN}$, where *AB* = 5.7, *CD* = 8.5, and the height, *h* = 3.5. Find the indicated values. (LESSON 3.7, 5.2)

44. *MN*　　　　　　　　　　**45.** Area of trapezoid *ABCD*

Look Beyond

46. Draw a scalene triangle, △*ABC*, and construct median $\overline{AM}$. Prove that the area of △*ABM* is equal to the area of △*ACM*.

47. Using △*ABC* from Exercise 46, construct the centroid (the intersection of the medians) of △*ABM* and label it *D*. Construct the centroid of △*ACM* and label it *E*. Draw $\overline{DE}$. What is the relationship between $\overline{AM}$ and $\overline{DE}$? What is the relationship between $\overline{BC}$ and $\overline{DE}$? If you are using geometry graphics software, try dragging the points to different locations to see if your results still hold. Otherwise, draw several different triangles to test your results.

Geometric Probability

5.8

Objective

● Develop and apply the basic formula for geometric probability.

Why *Some scientists believe that an asteroid or comet caused the extinction of the dinosaurs. What is the likelihood that one particular such object would have struck land? You can use geometric probability to help answer questions like this.*

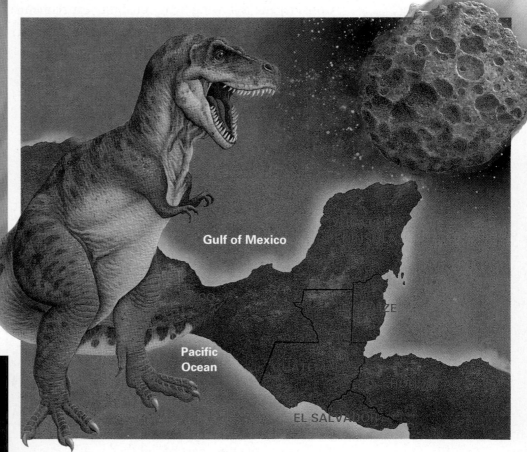

Gulf of Mexico

Pacific Ocean

EL SALVADOR

According to the fossil record, Earth experienced a sudden and drastic change about 65 million years ago that resulted in the extinction of dinosaurs like Tyrannosaurus rex. Some scientists believe that this upheaval was caused by an asteroid striking Earth near the Yucatán Peninsula.

The Probability of an Event

Consider the following question:

> The surface of Earth consists of about 30 percent land and 70 percent water. Assuming that a comet or asteroid would be equally likely to strike anywhere on Earth, what is the probability that such an object would strike land instead of water?

Mathematical intuition should tell you that there is a 30 percent chance, or *probability*, that the object would strike land. **Probability** is a number from 0 to 1 (or from 0 to 100 percent) that indicates how likely an event is to occur.
•A probability of 0 (or 0 percent) indicates that the event cannot occur.
•A probability of 1 (or 100 percent) indicates that the event must occur.

For many situations, it is possible to define and calculate the *theoretical probability* of an event with mathematical precision. The theoretical probability that an event will occur is a fraction whose denominator represents all equally likely outcomes and whose numerator represents the outcomes in which the event occurs.

Theoretical Probability

Consider the following two probability experiments:

a. There are 30 marbles in a bag, and 3 of them are red. A marble is drawn at random from the bag. The probability of drawing a red marble is:

$$P = \frac{3}{30} = 0.1$$

b. A point in the figure at right is selected at random. The probability that the selected point is from area A is:

$$P = \frac{\text{area } A}{\text{area } B}$$

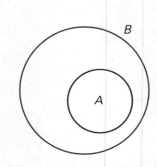

Figure B includes Figure A (the inner circle).

In cases like **a**, which you are probably already familiar with from your algebra studies, the number of possible outcomes can be counted. In cases like **b**, however, the number of possible outcomes is infinite. In these cases, the *areas* of the figures are used as measures of the sets of outcomes.

E X A M P L E ❶ If a dart lands in the blue part of the board below, a prize is given. What is the probability that a dart that hits the board at random will win?

SOLUTION

The area of the blue part of the board is 60 square units. The area of the entire board is 96 square units. The probability that the dart will land in the blue part is $\frac{60}{96}$, or $\frac{5}{8}$.

E X A M P L E ❷ In a game, pennies are tossed onto a grid of squares whose sides are equal to the diameter of a penny. To win, a penny must touch or cover an intersection of the grid. What is the probability of winning on a random toss?

SOLUTION

Imagine that circles are drawn around each intersection on the grid, as shown on the following page. If the *center* of a penny falls within one of these circles, the penny itself will touch or cover an intersection.

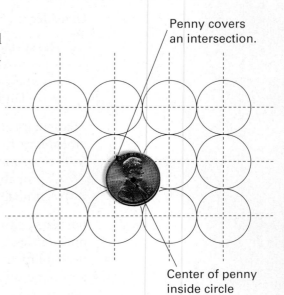

Penny covers an intersection.

Center of penny inside circle

The diagram at right represents a single square of the grid. If the center of the penny lands in a shaded region, the toss will be a winning toss.

Let a unit be defined as the radius of a penny. Then the width of a square (and the diameter of a penny) will be 2 units.

Area of a shaded region: $A_c = \pi \times 1^2 = \pi$ units^2

Area of a square: $A_s = 2 \times 2 = 4$ units^2

The area of each quarter of a circle is $\frac{\pi}{4}$ units^2, so the area of the shaded regions is π units^2. The theoretical probability, P, of the penny touching or covering a vertex—and of the player winning—is found below.

$$P = \frac{\text{area of shaded regions}}{\text{area of square}} = \frac{\pi}{4}$$

Experimental Probability

In the following Activity, you will use *experimental probability* and the result from Example 2 to estimate the value of π.

Activity

A "Monte Carlo" Method for Estimating π

YOU WILL NEED

grid paper with squares equal in width to the diameter of a penny

1. Toss a penny randomly onto the grid paper. Notice whether the coin touches or covers an intersection of two lines. Repeat for a total of 20 tosses. Record the number of tosses that touch or cover an intersection and the number that do not.

2. Share your results with the rest of your class. Find the totals for the entire class. To find the experimental proability that the penny will touch or cover an intersection, calculate the ratio of the number of tosses that touch or cover an intersection to the total number of tosses. Call this number R.

3. For a large number of tosses, you may assume that the value of the experimental probability, R, will be close to the theoretical probability, P, that you calculated in Example 2. Using this assumption, calculate an estimate for π.

$R \approx \frac{\pi}{4}$ (assumed to be true for large numbers)

$\pi \approx R \times 4$

4. Compare your estimate with the known value of π. Determine the relative error of your estimate by using the formula below.

$E = \frac{|V_t - V_e|}{V_t} \times 100$

where E = percent of error, V_t = true value, and V_e = estimated value

Exercises

Communicate

1. Why does a probability always have to be a number from 0 to 1 (or 0% to 100%)?

2. What does it mean for an outcome to have a probability of 0?

3. What does it mean for an outcome to have a probability of 1?

4. Assign a probability to each of the following words:

often	seldom	usually
never	maybe	frequently
sometimes	always	rarely

Rank the words in order from lowest probability to highest. Compare your list with those of your classmates. Which words do you agree on? Which words do you disagree on?

5. In the Activity, how did your estimate of π from 20 tosses compare with the estimate of π from the total tosses for the entire class? How do you think you could improve your estimate of π?

Guided Skills Practice

6. Find the probability that a dart tossed at random will land in the blue area of the figure at right. **(EXAMPLE 1)**

7. If the area of the shaded region at right is π square units and the area of the square is 4 square units, what is the area of the unshaded region? What is the probability that a penny tossed onto a grid with squares equal to the width of a penny will *not* touch or cover an intersection? **(EXAMPLE 2)**

8. Explain how you could use a grid with squares equal to the width of a penny to estimate π. **(ACTIVITY)**

Practice and Apply

Find the probability that a dart tossed at random onto each figure will land in the shaded area.

9.   10.   11.

For each spinner below, find the theoretical probability that the arrow will land on red.

12.

13.

14.

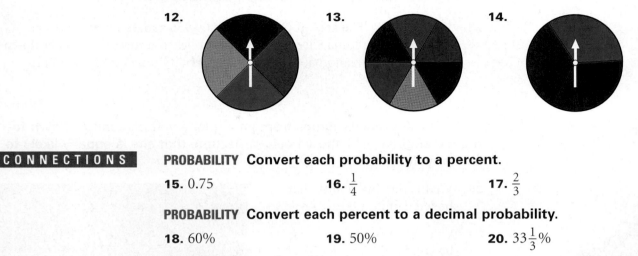

CONNECTIONS

PROBABILITY Convert each probability to a percent.

15. 0.75

16. $\frac{1}{4}$

17. $\frac{2}{3}$

PROBABILITY Convert each percent to a decimal probability.

18. 60%

19. 50%

20. $33\frac{1}{3}\%$

PROBABILITY Convert each percent to a fractional probability. Write your answers in lowest terms.

21. 45%

22. 80%

23. $66\frac{2}{3}\%$

For Exercises 24–27, refer to the spinner shown at left.

24. What is the probability that the arrow will land on 5?

25. What is the probability that the arrow will land on an odd number?

26. What is the probability that the arrow will land on an even number?

27. Add your results from Exercises 25 and 26. What does this result represent in terms of probability?

28. Design a dartboard in which the probability of a dart landing in a red circle is 0.5.

29. Design a dartboard in which the probability of a dart landing in a red triangle is $\frac{1}{3}$.

In the dartboard shown at left, the radius of the inner circle is half of the radius of the outer circle.

30. What is the probability that a dart thrown at random will hit a red region?

31. What is the probability that a dart thrown at random will hit a black region?

32. What is the probability that a dart thrown at random will hit a white region?

CHALLENGE

33. The squares in the grid at right are exactly the width of a penny. What is the probability that the center of a penny tossed at random onto the grid will land within a white square, with no part of the penny touching an intersection?

34. METEOROLOGY The weather forecaster predicts an 80% chance of rain. Express this as a probability in decimal and fractional forms.

35. METEOROLOGY The area of Oklahoma is 69,956 square miles. The area of its capital, Oklahoma City, is 625 square miles. If a tornado touches down at random in Oklahoma, what is the probability it will touch down in Oklahoma City?

SKYDIVING A skydiver jumps from an airplane and parachutes down to the rectangular field shown below. Assume that she is equally likely to land anywhere in the field.

36. What is the probability that she will land on target *A*?

37. What is the probability that she will land on target *B*?

38. What is the probability that she will land on target *C*?

39. What is the probability that she will miss all three of the targets?

40. TRANSPORTATION At a subway stop in New York City, a train arrives every 5 minutes, waits 1 minute, and then leaves. Suppose that you go to the stop at a random time. Use the diagram below to determine the probability that you will wait 2 minutes or less for a train to arrive.

Look Back

In Exercises 41–46, three lengths are given. Classify the lengths as sides of a right triangle, an acute triangle, an obtuse triangle, or no triangle. (LESSONS 4.8 AND 5.4)

41. 16, 30, 34   **42.** 27, 36, 50   **43.** 16, 63, 82

44. 11, 60, 61   **45.** 10, 24, 25   **46.** 8, 15, 23

Find the area of each figure described below. (LESSONS 5.2 AND 5.3)

47. a triangle with a base of 4 and a height of 7.5

48. a parallelogram with a base of 4 and a height of 7.5

49. a trapezoid with bases of 20 and 30 and a height of 12.6

50. a circle with a radius of 16 (Note: Use $\frac{22}{7}$ for π.)

Look Beyond

RECREATION The dimensions of a standard dartboard are given below.

outside edge of double ring to center = 170 mm
outside edge of triple ring to center = 117 mm
outer bull's-eye diameter = 31 mm
inner bull's-eye diameter = 12.7 mm
double and triple rings inside width = 8 mm

Double Ring

Triple Ring

Suppose that a thrown dart is equally likely to land anywhere on the board.

51. What is the probability that the dart will land in the double ring?

52. What is the probability that the dart will land in the triple ring?

Each sector of the board has a number from 1 to 20.

53. The highest scoring section of the board is in the 20 sector of the triple ring (60 points). What is the probability of hitting this section?

54. The inner bull's-eye (50 points) is the small red circle in the center of the board. What is the probability of hitting the inner bull's-eye?

55. The outer bull's-eye (25 points) is the black ring around the inner bull's-eye. What is the probability of hitting the outer bull's-eye?

56. When playing darts, why might the actual results be different from the theoretical probabilities?

TECHNOLOGY Most scientific and graphics calculators have random-number generators, which generate a number between 0 and 1.

1. Use your calculator to generate two random numbers. Multiply each number by 10 to get a number between 0 and 10, and use the results to create an ordered pair. Round to three decimal places.

For example: 0.6984024365 and 0.3306812911 → (6.984, 3.307)

2. Use this method to generate 20 ordered pairs.

If a point chosen at random is on the 10 × 10 grid shown at right, what is the probability the point will be in the shaded region?

You will use the points generated in Steps 1 and 2 to test this probability.

3. Copy the 10 × 10 grid at right, including the shaded region. Plot the ordered pairs on your grid.

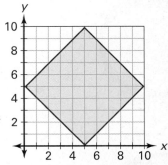

4. How many points are inside the shaded region? Divide this number by 20 to estimate the probability.

5. Was your estimate close to the probability you calculated? Compute the relative error by using the formula $E = \dfrac{|V_e - V_t|}{V_t} \times 100$, where V_e is the estimated (experimental) value and V_t is the true (theoretical) value.

WORKING ON THE CHAPTER PROJECT

You should now be able to complete Activity 2 of the Chapter Project.

PROJECT FIVE

Area of a Polygon

This project challenges you to find a formula for the areas of polygons drawn on dot paper with all of the vertices on dots—surprisingly, one formula will work for all of them! Work with a team of three or four classmates. You will probably find it helpful to divide the work among your team members.

Activity 1

Compute the areas for the figures in each group below. Then, on a separate piece of paper, complete a table like the ones given for each group. For each figure, N_b is the number of dots on the boundary, N_i is the number of dots in the interior, and A is the area.

GROUP A

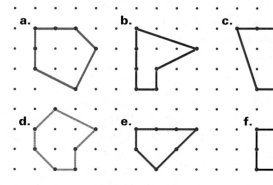

Group A			
	N_b	N_i	A
a.	7	4	$6\frac{1}{2}$
b.	?	?	?
c.	?	?	?
d.	?	?	?
e.	?	?	?
f.	?	?	?

GROUP B

Group B			
	N_b	N_i	A
a.	?	?	?
b.	?	?	?
c.	?	?	?
d.	?	?	?
e.	?	?	?
f.	?	?	?

GROUP C

	Group C		
	N_b	N_i	A
a.	?	?	?
b.	?	?	?
c.	?	?	?
d.	?	?	?
e.	?	?	?
f.	?	?	?

 Activity 2

To help you determine the formula, notice that all the figures in each group of Activity 1 have the same number of boundary points.

1. What is the pattern for calculating the area?

2. Write the pattern as a formula in terms of N_b and N_i.

3. Test your formula by creating new figures on dot paper and calculating their areas.

The formula you have just discovered was originally found by George Pick in 1899.

Activity 3

Refer to the figure at right.

1. Calculate the probability that a point chosen at random in the grid will be inside the shaded region.

2. Use a random-number generator to create 20 ordered pairs. Copy the figure onto graph paper and plot the ordered pairs on the figure. Divide the number of points that fall inside the shaded region by 20. Is your result close to the probability you calculated in Step 1?

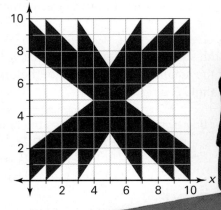

Chapter Review and Assessment

POSTULATES, FORMULAS, AND THEOREMS

Lesson	Number	Postulate or Theorem
5.1	5.1.3 Postulate: Sum of Areas	If a figure is composed of non-overlapping regions A and B, then the area of the figure is the sum of the areas of regions A and B.
	5.1.4 Perimeter of a Rectangle	The perimeter of a rectangle with base b and height h is $P = 2b + 2h$.
	5.1.5 Postulate: Area of a Rectangle	The area of a rectangle with base b and height h is $A = bh$.
5.2	5.2.1 Area of a Triangle	The area of a triangle with base b and height h is $A = \frac{1}{2}bh$.
	5.2.2 Area of a Parallelogram	The area of a parallelogram with base b and height h is $A = bh$.
	5.2.3 Area of a Trapezoid	The area of a trapezoid with bases b_1 and b_2 and height h is $A = \frac{1}{2}(b_1 + b_2)h$.
5.3	5.3.2 Circumference of a Circle	The circumference of a circle with diameter d and radius r is $C = \pi d$ or $C = 2\pi r$.
	5.3.3 Area of a Circle	The area of a circle with radius r is $A = \pi r^2$.
5.4	5.4.1 Pythagorean Theorem	For any right triangle, the square of the length of the hypotenuse is equal to the sum of the squares of the lengths of the legs; that is, $c^2 = a^2 + b^2$.
	5.4.2 Converse of the Pythagorean Theorem	If the square of the length of one side of a triangle equals the sum of the squares of the lengths of the other two sides, then the triangle is a right triangle.
	5.4.3 Pythagorean Inequalities	For any triangle ABC, with c as the length of the longest side: If $c^2 = a^2 + b^2$, then $\triangle ABC$ is a right triangle. If $c^2 > a^2 + b^2$, then $\triangle ABC$ is an obtuse triangle. If $c^2 < a^2 + b^2$, then $\triangle ABC$ is an acute triangle.

Lesson	Number	Postulate or Theorem
5.5	5.5.1 45-45-90 Triangle Theorem	In any 45-45-90 triangle, the length of the hypotenuse is $\sqrt{2}$ times the length of a leg.
	5.5.2 30-60-90 Triangle Theorem	In any 30-60-90 triangle, the length of the hypotenuse is 2 times the length of the shorter leg, and the length of the longer leg is $\sqrt{3}$ times the length of the shorter leg.
	5.5.3 Area of a Regular Polygon	The area of a regular polygon with apothem a and perimeter p is $A = \frac{1}{2}ap$.
5.6	5.6.1 Distance formula	On a coordinate plane, the distance between two points (x_1, y_1) and (x_2, y_2) is $d = \sqrt{(x_2 - x_1)^2 + (y_2 - y_1)^2}$.

Key Skills & Exercises

LESSON 5.1

Key Skills

Find the perimeter of a polygon.

Find the perimeter of polygon $ABFCDE$ below.

Polygon $ABCDEF$ has a perimeter of
$3 + 4 + 5 + 5 + 4 + 3 = 24$ cm.

Find the area of a rectangle.

Find the area of rectangle $ABCD$ above.

The base of the rectangle is 4 and the height is 2.3, so the area is $4 \times 2.3 = 9.2$ cm^2.

Exercises

Find the perimeter of each polygon.

1. 23, 30, 37

2. 1.3, 3.8, 2.0, 2.5, 2.4, 2.8

Find the area of each rectangle.

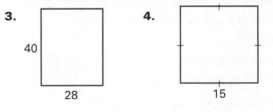

3. 40, 28

4. 15

LESSON 5.2

Key Skills

Find the area of a triangle.

Find the area of a triangle with a base of 6 and a height of 8.

The area is $\frac{1}{2} \times 6 \times 8 = 24$.

Find the area of a parallelogram.

Find the area of a parallelogram with a base of 6 and a height of 8.

The area is $6 \times 8 = 48$.

Exercises

For Exercises 5–8, find the area of the given figure.

5. a triangle with a base of 4 and a height of 12

6. a triangle with a base of 7 and a height of 5

7. a parallelogram with a base of 4 and a height of 9

Find the area of a trapezoid.

Find the area of a trapezoid with bases of 6 and 10 and a height of 8.

The area is $\frac{1}{2}(6 + 10) \times 8 = 64$.

8. a trapezoid with bases of 3 and 5 and a height of 6

LESSON 5.3
Key Skills

Exercises

Find the circumference of a circle.

Find the circumference of a circle with a diameter of 8 inches.

The circumference is $\pi \times 8 \approx 25.12$ inches.

Find the area of a circle.

Find the area of a circle with a radius of 3 meters.

The area is $\pi \times 3^2 \approx 28.26$ square meters.

9. Find the circumference of a circle with a diameter of 11.

10. Find the circumference of a circle with a radius of 2.5.

11. Find the area of a circle with a radius of 10.

12. Find the area of a circle with a diameter of 1.

LESSON 5.4
Key Skills

Exercises

Find the lengths of the sides of a right triangle by using the Pythagorean Theorem.

Find the missing leg of a right triangle with a leg of 40 and a hypotenuse of 41.

If the missing leg is a, then $a^2 + 40^2 = 41^2$, or $a^2 + 1600 = 1681$. Thus, $a^2 = 81$, and $a = 9$.

Determine whether a triangle is right, acute, or obtuse by using the Pythagorean inequalities.

A triangle has side lengths of 28, 45, and 50. Classify the triangle as right, acute, or obtuse.

$50^2 = 2500$ and $28^2 + 45^2 = 2809$
Since $2500 < 2809$, the triangle is acute.

Two side lengths of a triangle with hypotenuse c are given. Find the third side.

13. $a = 20$ $b = 21$ $c = \underline{\ ?\ }$

14. $a = \underline{\ ?\ }$ $b = 99$ $c = 101$

Three side lengths of a triangle are given. Classify the triangle as right, acute, or obtuse.

15. 48, 63, and 65

16. 48, 55, and 73

LESSON 5.5
Key Skills

Exercises

Find the side lengths of 45-45-90 triangles.

The hypotenuse of a 45-45-90 triangle is 18. Find the length of the legs.

The hypotenuse is $\sqrt{2}$ times the length of a leg, so the legs have lengths of
$\frac{18}{\sqrt{2}} = \frac{18\sqrt{2}}{2} = 9\sqrt{2}$.

17. The hypotenuse of a 45-45-90 triangle is 100. Find the length of the legs.

18. The length of each leg of a 45-45-90 triangle is 34. Find the hypotenuse.

19. The longer leg of a 30-60-90 triangle is 27. Find the lengths of the shorter leg and the hypotenuse.

Find the side lengths of 30-60-90 triangles.

The longer leg of a 30-60-90 triangle is 12. Find the shorter leg and the hypotenuse.

The longer leg is $\sqrt{3}$ times the shorter leg, so the shorter leg has a length of $\frac{12}{\sqrt{3}} = \frac{12\sqrt{3}}{3} = 4\sqrt{3}$.
The hypotenuse is twice the length of the shorter leg, or $8\sqrt{3}$.

20. The hypotenuse of a 30-60-90 triangle is 16. Find the lengths of the shorter leg and the longer leg.

Key Skills

Exercises

Determine the distance between two points on a coordinate plane.

Find the distance between the points $(4, 1)$ and $(2, 3)$.

$$d = \sqrt{(2-4)^2 + (3-1)^2} = \sqrt{8} = 2\sqrt{2}$$

Estimate the area under a curve by the method of quadrature.

Use the method of quadrature to estimate the area between the curve $y = x^2$ and the x-axis from $x = 1$ to $x = 4$.

The area can be estimated by three rectangles with a base of 1 and the following heights:

$$1^2 = 1 \qquad 2^2 = 4 \qquad 3^2 = 9$$

Thus, the area is approximately $1 + 4 + 9 = 14$.

Find the distance between each pair of points.

21. $(0, 0)$ and $(6, 8)$

22. $(3, 3)$ and $(7, 2)$

23. $(-2, 4)$ and $(3, -1)$

24. Use the method of quadrature to estimate the area between the curve $y = x^3$ and the x-axis from $x = 1$ to $x = 4$.

Key Skills

Exercises

Prove theorems by using coordinate geometry.

Prove that if the diagonals of a rectangle are perpendicular to each other, then the rectangle is a square.

A general rectangle can be drawn with vertices at $(0, 0)$, $(a, 0)$, (a, b), and $(0, b)$. The slopes of the diagonals are $\frac{b}{a}$ and $-\frac{b}{a}$. If the diagonals are perpendicular to each other, then $\left(\frac{b}{a}\right)\left(-\frac{b}{a}\right) = -1$, so $\frac{b^2}{a^2}$, which means that $b = a$ (or $b = -a$). Thus, the lengths of the sides of the rectangle are the same, so the rectangle is a square.

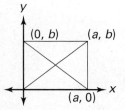

In Exercises 25–28, you will prove that if the diagonals of a parallelogram are equal, the parallelogram is a rectangle.

25. Draw a general parallelogram in a coordinate plane with three vertices at $A(0, 0)$, $B(a, 0)$, and $D(b, c)$. Give the coordinates of the fourth vertex, C.

26. Write expressions for the lengths of the diagonals of the parallelogram you drew in Exercise 25.

27. Assume the diagonals are the same length. Set the expressions from Exercise 26 equal to each other and prove that a or b must equal 0.

28. Use your result from Exercise 27 to complete the proof.

Find the probability of an event.

What is the probability that a randomly generated point with $0 \le x \le 5$ and $0 \le y \le 5$ will land in a circle with radius 2 centered at (3, 3)?

The diagram below represents the points with $0 \le x \le 5$ and $0 \le y \le 5$ and the circle with a radius of 2.

The area of the circle is 4π, and the area of the square is 25, so the probability is $\frac{4\pi}{25} \approx 0.50$.

Exercises

Find the probability that a dart tossed randomly onto each figure will land in the shaded area.

29.

30.

31.

32.

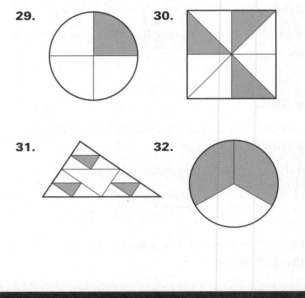

Applications

33. **SAFETY** You have a 20-foot ladder to climb onto a wall that is 16 feet high. Due to safety considerations, the ladder must extend at least 3 feet past the top of the wall, and the bottom of the ladder must be placed away from the wall at a distance of $\frac{1}{4}$ of the length of the ladder from the ground to where the ladder rests against the roof. Use the diagram at right to find x, the length of the ladder from the ground to the wall. Is the ladder long enough? How far will the bottom of the ladder be from the wall?

at least 3 ft

x ft   16 ft

$\frac{x}{4}$ ft

34. **SPORTS** A running track has straight sides and semicircular ends. The outer radius of the semicircles is 80 meters, the inner radius is 40 meters, and the straight sides are 160 meters long. What is the perimeter of the outside edge of the track? What is the perimeter of the inside edge of the track? What is the area of the track?

Alternative Assessment

Performance Assessment

The diagram below shows a set of *concentric circles*, which are circles that have the same center. The innermost circle has a radius of 1. As you move outward, the radius of each circle increases by 1.

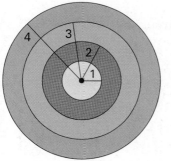

The region between two concentric circles is called an *annulus* (plural, *annuli*). The area of an annulus can be found by subtracting the area of the outer circle from the area of the inner circle.

Annulus

1. Find the area of the annulus with inner radius 1 and outer radius 2. Leave your answer in terms of π.

2. Repeat for the annulus with inner radius 2 and outer radius 3, and for the annulus with inner radius 3 and outer radius 4. Compare these areas, and make a conjecture about the area of an annulus with inner radius n and outer radius $n + 1$.

3. Verify your conjecture algebraically by subtracting the area of a circle with radius n from the area of a circle of radius $n + 1$, and simplifying your result.

Portfolio Projects

CULTURAL CONNECTION: ASIA A tangram is an ancient Chinese puzzle made from a large square.

A basic tangram consists of seven pieces: two small triangles, one medium triangle, two large triangles, a square, and a parallelogram, as shown.

The area of the entire puzzle is 16 square units.

1. Find the area of each piece.

2. Find all of the side lengths of each piece.

3. Make your own set of tangram pieces by copying the figure above onto graph paper and cutting out the pieces. Arrange all seven pieces to form a rectangle that is not a square. What is the perimeter of the rectangle?

4. Make another set of tangram pieces. Arrange all the pieces of both sets together to make a square. Prove that the figure you made is a square.

College Entrance Exam Practice

QUANTITATIVE COMPARISON For Items 1–3, write
A if the quantity in Column A is greater than the quantity in Column B;
B if the quantity in Column B is greater than the quantity in Column A;
C if the two quantities are equal; or
D if the relationship cannot be determined from the information given.

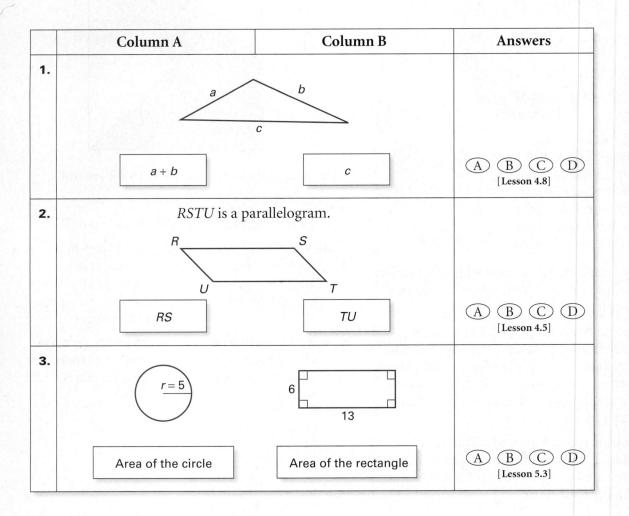

Column A	Column B	Answers
1.		
$a + b$	c	Ⓐ Ⓑ Ⓒ Ⓓ [Lesson 4.8]
2. *RSTU* is a parallelogram.		
RS	TU	Ⓐ Ⓑ Ⓒ Ⓓ [Lesson 4.5]
3. $r = 5$	6 / 13	
Area of the circle	Area of the rectangle	Ⓐ Ⓑ Ⓒ Ⓓ [Lesson 5.3]

Give the most specific name for each figure. It may help to draw a possible example. (LESSONS 3.1, 4.6, AND 5.3)

4. quadrilateral $ABCD$, in which $AB = CD$ and $AD = BC$

5. polygon $MNOPQR$

6. quadrilateral $FGHI$, in which $FG = HI$, $\overline{FG} \nparallel \overline{HI}$, and $\overline{FI} \parallel \overline{GH}$

7. a plane figure in which all points are equidistant from point P

For Items 8–12, refer to the diagram below. Complete the proof that △ABD is isosceles. (LESSON 4.4)

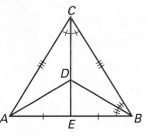

8. ∠CAB ≅ ∠CBA by ___?___.
  a. CPCTC
  b. ASA
  c. Isosceles Triangle Theorem
  d. Converse of the Isosceles Triangle Theorem

9. △ACD ≅ △BCD by ___?___.
  a. Isosceles Triangle Theorem
  b. SAS
  c. ASA
  d. Angle Addition Postulate

10. ∠CAD ≅ ∠CBD by ___?___.
  a. CPCTC
  b. ASA
  c. Isosceles Triangle Theorem
  d. definition of angle bisector

11. ∠DAE ≅ ∠DBE by ___?___.
  a. Isosceles Triangle Theorem
  b. Converse of the Isosceles Triangle Theorem
  c. Angle Addition Postulate
  d. SAS

12. △ABD is isosceles by ___?___.
  a. definition of isosceles triangle
  b. CPCTC
  c. Isosceles Triangle Theorem
  d. Converse of the Isosceles Triangle Theorem

In Items 13–16, suppose that △ABC has vertices at A (–1, 8), B (4, 3), and C (1, 2).

13. Find the slope of each side of △ABC. **(LESSON 3.8)**

14. Prove that △ABC is a right triangle. **(LESSON 3.8)**

15. Find the midpoint of each side of △ABC. **(LESSON 3.8)**

16. Find the length of each side of △ABC. Round your answers to the nearest hundredth. **(LESSON 5.6)**

FREE-RESPONSE GRID
Items 17–20 may be answered by using a free-response grid such as that commonly used by standardized-test services.

For Items 17–19, refer to the figure below. (LESSONS 5.2 AND 5.4)

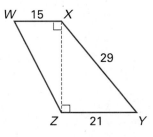

17. XZ = ___?___

18. WZ = ___?___

19. area of trapezoid WXYZ = ___?___

20. Find the length of a side of a square with the same area as a circle with a radius of 10. Round your answer to the nearest tenth. **(LESSONS 5.1 AND 5.3)**

Shapes in Space

WHAT GIVES THE PICTURE ON THE FACING page such a dramatic effect? Even though it is printed on a flat, two-dimensional surface, the French Concorde jet has the illusion of depth. In this chapter, you will find out how to give drawings this three-dimensional quality.

As you study geometric figures in space, you will discover that three-dimensional figures have properties similar to those of two-dimensional figures. In this chapter, you will apply what you already know about points and lines in a plane to points, lines, and planes in space.

DNA molecule

Three-dimensional computer illustrations often begin with a "wire frame."

About the Chapter Project

Solid figures that are bounded by polygons are called polyhedra. Mathematicians have long been interested in regular polyhedra, which are bounded by congruent regular polygons.

The last book of Euclid's *Elements* contains a proof that only five regular polyhedra exist. The earlier Greek philosopher Plato also knew this fact, and the five regular polyhedra are often called Platonic solids after him.

In the Chapter Project, you will explore why there are exactly five regular polyhedra.

After completing the Chapter Project you will be able to do the following:

- Visualize and understand the relationships among polygons in space.

About the Portfolio Activities

Throughout the chapter, you will be given opportunities to complete Portfolio Activities that are designed to support your work on the Chapter Project.

The theme of each Portfolio Activity and of the Chapter Project is polyhedra.

- In the Portfolio Activity on page 387, you will build the five Platonic solids from patterns called nets.

- In the Portfolio Activity on page 395, you will create two nets that "pop up" to form a dodecahedron with the help of a rubber band.

- In the Portfolio Activity on page 408, you will create semiregular polyhedra, also called Archimedean solids, from nets.

Solid Shapes

Objectives

- Use isometric dot paper to draw three-dimensional shapes composed of cubes.

- Develop an understanding of orthographic projection.

- Develop a basic understanding of volume and surface area.

Why *Architects, engineers, and many other professionals must be skilled in the art of drawing three-dimensional shapes. In this lesson, you will learn some of the fundamental concepts of this art.*

APPLICATION

MECHANICAL DRAWING

An **isometric drawing** is one in which the horizontal lines of an object are represented by lines that form 30° angles with a horizontal line in the picture.

30°   30°

Drawing Cubes

A type of graph paper called isometric dot paper can be helpful in drawing solid shapes such as cubes. This paper has diagonal rows of dots that form a 30° angle with horizontal lines.

Connect the dots as shown to draw a cube. On a solid cube, at most three faces are visible at a time; however, sometimes you may wish to represent the hidden faces of a cube. This is done by using dashed lines, as shown in the second drawing.

In the first and second drawings, the cube is viewed from above. In the third drawing, the cube is viewed from below.

Activity 1

Using Isometric Grid Paper

1. Draw each solid figure below on isometric dot paper. Then redraw the figures with the red cube or cubes removed.

2. Draw each solid figure below on isometric dot paper. Then add a cube at each red face and draw the new figure.

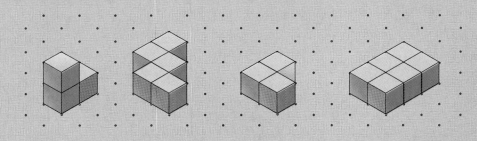

CHECKPOINT ✔ **3.** Describe how the drawings change as you add or subtract cubes.

Activity 2

Using Unit Cubes

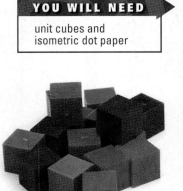

1. Use unit cubes to build the three figures below. Are your figures the same as those of your classmates? If not, how are they different?

2. Build three solids of your own with at least six unit cubes in each. Draw your solids on isometric dot paper.

CHECKPOINT ✔ **3.** Could any of the drawings you made in Step 2 represent more than one solid figure? Explain your reasoning.

Orthographic Projections

An **orthographic projection** is a view of an object in which points of the object are "projected" onto the picture plane along lines perpendicular to the picture plane.

Typically, a solid may be drawn from six different views: front, back, left, right, top. and bottom. For the solid at right, all six orthographic views are shown below. Edges that cannot be seen in a particular view are represented by dashed lines.

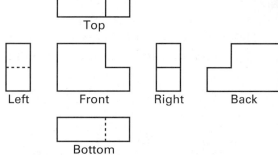

Top

Left   Front   Right   Back

Bottom

You may find it helpful to think of the different views of an object on an unfolded box that has a different view on each face.

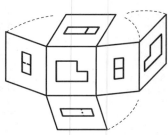

Activity 3

Volume and Surface Area

YOU WILL NEED

unit cubes and graph paper

1. Use 7 unit cubes to build a model of the solid shown at right. Then draw six orthographic projections of the object on graph paper, representing the views from the back, front, left, right, top, and bottom.

CHECKPOINT ✔

2. The volume of a solid figure is the number of unit cubes that it takes to completely fill it. How can you determine the volume of the figure that you built?

Left   Front

CHECKPOINT ✔

3. The total area of the exposed surfaces of an object is called its surface area. How can you use the orthographic projections of the object to determine its surface area?

CRITICAL THINKING   In the orthographic projections you drew in Activity 3, each exposed face of the object appeared in exactly one view. Do you think this would be true for any object you might build? Use drawings or models to explain your reasoning.

Exercises

Communicate

1. A solid built from cubes has the top view shown at right. Draw two possibilities for this solid.

2. A solid has the right view and front view shown at right. What is a possible top view for this solid?

   Front   Right

3. Why do you think an isometric drawing uses angles of 30° from a horizontal line? (Hint: Draw a cube on isometric grid paper and measure its edges.)

4. Describe the difference between volume and surface area. Include units in your description.

Guided Skills Practice

Use isometric dot paper for Exercises 5 and 6. *(ACTIVITY 1)*

5. Draw the solid below with the red cube removed.

6. Draw the solid below with a cube added at the red face.

Use unit cubes for Exercises 7–10. Refer to the figure below.
(ACTIVITIES 2 AND 3)

7. Build a model of the figure out of unit cubes. Then turn the figure and draw a different view on isometric dot paper.

8. Draw six orthographic projections of the solid.

9. Find the solid's volume in cubic units.

10. Find the solid's surface area in square units.

Front

Practice and Apply

For Exercises 11–16, refer to the isometric drawing of a cube shown below. Identify the letter of the indicated face.

11. front

12. right

13. top

14. left

15. bottom

16. right

Front   f

17. What lines would you add to the drawing at right to indicate the hidden faces?

18. Draw an orthographic projection of the cube.

For Exercises 19–23, refer to the isometric drawing below.

19. Give the solid's volume in cubic units.

20. Give the solid's surface area in square units.

21. Draw six orthographic views of the solid.

22. Draw the solid on isometric dot paper from a different viewpoint.

23. Draw the solid on isometric dot paper with a cube added at each red face.

Front

Refer to the drawing below for Exercises 24–27. Assume that there are no hidden cubes.

24. Each flat area of the solid is a face. How many faces does this solid have?

25. Draw the front-right face.

26. Is any other face of the solid congruent to the front-right face?

27. Are any two faces of the solid congruent? Explain your answer.

Front

The three solids below each have a volume of 4 cubic units.

a.

Front

b.

c.

Front

28. Find each solid's surface area in square units. Which has the smallest surface area?

29. Draw six orthographic projections of each solid.

30. Draw two other solids with a volume of 4 cubic units that are different solids from those above (not just different views).

Refer to the solid below for Exercises 31 and 32.

31. Draw six orthographic projections of the solid.

32. Are there any exposed faces of the solid that do not appear in any of the orthographic views you drew in Exercise 31? Explain.

33. Use isometric dot paper to draw two solids that have the same volume but different surface areas.

34. Use isometric dot paper to draw two solids that have the same surface area but different volumes.

35. The solid at right is made up of 8 unit cubes. Draw a solid with a smaller volume but a larger surface area.

36. Draw two different solids that have the same orthographic projections from the top and bottom views.

37. Draw two different solids that have the same orthographic projections from the front and back views.

38. How many unit cubes does it take to build a $2 \times 2 \times 2$ cube? a $3 \times 3 \times 3$ cube? a $4 \times 4 \times 4$ cube? an $n \times n \times n$ cube?

39. What is the surface area in square units of a $2 \times 2 \times 2$ cube? a $3 \times 3 \times 3$ cube? a $4 \times 4 \times 4$ cube? an $n \times n \times n$ cube?

CHALLENGES

40. The figure at right shows a $3 \times 3 \times 3$ cube built from unit cubes. Suppose that the exposed faces of the cube were painted red and then the solid was disassembled into unit cubes.

How many unit cubes would have
a. 3 red faces?
b. 2 red faces?
c. 1 red face?
d. no red faces?

41. Repeat Exercise 40 for a $4 \times 4 \times 4$ cube.

APPLICATIONS

42. ARCHITECTURE Draw orthographic projections that represent the top, front, and right views of the building shown at right. Assume that the front faces are the ones on the right, in sunlight.

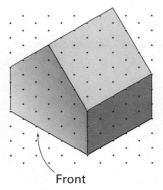

Front

43. RECREATION The isometric drawing of a tent, shown at left, can be used to show how the tent will look when it is assembled.
Draw six orthographic views of the tent. What might these views be used for?

Find the area of each figure. *(LESSONS 5.2 AND 5.3)*

44.

45.

2.9 m   2.3 m
10.7 m

46.

5 ft
3.7 ft
8 ft

47.

$x = 10$ m

Find the unknown length in each right triangle. *(LESSON 5.4)*

48.

3 cm   ?
6 cm

49.

?   315 yd
265 yd

50. ARCHAEOLOGY The top of a pyramid has been destroyed. Use the diagram of a cross section at left to estimate its original height. *(LESSONS 3.7 AND 5.4)*

350 ft
300 ft
600 ft

Use the orthographic projections at right for Exercises 51–54.

51. Use isometric dot paper to draw two different views of the solid.

52. Are any cubes hidden in your drawings?

53. How many faces does the solid have?

54. How many congruent faces does the solid have? Explain your answer.

Top

Left   Front   Right   Back

Bottom

Spatial Relationships

Objectives

- Define *polyhedron*.
- Identify the relationships among points, lines, segments, planes, and angles in three-dimensional space.
- Define *dihedral angle*.

Why *The interplay of lines and planes is apparent in real-world objects such as this quartz crystal. Understanding the relationships among lines and planes in three dimensions is essential to understanding the structure of matter.*

Figures in Space

A closed spatial figure made up of polygons is called a *polyhedron* (plural, *polyhedra* or *polyhedrons*). Closed spatial figures are also known as **solids**.

Definition: Polyhedron

A **polyhedron** is a closed spatial figure composed of polygons, called the **faces** of the polyhedron. The intersections of the faces are the **edges** of the polyhedron. The vertices of the faces are the **vertices** of the polyhedron.

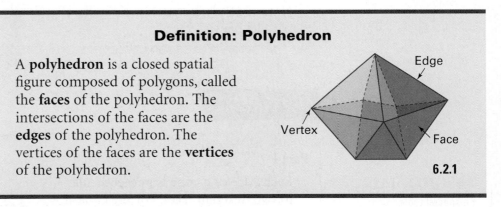

6.2.1

The polyhedron you are probably most familiar with is a cube. A cube is an example of a regular polyhedron. In a *regular polyhedron*, all of the faces are congruent regular polygons, and the same number of polygons meet at each vertex.

Lines and Planes in Space: A Step-by-Step Procedure

In the Activities that follow, you will discover and develop ideas about how lines and planes relate to each other in space.

Activity 1

Parallel Lines and Planes in Space

Part I

1. Sketch a cube and label its vertices as shown. Identify the segments that form the vertical edges.

   Do $\overline{AE}$ and $\overline{CG}$ seem to be coplanar? Do they seem to be parallel? Do you think that they would meet if they were extended infinitely?

2. Which edges of the cube seem to be parallel?

CHECKPOINT ✔

3. Are there edges of the cube that are not parallel and yet would never meet if they were extended infinitely in either direction? Explain how this can be. These segments (or lines or rays) are said to be **skew**. Make a list of four pairs of skew edges in the cube.

Part II

1. How many faces does a cube have? Which faces seem to be parallel to each other?

CHECKPOINT ✔

2. Write your own definition of **parallel planes** by completing the sentence below.

Definition: Parallel Planes

Two planes are parallel if and only if ____?____ .

6.2.2

Note: Two plane figures are parallel if and only if they are contained in parallel planes.

Activity 2

Segments and Planes

Part I

1. Each edge of a cube is perpendicular to two different faces. For the cube at right, make a list of edges and the faces to which they are perpendicular.

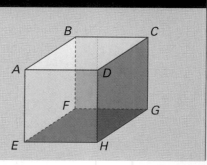

2. What do you think it means for a segment or a line to be perpendicular to a plane? Draw a line, ℓ, on a piece of paper and label a point, P, on it. Hold your pencil so that it is perpendicular to the paper, and place the tip on point P. Is the pencil perpendicular to line ℓ?

3. Is it possible to tilt your pencil so that it is still perpendicular to line ℓ but not to the paper? Make a sketch illustrating your answer.

4. Draw a new line, m, through point P. Place your pencil on P so that it is perpendicular to ℓ and m at point P. What is the relationship between the pencil and the plane of the paper?

5. Draw several other lines through point P. When you place your pencil on P so that it is perpendicular to the paper, is it also perpendicular to these other lines?

CHECKPOINT ✔   **6.** Write your own definition of a **line perpendicular to a plane** by completing the sentence below.

Definition: A Line Perpendicular to a Plane

A line is perpendicular to a plane at a point P if and only if it is perpendicular to every line in the plane that _____?_____.   **6.2.3**

Part II

1. Each edge of a cube (see Step 1 of Part I) is parallel to two different faces. Make a list of edges and the faces to which they are parallel. What do you think it means for a segment or a line to be parallel to a plane?

2. Draw a line, ℓ, on a piece of paper. Hold your pencil above ℓ so that it is parallel to the line. Does the pencil seem to be parallel to the plane of the paper?

3. Turn your pencil so that it is still parallel to the paper but not to the line. Do you think that you could draw another line on the paper that is parallel to the pencil in this position?

CHECKPOINT ✔   **4.** Write your own definition of a **line parallel to a plane** by completing the sentence below.

Definition: A Line Parallel to a Plane

A line that is not contained in a given plane is parallel to the plane if and only if it is parallel to _____?_____.   **6.2.4**

Angles Formed by Planes

A **half-plane** is the portion of a plane that lies on one side of a line in the plane and that includes the line. The angle between two half-planes is known as a *dihedral angle*.

Definition: Dihedral Angle

A **dihedral angle** is the figure formed by two half-planes with a common edge. Each half-plane is called a **face** of the angle, and the common edge of the half-planes is called the **edge** of the angle.

Edge

Faces

Half-plane

Dihedral angle

6.2.5

Activity 3

Measuring Dihedral Angles

1. Some of the faces of a cube form right dihedral angles (that is, the faces are perpendicular to each other). Each face of the cube is perpendicular to how many other faces of the cube?

2. Draw a horizontal line, ℓ, on a piece of paper. Mark and label points A, B, and C on the line, with point B between A and C. Make a crease through point B so that line ℓ folds onto itself. What is the relationship between line ℓ and the line of the crease?

3. Open the paper slightly. The angle formed by the sides of the paper is a dihedral angle. The measure of the dihedral angle is the measure of $\angle ABC$.

CHECKPOINT ✔

4. Write your own definition of the **measure of a dihedral angle** by completing the sentence below.

Definition: Measure of a Dihedral Angle

The measure of a dihedral angle is the measure of an angle formed by two rays that are on the faces and that are ____?____ to the edge.

6.2.6

5. Open the paper and flatten it. Add points X and Y on the line of the crease. Draw two rays from each point, with one ray on each side of the crease, as shown.

6. Fold the paper again along the same crease. Cut out pieces of paper that fit neatly into the angles formed by the different rays. Compare the shapes of the pieces of paper. Do the angles that are formed by the new rays have the same or different measures than that of $\angle ABC$?

CHECKPOINT ✔ **7.** By measuring the angles formed by different rays in Step 6, you can get a variety of results. What is the smallest angle that you could measure? What is the largest? Use your results to explain why a dihedral angle is measured along rays that are perpendicular to the edge.

Exercises

● *Communicate*

1. If two lines in space are perpendicular to the same line, are the two lines parallel to each other? Why or why not?

2. If a line not in a plane is perpendicular to a line in the plane, is the first line perpendicular to the plane? Why or why not?

3. If a line is perpendicular to two intersecting lines in a plane, is the first line perpendicular to the plane? Why or why not?

APPLICATION

CHEMISTRY Use the model of a sodium chloride crystal, which has a cubic structure, for Exercises 4–6.

4. Describe a pair of parallel segments, and explain why the segments are parallel.

5. Describe a pair of parallel planes, and explain why the planes are parallel.

6. Describe a pair of perpendicular planes, and explain why the planes are perpendicular.

Crystal lattice of sodium chloride (table salt)

Use the figure of the cube at right for Exercises 7–11. (ACTIVITY 1)

7. Name two pairs of parallel edges.

8. Name two pairs of skew edges.

9. Name two pairs of parallel faces.

10. List two edges and the planes to which they are perpendicular.

11. List two pairs of parallel edges that do not lie on the same face of the cube.

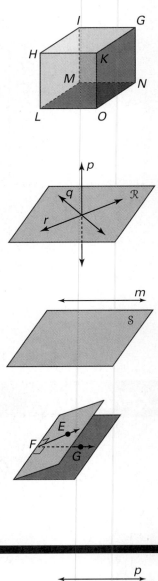

12. In the figure at right, line p is perpendicular to plane $\mathcal{R}$. What is the relationship between line p and line q? line p and line r? **(ACTIVITY 2)**

13. In the figure at right, line m is parallel to plane $\mathcal{S}$. What can you conclude about a certain line in plane $\mathcal{S}$? **(ACTIVITY 2)**

14. What is the measure of the dihedral angle in the figure at right? **(ACTIVITY 3)**

Practice and Apply

15. In the figure at right, line p is parallel to line q. What can you conclude about the relationship between line p and plane $\mathcal{M}$?

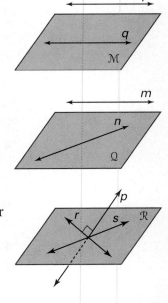

16. In the figure at right, line m is not parallel to line n. Can you draw a conclusion about the relationship between line m and plane $\mathcal{Q}$? Explain your answer.

17. In the figure at right, line p is perpendicular to line r but not to line s. What can you conclude about the relationship between line p and plane $\mathcal{R}$?

For Exercises 18–22, indicate whether each statement is true or false for a figure in space. Explain your answers by making sketches.

18. If two lines are parallel to a third line, then the two lines are parallel.

19. If two planes are parallel to a third plane, then the two planes are parallel.

20. If two planes are perpendicular to a third plane, then the two planes are parallel.

21. If two planes are perpendicular to the same line, then the planes are parallel.

22. If two lines are perpendicular to the same plane, then the lines are parallel.

Use your model from Exercise 23 to answer Exercises 24–26.

23. Fold an index card in half and mark it with two segments, $\overline{AB}$ and $\overline{CD}$, so that $\overline{AB}$ is perpendicular to the folded edge and $\overline{CD}$ is not. Cut along each segment from the folded edge. Insert two more index cards to model intersecting planes, as shown at right.

24. Which inserted card forms a plane that is not perpendicular to the edge?

25. Which inserted card can be used to measure the dihedral angle formed by the folded card?

26. Which of the two angles formed by the inserted cards is larger?

APPLICATIONS

Ships navigate the surface of the ocean, which can be modeled by a plane. Airplanes navigate in space. Exercise 27–30 concern the navigation of ships and airplanes.

27. NAVIGATION Ship *A* is traveling south. Ship *B* in the same vicinity is traveling southeast. Their lines of travel _____?_____.

28. NAVIGATION Airplane *A* is traveling south at an altitude of 23,000 feet. Nearby, airplane *B* is traveling southeast at an altitude of 18,000 feet. Their lines of travel are _____?_____.

29. NAVIGATION Two airplanes flying horizontally at the same altitude are flying in the same _____?_____.

30. NAVIGATION An airplane takes off from a runway. The airplane's nose and wingtips define the *plane of flight*. Sketch the dihedral angle formed by the airplane's plane of flight and the surface of the runway. Is the angle acute, right, or obtuse? Explain.

In Exercises 31–34, use a compass and straightedge to construct each figure. *(LESSON 4.7)*

31. Draw an angle and label it ∠ABC. Construct a copy of ∠ABC.

32. Draw a segment and label it $\overline{DE}$. Construct its perpendicular bisector.

33. Draw segment and label it $\overline{FG}$. Mark a point, H, not on $\overline{FG}$. Construct a line through point H parallel to $\overline{FG}$.

34. Draw a triangle and label it △JKL. Construct a copy of △JKL.

Find the area of each polygon below. *(LESSONS 5.1, 5.2, AND 5.5)*

35. square

14

36. trapezoid

22
16
32

37. parallelogram

21
18
28

38. right triangle

28
53

39. equilateral triangle

15

40. regular hexagon

9

CHALLENGES

41. Arrange the figure below into exactly three triangles by moving only two toothpicks.

42. Arrange the figure below into exactly three squares by moving only three toothpicks.

BUILDING PLATONIC SOLIDS

A **regular polyhedron** has congruent regular polygons as faces, with the same number of faces meeting at each vertex. There are exactly five different convex regular polyhedra, shown below. Because the Greek mathematician Plato proved that only these five exist, they are also known as the **Platonic solids**. Polyhedra are named for the number of faces they have.

Tetrahedron
(4 faces)

Hexahedron (cube)
(6 faces)

Octahedron
(8 faces)

Dodecahedron
(12 faces)

Icosahedron
(20 faces)

tetrahedron

cube

octahedron

dodecahedron

icosahedron

1. The patterns at right are nets for the Platonic solids. Enlarge and copy the nets onto a sturdy material such as poster board or cardboard.

Tetrahedron

Hexahedron (cube)

2. Cut out the nets and fold along all of the dotted lines. You may wish to color the faces or decorate them in some way.

Octahedron

3. Use tape or glue to assemble each polyhedron.

Dodecahedron

WORKING ON THE CHAPTER PROJECT

You should now be able to complete Activity 1 of the Chapter Project.

Icosahedron

Prisms

Why *Many familiar objects are shaped like polyhedra known as prisms. As you study prisms, you will apply your knowledge of congruent segments, angles, and polygons.*

Objectives

- Define *prism, right prism,* and *oblique prism.*

- Examine the shapes of lateral faces of prisms.

- Solve problems by using the diagonal measure of a right prism.

A laser beam is bent as it passes through the triangular glass prism in this experiment. Laser light consists of one pure color, so it is not refracted into a spectrum as ordinary light would be.

Prisms

The figures shown below are prisms. A prism is named by the shape of its base.

Triangular prism   Rectangular prism   Pentagonal prism   Hexagonal prism

A **prism** is a polyhedron that consists of a polygonal region and its translated image on a parallel plane, with quadrilateral faces connecting corresponding edges.

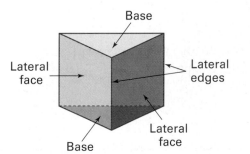

The faces formed by the polygonal region and its image are each called a **base** of the prism. The remaining faces, which are quadrilaterals, are called **lateral faces** of the prism. The edges of the lateral faces that are not edges of either base are called **lateral edges** of the prism.

CRITICAL THINKING   What type of quadrilateral are the lateral faces of a prism? (Hint: Which edges are parallel? Which edges are congruent?)

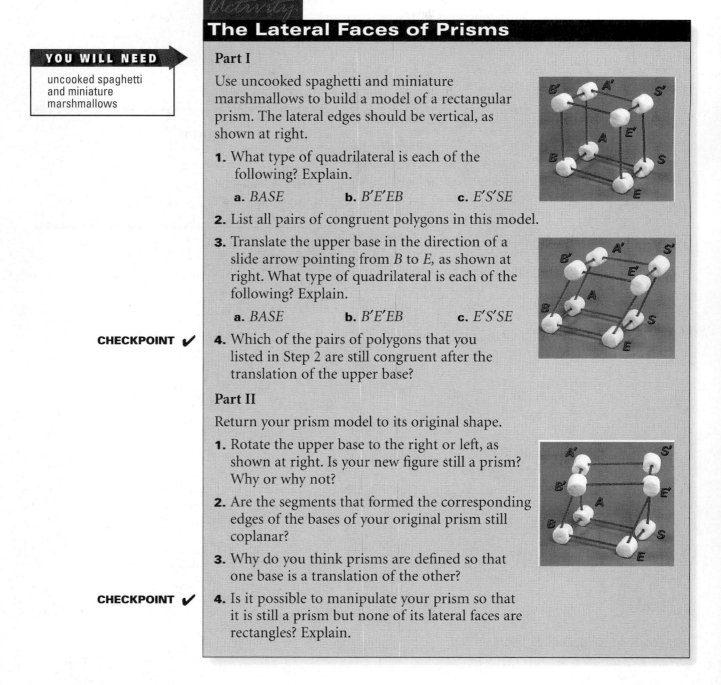

Activity
The Lateral Faces of Prisms

Part I

Use uncooked spaghetti and miniature marshmallows to build a model of a rectangular prism. The lateral edges should be vertical, as shown at right.

1. What type of quadrilateral is each of the following? Explain.

   a. *BASE*   **b.** *B'E'EB*   **c.** *E'S'SE*

2. List all pairs of congruent polygons in this model.

3. Translate the upper base in the direction of a slide arrow pointing from *B* to *E*, as shown at right. What type of quadrilateral is each of the following? Explain.

   a. *BASE*   **b.** *B'E'EB*   **c.** *E'S'SE*

CHECKPOINT ✔   **4.** Which of the pairs of polygons that you listed in Step 2 are still congruent after the translation of the upper base?

Part II

Return your prism model to its original shape.

1. Rotate the upper base to the right or left, as shown at right. Is your new figure still a prism? Why or why not?

2. Are the segments that formed the corresponding edges of the bases of your original prism still coplanar?

3. Why do you think prisms are defined so that one base is a translation of the other?

CHECKPOINT ✔   **4.** Is it possible to manipulate your prism so that it is still a prism but none of its lateral faces are rectangles? Explain.

In the Activity, you discovered that all lateral faces of a prism are rectangles or nonrectangular parallelograms. This gives an additional classification for prisms.

A **right prism** is a prism in which all of the lateral faces are rectangles.

An **oblique prism** has at least one nonrectangular lateral face.

The Diagonals of a Right Rectangular Prism

A **diagonal** of a polyhedron is a segment whose endpoints are vertices of two different faces of the polyhedron.

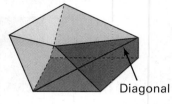

Diagonal

Diagonal of a Right Rectangular Prism

In a right rectangular prism with dimansions $\ell \times w \times h$, the length of a diagonal is given by

$$d = \sqrt{\ell^2 + w^2 + h^2}.$$

6.3.1

Given: rectangular prism with dimensions $\ell \times w \times h$

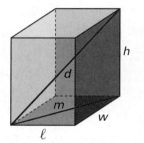

Prove: $d = \sqrt{\ell^2 + w^2 + h^2}$

Proof:

Statements	Reasons
$m^2 = \ell^2 + w^2$	Pythagorean Theorem
$d^2 = m^2 + h^2$	Pythagorean Theorem
$d^2 = \ell^2 + w^2 + h^2$	Substitution Property of Equality
$d = \sqrt{\ell^2 + w^2 + h^2}$	Take the square root of each side.

E X A M P L E ●

APPLICATION
CHEMISTRY

Iron atoms form *body-centered cubic* structures. The atoms form cubes with one atom in the center and one at each vertex, as shown at right. The cube, called a *unit cell*, is the smallest unit of a repeating pattern.

The atoms along the diagonal may be thought of as touching each other. If a unit cell measures 291 picometers (1 picometer = 1×10^{-12} meters) on each edge, what is the approximate radius of an iron atom?

Each vertex of the dashed cube is at the center of one of the atoms.

First find the length of the diagonal of the unit cell.

$$d = \sqrt{\ell^2 + w^2 + h^2}$$
$$d = \sqrt{291^2 + 291^2 + 291^2} = \sqrt{3(291^2)} = 291\sqrt{3}$$
$$d \approx 504 \text{ picometers}$$

The diagonal is the length of four radii, so divide by 4.

$$r \approx 504 \div 4 \approx 126 \text{ picometers}$$

The radius of an iron atom is approximately 126 picometers, or 0.000000000126 meter.

Exercises

1. Why is a prism with rectangular lateral faces called a "right" prism?

2. In the right rectangular prism shown at right, which faces are congruent? Would they still be congruent if the prism were oblique? Explain your reasoning.

3. In the rectangular prism shown at right above, could more than one pair of faces be called the bases? Explain your reasoning.

4. Identify some real-world objects that are prisms, including oblique prisms and prisms that do not have rectangular bases.

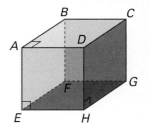

A rare example of an oblique prism

Refer to the oblique rectangular prism with base *EFGH* shown at right. Classify each quadrilateral named below. *(ACTIVITY)*

5. *EFGH*   6. *ADHE*   7. *ABFE*

8. **CHEMISTRY** Sodium has a body-centered cubic structure with a unit cell that is 430 picometers wide. Estimate the radius of a sodium atom. *(EXAMPLE)*

● *Practice and Apply*

Which of the figures below appear to be prisms? Give the name for each prism. If the figure is not a prism, explain why not.

9.

10.

11.

12.

13.

14.

Use the right triangular prism below for Exercises 15–20.

15. Which face is congruent to △*ABC*? Explain.

16. Name all segments congruent to $\overline{BE}$.

17. What type of quadrilateral is *ACFD*?

18. Name all segments congruent to $\overline{EF}$.

19. Name two congruent lateral faces.

20. List all right angles in the prism.

Use the oblique rectangular prism below for Exercises 21–25. In the prism, m∠*GKN* = 60° and m∠*GKL* = 80°.

21. Name all segments congruent to $\overline{GN}$.

22. What type of quadrilateral is *GJNK*?

23. What type of quadrilateral is *JIMN*?

24. Name all pairs of congruent faces.

25. List all obtuse angles in the prism.

Find the length of a diagonal of a right rectangular prism with the given dimensions.

26. $\ell = 4, w = 12, h = 3$

27. $\ell = 10, w = 5, h = 12$

28. $\ell = 7.5, w = 8, h = 8.5$

29. $\ell = a, w = a, h = a$

For Exercises 30–33, refer to the right rectangular prism shown at right.

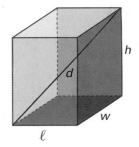

30. $\ell = $ ____?____ , $w = 8$, $h = 24$, $d = 26$

31. $\ell = 7$, $w = $ ____?____ , $h = 10$, $d = 15$

32. $\ell = a$, $w = a$, $h = 2a$, $d = $ ____?____

33. $\ell = x$, $w = x$, $h = x$, $d = 10$, $x = $ ____?____

For Exercises 34–38, complete the table below.

	Type of prism	Number of faces	Number of vertices	Number of edges
34.	triangular	?	?	?
35.	rectangular	?	?	?
36.	pentagonal	?	?	?
37.	hexagonal	?	?	?
38.	*n*-gonal	?	?	?

39. Use the pattern from the table above to determine how many faces, vertices, and edges a 20-gonal prism has.

40. **CULTURAL CONNECTION: EUROPE** Swiss mathematician Leonhard Euler (1707–1783) proved a relationship between the faces, vertices, and edges of a polyhedron. Let F represent the number of faces, V represent the number of vertices, and E represent the number of faces of a prism. For each prism in the table above, calculate $V - E + F$. What do you notice?

Leonhard Euler

CHALLENGE

41. The base of the right hexagonal prism shown at right is a regular hexagon with a side length of a and a height of h. Find the length of the indicated diagonal in terms of a and h.

APPLICATION

42. **PACKAGING** A manufacturer wishes to package a candy bar in a box shaped like a right triangular prism with equilateral triangles for the bases. The unfolded pattern for the box, called a *net*, is shown at right. Explain how the net should be folded in order to produce the desired shape. Draw a net for a box shaped like a right hexagonal prism.

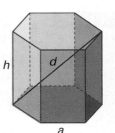

43. CHEMISTRY A given substance has a body-centered cubic structure, like that of iron and sodium. If the radius of one atom of this substance is a, find the length of a side, in terms of a, of a unit cell of the substance.

44. CHEMISTRY Another type of atomic structure is called *face-centered cubic*. This structure is a cube with one atom at each vertex and one in the center of each face, as shown at right. Calcium has a face-centered cubic structure and has an atomic radius of 197 picometers. Find the dimensions, including the diagonal, of a unit cell of calcium.

Each vertex of the dashed cube is at the center of one of the atoms.

 Look Back

Classify each statement as true or false and explain your reasoning. (LESSON 4.6)

45. The diagonals of a rectangle that is not a square bisect each other.

46. The diagonals of a rectangle that is not a square bisect the angles of the rectangle.

47. The diagonals of a rectangle that is not a square are perpendicular to each other.

48. The diagonals of a rectangle that is not a square divide the rectangle into four congruent triangles.

COORDINATE GEOMETRY For Exercises 49–51, find the length and midpoint of the segment connecting each pair of points. *(LESSON 5.6)*

49. $(3, 4), (3, -4)$ **50.** $(4, -2), (-2, 3)$ **51.** $(2, -2), (5, 6)$

52. OPTICS When white light passes through a glass prism, it is separated into a spectrum of different wavelengths, as shown at right. In the diagram, the light hits the prism at an angle of 36°. Use the information in the diagram to determine the angle between the red and violet light rays leaving the prism.

The Spider and the Fly Suppose that a spider and a fly are in a room shaped like a right rectangular prism. The fly can travel directly to any point in the room by flying, but the spider must walk on the walls, floor, or ceiling.

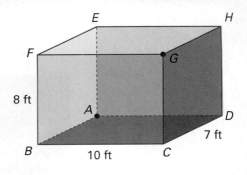

53. What is the shortest distance that the fly can travel to get from point *A* to point *G*?

54. What is the shortest distance that the spider can travel to get from point *A* to point *G*? (Hint: Think of the room as a box. Unfold the box so that it lies flat, and the path traveled by the spider will be a straight line.)

POP-UP DODECAHEDRON

Follow the steps below to create a model of a regular dodecahedron.

1. Make two copies of the pattern at right out of cardboard. Fold along the dotted lines.

2. Holding the two pattern pieces together, place a rubber band around them as shown.

3. Release your hold on the pieces to form the dedecahedron.

Coordinates in Three Dimensions

6.4

Objectives

- Identify the features of a three-dimensional coordinate system, including the axes, octants, and coordinate planes.

- Solve problems by using the distance formula in three dimensions.

Why *By using a three-dimensional coordinate system, it is possible to locate objects anywhere in space. One method used by astronomers to identify points in space is based on this system.*

By using the x- and y-axes in a coordinate plane, you can give the location of any point on the plane. Only two numbers are required to do this, so a plane is said to be *two-dimensional*. By adding a third axis, called the z-axis, that intersects the x- and y-axes at right angles, you can give the location of any point in space. Three numbers are required for this, so space is said to be *three-dimensional*.

The Arrangement of the Axes

The arrows on the axes point in the positive direction.

Examine the x- and y-axes at left. There are two ways to add a third axis. The positive direction of the z-axis can point either straight up (out of the page) from the origin or straight down (into the page). In the most common system, the z-axis points straight up. To represent the three axes on a page, turn the axes so that the x-axis points down and to the left, the y-axis points to the right, and the z-axis points straight up, as shown at right. Imagine that the x-axis points straight out from the page.

This arrangement of the axes is called a **right-handed system**. With your right hand, let your index finger represent the x-axis, your middle finger, the y-axis, and your thumb, the z-axis. Hold these fingers at right angles to each other, as shown. Each finger points in the positive direction of the axis.

View your right hand from different perspectives while keeping your fingers in the same positions to see different views of the right-handed system. Do the axes shown at right represent right-hand systems?

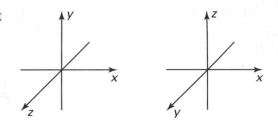

EXAMPLE 1 Locate the point $P(1, 2, 3)$ in a three-dimensional coordinate system.

● **SOLUTION**

1. Starting at the origin, count 1 unit in the positive direction along the x-axis. Make a mark on the x-axis at this position.

2. From your mark on the x-axis, count 2 units in the positive direction along the y-axis, drawing a dashed line to represent the distance. Make a mark at the new position.

3. From the new position, count 3 units in the positive direction along the z-axis, drawing a dashed line to represent the distance. Label your final position as point $P(1, 2, 3)$.

The Octants and the Coordinate Planes

Just as the x- and y-axes divide the plane into four quadrants, the x-, y- and z-axes divide space into eight **octants**. The octant in which all three coordinates of a point are positive, abbreviated $(+, +, +)$, is called the *first octant*.

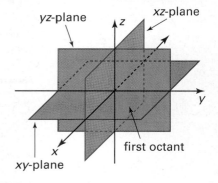

The remaining octants are described by the words *top*, *bottom*, *front*, *back*, *left*, and *right*. For example, the top-front-left octant is the octant in which each point has positive x- and z-coordinates and a negative y-coordinate, or $(+, -, +)$.

CHECKPOINT ✔ What are the signs for the coordinates of points in each of the other six octants?

Each pair of axes also determines a **coordinate plane**. There are three coordinate planes, each named by the pair of axes that determines the plane: the xy-plane, the xz-plane, and the yz-plane.

• In the xy-plane, the z-coordinate of every point is 0.
• In the xz-plane, the y-coordinate of every point is 0.
• In the yz-plane, the x-coordinate of every point is 0.

CHECKPOINT ✔ What can you say about the coordinates of a point on the x-axis? on the y-axis? on the z-axis?

EXAMPLE 2 Draw a right rectangular prism in the first octant so that three of its faces are in the three coordinate planes. Label the coordinates of each vertex.

SOLUTION

One possible solution is shown in the figure at right.

The Distance Formula in Three Dimensions

Recall from Lesson 6.3 that the length of the diagonal of a right rectangular prism with dimensions $\ell \times w \times h$ is given by $d = \sqrt{\ell^2 + w^2 + h^2}$.

To find the distance between two points in three-dimensions, construct a right rectangular prism with faces parallel to the coordinate planes and with the given points as vertices. The segment joining the two points is a diagonal of this prism. The dimensions of the prism are $\ell = |x_2 - x_1|$, $w = |y_2 - y_1|$, and $h = |z_2 - z_1|$.

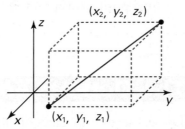

Therefore, the distance between two points in space is given by the following formula:

Distance Formula in Three Dimensions

The distance, d, between the points (x_1, y_1, z_1) and (x_2, y_2, z_2) is given by

$$d = \sqrt{(x_2 - x_1)^2 + (y_2 - x_1)^2 + (z_2 - z_1)^2}.$$ **6.4.1**

EXAMPLE 3 Find the distance between the points $R(4, 6, -9)$ and $S(-3, 2, -6)$.

SOLUTION

$$
\begin{aligned}
RS &= \sqrt{(-3 - 4)^2 + (2 - 6)^2 + [-6 - (-9)]^2} \\
&= \sqrt{(-7)^2 + (-4)^2 + (3)^2} \\
&= \sqrt{49 + 16 + 9} \\
&= \sqrt{74} \approx 8.6
\end{aligned}
$$

TRY THIS Find the distance between the points $C(3, -4, -5)$ and $D(2, 0, -1)$.

Exercises

Communicate

1. What can you say about the location of a point in space that has one coordinate equal to 0? two coordinates equal to 0?

Exercises 2 and 3 refer to the coordinate axes and point *P* at right.

2. Is it possible to determine the coordinates of point *P* from the diagram? Explain your reasoning. If not, what information is needed?

3. Is it possible to determine the octant in which point *P* is located from the diagram? If not, is it possible to determine an octant in which *P* could not be located?

4. In the distance formula in three dimensions, why is it not necessary to use absolute-value signs on the lengths of the sides?

5. Imagine that your classroom is in a three-dimensional coordinate system. Choose a point in the room to represent the origin, and describe the coordinates of some objects in the room.

Guided Skills Practice

Locate each point in a three-dimensional coordinate system. (EXAMPLE 1)

6. $(2, 0, -1)$ 7. $(-1, -2, -3)$ 8. $(3, -1, 4)$

9. A right rectangular prism in the first octant of a three-dimensional coordinate system is positioned so that three of its faces are in the coordinate planes. The prism has a length of 2, width of 7, and height of 6. Find the coordinates of each vertex. **(EXAMPLE 2)**

Find the distance between each pair of points. (EXAMPLE 3)

10. $(0, 0, 0)$ and $(3, 6, 1)$ 11. $(3, 1, 0)$ and $(5, -3, 1)$

12. $(-4, 7, -2)$ and $(2, 3, -5)$

Practice and Apply

Locate each point in a three-dimensional coordinate system.

13. $(4, 1, 5)$ 14. $(0, -3, 0)$ 15. $(0, -5, 2)$ 16. $(-3, -1, 3)$

Name the octant, coordinate plane, or axis in which each point is located.

17. $(3, 1, 7)$ 18. $(-2, 10, -4)$ 19. $(4, 0, -2)$ 20. $(0, -5, 0)$

21. $(-3, 1, 7)$ 22. $(-1, 3, 0)$ 23. $(4, 0, 0)$ 24. $(-8, -1, -5)$

Give the coordinates of a point in the given octant.

25. first octant

26. bottom-front-right

27. top-back-left

28. bottom-back-left

Algebra

Find the distance between each pair of points.

29. $(1, 1, 1)$ and $(-1, -1, -1)$

30. $(2, 1, 3)$ and $(5, -2, 7)$

31. $(2, 0, 1)$ and $(-5, -6, -5)$

32. $(7, -6, 5)$ and $(6, 4, 3)$

For Exercises 33–42, refer to the diagram below of a right rectangular prism.

Determine the coordinates of each point.

33. point F

34. point H

35. point C

36. point E

Find each measure.

37. AD

38. AG

39. AH

40. DF

41. area of $EGDA$

42. area of $ABCD$

Algebra

The midpoint of a segment in a three dimensional coordinate system is given by the following formula:

Midpoint Formula in Three Dimensions

The midpoint of a segment with endpoints at (x_1, y_1, z_1) and (x_2, y_2, z_2) is the point $\left(\dfrac{x_1 + x_2}{2}, \dfrac{y_1 + y_1}{2}, \dfrac{z_1 + z_2}{2} \right)$.

6.4.2

Find the midpoint of the segment with the given endpoints.

43. $(5, -2, 3)$ and $(6, -7, 4)$

44. $(3, 2, 1)$ and $(1, 2, 3)$

45. $(-1, -4, -5)$ and $(6, 1, 0)$

46. $(2, -1, 0)$ and $(0, 0, 1)$

47. $(1, 1, 1)$ and $(-1, -1, -1)$

48. (a, b, c) and $(-a, -b, -c)$

CHALLENGE

49. Give the coordinates of a point that is not on an axis or in a coordinate plane and that is a distance of 1 unit from the origin.

APPLICATION

50. GRAPHICS A white sphere is illuminated by two different light sources, one red and one blue. Where the red and blue light mix, magenta light is produced. Examine the illustration at right. Where are the light sources located with respect to the coordinate axes?

Look Back

Find the slope of the segment with the given endpoints. *(LESSON 3.8)*

51. (0, 0) and (6, 4)

52. (1, 5) and (3, 3)

53. (3, 1) and (2, −5)

54. (−3, 1) and (6, −4)

Graph each line and find the *x*- and *y*-intercepts.

55. $y = 2x + 1$

56. $2x + 3y = 6$

57. $y − 3 = 4(x + 1)$

58. $3x − y = 5$

APPLICATION

59. PHYSICAL SCIENCE The relationship between Celsius and Fahrenheit temperatures can be represented by a straight line. Suppose that the *x*-coordinate represents Celsius temperature and the *y*-coordinate represents Fahrenheit temperature. Then the freezing point of water is the point (0, 32) and the boiling point is (100, 212).

  a. Find the slope of the line through the given points.

  b. Find the equation of the line through the given points.

Look Beyond

APPLICATION

ASTRONOMY **The table below gives the *x*-, *y*-, and *z*-coordinates of points in a three-dimensional coordinate system with the Sun at the origin.**

The *x*-axis points in the direction of Earth's position at the first moment of the spring, or vernal, equinox. The *xy*-plane is the plane of Earth's equator at the time of the equinox.

An astronomical unit is a measure of length that is equal to the mean distance from Earth to the Sun, about 92,960,000 miles or 149,600,000 kilometers.

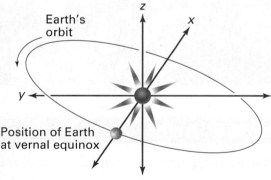

Sun-Centered Coordinates of Major Planets on August 17, 1990 (in astronomical units)			
Planets	*x*	*y*	*z*
Earth	0.819	−0.546	−0.237
Mars	1.389	0.145	0.029
Jupiter	−2.238	4.328	1.910
Saturn	3.935	−8.443	−3.656

Source: *The Astronomical Almanac for the year 1990*

60. Calculate the distance from Earth to Jupiter on August 17, 1990. Then convert from astronomical units to miles.

61. Calculate the distance from Mars to Saturn on August 17, 1990. Then convert from astronomical units to kilometers.

62. Calculate the distance from the Sun to Mars on August 17, 1990. The mean distance from the Sun to Mars is 227,940,000 kilometers. How close to this value is your calculation? Why do you think there is a difference?

63. Of the planets in the table, which is closest to the Sun?

Lines and Planes in Space

Objectives

- Define the equation of a line and the equation of a plane in space.

- Solve problems by using the equations of lines and planes in space.

Why *Planes in space can form three-dimensional shapes, as in these computer drawings. Each plane in the drawings can be described mathematically.*

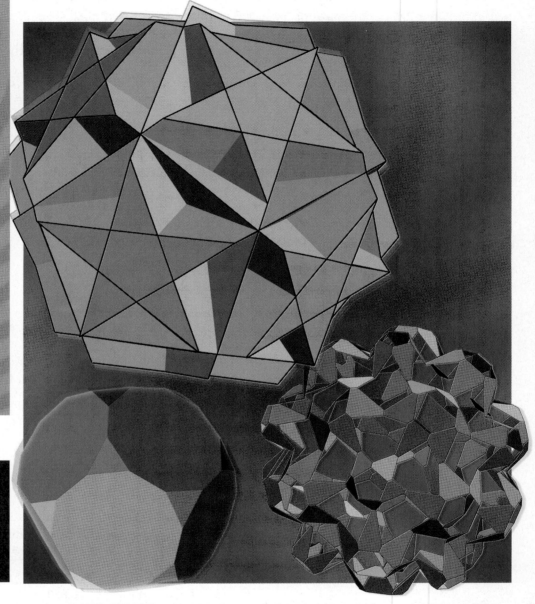

The Equation of a Plane

Recall from algebra that the standard form of the equation of a line in a plane is

$$Ax + By = C,$$

where A, B, and C are real numbers, A is nonnegative, and A, B, and C are not all zero. The standard equation of a plane in space resembles the equation of a line, with an extra variable for the added dimension of space:

$$Ax + By + Cz = D,$$

where A, B, C, and D are real numbers, A is nonnegative, and A, B, C, and D are not all zero. For example,

$$2x + 5y - 2z = 9$$

is an equation of a plane where $A = 2$, $B = 5$, $C = -2$, and $D = 9$.

Using Intercepts in Graphing

In the coordinate plane, a line that does not lie entirely on an axis crosses the x- and y-axes at one or two points called **intercepts**. In coordinate space, a plane that does not lie entirely on an axis has one, two, or three intercepts.

EXAMPLE ❶ Sketch the graph of the plane defined by the equation $2x + y + 3z = 6$.

Algebra

SOLUTION

Find the intercepts where the plane crosses each axis. The x-intercept, for example, is the x-coordinate of the point where the plane crosses the x-axis. At this point, the y- and z-coordinates are 0. To find the x-intercept, set $y = 0$ and $z = 0$ and solve the equation for x.

$$2x + 0 + 3(0) = 6$$
$$2x = 6$$
$$x = 3$$

The x-intercept is $(3, 0, 0)$. Similarly, the y-intercept is $(0, 6, 0)$, and the z-intercept is $(0, 0, 2)$. Plot the intercepts as points on the axes. These three noncollinear points determine the plane. To sketch the plane, connect the points with segments and add shading. NOTE: Intercepts are often written as single numbers.

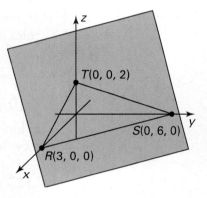

What Happens to the Equation of a Line in Space?

In a coordinate plane, $2x + 4y = 8$ is the equation of a line.

In a three dimensional coordinate space, the equation $2x + 4y = 8$ is the equation of a plane in which the coefficient of z is equal to 0.

Notice that this equation of a plane is unaffected by the values of z. Thus for any values of x and y that satisfy the equation, any value of z will also work. Thus every point directly above and below the line is in the graph of the plane.

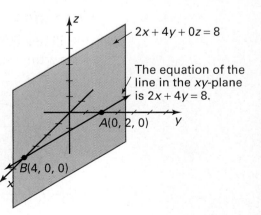

Lines and Planes in Space: A Step-by-Step Procedure

Imagine that you are able to plot one point of a graph in space every minute according to a given set of instructions. Your instructions have one rule for the *x*-coordinates, another for the *y*-coordinates, and another for the *z*-coordinates. These rules are called **parametric equations**.

Let $t = 1, 2, 3, \ldots$ represent the time (in minutes) at which you plot each point. Your rules for each coordinate would have the following form:

$$x = [\text{an expression involving } t]$$
$$y = [\text{an expression involving } t]$$
$$z = [\text{an expression involving } t]$$

Note: *t* can also be 0 or negative.

E X A M P L E 2 To plot a graph in a coordinate plane, you will need just two rules. Use the rules below to plot a line in a coordinate plane for $t = 1, 2, 3, 4 \ldots$

$$x = 2t$$
$$y = 3t + 1$$

SOLUTION

Fill in a table like the one below. Then plot the points in the *xy*-coordinate plane.

t	x	y
1	2	4
2	4	7
3	6	10
4	8	13

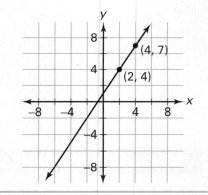

E X A M P L E 3 Use the rules below to plot a line in a coordinate system for $t = 1, 2, 3, 4 \ldots$

$$x = 2t + 1$$
$$y = 3t - 5$$
$$z = 4t$$

SOLUTION

Fill in a table like the one below. Then plot the graph in a three-dimensional coordinate system.

t	x	y	z
1	3	-2	4
2	5	1	8
3	7	4	12
4	9	7	16

Exercises

Communicate

1. What is the standard form for an equation of a plane? How is it similar to the standard equation of a line? How is it different?

2. Describe the characteristics of a plane defined by an equation in which the coefficient of x is 0.

3. Describe the characteristics of a plane defined by an equation in which the coefficient of z is 0.

4. Describe the characteristics of a plane defined by an equation in which the coefficients of y and z are 0.

5. Can a line in a three-dimensional coordinate system pass through exactly one octant? exactly two octants? exactly three octants? more than three octants? Explain your reasoning.

Guided Skills Practice

Sketch the graph of each plane. *(EXAMPLE 1)*

6. $3x + 6y + 4z = 12$

7. $2x + 5y - z = 2$

Using $t = 1, 2, 3, \ldots$, create a table of x- and y-values for the equations below. Then graph each line in a coordinate plane. *(EXAMPLE 2)*

8. $x = t + 2$
$y = 4t$

9. $x = t - 1$
$y = 2t + 3$

Using $t = 1, 2, 3, \ldots$, create a table of x-, y-, and z-values for the equations below. Then graph each line in a coordinate system. *(EXAMPLE 3)*

10. $x = t$
$y = 2t$
$z = t + 1$

11. $x = 3t - 2$
$y = t + 6$
$z = -t$

Practice and Apply

Use intercepts to sketch the plane defined by each equation below.

12. $3x + 2y + 7z = 4$

13. $2x - 4y + z = -2$

14. $x - 2y - 2z = -4$

15. $-3x + y = 4$

16. $x - 2y = 2$

17. $x = -4$

Algebra

Plot the line for each pair of parametric equations in a coordinate plane.

18. $x = t + 3$
$y = 1 - t$

19. $x = 2t - 1$
$y = -t$

20. $x = 3t$
$y = 5$

21. $x = 2$
$y = -4t + 6$

Plot the line for each set of parametric equations in a coordinate system.

22. $x = t$
$y = 2t$
$z = 3t$

23. $x = t + 1$
$y = 2t + 1$
$z = 0$

24. $x = t$
$y = t$
$z = 1 - t$

25. $x = 3t$
$y = 4$
$z = 4t - 7$

The *trace* of a plane is its intersection with the *xy*-plane. To find the equation of the trace, set the *z*-coordinate equal to 0 (all points in the *xy*-plane have a *z*-coordinate of 0). For example:

$$2x + y + 3z = 6 \quad \textit{Equation of the plane}$$
$$2x + y + 3(0) = 6 \quad \textit{Set z equal to 0.}$$
$$2x + y = 6 \quad \textit{Equation of the trace}$$

Find the equation of the trace for each plane defined below. Sketch the plane and indicate the trace.

26. $x + 3y - z = 7$

27. $5x - 2y + z = 2$

28. $2x + 7y + 3z = 2$

29. $-4x - 2y + 2z = 1$

30. Write parametric equations to describe the line that passes through the points $(0, 0, 0)$ and $(2, 4, -1)$. (Hint: Start with the rule for the *x*-coordinate. The equation will be of the form $x = at + b$, where a and b are constants.

31. Write parametric equations to describe the line that passes through the points $(-2, 1, 5)$ and $(6, -4, 7)$.

In a coordinate system, plot the pair of lines described by each set of parametric equations below. Determine whether the lines are parallel, intersecting, or skew. Explain your reasoning. (Note: Let $t = 1, 2, 3, \ldots$ and $s = 1, 2, 3, \ldots$)

32. $x = 1 \quad x = 1$
$y = 2 \quad y = s$
$z = t \quad z = 3$

33. $x = 1 \quad x = 3$
$y = 2 \quad y = 5$
$z = t \quad z = s$

34. $x = 1 \quad x = 3$
$y = 2 \quad y = s$
$z = t \quad z = s$

CHALLENGE

35. Do the lines described by the parametric equations below intersect? How can you tell? If they intersect, find the point of intersection. (Hint: Set the *x*-coordinates equal to each other to form an equation. Repeat with the *y*- and *z*-coordinates to get a system of three equations in *s* and *t*, and then solve the system.)

$$x = t - 1 \qquad x = s - 3$$
$$y = 2t + 1 \qquad y = s$$
$$z = -t \qquad z = 2s - 7$$

APPLICATION

36. HOUSE PAINTING You agree to pay a friend $5 per hour to help paint your house. The paint costs $25 per gallon, and the equipment (ladder, brushes, etc.) costs $200. An equation representing the cost, *z*, of painting for *x* hours and using *y* gallons of paint is $z = 5x + 25y + 200$. Write this equation in the standard form for an equation of a plane, and sketch a graph of the plane.

37. HOBBIES Delia rented a booth at a craft fair for $50. She is selling earrings at a profit of $7 per pair and bracelets at a profit of $4 each. Her profit, z, for selling x pairs of earrings and y bracelets is represented by the equation $z = 7x + 4y - 50$. Write this equation in the standard form for an equation of a plane, and sketch a graph of the plane. Find the trace of the plane. What is Delia's profit for points in the trace?

38. NAVIGATION Suppose that an airplane's initial path at takeoff is described by the following parametric equations:

$$x = -t$$
$$y = t$$
$$z = 0.5t$$

Plot the line described by the given equations. If the positive x-axis points north, in which direction did the airplane take off?

Look Back

COORDINATE GEOMETRY For each pair of lines below, state whether the lines are parallel, perpendicular, or neither. *(LESSON 3.8)*

39. $y = 3x + 5$
   $y = 3x - 7$

40. $y = x + 2$
   $y = 2 - x$

41. $y = 2x - 1$
   $y = -2x + 4$

42. $2x + 3y = 6$
   $3x - 2y = 6$

43. Given: $\angle 1 \cong \angle 2$ and $\overline{EH} \cong \overline{FG}$
   Prove: $\overline{EF} \cong \overline{HG}$ *(LESSON 3.3)*

44. Given rectangle $ABCD$ with diagonals $\overline{AC}$ and $\overline{BD}$, prove $\triangle ADC \cong \triangle BCD$. *(LESSON 4.6)*

45. Given rectangle $ABCD$ with diagonals $\overline{AC}$ and $\overline{BD}$ that intersect at point E, prove $\triangle AED \cong \triangle CEB$. *(LESSON 4.6)*

46. Given quadrilateral $PQRS$ in which $\overline{PQ} \cong \overline{RS}$ and $\overline{PS} \cong \overline{QR}$, prove that $PQRS$ is a parallelogram. *(LESSON 4.6)*

Algebra

Symmetric equations for a line in space are in the form $\frac{x - x_1}{a} = \frac{y - y_1}{b} = \frac{z - z_1}{c}$, where (x_1, y_1, z_1) is a point on the line and a, b, and c are constants.

The relationship between the parametric and symmetric forms is shown below.

$$\frac{x - x_1}{a} = \frac{y - y_1}{b} = \frac{z - z_1}{c} \longleftrightarrow \begin{array}{l} t = \dfrac{x - x_1}{a} \\[2mm] t = \dfrac{y - y_1}{b} \\[2mm] t = \dfrac{z - z_1}{c} \end{array} \longleftrightarrow \begin{array}{l} x = at + x_1 \\ y = bt + y_1 \\ z = ct + z_1 \end{array}$$

Write each set of symmetric equations in parametric form.

47. $\dfrac{x - 2}{3} = \dfrac{y - 5}{-1} = \dfrac{z + 1}{4}$

48. $x - 2 = \dfrac{y + 6}{2} = \dfrac{z - 3}{-5}$

Write each set of parametric equations in symmetric form.

49. $x = 2t + 1$
$y = -3t + 6$
$z = 4t$

50. $x = t - 4$
$y = 2t$
$z = 6t + 2$

PORTFOLIO ACTIVITY

ARCHIMEDEAN SOLIDS

Archimedean solids, or semiregular polyhedra, are named after Greek mathematician Archimedes. All of the faces of an Archimedean solid are regular polygons, but two or more types of polygons are used. In addition, every vertex must have the exact same arrangement of faces around it. There are exactly 13 Archimedean solids.

1. Enlarge and copy the net at right. You may wish to add tabs at some of the edges to help in assembling the polyhedron. This solid is called a *rhombicuboctahedron*. It has 8 equilateral triangular faces and 18 square faces. Three squares and one triangle meet at every vertex.

2. Enlarge and make two copies of the net at right. The net forms half of the solid known as a *truncated icosahedron*. If you truncate, or cut off, the vertices of a regular icosahedron (refer back to the Portfolio Activity on page 379), this shape is formed. A truncated icosahedron has 12 regular pentagonal faces and 20 regular hexagonal faces. You may recognize the truncated icosahedron as a common object—a soccer ball.

Perspective Drawing

Why
You know that objects that are far away appear smaller than they would if they were close to you. In studying perspective drawing you will learn the rules for making things appear in proper relationship to each other.

Objectives

- Identify and define the basic concepts of perspective drawing.

- Apply these basic concepts to create your own perspective drawings.

European Renaissance artists in the fourteenth through sixteenth centuries rediscovered, from classical Greek and Roman art, how to create the illusion of depth in drawings and paintings. Notice the relatively flat appearance of the pre-Renaissance work on the right compared with the Renaissance painting to the left.

Perspective Drawings: Windows to Reality

A picture-plane "window" containing a projected image

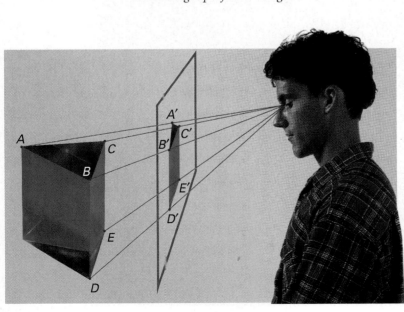

Modern perspective drawing methods, discovered by the Italian architect Fillip Brunelleschi (1377–1446), are based on the idea that a picture is like a window.

An artist creating a picture, or a person looking at the finished picture, is *thought* of as looking through the picture to the reality it portrays. (The word *perspective* comes from the Latin words meaning "looking through.")

When someone looks at an object, there is a line of sight from every point on the object to the eye. Imagine a plane, such as an empty canvas, that intersects the lines of sight. The points of intersection on the plane make up the image of the object, and the image is said to be *projected* onto the "picture plane."

Albrecht Dürer (1471–1528), a German artist, visited Italy to learn the techniques of perspective drawing. He produced a number of works that showed artists employing these techniques, as in the woodcut at right.

Two artists producing an image of a lute, one point at a time

CHECKPOINT ✔ How does the technique used by the artist in Dürer's woodcut illustrate the projection of an object onto a picture plane?

Parallel Lines and Vanishing Points

Have you ever noticed how the rails of a railroad track or the sides of a highway seem to meet as they recede into the distance? The point where parallel lines seem to meet, which is often on the horizon, is known in perspective drawing as the **vanishing point**.

$\overline{AB}$ and $\overline{CD}$ on the sidewalk in the picture below are actually the same length, but when they are projected onto the picture plane seen by the student, the image of $\overline{AB}$ is longer than the image of $\overline{CD}$.

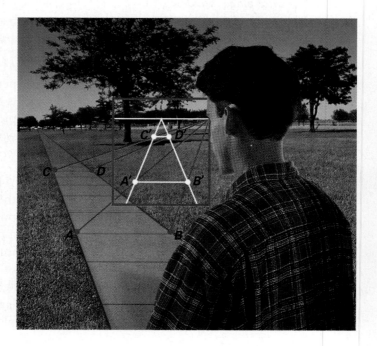

Principles of Perspective Drawing

The two theorems on this page provide the basis for an understanding of how perspective drawings are made. They can be demonstrated by studying the way parallel lines project onto the picture plane of a perspective drawing.

Theorem: Sets of Parallel Lines

In a perspective drawing, all lines that are parallel to each other, but not to the picture plane, meet at a single point known as a vanishing point. **6.6.1**

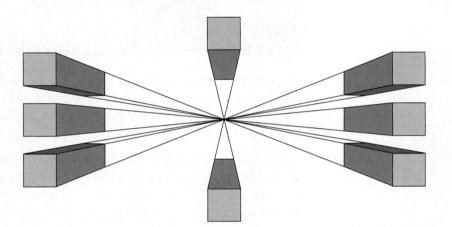

CRITICAL THINKING Do you think that the point where a set of parallel lines seem to meet has to be somewhere in the drawing? Use illustrations in your explanation.

Theorem: Lines Parallel to the Ground

In a perspective drawing, a line that is in the plane of the ground in the drawing and is not parallel to the picture plane will meet the horizon of the drawing. Any line parallel to this line will meet the horizon of the drawing at the same point. **6.6.2**

CRITICAL THINKING Parallel lines that are parallel to the picture plane of a perspective drawing are usually represented with no vanishing point. In most cases this procedure causes no problems. Can you think of situations in which it would result in unrealistic drawings?

In a perspective drawing, the concept of vanishing points applies even when no parallel lines are actually shown in the drawing. In the row of telescope dishes shown below, for example, there are imaginary lines that pass through points at the tops and at the bottoms of the reflectors. These lines meet at

CHECKPOINT ✔ the horizon. Explain why.

Much of the subject matter of early perspective drawings was architectural. Buildings and houses are ideally suited to the development of the theories of perspective drawing because they usually contain many lines that are parallel to each other and to the ground.

Eventually, perspective drawing began to influence architecture, as illustrated in the Church of San Lorenzo in Florence, Italy, shown below. Brunelleschi used the principles of perspective drawing in his design for this church. The vanishing point of the structural elements appears to be at the altar.

Church of San Lorenzo in Florence, Italy

Exercises

● Communicate

1. Explain what is meant by a vanishing point in a perspective drawing.

2. Assume that any two lines in a perspective drawing that are parallel to each other but not to the picture plane meet at a point. How is this statement extended to apply to sets of parallel lines in Theorem 6.6.1? What theorem or postulate justifies this extension?

3. Explain why a line that is in the plane of the ground and is not parallel to the picture plane must end at the horizon. Make a sketch to explain your answer.

4. Explain why drawings of buildings and houses are ideally suited to the development of perspective drawing. Make a sketch to explain your answer.

5. In a perspective drawing, the horizon is usually represented by a horizontal line that is assumed to be at eye level. Why do you think this assumption is made?

6. What is the difference between a perspective drawing and an isometric drawing?

CRITICAL THINKING

7. The images of the two polar bears in the picture at right are actually the same size (measure them), but one appears larger than the other. Explain why.

● Guided Skills Practice

8. The drawing at right is a perspective drawing of a cube. What do you know about the lines that contain $\overline{AB}$, $\overline{CD}$, and $\overline{EF}$? about the lines that contain $\overline{AC}$, $\overline{BD}$, and $\overline{GF}$? **(THEOREM 6.6.1)**

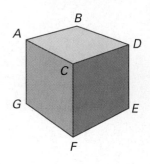

9. In the drawing of the cube, the lines that contain the vertical segments will *not* meet at a vanishing point. Use the principles of perspective drawing to explain why. **(THEOREM 6.6.1)**

10. Imagine that the drawing of the cube represents a cube-shaped building on level ground. Where will the lines that contain the nonvertical sides of the building meet? **(THEOREM 6.6.2)**

The exercises below give the steps that are used to produce various types of perspective drawings.

11. A drawing that has just one vanishing point is said to have one-point perspective. Follow the steps below to produce a one-point perspective drawing of a cube.

    a. Draw a square. Then draw a horizontal line to represent the horizon. Mark a vanishing point on the horizon.

    b. From each corner of the square, lightly draw dashed lines to the vanishing point you marked in part **a.**

    c. Lightly draw the sides of a smaller square whose vertices touch the lines you drew in part **b.**

    d. Erase the perspective lines that extend "behind" the smaller cube. The dashed lines that remain indicate the edges of the cube that are hidden from view.

12. Repeat the steps in Exercise 11, but place the vanishing point to the left or right of the square.

13. Repeat the steps in Exercise 11, but place the horizon line and vanishing point below the square.

14. What happens if you place the vanishing point in the interior or on an edge of the square?

For Exercises 15 and 16, trace the figure below onto your paper.

15. Locate the vanishing point for the figure.

16. Draw the horizon line.

17. A drawing that has two vanishing points is said to have **two-point perspective**. Follow the steps below to produce a two-point perspective drawing of a cube.

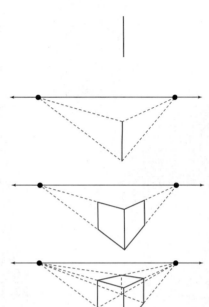

  a. Draw a vertical segment. This will be the front edge of your cube. Draw a horizon line above the segment. Place two vanishing points on the horizon line as shown, with one on either side of the vertical segment.

  b. Lightly draw lines back to each vanishing point from the endpoints of the vertical segment as shown.

  c. Draw vertical segments to complete the front sides of the cube.

  d. Lightly draw lines from the endpoints of the segments you drew in step **c** to each vanishing point. Draw a vertical dashed line between the two intersection points of the light perspective lines.

  e. Erase the perspective lines that extend beyond the edges of the cube. Use dashed lines to indicate the edges of the cube that are hidden from view.

18. Repeat the steps in Exercise 17, but place the horizon line and vanishing points below the vertical line.

19. Repeat the steps in Exercise 17, but place the horizon line so that it intersects the vertical line.

20. What happens to a two-point perspective drawing of a cube as the vanishing points are moved closer together? farther apart?

21. What happens to a two-point perspective drawing of a cube if both vanishing points are moved to the same side of the original vertical line? if one vanishing point is directly above the vertical line?

For Exercises 22 and 23, trace the figure below onto your paper.

22. Locate the vanishing points for the figure.

23. Draw the horizon line.

24. Follow the steps below to create a perspective drawing of your name in black letters.

  a. Draw "flat" block letters and a horizon line with a vanishing point, as shown.

  b. Draw lines from all corners and appropriate curved edges of the letters, as shown.

  c. Fill in the edges as shown. Erase the perspective lines to complete the drawing.

25. Use one-point or two-point perspective to draw a city view. Start with boxes for the buildings and then add details.

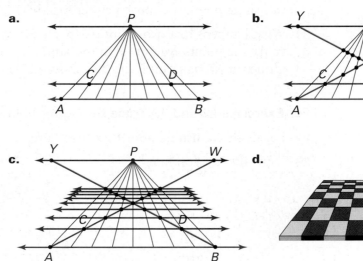

Tiling was particularly intriguing to the artists who first explored perspective studies. The pictures below suggest a technique for creating a tile pattern in a perspective drawing. Study the pictures to answer Exercises 26–28.

26. A method for finding the lines parallel to $\overline{AB}$ is suggested by the diagonals. Explain how the diagonals can be used to determine these parallel lines.

27. Explain how the tile pattern could be viewed from a corner by applying two-point perspective. How would the intersecting lines be determined?

28. Create your own one- or two-point perspective drawing of a square tile pattern.

Look Back

29. A solid composed of unit cubes has the top view shown at right. What are two possibilities for the solid? Make an isometric drawing of each. *(LESSON 6.1)*

30. A solid composed of unit cubes has the front and right views shown at right. What might the top view look like? *(LESSON 6.1)*

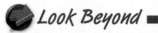

Front   Right

Locate and sketch each point in a three-dimensional coordinate system. (LESSON 6.4)

31. $(5, -1, -2)$ **32.** $(13, 0, 0)$ **33.** $(-2, 0, 5)$

Find the distance between each pair of points with the given coordinates in a three-dimensional coordinate system. (LESSON 6.4)

34. $(4, 3, 2)$ and $(-5, 2, -1)$ **35.** $(-1, 0, 1)$ and $(15, 5, -2)$

Find the midpoint of the segment with the given endpoints in a three-dimensional coordinate system. (LESSON 6.4)

36. $(5, 5, 5)$ and $(-3, -3, -3)$ **37.** $(0, 0, 0)$ and $(-1, 10, 9)$

Look Beyond

38. The artist in the photograph at right is using an image produced by a projector to create a mural. Draw a diagram showing how an image on a slide is projected onto a screen or wall. Assume that the light source is a single point of light. (The lens arrangement of the projector makes this assumption appropriate.)

39. How is the projection process of the slide projector like the projection in perspective drawing? How is it different?

PROJECT

Polyhedra

Activity 1

The Small Stellated Dodecahedron

A *stellated* polyhedron is formed from an existing polyhedron by extending the plane of each face until the planes meet, forming a new polyhedron, which is larger and has a star shape.

- Enlarge and copy the net below to create a regular dodecahedron out of cardboard.

- Enlarge and make 12 copies of the net below. The base should be the same length as an edge of the dodecahedron. The triangles are isosceles, with a vertex angle of 36° and a base angle of 72°. Color the triangles, if you wish.

- Fold the 12 nets above to form 12 pyramids that are open at the bottom. Use glue or tape to attach one pyramid to each face of the dodecahedron.

Why Are There Five?

The five convex regular polyhedra, known as the Platonic solids, have been studied extensively since the time of the ancient Greeks. To see why there are only five, consider a single vertex of each one. Since all vertices of a regular polyhedron are identical, the shape of the polyhedron is uniquely determined by one vertex.

1. The patterns below are possible arrangements of equilateral triangles at a vertex. Copy the patterns and cut them out. Glue or tape the red edges of each pattern to form a single vertex.

   Which Platonic solid has 3 triangles at each vertex? Which has 4 triangles at each vertex?

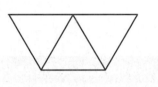

2. Try some other arrangements of equilateral triangles at a vertex. What is the minimum number possible? What is the maximum number possible? Explain your reasoning.

Johannes Kepler (1571–1630) believed that the orbits of the planets were described by nested regular polyhedra.

3. The pattern at right is a possible arrangement of squares at a vertex. Copy the pattern and cut it out. Glue or tape the red edges to form a single vertex.

   Which Platonic solid has 3 squares at each vertex? Are any other arrangements of squares possible at a single vertex?

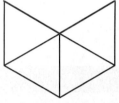

4. What arrangements of regular pentagons are possible at a single vertex? Which Platonic solid has an arrangement of pentagons at each vertex?

5. Can any other regular polygons be arranged at a single vertex? Explain your reasoning.

Extension

1. In each vertex arrangement above, what is the sum of the angles at the vertex?

2. As the sum of the angles increases, what happens to the sharpness of the vertex? What do you think is the maximum possible sum of the angles?

3. How do you think it would be possible to prove that there are exactly 13 Archimedean solids?

Chapter Review and Assessment

VOCABULARY

POSTULATES AND THEOREMS

Lesson	Number	Postulate or Theorem
6.3	6.3.1 Diagonal of a Right Rectangular Prism	The length of the diagonal, d, of a right rectangular prism is given by $d = \sqrt{\ell^2 + w^2 + h^2}$.
6.4	6.4.1 Distance Formula in Three Dimensions	The distance, d, between the points (x_1, y_1, z_1) and (x_2, y_2, z_2) in space is given by $d = \sqrt{(x_2 - x_1)^2 + (y_2 - y_1)^2 + (z_2 - z_1)^2}$.
	6.4.2 Midpoint Formula in Three Dimensions	The midpoint of a segment with endpoints at (x_1, y_1, z_1) and (x_2, y_2, z_2) in space is given by $\left(\frac{x_1 + x_2}{2}, \frac{y_1 + y_2}{2}, \frac{z_1 + z_2}{2} \right)$.
6.6	6.6.1 Theorem: Sets of Parallel Lines	In a perspective drawing, all lines that are parallel to each other, but not to the picture plane, will seem to meet at a single point known as a vanishing point.
	6.6.2 Theorem: Lines Parallel to the Ground	In a perspective drawing, a line that is on the plane of the ground and is not parallel to the picture plane will meet the horizon of the drawing. Any line parallel to this line will meet the horizon at the same point.

LESSON 6.1

Key Skills

Create isometric drawings of solid figures.

A solid figure is composed of seven cubes as shown.

Make a drawing of the solid on isometric dot paper.

Draw orthographic projections of solid figures.

Draw six orthographic projections of the solid above.

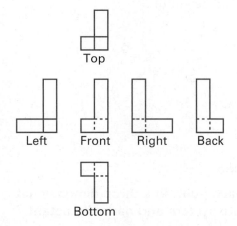

Find the volume and surface area of solids that are composed of cubes.

Find the volume and surface area of the solid above.

The solid is composed of seven cubes, so its volume is 7 cubic units. There are 30 exposed cube faces, so the surface area is 30 square units.

Exercises

Refer to the isometric drawing below. Assume that no cubes are hidden in the drawing.

1. Make an isometric drawing of the solid from a different angle. (You may wish to create a model of the solid by using unit cubes.)

2. Draw the six orthographic views of the solid.

3. Find the volume of the solid.

4. Find the surface area of the solid.

Key Skills

Identify relationships among lines and planes in space.

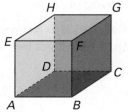

In the cube, name a set of parallel faces, perpendicular faces, and skew edges. Then name an edge and a face that are perpendicular.

Faces *ADHE* and *BCGF* are parallel. Faces *ADHE* and *ABFE* are perpendicular. $\overline{EH}$ and $\overline{BF}$ are skew edges. $\overline{AB}$ is perpendicular to *BCGF*.

Exercises

Refer to the right hexagonal prism below. *UVWXYZ* is a regular hexagon.

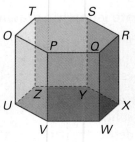

5. Name two pairs of parallel faces.

6. Name a pair of perpendicular faces.

7. Name a pair of skew edges.

8. Name an edge and a face that are perpendicular.

Key Skills

Identify parts of a prism.

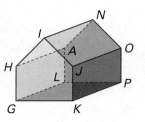

In the pentagonal prism at right, name a base, a lateral face, and a lateral edge.

Pentagon *GHIJK* is a base, parallelogram *GLPK* is a lateral face, and $\overline{IN}$ is a lateral edge.

Find the length of a diagonal of a right rectangular prism.

A right rectangular prism has a length of 20, a width of 10, and a height of 24. What is the length of its diagonal?

The length of a diagonal is given by the formula $d = \sqrt{\ell^2 + w^2 + h^2}$.

$d = \sqrt{20^2 + 10^2 + 24^2} = \sqrt{1076} \approx 32.8$

Exercises

Refer to the prism below.

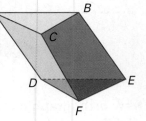

9. Name a base.

10. Name a lateral face.

11. Name a lateral edge.

12. What is the length of a diagonal of a right rectangular prism with a length of 24, width of 8, and height of 6?

Key Skills

Locate points in a three-dimensional coordinate system.

Locate the point $(3, -1, 7)$ in a three-dimensional coordinate system. Name the octant of the point.

(3, 1, 7) is in the top-front-left octant.

Exercises

Locate each point in a three-dimensional coordinate system and name its octant.

13. $(5, 3, 4)$

14. $(-6, 1, -2)$

Find the distance between two points in space.

Find the distance between the points $(3, -1, 7)$ and $(2, 0, -5)$.

$$\begin{aligned}
d &= \sqrt{(x_2 - x_1)^2 + (y_2 - y_1)^2 + (z_2 - z_1)^2} \\
&= \sqrt{(2-3)^2 + [0 - (-1)]^2 + (-5 - 7)^2} \\
&= \sqrt{1 + 1 + 144} = \sqrt{146} \approx 12.1
\end{aligned}$$

Find the distance between each pair of points in space.

15. $(2, 4, 5)$ and $(0, 7, -1)$

16. $(-1, 6, -2)$ and $(-5, 2, 4)$

LESSON 6.5

Key Skills

Sketch planes in space.

Sketch the plane defined by the equation $x + 2y - 3z = 18$.

First, find the intercepts.

x-intercept: $x + 2(0) + 3(0) = 18$, so $x = 18$
y-intercept: $0 + 2y + 3(0) = 18$, so $y = 9$
z-intercept: $0 + 2(0) + 3z = 18$, so $z = 6$

Plot the intercepts as points and sketch the plane.

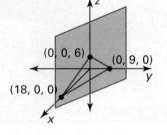

Plot lines in space.

Plot the line represented by the parametric equations below.

$$x = 2t$$
$$y = t + 1$$
$$z = -t$$

When $t = 0$, the point defined by the equations is $(0, 1, 0)$. When $t = 1$, the point defined by the equations is $(2, 2, -1)$. Locate these points and sketch the line.

Exercises

17. Sketch the plane defined by the equation $2x + y - z = 10$.

18. Sketch the plane defined by the equation $-x + 4y + z = 8$.

19. Plot the line represented by the parametric equations below.

$$x = t + 3$$
$$y = t - 1$$
$$z = 2t$$

20. Plot the line represented by the parametric equations below.

$$x = -t + 2$$
$$y = t - 4$$
$$z = -3t + 1$$

Key Skills

Use vanishing points to make perspective drawings.

Make a one-point and a two-point perspective drawing of the letter **T**.

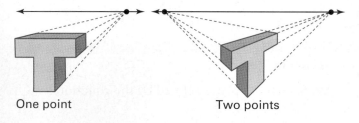

One point   Two points

Exercises

21. Make a one-point perspective drawing of a right rectangular prism with the vanishing point to the right of the prism.

22. Make a one-point perspective drawing of a right rectangular prism with the vanishing point below the prism.

23. Make a two-point perspective drawing of a right rectangular prism with the horizon above the prism.

24. Make a two-point perspective drawing of a right rectangular prism with the horizon below the prism.

Applications

25. GEOLOGY Crystals are classified by the different axes, called *crystallographic axes*, that passes through the center of the crystal. Topaz has a crystal structure called *orthorhombic*. In topaz, three axes of three different lengths all meet at right angles. Draw the three axes for topaz. Suppose that the endpoints of the axes are vertices of a crystal. Draw an outline of the crystal around your axes.

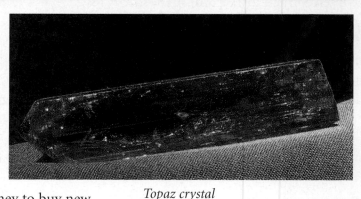

Topaz crystal

26. FUND-RAISING The school band is raising money to buy new uniforms. The band members will wash cars in a space donated by a local business. The supplies for the car wash cost $20. The charges are $3 for a regular car wash and $5 for a deluxe car wash. The amount of money, z, raised by x regular washes and y deluxe washes is given by $z = 3x + 5y - 20$.

Write this equation in the standard form and sketch the graph of the plane. Indicate the trace of the plane on your graph. What does the trace represent?

27. DESIGN The diagrams at right are three orthographic views of a table. Use the diagrams to make a perspective drawing of the table.

Top

Front   Side

Alternative Assessment

Performance Assessment

Curved surfaces in a three-dimensional coordinate system can be defined by equations.

1. *Level curves* are defined by letting z have fixed values, such as 0, 1, 2, 3, ..., and graphing the resulting equation in x and y. Some level curves with equations of the form $z = -x^2 + y$ are given below.

$z = 0 \longrightarrow -x^2 + y = 0$, or $y = x^2$
$z = 1 \longrightarrow -x^2 + y = 1$, or $y = x^2 + 1$
$z = 2 \longrightarrow -x^2 + y = 2$, or $y = x^2 + 2$
$z = 3 \longrightarrow -x^2 + y = 3$, or $y = x^2 + 3$

Graph the curves together on the xy-plane.

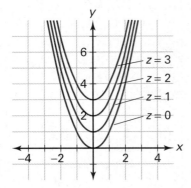

Imagine these curves at different heights, determined by the z-value, to visualize the curve in three dimensions.

Try graphing the level curves for $z = -x + y^2$ by letting $z = 0, 1, 2$, and 3.

2. **TECHNOLOGY** Some graphing software can graph in three dimensions. If you have access to this type of software, try graphing the two surfaces from Item 1. Below are some other interesting surfaces to try.

$z = x^2 - y^2$ $\qquad$ $z = x^2 + y^2$
$z = \sqrt{1 - x^2 - y^2}$ $\qquad$ $z = \sqrt{1 + x^2 + y^2}$

Portfolio Projects

Swiss mathematician Leonhard Euler made contributions to many areas of mathematics. One of his discoveries is the relationship among the numbers of vertices, edges, and faces of a polyhedron.

1. Find several models of polyhedra. They can be models you have made, everyday objects such as boxes, or drawings of polyhedra that show all of the hidden faces and edges, such as the one below. Count the vertices, edges, and faces for each of your models.

5 vertices
8 edges
5 faces

2. Let V represent the number of vertices, E represent the number of edges, and F represent the number of faces. Find $V - E + F$ for each polyhedron. Do you always get the same number? Can you find a polyhedron that gives a different number? (Hint: Find $V - E + F$ for the polyhedron below.)

---

internetconnect

(HRW) The HRW Web site contains many resources to reinforce and expand your knowledge of shapes in space. The Web site also provides Internet links to other sites where you can find information and real-world data for use in research projects, reports, and activities that involve shapes in space. Visit the HRW Web site at **go.hrw.com**, and enter the keyword **MG1 CH6**.

QUANTITATIVE COMPARISON For Items 1–4, write
A if the quantity in Column A is greater than the quantity in Column B;
B if the quantity in Column B is greater than the quantity in Column A;
C if the two quantities are equal; or
D if the relationship cannot be determined from the information given.

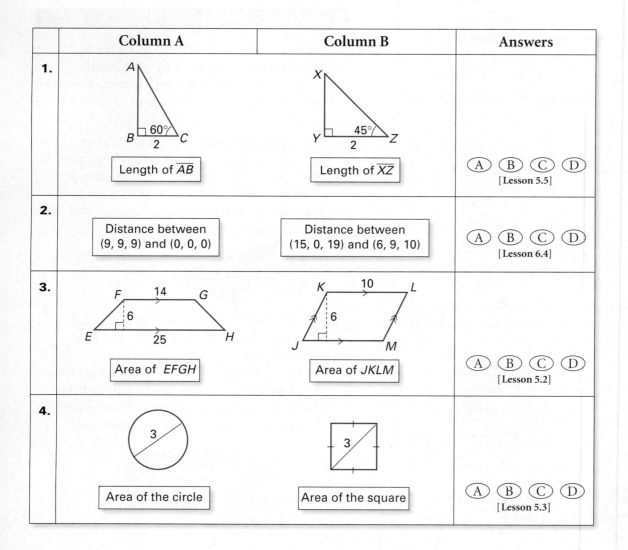

	Column A	Column B	Answers
1.	Length of $\overline{AB}$	Length of $\overline{XZ}$	Ⓐ Ⓑ Ⓒ Ⓓ [Lesson 5.5]
2.	Distance between (9, 9, 9) and (0, 0, 0)	Distance between (15, 0, 19) and (6, 9, 10)	Ⓐ Ⓑ Ⓒ Ⓓ [Lesson 6.4]
3.	Area of *EFGH*	Area of *JKLM*	Ⓐ Ⓑ Ⓒ Ⓓ [Lesson 5.2]
4.	Area of the circle	Area of the square	Ⓐ Ⓑ Ⓒ Ⓓ [Lesson 5.3]

5. What is the measure of an interior angle of a regular nonagon (9-gon)? *(LESSON 3.6)*

    a. 40°
    b. 100°
    c. 140°
    d. 160°

6. The area of a circle is 154 square inches. Approximate its circumference. *(LESSON 5.3)*

    a. 22 in.
    b. 44 in.
    c. 51 in.
    d. 77 in.

7. What is the midpoint of a segment with endpoints at (9, 12) and (−12, 9)? *(LESSON 5.6)*
  a. (−1.5, 10.5)
  b. (4.5, 6)
  c. (−6, −4.5)
  d. (10.5, −10.5)

8. What is the midpoint of a segment with endpoints at (2, 10, 0) and (11, 10, −6)? *(LESSON 6.4)*
  a. (6.5, 0, −3)
  b. (4.5, 10, 3)
  c. (6.5, 10, −3)
  d. (5.5, 5, 3)

9. In the oblique rectangular prism shown below, find x. *(LESSON 2.2)*
  a. 60
  b. 80
  c. 100
  d. 120

10. Find the surface area of the solid shown below. Assume that no cubes are hidden. *(LESSON 6.1)*
  a. 34 square units
  b. 45 square units
  c. 54 square units
  d. 66 square units

11. Is the statement below a definition? Explain your reasoning. *(LESSON 2.3)*

Skew lines are not parallel and do not intersect.

12. A circle and a square both have an area of 441 square inches. Find the circumference and perimeter, respectively. *(LESSONS 5.1 AND 5.3)*

13. A quadrilateral has vertices at (3, −1), (9, −5), (7, −8), and (1, −4). Use slopes to prove that the quadrilateral is a rectangle. *(LESSON 3.8)*

14. Find the area of the regular hexagon below. *(LESSON 5.5)*

15. Draw a rhombus and construct a reflection of it over a horizontal line. *(LESSON 4.5)*

FREE-RESPONSE GRID
Items 16–19 may be answered by using a free-response grid such as that commonly used by standardized-test services.

16. Find the area of △PEG. *(LESSON 5.2)*

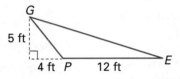

17. Find the total area of the three rectangles below. *(LESSON 5.1)*

18. Refer to the diagram below. Find the value of x. *(LESSON 3.6)*

19. A point, P, is selected at random on the segment below. What is the probability that $2 \leq P \leq 2.5$? *(LESSON 5.7)*

7

Surface Area and Volume

FOR SEVEN DAYS, 120 WORKERS AND 90 ROCK climbers worked to cover the Reichstag building in Berlin with shiny silver fabric. A million square feet of fabric and over 10 miles of plastic cord were required for the feat.

Why would anyone want to wrap a huge building? This is a question that conceptual artist Christo, who conceived of the project 24 years before he was allowed to do it, has heard many times. What do you think? Is it art?

In this chapter, you will investigate the surface area of solid objects. Christo certainly needed to determine the surface area of the Reichstag to prepare for the project.

Kugel Ball, at the Houston Museum of Natural Science

Quartet by Barry LeVa

Pine Circles, Cone Sphere by Chris Drury

Conceptual artist Christo explaining his project

CHAPTER PROJECT
PORTFOLIO ACTIVITIES

About the Chapter Project

Models of solid figures can be very useful in studying the properties of these figures. Throughout this chapter, you will use nets for solid figures to determine surface area.

In the Chapter Project, you will create a three-dimensional puzzle based on the tangram.

After completing the Chapter Project, you will be able to do the following:

- Assemble three-dimensional objects from nets.

- Understand the volume of a three-dimensional object in terms of the sum of its parts.

About the Portfolio Activities

Throughout the chapter, you will be given opportunities to complete Portfolio Activities that are designed to support your work on the Chapter Project.

The theme of each Portfolio Activity and of the Chapter Project is modeling solid figures.

- In the Portfolio Activity on page 452, you will build three oblique pyramids that fit together to form a cube.

- In the Portfolio Activity on page 459, you will examine the lateral surface of an oblique cylinder.

- In the Portfolio Activity on page 485, you will create a solid of revolution by using cardboard and thread.

Surface Area and Volume

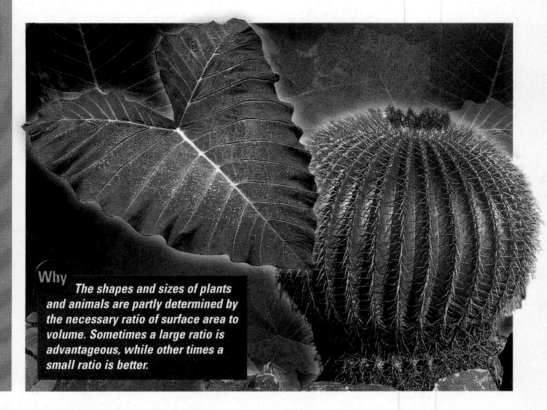

Objectives

- Explore ratios of surface area to volume.

- Develop the concepts of maximizing volume and minimizing surface area.

Why *The shapes and sizes of plants and animals are partly determined by the necessary ratio of surface area to volume. Sometimes a large ratio is advantageous, while other times a small ratio is better.*

Desert plants must conserve water but have a lot of light. Tropical plants receive a lot of water but must often compete for light. Leaves with a large surface area collect more light than leaves with a smaller surface area, but they also lose more water through evaporation.

Surface Area and Volume

The surface area of an object is the total area of all the exposed surfaces of the object. The volume of a solid object is the number of nonoverlapping unit cubes that will exactly fill the interior of the figure.

Surface Area and Volume Formulas

The surface area, S, and volume, V, of a right rectangular prism with length ℓ, width w, and height h are

$$S = 2\ell w + 2wh + 2\ell h \quad \text{and} \quad V = \ell wh.$$

7.1.1

The surface area, S, and volume, V, of a cube with side s are

$$S = 6s^2 \quad \text{and} \quad V = s^3.$$

7.1.2

In Activity 1 below, you will explore the relationship between the volume and surface area of right rectangular prisms.

Ratio of Surface Area to Volume

YOU WILL NEED

isometric dot paper or unit cubes, and spreadsheet software or a graphics calculator (optional)

Algebra

Part I

1. Draw or build several rectangular prisms with a width and height of 1 unit and a length $\ell = 1, 2, 3, \ldots$ Copy and complete the table below, or use a spreadsheet to generate a similar table.

Length, ℓ	Surface area = $2\ell w + 2\ell h + 2wh$	Volume = ℓwh	Ratio of $\frac{\text{surface area}}{\text{volume}}$
1	6	1	?
2	?	?	?
3	?	?	?
n	?	?	?

2. If $\ell = 100$, what is the ratio of surface area to volume? What happens to the ratio of surface area to volume as the value of ℓ increases? Is there a number that the ratio approaches?

Part II

Algebra

1. Draw or build several cubes with side $s = 1, 2, 3, \ldots$ Copy and complete the table below, or use a spreadsheet to generate a similar table.

Side, s	Surface area = $6s^2$	Volume = s^3	Ratio of $\frac{\text{surface area}}{\text{volume}}$
1	6	1	?
2	?	?	?
3	?	?	?
n	?	?	?

CHECKPOINT ✔

2. If $s = 100$, what is the ratio of surface area to volume? What happens to the ratio of surface area to volume as the value of s increases? What conclusions can you draw about the surface-area-to-volume ratios of smaller cubes compared to those of larger cubes?

By extending your tables from Activity 1, you can discover another important result: *For a given volume greater than 1*, the cube in Part II will have a smaller surface-area-to-volume ratio than the long rectangular prism in Part I. Compare, for example, the ratios for a cube and a long prism, each with a volume of 8. (For the cube, $s = 2$, and for the prism, $\ell = 8$.)

Small animals such as hummingbirds have high surface-area-to-volume ratios. They must maintain high metabolisms to compensate for the loss of body heat.

In general, a "bunched-up" object (like a cube or a sphere) will have a smaller surface-area-to-volume ratio than a long, thin object with the same volume. Also, a large object has a smaller surface-area-to-volume ratio than a small object of the same shape (compare the ratios for cubes with different side lengths).

These observations have important consequences in the sciences. In biology and physics, many important processes happen at the surface of a solid body. For example, an animal's loss of body heat to the atmosphere occurs at the surface. If its body is large and the surface area is small, the rate of heat loss will be low. If its body is small and the surface area is large, the rate of heat loss will be high.

On sunny days, a snake may stretch out in the sun to absorb heat. At night, a snake may coil its body to retain heat. Why does this strategy work?

CRITICAL THINKING

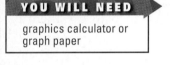

Maximizing Volume

YOU WILL NEED

graphics calculator or graph paper

1. Create an open box from a standard 8.5×11 inch piece of paper by cutting squares from the corners and folding up the sides, as shown. Copy and complete the table below.

Side of square, x	Length, ℓ	Width, w	Height, h	Volume, ℓwh
1	9	6.5	1	58.5
2	?	?	?	?
3	?	?	?	?
x	?	?	?	?

2. What side length of the square will maximize the volume of the box?

3. Use a graphics calculator or graph paper to graph the volume as a function of the side length of the square. Let x be the side length of the square, and let y be the volume of the open box.

CONNECTION

MAXIMUM/MINIMUM

4. Use the trace function of your calculator or estimate from your graph to find the following:

CHECKPOINT ✔

   a. the side length that produces the largest volume
   b. the largest volume

A cube has a larger volume for its surface area than any other rectangular prism. However, it is not possible to make a cubic box from an 8.5×11 inch piece of paper by using the given method. The shape you determined in Activity 2 is the closest shape to a cube that can be created by this method.

A cereal company is choosing between two box designs with the dimensions shown at right. Which design has the greater surface area and thus requires more material for the same volume?

Box A   Box B

8 in.

10 in.

4 in.   5 in.   2 in.   8 in.

SOLUTION

Both boxes have a volume of 160 cubic inches. The surface area of box A is $2(8)(5) + 2(4)(5) + 2(4)(8) = 184$ square inches.

The surface area of box B is $2(10)(8) + 2(2)(8) + 2(2)(10) = 232$ square inches.

Box B has the greater surface area.

CRITICAL THINKING   Most cereal boxes use a design that is more like box B than box A. What considerations other than surface area might be important in choosing a box design?

Exercises

Communicate

1. Explain why the surface area of a right rectangular prism is given by the formula $S = 2\ell w + 2\ell h + 2wh$.

2. What can you say about a solid whose surface-area-to-volume ratio is greater than 1? What can you say about a solid whose surface-area-to-volume ratio is less than 1?

3. Explain why the surface-area-to-volume ratio of a rectangular prism with dimensions of $\ell \times 1 \times 1$ approaches 4 as ℓ increases. (Hint: Which faces of the prism have the same area as ℓ increases?)

4. In Activity 2, you discovered that the surface-area-to-volume ratio of a cube approaches 0 as the side length increases. Can this ratio ever equal 0? Why or why not?

5. **BIOLOGY** Single-celled organisms are able to absorb all of their oxygen and food through their surface. Use surface-area-to-volume ratios to explain why this is not possible for larger animals.

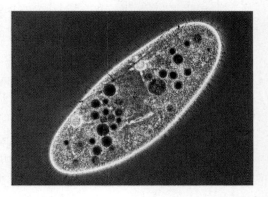

Determine the surface-area-to-volume ratio for a rectangular prism with the given dimensions. *(ACTIVITY 1)*

6. $5 \times 1 \times 1$　　**7.** $10 \times 1 \times 1$　　**8.** $5 \times 5 \times 5$　　**9.** $10 \times 10 \times 10$

APPLICATIONS

10. HOBBIES Teresa is selling vegetables from her garden at a farmer's market. She has some rectangular pieces of cardboard that are 42×48 inches. What is the maximum volume for a box made by cutting squares out of the corners of these pieces and folding up the edges? What size squares should be cut out to make a box with this volume? *(ACTIVITY 2)*

11. MANUFACTURING A company is choosing between two box designs. The dimensions (in inches) of box A are $4 \times 2 \times 6$ and of box B are $3 \times 2 \times 8$. Which has the smaller surface area? *(EXAMPLE)*

● *Practice and Apply*

Algebra

Copy and complete the following table for rectangular prisms:

Length	Width	Height	Surface area	Volume	Surface area / volume
2	2	1	**12.** ?	**13.** ?	**14.** ?
4	4	1	**15.** ?	**16.** ?	**17.** ?
7	3	5	**18.** ?	**19.** ?	**20.** ?
4	**21.** ?	3	**22.** ?	84	**23.** ?
2	5	**24.** ?	104	**25.** ?	**26.** ?

Find the surface-area-to-volume ratio of each of the following:

27. a cube with a volume of 64 cubic units

28. a cube with a volume of 1000 cubic units

29. a rectangular prism with dimensions of $n \times n \times 1$

30. a prism with a square base, where the lateral edge is twice the length of a base edge

CONNECTIONS

MAXIMUM/MINIMUM For each situation below, determine whether you should maximize the volume or minimize the surface area. Explain your reasoning.

31. building a storage bin with a limited amount of lumber

32. designing soup cans that will hold 15 ounces of soup

33. mailing a package with dimensions whose sum is less than 72 inches

34. building a tank that must hold 100 gallons of water

35. MAXIMUM/MINIMUM Compare the surface-area-to-volume ratio of an $n \times n \times 1$ rectangular prism with that of an $n \times n \times n$ rectangular prism as n increases.

36. MAXIMUM/MINIMUM Based on your answer to Exercise 35, make a conjecture about what type of rectangular prism has the smallest surface area for a given volume.

37. Find the side length of a cube with a surface-area-to-volume ratio of 1.

38. A prism with a square base has a lateral edge that is twice as long as the base edges. If its surface-area-to-volume ratio is 1, find the length of a base edge.

39. BIOLOGY Flatworms do not have gills or lungs. They absorb oxygen through their skin. Use surface-area-to-volume ratios to explain how the shape of a flatworm helps it absorb enough oxygen.

Marine flatworm

40. PHYSIOLOGY Human lungs are subdivided into thousands of air sacs. The total surface area of an average person's lungs is about 100 square yards. Use surface-area-to-volume ratios to explain why such a large surface area is necessary.

41. BOTANY Use surface-area-to-volume ratios to explain why tall trees with large, broad leaves do not usually grow in deserts.

42. COOKING A roast is finished cooking when the internal temperature is 160°F. Use surface-area-to-volume ratios to explain why a large roast takes longer to cook than a small one of the same shape.

43. CHEMISTRY Use surface-area-to-volume ratios to explain why a large block of ice will melt more quickly if it is broken up into smaller pieces.

44. PHYSIOLOGY Use surface-area-to-volume ratios to explain why chewing your food thoroughly makes digestion easier.

45. BOTANY A barrel cactus has a short, squatty form. Use surface-area-to-volume ratios to explain how this form helps the cactus conserve water in desert conditions.

Look Back

46. Show that AAA information cannot be used to prove that triangles are congruent. *(LESSON 4.3)*

Find the area of each figure below. *(LESSONS 5.1 AND 5.2)*

47.
15
29

48.
45
28
53

49.
5.5
7.3
12.2

50.
33
16
46

51. If two triangles have the same area, are they necessarily congruent? If not, give a counterexample. **(LESSON 5.2)**

52. If the perimeter of a rectangle is 20 inches, give five possible areas and the dimensions for each. **(LESSON 5.2)**

53. What is the area of a circle with a circumference of 10π? **(LESSON 5.3)**

54. The legs of an isosceles triangle measure 15 feet and the base is 16 feet. Find the area. **(LESSON 5.4)**

55. The legs of an isosceles right triangle measure 6 centimeters. Find the length of the hypotenuse. **(LESSONS 5.4 AND 5.5)**

Look Beyond

Ratios of surface area to volume play a part in determining how big a person can be. As the size of a person increases, the weight increases in proportion to the volume, while certain other measurements increase in proportion to the surface area.

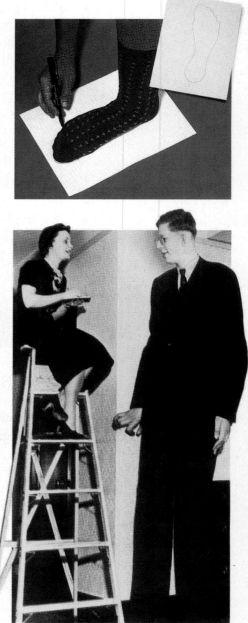

56. Estimate the total area of the bottom of both of your feet in square inches. One way to do this is to trace your feet onto a piece of graph paper and count the squares.

57. Divide your weight in pounds by the area of the bottom of both of your feet. This gives the pressure, in pounds per square inch, on the bottom of your feet.

58. Suppose that the size of your body increased while keeping the same shape. If the length and width of your feet were doubled, the area of your feet would increase by a factor of ____?____. Estimate the new area of the bottom of your feet. If your length, width, and height were doubled, your weight would increase by a factor of ____?____. What would be your new weight?

59. Divide the new weight by the new area to find the new pressure on the bottoms of your feet. Explain what happens to the pressure on your feet as your size increases. How does this limit the size of the human body?

The tallest man in history was Robert Wadlow (1918–1940), who stood 8 feet and 11.1 inches tall. He is shown here at 19 years old, being interviewed by a reporter.

Surface Area and Volume of Prisms

7.2

Why Many products are packaged in boxes that are prisms. You can use surface area to estimate the amount of material used in each box or use volume to estimate the amount of the product in each box.

Objectives

- Define and use a formula for finding the surface area of a right prism.

- Define and use a formula for finding the volume of a right prism.

- Use Cavalieri's Principle to develop a formula for the volume of a right or oblique prism.

You can use the surface area of a box to estimate the amount of wrapping paper you will need to cover the box.

Recall from Lesson 6.3 that a prism has two parallel congruent faces called bases and that the remaining faces, called lateral faces, are parallelograms. In a right prism, all of the lateral faces are rectangles.

An **altitude** of a prism is a segment that has endpoints in the planes containing the bases and that is perpendicular to both planes. The **height** of a prism is the length of an altitude.

Altitudes

Right prism   Oblique prism

Surface Area of Right Prisms

The surface area of a prism may be broken down into two parts: the area of the bases, or base area, and the area of the lateral faces, or lateral area.

Since the bases are congruent, the base area is twice the area of one base, or 2*B*, where *B* is the area of one base.

To find the lateral area of a right prism, you may find it helpful to use a net. Because the net folds up to form the prism, the area of the net is equal to the surface area of the prism.

If the sides of the base are s_1, s_2, and s_3 and the height is h, then the lateral area is given by the following formula:

$$L = s_1 h + s_2 h + s_3 h = h(s_1 + s_2 + s_3)$$

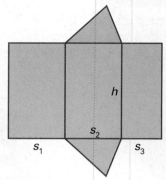

Because $s_1 + s_2 + s_3$ is the perimeter of the base, we can write the lateral area as $L = hp$, where p is the perimeter of the base.

CRITICAL THINKING How can you show that the formula for lateral area, $L = hp$, is true for right prisms with bases that are not triangles?

The surface area of a prism is the sum of the base area and the lateral area.

Surface Area of a Right Prism

The surface area, S, of a right prism with lateral area L, base area B, perimeter p, and height h is

$$S = L + 2B \quad \text{or} \quad S = hp + 2B.$$

7.2.1

E X A M P L E **1** The net for a right triangular prism is below. What is its surface area?

SOLUTION

The area of each base is $B = \frac{1}{2}(2)(21) = 21$.

The perimeter of each base is $p = 10 + 21 + 17 = 48$, so the lateral area is $L = hp = 30(48) = 1440$.

Thus, the surface area is $S = L + 2B = 1440 + 2(21) = 1440 + 42 = 1482$.

Volumes of Right Prisms

Recall from Lesson 7.1 that the volume of a right rectangular prism with length ℓ, width w, and height h is given by $V = \ell wh$. Because the base area, B, of this type of prism is equal to ℓw, you can also write the formula for the volume as $V = Bh$.

Does this formula work for right prisms with bases that are not rectangles? The discussion that follows begins by considering a triangular prism with a height of 1.

The prism at right has a height of 1. The number of unit cubes needed to fill the prism is equal to the number of unit squares needed to cover the base, which is equal to the area of the base.

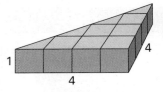

A prism with a height of h can be made by stacking h prisms with a height of 1 on top of each other. The volume of this prism is h times the volume of the prism with a height of 1, or h times the area of the base. Thus, the volume of any right prism with a base area of B and height of h is $V = Bh$.

EXAMPLE ❷

APPLICATION
AQUARIUMS

An aquarium in the shape of a right rectangular prism has dimensions of $110 \times 50 \times 7$ feet. Given that 1 gallon ≈ 0.134 cubic feet, how many gallons of water will the aquarium hold? Given 1 gallon of water ≈ 8.33 pounds, how much will the water weigh?

● **SOLUTION**

The volume of the aquarium is found by using the volume formula.
$$V = Bh = \ell wh = (110)(50)(7) = 38{,}500 \text{ cubic feet}$$

To approximate the volume in gallons, divide by 0.134.
$$V = 38{,}500 \div 0.134 \approx 287{,}313 \text{ gallons}$$

To approximate the weight, multiply by 8.33.
$$\text{weight} \approx (287{,}313)(8.33) \approx 2{,}393{,}317 \text{ pounds}$$

EXAMPLE ❸

An aquarium has the shape of a right regular hexagonal prism with the dimensions shown at right. Find the volume of the aquarium.

14 in. $7\sqrt{3}$ in.

48 in.

● **SOLUTION**

The base of the aquarium has a perimeter of $(14)(6)$, or 84, inches and an apothem of $7\sqrt{3}$ inches, so the base area is found as follows:

$$B = \tfrac{1}{2}ap = \tfrac{1}{2}(84)(7\sqrt{3}) = 294\sqrt{3} \approx 509.22 \text{ square inches}$$

The volume is $V = Bh = (294\sqrt{3})(48) = 14112\sqrt{3} \approx 24{,}443$ cubic inches.

Volumes of Oblique Prisms

In an oblique prism, the lateral edges are not perpendicular to the bases, and there is no simple general formula for surface area. However, the formula for the volume is the same as that for a right prism. To understand why this is true, consider the explanation below.

Stack a set of index cards in the shape of a right rectangular prism. If you push the stack into the shape of an oblique prism, the volume of the solid does not change because the number of cards does not change.

Both stacks have the same number of cards, and each prism is the same height. Also, because every card has the same size and shape, they all have the same area. Any card in either stack represents a cross section of each prism.

The prisms above illustrate an important geometry concept.

Cavalieri's Principle

If two solids have equal heights and the cross sections formed by every plane parallel to the bases of both solids have equal areas, then the two solids have equal volumes.

7.2.2

The solids will have equal volumes if they are the same height and all cross-sectional areas are equal.

Every oblique prism can be compared with a right prism with the same base and height. As the stacks of cards suggest, every cross section of a prism is congruent to its bases, so all of the cross sections of the oblique prism are equal to the cross sections of the right prism. Thus, by Cavalieri's Principle, they have equal volumes.

The formula for the volume of any prism is given below.

Volume of a Prism

The volume, V, of a prism with height h and base area B is
$$V = Bh.$$

7.2.3

● *Communicate*

1. Explain how to find the surface area of a right prism.

2. Explain the formula for the volume of a right prism.

3. Explain Cavalieri's Principle and how it can be used to find the volume of an oblique prism.

4. Can Cavalieri's Principle be used for two prisms of the same height if the base of one is a triangle and the base of the other is a hexagon? Why or why not?

● *Guided Skills Practice*

Determine the surface area of the prism formed by each net.
(EXAMPLE 1)

5.

6.

Find the volume of each prism. *(EXAMPLES 2 AND 3)*

7. right rectangular prism

8. right regular octagonal prism

Draw a net for each figure named below.

9. cube

10. right rectangular prism

11. right equilateral triangular prism

12. right regular hexagonal prism

Find the volume of a prism with the given dimensions.

13. $B = 7$ sq cm, $h = 5$ cm

14. $B = 9$ sq m, $h = 6$ m

15. $B = 17$ sq in., $h = 23$ in.

16. $B = 32$ sq ft, $h = 17$ ft

Find the surface area and volume of a right rectangular prism with the given dimensions.

17. $\ell = 5$, $w = 7$, $h = 2$

18. $\ell = 16$, $w = 9$, $h = 10$

19. $\ell = \frac{1}{2}$, $w = \frac{2}{3}$, $h = 1$

20. $\ell = 1.3$, $w = 4$, $h = 0.5$

Algebra

21. Find the height of a rectangular prism with a surface area of 286 in.^{2} and a base measuring 7×9 in.

22. The height of a right regular hexagonal prism is 25 cm, and the sides of its base measure 18 cm. Find its surface area.

23. The height of a right regular hexagonal prism is 20 cm, and the apothem of its base is $4\sqrt{3}$ cm. Find its surface area.

24. If a cube has a volume of 343 yd^3, what is its surface area?

25. The figure at right is a net for an oblique rectangular prism. The 3×2 rectangular faces are the bases of the prism. Find its volume and surface area. It may help to copy the net and fold it to form the prism.

26. The figure at right is a net for an oblique triangular prism. Find its volume and surface area. It may help to copy the net and fold it to form the prism.

27. Find the volume of a right trapezoidal prism if the bases of the trapezoid measure 6 m and 8 m, the height of the trapezoid is 7 m, and the height of the prism is 18 m.

28. Find the volume of a right triangular prism whose base is an isosceles right triangle with a hypotenuse of 10 cm and whose height is 23 cm.

29. Find the surface area and volume of a right triangular prism whose base is an isosceles triangle with legs measuring 4 in. and a base of 6 in. and whose height is 13 in.

30. If the height of a prism is doubled and the bases are unchanged, what happens to the volume of the prism?

31. If the edges of a cube are doubled, what happens to the surface area? What happens to the volume?

32. If the edges of a cube are tripled, what happens to the surface area? What happens to the volume?

33. The height of a right regular hexagonal prism is equal to the side length of its base. If these side lengths are doubled, what happens to the volume? What happens to the surface area?

CHALLENGE

34. What happens to the side length of a cube if its volume is doubled? What happens to the surface area?

APPLICATIONS

35. **MANUFACTURING** Find the surface area and volume of each box of cat food shown below.

36. **RECREATION** The swimming pool shown below is a right prism with concave hexagonal bases. Use the given dimensions to find the volume in cubic feet and in gallons (1 gallon ≈ 0.134 cubic foot).

37. **RECREATION** Find the surface area of the sides and bottom of the swimming pool shown above. If 1 gallon of paint covers 400 square feet, how many gallons of paint will be needed to paint the inside of the pool?

38. RECREATION The tent shown at right is a right triangular prism. The bases of the prism are isosceles triangles. Find the surface area of the tent, including the floor.

3.5 ft

4.5 ft

7.0 ft

39. ARCHITECTURE Use the lateral area of the building shown at right to estimate how much glass was used to cover the outside walls of the building. Each story is 12 feet high, and the base of the building is a square with sides of 48 feet.

Look Back

Simplify each radical expression. Give answers in simplified radical form. *(ALGEBRA REVIEW)*

40. $\sqrt{20}$

41. $(\sqrt{18})(\sqrt{2})$

42. $(5\sqrt{7})^2$

43. $(2\sqrt{2})(3\sqrt{8})$

44. $\dfrac{2}{\sqrt{2}}$

45. $\sqrt{27} + \sqrt{12}$

Exercises 46 and 47 refer to a 30-60-90 triangle with a hypotenuse of 10 inches. *(LESSON 5.5)*

46. Find the length of the shorter leg.

47. Find the length of the longer leg.

Look Beyond

COORDINATE GEOMETRY A prism in a three-dimensional coordinate system has one base with its vertices at A(0, 0, 0), B(0, 5, 0), C(4, 5, 0), and D(4, 0, 0). The second base has its vertices at E(2, 2, 2), F(2, 7, 2), G(6, 7, 2), and H(6, 2, 2).

48. Draw the prism on a set of coordinate axes. Name the type of prism.

49. Find the volume of the prism.

50. Find the surface area of the prism.

Surface Area and Volume of Pyramids

7.3

Objectives

- Define and use a formula for the surface area of a regular pyramid.
- Define and use a formula for the volume of a pyramid.

Why *Across cultures, the pyramid has an enduring appeal. From ancient Egyptian tombs to the Transamerica Pyramid in San Francisco, a pyramid's shape has strength and simple beauty.*

Pyramids

A **pyramid** is a polyhedron consisting of a **base**, which is a polygon, and three or more **lateral faces**. The lateral faces are triangles that share a single vertex, called the **vertex of the pyramid**. Each lateral face has one edge in common with the base, called a **base edge**. The intersection of two lateral faces is a **lateral edge**.

The **altitude** of a pyramid is the perpendicular segment from the vertex to the plane of the base. The **height** of a pyramid is the length of its altitude.

A **regular pyramid** is a pyramid whose base is a regular polygon and whose lateral faces are congruent isosceles triangles. In a regular pyramid, all of the lateral edges are congruent, and the altitude intersects the base at its center. The length of an altitude of a lateral face of a regular pyramid is called the **slant height** of the pyramid.

Pyramids, like prisms, are named by the shape of their base.

Triangular pyramid

Rectangular pyramid

Pentagonal pyramid

Hexagonal pyramid

The Surface Area of a Pyramid

To analyze the surface area of a pyramid, it is helpful to use a net. The area of the net is the same as the surface area of the pyramid. For pyramids that are not regular, the area of each face must be calculated separately, and then the areas must be added together. For regular pyramids, however, there is an easier way.

EXAMPLE ❶ Find the surface area of a regular square pyramid whose slant height is ℓ and whose base edge length is s.

Area of each triangle $= \frac{1}{2}s\ell$

SOLUTION

The surface area is the sum of the lateral areas and the base area.

$$S = L + B$$

$$S = 4\left(\frac{1}{2}s\ell\right) + s^2$$

This can be rewritten as follows:

$$S = \frac{1}{2}\ell(4s) + s^2$$

Because $4s$ is the perimeter of the base,

$$S = \frac{1}{2}\ell p + s^2.$$

The lateral area of any regular pyramid is equal to $\frac{1}{2}\ell p$. The surface area is found by adding the base area, B, to this value.

Surface Area of a Regular Pyramid

The surface area, S, of a regular pyramid with lateral area L, base area B, perimeter of the base p, and slant height ℓ is

$$S = L + B \quad \text{or} \quad S = \frac{1}{2}\ell p + B.$$

7.3.1

EXAMPLE ❷ The roof of a gazebo is a regular octagonal pyramid with a base edge of 4 feet and a slant height of 6 feet. Find the area of the roof. If roofing material costs $3.50 per square foot, find the cost of covering the roof with this material.

SOLUTION

The area of the roof is the lateral area of the pyramid.

$$L = \frac{1}{2}\ell p = \frac{1}{2}(6)(8 \times 4) = 96 \text{ square feet}$$

96 square feet $\times$ $3.50 per square foot = $336.00

The Volume of a Pyramid

The volume of a pyramid has an interesting relationship to the volume of a prism with the same base and height as the pyramid.

Activity

Pyramids and Prisms

1. Using construction paper, make a right square prism and a regular square pyramid with the same base and the same height. Seal the lateral edges with tape. (Do not seal the base edges.)

2. Fill the pyramid with dry cereal or packing material. Pour the contents into the prism. Repeat as necessary to fill the prism completely. How many times did you have to fill the pyramid in order to fill the prism?

CHECKPOINT ✔

3. Make a conjecture about the relationship of the volume of a pyramid to the volume of a prism with the same base and height. Express your conjecture as a formula for the volume of a pyramid.

This relationship between triangular prisms and pyramids can be verified by dividing a triangular prism into three triangular pyramids, as shown below. The pyramids are not congruent to each other, but they each have the same volume.

Triangular prism

Pyramid I

Pyramid II

Pyramid III

Consider the pyramids in pairs, as follows:

Pair A: Pyramid I has base *DEF* and height *h*.
Pyramid II has base *ABC* and height *h*.

Notice that the bases are congruent and the heights are equal. It can be shown, using Cavelieri's principle, that pyramids with congruent bases and equal heights have equal volumes.

Pair B: Pyramid II has base *EBC*.
Pyramid III has base *EFC*.

The bases of these pyramids are congruent because $\overline{EC}$ is a diagonal of *BEFC*. The heights are also equal because both pyramids include vertex *A*. Therefore, these two pyramids have equal volumes.

By the Transitive Property, the three pyramids all have equal volume, so each must be one-third of the original prism. This suggests the following formula:

Volume of a Pyramid

The volume, V, of a pyramid with height h and base area B is
$$V = \tfrac{1}{3}Bh.$$

7.3.2

CRITICAL THINKING

Once you know that the above formula works for all triangular pyramids, how would you show that it works for pyramids with other bases?

EXAMPLE **3**

APPLICATION
ARCHAEOLOGY

CULTURAL CONNECTION: AFRICA The pyramid of Khufu is a regular square pyramid with a base edge of approximately 776 feet and an original height of 481 feet. The limestone used to construct the pyramid weighs approximately 167 pounds per cubic foot. Estimate the weight of the pyramid of Khufu. (Assume the pyramid is solid.)

SOLUTION

The volume of the pyramid is is found as follows:

$$V = \tfrac{1}{3}Bh$$
$$\approx \tfrac{1}{3}(776^2)(481)$$
$$\approx 96{,}548{,}885 \text{ cubic feet}$$

The weight in pounds is 96,548,885 cubic feet × 167 pounds per cubic foot ≈ 16,123,663,850 pounds, or 8,061,831 tons.

Exercises

Communicate

1. Define *pyramid*. Is there any type of pyramid in which more than one face could be considered the base?

2. Explain how to find the surface area of a regular pyramid.

3. Explain how to find the volume of a pyramid.

4. In a regular pyramid, which is larger, the height or the slant height? Explain your reasoning.

5. Explain the relationship between the volume of a pyramid and the volume of a prism with the same base and height.

These strings of lights form the lateral edges of a 39-gonal pyramid.

Guided Skills Practice

Determine the surface area of each regular pyramid. *(EXAMPLE 1)*

6.

10
7

7.

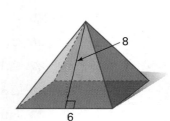

8
6

APPLICATION

8. CONSTRUCTION The roof of a gazebo is a regular decagonal pyramid with a base edge of 4 ft and a slant height of 7 ft. Find the surface area of the roof. *(EXAMPLE 2)*

9. CULTURAL CONNECTION: AFRICA The pyramid of Khafre is a regular square pyramid with a base edge of 708 ft and a height of 471 ft. It is constructed of limestone, which weighs approximately 167 pounds per cubic foot. Estimate the weight of the pyramid. *(EXAMPLE 3)*

Practice and Apply

Draw a net for each regular pyramid named below.

10. a square pyramid

11. a triangular pyramid

12. a pentagonal pyramid

13. a hexagonal pyramid

Find the surface area of each regular pyramid with side length s and slant height ℓ given below. The number of sides of the base is given by n.

14. $s = 8$
$\ell = 9$
$n = 3$

15. $s = 6$
$\ell = 7$
$n = 4$

16. $s = 10$
$\ell = 12$
$n = 6$

Find the volume of each rectangular pyramid.

17.

7.6 m
3 m
4 m

18.

8 m
7 m
9 m

19.

9 in.
13 in.
12 in.

In Exercises 20–23, draw a diagram of each pyramid and label the base edges and the altitude. Find the volume of each pyramid. Give exact answers.

20. a rectangular pyramid with a 5 × 7 base and a height of 11

21. an octagonal pyramid with a base area of 16 and a height of 14

22. a right triangular pyramid with base edges of 5, 12, and 13, and a height of 10

23. a square pyramid with a base edge of 4 a height equal to the diagonal of the base

Use the diagram of the pyramid below for Exercises 24–30.

24. Find the area of the base, *BCDE*.

Find the area of each lateral face.

25. △*ABC*

26. △*ACD*

27. △*ADE*

28. △*AEB*

29. Find the surface area of the pyramid.

30. Find the volume of the pyramid.

Find the height of each pyramid described below.

31. a pentagonal pyramid with a base area of 24 square units and a volume of 104 cubic units

32. a regular square pyramid with a base edge of 10 units and a volume of 500 cubic units

33. a regular triangular pyramid with a base perimeter of 12 units and a volume of 8 cubic units

34. a regular hexagonal pyramid with a base edge of 2 units and a slant height of 2 units

Copy and complete the table below for the vertices, edges, and faces of pyramids. For each entry in the last row, explain your reasoning.

Number of sides of base, *n*	Number of vertices, *V*	Number of edges, *E*	Number of faces, *F*
3	4	6	4
4	**35.** ?	**36.** ?	**37.** ?
5	**38.** ?	**39.** ?	**40.** ?
n	**41.** ?	**42.** ?	**43.** ?

TABLE PROOF

44. Use the last row of the table above to prove that $V - E + F = 2$ for all pyramids.

45. Find the volume and surface area of the rectangular pyramid below.

46. CONSTRUCTION The entrance of the Louvre museum in France is a square pyramid with a base area of 225 m² and a height of 15 m. What is the volume of the pyramid?

47. CONSTRUCTION How much glass would it take to cover the pyramidal entrance of the Louvre? (Ignore the trapezoidal doorway and do not include the floor.)

Look Back

48. An angle measures 51°. What is the measure of its complement? What is the measure of its supplement? *(LESSON 1.3)*

Quadrilateral *ABCD* has vertices at *A*(0, 0), *B*(5, 0), *C*(7, 6), and *D*(2, 6).
(LESSONS 3.8 AND 5.6)

49. What type of special quadrilateral is *ABCD*? Prove your answer.

50. Find the perimeter of the quadrilateral.

51. Find the area of the quadrilateral.

52. Give the vertices of a quadrilateral with the same area as *ABCD* but a different perimeter.

Find the area of the shaded region in each figure. *(LESSONS 5.1 AND 5.3)*

53.

54.

55.

In Exercises 56–60, write a conjecture for each question. If a theorem you know supports your reasoning, state the theorem.

In the proof of the formula for the volume of a pyramid, it was stated that if two pyramids have the same height and bases of the same area, then they have equal volumes. Examine the illustration below.

56. In the illustration, the intersecting plane is parallel to the plane of the bases of the pyramid, and it intersects each altitude at its midpoint. How does the intersecting plane seem to divide the lateral edges of the pyramids?

57. How do you think the lengths of the sides of the red triangles compare with the lengths of the sides of the bases in each pyramid?

58. How do you think the areas of the red triangles compare with the areas of the bases? How do you think they compare with each other?

59. Do you think the results from Exercise 58 would be the same if the intersecting plane intersected the altitudes at points other than their midpoints?

60. How do your results show that the pyramids have equal volumes?

NETS FOR OBLIQUE PYRAMIDS

The net at right is for an oblique square pyramid.

1. Find *BE*. Which sides are congruent to $\overline{BE}$?

2. Find *FC*. Which side is congruent to $\overline{FC}$?

3. Make three copies of the net and fold them into three congruent square pyramids. Fit the three pyramids together to form a cube. How does this illustrate the formula for the volume of a pyramid?

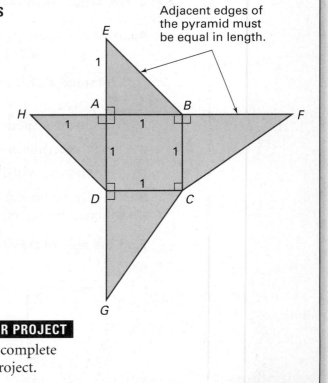

Adjacent edges of the pyramid must be equal in length.

WORKING ON THE CHAPTER PROJECT

You should now be able to complete Activity 1 of the Chapter Project.

Surface Area and Volume of Cylinders

Objectives

- Define and use a formula for the surface area of a right cylinder.

- Define and use a formula for the volume of a cylinder.

Why *Many everyday objects are cylindrical in shape. You can use the volume of these objects to find how much liquid they will hold.*

APPLICATION
ENGINEERING

The gasoline you buy at a pump is stored in underground tanks. How could you use the dimensions of an underground gasoline tank to estimate the number of car tanks that could be filled from it? (See Example 2, page 455.)

Cylinders

A **cylinder** is a solid that consists of a circular region and its translated image on a parallel plane, with a **lateral surface** connecting the circles.

Bases

The faces formed by the circular region and its translated image are called the **bases** of the cylinder.

An **altitude** of a cylinder is a segment that has endpoints in the planes containing the bases and is perpendicular to both planes. The **height** of a cylinder is the length of an altitude.

Altitude

Axis

The **axis** of a cylinder is the segment joining the centers of the two bases.

If the axis of a cylinder is perpendicular to the bases, then the cylinder is a **right cylinder**. If not, it is an **oblique cylinder**.

Cylinders and Prisms

As the number of sides of a regular polygon increases, the figure becomes more and more like a circle.

Similarly, as the number of lateral faces of a regular polygonal prism increases, the figure becomes more and more like a cylinder.

This fact suggests that the formulas for surface areas and volumes of prisms and cylinders are similar.

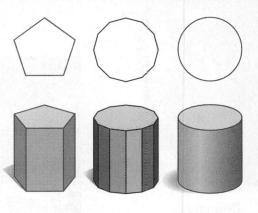

The Surface Area of a Right Cylinder

The surface area of a right cylinder with a radius of r and a height of h can be found by using a net. The net for a right cylinder, shown at right, includes the two circular bases and the lateral surface, which becomes a rectangle. The length of this rectangle is the circumference of the base of the cylinder, or $2\pi r$. The height of the rectangle is the height of the cylinder, or h.

Thus, the lateral area of the cylinder is $2\pi rh$, and the area of each base is πr^2. The surface area of a cylinder is the sum of the lateral area and the base areas.

Surface Area of a Right Cylinder

The surface area, S, of a right cylinder with lateral area L, base area B, radius r, and height h is
$$S = L + 2B \quad \text{or} \quad S = 2\pi rh + 2\pi r^2.$$

7.4.1

E X A M P L E ❶ A penny is a right cylinder with a diameter of 19.05 millimeters and a thickness of 1.55 millimeters. Ignoring the raised design, estimate the surface area of a penny.

APPLICATION
COINS

● **SOLUTION**

The radius of a penny is half of the diameter, or 9.525 millimeters. Use the formula for the surface area of a right cylinder.

Pennies, at one time made of pure copper, are now made of copper-plated zinc.

$$S = 2\pi rh + 2\pi r^2$$
$$S = 2\pi(9.525)(1.55) + 2\pi(9.525)^2 \approx 663.46 \text{ square millimeters}$$

Volumes of Cylinders

For the following Activity, recall the method you used in Lesson 5.3 to find the area of a circle.

Activity
Analyzing the Volume of a Cylinder

YOU WILL NEED

no special tools

The formula for the area of a circle was found by dividing the circle into sectors and fitting them together to form a shape that was close to that of a rectangle. The same idea can be used to find the volume of a cylinder.

1. Refer to the figure above. What geometric solid does the cylinder approximate when the sections are arranged as they are at right above?

CHECKPOINT ✔ 2. Use the length, width, and height of this arrangement to write a formula for the volume of a cylinder in terms of its radius and height.

CRITICAL THINKING   How would you show that the formula for the volume of an oblique cylinder is the same as the formula for the volume of a right cylinder? Use sketches to illustrate your answer.

Volume of a Cylinder

The volume, V, of a cylinder with radius r, height h, and base area B is

$$V = Bh \quad \text{or} \quad V = \pi r^2 h.$$

7.4.2

EXAMPLE ② The tank in the illustration on page 453 has a length of 31 feet $6\frac{1}{2}$ inches and an outer diameter of 8 feet 0 inches. Assuming a wall thickness of about 2 inches, what is the volume of the tank? At 15 gallons per car, how many car tanks could be filled from the storage tank if it starts out completely full of gasoline?

APPLICATION
ENGINEERING

● **SOLUTION**

The tank is not perfectly cylindrical, because of its hemispherical heads, but you can approximate its volume by a slightly shorter cylindrical tank, say, 29 feet long. Subtracting the wall thickness from the dimensions of the tank,

$$V = \pi r^2 h \approx \pi (3.833)^2 (28.667) \approx 1323 \text{ cubic feet}$$

Convert from cubic feet to gallons.

Note: 1 cubic foot ≈ 7.48 gallons.

1323 cubic feet × 7.48 gallons per cubic foot ≈ 9896 gallons

Thus the tank could deliver about $\frac{9896}{15}$, or ≈ 660, 15-gallon fill-ups.

Exercises

Communicate

1. Explain the difference between an altitude and the axis of a cylinder.

2. Is a cylinder a polyhedron? Why or why not?

3. Explain how to find the surface area of a right cylinder.

4. Write the formula for the surface area of a right cylinder in factored form. Which form do you prefer, and why?

5. How are cylinders and prisms alike? How are they different?

Guided Skills Practice

COINS The dimensions of various coins are given in the table below.

Coin	Diameter	Thickness
nickel	21.21 mm	1.95 mm
dime	17.91 mm	1.35 mm
quarter	24.26 mm	1.75 mm

Find the surface area of each coin. *(EXAMPLE 1)*

6. nickel   7. dime   8. quarter

Find the volume of each coin. *(EXAMPLE 2)*

9. nickel   10. dime   11. quarter

Practice and Apply

Draw a net for the right cylinders shown below. Label the dimensions of the net.

12.

13.

Algebra

Find the unknown value for a right cylinder with radius r, height h, and surface area S. Round your answers to the nearest tenth.

14. $r = 5, h = 4, S = $ ___?___

15. $r = 4, h = 15, S = $ ___?___

16. $r = \frac{1}{2}, h = 1, S = $ ___?___

17. $r = 3, h = $ ___?___ $, S = 72\pi$

18. $r = 7, h = $ ___?___ $, S = 550$

19. $r = $ ___?___ $, h = 2, S = 70\pi$

Find the unknown value for a right cylinder with radius *r*, height *h*, and volume *V*. Give your answers in exact form.

20. $r = 5, h = 4, V = \underline{\ ?\ }$

21. $r = 4, h = 15, V = \underline{\ ?\ }$

22. $r = 2, h = \underline{\ ?\ }, V = 12\pi$

23. $r = 8, h = \underline{\ ?\ }, V = 1536$

24. $r = \underline{\ ?\ }, h = 6, V = 54\pi$

25. $r = \underline{\ ?\ }, h = 9, V = 80\pi$

26. The surface area of a cylinder is 200 cm². The diameter is equal to the height. Find the radius.

27. The volume of a cylinder is 360π mm³ and the height is 10 mm. Find the circumference of the base.

28. A semicircular cylinder is formed by cutting a solid cylinder with a radius of 8 ft and height of 10 ft in half along a diameter. Find the volume and surface area of the semicircular cylinder.

29. How does doubling the height of a cylinder affect the volume?

30. How does doubling the radius of a cylinder affect the volume?

31. How does doubling the height and radius of a cylinder affect the volume?

32. How does doubling the height and radius of a cylinder affect the surface area?

CHALLENGE

33. The volume of a cylinder is equal to its surface area. Prove that the radius and height must both be greater than 2.

CONNECTION

34. MAXIMUM/MINIMUM A right cylinder has a volume of 16π cubic units. To find the minimum possible surface area, first solve the volume equation $\pi r^2 h = 16\pi$ for *h* and substitute the expression for *h* in the surface area formula.

Graph this formula for surface area on a graphics calculator, using *x* for the radius. (Hint: Use a viewing window with $0 \le x \le 10$ and $0 \le y \le 200$.)

Use the trace function of the calculator to estimate the radius for a cylinder with the minimum surface area. What is the height of the cylinder? What is the surface area?

35. CULTURAL CONNECTION: ASIA In ancient Mesopotamia, *cylinder seals* were used to make impressions in clay tablets. These seals, which were shaped like cylinders, had carved designs and were rolled on soft clay to create a repeating pattern. Suppose that an archaeologist discovers an impression from a cylinder seal that is 4.7 cm wide. What was the radius of the cylinder seal?

36. MARINE BIOLOGY The Giant Ocean Tank in Boston's New England Aquarium, shown at right, is a cylinder that is 23 ft high and has a volume of 200,000 gal. Find the diameter of the tank. (Note: 1 gal ≈ 0.134 ft^3)

37. PUBLIC HEALTH A scientist researching the health risks of cigars and filtered cigarettes wishes to compare the amount of tobacco in each. A cigarette is a cylinder with a diameter of 0.16 in. and a length (without the filter) of 2.56 in. A cigar is a cylinder with a diameter of 0.75 in. and a length of 7 in. How many times more tobacco is contained in the cigar than in the filtered cigarette?

38. MANUFACTURING A processing plant needs storage tanks to hold at least half a million gallons of waste water. How many cylindrical tanks must be built to hold the water if each tank has a diameter of 50 ft and a height of 25 ft? (Note: 1 ft^3 = 7.48 gal)

39. PRODUCT PACKAGING A manufacturer is designing a can that will hold 64 fluid ounces, or 115.5 in.^{3}, and that uses the least amount of materials. Use the method described in Exercise 34 to find the can with the minimum surface area.

Look Back

Copy each symbol. Draw all of the lines of symmetry and describe all rotational symmetries of each symbol. *(LESSON 3.1)*

40. ⊗   **41.** ⟺   **42.** ↖   **43.** §

Find the surface area and volume of each solid. *(LESSONS 7.2 AND 7.3)*

44. right rectangular prism   **45.** cube

46. right triangular prism   **47.** regular square pyramid

APPLICATION

CARPENTRY **The cross section of the strongest beam that can be cut from a cylindrical log is shown at right.**

The diameter is divided into three equal segments, and perpendicular segments are drawn as shown. The points where these segments intersect the circle are the vertices of the beam.

48. Use the Pythagorean Theorem to write an equation relating the quantities h, w, and d for the shaded triangle at right.

49. Use the Pythagorean Theorem to write an equation relating the quantities w, d, and x for the shaded triangle at right.

50. Use the Pythagorean Theorem to write an equation relating the quantities h, d, and x for the shaded triangle at right.

Algebra

51. Use the equations you found in Exercises 48–50 to find the height, h, and width, w, of the strongest beam that can be cut from a log with a diameter of 18 inches. Round your answers to the nearest tenth.

PORTFOLIO ACTIVITY

NETS FOR OBLIQUE CYLINDERS

A net for the lateral surface of an oblique cylinder has an interesting shape. To examine this, construct an oblique cylinder from a right cylinder.

- Create a right cylinder by rolling up a rectangle of heavy paper. Tape the edges together. Your diameter should be at least 2 inches.

- Use a pencil to mark two parallel circles at an angle to the bases. Cut along the circles to form an oblique.

- Unroll the paper to see the net. What do you notice about the sides of the net?

Surface Area and Volume of Cones

Objectives

- Define and use the formula for the surface area of a cone.

- Define and use the formula for the volume of a cone.

Why *The properties of cones can be used to model the physical properties of real-world objects that have approximately conic shapes.*

As a volcano erupts and deposits lava and ash over a period of time, it forms a cone. Volcanic cones may be different shapes and sizes, depending on factors such as the rate at which the lava and ash are deposited, how fast the lava cools, etc.

Cones

A **cone** is a three-dimensional figure that consists of a circular **base** and a curved **lateral surface** that connects the base to a single point not in the plane of the base, called the **vertex**.

The **altitude** of a cone is the perpendicular segment from the vertex to the plane of the base. The **height** of the cone is the length of the altitude.

If the altitude of a cone intersects the base of the cone at its center, the cone is a **right cone**. If not, it is an **oblique cone**.

Right cone

Oblique cone

Just as a cylinder resembles a prism, a cone resembles a pyramid. As the number of sides of the base of a regular pyramid increases, the figure becomes more and more like a right cone. The illustrations below suggest that the formulas for the surface areas and volumes of prisms and cylinders are similar.

The Surface Area of a Right Cone

Activity 1

The Surface Area of a Right Cone

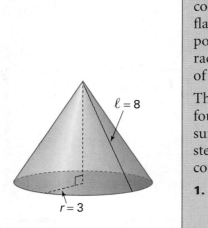

$\ell = 8$

$r = 3$

The surface area of a right cone is found by using a method similar to the one used for a right pyramid. The net for a right cone includes the circular base and the flattened lateral surface, which becomes a portion of a circle known as a *sector*. The radius, ℓ, of the sector is the **slant height** of the cone.

Slant height, ℓ Base

r

ℓ

Lateral surface

The surface area of a right cone can be found by adding the area of the lateral surface and the area of the base. Use the steps below to find the surface area of the cone at left.

1. The lateral surface of a right cone occupies a part of a circle. The arc from A to B matches the circumference of the base of the cone. The length, c, of this arc is equal to the circumference of the base. Find c for the given cone by using the radius of the base ($r = 3$).

Circumference, c (small circle)

A r

ℓ

Length of arc from A to B, c

Lateral surface

B

Circumference, C (large circle)

2. Find C, the circumference of the larger circle, for the given cone by using the slant height ($\ell = 8$) as the radius.

3. Divide c by C. This number tells you what fractional part the lateral surface occupies in the larger circle.

4. Find the area of the larger circle for the given cone by using the slant height ($\ell = 8$). Multiply this number by the fraction from Step 3. The result is the lateral area, L, of the cone.

CHECKPOINT ✔

5. Find B, the area of the base, for the given cone by using the radius ($r = 3$). Add this to the lateral area, L, of the cone from Step 4. What does your answer represent?

EXAMPLE **1** Find the surface area of a right cone with the indicated measurements.

SOLUTION

The circumference of the base is $c = 2\pi r = 14\pi$.

The lateral area is a sector of a circular region with circumference $C = 2\pi\ell = 30\pi$.

The portion of the circular region occupied by the sector is $\frac{c}{C} = \frac{14\pi}{30\pi} = \frac{7}{15}$.

Calculate the area of the sector (lateral area).

$$\pi\ell^2 = 225\pi$$

$$L = \frac{7}{15} \cdot 225\pi = 105\pi$$

Calculate the base area and add the lateral area.

$$B = \pi r^2 = 49\pi$$

$$B + L = 49\pi + 105\pi = 154\pi \approx 483.8$$

ℓ $\ell = 15$
$r = 7$

Activity 2

The Surface Area Formula for a Right Cone

1. Follow the steps in Example 1 above and in Activity 1 on the previous page to find the surface area of a general right cone. That is, use the variables r, h, and ℓ instead of numerical measures. Your answer will be a formula for the surface area of a right cone.

2. Use algebra to simplify your formula. Show your steps in an organized way and save your work in your notebook.

CHECKPOINT ✔ 3. Compare your results with the formula given below. Are they the same? Explain.

Surface Area of a Right Cone

The surface area, S, of a right cone with lateral area L, base of area B, radius r, and slant height ℓ is

$$S = L + B \quad \text{or} \quad S = \pi r\ell + \pi r^2. \qquad \textbf{7.5.1}$$

Many families celebrate special occasions by placing lateral surfaces of cones on their heads.

The Volume of a Cone

In Lesson 7.3, you found that a pyramid had one-third of the area of a prism with a congruent base and equal height. Now imagine performing a similar experiment with a cone and a cylinder. The result of the experiment should be the same—you can try it—because cones and cylinders are like many-sided pyramids and prisms.

This similarity leads to the formula for the volume of a cone.

Volume of a Cone

The volume, V, of a cone with radius r, height h, and base area B is

$$V = \tfrac{1}{3}Bh \quad \text{or} \quad V = \tfrac{1}{3}\pi r^2 h.$$

7.5.2

CRITICAL THINKING

Is the formula for the volume of an oblique cone the same as the formula for the volume of a right cone? What principle justifies your answer? Illustrate your answer with sketches.

E X A M P L E ❷

APPLICATION
GEOLOGY

A volcanologist is studying a violent eruption of a cone-shaped volcano. The original volcanic cone had a radius of 5 miles and a height of 2 miles. The eruption removed a cone-shaped area from the top of the volcano. This cone had a radius of 1 mile and a height of $\frac{1}{2}$ mile. What percent of the total volume of the original volcano was removed by the eruption?

● **SOLUTION**

Find the volume of the original volcano.

$$V = \tfrac{1}{3}\pi r^2 h = \tfrac{1}{3}(5^2)(2) \approx 52.4 \text{ cubic miles}$$

July 22nd, 1980 Eruption of Mount Saint Helens.

Find the volume of the destroyed cone.

$$V = \tfrac{1}{3}\pi r^2 h = \tfrac{1}{3}(1^2)(0.5) \approx 0.52 \text{ cubic miles}$$

Find the percent of the original volcano removed by the eruption.

$$\left(\frac{0.52}{52.4}\right)100 \approx 1\%$$

Exercises

● Communicate

1. Explain how to find the surface area of a right cone.

2. Explain how to find the volume of a right cone.

3. In a right cone, which is longer, the altitude or the slant height? Explain your reasoning.

4. Slant height is not defined for an oblique cone. Explain why.

5. What happens to the volume of a cone if the radius is doubled? if the height is doubled? if both are doubled?

● Guided Skills Practice

6. The diagram below represents a net for a right cone. Find the surface area of the cone. **(ACTIVITY 1)**

$r = 5$

$\ell = 13$

7. Find the surface area of a right cone with the measures shown below. **(EXAMPLE 1)**

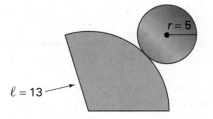

17

15

8

8. A formula for the surface area of a right cone is
   S (surface area) = L (lateral area) + B (base area).

   Explain how to rewrite the formula by substituting values for L and B in terms of π, r, and ℓ, where r is the radius of the base of the cone and ℓ is the slant height of the cone. **(ACTIVITY 2)**

9. Find the volume of the oblique cone shown below. **(EXAMPLE 2)**

27

8

Practice and Apply

Find the surface area of each right cone.

10. 5 cm, 4 cm, 3 cm

11. 10 cm, 8 cm, 6 cm

12. 2 in., 1.6 in., 1.2 in.

13. 1.3, 1.2, 0.5

14. 1.7, 1.5, 0.8

15. 5.3, 2.8, 4.5

16. A right cone has a radius of 3 in. and a height of 4 in. What is the slant height of the cone?

17. A right cone has a radius of 5 in. and a height of 6 in. What is the slant height of the cone?

Find the surface area of each right cone.

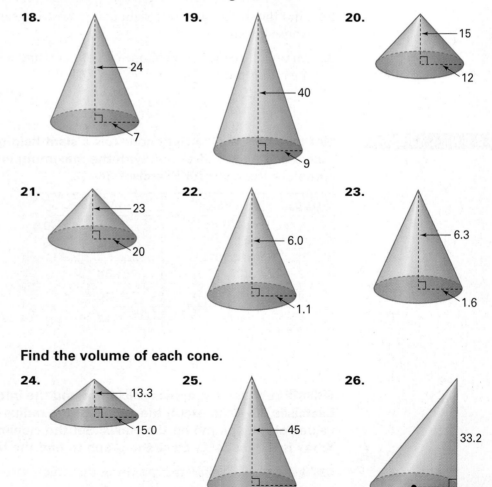

18. 24, 7

19. 40, 9

20. 15, 12

21. 23, 20

22. 6.0, 1.1

23. 6.3, 1.6

Find the volume of each cone.

24. 13.3, 15.0

25. 45, 7

26. 33.2, 6.0

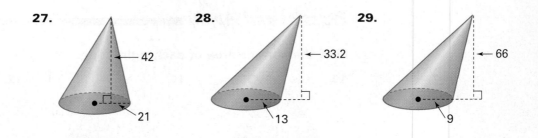

27.

28.

29.

30. If trianguar region *ABC* below is rotated in three dimensions about $\overleftrightarrow{AB}$, what solid figure is formed? Find its surface area.

31. A right triangle has sides of 10, 24, and 26. Rotate the triangle about each leg, and find the volume of the figures formed.

32. A regular square pyramid is inscribed in a right cone of the same height. The radius of the base and the height of the cone are each 10 cm. What is the ratio of the volume of the pyramid to the volume of the cone?

33. A right cone has a radius of 5 in. and a surface area of 180 in.^{2} What is the slant height of the cone?

34. Find the surface area of a right cone whose base area is 25π cm^2 and whose height is 13 cm.

35. An oblique cone has a volume of 1000 cm^3 and a height of 10 cm. What is the radius of the cone?

CONNECTIONS

MAXIMUM/MINIMUM A right cone has a slant height of 10. What are the radius and height of a cone with the maximum volume? Copy and complete the chart for Exercises 36–42.

Radius	Height	Volume
1	$\sqrt{99}$	$\frac{1}{3}\pi \times 1^2 \times \sqrt{99} \approx 10.4$
2	$\sqrt{96}$	**36.** ?
3	**37.** ?	**38.** ?
4	**39.** ?	**40.** ?
5	**41.** ?	**42.** ?

TECHNOLOGY Use a graphics calculator and the information from Exercises 36–42 to graph the values of the radius on the *x*-axis and the values of the volume on the *y*-axis. Set the viewing window so that Ymax is at least 410. Trace the graph to find the following:

43. the radius (*x*-value) that produces the largest volume

44. the largest volume

45. RECREATION An oblique conical tower is to be built at an amusement park, as shown in the architect's sketch below. The vertex is to be directly above the edge of the circular base. The tower is 150 ft tall, and the diameter of the base is 60 ft. What is the volume of the cone?

46. SMALL BUSINESS A businesswoman is selecting glasses for her new ice-cream parlor. Estimate the volumes of the two glasses shown at right.

MANUFACTURING Cone-shaped paper cups can be manufactured from patterns shaped like sectors of a circle. The figures below show two patterns. The first has a straight angle (180°). The second has an angle of 120°. Use the information provided in the diagrams for Exercises 47–49.

47. If each sector has a radius of 6 cm, what will be the area of each sector? (Hint: What fraction of the circle is included in the sector?)

48. Find the volume of the cup formed by the sector with the straight angle. (Hint: The length of the arc of the sector is equal to the circumference of the base of the cone.)

49. Find the volume of the cup formed by the sector with a 120° angle.

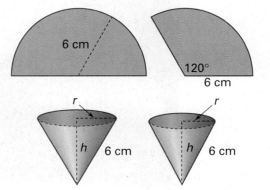

Refer to the graph below for Exercises 50 and 51. *(LESSON 3.8)*

50. If $\overleftrightarrow{AB}$ is perpendicular to $\overleftrightarrow{CD}$, what is the slope of $\overleftrightarrow{AB}$?

51. If $\overleftrightarrow{AB}$ is parallel to $\overleftrightarrow{CD}$, what is the slope of $\overleftrightarrow{AB}$?

52. Prove that a diagonal divides a parallelogram into two congruent triangles. *(LESSON 4.6)*

53. Find the distance between $(2, 4)$ and $(-5, 8)$ in a coordinate plane. *(LESSON 5.6)*

CONNECTION

54. **MAXIMUM/MINIMUM** A manufacturing company makes cardboard boxes of varying sizes by cutting out square corners from rectangular sheets of cardboard that are 12 in. wide and 18 in. long. The cardboard is then folded to form a box. If a customer wants a box with the greatest possible volume for the given sheet of cardboard, what should be the dimensions of the squares that are cut from the corners? What will be the dimensions of the box? What will be the volume of the box? *(LESSON 7.1)*

Look Beyond

APPLICATION

55. **SCALE MODELS** An "HO gauge" model train engine is $\frac{1}{87}$ of the size of a real train engine. If the model's width is 1.5 in. and its length is 5.5 in., what is the length of the real engine?

Surface Area and Volume of Spheres

7.6

Objectives

- Define and use the formula for the surface area of a sphere.

- Define and use the formula for the volume of a sphere.

Why *How much cloth do you think is needed to make a hot-air balloon? You can use the formula for a sphere to estimate the balloon's surface area.*

The balloon is being inflated by fanning hot air into the cloth "envelope."

The Volume of a Sphere

A **sphere** is the set of all points in space that are the same distance, r, from a given point known as the center of the sphere. To find the formula for the volume of a sphere, we first show that a sphere has the same volume as a cylinder with a double cone cut out of it. Then, by using the formulas for cones and cylinders, we derive the formula for the volume of each figure. The discussion that follows begins with a numerical calculation before moving on to the general case.

The height and diameter of the cylinder are the same as the sphere's diameter.

Suppose that the sphere has a diameter of 50 units and that the cylinder has a height and diameter of 50 units. Assume that a blue plane cuts through the sphere and the cylinder 10 units above their centers. You can prove that the two red cross sections—the circular region in the sphere and the **annulus** (the ring-shaped figure) in the cylinder—are both equal to 525π square units.

There are three red right triangles in the figure below: one in the sphere and two in the cylinder with the cones removed. The triangle in the sphere can be solved for b, the radius of the disc. The two triangles in the cylinder are isosceles and have known side lengths, as you can convince yourself by studying the figure. (The blue cutting plane is parallel to the bases of the hollow cones in the cylinder.) The information from the triangles is used to calculate the areas of the cross sections.

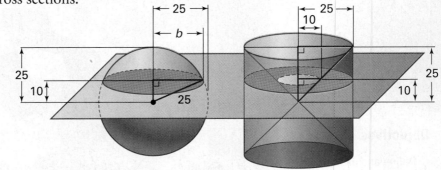

> In the cylinder, the acute angles of the large red triangle are congruent. You can show that the acute angles of the small triangle are congruent, as well.

Area of the circle in the sphere

$b^2 + 10^2 = 25^2$

$b^2 + 100 = 625$

$b^2 = 525$

$b = \sqrt{525}$

$\text{area} = \pi b^2 = \pi\left(\sqrt{525}\right)^2$

$= 525\pi$ square units

Area of the annulus

area of large circle − area of small circle

$= \pi 25^2 - \pi 10^2$

$= \pi(25^2 - 10^2)$

$= \pi(625 - 100)$

$= \pi(525)$

$\text{area} = 525\pi$ square units

PROOF

Thus, the areas are equal. Next, *generalize* the procedure to get the formulas for a sphere and a cylinder of radius r. Let r be the radius of the sphere and y be the distance from the centers of the figures to the blue cutting plane.

Area of the circle in the sphere

$b^2 + y^2 = r^2$

$b^2 = r^2 - y^2$

$b^2 = \sqrt{r^2 - y^2}$

$\text{area} = \pi\left(\sqrt{r^2 - y^2}\right)^2 = \pi(r^2 - y^2)$

Area of the annulus

area of large circle − area of small circle

$= \pi r^2 - \pi y^2$

$\text{area} = \pi(r^2 - y^2)$

This shows that the formulas are true for all planes parallel to the bases of each figure and for all values of y.

> The corresponding cross sections have equal areas.

Therefore, the conditions of Cavalieri's Principle are satisfied. The volume of the sphere equals the volume of the cylinder with the double cones removed.

$V(\text{sphere}) = V(\text{cylinder}) - V(\text{cones})$

$V(\text{sphere}) = \pi r^2(2r) - 2\left(\frac{1}{3}\pi^2\right)(r)$

$= 2\pi r^3 - \frac{2}{3}\pi r^3$

$= \frac{4}{3}\pi r^3$

Volume of a Sphere

The volume, V, of a sphere with radius r is

$$V = \tfrac{4}{3}\pi r^3.$$

7.6.1

CRITICAL THINKING

When the cutting plane in the proof of the volume formula cuts through the center of the sphere and the center of the cylinder, what happens to the annulus? How does this affect the calculations? Explain.

E X A M P L E ❶ The envelope of a hot-air balloon has a radius of 27 feet when fully inflated. Approximately how many cubic feet of hot air can it hold?

● **SOLUTION**

$V = \tfrac{4}{3}\pi r^3$

$\quad = \tfrac{4}{3}\pi(27)^3$

$\quad = \tfrac{4}{3}(19{,}683)\pi$

$\quad = 26{,}244\pi$ cubic feet $\approx 82{,}488$ cubic feet

The Surface Area of a Sphere

APPLICATION

CARTOGRAPHY

In previous lessons, you analyzed the surface areas of three-dimensional figures by unfolding them to form a net on a flat surface. But, as mapmakers know, a sphere cannot be unfolded smoothly onto a flat surface.

The most common map of the world uses a Mercator projection of Earth's surface onto a flat plane. On this kind of map, the landmasses near the North and South Poles, such as Greenland and Antartica, appear to be much larger in relation to other landmasses than they actually are.

This map is a Mercator projection. You will learn more about projecting a sphere onto a flat surface in Lesson 11.7.

The formula for the surface area of a sphere can nevertheless be derived by using some clever techniques.

To derive the formula for the surface area of a sphere, begin with an approximation. Imagine the surface of a sphere as a large number of polygons, as in the geodesic dome in the photo at right. The smaller these polygons are, the more closely they approximate the surface of a sphere.

Consider each polygon to be the base of a pyramid with its vertex at the center of the sphere. The volumes of these pyramids added together will approximate the volume of the sphere. The height of each pyramid is the radius of the sphere. Therefore, the volume of each pyramid is $\frac{1}{3}Br$, where B is the area of the base of the pyramid and r is the radius of the sphere.

Doppler radar tower

$$\text{Volume of a sphere} \approx \frac{1}{3}B_1r + \frac{1}{3}B_2r + \cdots + \frac{1}{3}B_nr$$
$$\approx \frac{1}{3}r(B_1 + B_2 + \cdots + B_n)$$

If the total area of the bases of the pyramids is assumed to equal the surface area of the sphere, S, then the volume, V, of the sphere can be written as

> The larger the number of pyramids, the closer the approximation will be to the actual volume of the sphere. The approximation is said to be exact "in the limit."

$$V = \frac{1}{3}r(S).$$

Solve for S to get

$$S = \frac{3V}{r}.$$

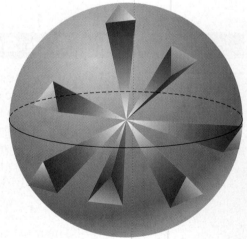

You now have a formula for the surface area of a sphere in terms of its volume. Now substitute in the formula for volume from the previous page.

$$S = \frac{3V}{r}$$
$$= \frac{3\left(\frac{4}{3}\pi r^3\right)}{r}$$
$$= 4\pi r^2$$

Surface Area of a Sphere

The surface area, S, of a sphere with radius r is
$$S = 4\pi r^2.$$

7.6.2

EXAMPLE 2

APPLICATION
HOT-AIR BALLOONING

The envelope of a hot-air balloon is 54 feet in diameter when inflated. The cost of the fabric used to make the envelope is $1.31 per square foot. Estimate the total cost of the fabric for the balloon envelope.

● **SOLUTION**

First estimate the surface area of the inflated balloon envelope. The balloon is approximately a sphere with a diameter of 54 feet, so the radius is 27 feet.

$$S = 4\pi r^2$$
$$= 4\pi(27)^2$$
$$= 4(729)\pi$$
$$= 2916\pi \approx 9160.9 \text{ square feet}$$

Now multiply the surface area of the fabric by the cost per square foot to find the approximate cost of the fabric.

9160.9 square feet × $1.31 per square foot ≈ $12,000

Exercises

● **Communicate**

1. How does the formula for the surface area of a sphere with a radius of r relate to the formula for the area of a circle with a radius of r?

2. If the area of a circle is approximately 39 ft^2, what is the surface area of a sphere with the same radius as the circle? Explain your answer.

3. What happens to the area of a circle when the radius is doubled? when it is tripled?

4. What happens to the surface area of a sphere when the radius is doubled? when it is tripled?

5. What happens to the volume of a sphere when the radius is doubled? when it is tripled?

● **Guided Skills Practice**

6. Find the approximate volume of a sphere that has a radius of 40 ft. Round your answer to the nearest whole number. **(EXAMPLE 1)**

7. Find the surface area of a sphere that has a diameter of 48 ft. Find the cost of constructing the sphere if the cost for materials and labor is $1.50 per square foot of surface area. Round your answer to the nearest whole dollar. **(EXAMPLE 2)**

Practice and Apply

In Exercises 8–19, find the surface area and volume of each sphere, where *r* is the radius of the sphere and *d* is the diameter. Express your answers in two ways: (a) as an exact answer in terms of π, and (b) as an approximate answer rounded to the nearest hundredth.

8. $r = 4$ **9.** $r = 8$ **10.** $r = 41$

11. $r = 33$ **12.** $r = 13.41$ **13.** $r = 12.22$

14. $d = 18$ **15.** $d = 16$ **16.** $d = 22.34$

17. $d = 11.48$ **18.** $r = 12.33$ **19.** $r = 99.98$

In Exercises 20–25, find the area and volume of each sphere with the given radius or diameter. Give exact answers in terms of π and a variable.

20. $r = x$ **21.** $r = 2y$ **22.** $d = 12x$

23. $d = 4y$ **24.** $r = \frac{x}{2}$ **25.** $r = \frac{y}{4}$

Area of cross section, *A*

In Exercises 26–31, find the surface area of the sphere at left based on the area, *A*, of a cross section through its center. (Hint: See Exercise 1.)

26. $A = 225$ **27.** $A = 125$ **28.** $A = 32.30$

29. $A = 11.22$ **30.** $A = 16\pi$ **31.** $A = 225\pi$

32. What is the volume of the largest ball that will fit into a cubical box with edges of 12 in.?

33. A cube and a sphere both have a volume of 1000 cubic units. What are their surface areas?

34. A cube and a sphere both have a surface area of 864 square units. What are their volumes?

CHALLENGE

35. The figures shown at right are a hemisphere, a right circular cone, and a right circular cylinder. Each has the same volume, and each has a base with a radius of 10 in. Find the altitude of each.

Radius of bases = 10 in.

APPLICATIONS

In Exercises 36 and 37, round your answers to the nearest tenth.

36. **SPORTS** Find the surface area and volume of the softball at right.

37. **SPORTS** Find the surface area and volume of the baseball at right.

3.5 in.   2.9 in.

softball   baseball

9 in.

38. SPORTS A can of tennis balls has 3 balls stacked tightly in it. The approximate height of the stacked balls and of the interior of the can is 9 in. How much space to do the tennis balls occupy? Approximately what percent of the space inside the can do the balls occupy? Round your answers to the nearest tenth.

39. GEOGRAPHY Earth's radius is approximately 4000 mi. If two-thirds of Earth's surface is covered by water and one-third is land, estimate the land area on Earth.

FOOD Hosea buys an ice-cream cone. The ice cream is a sphere with a radius of 1.25 in. The cone has a height of 8 in. and a diameter of 2.5 in. Round your answers for Exercises 40–42 to the nearest hundredth.

40. What is the volume of the ice cream?

41. What is the volume of the cone?

42. If the ice cream sits in the cone so that it forms a hemisphere (a half of a sphere) above the rim of the cone, what is the total surface area of the hemisphere of ice cream and the cone?

43. METALWORK A metal sculptor has a solid bronze sphere with a radius of 10 in. She melts the sphere and casts a hollow sphere with an inner radius of 10 in. What is the thickness of the shell of the hollow sphere?

Pouring molten metal into a mold

Look Back

Find the surface area of each figure. (LESSONS 7.2, 7.4, AND 7.5)

44. right rectangular prism: $\ell = 3$ in., $w = 10$ in., $h = 5$ in.

45. right cone: $r = 15$ cm, slant height $= 45$ cm

46. right cylinder: $r = 9$ m, $h = 10$ m

Find the volume of each figure. (LESSONS 7.3, 7.4, AND 7.5)

47. rectangular pyramid: base length $= 15$ ft, base width $= 7$ ft, altitude $= 12$ ft

48. right cone: $r = 5$ in., $h = 10$ in.

49. oblique cylinder: $r = 7.5$ m, $h = 20$ m

Look Beyond

Algebra

50. What is the ratio of the volume of a sphere to the surface area of the same sphere? Express your answer in terms of the radius, r, of the sphere.

51. Write a formula for the volume of a sphere in terms of its surface area, S.

52. If the radius of a sphere is 4.5 units, by what number can you multiply the surface area of the sphere to find the volume?

53. A sphere with a diameter of $2r$ is contained in a cube with edges of $2r$. What is the ratio the volume of the cube to the volume of the sphere?

TREASURE
of King Tut's Tomb

GEM-STUDDED RELICS AMAZE EXPLORERS
Special Cable to the New York Times

LONDON, Nov. 30, 1922—The Cairo correspondent of the London Times, in a dispatch to his paper, describes how Lord Carnarvon and Howard Carter unearthed below the tomb of Ramses VI, near Luxor, two rooms containing the funeral paraphernalia of King Tutankhamen, who reigned about 1350 b.c.

A sealed outer door was carefully opened, then a way was cleared down some sixteen steps and along a passage of about twenty-five feet. A door to the chambers was found to be sealed as the outer door had been and as on the outer door, there were traces of re-closing.

The excitement of unearthing the tomb was followed by three years of hard work. Unpacking a box of priceless relics took three painstaking weeks. The items, which had been stored for thousands of years, could have been easily damaged if not handled with great care.

A crumpled robe in Box 21 presented a dilemma for Howard Carter. Should he preserve the garment as is and lose the chance to learn its full design? Or should he handle it and sacrifice the cloth in order to examine the complete robe?

Excerpt from Howard Carter's notes

. . . by sacrificing the cloth, picking it carefully away piece by piece, we could recover, as a rule, the whole scheme of decoration. Later, in the museum, it will be possible to make a new garment of the exact size, to which the original ornamentation—beadwork, gold sequins, or whatever it may be—can be applied. Restorations of this kind will be far more useful, and have much greater archaeological value, than a few irregularly shaped pieces of cloth and a collection of loose beads and sequins.

Men build a single track railway through the Valley of Kings for transporting cases of relics from Tutankhamen's tomb in Luxor, 1922.

Cooperative Learning

The robe in Box 21 had the shape of the lateral surface of a cylindrical shape, which was laid flat to form two rectangular layers. By estimating the height and width of the rectangles, Howard Carter was able to determine the size of the garment. The robe was covered by a repeating pattern of glass beads and gold sequins, with a lower band of beaded fringe, or pendant strings. In the following Activity, some of the data has been removed from his notes.

We know the distance between the pendant strings of the lower band to have averaged 8 mm. There are 137 of these pendants. Therefore the circumference of the lower part of the garment must have been _____?_____, which would make one width about _____?_____.

Now there are 3054 gold sequins. Three sequins in the pattern require 9 square centimetres. Therefore total area of network = _____?_____. As circumference was _____?_____, this would make the height work out at about _____?_____.

Thus size of garment would work out at about _____?_____ × _____?_____.

1. Determine the missing information in Howard Carter's notes, and explain how you found each value. Include units in your answer. You may wish to convert all the measurements to the same units first.

2. Use a tape measure or ruler and string to compare the size of the robe to that of your own clothes. How do they compare?

British archeologist and Egyptologist Howard Carter (left) and Callender (right) opening the entrance to the fourth chamber of Tutankhamen's tomb, 1923.

Three-Dimensional Symmetry

7.7

Objectives

● Define various transformations in three-dimensional space.

● Solve problems by using transformations in three-dimensional space.

Why
So far, the definitions of symmetry have been limited to a plane. But three-dimensional figures, like the tiger in the photo, may also have symmetry.

Tiger! Tiger! burning bright
In the forests of the night,
What immortal hand or eye
Could frame thy fearful symmetry?
—William Blake

Three-Dimensional Reflections

A three-dimensional figure may be reflected across a plane, just as a two-dimensional figure can be reflected across a line. What happens to each point in the preimage of a figure as a result of a reflection across a plane? You will investigate this question in Activity 1.

Activity 1

Reflections in Coordinate Space

Part I

1. Graph the point $A(1, 1, 1)$ in a three-dimensional coordinate space. Use dashed lines to make the location of the point in space evident.

2. Multiply the x-coordinate of point A by -1 and graph the resulting point. Label the new point A'.

3. Point A' is the image of point A reflected across a plane. Name this plane. If you connect points A and A' to form $\overline{AA'}$, what is the relationship between this segment and the plane of reflection?

4. Write your definition for the reflection of a point across a plane in a three-dimensional coordinate space.

5. Experiment with other reflections of point A. What happens if you multiply the x-coordinates by -1? the z-coordinates?

6. Now study the reflection of an entire segment, such as $\overline{AB}$, across a plane. (For example, multiply the x-coordinates of points A and B by -1 and connect the resulting points.) Experiment with the reflections of other figures, such as cubes.

CHECKPOINT ✔ 7. Write your own definition for the reflection of a figure in a three-dimensional coordinate space across a plane.

Part II

1. Fill in the table below. Use the terms *front*, *back*, *left*, *right*, *top*, and *bottom* to describe the octant of a point.

		Octant of image	Coordinates of image
$A(2, 3, 4)$	Reflection across the xy-plane	front-right-bottom	$(2, 3, -4)$
	Reflection across the xz-plane	?	?
	Reflection across the yz-plane	?	?
$B(-4, 5, 6)$	Reflection across the xy-plane	?	?
	Reflection across the xz-plane	?	?
	Reflection across the yz-plane	?	?

CHECKPOINT ✔ 2. Generalize your findings for point $P(x, y, z)$.
   a. What are the coordinates of the reflection of P across the xy-plane?
   b. What are the coordinates of the reflection of P across the xz-plane?
   c. What are the coordinates of the reflection of P across the yz-plane?

✔ The three-dimensional figures below have reflectional symmetry. Explain why. Where is the reflection "mirror" in each case? Create your own definition of reflectional symmetry in space. (Use the earlier definition of reflectional symmetry in a plane as a model.)

Activity 2
Rotations in Coordinate Space

YOU WILL NEED

no special tools

1. Graph $\overline{AB}$ in a three-dimensional coordinate space. Use dashed lines to make the location of the segment evident.

2. Multiply the *x*- and *y*-coordinates of points *A* and *B* by −1 and graph the resulting points, *A′* and *B′*. Connect the new points to form $\overline{A'B'}$.

3. $\overline{A'B'}$ is a rotation image of $\overline{AB}$ about the *z*-axis. About what point on the *z*-axis has point *A* been rotated?

4. Imagine the segment rotating about the *z*-axis, as suggested by the picture. Does it seem to you that each of the rotation images of point *A* is in the same plane? of point *B*? What is the relationship between these planes and the *z*-axis?

CHECKPOINT ✔ 5. Write your own definition of the rotation of a figure about an axis in coordinate space.

The three-dimensional figures below have rotational symmetry. Explain why. What is the axis of rotation in each case? Create your own definition of rotational symmetry in space. (Use the earlier definition of rotational symmetry in the plane as a model.)

Revolutions in Coordinate Space

If you rotate a figure about an axis, a spatial figure is formed. The spatial figure is the set of points through which the original figure passes in one complete revolution.

EXAMPLE ● You are given $\overline{AB}$ with endpoints $A(0, 5, 0)$ and $B(0, 5, 5)$. Sketch, describe, and give the dimensions of the figure that results when

    a. $\overline{AB}$ is rotated about the z-axis.

    b. $\overline{AB}$ is rotated about the y-axis.

● **SOLUTION**

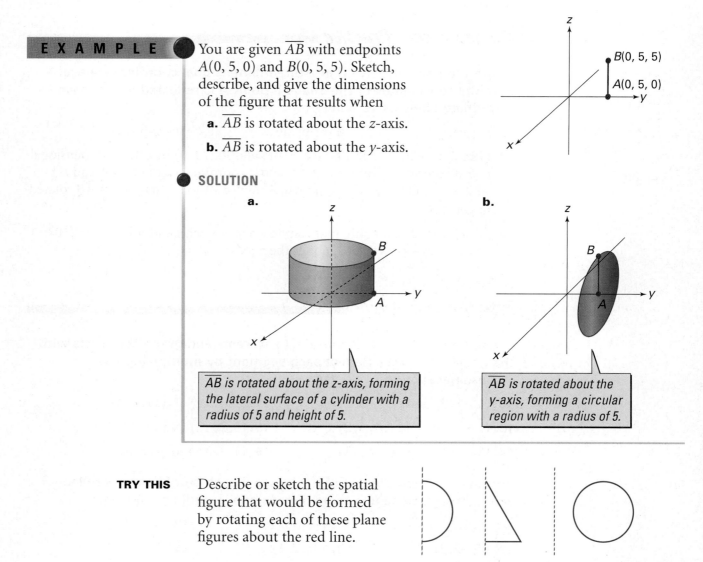

a.

$\overline{AB}$ is rotated about the z-axis, forming the lateral surface of a cylinder with a radius of 5 and height of 5.

b.

$\overline{AB}$ is rotated about the y-axis, forming a circular region with a radius of 5.

TRY THIS   Describe or sketch the spatial figure that would be formed by rotating each of these plane figures about the red line.

Exercises

Communicate

1. Describe the similarities and differences betweeen three-dimensional reflectional symmetry and two-dimensional reflectional symmetry.

2. What spatial figure is formed by rotating rectangle *ABCD* about $\overline{AB}$?

3. What spatial figure is formed by rotating △*EFG* about $\overline{EG}$?

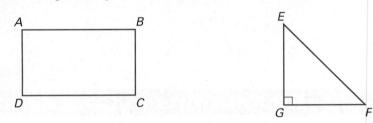

4. List some objects in your classroom that have three-dimensional rotational symmetry.

Guided Skills Practice

Graph the point *A*(2, 1, –1) in a three-dimensional coordinate space. Give the coordinates of the image if point *A* is reflected across each coordinate plane listed below. (ACTIVITY 1)

5. *xy*-plane

6. *xz*-plane

7. *yz*-plane

8. Graph $\overline{AB}$ with endpoints *A*(1, 3, –2) and *B*(1, 3, 0) in a three-dimensional coordinate space. Multiply the *y*- and *z*-coordinates of points *A* and *B* by –1 and graph the resulting points, *A′* and *B′*. About what axis is $\overline{AB}$ rotated to get $\overline{A'B'}$? (ACTIVITY 2)

9. Sketch the spatial figure that results when $\overline{AB}$, with endpoints *A*(1, 2, 0) and *B*(1, 2, –4), is rotated about the *z*-axis. (EXAMPLE)

Practice and Apply

Draw three-dimensional coordinate systems and graph segments with the given endpoints. Reflect each segment by multiplying the *y*-coordinates by –1.

10. (4, –2, 3) and (–2, –3, 2)

11. (–5, 2, 1) and (1, 1, 1)

12. (1, –2, –3) and (–1, 5, 2)

13. (3, 2, –3) and (–4, 3, –2)

14. (5, 3, 2) and (–1, –1, –1)

15. (1, 2, 3) and (1, 3, 3)

What are the coordinates of the image if each point below is reflected across the *xy*-plane in a three-dimensional coordinate system?

16. (6, 5, 8)

17. (–2, 3, –1)

18. (1, 1, 1)

19. (4, –2, 3)

20. (–5, 2, 1)

21. (1, –2, –3)

What are the octants and coordinates of the image if each point below is reflected across the *xz*-plane in a three-dimensional coordinate system?

22. $(6, -2, 8)$ **23.** $(-4, -4, -1)$ **24.** $(1, 0, 1)$

25. $(1, 1, 0)$ **26.** $(4, -4, 1)$ **27.** $(2, 2, -8)$

In Exercises 28–31, $\overline{AB}$ has endpoints A(5, 0, 10) and B(5, 0, 0).

28. What spatial figure is formed by rotating $\overline{AB}$ about the *x*-axis?

29. What is the area of the figure formed by rotating $\overline{AB}$ about the *x*-axis?

30. What spatial figure is formed by rotating $\overline{AB}$ about the *z*-axis?

31. What is the volume of the figure formed by rotating $\overline{AB}$ about the *z*-axis?

In Exercises 32 and 33, $\overline{CD}$ has endpoints C(4, 0, 0) and D(0, 0, 4).

32. What spatial figure is formed by rotating $\overline{CD}$ about the *z*-axis?

33. What is the volume of the figure formed by rotating $\overline{CD}$ about the *z*-axis?

CONNECTION

CHALLENGE

34. **MAXIMUM/MINIMUM** The area of a right triangle with a fixed perimeter is maximized when the triangle is a 45-45-90 triangle. Suppose that you rotate a 45-45-90 triangle about one leg to create a cone. Does the cone have the maximum volume for the given slant height? Why or why not? (See Exercises 43–44 in Lesson 7.5, page 466.)

APPLICATION

POTTERY A potter is making pots according to certain patterns. For each half of a pattern given below, sketch what the complete pot will look like. (Rotate about the dashed red line.)

35.   **36.**   **37.**

38.   **39.**   **40.**

Look Back

Find the volume and surface area of each prism. *(LESSON 7.2)*

41.   **42.**

7
11
15

Right-triangular prism

6
12

Regular hexagonal prism

43. TRANSPORTATION A truck's storage space is shaped like a rectangular prism that measures 7 ft 10 in. × 8 ft 2 in. × 24 ft. What is the volume of the storage space? *(LESSON 7.2)*

44. Find the surface area and volume of a regular square pyramid with a base area, *B*, of 36 cm² and a height, *h*, of 5 cm. *(LESSON 7.3)*

45. HOBBIES Seeing barbecue smokers displayed next to bags of potting soil, Dolores decides to change her old smoker at home into a planter. The center of her smoker comes up to her waist and is thus half of her height of 5 ft. Use this information to estimate other dimensions in the photo, and then determine how much soil it will take to fill the smoker to the centerline of the barrel from which it is made.
(LESSON 7.4)

Look Beyond

OPTICS A spinning fan creates the illusion of being solid. A strobe light, which emits a flash at certain time intervals, can "freeze" the motion of the fan.

46. Suppose that a fan with four equally spaced, identical blades is turning at 12 revolutions per second. How often will the strobe need to flash to make the fan appear to be frozen?

47. If the strobe flashes 36 times per second, at what speeds (in revolutions per second) can the fan move and still appear to be frozen?

CREATING SOLIDS OF REVOLUTION

If a region in a plane is rotated about an axis that lies in the same plane, the resulting figure is called a *solid of revolution*. Follow the directions below to create your own solids of revolution with cardboard, foam core board, a needle and thread.

1. On a piece of graph paper, draw a shape that has at least one vertical edge, such as the one shown at right. This edge will be the axis of revolution of your solid. Your shape should be at least 10 units high.

2. Number the intersections of the grid lines with the edge of your figure, as shown. For each intersection, set your compass to the width of your figure, and draw a circle on a piece of cardboard. Number the circle according to the number of the intersection.

3. Cut the circles out of the cardboard. These circles represent cross-sections of your solid of revolution. Cut 10 small pieces of foam core board that are approximately the thickness of the squares on your graph paper to use as spacers.

4. Using a small needle, thread the circles of cardboard in order, alternating with the pieces of foam core. Secure the string at the bottom with a small piece of tape. Tie a loop in the string at the top for a hanger.

A THREE-DIMENSIONAL PUZZLE

This puzzle is a three-dimensional version of the ancient Chinese puzzle known as the tangram (see page 359).

Activity 1

THE PUZZLE Copy the nets shown below and on the next page onto heavy paper and assemble them to form six solids. Then see if you can fit them together to make a cube. The measurements are given in centimeters.

Make two copies of this piece.

4

4

4

2

4

2

2

2

2

2

2

2

2

2

Activity 2

Describe each piece in as much detail as possible. What shape is it? What are the lateral faces and bases? Find the lengths of all edges that are not given. Then find the volume of each piece.

Extension

1. What are the dimensions of the cube that is formed when the pieces are assembled? Does the sum of the volumes of the individual pieces equal the volume of the cube?

2. Can you assemble the pieces to form any other interesting solids?

7 Chapter Review and Assessment

POSTULATES AND THEOREMS

Lesson	Number	Postulate or Theorem
7.1	7.1.1 Surface Area and Volume of a Right Rectangular Prism	The surface area, S, and volume, V, of a right rectangular prism with length ℓ, width w, and height h, are $S = 2\ell w + 2wh + 2\ell h$ and $V = \ell wh$.
	7.1.2 Surface Area and Volume of a Cube	The surface area, S, and volume, V, of a cube with side length s are $S = 6s^2$ and $V = s^3$.
7.2	7.2.1 Surface Area of a Right Prism	The surface area, S, of a right prism with lateral area L, base area B, perimeter p, and height h is $S = L + 2B$ or $S = hp + 2B$.
	7.2.2 Cavalieri's Principle	If two solids have equal heights and the cross sections formed by every plane parallel to the bases of both have equal areas, then the two solids have equal volumes.
	7.2.3 Volume of a Prism	The volume, V, of a prism with height h and base area B is $V = Bh$.
7.3	7.3.1 Surface Area of a Regular Pyramid	The surface area, S, of a regular pyramid with lateral area L, base area B, perimeter of the base p, and slant height ℓ is $S = L + B$ or $S = \frac{1}{2}\ell p + B$.
	7.3.2 Volume of a Pyramid	The volume, V, of a pyramid with height h and base area B is $V = \frac{1}{3}Bh$.
7.4	7.4.1 Surface Area of a Right Cylinder	The surface area, S, of a right cylinder with lateral area L, base area B, radius r, and height h is $S = L + 2B$ or $S = 2\pi rh + 2\pi r^2$.
	7.4.3 Volume of a Cylinder	The volume, V, of a cylinder with radius r, height h, and base area B is $V = Bh$ or $V = \pi r^2 h$.

Lesson	Number	Postulate or Theorem
7.5	7.5.1 Surface Area of a Right Cone	The surface area, S, of a right cone with lateral area L, base area B, radius r, and slant height ℓ is $S = L + B$ or $S = \pi r \ell + \pi r^2$.
	7.5.2 Volume of a Cone	The volume, V, of a cone with radius r, height h, and base area B is $V = \frac{1}{3}Bh$ or $V = \frac{1}{3}\pi r^2 h$.
7.6	7.6.1 Volume of a Sphere	The volume, V, of a sphere with radius r is $V = \frac{4}{3}\pi r^3$.
	7.6.2 Surface Area of a Sphere	The surface area, S, of a sphere with radius r is $S = 4\pi r^2$.

Key Skills & Exercises

LESSON 7.1

Key Skills

Solve problems by using the ratio of surface area to volume.

A cube has a volume of 27,000 cubic millimeters. What is its ratio of surface area to volume?

The length of each edge of the cube is $\sqrt[3]{27{,}000} = 30$ millimeters, so the area of each face is $30 \times 30 = 900$ square millimeters. There are 6 faces, so the surface area is $6(900) = 5400$ square millimeters. Thus, the ratio of surface area to volume is $\frac{5400}{27{,}000} = 0.2$.

Exercises

1. Find the ratio of surface area to volume for a cube with a volume of 64 cubic inches.

2. Find the ratio of surface area to volume for a cube with a volume of 100 cubic inches.

3. Compare the ratios of surface area to volume for a cube with an edge length of 4 and for a right rectangular prism with the dimensions $4 \times 7 \times 3$.

4. A right rectangular prism has a square base, and its height is triple the base edge. Find the ratio of its surface area to volume.

LESSON 7.2

Key Skills

Find the surface area of a right prism.

Find the surface area of the right triangular prism below.

The lateral area is hp, or
$100(36 + 77 + 85) = 19{,}800$.

The area of each base is
$\frac{1}{2}(36)(77) = 1386$.

The surface area is
$S = L + 2B$, or
$19{,}800 + 2(1386) = 22{,}572$.

Exercises

Refer to the right hexagonal prism with a regular hexagonal base below.

5. Find the lateral area of the prism.

6. Find the surface area of the prism.

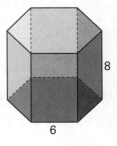

Find the volume of a prism.

Find the volume of the prism on the previous page.

The area of the base is 1386, so the volume is $V = Bh$, or $1386(100) = 138{,}600$.

7. Find the volume of the prism on the previous page.

8. Find the volume of the prism on the previous page if the dimensions are doubled.

LESSON 7.3
Key Skills

Find the surface area of a regular pyramid.

Find the surface area of the regular square pyramid below.

The lateral area is $\frac{1}{2}\ell p$, or $\frac{1}{2}(25)(56) = 700$.

The area of the base is $14^2 = 196$.

The surface area is $S = L + B$, or $700 + 196 = 896$.

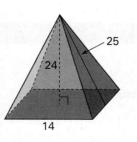

Find the volume of a pyramid.

Find the volume of the pyramid above.

The area of the base is 700, so the volume is $V = \frac{1}{3}Bh$, or $\frac{1}{3}(700)(24) = 5600$.

Exercises

Refer to the regular pyramid below.

9. Find the lateral area of the pyramid.

10. Find the surface area of the pyramid.

11. Find the volume of the pyramid.

12. Find the volume of the pyramid if the height is doubled.

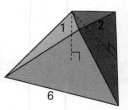

LESSON 7.4
Key Skills

Find the surface area of a right cylinder.

Find the surface area of the right cylinder below.

The lateral area is $2\pi rh$, or $2\pi(12)(7) = 168\pi$.

The area of each base is πr^2, or $\pi(12^2) = 144\pi$.

The surface area is $S = L + 2B$, or $168\pi + 2(144\pi) = 456\pi$.

Find the volume of a cylinder.

Find the volume of the right cylinder above.

The area of the base is 144π, so the volume is $V = Bh$, or $144\pi(7) = 1008\pi$.

Exercises

A right cylinder has a radius of 4 and a height of 9.

13. Find the lateral area of the cylinder.

14. Find the surface area of the cylinder.

15. Find the volume of the cylinder.

16. Which increases the volume of a cylinder more, doubling the height or doubling the radius? Explain your reasoning.

Key Skills

Find the surface area of a right cone.

Find the surface area of the right cone below.

The lateral area, L, is $\pi r\ell$, or
$\pi(6)(10.9) = 65.4\pi$.

The area of the base, B, is πr^2, or $\pi(6^2) = 36\pi$.

The surface area is
$S = L + B$, or
$65.4\pi + 36\pi = 101.4\pi$.

Find the volume of a cone.

Find the volume of the cone above.

The area of the base is 36π, so the volume is
$V = \frac{1}{3}Bh$, or $\frac{1}{3}36\pi(9.1) = 109.2\pi$.

Exercises

Refer to the right cone below.

17. Find the lateral area of the cone.

18. Find the surface area of the cone.

19. Find the volume of the cone.

20. Find the volume of the cone if the height is doubled. Round your answer to the nearest hundredth.

Key Skills

Find the surface area of a sphere.

Find the surface area of a sphere with a radius of 9.

The surface area is $S = 4\pi r^2$, or $4\pi 9^2 = 324\pi$.

Find the volume of a sphere.

Find the volume of a sphere with a radius of 9.

The volume is $V = \frac{4}{3}\pi r^3$, or $\frac{4}{3}\pi 9^3 = 972\pi$.

Exercises

21. Find the surface area of a sphere with a radius of 5.

22. Find the volume of a sphere with a radius of 5.

23. Find the radius of a sphere with a volume of 36π.

24. What is the side length of a cube that has the same volume as a sphere with a radius of 1?

Key Skills

Reflect a figure in a three-dimensional coordinate system.

Reflect the segment with endpoints at $(5, 5, 2)$ and $(5, 5, -2)$ across the xz-plane.

The image has endpoints at $(5, -5, 2)$ and $(5, -5, -2)$.

Exercises

A segment has endpoints at $A(0, 0, 0)$ and $B(0, 4, 6)$.

25. Draw a reflection of the segment across the xy-plane. Give the coordinates of the endpoints of the image.

26. Draw a reflection of the segment across the yz-plane. Give the coordinates of the endpoints of the image.

Sketch the spatial figure formed by rotating a figure about an axis.

Sketch the spatial figure formed by revolving the segment with endpoints at (5, 5, 2) and (5, 5, −2) about the z-axis.

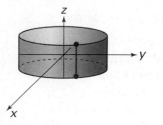

A triangle has vertices at (0, 0, 0), (2, 0, 0), and (0, 0, 5).

27. Sketch the spatial figure formed by rotating the triangle about the x-axis.

28. Sketch the spatial figure formed by rotating the triangle about the z-axis.

Applications

29. PRINTING A roll of paper used in printing has a diameter of 3 feet, a hollow core of 4 inches, and a width of 5 feet. If the thickness of the paper is 0.0015 inch, what is the length of the paper on the roll?

30. SPORTS A basketball has a radius of approximately 4.75 inches when filled. How much material is needed to make one? How much air will it hold? If the basketball is stored in a cubic box whose edges are 9.5 inches long, what percent of the box is not filled by the basketball? (Round your answers to the nearest hundredth.)

31. PHYSICS A spherical soap bubble with a radius of 5 centimeters lands on a flat surface and becomes a hemisphere. What is the radius of the hemisphere?

Alternative Assessment

Performance Assessment

1. A shape with an area of B and a perimeter of p is translated from one plane to a parallel plane, and each point is connected with its image to form a solid.

The distance between the parallel planes is h, and the distance from each point on the shape to its image in the parallel plane is ℓ. What is the volume of this solid? What principle justifies your answer?

2. The solid can be called *right* if the height, h, is equal to the distance, ℓ. What is the surface area of a right solid like the one above with a base area of B, base perimeter of p, and height of h? Explain your reasoning.

3. A point is selected in the parallel plane used to form the solid above, and every point on the original shape is connected to this point to form a new solid. Make a conjecture about the volume of this solid. Explain your reasoning.

Portfolio Projects

Suppose that you plan to manufacture and sell a food product and that you need to design a container with a volume of 100 cubic inches. Your container can be any of the following: a prism, a pyramid, a cylinder, or a cone.

1. Decide on the type of food product you plan to sell, and choose at least three different container shapes to investigate. Create several patterns for containers of each shape with different heights, widths, and radii. Your pattern should include any flaps or tabs that will be needed to assemble the container. Each container should have a volume of 100 cubic inches.

2. Compare the surface areas of the different containers you created. Which shape seems to require the least amount of materials to create? What are the dimensions of the container of that shape that requires the least amount of materials to create?

3. Decide on a container for your package. Some of the factors besides surface area that you should consider are
   • the look of the container,
   • stacking and storing your product,
   • printing or labeling your package, and
   • convenience for the consumer.

🔗 internet connect

HRW The HRW Web site contains many resources to reinforce and expand your knowledge of surface area and volume of solids. The Web site also provides Internet links to other sites where you can find information and real-world data for use in research projects, reports, and activities that involve surface area and volume of solids. Visit the HRW Web site at **go.hrw.com**, and enter the keyword **MG1 CH7** to access the resources for this chapter.

QUANTITATIVE COMPARISON For Items 1–4, write
A if the quantity in Column A is greater than the quantity in Column B;
B if the quantity in Column B is greater than the quantity in Column A;
C if the two quantities are equal; or
D if the relationship cannot be determined from the information given.

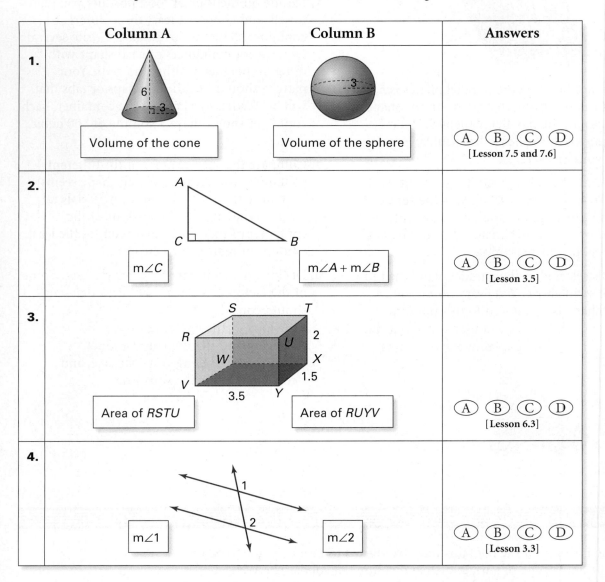

	Column A	Column B	Answers
1.	Volume of the cone	Volume of the sphere	Ⓐ Ⓑ Ⓒ Ⓓ [Lesson 7.5 and 7.6]
2.	m∠C	m∠A + m∠B	Ⓐ Ⓑ Ⓒ Ⓓ [Lesson 3.5]
3.	Area of *RSTU*	Area of *RUYV*	Ⓐ Ⓑ Ⓒ Ⓓ [Lesson 6.3]
4.	m∠1	m∠2	Ⓐ Ⓑ Ⓒ Ⓓ [Lesson 3.3]

5. Vertical angles are ____?____. *(LESSON 2.5)*
　　a. complementary　　**b.** supplementary
　　c. congruent　　**d.** obtuse

6. If lines are parallel, then same-side interior angles are ____?____. *(LESSON 3.3)*
　　a. complementary　　**b.** supplementary
　　c. congruent　　**d.** obtuse

7. Which of the following is valid for proving that triangles are congruent?
(LESSONS 4.2 AND 4.3)

a. AAA

b. ASA

c. SSA

d. none of the above

8. The diagonals of a rhombus are _____?_____.
(LESSON 4.5)

a. congruent

b. parallel

c. perpendicular

d. convex

9. The diagonals of a rectangle are _____?_____.
(LESSON 4.5)

a. congruent

b. parallel

c. perpendicular

d. convex

10. Find x in the diagram below. **(LESSON 3.7)**

a. 12

b. 3.6

c. 6

d. 5.4

11. Write the conclusion, if any, that follows from the given statements. **(LESSON 2.3)**

If a person exercises regularly, then that person is in shape.

Vanessa is in shape.

12. Complete the Euler diagram below.
(LESSON 3.2)

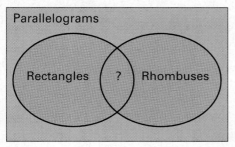

13. What is the area of a circle with a radius of 7? Give an exact answer. **(LESSON 5.3)**

14. What is the radius of a circle with an area of 7? Give an exact answer. **(LESSON 5.3)**

15. Draw six orthographic views of the solid below. **(LESSON 6.1)**

FREE-RESPONSE GRID
Items 16–18 may be answered by using a free-response grid such as that commonly used by standardized-test services.

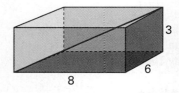

16. Find the length of the diagonal of the right rectangular prism below. Round your answer to the nearest tenth. **(LESSON 6.3)**

17. Find the surface area of the right rectangular prism above. **(LESSON 7.1)**

18. Find the volume of the regular hexagonal pyramid below. Round your answer to the nearest unit. **(LESSON 7.3)**

Similar Shapes

IN 1948 KORCZAK ZIOLKOWSKI (1908–1982) began work on the world's largest sculpture, the Crazy Horse Memorial, in the Black Hills of South Dakota. The sculpture is 563 feet high, more than 9 times the height of the sculptures on Mount Rushmore. The completed face of Crazy Horse is nine stories tall.

In the photo on the page facing, a scale model of the finished statue stands in front of the actual work, which is 1 mile behind it. The model is $\frac{1}{34}$ the height of the mountain carving in progress. Such figures are known as *similar* figures. You will be studying similar figures throughout this chapter.

Korczak Ziolkowski with scale model

Precision explosives were used to create the outline of the form

Crazy Horse 1/34ᵗʰ Scale Model
© KORCZAK, Sculptor

About the Chapter Project

Similarity is a term used to describe objects that are the same shape, but not necessarily the same size. One application of similarity is a scale model, which is an object built in the exact shape of another object, but usually smaller. Scale models are often used to get an overall view of a large object or area.

In the Chapter Project, you and your classmates build a scale model of your or school, starting with a map.

After completing the Chapter Project you will be able to do the following:

- Work as a team to apply the techniques of meausrement to real-world objects.

- Create maps and models of large structues that cannot be measured directly.

About the Portfolio Activities

Throughout the chapter, you will be given opportunities to complete Portfolio Activities that are designed to support your work on the Chapter Project.

The theme of each Portfolio Activity and of the Chapter Project is scale models.

- In the Portfolio Activity on page 506, you will enlarge a design by using grids.

- In the Portfolio Activity on page 542, you will use indirect measurement to find the dimensions of a building.

Dilations and Scale Factors

Objectives

- Construct a dilation of a segment and a point by using a scale factor.

- Construct a dilation of a closed plane figure.

A camera obscura projects an inverted image through a small opening into a dark room. Dilations can be used to explain the process.

The principle of the camera obscura, the forerunner of the modern camera, was discovered by physicist Ibn al-Haitham (965–1039 C.E.).

APPLICATION

PHOTOGRAPHY

The camera obscura was used to draw accurate images of an object or a scene. The artist traced the image that was projected through a pinhole onto a drawing surface. In a modern camera, the pinhole has been replaced by a lens and the drawing surface has been replaced by light-sensitive photographic film.

Dilations

So far you have studied three types of transformations: translations, rotations, and reflections. These transformations are called rigid because they preserve size and shape.

A **dilation** is an example of a transformation that is not rigid. Dilations preserve the shape of an object, but they may change its size.

The size of the figure is changed by a dilation.

A dilation of a point in a coordinate plane can be found by multiplying the x- and y-coordinates of a point by the same number, n.

$$D(x, y) = (nx, ny)$$

The number n is called the **scale factor** of the transformation.

1 What is the image of the point $(2, 3)$ transformed by the dilation $D(x, y) = (4x, 4y)$? What is the scale factor?

● **SOLUTION**

The image is the point $D(2, 3) = (4 \cdot 2, 4 \cdot 3) = (8, 12)$. The scale factor is the multiplier, 4.

Activity 1

Dilations in the Coordinate Plane

YOU WILL NEED

graph paper, ruler, and calculator

OR

geometry graphics software

1. Plot $A(3, 4)$ on a coordinate plane. Use the distance formula or a ruler to find the distance, OA, from $O(0, 0)$ to A. Use dilations with the scale factors given below to transform point A, and then copy and complete the table.

A	OA	Scale factor	Image, A′	OA′	Ratio $\frac{OA′}{OA}$
(3, 4)	?	2	(?, ?)	?	?
(3, 4)	?	0.5	(?, ?)	?	?
(3, 4)	?	−1	(?, ?)	?	?
(3, 4)	?	n	(?, ?)	?	?

2. Plot point A and its image, A', for each dilation above. What is the simplest geometric figure that contains all of these points? Add this figure to your graph.

3. Complete the following conjecture:

CHECKPOINT ✔

Conjecture

The distance from the origin to the image of a point transformed by a dilation with scale factor *n* is ____?____ the distance from the origin to the preimage.

4. Plot point A again and a new point, $B(5, 6)$, on a new coordinate plane. Use the distance formula or a ruler to find the length of $\overline{AB}$. Use dilations with the scale factors given below to transform points A and B, and then copy and complete the table.

B	AB	Scale factor	Image, B′	A′B′	Ratio $\frac{A′B′}{AB}$
(5, 6)	?	2	(?, ?)	?	?
(5, 6)	?	0.5	(?, ?)	?	?
(5, 6)	?	−1	(?, ?)	?	?
(5, 6)	?	n	(?, ?)	?	?

5. Find the slope of each dilated segment, $\overline{A'B'}$.

6. Complete the following conjecture:

CHECKPOINT ✔

Conjecture

The image of a segment transformed by a dilation with a scale factor of *n* is ____?____ the length of the preimage, and the slopes of the image and preimage are ____?____ .

In a dilation, each point and its image lie on a straight line that passes through a point known as the **center of dilation**. In Activity 1, that point was the origin. Every dilation has a center of dilation.

APPLICATION

OPTICS

The colored portion of the eye, known as the *iris*, changes shape to let in more or less light. When the iris is relatively open it is said to be *dilated*. A similar device in a camera, also known as the *iris*, is placed behind the lens to let in more or less light to accommodate different lighting conditions.

In the photo on the left, the eye is contracted for relatively bright light.
In the photo on the right, it is dilated for less bright conditions.

You may have noticed that the size of images of segments that have been dilated varies according to the scale factor. If the size of a figure is reduced by a dilation, the dilation is called a **contraction**. If the size of a figure is enlarged by a dilation, the dilation is called an **expansion**. For a dilation with a scale factor of *n*:

If $|n| < 1$, the dilation is a contraction.

If $|n| > 1$, the dilation is an expansion.

CRITICAL THINKING   How is the image of a point or segment affected by a negative scale factor?

Activity 2
Drawing a Dilation

YOU WILL NEED

ruler and calculator
OR
geometry graphics software

1. Draw a polygon, such as a triangle, and a center of dilation. Draw lines that pass through each vertex of the figure and the center of dilation.

2. Decide on a scale factor *n* for your dilation. Choose one of the vertices and measure the distance, *x*, from the center of dilation to the vertex. Multiply this distance by *n* to obtain a new value, *x'*. On the line containing the chosen vertex, plot a point that is a distance of *x'* from the center of dilation. This new point is the image of the chosen vertex.

3. Repeat for each of the other vertices of your figure.

CHECKPOINT ✔   4. Connect the images of the vertices to form the dilated image of the figure.

CRITICAL THINKING   How could you draw an approximate dilation of a curved plane figure?

EXAMPLE **2**

APPLICATION
ASTRONOMY

The students shown below are using a pinhole to observe a solar eclipse. The process may be understood as a dilation, with the pinhole as the center of dilation. The diameter of the Sun is approximately 870,000 mi. If the image is 0.25 inch in diameter, what is the scale factor of the dilation?

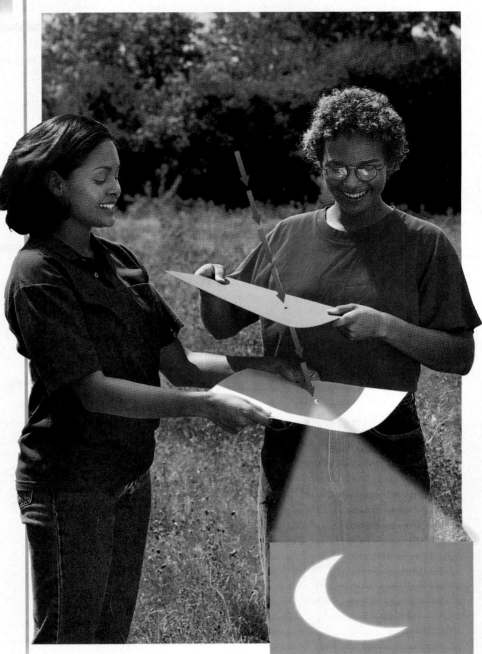

SOLUTION

The diagram at right (not to scale) shows the path of rays of light from the edges of the Sun as seen from Earth. The scale factor is negative because the image is on the opposite side of the center of dilation (the pinhole). The ratio of the image to the preimage is
$$\frac{0.25 \text{ inch}}{55,123,200,000 \text{ inch}} \approx 4.5 \times 10^{-12}, \text{ so the}$$
scale factor is approximately -4.5×10^{-12}.

Sun →

Pinhole →

Image →

Exercises

Communicate

1. What is a dilation? How is a dilation different from other transformations you have studied?

2. What is a scale factor? How can you determine the scale factor of a dilation by looking at a segment and its image?

Explain how the image of a figure transformed by a dilation would be affected by the following scale factors:

3. 2   **4.** 0.5   **5.** −1   **6.** 1

● **Guided Skills Practice**

Find the image of each point transformed by the given dilation. Plot the point and its image on a coordinate plane. *(EXAMPLE 1 AND ACTIVITY 1)*

7. $(1, 5)$; $D(x, y) = (3x, 3y)$ **8.** $(-1, 4)$; $D(x, y) = (2x, 2y)$

9. $(6, -2)$; $D(x, y) = (0.25x, 0.25y)$ **10.** $(2, 3)$; $D(x, y) = (-2x, -2y)$

Copy each figure below and draw a dilation with the given scale factor. *(ACTIVITY 2)*

11. scale factor of 2   **12.** scale factor of −1

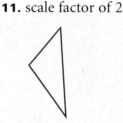

13. The figure at right shows a segment and its image under a dilation. What is the scale factor of the dilation? *(EXAMPLE 2)*

● **Practice and Apply**

For Exercises 14–17, the vertices of a figure and a scale factor, *n,* are given. Use the dilation $D(x, y) = (nx, ny)$ to transform each figure, and plot the preimage and image on a coordinate plane.

14. $(1, 3), (2, 5), (4, 3)$ **15.** $(-3, 5), (8, 9), (2, -6)$

    $n = 2$ $n = \dfrac{1}{3}$

16. $(0, 0), (6, 0), (4, 4), (2, 3)$ **17.** $(1,1), (3, -1), (-2, -3)$

    $n = -\dfrac{1}{2}$ $n = 1.6$

In Exercises 18–21, the blue figures represent the preimages of dilations, and the red figures represent the images. Find the scale factor of each dilation.

18.

(3, 3)

(1, 1)

y

x

19.

y

(8, 2)

(4, 1)

x

20.

y

(6, 4)

x

(−6, −4)

21.

y

(6, 0)

(−2, 0)

x

Copy each figure below and draw a dilation with the given scale factor, *n*.

22. $n = 3$ **23.** $n = \dfrac{3}{4}$ **24.** $n = -2$

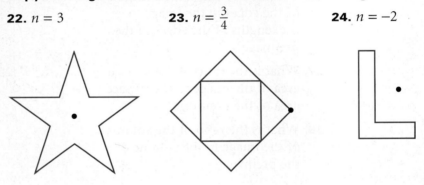

25. Copy the figure below. Draw three dilations of the triangle with a scale factor of 2, using the three given points as centers of dilation. How are the images alike? How are they different?

A   *B*   *C*

26. Copy the figure below. Locate the center of dilation and find the scale factor.

Preimage

Image

For Exercises 27–30, the endpoints of a segment and a scale factor _n_ are given. Show that the dilation image of the segment has the same slope as the preimage.

27. $(1, 0)$ and $(5, 3)$; $n = 2$ **28.** $(-2, 3)$ and $(3, 1)$; $n = 5$

29. $(-2, 4)$ and $(4, 8)$; $n = -\dfrac{1}{2}$ **30.** $(1, 1)$ and $(2, 4)$; $n = 1.7$

For Exercises 31–34, a point and a scale factor are given. Find the line that passes through the preimage and image, and show that this line contains the origin.

31. $(5, 1)$; $n = 4$ **32.** $(-2, 3)$; $n = \dfrac{5}{6}$

33. $(3, -5)$; $n = -3$ **34.** $(4, 7)$; $n = 2.5$

The figure below shows a right rectangular prism in a three-dimensional coordinate system.

35. Draw the image of the prism transformed by the dilation $D(x, y, z) = (2x, 2y, 2z)$. Label the coordinates of the image's vertices.

36. What is the ratio of the lengths of the edges in the image to the lengths of the edges in the preimage?

37. What is the ratio of the surface area of the image to the surface area of the preimage?

38. What is the ratio of the volume of the image to the volume of the preimage?

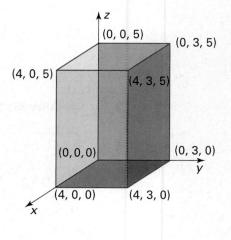

CHALLENGE

39. An example of a dilation that is not centered at the origin of a coordinate plane is $D(x, y) = (2x - 4, 2y - 3)$. Use this dilation to transform the segment with endpoints $(2, 3)$ and $(5, 5)$. Plot the preimage and the image on a coordinate plane. Determine the scale factor and locate the center of dilation. Write the rule for a dilation with a center of $(2, 1)$ and a scale factor of 4.

APPLICATION

Cubed lattice

40. HOBBIES A quilter has a pattern for a 4-in. square quilt block and wishes to enlarge it to a 12-in. block. What is the scale factor of the enlargement? The pattern is traced onto a grid of 1-in. squares, as shown at left. One of the shapes in the pattern has vertices at $(0, 1)$, $(1, 1)$, $(2, 2)$, and $(1, 2)$. What are the coordinates of the image of this shape in the 12-in. block?

APPLICATIONS

OPTICS The diagram below shows a part of a camera obscura. The image is projected through a small hole into a dark room or chamber. This projection is an example of a dilation.

41. What part of the camera obscura acts as the center of dilation?

42. Is the scale factor positive or negative? Explain your answer.

43. Explain why the projected image is inverted.

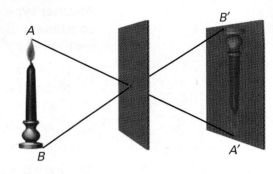

44. GRAPHIC ARTS An artist is using a photocopier to reduce a design. The original design is 5 in. wide. The copy should be 2 in. wide. What is the scale factor of the dilation?

Look Back

45. The base of an isosceles triangle is 6 m and the legs are 8 m each. Find the perimeter and area of the triangle. *(LESSONS 5.2 AND 5.4)*

46. A leg of a 45-45-90 triangle is 7 cm long. What is the length of the hypotenuse? *(LESSON 5.5)*

APPLICATIONS

47. ENGINEERING A spherical gas tank has an outer diameter of 40 ft. The tank is made with 1-in. thick steel. Find the difference between the surface area of the outside and of inside of the tank. *(LESSON 7.6)*

48. ENGINEERING Suppose that 1 gal of paint covers 400 ft^2. How many gallons of paint are needed to paint the inside and outside of the tank described in Exercise 47? *(LESSON 7.6)*

49. EARTH SCIENCE The circumference of a great circle of Earth is about 40,000 km. What is the radius of Earth? *(LESSON 7.6)*

50. EARTH SCIENCE The height of Earth's atmosphere is about 550 km. Use this information and your answer to Exercise 49 to find the volume of Earth and its atmosphere. *(LESSON 7.6)*

Earth's atmosphere, as photographed by a Russian cosmonaut

Another type of transformation that is not rigid can be described in a coordinate plane by $T(x, y) = (mx, ny)$, where the x- and y-values are multiplied by two different scale factors.

51. Draw a triangle with vertices at $(1, 3)$, $(5, 1)$, and $(5, 3)$ in a coordinate plane. Transform the figure by using the transformation $T(x, y) = (2x, 4y)$.

52. How are the preimage and image alike? How are they different? How is this type of transformation like a dilation? How is it different?

53. Given a segment with endpoints $(1, 3)$ and $(5, 1)$, what is the ratio of its slope to the slope of its image?

54. Use the transformation $T(x, y) = (3x, -y)$ to transform the figure at left.

Calvin and Hobbes   by Bill Watterson

USING GRIDS TO ENLARGE A DESIGN

Choose a design, cartoon, or photo that you wish to enlarge. You may wish to photocopy the design so that you do not damage the original.

1. Using a ruler, draw a grid of $\frac{1}{2}$-cm squares on the design.

2. Draw a grid with the same number of squares on a larger piece of paper. These squares should be at least twice as big.

3. Copy the pattern that appears in each square of the first grid onto the corresponding square in the larger grid. It may help to cover all but one square of the design as you copy it.

Be sure to copy one square at a time—do not try to draw larger parts of the design.

Once all the squares are copied, you will have an accurate enlargement of the design.

WORKING ON THE CHAPTER PROJECT

You should now be able to complete Activity 1 of the Chapter Project.

Similar Polygons

Why
Scale models are used by architects, city planners, movie set designers, and hobbyists. To create a scale model, properties of similar figures are needed.

Similar Polygons

When a figure undergoes a dilation (Lesson 8.1), the preimage and image have the same shape but are not necessarily the same size. They are said to be *similar*.

Definition: Similar Figures

Two figures are **similar** if and only if one is congruent to the image of the other by a dilation.   **8.2.1**

In the dilation at right, the corresponding angles of the triangles are congruent, and the ratios of the lengths of the corresponding sides are all equal to the absolute value of the scale factor of the dilation—which is 2 in this case.

The symbol ~ means similar.
$\triangle ABC \sim \triangle A'B'C'$

$\angle A \cong \angle A' \quad \angle B \cong \angle B' \quad \angle C \cong \angle C'$

$$\frac{A'B'}{AB} = \frac{B'C'}{BC} = \frac{A'C'}{AC} = 2 \quad \text{Scale factor}$$

When the ratios of corresponding sides of two polygons are equal (as in the illustration on the previous page), the sides are said to be **proportional**. A statement of the equality of two ratios is called a **proportion**. The concept of proportion is important in the Polygon Similarity Postulate.

Polygon Similarity Postulate

Two polygons are similar if and only if there is a way of setting up a correspondence between their sides and angles such that the following conditions are met:

- Each pair of corresponding angles is congruent.
- Each pair of corresponding sides is proportional.

8.2.2

In a similarity statement, as in a congruence statement, the letters of the vertices must be written in corresponding order.

$\triangle ABC \sim \triangle DFE$

CRITICAL THINKING

Is either condition in the Polygon Similarity Postulate, taken separately, enough to guarantee that two polygons are similar? Use the figures at right to explain your answer.

EXAMPLE ❶ Are the triangles at right similar?

SOLUTION

It is given that $\angle P \cong \angle S$, $\angle Q \cong \angle T$, and $\angle R \cong \angle U$, so the first condition of the Polygon Similarity Postulate is satisfied. To see whether the second condition is satisfied, check each ratio of corresponding sides.

$$\frac{ST}{PQ} = \frac{57}{38} = \frac{3}{2} \qquad \frac{TU}{QR} = \frac{30}{30} = \frac{3}{2} \qquad \frac{SU}{PR} = \frac{39}{26} = \frac{3}{2}$$

Because the ratios are equal, the corresponding sides are proportional, and the second condition of the Polygon Similarity Postulate is satisfied. Therefore, $\triangle PQR \sim \triangle STU$.

Properties of Proportions

When working with similar figures, it is often helpful to know the following Properties of Proportions:

Properties of Proportions

Let a, b, c, and d be any real numbers.

Cross-Multiplication Property

If $\frac{a}{b} = \frac{c}{d}$ and b and $d \neq 0$, then $ad = bc$.   **8.2.3**

Reciprocal Property

If $\frac{a}{b} = \frac{c}{d}$ and a, b, c, and $d \neq 0$, then $\frac{b}{a} = \frac{d}{c}$.   **8.2.4**

Exchange Property

If $\frac{a}{b} = \frac{c}{d}$ and a, b, c, and $d \neq 0$, then $\frac{a}{c} = \frac{b}{d}$.   **8.2.5**

"Add-One" Property

If $\frac{a}{b} = \frac{c}{d}$ and b and $d \neq 0$, then $\frac{a+b}{b} = \frac{c+d}{d}$.   **8.2.6**

TRY THIS   Verify each property for the proportion $\frac{1}{3} = \frac{2}{6}$.

CRITICAL THINKING   Why do you think the last property in the list above is called the "Add-One" Property? (Hint: Separate each side of the equation into two separate fractions.)

E X A M P L E ➋ In the figure at right, pentagon $ABCDE \sim$ pentagon $FGHIJ$. Find AB.

● **SOLUTION**

Because the pentagons are similar, the Polygon Similarity Postulate states that the sides are proportional; thus,

$\dfrac{AB}{FG} = \dfrac{BC}{GH} = \dfrac{CD}{HI} = \dfrac{DE}{IJ} = \dfrac{EA}{JF}$.

Because all the ratios of the sides are equal, any two ratios are equal. For example:

$$\frac{AB}{FG} = \frac{CD}{HI}$$

$$\frac{AB}{16} = \frac{15}{20} \qquad \textit{Substitute the known lengths into the proportion.}$$

$$16 \cdot \frac{AB}{16} = \frac{15}{20} \cdot 16 \qquad \textit{To solve, multiply both sides by 16.}$$

$$AB = \frac{240}{20} = 12$$

Proportionality Within a Figure

Consider the two similar rectangles shown at right. By the Polygon Similarity Postulate,

$$\frac{KL}{WX} = \frac{LM}{XY} = \frac{MN}{YZ} = \frac{NK}{ZW}.$$

You can form a proportion with any two of these ratios, such as $\frac{MN}{YZ} = \frac{NK}{ZW}$.

Notice that each ratio in the proportion contains a side length from each rectangle. By using the Exchange Property, you get $\frac{MN}{NK} = \frac{YZ}{ZW}$.

Thus, the ratio of the long side to the short side is the same in each rectangle. This provides another way of thinking about similarity: the ratio of any two sides in one polygon is the same as the ratio of the corresponding sides in a similar polygon.

E X A M P L E ③

APPLICATION
ARCHITECTURE

Amber and Adrianne are making a scale model of a building with a rectangular foundation, as shown below. If the long sides of the model are 24 inches, how long are the short sides?

● **SOLUTION**

Because the scale model is similar to the original building, the ratio of the shorter side to the longer side is the same in each rectangle.

$$\frac{x}{24} = \frac{18}{32}$$

$$24 \cdot \frac{x}{24} = \frac{18}{32} \cdot 24$$

$$x = \frac{432}{32}, \text{ or } 13.5 \text{ inches.}$$

Exercises

● *Communicate*

Classify each statement as true or false and explain your reasoning.

1. If △ABC ~ △DEF, then △DEF ~ △ABC.

2. If △ABC ~ △DEF, then △ABC ~ △EFD.

3. If two figures are congruent, then they are similar.

4. If two figures are similar, then they are congruent.

5. Any two regular polygons with the same number of sides are similar.

● *Guided Skills Practice*

Determine whether each pair of figures is similar. Explain your reasoning. (EXAMPLE 1)

6.

7.

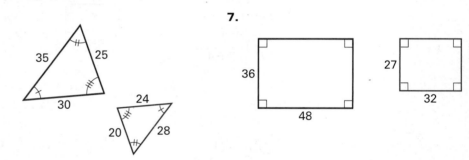

In Exercises 8 and 9, the polygons in each pair are similar. Find the missing length. (EXAMPLE 2)

8. *EH*

9. *JK*

APPLICATION

10. ARCHITECTURE A scale model of a building has the dimensions shown below. Find the length of the actual building.

Model 3.5 cm
7.5 cm

Building 42 ft
?

11. Given the proportionality statement $\frac{SG}{MW} = \frac{GT}{WR} = \frac{TS}{RM}$ for two similar triangles, write a similarity statement that shows the correct correspondence.

12. Given the proportionality statement $\frac{JD}{LE} = \frac{DC}{EB} = \frac{CP}{BH} = \frac{PJ}{HL}$ for two similar rectangles, write a similarity statement that shows the correct correspondence.

13. Given $\triangle ABC \sim \triangle XYZ$, write a proportionality statement for the ratios between the sides.

14. Given quadrilateral $EFGH \sim$ quadrilateral $VWXY$, write a proportionality statement for the ratios between the sides.

15. Given pentagon $JKLMN \sim$ pentagon $PQRST$, write a proportionality statement for the ratios between the sides.

For Exercises 16–19, determine whether the polygons are similar. Explain your reasoning.

16.   **17.**

18.   **19.**

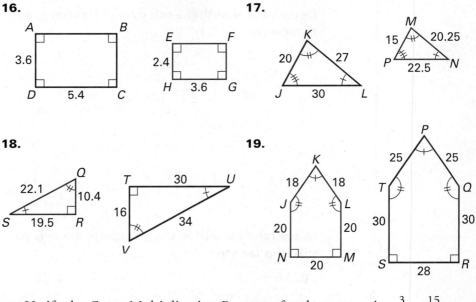

20. Verify the Cross-Multiplication Property for the proportion $\frac{3}{4} = \frac{15}{20}$.

21. Verify the Reciprocal Property for the proportion $\frac{5}{2} = \frac{10}{4}$.

22. Verify the Exchange Property for the proportion $\frac{6}{9} = \frac{2}{3}$.

23. Given $\frac{x}{4} = \frac{y}{8}$, find $\frac{x}{y}$.

For each pair of similar figures below, compare the ratio of the sides of the figures with the ratio of the areas of the figures.

24.   **25.**

Algebra

For Exercises 26–29, the given polygons are similar. Find _x_.

26.

27.

28.

29.

30. Use the diagram below to determine whether quadrilateral *GHIJ* ~ quadrilateral *KLMN*. Explain your answer.

Algebra

Solve each proportion for _x_.

31. $\frac{6x}{24} = \frac{27}{9}$

32. $\frac{4.8}{x} = \frac{6}{8.4}$

33. $\frac{\frac{2}{5}}{8} = \frac{\frac{7}{10}}{x}$

34. $\frac{6}{x} = \frac{x}{150}$

35. $\frac{3}{x-4} = \frac{7}{x+4}$

36. $\frac{5-2x}{8} = \frac{3x+1}{4}$

37. In general, what do you think is true about the ratio of the sides and the ratio of the areas of two similar polygons?

Algebra

For real numbers _a_, _b_, _c_, and _d_, where $b \neq 0$ and $d \neq 0$, $\frac{a}{b} = \frac{c}{d}$.
Determine whether each proportion is true for all values of the variables for which the proportion is defined. If a proportion is not true, give a numerical counterexample.

38. $\frac{a+x}{b} = \frac{c+x}{d}$

39. $\frac{a+x}{b+x} = \frac{c+x}{d+x}$

40. $\frac{a+b}{b} = \frac{c+d}{d}$

41. $\frac{a}{a+b} = \frac{c}{c+d}$

CHALLENGE

42. Verify that the following proportion is true for all real numbers *a*, *b*, *c*, *d*, *e*, and *f*, where *b*, *d*, and $f \neq 0$: If $\frac{a}{b} = \frac{c}{d} = \frac{e}{f}$, then $\frac{a}{b} = \frac{a+c+e}{b+d+f}$.

43. WILDLIFE MANAGEMENT A method used to estimate the size of wild-animal populations uses proportions. Suppose that a scientist catches 300 fish from a lake and then tags and releases them. After a short time, the scientist comes back, catches 100 fish, and finds that 8 of them are already tagged. Assuming that the proportion of tagged fish in this catch is equal to the proportion of tagged fish in the entire population, estimate the number of fish in the lake.

44. WILDLIFE MANAGEMENT On the map of the Aransas National Wildlife Refuge shown below, the scale is 1 cm = 3.5 mi. Estimate the area of the refuge in square centimeters, and then use that value to estimate the area of the refuge in square miles. How many acres are in the refuge? (Note: 1 square mile = 640 acres)

The Aransas National Wildlife Refuge on the Texas coast contains important nesting sites for endangered whooping cranes. The delicate salt-marsh environment must be carefully maintained in order to help this and other species of wildlife survive.

45. LANDSCAPING In the landscaping diagram below, the scale is 1 cm = 15 ft. Measure the diagram and use the given scale to determine how far apart the trees should be planted.

46. FINE ART Brenda is attempting to paint a reproduction of the *Mona Lisa* from a print that is 16 in. by 24 in. If her canvas is 15 in. wide, how tall should it be in order for the reproduction to be proportional to the print?

47. FINE ART In the print of the *Mona Lisa* described in Exercise 46, the face is 6 in. tall and 4 in. wide. What should be the dimensions of the face in Brenda's reproduction?

Mona Lisa (1503-6) by Leonardo da Vinci (1452-1519). Louvre, Paris, France

48. INTERIOR DECORATING Fernando is drawing a floor plan of his house to help him in arranging furniture. His dining room is a 12 ft × 15 ft rectangle, and his table is a regular octagon with 2 ft 3 in. sides. If the room on the floor plan is 8 in. × 10 in., how long should the sides of the table in the floor plan be?

Designers use floor plans to help in determining the layout of a room.

49. The angles of a triangle measure $(x + 5)°$, $(5x + 12)°$, and $(2x + 3)°$. Find the measures of the angles. *(LESSON 3.5)*

Which of the following can be used to prove triangle congruence?
(LESSONS 4.2 AND 4.3)

50. ASA   **51.** AAA   **52.** SAA   **53.** SAS

54. Find the measure of a base angle of an isosceles triangle whose vertex angle is 92°. *(LESSON 4.4)*

55. A right triangle has legs of 5 cm and 7 cm. Find the length of the hypotenuse.

56. A right triangle has a leg of 5 cm and a hypotenuse of 7 cm. Find the length of the other leg.

Look Beyond

INDIRECT MEASUREMENT Anthony uses the following method to estimate the width of a river:

Anthony stands at point *N* and adjusts the visor of his cap until it is in his line of sight to point *R* on the opposite shore. Without changing the position of his cap, he turns and sights along the visor to point *A* on his side of the river.

57. Explain how Anthony can find the width of the river.

58. Which segment in the figure has the same length as $\overline{NR}$?

59. Which postulate can be used to prove that $\triangle SNR \cong \triangle SNA$? Explain your answer.

60. What are some possible problems with using this method of indirect measurement?

Triangle Similarity

8.3

Objective

- Develop the AA Triangle Similarity Postulate and the SSS and SAS Triangle Similarity Theorems.

Why *Similar triangles have interesting mathematical properties.*

Triangle Similarity

Activities 1, 2, and 3 suggest some shortcuts for determining triangle similarity.

Activity 1

AA Triangle Similarity Postulate

YOU WILL NEED

ruler and protractor
OR
geometry graphics software

1. Draw $\triangle ABC$ with m$\angle A = 45°$ and m$\angle B = 65°$. What is m$\angle C$? Measure the sides of $\triangle ABC$.

2. Draw $\triangle DEF$ with m$\angle D = 45°$ and m$\angle E = 65°$ such that DE is longer than AB. What is m$\angle F$? Measure the sides of $\triangle DEF$.

3. Use your measurements to complete the table below.

	Sides			Angles		
$\triangle ABC$	$AB = ?$	$BC = ?$	$AC = ?$	m$\angle A = 45°$	m$\angle B = 65°$	m$\angle C = ?$
$\triangle DEF$	$DE = ?$	$EF = ?$	$DF = ?$	m$\angle D = 45°$	m$\angle E = 65°$	m$\angle F = ?$
Ratio	$\frac{DE}{AB} = ?$	$\frac{EF}{BC} = ?$	$\frac{DF}{AC} = ?$			

4. What is the relationship between corresponding sides? between corresponding angles? Are the triangles similar? Explain your reasoning.

CHECKPOINT ✔

5. Based on your results, complete the postulate below.

AA (Angle-Angle) Similarity Postulate

If two ____?____ of one triangle are congruent to two ____?____ of another triangle, then the triangles are ____?____.   **8.3.1**

EXAMPLE 1 Are the triangles at right similar?

● SOLUTION

By the Triangle Sum Theorem,
$m\angle J = 180° − 90° − 35° = 55°$, so
$m\angle J = m\angle P$ ($\angle J \cong \angle P$) and
$m\angle K = m\angle M$ ($\angle K \cong \angle M$).

Thus, by the AA Similarity Postulate, $\triangle JKL \sim \triangle PMN$.

Activity 2

SSS Triangle Similarity Theorem

YOU WILL NEED

ruler,
protractor, and
compass
OR
geometry
graphics
software

1. Draw $\triangle ABC$ with $AB = 2$ cm, $BC = 3$ cm, and $AC = 4$ cm. Measure the angles of $\triangle ABC$ to the nearest degree.

2. Draw $\triangle DEF$ with $DE = 6$ cm, $EF = 9$ cm, and $DF = 12$ cm. Measure the angles of $\triangle DEF$ to the nearest degree.

3. Use your measurements to complete the table below.

	Sides			Angles		
$\triangle ABC$	$AB = 2$	$BC = 3$	$AC = 4$	$m\angle A = ?$	$m\angle B = ?$	$m\angle C = ?$
$\triangle DEF$	$DE = 6$	$EF = 9$	$DF = 12$	$m\angle D = ?$	$m\angle E = ?$	$m\angle F = ?$
Ratio	$\frac{DE}{AB} = ?$	$\frac{EF}{BC} = ?$	$\frac{DF}{AC} = ?$			

4. What is the relationship between corresponding sides? between corresponding angles? Are the triangles similar? Explain your reasoning.

CHECKPOINT ✔

5. Based on your results, complete the theorem below, which you will be asked to prove in Exercises 25–27.

SSS (Side-Side-Side) Similarity Theorem

If the three ____?____ of one triangle are proportional to the three ____?____ of another triangle, then the triangles are ____?____.

8.3.2

EXAMPLE 2 Are the triangles at right similar?

● SOLUTION

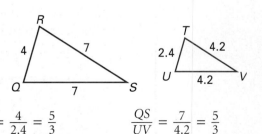

The ratios of the three sides are as follows:

$$\frac{QR}{UT} = \frac{4}{2.4} = \frac{5}{3} \qquad \frac{QR}{UT} = \frac{4}{2.4} = \frac{5}{3} \qquad \frac{QS}{UV} = \frac{7}{4.2} = \frac{5}{3}$$

Thus, the sides of the triangle are proportional and, by the SSS Similarity Theorem, $\triangle QRS \sim \triangle UTV$.

SAS Triangle Similarity Theorem

1. Draw $\triangle ABC$ with $AB = 3$ cm, $BC = 4$ cm, and m$\angle B = 60°$. Measure the sides and angles of $\triangle ABC$.

2. Draw $\triangle DEF$ with $DE = 6$ cm, $EF = 8$ cm, and m$\angle E = 60°$. Measure the sides and angles of $\triangle DEF$.

3. Use your measurements to complete the table below.

	Sides			Angles		
$\triangle ABC$	$AB = 3$	$BC = 4$	$AC = ?$	m$\angle A = ?$	m$\angle B = 60°$	m$\angle C = ?$
$\triangle DEF$	$DE = 6$	$EF = 8$	$DF = ?$	m$\angle D = ?$	m$\angle E = 60°$	m$\angle F = ?$
Ratio	$\frac{DE}{AB} = ?$	$\frac{EF}{BC} = ?$	$\frac{DF}{AC} = ?$			

4. What is the relationship between corresponding sides? between corresponding angles? Are the triangles similar? Explain your reasoning.

CHECKPOINT ✔

5. Based on your results, complete the theorem below, which you will be asked to prove in Exercises 28–30.

SAS (Side-Angle-Side) Similarity Theorem

If two _____?_____ of one triangle are proportional to two _____?_____ of another triangle and their _____?_____ _____?_____ are congruent, then the triangles are _____?_____.

8.3.3

E X A M P L E **3** Are the triangles shown below similar?

SOLUTION

The ratios of the given sides are as follows:

$$\frac{GF}{XY} = \frac{3}{2} \qquad \frac{GH}{XZ} = \frac{5.4}{3.6} = \frac{3}{2}$$

Thus, the sides of the triangle are proportional, and the included angles of these sides are congruent. By the SAS Similarity Theorem, $\triangle QRS \sim \triangle UTV$.

CRITICAL THINKING Why are ASA and AAS not included in a list of triangle similarity theorems?

Communicate

1. In the figure below, the midsegments divide the large triangle into four smaller triangles. How would you show that each of the smaller triangles is similar to the large triangle?

2. Recall the HL Congruence Theorem from Lesson 4.3. Could there be an HL Similarity Theorem? Why or why not?

3. Use a counterexample to explain why there is no AAA Similarity Postulate for quadrilaterals. That is, if three angles of one quadrilateral are congruent to three corresponding angles of another quadrilateral, explain why the quadrilaterals are not necessarily similar.

4. Use a counterexample to explain why there is no SSSS Similarity Postulate for quadrilaterals. That is, if four sides of one quadrilateral are congruent to four corresponding sides of another quadrilateral, explain why the quadrilaterals are not necessarily similar.

Guided Skills Practice

Each pair of triangles below can be proven similar by using AA, SSS, or SAS information. Write a similarity statement for each pair, and identify the postulate or theorem used. (EXAMPLES 1, 2, AND 3)

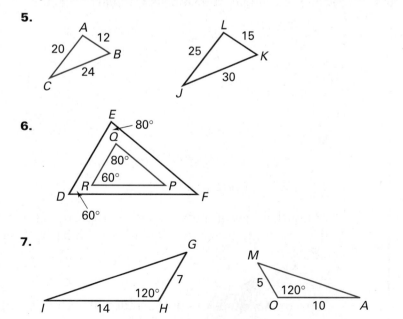

5.

6.

7.

Determine whether each pair of triangles can be proven similar by using AA, SSS, or SAS similarity. If so, write a similarity statement, and identify the postulate or theorem used.

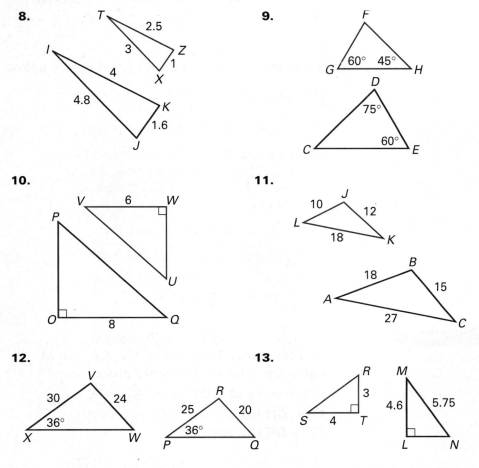

8.

9.

10.

11.

12.

13.

In Exercises 14 and 15, can the pairs of triangles be proven similar? Why or why not?

14.

15.

In Exercises 16 and 17, indicate which figures are similar. Explain your reasoning.

16.

17.

For Exercises 18–21, refer to the diagram below. Is the given information enough to prove that △*ABC* ~ △*ADE*? Explain your reasoning.

18. $\overline{BC} \| \overline{DE}$

19. $\dfrac{AB}{AD} = \dfrac{AC}{AE}$

20. $AD = DB$ and $AE = EC$

21. $\dfrac{AB}{AD} = \dfrac{BC}{DE}$

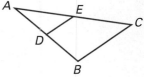

For Exercises 22–24, refer to the diagram below.

22. Draw △*NOP* in which each side is twice as long as each side of △*KLM*. Is △*NOP* ~ △*KLM*? Why or why not?

23. Draw △*QRS* in which each side is 1 cm longer than the corresponding side of △*KLM*. Is △*QRS* ~ △*KLM*? Why or why not?

24. Draw △*TUV* ~ △*KLM* such that $\dfrac{TU}{KL} = \dfrac{UV}{LM} = \dfrac{VT}{MK} = \dfrac{3}{2}$. What is the ratio of the perimeter of △*TUV* to the perimeter of △*KLM*?

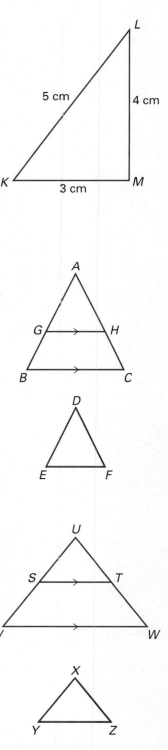

PROOF

In Exercises 25–27 you will prove the SSS Similarity Theorem. In the diagram at right, the sides of △*ABC* are proportional to the sides of △*DEF*, and $\dfrac{AB}{DE} = \dfrac{BC}{EF} = \dfrac{CA}{FD}$. Also, $\overline{GH}$ has been added such that $AG = DE$ and $\overline{GH} \| \overline{BC}$.

25. Use the AA Similarity Postulate to prove that △*AGH* ~ △*ABC*.

26. Use the result of Exercise 25 to prove that △*AGH* ≅ △*DEF*.

27. Use the results of Exercises 25 and 26 to prove that △*ABC* ~ △*DEF*.

PROOF

In Exercises 28–30 you will prove the SAS Similarity Theorem. In the diagram at right, two sides of △*UVW* are proportional to two sides of △*XYZ*, $\dfrac{UV}{XY} = \dfrac{UW}{XZ}$, and ∠*U* ≅ ∠*X*. Also $\overline{ST}$ has been added such that $US = XY$ and $\overline{ST} \| \overline{VW}$.

28. Use the AA Similarity Postulate to prove that △*UST* ~ △*UVW*.

29. Use the result of Exercise 28 to prove that △*UST* ≅ △*XYZ*.

30. Use the results of Exercises 28 and 29 to prove that △*UVW* ~ △*XYZ*.

Recall from Lesson 2.4 that a relationship is called an *equivalence relation* if it satisfies the Reflexive, Symmetric, and Transitive Properties. Exercises 31–33 establish the fact that similarity is an equivalence relation.

31. Prove that △*ABC* ~ △*ABC* (Symmetric Property).

32. Prove that if △*ABC* ~ △*DEF*, then △*DEF* ~ △*ABC* (Reflexive Property).

33. Prove that if △*ABC* ~ △*DEF* and △*DEF* ~ △*GHI*, then △*ABC* ~ △*GHI* (Transitive Property).

CHALLENGE

34. Refer to the diagram at right. Suppose that there is a quadrilateral *STUV* with ∠*T* ≅ ∠*X* and ∠*V* ≅ ∠*Z* and that all four sides of *STUV* are proportional to all four sides of *WXYZ*. Show that the two quadrilaterals are similar. Generalize your results from this case to make a conjecture about a set of conditions that can be used to prove similarity in quadrilaterals.

Artists creating a mural by enlarging smaller images.

In Exercises 35–38, refer to the following situation: Four artists are copying a design to a larger scale. The design contains a triangle, and the artists need to make a triangle that is similar to it.

APPLICATIONS

35. GRAPHIC DESIGN Tony first draws one side of his triangle and then copies the angles from the original triangle at the endpoints of that side. He then extends the angles until they meet. Will Tony's triangle be similar to the original? Explain your reasoning.

36. GRAPHIC DESIGN Biata measures the sides of the original triangle and then multiplies each length by 5 to get the sides of the triangle in her design. Will Biata's triangle be similar to the original? Explain your reasoning.

37. GRAPHIC DESIGN Miki copies one angle from the original triangle, then measures the two sides adjacent to that angle, and multiplies the lengths by 5. She then draws sides with these lengths adjacent to the copied angle and connects the endpoints to form a triangle. Will Miki's triangle be similar to the original? Explain your reasoning.

38. GRAPHIC DESIGN George measures two sides of the triangle and multiplies the lengths by 5. He draws one of the sides and copies the nonincluded angle at one of the endpoints. He extends the angle and then draws the remaining side from the other endpoint so that it intersects the side of the copied angle. Will George's triangle be similar to the original? Explain your reasoning.

Refer to the diagram at right.
$\overline{AB} \cong \overline{BC}$ and line *j* ∥ line *k*.
Find each angle listed below.
(LESSON 3.3)

39. *x* **40.** *y*

41. *z* **42.** *v*

Algebra

Plot point *A*(3, 5) on a coordinate plane and draw a line through this point and the origin.

43. What is the slope of this line? *(LESSON 3.8)*

44. What is the distance from the origin to point *A*? *(LESSON 5.6)*

45. Draw a right triangle with point *A* as one vertex, the origin as one vertex, and one side on the *x*-axis. Find the ratio of the length of the longer leg to the length of the the shorter leg. *(LESSON 8.2)*

46. Choose another point anywhere on the original line and label it *B*. Draw a right triangle with point *B* as one vertex, the origin as one vertex, and one side on the *x*-axis. Find the ratio of the length of the longer leg to the length of the shorter leg. *(LESSON 8.2)*

47. Are the triangles in Exercises 45 and 46 similar? Explain your reasoning. *(LESSON 8.2)*

 Look Beyond

In the diagram below, the three triangles are similar, and $\overline{AC}$, $\overline{DF}$, and $\overline{GI}$ are parallel.

48. Copy the diagram and draw lines $\overleftrightarrow{AD}$, $\overleftrightarrow{BE}$, and $\overleftrightarrow{CF}$. Label the point of intersection *X*.

49. Draw lines $\overleftrightarrow{AG}$, $\overleftrightarrow{BH}$, and $\overleftrightarrow{CI}$. Label the point of intersection *Y*.

50. Draw lines $\overleftrightarrow{DG}$, $\overleftrightarrow{EH}$, and $\overleftrightarrow{FI}$. Label the point of intersection *Z*.

51. What do you notice about points *X*, *Y*, and *Z*?

The Side-Splitting Theorem

8.4

Objectives

- Develop and prove the Side-Splitting Theorem.

- Use the Side-Splitting Theorem to solve problems.

Why *When you hear the phrase "side splitting," you may think of laughter. In geometry, it refers to a useful theorem.*

The capital letter A may be embellished in endless ways for reasons of beauty and style. But in its most basic form, it suggests a geometry theorem.

The Side-Splitting Theorem

As you will see in Example 2, people have been solving problems about proportions in triangles since ancient times. One useful result is the Side-Splitting Theorem.

Recall from the Triangle Midsegment Theorem in Lesson 4.6 that the midsegment of a triangle is parallel to one side of the triangle. The following theorem applies to any segment that is parallel to one side of a triangle.

Side-Splitting Theorem

A line parallel to one side of the triangle divides the other two sides proportionally.

8.4.1

Given: $\overline{DE} \parallel \overline{BC}$

Prove: $\dfrac{DB}{AD} = \dfrac{EC}{AE}$

Proof:

Statements	Reasons
1. $\overline{DE} \parallel \overline{BC}$	Given
2. $m\angle B = m\angle 1$ $m\angle C = m\angle 2$	If ∥ lines are cut by a transversal, corresponding angles are ≅.
3. $\triangle ABC \sim \triangle ADE$	AA Similarity Postulate
4. $\dfrac{AB}{AD} = \dfrac{AC}{AE}$	Polygon Similarity Postulate
5. $AD + DB = AB$ $AE + EC = AC$	Segment Addition Postulate
6. $\dfrac{AD + DB}{AD} = \dfrac{AE + EC}{AE}$	Substitution Property
7. $\dfrac{AD}{AD} + \dfrac{DB}{AD} = \dfrac{AE}{AE} + \dfrac{EC}{AE}$	Addition of fractions
8. $1 + \dfrac{DB}{AD} = 1 + \dfrac{EC}{AE}$	Simplify.
9. $\dfrac{DB}{AD} = \dfrac{EC}{AE}$	Subtraction Property of Equality

CRITICAL THINKING

What other proportions can you find in $\triangle ABC$ by using the Side-Splitting Theorem?

The diagram at right may help you remember the proportions in a triangle with a segment parallel to one side. The following are some of the proportions resulting from the Side-Splitting Theorem:

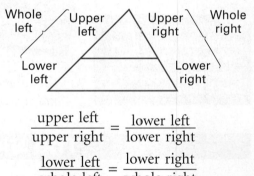

$$\frac{\text{upper left}}{\text{lower left}} = \frac{\text{upper right}}{\text{lower right}}$$

$$\frac{\text{upper left}}{\text{upper right}} = \frac{\text{lower left}}{\text{lower right}}$$

$$\frac{\text{upper left}}{\text{whole left}} = \frac{\text{upper right}}{\text{whole right}}$$

$$\frac{\text{lower left}}{\text{whole left}} = \frac{\text{lower right}}{\text{whole right}}$$

E X A M P L E ❶ Use the Side-Splitting Theorem to find x in the triangle below.

● **SOLUTION**

Choose a proportion that includes x as a single term.

$$\frac{\text{upper left}}{\text{lower left}} = \frac{\text{upper right}}{\text{lower right}}$$

$$\frac{12}{16} = \frac{15}{x}$$

$$12x = 240$$

$$x = 20$$

The following is a corollary of the Side-Splitting Theorem:

Two-Transversal Proportionality Corollary

Three or more parallel lines divide two intersecting transversals proportionally.

8.4.2

In the diagram at right, lines ℓ, m, and n are parallel, with transversals s and t.

One proportion that results from Corollary 8.4.2 is $\frac{AB}{BC} = \frac{XY}{YZ}$.

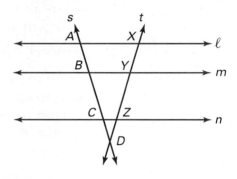

CRITICAL THINKING   How can you tell that $\triangle ADX \sim \triangle BDY$?

EXAMPLE ②

CULTURAL CONNECTION: AFRICA Students in ancient Egypt studied geometry to solve practical problems involving the pyramids. This problem is based on a problem in a papyrus copied in 1650 B.C.E. by the scribe Ahmes from a source that may date back to 2000 B.C.E.

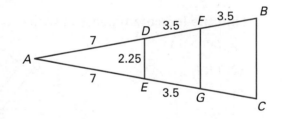

In the diagram above, $\overline{DE}$, $\overline{FG}$, and $\overline{BC}$ are parallel, $AD = AE = 7$ cubits, $DF = FB = 3.5$ cubits, and $DE = 2.25$ cubits. Find the remaining lengths.

● **SOLUTION**

Use the Two-Transversal Proportionality Corollary:

$$\frac{AD}{DF} = \frac{EG}{GC} \quad \Rightarrow \quad \frac{7}{3.5} = \frac{7}{EG} \quad \Rightarrow \quad EG = 3.5$$

and

$$\frac{DF}{FB} = \frac{EG}{GC} \quad \Rightarrow \quad \frac{3.5}{3.5} = \frac{3.5}{GC} \quad \Rightarrow \quad GC = 3.5$$

Use the definition of similar triangles:

$$\frac{AD}{AF} = \frac{DE}{FG} \quad \Rightarrow \quad \frac{7}{10.5} = \frac{2.25}{GF} \quad \Rightarrow \quad GF = 3.375$$

$$\frac{AD}{AB} = \frac{DE}{BC} \quad \Rightarrow \quad \frac{7}{14} = \frac{2.25}{BC} \quad \Rightarrow \quad BC = 4.5$$

Exercises

Communicate

1. Use the examples on page 526 to make a list of proportions in the figure at right.

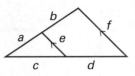

2. Are all isosceles triangles similar? Explain or give a counterexample.

3. Are all equilateral triangles similar? Explain or give a counterexample.

4. Are all right triangles similar? Explain or give a counterexample.

5. How does the capital letter A relate to the Side-Splitting Theorem? If the cross bar is horizontal, what is true of the places where it intersects the sides of the letter? Would this be true of an italic (slanted) A, as well?

Guided Skills Practice

Use the Side-Splitting Theorem to find x. (EXAMPLE 1)

6.

7.

8.

Find the indicated measurements. (EXAMPLE 2)

9. DF

10. FB

11. FG

12. BC

Practice and Apply

In Exercises 13–20, use the Side-Splitting Theorem to find *x*. In some exercises, there may be more than one possible value for *x*.

13.

14.

15.

16.

17.

18.

19.

20.

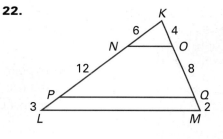

Name all similar triangles in each figure. State the postulate or theorem that justifies each similarity.

21.

22.

Algebra

Find x and y in each figure below.

23. *QRST* is a parallelogram.

24.

25.

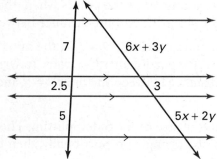

The Side-Splitting Theorem can be used to divide a segment into any number of congruent parts using a compass and straightedge.

26. Draw a segment and label its endpoints A and B. Using your straightedge, draw a ray extending from point A to form an acute angle with $\overline{AB}$.

27. Set your compass to some small length (such as 1 cm) and mark off several lengths along the ray, as many as the number of parts you wish to divide the segment into.

28. Connect the last mark to point B with a segment. Construct lines parallel to this segment through each mark of the compass.

29. Explain why the parallel lines in the last figure divide $\overline{AB}$ into five congruent segments.

30. Draw a segment that is 15 cm long. Use the method described above to divide the segment into seven congruent segments.

CONNECTION

COORDINATE GEOMETRY The distance between two parallel lines is measured along a line perpendicular to both. In Exercises 31–34, you will explore the distances between parallel lines in a coordinate plane.

31. Graph the lines $y = x$, $y = x + 2$, and $y = x - 3$ in a coordinate plane. How can you verify that these lines are are parallel?

32. Graph the line $y = -x$ in the same plane. How can you verify that this line is perpendicular to the lines $y = x$, $y = x + 2$, and $y = x - 3$?

33. Give the coordinates of the points where the line $y = -x$ intersects the lines $y = x$ and $y = x + 2$. Find the distance between these parallel lines.

34. Graph the horizontal line $y = 3$ in the same plane. Use the Two-Transversal Proportionality Corollary to write a proportion, and solve it to find the distance between the lines $y = x$ and $y = x - 3$.

PROOF

35. Prove the converse of the Side-Splitting Theorem: If a segment divides two sides of a triangle proportionally, then the segment is parallel to the third side.

36. In the diagram at right, $\frac{AE}{AG} = 1.3$, $\frac{AB}{AE} = 2.2$, and $AD = 10$ cm. Find AB.

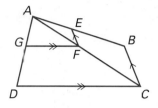

37. MUSIC A simple stringed instrument can be created with parallel strings on an isosceles trapezoidal frame, as shown. If the note for the first (lowest) string, with a length of 40 cm, is an F, then a string with a length of 20 cm would be one *octave* higher, also an F.

If a string for the note C is $\frac{2}{3}$ of the length of the string for the lower F, where would you place the string on the frame?

38. MUSIC For the instrument above, the string for the note G is $\frac{8}{9}$ of the length of the string for the lower F. Where would you place the string on the frame?

The diagram represents one sector of the web.

39. BIOLOGY An orb spider web consists of a spiral of straight segments attached to a series of radial segments that meet at the center. Each sector of the web can be modelled by nested similar triangles. Suppose the outside of a spider web is approximately a regular 16-gon with a side length of 19.5 cm and the 9 inner segments are spaced at equal intervals. Use the diagram below to estimate the length of silk used to build the web.

Notice the spiral of straight segments in this orb spider web. Though spider webs in nature are not perfectly symmetrical, or the spacing of their spirals even, you can model a web mathematically and get a good estimate of the length of the silk in its spiral.

APPLICATION

40. ENGINEERING Find the height of the stack of pipes shown at right. The diameter of each pipe is 8 ft. (Hint: Connect the centers of the circles to form an equilateral triangle.)

CONNECTION

COORDINATE GEOMETRY For Exercises 41–47, refer to △*ABC* with vertices *A*(3, 5), *B*(6, 1), and *C*(–2, –7) in a coordinate plane. *(LESSON 5.6)*

41. Draw △*ABC* in a coordinate plane.

42. Find the perimeter of △*ABC*.

43. Find the midpoint of each side. Label them *D*, *E*, and *F*.

44. Find the perimeter of △*DEF*.

Recall that the median of a triangle is a segment that connects a vertex to the midpoint of the opposite side. *(LESSON 5.6)*

45. Draw the median of △*ABC* from *A*, and find its length.

46. Draw the median of △*ABC* from *B*, and find its length.

47. Draw the median of △*ABC* from *C*, and find its length.

Refer to the figures below for Exercises 48–49.

48. Are the triangles similar? Why or why not?

49. Find the ratio of the length of the shorter leg to the length of the hypotenuse for each triangle. What do you notice?

50. The shorter leg of another right triangle with a 30° angle has a length of 17. What is the ratio of the length of the shorter leg to the length of the hypotenuse for this triangle?

Indirect Measurement and Additional Similarity Theorems

8.5

Monument Valley in Arizona

Objectives

- Use triangle similarity to measure distances indirectly.

- Develop and use similarity theorems for altitudes and medians of triangles.

Why *Directly measuring the distance across a lake or the height of a mountain can be difficult or even impossible. A more practical method uses similar triangles to estimate the desired measurement.*

Using Similar Triangles to Measure Distances

Inaccessible distances can often be measured by using similar triangles.

EXAMPLE ① Suppose that a military engineer needs to know the distance across a river in order to build a temporary bridge. Using a point on the opposite side of the river as a reference point, the engineer sets up right triangles along the bank of the river. Use the diagram at right to find the distance across the river.

● **SOLUTION**

Using the right angles and vertical angles, $\triangle ABC \sim \triangle DEC$ by AA Similarity, so $\frac{AB}{DE} = \frac{BC}{EC}$. Therefore:

$$\frac{x}{40} = \frac{72}{30}, \text{ or } \frac{x}{40} = \frac{12}{5}$$

$$x = \frac{12}{5} \cdot 40 = 96 \text{ meters}$$

EXAMPLE ❷ The diagram at right illustrates another method of setting up right triangles along the river bank. Use this diagram to find the distance across the river.

● **SOLUTION**

Because $\angle ABE \cong ACD$ and $\angle A \cong \angle A$, $\triangle ABE \sim \triangle ACD$ by AA Similarity, and so $\frac{AB}{AC} = \frac{BE}{CD}$. Therefore:

$$\frac{x}{x+32} = \frac{60}{80}, \text{ or } \frac{x}{x+32} = \frac{3}{4}$$

$$4x = 3x + 96$$

$$x = 96 \text{ meters}$$

CRITICAL THINKING   Why do you think the engineer used 90° angles? Could other angles be used?

EXAMPLE ❸ Kim wants to know the height of a basketball hoop. The regulation height is 10 feet. Use Kim's height and the length of the shadows in the diagram at right to find the height of the basketball hoop. Is the hoop at regulation height?

● **SOLUTION**

Because the Sun is so far away, the sun's rays are essentially parallel, and $\angle 1 = \angle 2$. Both Kim and the post of the basketball hoop are perpendicular to the ground, so the two triangles are similar by the AA Similarity Postulate.

$$\frac{x}{70} = \frac{85}{50}$$

$$70 \cdot \frac{x}{70} = \frac{85}{50} \cdot 70$$

$$x = 119 \text{ inches, or 9 feet} \atop \text{and 11 inches.}$$

The hoop is 1 inch below regulation height.

Additional Similarity Theorems

As you know, the sides of similar triangles are proportional. As you will see below, other parts of similar triangles are also proportional.

Proportional Altitudes Theorem

If two triangles are similar, then their corresponding altitudes have the same ratio as their corresponding sides.   **8.5.1**

Recall that an altitude of a triangle may be inside the triangle, outside the triangle, or a side of the triangle.

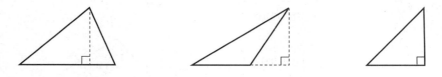

The following proof shows the case of an altitude inside the triangle. You will be asked to prove the case of an altitude outside the triangle in Exercise 26.

PARAGRAPH PROOF

Given: $\triangle ABC \sim \triangle XYZ$; $\overline{AD}$ is an altitude of $\triangle ABC$ and $\overline{XW}$ is an altitude of $\triangle XYZ$.

Prove: $\dfrac{AD}{XW} = \dfrac{AB}{XY}$

Proof:

Because $\triangle ABC \sim \triangle XYZ$, $\angle B \cong \angle Y$ by the Polygon Similarity Postulate, and $\angle BDA$ and $\angle YWX$ are right angles by definition of altitude. Thus, $\triangle ABD \sim \triangle XYW$ by the AA Similarity Postulate, and $\dfrac{AD}{XW} = \dfrac{AB}{XY}$ by the Polygon Similarity Postulate.

CRITICAL THINKING

How would you prove the case of an altitude that is a side of the triangle?

EXAMPLE ④ The triangles in the diagram are similar. Solve for x.

● **SOLUTION**

$$\frac{x}{12} = \frac{16}{20}$$

$$12 \cdot \frac{x}{12} = \frac{16}{20} \cdot 12$$

$$x = 9.6$$

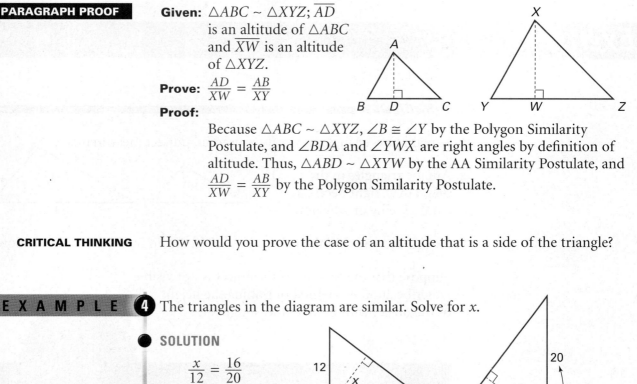

These are corresponding sides.

Recall that a *median* of a triangle is a segment that joins a vertex to the midpoint of the opposite side.

Proportional Medians Theorem

If two triangles are similar, then their corresponding medians have the same ratio as their corresponding sides.

8.5.2

You will be asked to prove this theorem in Exercises 35–41.

5 The triangles in the diagram are similar. Solve for x.

SOLUTION

$$\frac{4}{6} = \frac{3.2}{x}$$

$$4x = 19.2$$

$$x = 4.8$$

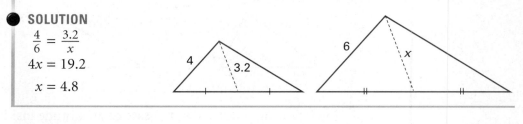

Exercises

Communicate

1. Explain how triangle similarity is used in indirect measurement.

2. Are the triangles in the diagram at right necessarily similar? Why or why not?

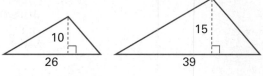

3. How could you find the height of an object from the length of its shadow?

4. Suppose that the shadow of the object is not visible. Describe another method of finding the height of the object with indirect measurement.

5. Use the diagram at right to estimate the width of the lake. Justify your answer. **(EXAMPLE 1)**

5.4 km
3 km
10 km

6. Use the diagram at right to estimate the width of the lake. Justify your answer. **(EXAMPLE 2)**

10 km
9 km
14 km

7. Tranh wants to know the height of a streetlight in his neighborhood. He measures his shadow and the pole's shadow at the same time of day. Tranh is 170 cm tall. Find the height of the streetlight. **(EXAMPLE 3)**

840 cm   310 cm

In Exercises 8 and 9, the triangles are similar. Find x. **(EXAMPLES 4 AND 5)**

8.

2.4
x
3.6
7.2

9.

1.5   1
3.3
x

For Exercises 10–13, use the diagrams to find the height of each building.

10.

22.5 ft
81 ft   27 ft

11.

12 ft
10 ft
20 ft

12.

18 ft
12 ft   20 ft

13.

9 ft
6 ft   14 ft

For Exercises 14–17, the triangles are similar. Find x.

14. **15.**

16. **17.**

In the diagram at right, △ADE ~ △RVW, AC = CD, and RT = TV.

18. $RW =$ _____?_____ **19.** $EB =$ _____?_____

20. $WV =$ _____?_____

For Exercises 21–25, refer to the triangle below.

21. Are △JKL and △MKJ similar? Explain.

22. Find the length of the hypotenuse of △JKL and of △MKJ.

23. Find the area of △JKL and of △MKJ.

24. Find the lengths of the dashed altitudes of △JKL and of △MKJ.

25. Is the ratio of the altitudes of △JKL and △MKJ the same as that of the corresponding sides of △JKL and △MKJ? Explain your reasoning.

PROOF

26. Use the diagram below to prove the Proportional Altitudes Theorem for the case when the altitude is outside the triangle.

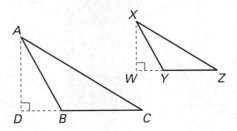

Complete the proof of the following theorem:

Proportional Angle Bisectors Theorem

If two triangles are similar, then their corresponding angle bisectors have the same ratio as the corresponding sides.   **8.5.3**

Given: $\triangle ABC \sim \triangle XYZ, \overline{AD}$
bisects $\angle BAC$, and
$\overline{XW}$ bisects $\angle YXZ$.

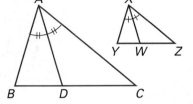

Prove: $\dfrac{AD}{XW} = \dfrac{AB}{XY}$

Proof:

Statements	Reasons
$\triangle ABC \sim \triangle XYZ$	27. ___?___
$\angle B \cong \angle Y$	28. ___?___
$m\angle BAC = m\angle YXZ$	29. ___?___
$\overline{AD}$ bisects $\angle BAC$; $\overline{XW}$ bisects $\angle YXZ$.	Given
30. ___?___ $= 2 \cdot m\angle BAD$ ___?___ $= 2 \cdot m\angle YXW$	Definition of angle bisector
31. ___?___	Substitution Property and Division Property
32. ___?___	AA Similarity Postulate
33. ___?___	Polygon Similarity Postulate

34. Prove the following theorem:

Proportional Segments Theorem

An angle bisector of a triangle divides the opposite side into two segments that have the same ratio as the other two sides.   **8.5.4**

Given: In $\triangle JKL, \overline{JM}$ bisects $\angle KJL$.

Prove: $\dfrac{KM}{LM} = \dfrac{JK}{JL}$

Hint: First construct a line parallel to $\overline{JM}$ through K. Extend $\overline{JL}$ to intersect this line at point N.

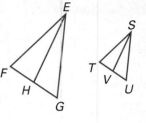

FLOWCHART PROOF

For Exercises 35–41, complete the proof of the Proportional Medians Theorem.

Given: $\triangle EFG \sim \triangle STU$, $\overline{EH}$ is the median of $\triangle EFG$, and $\overline{SV}$ is the median of $\triangle STU$.

Prove: $\dfrac{EH}{SV} = \dfrac{EF}{ST}$

Proof:

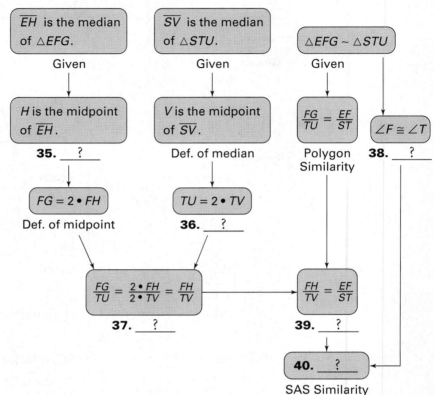

| $\overline{EH}$ is the median of $\triangle EFG$. | $\overline{SV}$ is the median of $\triangle STU$. | $\triangle EFG \sim \triangle STU$ |

Given   Given   Given

H is the midpoint of $\overline{EH}$. V is the midpoint of $\overline{SV}$. $\dfrac{FG}{TU} = \dfrac{EF}{ST}$ $\angle F \cong \angle T$

35. ___?___   Def. of median   Polygon Similarity   **38.** ___?___

$FG = 2 \cdot FH$ $TU = 2 \cdot TV$

Def. of midpoint   **36.** ___?___

$\dfrac{FG}{TU} = \dfrac{2 \cdot FH}{2 \cdot TV} = \dfrac{FH}{TV}$ $\dfrac{FH}{TV} = \dfrac{EF}{ST}$

37. ___?___   **39.** ___?___

40. ___?___

SAS Similarity

41. ___?___

Polygon Similarity

APPLICATION

42. PALEONTOLOGY You can use a mirror to estimate the height of an object, as shown in the photo. According to the laws of optics, the light reflects off a mirror at the same angle from which it strikes the mirror. Use the diagram to estimate the height of the dinosaur skeleton.

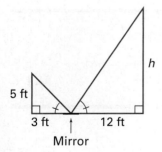

The laws of optics guarantee that the indicated angles are congruent.

OPTICS The diagram at right shows a *convex lens*, which is thicker in the middle than at the edges. A convex lens bends light rays to form an image of an object on the opposite side of the lens. Light rays through the center of the lens are not bent, so $\overline{BB'}$ and $\overline{CC'}$ are straight segments. Thus, $\triangle ABC$ and $\triangle AB'C'$ are similar and $\dfrac{BC}{B'C'} = \dfrac{AC}{AC'}$, or $\dfrac{\text{object size}}{\text{image size}} = \dfrac{\text{object distance}}{\text{image distance}}$.

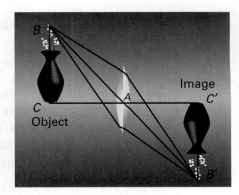

43. Prove that $\triangle ABC \sim \triangle AB'C'$.

44. If an object 6 cm from a convex lens forms an image 15 cm from the lens on the opposite side, what is the ratio of the size of the object to the size of the image?

45. Rosa placed a lens 25 cm from an object 10 cm tall. An image was formed 5 cm from the lens on the opposite side. How tall was the image?

46. How could you arrange an object, a lens, and an image so that the object and the image are the same size?

47. How could you arrange an object, a lens, and an image so that the image is 20 times taller than the object?

Look Back

Find the shaded area of each figure. *(LESSONS 5.2, 5.3, 5.4, AND 8.4)*

48.

49.

50.

51.

Find the volume of each solid. *(LESSONS 7.3 AND 7.6)*

52.

53.

Look Beyond

CULTURAL CONNECTION: AFRICA Eratosthenes, a Greek astronomer who lived in northern Africa, used indirect measurement to estimate Earth's circumference in approximately 200 B.C.E. Eratosthenes knew that the Sun was directly above the city of Aswan at noon on the summer solstice, so a vertical rod would cast no shadow. He measured the angle that the Sun's rays formed with a vertical rod in Alexandria at the same time and found it to be 7.5°.

54. Explain why $\angle 1 \cong \angle 2$ in the diagram below. Remember that the Sun's rays are considered to be parallel.

55. The distance from Aswan to Alexandria is 5000 *stades*, or about 575 mi. Complete the proportion $\frac{7.5°}{360°} = \frac{?}{?}$ and solve it to find Eratosthenes' estimate of Earth's circumference.

56. Given that Earth's circumference is approximately 24,900 mi, find the relative error in Eratosthenes' estimate.

TECHNIQUES FOR INDIRECT MEASUREMENT

Use proportions and at least two of the methods listed below to find the dimensions of a building or other structure at your school or in your neighborhood.

- Measure the shadow of the building and the shadow of a person or object with a known height.
- Use a mirror to create similar triangles, as shown in Exercise 43.
- Take a photograph of the building with a person or object of known height standing in front of it. Measure the building and the person in the photograph.

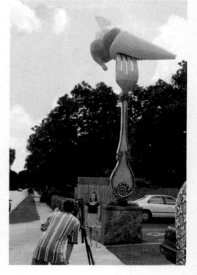

WORKING ON THE CHAPTER PROJECT

You should now be able to complete Activity 2 of the Chapter Project.

Area and Volume Ratios

Objectives

- Develop and use ratios for areas of similar figures.

- Develop and use ratios for volumes of similar solids.

- Explore relationships between cross-sectional area, weight, and height.

Why If one spherical tank has twice the diameter of another, how much more will the larger tank hold? This lesson will show you how to answer questions like these.

Spherical tanks are best for storing contents under pressure, such as liquid butane and related chemicals like butylene and butadiene.

Changing Dimensions of Figures

If the diameter of the spherical container in the photo is doubled, what would happen to the volume? In the Activities that follow, you will answer this and related questions.

Activity 1

Ratios of Areas of Similar Figures

YOU WILL NEED
calculator

For each pair of similar figures below, the ratio of a pair of corresponding measures is given. Find the ratio of the areas for each pair.

1. Two squares

$$\frac{\text{side of square A}}{\text{side of square B}} = \frac{3}{1}$$

$$\frac{\text{area of square A}}{\text{area of square B}} = \frac{?}{?}$$

2. Two triangles

$$\frac{\text{side of triangle A}}{\text{side of triangle B}} = \frac{2}{1}$$

$$\frac{\text{area of triangle A}}{\text{area of triangle B}} = \frac{?}{?}$$

3. Two rectangles

$$\frac{\text{side of rectangle A}}{\text{side of rectangle B}} = \frac{3}{2}$$

$$\frac{\text{area of rectangle A}}{\text{area of rectangle B}} = \frac{?}{?}$$

4. Two circles

$$\frac{\text{radius of circle A}}{\text{radius of circle B}} = \frac{4}{3}$$

$$\frac{\text{area of circle A}}{\text{area of circle B}} = \frac{?}{?}$$

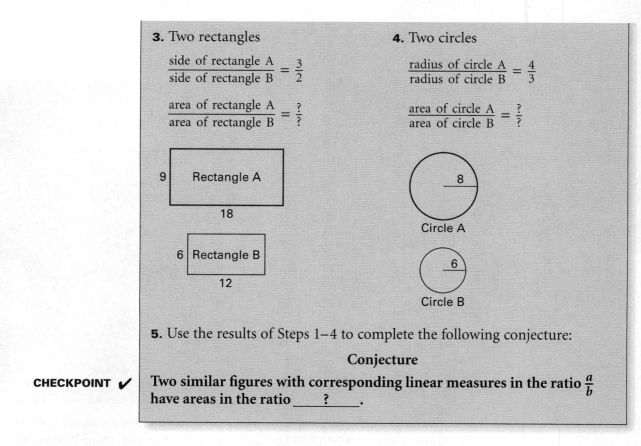

5. Use the results of Steps 1–4 to complete the following conjecture:

Conjecture

CHECKPOINT ✔ Two similar figures with corresponding linear measures in the ratio $\frac{a}{b}$ have areas in the ratio ____?____.

Solids are similar if they are the same shape and all corresponding linear dimensions are proportional. For example, two rectangular prisms are similar if the lengths and widths of the corresponding faces and bases are proportional. In Activity 2, you will explore the ratios of the volumes of similar solids.

Activity 2

Ratios of Volumes of Similar Figures

YOU WILL NEED

calculator

For each pair of similar solids below, the ratio of a pair of corresponding measures is given. Find the ratio of the volumes for each pair.

1. Two cubes

$$\frac{\text{edge of cube A}}{\text{edge of cube B}} = \frac{3}{1}$$

$$\frac{\text{volume of cube A}}{\text{volume of cube B}} = \frac{?}{?}$$

2. Two rectangular prisms

$$\frac{\text{edge of prism A}}{\text{edge of prism B}} = \frac{3}{2}$$

$$\frac{\text{volume of prism A}}{\text{volume of prism B}} = \frac{?}{?}$$

3. Two spheres

$$\frac{\text{radius of sphere A}}{\text{radius of sphere B}} = \frac{2}{1}$$

$$\frac{\text{volume of cube A}}{\text{volume of cube B}} = \frac{?}{?}$$

4. Two cylinders

$$\frac{\text{radius of cylinder A}}{\text{radius of cylinder B}} = \frac{2}{1}$$

$$\frac{\text{volume of cylinder A}}{\text{volume of cylinder B}} = \frac{?}{?}$$

Sphere A   Sphere B   Cylinder A   Cylinder B

5. Use the results of Steps 1–4 to complete the following conjecture:

Conjecture

CHECKPOINT ✔ **Two similar solids with corresponding linear measures in the ratio $\frac{a}{b}$ have volumes in the ratio ____?____.**

CRITICAL THINKING   Are all cubes similar? Why or why not? Area all spheres similar? Why or why not? Are all cylinders similar? Why or why not?

TRY THIS

a. The corresponding sides of two similar triangles are in the ratio $\frac{5}{9}$. What is the ratio of their areas?

b. The surface areas of two similar rectangular prisms are in the ratio $\frac{9}{4}$. What is the ratio of their corresponding sides?

c. One sphere has a radius of 5 meters. Another sphere has a radius of 15 meters. What is the ratio of their volumes?

Cross-Sectional Areas, Weight, and Height

The amount of weight that a structure can support is proportional to its cross-sectional area. For example, column A, whose radius is 3 times the radius of column B, can support 9 times more weight than column B. This is because the cross-sectional area of column A is 9 times that of column B.

Column A   Column B

The load-bearing capacities of the columns depend on their cross-sectional areas.

In Example 1 and Activity 3, you will investigate the mathematical consequences of some animals' support requirements.

A young elephant that is 4 ft tall at the shoulder has a leg bone whose circular cross section has a radius of 4 cm. How much thicker would the leg bone need to be to give the same support to an elephant that is 8 ft tall at the shoulder? By what scale factor does the original radius need to be multiplied to provide a sufficient cross-sectional area?

● **SOLUTION**

The height of the elephant increases by a factor of 2. However, the volume of the elephant increases by a factor of 8. Thus, the larger elephant would need a leg bone with a cross-sectional area 8 times larger than that of the original elephant.

Thus, $8\pi r^2$ is the required cross-sectional area.

Let R represent the radius of the leg bone of the larger elephant.

$$\pi R^2 = 8\pi r^2$$
$$R^2 = 8r^2$$
$$R = \sqrt{8}r$$

Thus, the radius of the larger bone must be $\sqrt{8}$, or about 2.8, times the radius of the smaller bone.

Have you read stories of giants hundreds of feet tall? Do you think giants like this could be real? The elephant example shows how increased linear dimensions affect the structures that must support the increased bulk. Think about the question of giants as you do the next Activity.

Activity 3

Increasing Height and Volume

1. Complete the chart to determine the necessary scale factors for the cross-sectional radius of leg bones of animals.

Height scale factor	Volume scale factor	Crosss-sectional radius scale factor
2	8	$\sqrt{8}$
3	27	$\sqrt{27}$
4	64	?
5	?	?
100	?	?

2. If a normal-sized horse requires a leg bone with a cross-sectional radius of 2.5 cm, what must the cross-sectional radius of a leg bone be for a horse that is 20 times taller? 50 times taller? 100 times taller?

CHECKPOINT ✔ **3.** What effect does increasing the size of the horse have on the relative proportions of the legs? Would the legs of a larger horse be proportionally thicker or thinner than those of a normal-sized horse?

Communicate

1. What is the relationship between the ratio of the edges of two cubes and the ratio of their volumes?

2. Why are all spheres similar? Why are all cubes similar? Is this true for other three-dimensional figures? If so, name or describe them.

3. If you know the surface areas of two similar prisms, explain how you can find the ratio of their volumes.

4. What happens to the volume of a cylinder if the radius is doubled but the height stays the same?

5. How does the cross-sectional area of a bone relate to an animal's weight and height?

Guided Skills Practice

6. Of two similar parallelograms, one has a base of 8 and a height of 3. The linear dimensions of the other parallelogram are one-half the linear dimensions of the first. Find the ratio of the areas. *(ACTIVITY 1)*

7. Of two similar pyramids, one has a square base with each side equal to 6 and a height of 4. The linear dimensions of the other pyramid are one-half the linear dimensions of the first. Find the ratio of the volumes. *(ACTIVITY 2)*

8. The leg bone of an animal has a circular cross section with an area of about 4π cm^2. If the animal's height were tripled without changing the proportions of its measurements, how would the volume of the animal be affected? How would the cross-sectional area of the leg bone be affected? *(EXAMPLE 1 AND ACTIVITY 3)*

Practice and Apply

In the figure at right, $\overline{EC} \parallel \overline{AB}$.

9. Find the ratio of the perimeters of $\triangle ABD$ and $\triangle ECD$.

10. Find the ratio of the areas of $\triangle ABD$ and $\triangle ECD$.

In the figure at right, JKLMN ~ VWXYZ. JKLMN has a perimeter of 24 m and an area of 50 m².

11. Find the perimeter of $VWXYZ$.

12. Find the area of $VWXYZ$.

13. In the figure at right, $ABCDEF \sim PQRSTU$. $ABCDEF$ has a perimeter of 42 m and an area of 96 m². Find the perimeter and area of $PQRSTU$.

The ratio of the corresponding edges of two similar pyramids is $\frac{7}{5}$. Find the ratio of the following:

14. the perimeters of their bases

15. the areas of their bases

16. their volumes

Two spheres have radii of 5 cm and 7 cm. Find the ratio of the following:

17. the circumferences of their great circles

18. their surface areas

19. their volumes

The ratio of the heights of two similar cones is $\frac{7}{9}$. Find the ratio of the following:

20. their radii

21. their volumes

22. the areas of their bases

The ratio of the base areas of two similar cylinders is $\frac{16}{25}$. Find the ratio of the following:

23. their heights

24. the circumferences of their bases

25. their volumes

The ratio of the surface areas of two spheres is $\frac{144}{169}$. Find the ratio of the following:

26. their radii

27. their volumes

28. the circumferences of their great circles

29. Two similar cylinders have base areas of 16 cm² and 49 cm². If the larger cylinder has a height of 21 cm, find the height of the smaller cylinder.

Two similar cones have surface areas of 225 cm² and 441 cm².

30. If the height of the larger cone is 12 cm, find the height of the smaller cone.

31. If the volume of the smaller cone is 250 cm³, find the volume of the larger cone.

The ratio of the volumes of two prisms is $\frac{64}{125}$. Find the ratio of the following:

32. their surface areas

33. the perimeters of their bases

34. corresponding base diagonals

35. the areas of their bases

The small cone is formed by cutting off the lower part of the larger cone. Use the figure at right for Exercises 36 and 37.

2 cm
6 cm

36. If the larger cone has a volume of 288 cm³ find the volume of the smaller cone.

37. If the base of the smaller cone has an area of 3.6 cm², find the area of the base of the larger cone.

38. A cross section of the leg bone of a horse is a circle with an area of 19 cm². What is the diameter of the leg bone?

39. What should the diameter of the cross section in Exercise 38 be to support a horse twice as tall?

40. A 100-ft tall cylindrical tower has a cross-sectional radius of 26 ft. What would be the radius of a 350-ft tall tower that is similar to the 100-ft tall tower?

41. ASTRONOMY The circumferences of Earth and the Moon are about 40,200 km and 10,000 km, respectively. Find the ratio of their diameters and the ratio of their volumes.

Earthrise, viewed from the moon

42. FOOD A new pizzeria sells an 8-in. diameter pizza for $4.00 and a 16-in. diameter pizza for $8.00. Which pizza is the better deal? Explain.

43. PACKAGING A 12-oz box of noodles is a rectangular prism that measures 15 cm × 20 cm × 4 cm. To make 24-oz box, should the company double all of the dimensions of the 12-oz box? Explain.

44. PACKAGING A toothpaste company packages its product in a tube that has a circular opening with a radius of 2 mm. The company increases the radius of the opening to 3 mm. Predict what will happen to the amount of toothpaste used if people continue to use the same length of toothpaste on their toothbrushes.

45. BUSINESS The area of the parking lot at Jerome's Restaurant is 400 m². Jerome buys some adjoining land and expands the parking lot to 1.5 times as wide and 1.75 times as long as the original lot. Find the area of the expanded lot.

46. SPORTS The diameter of a standard basketball is about 9.5 in. A company that makes basketballs is contracted to make promotional basketballs with a diameter of 5 in. The materials for a standard-sized basketball cost $1.40. How much will the materials cost for a promotional basketball made of the same materials?

47. STORAGE A city stores rock salt for winter road maintenance in a dome-shaped building that is 82 ft in diameter at the base. The building holds 3366 tons of salt. Because the city is growing, the city planners decide to build a second, smaller dome. The linear dimensions of the new dome will be three-fourths the linear dimensions of the old dome. Estimate the storage capacity of the new dome.

Look Back

48. The measure of each interior angle of a regular polygon is 165°. How many sides does the polygon have? **(LESSON 3.6)**

49. The exterior angles of a regular polygon each measure 40°. How many sides does the polygon have? **(LESSON 3.6)**

Algebra

50. Which of the lines below are parallel? **(LESSON 3.8)**
   a. $5x + 4y = 18$
   b. $-10x + 8y = 21$
   c. $30x - 24y = 45$

51. Write an equation for the line that is parallel to $5x + 4y = 18$ and passes through the point $(3, 8)$. **(LESSON 3.8)**

52. Write an equation of the line that is perpendicular to $5x + 4y = 18$ and passes through the point $(3, 8)$. **(LESSON 3.8)**

Solve each proportion. Give all possible values of x. (LESSON 8.2)

53. $\frac{x + 1}{4} = \frac{3x}{8}$

54. $\frac{2}{x + 2} = \frac{x - 1}{2}$

55. $\frac{x - 1}{x + 1} = \frac{x + 2}{2x + 2}$

56. In a rectangle, the ratio of the length of the long side to the length of the short side is $\frac{5}{3}$. If the short side has a length of 9, what is the length of the long side? **(LESSON 8.2)**

57. In the triangle below, $\overline{DE} \| \overline{BC}$. Solve for x. **(LESSON 8.4)**

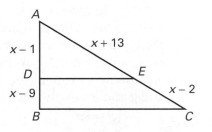

Look Beyond

CHALLENGE

ABCD is a square with side lengths of 1. The squares inside ABCD are formed by connecting the midpoints of the larger square.

Algebra

58. What happens to the perimeter of each successive square? Using a calculator, add the perimeters of the first eight squares. Keep track of the sums as each perimeter is added. Do you think the sum of the perimeters of all the squares generated this way will ever reach 14? Explain.

PROJECT
CHAPTER EIGHT

Indirect Measurement

In this project, you will build a scale model of your school and possibly the area around it. You and your class should use any methods you know, both direct and indirect, to find the dimensions you will need to build your model. Be creative!

Direct Measurement　　　*Photographic Method*　　　*Measuring a student's stride*

Activity 1

THE MAP To give context to your model, you will need a map of the area you are modeling. One possibility is to enlarge the portion of your city map in which your school is located. Draw the map on a coordinate grid and locate one or more physical landmarks on it to serve as reference points for the measurements you make. Make as many measurements as you will need to model and locate the features of your school.

Activity 2

Lay out a coordinate grid on the area where your model will be built. Use the grid to give organization and structure to the measurements you took in Activity 1. Create physical models of the buildings and other structures on your campus and place them on the grid.

Chapter Review and Assessment

center of dilation 500	expansion 500	scale factor 498
contraction 500	proportion 508	similar 507
dilation 498	proportional 508	

POSTULATES AND THEOREMS

Lesson	Number	Postulate or Theorem
8.2	8.2.2 Polygon Similarity Postulate	Two polygons are similar if and only if there is a way of setting up a correspondence between their sides and angles such that: • Each pair of corresponding angles is congruent. • Each pair of corresponding sides is proportional.
	8.2.3 Cross-Multiplication Property	For real numbers a, b, c, and d such that b, $d \neq 0$, if $\frac{a}{b} = \frac{c}{d}$, then $ad = bc$.
	8.2.4 Reciprocal Property	For real numbers a, b, c, and d such that a, b, c, $d \neq 0$, if $\frac{a}{b} = \frac{c}{d}$, then $\frac{b}{a} = \frac{d}{c}$.
	8.2.5 Exchange Property	For real numbers a, b, c, and d such that a, b, c, $d \neq 0$, if $\frac{a}{b} = \frac{c}{d}$, then $\frac{a}{c} = \frac{b}{d}$.
	8.2.6 "Add-One" Property	For real numbers a, b, c, and d such that b, $d \neq 0$, if $\frac{a}{b} = \frac{c}{d}$, then $\frac{a+b}{b} = \frac{c+d}{d}$.
8.3	8.3.1 AA (Angle-Angle) Similarity Postulate	If two angles of one triangle are congruent to two angles of another triangle, then the triangles are similar.
	8.3.2 SSS (Side-Side-Side) Similarity Theorem	If the three sides of one triangle are proportional to the three sides of another triangle, then the triangles are similar.
	8.3.3 SAS (Side-Angle-Side) Similarity Theorem	If two sides of one triangle are proportional to two sides of another triangle and their included angles are congruent, then the triangles are similar.
8.4	8.4.1 Side-Splitting Theorem	A line parallel to one side of a triangle divides the other two sides proportionally.
	8.4.2 Two-Transversal Proportionality Corollary	Three or more parallel lines divide two intersecting transversals proportionally.
8.5	8.5.1 Proportional Altitudes Theorem	If two triangles are similar, then their corresponding altitudes have the same ratio as their corresponding sides.
	8.5.2 Proportional Medians Theorem	If two triangles are similar, then their corresponding medians have the same ratio as their corresponding sides.
	8.5.3 Proportional Angle Bisectors Theorem	If two triangles are similar, then their corresponding angle bisectors have the same ratio as their corresponding sides.
	8.5.4 Proportional Segments Theorem	An angle bisector of a triangle divides the opposite side into two segments that have the same ratio as the other two sides.

Key Skills & Exercises

Key Skills

Draw a dilation on a coordinate plane.

Find the image of a segment with endpoints at (1, 3) and (4, 0) that is transformed by the dilation $D(x, y) = (2x, 2y)$.

The endpoints of the image are (2, 6) and (8, 0).

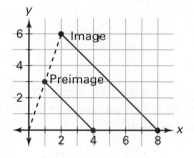

Draw a dilation in a plane.

Draw an equilateral triangle and dilate it about one vertex by a scale factor of 2.

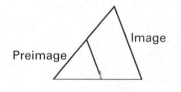

Exercises

Find the image of a segment with endpoints at (–2, 1) and (3, 4) that is transformed by the given dilation.

1. $D(x, y) = (3x, 3y)$

2. $D(x, y) = (-x, -y)$

Copy each figure below and dilate the figure about the given point by the given scale factor.

3. $n = 3$

4. $n = \frac{1}{2}$

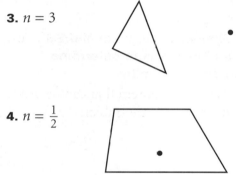

Key Skills

Determine whether polygons are similar.

Are the rectangles in the diagram below similar? Why or why not?

If the rectangles were similar, the proportion $\frac{18}{30} = \frac{48}{72}$ would be true. Cross-multiplying gives $18 \cdot 72 = 30 \cdot 48$, or $1296 = 1440$, so the rectangles are not similar.

Exercises

Determine whether the polygons in each pair are similar. Explain your reasoning.

5.

6.

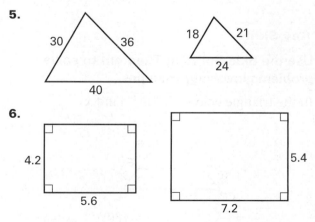

Use proportions to find the side lengths of similar figures.

The triangles in the diagram below are similar. Find x.

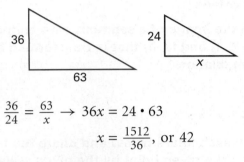

$$\frac{36}{24} = \frac{63}{x} \rightarrow 36x = 24 \cdot 63$$

$$x = \frac{1512}{36}, \text{ or } 42$$

The figures in each pair below are similar. Find x.

7.

8.

LESSON 8.3
Key Skills

Use the AA Similarity Postulate and SSS and SAS Similarity Theorems to determine whether triangles are similar.

Name the postulate or theorem that can be used to prove that the triangles are similar.

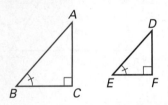

$\angle B \cong \angle E$ and $\angle C \cong \angle F$, so the triangles are similar by the AA Similarity Postulate.

Exercises

For each pair of triangles, name the postulate or theorem that can be used to prove that the triangles are similar.

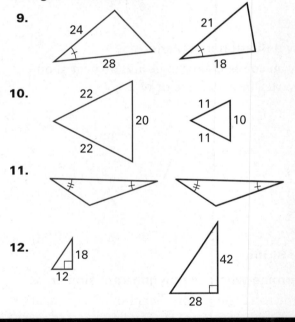

9.

10.

11.

12.

LESSON 8.4
Key Skills

Use the Side-Splitting Theorem to solve problems involving triangles.

In the triangle below, $\overline{DE} \| \overline{BC}$. Find x.

$$\frac{20}{10} = \frac{18}{x} \qquad 20x = 180 \qquad x = 9$$

Exercises

Use the Side-Splitting Theorem to find x.

13.

14.

15.

16.

LESSON 8.5

Key Skills

Use similar triangles to measure distance indirectly.

A right square pyramid has a base edge of 40 ft. If the pyramid casts a shadow 60 ft long at the same time that a yardstick casts a shadow 6 ft long, what is the height of the pyramid?

Because the sun's rays are considered parallel, the triangles formed by the pyramid and the yardstick are similar. The horizontal side of the large triangle is 60 ft plus 20 ft, or 80 ft.

$$\frac{6}{80} = \frac{3}{x} \qquad 6x = 240 \qquad x = 40$$

The height of the pyramid is 40 ft.

Use similarity theorems to solve problems involving altitudes and medians of triangles.

The triangles below are similar. Find x.

Because the triangles are similar, their altitudes are proportional.

$$\frac{x}{1.2} = \frac{5.6}{6.4}$$

$$1.2 \cdot \frac{x}{1.2} = \frac{5.6}{6.4} \cdot 1.2$$

$$x = 1.05$$

Exercises

17. Catherine is 1.6 m tall and casts a shadow of 3.5 m. At the same time, a house casts a shadow of 17.5 m. Find the height of the house.

Refer to the diagram below.

18. A surveyor made the measurements shown below. $\overline{ZY}$ is parallel to $\overline{WX}$. Find the distance across the base of the hill.

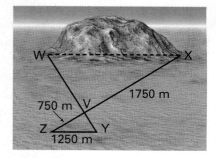

The triangles in each pair below are similar. Find x.

19.

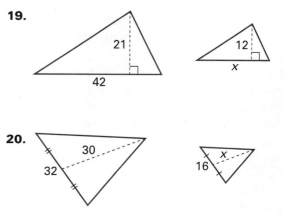

20.

LESSON 8.6

Key Skills

Find the ratio of the areas of similar figures.

The ratio of the sides of two similar quadrilaterals is $\frac{2}{3}$. Find the ratio of their areas. The ratio of their areas is $\frac{2^2}{3^2} = \frac{4}{9}$.

Exercises

21. The ratio of the sides of two similar triangles is $\frac{5}{1}$. What is the ratio of their areas?

22. The ratio of the sides of two similar parallelograms is $\frac{1}{2}$. What is the ratio of their areas?

Find the ratio of the volumes of similar solids.

The ratio of the edges of two similar pyramids is $\frac{2}{3}$. Find the ratio of their volumes.

The ratio of their volumes is $\frac{2^3}{3^3} = \frac{8}{27}$.

23. The ratio of the edges of two similar prisms is $\frac{1}{10}$. What is the ratio of their volumes?

24. The ratio of the radii of two similar spheres is $\frac{1}{2}$. What is the ratio of their volumes?

Applications

25. GRAPHIC DESIGN An artist uses a photocopier to enlarge a design by 120%. If the area of the original design is 10 in.², what will be the area of the enlarged design?

26. LANDSCAPING Brian needs to cut down a tree that is 30 m away from his house. To estimate the height of the tree, he places a mirror on the ground 15 m from the base of the tree and stands 1 m away from the mirror, in which he can see the reflection of the top of the tree. Brian's eyes are 1.8 m above the ground. If the tree falls toward Brian's house, will it hit the house?

1.8 m

15 m   1 m

27. FINE ARTS The sculpture *Batcolumn* by Claes Oldenburg and Coosje van Bruggen is a giant replica of a baseball bat and is 100 ft and 9 in. tall. What is the ratio of the volume of this sculpture to the volume of a real baseball bat that is 35 in. long?

Alternative Assessment

Performance Assessment

1. Use the figures below to write a test for similar quadrilaterals. Try to use the smallest possible number of conditions needed to ensure that the quadrilaterals are similar. Prove that your test is valid.

2. Generalize your results from Exercise 1 to include all polygons. What is the fewest number of sides and angles needed to prove that two polygons with *n* sides are similar?

3. Describe a method for deciding whether two curved figures are similar. Is this method an approximate method or an exact method?

Portfolio Projects

Suppose you are building a scale model of the solar system. Use the table below to determine the model sizes of the planets and the model distances of the planets from the Sun.

	Diameter	Distance from the Sun
Sun	1,391,900 km	0 km
Mercury	4866 km	57,950,000 km
Venus	12,106 km	108,110,000 km
Earth	12,742 km	149,570,000 km
Mars	6760 km	227,840,000 km
Jupiter	139,516 km	778,140,000 km
Saturn	116,438 km	1,427,000,000 km
Uranus	46,940 km	2,870,300,000 km
Neptune	45,432 km	4,499,900,000 km
Pluto	3400 km	5,913,000,000 km

1. Select a convenient size to make the model of the Sun. What is the scale factor for your model?

2. Using your scale factor from Step 1, find the sizes of the planets in your model and the model distances of the planets from the Sun.

3. What is the ratio of the volume of the Sun to the volume of your model of the Sun?

4. What problems would you have in making your model? How big a space would you need for it? Would the smaller planets be big enough for you to see them well?

College Entrance Exam Practice

QUANTITATIVE COMPARISON For Items 1–4, write
A if the quantity in Column A is greater than the quantity in Column B;
B if the quantity in Column B is greater than the quantity in Column A;
C if the two quantities are equal; or
D if the relationship cannot be determined from the information given.

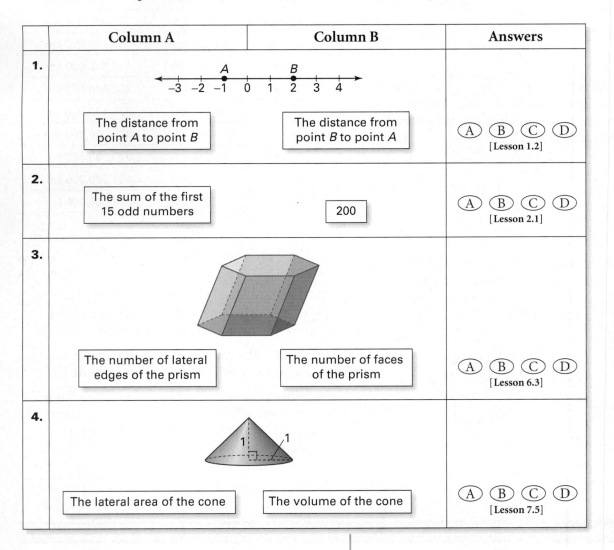

	Column A	Column B	Answers
1.	The distance from point A to point B	The distance from point B to point A	Ⓐ Ⓑ Ⓒ Ⓓ [Lesson 1.2]
2.	The sum of the first 15 odd numbers	200	Ⓐ Ⓑ Ⓒ Ⓓ [Lesson 2.1]
3.	The number of lateral edges of the prism	The number of faces of the prism	Ⓐ Ⓑ Ⓒ Ⓓ [Lesson 6.3]
4.	The lateral area of the cone	The volume of the cone	Ⓐ Ⓑ Ⓒ Ⓓ [Lesson 7.5]

5. If line ℓ has a slope of 5 and line m has a slope of 0.2, then the lines are ___?___. *(LESSON 3.8)*
 a. parallel
 b. perpendicular
 c. vertical
 d. none of the above

6. If pentagon $ABCDE \cong$ pentagon $MNJKL$, which of the following is true? *(LESSON 4.1)*
 a. $\angle C \cong \angle K$
 b. $\angle A \cong \angle M$
 c. $\overline{AE} \cong \overline{MN}$
 d. cannot be determined from the given information

7. Which of the following quadrilaterals has congruent diagonals? *(LESSON 4.5)*
  a. trapezoid
  b. rectangle
  c. parallelogram
  d. rhombus

8. Which of the following quadrilaterals can have exactly two right angles? *(LESSON 4.5)*
  a. trapezoid
  b. rectangle
  c. parallelogram
  d. rhombus

9. Find the midpoint of a segment with endpoints at $(5, 1)$ and $(1, -3)$. *(LESSON 5.6)*
  a. $(6, -2)$
  b. $(2, 2)$
  c. $(3, -1)$
  d. $(4, -2)$

10. The point $(3, -1, 2)$ is in which octant of a three-dimensional coordinate system? *(LESSON 6.4)*
  a. first octant
  b. top-right-back
  c. top-front-left
  d. bottom-front-right

11. Find the y-intercept of the plane defined by the equation $5x + 2y - z = 6$. *(LESSON 6.5)*

12. Solve the proportion for x. *(LESSON 8.2)*
$$\frac{2 + x}{3} = \frac{x}{2}$$

13. Which of the following is not valid for proving triangles similar? *(LESSON 8.3)*
  a. SSS
  b. AA
  c. SSA
  d. SAS

14. Find the ratio of the volumes of a sphere with a radius of 3 and a sphere with a radius of 6. *(LESSON 8.6)*

FREE-RESPONSE GRID
Items 15–18 may be answered by using a free-response grid such as that commonly used by standardized-test services.

15. What is the sum of the interior angles of the figure below? *(LESSON 3.6)*

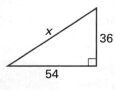

16. What is the sum of the exterior angles of the figure above? *(LESSON 3.6)*

17. Find the value of x in the figure below. Round your answer to the nearest tenth. *(LESSON 5.4)*

18. In the figure below, $\overline{BC} \parallel \overline{DE}$. Find BD. *(LESSON 8.4)*

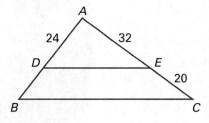

Circles

GREEK GEOMETERS AT THE TIME OF EUCLID believed that circles have a special perfection. With the rediscovery of Euclid's *Elements* by English philosopher Adelard (twelfth century), this way of thinking made its way into the European world. The designs in many early churches were based on geometric principles learned from Euclid.

Among the most famous works of art based on the geometry of the circle is the north rose window in the Notre Dame Cathedral in Paris (thirteenth century).

The circular structures in the aerial photo of the Pueblo ruins at Chaco Canyon are known as *kivas*. The circular design of these ceremonial structures reflects the Pueblo belief, suggested by the recurrence of celestial phenomena, that time is circular in nature.

Rose window, Notre Dame Cathedral

Chaco Canyon

From "Japanese Temple Geometry," by Tony Rothman, with the cooperation of Hidetoshi Fukagawa. Copyright © 1998 by Scientific American, Inc. All rights reserved.

562

Natural well or cenote at Chichén Itzá

About the Chapter Project

The art of constructing elegant egg shapes from curves of different radii is an ancient art that goes back to prehistoric times. Certain principles of geometry in this chapter will enable you to construct objects like the egg shape shown at right.

After completing the Chapter Project, you will be able to do the following:

- Construct artistic eggs from given examples.

- Create your own designs by using reverse curves.

About the Portfolio Activities

Throughout this chapter, you will be given opportunities to complete Portfolio Activities that are designed to support your work on the Chapter Project.

- The basic construction for circle flowers is the topic of the Portfolio Activity on page 572.

- A method for creating a single smooth curve from two curves of different radii is shown in the Portfolio Activity on page 579. With this method, you can construct the egg shapes in the Chapter Project.

- The method of creating a reverse curve, or "S-curve," by connecting two arcs smoothly is given in the Portfolio Activity on page 609. This principle, which is important in art, engineering, and architecture, will enable you to construct the figures in the Chapter Project, as well as figures of your own design.

Chairs and Arcs

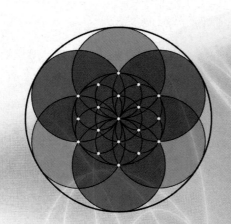

Objectives

- Define a circle and its associated parts, and use them in constructions.

- Define and use the degree measure of arcs.

- Define and use the length measure of arcs.

- Prove a theorem about chords and their intercepted arcs.

Why To study the geometry of circles, you will need to know some basic definitions and ideas. With these, you can quickly learn to make interesting constructions.

The mathematical features of these attractive designs, which were created with geometry graphics software, can easily be understood from the ideas in this lesson.

Circles: A Formal Definition

A circle can be named by using the symbol ⊙ and the center of the circle. The circle in the illustration below is ⊙P, or circle P.

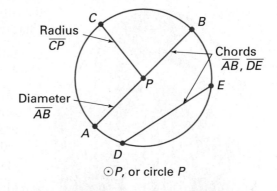

⊙P, or circle P

Definition: Circle

A circle is the set of all points in a plane that are equidistant from a given point in the plane known as the center of the circle. A **radius** (plural, *radii*) is a segment from the center of the circle to a point on the circle. A **chord** is a segment whose endpoints line on a circle. A **diameter** is a chord that contains the center of a circle. **9.1.1**

In Activity 1 below, you will use concepts from the definition of a circle to perform and analyze a construction.

Activity 1

Constructing a Hexagon in a Circle

YOU WILL NEED

compass and
straightedge

1. Draw a circle with a compass. Label the center *P*. Choose a point on the circle and label it *A* (figure **a**).

2. Without changing your compass setting, place the point of your compass on point *A*. Draw an arc that intersects the circle at a new point. Label the new point *B* (figure **b**).

3. Draw chord $\overline{AB}$. Draw radii $\overline{PA}$ and $\overline{PB}$ (figure **c**).

CHECKPOINT ✔ 4. What kind of triangle is △*ABP*? What are the measures of its angles? Explain your reasoning.

5. Without changing your compass setting, place the point of your compass on point *B*. Draw an arc that intersects the circle at a new point. Label the new point *C* (figure **d**).

6. Draw chord $\overline{BC}$ and the new radius $\overline{PC}$.

7. Continue drawing new points, chords, and radii until you have completed a figure like the one shown in figure **e**.

8. Is polygon *ABCDEF* a regular hexagon? Explain your answer.

CHECKPOINT ✔ 9. An angle such as ∠*APB* is known as a *central angle* of a circle. (See page 566.) Are all the central angles in the figure congruent? Does the sum of their measures equal 360°? Explain your reasoning.

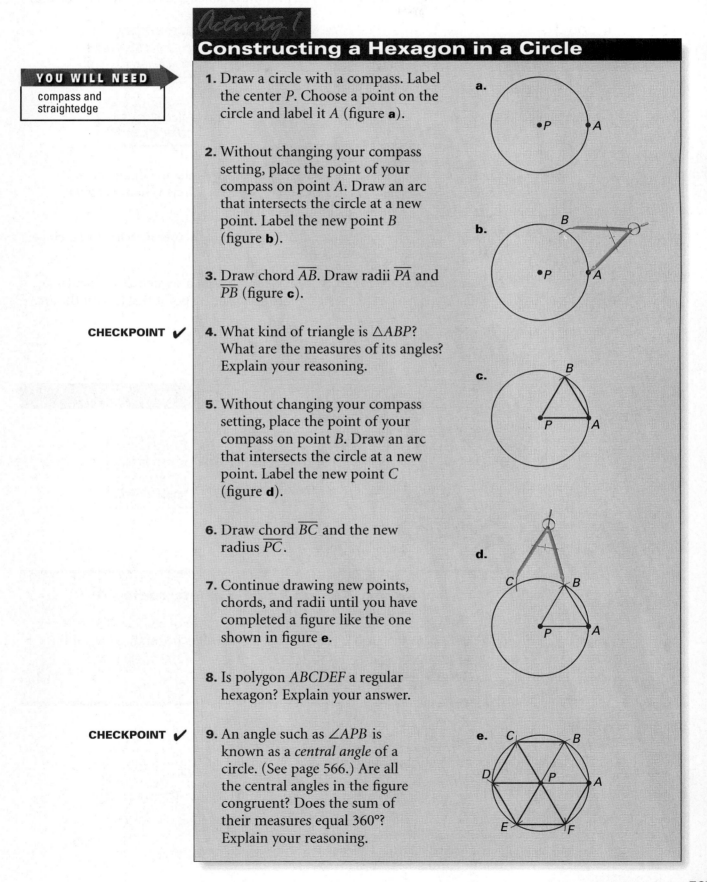

Major and Minor Arcs

An **arc** is an unbroken part of a circle. Any two distinct points on a circle divide the circle into two arcs. The points are called the **endpoints** of the arcs.

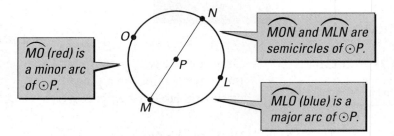

$\overgroup{MO}$ (red) is a minor arc of ⊙P.

$\overgroup{MON}$ and $\overgroup{MLN}$ are semicircles of ⊙P.

$\overgroup{MLO}$ (blue) is a major arc of ⊙P.

A **semicircle** is an arc whose endpoints are endpoints of a diameter. A semicircle is informally called a half-circle. A semicircle is named by its endpoints and another point that lies on the arc.

A **minor arc** of a circle is an arc that is shorter than a semicircle of that circle. A minor arc is named by its endpoints.

A **major arc** of a circle is an arc that is longer than a semicircle of that circle. A major arc is named by its endpoints and another point that lies on the arc.

Degree Measures of Arcs

Central angles of circles are used to find the measures of arcs.

$\overgroup{RS}$ is the intercepted arc of central angle ∠RPS.

Definitions: Central Angle and Intercepted Arc

A **central angle** of a circle is an angle in the plane of a circle whose vertex is the center of the circle. An arc whose endpoints lie on the sides of the angle and whose other points lie in the interior of the angle is the **intercepted arc** of the central angle.　　9.1.2

Central angle

Intercepted arc

An orange may consist of nine wedges, seen in cross section here. Thus, an average wedge would form a central angle of about one-ninth of the full circle, or 40°. (When you look at a typical orange wedge, does it seem to be about 40°?)

Central angle

Intercepted arc

Arcs can be measured
in terms of degrees.

$m\widehat{AB} = 50°$

$m\widehat{ACB} = 360° - 50° = 310°$

Definition: Degree Measure of Arcs

The degree measure of a minor arc is the measure of its central angle.
The degree measure of a major arc is 360° minus the degree measure
of its minor arc. The degree measure of a semicircle is 180°.   **9.1.3**

EXAMPLE **1** Find the measures of $\widehat{RT}$, $\widehat{TS}$, and $\widehat{RTS}$.

SOLUTION

The measures of $\widehat{RT}$ and $\widehat{TS}$ are found
from their central angles.

$m\widehat{RT} = 100°$ $m\widehat{TS} = 90°$

$\widehat{RT}$ and $\widehat{TS}$, which have just one endpoint in common, are called adjacent
arcs. Add their measures to find the measure of $\widehat{RTS}$.

$m\widehat{RTS} = m\widehat{RT} + m\widehat{TS} = 100° + 90° = 190°$

Arc Length

A second way to measure an arc is in terms of its length. To find the length of
an arc, you need to know the radius of the circle of which the arc is a part.

EXAMPLE **2** Find the length of the indicated arc.
Express your answer to the nearest
millimeter. (There are 20 equal
sectors on a dartboard.)

SOLUTION

The length of the arc is $\frac{1}{20}$ of the
circumference of the circle.
Remember that $C = 2\pi r$.

Length of arc $= \frac{1}{20}(2\pi \times 170)$
$= 17\pi \approx 53.4 \approx 53$ mm

r = 170 mm

In general, the length of an arc can be found by using the formula below.

Arc Length

If r is the radius of a circle and M is the degree measure of an arc of the circle, then the length, L, of the arc is given by the following:

$$L = \frac{M}{360°}(2\pi r)$$

9.1.4

TRY THIS   Find the length of $\widehat{RS}$. Round your answer to the nearest hundredth.

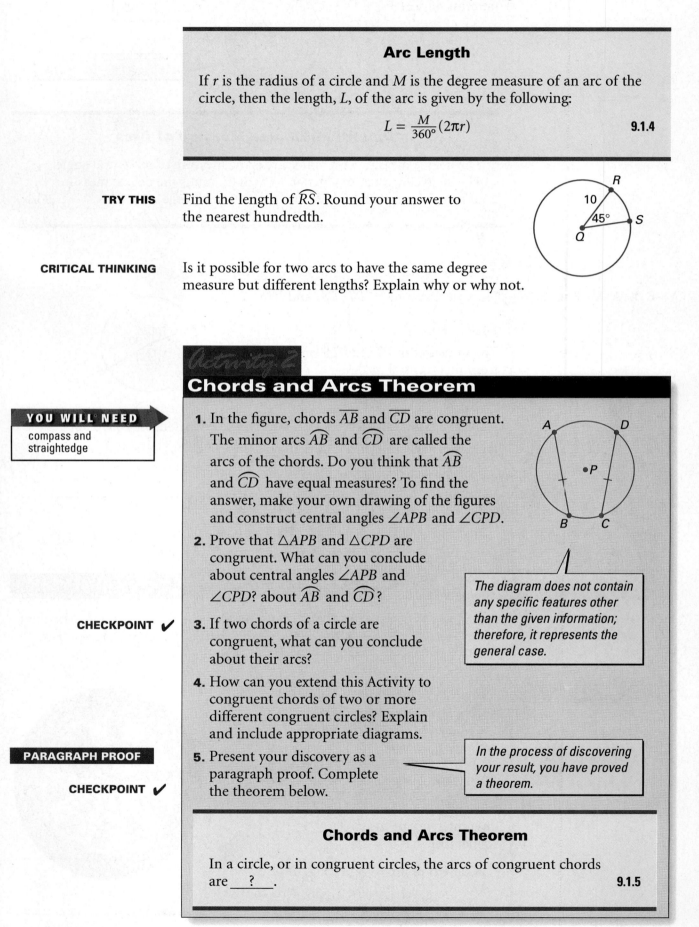

CRITICAL THINKING   Is it possible for two arcs to have the same degree measure but different lengths? Explain why or why not.

Activity 2

Chords and Arcs Theorem

YOU WILL NEED
compass and straightedge

1. In the figure, chords $\overline{AB}$ and $\overline{CD}$ are congruent. The minor arcs $\widehat{AB}$ and $\widehat{CD}$ are called the arcs of the chords. Do you think that $\widehat{AB}$ and $\widehat{CD}$ have equal measures? To find the answer, make your own drawing of the figures and construct central angles $\angle APB$ and $\angle CPD$.

2. Prove that $\triangle APB$ and $\triangle CPD$ are congruent. What can you conclude about central angles $\angle APB$ and $\angle CPD$? about $\widehat{AB}$ and $\widehat{CD}$?

   The diagram does not contain any specific features other than the given information; therefore, it represents the general case.

CHECKPOINT ✔   3. If two chords of a circle are congruent, what can you conclude about their arcs?

4. How can you extend this Activity to congruent chords of two or more different congruent circles? Explain and include appropriate diagrams.

PARAGRAPH PROOF
CHECKPOINT ✔   5. Present your discovery as a paragraph proof. Complete the theorem below.

   In the process of discovering your result, you have proved a theorem.

Chords and Arcs Theorem

In a circle, or in congruent circles, the arcs of congruent chords are ___?___.

9.1.5

Exercises

Communicate

In Exercises 1–5, classify each statement as true or false and explain your reasoning.

1. Every diameter of a circle is also a chord of the circle.

2. Every radius of a circle is also a chord of the circle.

3. Every chord of a circle contains exactly two points of the circle.

4. If two chords of a circle are congruent, then their arcs are also congruent.

5. If two arcs of a circle are congruent, then their chords are also congruent.

6. How can you show that two arcs of a circle are congruent? Is there more than one way? Discuss.

7. Are all semicircles of a circle congruent? Explain.

Guided Skills Practice

Use the figure below for Exercises 8 and 9.

8. Find the degree measures of $\overset{\frown}{AB}$, $\overset{\frown}{BC}$, and $\overset{\frown}{CA}$. *(EXAMPLE 1)*

9. Find the length of $\overset{\frown}{AB}$. Round your answer to the nearest hundredth. *(EXAMPLE 2)*

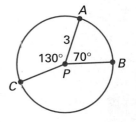

PARAGRAPH PROOF

10. Write a paragraph proof of the Chords and Arcs Theorem for the case in which the congruent chords are in different congruent circles. Use the diagram below to begin your proof. *(ACTIVITY 2 AND THEOREM 9.1.5)*

Practice and Apply

Use the figure of ⊙P below for Exercises 11–22.

11. Name the center of the circle.

12. Name a radius of the circle.

13. Name a chord of the circle.

14. Name a diameter of the circle.

15. Name a central angle of the circle.

16. Name a semicircle of the circle.

17. Name two minor arcs of the circle.

18. Name two major arcs of the circle.

Identify the given part of ⊙P.

19. $\overline{AP}$ **20.** $\overline{AC}$ **21.** $\overline{ED}$ **22.** $\angle APB$

Find the degree measure of each arc by using the central angle measures given in ⊙Q at right.

23. $\overset{\frown}{TU}$ **24.** $\overset{\frown}{TSU}$ **25.** $\overset{\frown}{RT}$

26. $\overset{\frown}{UR}$ **27.** $\overset{\frown}{VS}$ **28.** $\overset{\frown}{US}$

29. $\overset{\frown}{SUV}$ **30.** $\overset{\frown}{VTR}$

Determine the length of an arc with the given central angle measure, m∠P, in a circle with the given radius, r. Round your answer to the nearest hundredth.

31. m∠P = 90°; r = 10   **32.** m∠P = 60°; r = 3   **33.** m∠P = 30°; r = 120

Algebra

Determine the degree measure of an arc with the given length, L, in a circle with the given radius, r.

34. L = 14; r = 70   **35.** L = 20; r = 100

36. L = 3; r = 15   **37.** L = 5; r = 25

38. In ⊙R at right, if m∠ARB = 43° and AB = 5, find XY. Explain your reasoning.

Suppose that ABCDE is a regular pentagon inscribed in ⊙Q and that AQ = 2. Find the following:

39. m∠AQB   **40.** m$\overset{\frown}{AE}$ **41.** m$\overset{\frown}{ACE}$

42. length of $\overset{\frown}{AE}$ **43.** length of $\overset{\frown}{ACE}$

PROOF

44. Complete the converse of the Chords and Arcs Theorem below and prove your result.

The Converse of the Chords and Arcs Theorem

In a circle or in congruent circles, the chords of congruent arcs are ___?___ .

9.1.6

---

45. LANDSCAPING Sixty tulips are planted around the base of a circular fountain. If the fountain is 20 ft in diameter and the tulips are placed 1 ft away from the base, what is the length of the arc between consecutive tulips?

46. DEMOGRAPHICS Smith High School has 450 freshmen, 375 sophomores, 400 juniors, and 325 seniors. Create a pie chart that shows the distribution of students. First find the percent of the total student body for each class. Then multiply each percent by 360° to find the measure of the central angle for each section of the chart.

47. CIVIL ENGINEERING A cloverleaf intersection consists of four ramps that are roughly circular. Use the measurements and objects in the photo to estimate the length of the section of road indicated by the dashed arc.

To estimate, use 1 car ≈ 16 feet.

Look Back

48. What effect does doubling the radius of a sphere have on its volume? **(LESSONS 7.1 AND 7.6)**

SPORTS Find the volume of each object. **(LESSONS 7.2 AND 7.6)**

49.

1.45 in.

50.

$\ell = 59$ m
$w = 25$ m
$h = 2$ m

SPORTS A bicycle tire has a diameter of 26 in. **(LESSON 5.3)**

51. Find the circumference of the tire.

52. If the bicycle travels so that the wheel makes 100 complete turns, how far does the bicycle travel?

Look Beyond

CONNECTION

TRIGONOMETRY The circle with its center at the origin and a radius of 1 is called the unit circle. You will study functions defined on the unit circle in trigonometry.

53. What are the coordinates of A, B, C, and D on the unit circle shown at right?

54. What is the circumference of the unit circle?

Use the graph of the unit circle to find the following:

55. $m\overarc{AB}$ and length of $\overarc{AB}$

56. $m\overarc{ABC}$ and length of $\overarc{ABC}$

57. $m\overarc{ABD}$ and length of $\overarc{ABD}$

PORTFOLIO ACTIVITY

CONSTRUCTING A CIRCLE FLOWER A circle flower is formed when three or more circles meet at the center of another circle, forming "petals." A six-petal circle flower has a construction that is similar to that of the regular hexagon in Activity 1.

1. First draw a circle with a compass or geometry graphics software. Choose a point on the circle and label it A. Draw another circle congruent to the first with point A as its center.

2. Choose one of the points where circle A intersects the original circle and label it B. Using this point as a center, draw another circle congruent to the first. Repeat, going around the circle in one direction, until you have completed the flower.

3. How is the flower related to the hexagon in Activity 1? Think about why the petals meet at the center of the flower. You may wish to consider the measures of the arcs and central angles in the figure.

WORKING ON THE CHAPTER PROJECT

You should now be able to complete Activity 1 of the Chapter Project.

Many other interesting circle patterns are possible. This "quilt" was assembled by a computer from a pattern on page 223 of this book.

Tangents to Circles

Shuttle's orbit

Momentum of shuttle

Pull of
the Earth

Why *The momentum of an object revolving around Earth in a circular orbit is in the direction of a line tangent to the orbit of the object. You will encounter such lines frequently, both in the sciences and the arts.*

Objectives

- Define *tangents* and *secants* of circles.

- Understand the relationship between tangents and certain radii of circles.

- Understand the geometry of a radius perpendicular to a chord of a circle.

This photograph was taken by an astronaut during a space walk. The space shuttle is orbiting Earth from right to left. If it were not for the gravitational pull of the Earth, the shuttle would continue in a straight line in the direction of its momentum. It would quite literally "go off on a tangent."

Secants and Tangents

A line in the plane of a circle may or may not intersect the circle. There are three possibilities.

Secant line

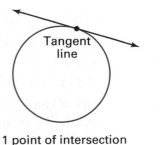

Tangent line

2 points of intersection   1 point of intersection   0 points of intersection

Secants and Tangents

A **secant** to a circle is a line that intersects the circle at two points. A **tangent** is a line in the plane of the circle that intersects the circle at exactly one point, which is known as the **point of tangency**.   **9.2.1**

CRITICAL THINKING   The word *tangent* comes from the Latin word meaning "to touch." The word *secant* comes from the Latin word meaning "to cut." Why are these words appropriate names for the lines they describe?

Activity 1

Radii and Tangents

1. Draw ⊙*P* with radius $\overline{PQ}$.

2. Locate a point *R* on the circle and draw $\overleftrightarrow{QR}$. Measure ∠*PQR*.

3. Repeat Step 2 with point *R* closer to point *Q*, but still on the circle. If you are using geometry graphics software, drag point *R* toward point *Q*. What do you observe about m∠*PQR* as point *R* moves closer to *Q*? What happens when *R* coincides with *Q*?

CHECKPOINT ✔

4. Make a conjecture about the relationship between a tangent to a circle and a radius drawn to the point of tangency. Based on your observations, complete the theorem below, which you will be asked to prove in Exercises 31–34.

Tangent Theorem

If a line is tangent to a circle, then the line is _____?_____ to a radius of the circle drawn to the point of tangency.   **9.2.2**

Activity 2

Radii Perpendicular to Chords, Part 1

1. Draw ⊙*P* with chord $\overline{AB}$.

2. Construct a radius perpendicular to chord $\overline{AB}$. Label the point of intersection *X*.

3. Measure $\overline{AX}$ and $\overline{BX}$. What do you observe?

4. Repeat Steps 1–3 with different circles and chords. If you are using geometry graphics software, experiment by changing the size of the circle and by dragging the chords to different locations. Make a conjecture about radii that are perpendicular to chords in circles. Based on your conjecture, complete the theorem below.

CHECKPOINT ✔

Radius and Chord Theorem

A radius that is perpendicular to a chord of a circle _____?_____ the chord.   **9.2.3**

TWO-COLUMN PROOF

5. Draw $\overline{PA}$ and $\overline{PB}$ in one of your circles. Using this diagram, write a two-column proof of the Radius and Chord Theorem.

⊙P has a radius of 5 in. and PX is 3 in. $\overline{PR}$ is perpendicular to $\overline{AB}$ at point X. Find AB.

Algebra

SOLUTION

By the Pythagorean Theorem:

$$(AX)^2 + 3^2 = 5^2$$
$$(AX)^2 = 5^2 - 3^2$$
$$(AX)^2 = 16$$
$$AX = 4$$

By the Radius and Chord Theorem, $\overline{PR}$ bisects $\overline{AB}$, so $BX = AX = 4$. Therefore, $AB = AX + BX = 4 + 4 = 8$.

Activity 3

Radii Perpendicular to Chords, Part 2

YOU WILL NEED

compass, ruler, and protractor

OR

geometry graphics software

1. Explain how this diagram is like the one in Activity 2.

2. Imagine moving point X to point R. What happens to intersection points A and B as X gets closer to R? What happens when X touches R? What conjecture does this suggest? Use your answer to complete the theorem below, which is proved on the following page.

CHECKPOINT ✔

Converse of the Tangent Theorem

If a line is perpendicular to a radius of a circle at its endpoint on the circle, then the line is ___?___ to the circle.   **9.2.4**

CRITICAL THINKING

The proof on the following page uses the fact that the hypotenuse is the longest side of a right triangle. How could you argue that this must be true?

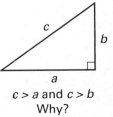

$c > a$ and $c > b$
Why?

The Converse of the Tangent Theorem

The following proof uses the definition of a circle in an interesting way. When you understand the proof, you should be able to summarize it quickly in your own words (see Exercise 18).

PARAGRAPH PROOF

Given: Point P is on $\odot O$, and $\overline{OP}$ is perpendicular to $\overleftrightarrow{AB}$.

Prove: $\overline{AB}$ is tangent to $\odot O$ at point P.

Proof: Choose any point on $\overleftrightarrow{AB}$ other than point P and label it Q. Draw right triangle OPQ. Since $\overline{OQ}$ is the hypotenuse of a right triangle, it is longer than $\overline{PO}$, which is a radius of the circle. Therefore, point Q does not lie on the circle. This is true for all points on $\overleftrightarrow{AB}$ except point P, so $\overleftrightarrow{AB}$ touches the circle at just one point. By definition, $\overleftrightarrow{AB}$ is tangent to $\odot O$ at point P.

Exercises

Communicate

1. Explain the three possible relationships between a line and a circle in a plane.

2. Explain how a secant intersects a circle.

3. How many lines are tangent to a circle? Explain.

4. How many lines are tangent to a circle at a given point? Explain.

5. Describe a point of tangency in the photo at right.

Guided Skills Practice

6. $\overleftrightarrow{KL}$ is tangent to $\odot M$ at K. If $KM = 1$ and $LM = 2$, find KL.
   (ACTIVITY 2 AND THEOREM 9.2.2)

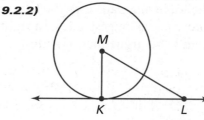

7. In $\odot P$, $QR = 3$. Find RS.
   (ACTIVITY 2 AND THEOREM 9.2.3)

8. ⊙X has a radius of 13, XW = 5, and $\overline{XV} \perp \overline{YZ}$. Find YZ. *(EXAMPLE AND THEOREM 9.2.3)*

9. Verify that $\overleftrightarrow{AB}$ is tangent to ⊙C at B. *(ACTIVITY 3 AND THEOREM 9.2.4)*

Practice and Apply

For Exercises 10–12, refer to ⊙R, in which $\overline{RY} \perp \overline{XZ}$ at W.

10. $\overline{XW} \cong$ ___?___

11. If RY = 7 and RW = 2, what is XW? What is WZ?

12. If RY = 3 and RW = 2, what is XW? What is WZ?

13. In the diagram at right, what is the radius of ⊙O?

14. In ⊙N, verify that line ℓ is a tangent by using the Converse of the Tangent Theorem.

PARAGRAPH PROOF

15. In Activity 2 you proved a theorem about a radius that is perpendicular to a chord. Write a paragraph proof of the following related theorem:

Theorem
The perpendicular bisector of a chord passes through the center of the circle.   **9.2.5**

CONSTRUCTION

16. Use Theorem 9.2.5 above to construct a circle through any three noncollinear points. First draw three points not on a straight line. Label them A, B, and C. Draw $\overline{AB}$ and $\overline{BC}$. Construct the perpendicular bisector of each segment. Where is the center of the circle that contains A, B, and C? Complete the construction. How does this construction relate to Activity 2 in Lesson 1.5?

17. Use the Converse of the Tangent Theorem to construct a tangent to a circle at a given point. First draw a circle and label the center P. Choose any point on the circle and label it A. Draw $\overline{AP}$. How is the tangent line at A related to $\overline{AP}$? Complete the construction.

18. Summarize the proof of the Converse of the Tangent Theorem from page 576 in your own words.

Algebra

Use the diagram of ⊙B to find the indicated lengths. Line *r* is tangent to ⊙B at *T*, *BT* = 2, *BS* = 1, and *WT* = 5. Round your answers to the nearest hundredth.

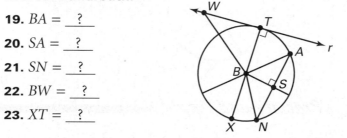

19. $BA =$ ___?___

20. $SA =$ ___?___

21. $SN =$ ___?___

22. $BW =$ ___?___

23. $XT =$ ___?___

APPLICATIONS

24. COMMUNICATIONS A radio station installs a VHF radio tower that stands 1500 ft tall. What is the maximum effective signal range of the tower? The diagram suggests a way to use tangents to solve the problem. Use the Pythagorean Theorem to find *d*. (The diameter of Earth is approximately 8000 mi.)

25. SPACE FLIGHT The space shuttle orbits at 155 mi above Earth. How far is it from the shuttle to the horizon? (The diameter of Earth is approximately 8000 mi.)

Space shuttle

155 mi

Orbit

26. DESIGN An artist wants to draw the largest circle that will fit into a square. She uses the following method:

Draw square *ABCD*. Connect the midpoints of sides $\overline{AB}$ and $\overline{CD}$ with a segment. Connect the midpoints of sides $\overline{AD}$ and $\overline{BC}$ with another segment.

Construct the desired circle. How can you prove that this is the largest circle that will fit in the square? How can you prove that no part of the circle lies outside the square?

Look Back

Algebra

27. A triangle has a perimeter of 24 cm and an area of 24 cm². What are the perimeter and area of a larger similar triangle if the scale factor is $\frac{2}{1}$? **(LESSONS 8.1 AND 8.6)**

28. A rectangle has a perimeter of 22 ft and an area of 22 ft². What are the perimeter and area of a larger similar rectangle if the scale factor is $\frac{8}{3}$? **(LESSONS 8.1 AND 8.6)**

29. A rectangular prism has dimensions $\ell = 12$ in., $w = 8$ in., and $h = 15$ in. What is the volume of a larger similar rectangular prism if the scale factor is $\frac{5}{3}$? *(LESSONS 8.1 AND 8.6)*

APPLICATION

30. ENGINEERING A cylindrical water tower has a radius of 30 ft and a height of 100 ft. What is the volume of a larger similar water tower if the scale factor is $\frac{7}{5}$? *(LESSONS 8.1 AND 8.6)*

Look Beyond

CHALLENGE

PARAGRAPH PROOF

Answer the questions below to prove the Tangent Theorem.

31. Suppose that the Tangent Theorem is *false.* That is, suppose that line m is tangent to $\odot O$ at point A, but that line m is *not* perpendicular to $\overline{OA}$. If this is true, then there *is* some segment with endpoint O, different from $\overline{OA}$, that is perpendicular to line m. Call that segment $\overline{OB}$. Then $\triangle OBA$ is a right triangle. What is the hypotenuse of $\triangle OBA$? Which segment is longer, $\overline{OA}$ or $\overline{OB}$?

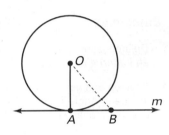

32. Point B must be in the exterior of the circle because m is a tangent line. What does this imply about the relative lengths of $\overline{OA}$ and $\overline{OB}$? Explain.

33. Compare your answers to Exercises 31 and 32. What do you observe?

34. If an assumption leads to a contradiction, it must be rejected. This is the basis for a type of proof known as an *indirect proof* or a *proof by contradiction.* Explain how the argument above leads to the desired conclusion.

CONSTRUCTING SMOOTH CURVES A curve may be made up of arcs of more than one circle. In order for arcs from two different circles to join smoothly at a point, they must have the same tangent at that point.

1. Using a compass and straightedge or geometry graphics software, draw $\odot P$ with radius $\overline{PR}$. Construct line ℓ tangent to $\odot P$ at R; that is, construct a line perpendicular to $\overline{PR}$ at R (refer to Exercise 17). For another circle to have the same tangent, ℓ, its center must be on the line $\overleftrightarrow{PR}$. Why?

2. Choose a point on $\overline{PR}$ that is not P or R and label it Q. Construct a circle centered at Q with radius $\overline{QR}$. $\odot P$ and $\odot Q$ will have the same tangent at R.

3. Try tracing part of your construction in a different color. Starting on $\odot P$ near R, trace until you get to R, and then continue tracing on $\odot Q$. The curves should join smoothly at R.

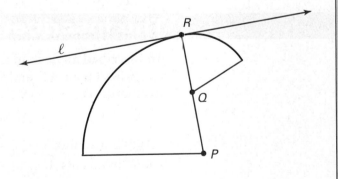

WORKING ON THE CHAPTER PROJECT

You should now be able to complete Activity 2 of the Chapter Project.

Inscribed Angles and Arcs

Objectives

- Define *inscribed angle* and *intercepted arc*.

- Develop and use the Inscribed Angle Theorem and its corollaries.

Why *You can find the center of a circular object, such as the tabletop shown here, by using a carpenter's square. The process involves inscribed angles.*

APPLICATION

CARPENTRY

A carpenter's square is used both for creating right angles in structures and for testing whether angles in structures are right angles. It can also be used to find the center of a circle. The method for finding the center of a circle relies on an important geometry corollary (see the application on page 584).

The Inscribed Angle Theorem

An **inscribed angle** is an angle whose vertex lies on a circle and whose sides are chords of the circle.

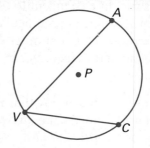

$\angle AVC$ is inscribed in $\odot P$.
$\angle AVC$ intercepts $\overset{\frown}{AC}$.

If you know the measure of the inscribed angle, you can determine the measure of its intercepted arc. How do you think this can be done? In the activities on the following pages, you will discover the answer.

CRITICAL THINKING   In the illustration above, what is $m\overset{\frown}{AVC} + m\overset{\frown}{AC}$? Would a similar result be true for all possible inscribed angles in a circle? Explain.

Activity 1

Angles and Intercepted Arcs: A Conjecture

YOU WILL NEED

compass, straightedge, and protractor

OR

geometry graphics software

1. Draw three different figures in which inscribed angle ∠AVC intercepts an arc of the circle. Include one minor arc, one major arc, and one semicircle.

2. Measure the inscribed angle and the intercepted arc in each figure. (You will need to draw central angles to determine the measures of the arcs.)

3. Compare the measures of the inscribed angle and its intercepted arc in each case.

CHECKPOINT ✔

4. Make a conjecture about the relationship between the measure of an inscribed angle and the measure of its intercepted arc.

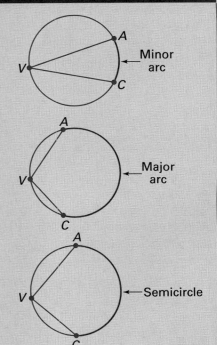

Activity 2

Proving the Conjecture

YOU WILL NEED

no special tools

To prove your conjecture from Activity 1, you will need to consider three separate cases.

Part I

1. In the figure at right, one side of the inscribed angle contains the center of the circle. What is the relationship between m∠1 and m∠2?

2. Notice that ∠3 is an exterior angle of △AVP. What is the relationship among m∠3, m∠1, and m∠2?

The center of the circle is on one side of the angle.

TABLE PROOF

3. Complete the table below. For each entry in the last row, give a reason.

m∠1	m∠2	m∠3	m$\widehat{AC}$
20°	?	?	?
30°	?	?	?
40°	?	?	?
x°	?	?	?

As you justify each entry of the table for the general case, you are also proving a theorem.

CHECKPOINT ✔

4. What does your table show about the relationship between m∠1 and m$\widehat{AC}$?

Part II

1. In the figure at right, the center of the circle is in the interior of the inscribed angle. What is the relationship between m∠1 and m$\overset{\frown}{AX}$? between m∠4 and m$\overset{\frown}{XC}$?

The center of the circle is in the interior of the angle.

2. Complete the table below. For each entry in the last row of the table, give a reason.

TABLE PROOF

> As you justify each entry of the table for the general case, you are also proving a theorem.

m∠1	m$\overset{\frown}{AX}$	m∠4	m$\overset{\frown}{CX}$	m∠AVC	m$\overset{\frown}{AXC}$
20°	?	20°	?	?	?
30°	?	20°	?	?	?
40°	?	50°	?	?	?
x°	?	y°	?	?	?

CHECKPOINT ✔

3. What does your table show about the relationship between m∠AVC and m$\overset{\frown}{AXC}$?

Part III

One more case remains to be proved. In this case, the center of the circle is in the *exterior* of the inscribed angle. Draw your own figure for this case and prove your conjecture for it. (You may want to make a table.)

The center of the circle is in the exterior of the angle.

Create your own figure.

Conclusion

CHECKPOINT ✔

Complete the theorem below, which your work in this Activity has proven.

Inscribed Angle Theorem

The measure of an angle inscribed in a circle is equal to ____?____ the measure of the intercepted arc.

9.3.1

E X A M P L E ❶ Find the measure of ∠XVY.

● **SOLUTION**

∠XVY is inscribed in ⊙P and intercepts $\overset{\frown}{XY}$. By the Inscribed Angle Theorem:

$$m\angle XYZ = \tfrac{1}{2}m\overset{\frown}{XY} = \tfrac{1}{2}(45°) = 22\tfrac{1}{2}°$$

Two Results of the Inscribed Angle Theorem

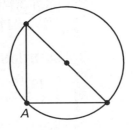

In the figure at right, $\angle A$ is inscribed in a semicircle. Therefore, the arc it intercepts is also a semicircle. Thus, the measure of the intercepted arc is 180° (why?), and the measure of $\angle A$ is $\frac{1}{2} \times 180°$, or 90°.

This calculation illustrates the following important corollary to the Inscribed Angle Theorem:

Right-Angle Corollary

If an inscribed angle intercepts a semicircle, then the angle is a right angle.   **9.3.2**

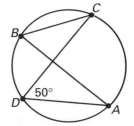

In the figure at right, the measure of $\overset{\frown}{AC}$ is twice the measure of $\angle D$, or $2 \times 50° = 100°$. The measure of $\angle B$ is one-half the measure of $\overset{\frown}{AC}$, or $\frac{1}{2} \times 100° = 50°$.

These calculations illustrate an important principle. Notice that $\angle D$ and $\angle B$ intercept the same arc. As you found, they have the same angle measure. This leads to another corollary of the Inscribed Angle Theorem.

Arc-Intercept Corollary

If two inscribed angles intercept the same arc, then they have the same measure.   **9.3.3**

E X A M P L E ❷

A person's effective field of vision is about 30°. In the diagram of the amphitheater, a person sitting at point A can see the entire stage. What is the measure of $\angle B$? Can the person sitting at point B view the entire stage?

SOLUTION

Angles A and B intercept the same arc. By Theorem 9.3.4, the angles must have the same measure, so $m\angle A = m\angle B = 30°$. The person sitting at point B can view the entire stage.

EXAMPLE ③ In ⊙*P* with diameter $\overline{AB}$, m$\widehat{CB}$ = 110°, and m$\widehat{BD}$ = 130°. Find the measures of ∠1, ∠2, ∠3, ∠4, ∠*APC*, ∠*ADB*, and ∠*CAD*.

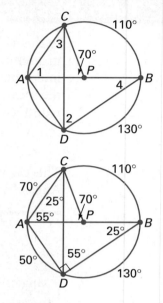

● **SOLUTION**

Copy the figure and add information as you work through the solution.

Begin by labeling as many arc measures as you can. Since $\overline{AB}$ is a diameter, $\widehat{ACB}$ and $\widehat{ADB}$ have a measure of 180°. Therefore, m$\widehat{AC}$ = 70° and m$\widehat{AD}$ = 50°.

∠1 and ∠2 intercept $\widehat{CB}$. Thus, m∠1 = m∠2 = $\frac{1}{2}$(110) = 55°.

∠3 and ∠4 intercept $\widehat{AD}$. Thus, m∠3 = m∠4 = $\frac{1}{2}$(50) = 25°.

∠*APC* is a central angle. Thus, m∠*APC* = m$\widehat{AC}$ = 70°.

∠*ADB* is inscribed in a semicircle. Thus, m∠*ADB* = 90°.

∠*CAD* intercepts $\widehat{CBD}$. Thus, m∠*CAD* = $\frac{1}{2}$(110 + 130) = 120°.

APPLICATION

CARPENTRY

A carpenter needs to find the center of a small tabletop. How can she do this by using a carpenter's square?

The carpenter inscribes two right angles, ∠*ABC* and ∠*DEF*, in the circle. She then draws the hypotenuses of the triangles by connecting points *A* and *C* and points *D* and *F*. The intersection of $\overline{AC}$ and $\overline{DF}$ is the center of the table.

You will be asked to perform a similar construction in Exercise 49.

Exercises

● *Communicate*

Refer to ⊙O in the photo for Exercises 1–5.

1. Name an inscribed angle in ⊙O.

2. Explain how to find m$\widehat{BD}$.

3. Explain how to find m∠1.

4. Explain how to find m$\widehat{AC}$.

5. Explain how to find m∠2.

6. Explain why two inscribed angles that intercept the same arc have the same measure.

● *Guided Skills Practice*

APPLICATION

PHOTOGRAPHY A person standing at the edge of a circular gallery takes a photograph. Suppose that $\overset{\frown}{OP}$ represents everything that is included in the photograph and that m$\overset{\frown}{OP}$ = 70°.

7. What is m∠OMP, known as the *picture angle* of the camera? *(ACTIVITIES 1 AND 2, THEOREM 9.3.2, AND EXAMPLE 1)*

8. What is m∠ONP? *(THEOREM 9.3.4 AND EXAMPLE 2)*

In ⊙F, $\overline{HK}$ is a diameter, m$\widehat{GH}$ = 50°, and m$\widehat{JK}$ = 90°. (THEOREM 9.3.1 AND EXAMPLE 3)

9. Find the measures of ∠GFH, ∠GKH, ∠JFK, and ∠KHJ.

10. Find the measure of ∠KGH.

● *Practice and Apply*

In ⊙W, m$\widehat{XZ}$ = 60°, m∠VYZ = 40°, and $\overleftrightarrow{YZ}$ is a diameter. Find the following:

11. m∠XYW

12. m∠WXY

13. m∠XWY

14. m∠XWZ

15. m$\widehat{YXZ}$

16. m∠YVZ

17. m$\widehat{XY}$

18. m$\widehat{VZ}$

19. m$\widehat{VY}$

20. m∠VZY

For Exercises 21–28, refer to ⊙P with diameter $\overline{AC}$. Find the following:

21. m∠A

22. m∠B

23. m∠BCA

24. m$\overparen{AB}$

25. m∠PCD

26. m∠CPD

27. m$\overparen{DC}$

28. m$\overparen{AD}$

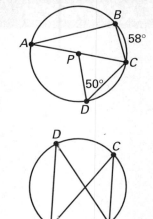

For Exercises 29–32, refer to the circle below.

29. m$\overparen{AB}$ = 68°   m∠C = ___?___   m∠D = ___?___

30. m∠D = 30°   m$\overparen{AB}$ = ___?___   m∠C = ___?___

31. m$\overparen{CD}$ = 87°   m∠B = ___?___   m∠A = ___?___

32. m∠B = a°   m$\overparen{CD}$ = ___?___   m∠A = ___?___

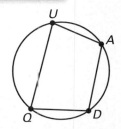

TABLE PROOF

Quadrilateral *QUAD* is inscribed in a circle, as shown below. Copy and complete the following table. For each entry in the last row, give a reason.

As you justify each entry of the table for the general case, you are also proving a theorem.

m∠U	m$\overparen{QDA}$	m$\overparen{QUA}$	m∠D	m∠U + m∠D	m∠Q + m∠A
100°	**33.** ?	**34.** ?	**35.** ?	**36.** ?	**37.** ?
38. ?	160°	**39.** ?	**40.** ?	**41.** ?	**42.** ?
x°	**43.** ?	**44.** ?	**45.** ?	**46.** ?	**47.** ?

48. Based on the information in the table, state a theorem about the angles of a quadrilateral that is inscribed in a circle.

APPLICATIONS

In Exercise 49, you will use a piece of paper with right-angle corners to model the use of a carpenter's square to find the center of a circular object, as described on page 584.

49. CARPENTRY Draw a large circle on a sheet of paper. Place one right-angle corner of a sheet of paper on the circle so that both sides of the angle cross the circle. Mark the three points where the right angle at the corner of the paper touches the circle, and connect them to form a triangle. Repeat these steps to form another triangle in the circle. Identify the point where the longest sides of your triangles intersect. Explain why this point is the center of the circle.

50. STAINED GLASS An artist is creating a circular stained-glass window with the design shown at right. The artist wants the arc intercepted by ∠A and ∠B to measure 80°. What should the measures be of ∠A and ∠B?

Center of circle

FLOWCHART PROOF

For Exercises 51–61, complete the flowchart proof below.

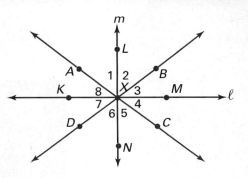

Given: Line ℓ is the angle bisector of $\angle BXC$, and line m is the angle bisector of $\angle CXD$.

Prove: Line ℓ and m are perpendicular.

Proof:

$\angle 3 \cong \angle 4$
$(m\angle 3 = m\angle 4)$

ℓ is the angle bisector of $\angle BXC$.

$\angle 6 \cong \angle 5$
$(m\angle 6 = m\angle 5)$

51. ?

$\angle 2 \cong \angle 6$
$(m\angle 2 = m\angle 6)$

52. ?

53. ?

Transitive Property of Equality

$m\angle 2 + m\angle 3 = m\angle 4 + m\angle 5$

54. ?

$m\angle 2 + m\angle 3 = m\angle MXL$
$m\angle 4 + m\angle 5 = m\angle MXN$

55. ?

$m\angle MXL = m\angle MXN$

56. ?

$m\angle MXL + m\angle MXN = 180°$

57. ?

58. ?

59. ?

60. ?

61. ?

Look Beyond

APPLICATION

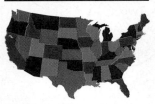

This map of the United States uses only four colors.

CARTOGRAPHY A famous theorem in mathematics is related to mapmaking. The *four-color theorem* states that any map in a plane can be colored with a maximum of four colors so that no two adjacent areas are the same color. The first "proof" of this theorem, by Appel and Haken in 1976, was controversial because parts of the proof used a computer and could not be verified by hand.

62. Explain why the "map" below cannot be colored with fewer than four colors.

63. Make up your own map and try to color it with only four colors.

Angles Formed by Secants and Tangents

Objectives

- Define angles formed by secants and tangents of circles.

- Develop and use theorems about measures of arcs intercepted by these angles.

Why *The principles of circle geometry are used by navigators at sea. For example, the "horizontal angle of danger" enables a navigator to stay a safe distance from a dangerous region.*

A navigator on board a ship can measure the angle between two lines of sight to lighthouses on the coastline. This measurement tells the navigator whether the ship is inside or outside a *circle of danger* that contains hazardous rocks or shallows. The technique is based on secant lines that intersect the circle of danger.

Classification of Angles With Circles

Angles formed by pairs of lines that intersect a circle in two or more places can be studied systematically. There are three cases to consider, according to the placement of the vertex of the angles.

Case 1: Vertex is on the circle.   **Case 2: Vertex is inside the circle.**

a. Secant and tangent   **b.** Two secants   Two secants

Case 3: Vertex is outside the circle.

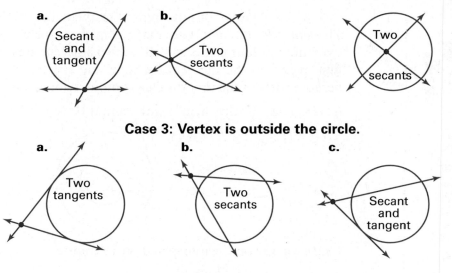

a. Two tangents   **b.** Two secants   **c.** Secant and tangent

Vertex on Circle—Secant and Tangent (Case 1a)

In this Activity, you will examine three configurations of secant-tangent angles.

1. The secant-tangent angle is a right angle. (The secant contains the center of the circle.)

$$m\angle AVC = \underline{\quad ? \quad} \qquad m\widehat{AV} = \underline{\quad ? \quad}$$

CHECKPOINT ✔

How does this relationship compare with the one between an inscribed angle and its intercepted arc?

TABLE PROOFS

2. The secant-tangent angle is acute. Copy and complete the following table:

m$\widehat{AV}$	m∠1	m∠2	m∠PVC	m∠AVC
120°	120°	30°	?	60°
100°	?	?	?	?
80°	?	?	?	?
x°	?	?	?	?

> *As you justify each entry of the table for the general case, you are also proving a theorem.*

CHECKPOINT ✔

Complete the following statement:

The measure of an acute secant-tangent angle with its vertex on a circle is __?__ the measure of its intercepted arc.

3. The secant-tangent angle is obtuse. Copy and complete the following table:

m$\widehat{AXV}$	m∠1	m∠2	m∠PVC	m∠AVC
200°	160°	10°	?	100°
220°	?	?	?	?
240°	?	?	?	?
x°	?	?	?	?

CHECKPOINT ✔

Complete the following statement:

The measure of an obtuse secant-tangent angle with its vertex on a circle is __?__ the measure of its intercepted arc.

CHECKPOINT ✔

4. Based on your results, complete the following theorem:

Theorem

If a tangent and a secant (or a chord) intersect on a circle at the point of tangency, then the measure of the angle formed is __?__ the measure of its intercepted arc.   **9.4.1**

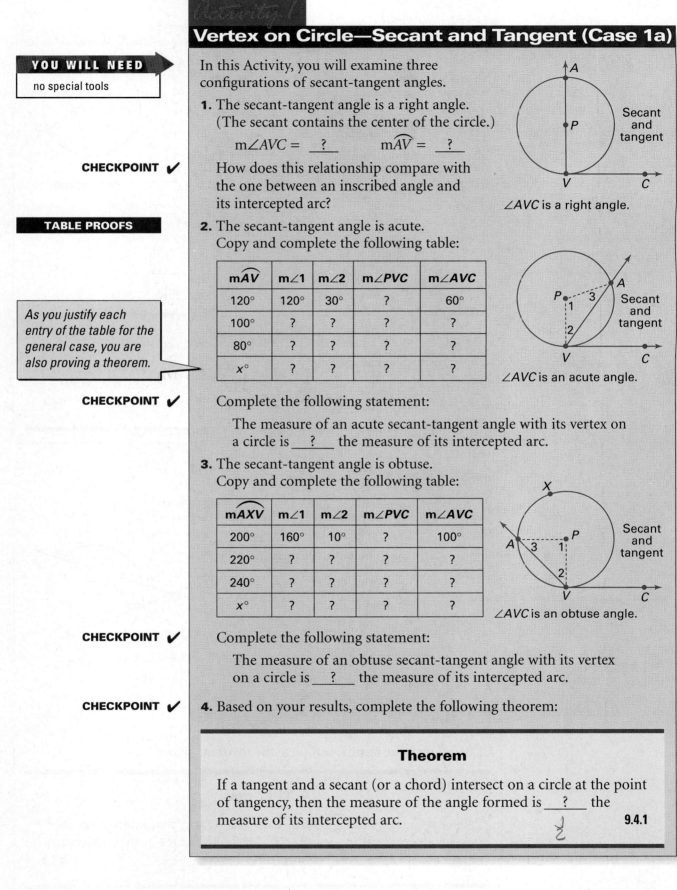

∠AVC is a right angle.

Secant and tangent

∠AVC is an acute angle.

Secant and tangent

∠AVC is an obtuse angle.

Secant and tangent

CRITICAL THINKING   Case 1b (vertex on circle, two secants) has already been studied in this book. Where? What theorem corresponds to this case?

Vertex Inside Circle—Two Secants (Case 2)

TABLE PROOF

1. ∠AVC is an exterior angle of △ADV. What is the relationship between the measure of ∠AVC and the measures of ∠1 and ∠2?

2. Copy and complete the following table:

> As you justify each entry of the table for the general case, you are also proving a theorem.

m$\widehat{AC}$	m$\widehat{BD}$	m∠1	m∠2	m∠AVC	m∠DVB
160°	40°	80°	20°	100°	100°
180°	70°	?	?	?	?
200°	60°	?	?	?	?
$x_1°$	$x_2°$	?	?	?	?

Two secants

3. Based on your results, complete the theorem below.

Theorem

CHECKPOINT ✔

The measure of an angle formed by two secants or chords that intersect in the interior of a circle is $\frac{1}{2}$? the ? *sum* of the measures of the arcs intercepted by the angle and its vertical angle. **9.4.2**

Vertex Outside Circle—Two Secants (Case 3b)

1. ∠1 is an exterior angle of △BVC. What is the relationship between the measure of ∠1 and the measures of ∠2 and ∠AVC?

TABLE PROOF

2. Copy and complete the following table:

> As you justify each entry of the table for the general case, you are also proving a theorem.

m$\widehat{BD}$	m$\widehat{AC}$	m∠1	m∠2	m∠AVC
200°	40°	100°	20°	80°
250°	60°	?	?	?
100°	50°	?	?	?
$x_1°$	$x_2°$	?	?	?

Two secants

3. Based on your results, complete the theorem below.

Theorem

CHECKPOINT ✔

The measure of an angle formed by two secants that intersect in the exterior of a circle is ? the ? of the measures of the intercepted arcs. **9.4.3**

$\frac{1}{2}$ *difference*

You will explore cases 3a and 3c in Exercises 27–37.

The illustration below shows a ship at point *F* and two lighthouses at points *A* and *B*. The circle encloses a region of dangerous rocks. The measure of an angle inscribed in $\overset{\frown}{AGB}$, such as ∠*C*, is known as the horizontal angle of danger. Navigation charts contain information about horizontal angles of danger for different regions.

The ship's navigator measures ∠*F*, the angle formed by his lines of sight to the two lighthouses. He knows that he is outside the circle of danger because m∠*F* is less than the horizontal angle of danger for the area. Why does this method work?

Notice that ∠*C* is an inscribed angle and that it intercepts $\overset{\frown}{AB}$. Thus, if the ship were on the circle, ∠*F* would equal ∠*C*. Therefore, it should be obvious that if m∠*F* is less than m∠*C*, the ship is outside the circle.

CRITICAL THINKING

How can you use theorems from this lesson to prove the "obvious" fact mentioned above? How can you prove that if m∠*F* were greater m∠*C*, then the ship would be *inside* the circle?

EXAMPLE **1** Find m∠*AVC* in each figure.

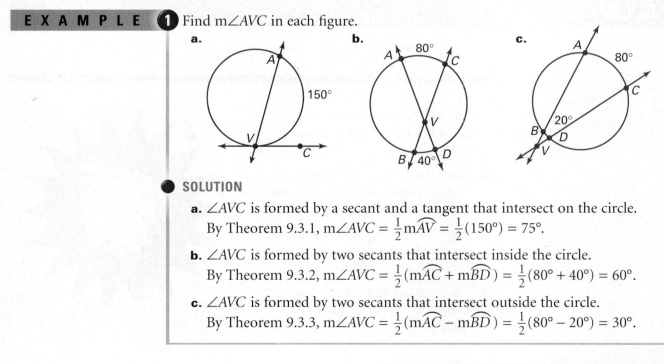

SOLUTION

a. ∠*AVC* is formed by a secant and a tangent that intersect on the circle. By Theorem 9.3.1, m∠*AVC* = $\frac{1}{2}$m$\overset{\frown}{AV}$ = $\frac{1}{2}$(150°) = 75°.

b. ∠*AVC* is formed by two secants that intersect inside the circle. By Theorem 9.3.2, m∠*AVC* = $\frac{1}{2}$(m$\overset{\frown}{AC}$ + m$\overset{\frown}{BD}$) = $\frac{1}{2}$(80° + 40°) = 60°.

c. ∠*AVC* is formed by two secants that intersect outside the circle. By Theorem 9.3.3, m∠*AVC* = $\frac{1}{2}$(m$\overset{\frown}{AC}$ − m$\overset{\frown}{BD}$) = $\frac{1}{2}$(80° − 20°) = 30°.

2 **Given:** $\overleftrightarrow{TU}$ is tangent to $\odot P$ at point T.

$m\widehat{QR} = 90°$

$m\widehat{RT} = 150°$

$m\widehat{QS} = 50°$

Find the following:

  a. $m\angle STU$ **b.** $m\angle 1$ **c.** $m\angle 2$

● **SOLUTION**

Make your own sketch of the figure. As you obtain new information, add it to the figure.

  a. Because the vertex of $\angle STU$ is on $\odot P$ and $\angle STU$ is formed by a tangent and a secant line, $m\angle STU = \frac{1}{2}m\widehat{ST}$. To find $m\widehat{ST}$, note that $50° + 90° + 150° + m\widehat{ST} = 360°$. Thus, $m\widehat{ST} = 70°$ and $m\angle STU = \frac{1}{2}(70°) = 35°$.

  b. Because the vertex of $\angle 1$ is in the interior of $\odot P$ and is formed by two intersecting chords, $m\angle 1 = \frac{1}{2}(m\widehat{QR} + m\widehat{ST}) = \frac{1}{2}(90° + 70°) = 80°$.

  c. Because the vertex of $\angle 2$ is outside the circle and is formed by two secants, $m\angle 2 = \frac{1}{2}(m\widehat{RT} - m\widehat{QS}) = \frac{1}{2}(150° - 50°) = 50°$.

● *Communicate*

  1. The measure of an acute secant-tangent angle with its vertex on a circle is ___?___.

  2. The measure of an obtuse secant-tangent angle with its vertex on a circle is ___?___.

  3. The measure of a secant-secant (or chord-chord) angle with its vertex inside a circle is ___?___.

  4. The measure of a secant-secant angle with its vertex outside a circle is ___?___.

In this figure, every chord of the outer circle is tangent to the inner circle.

5. In ⊙Z, m$\widehat{WY}$ = 138°. If $\overleftrightarrow{XY}$ is tangent to ⊙Z at Y, find m∠WYX. *(THEOREM 9.4.1 AND EXAMPLE 1)*

6. In ⊙Q, m$\widehat{MN}$ = 60° and m$\widehat{OP}$ = 150°. Find m∠1. *(THEOREM 9.4.2 AND EXAMPLE 1)*

7. In ⊙A, m$\widehat{BC}$ = 84° and m$\widehat{DE}$ = 40°. Find m∠BFC. *(THEOREM 9.4.3 AND EXAMPLE 1)*

8. $\overleftrightarrow{AB}$ is tangent to ⊙G at B, m$\widehat{BC}$ = 70°, m$\widehat{BF}$ = 100°, and m$\widehat{EF}$ = 130°. Find the following. *(EXAMPLE 2)*
  a. m∠ABC
  b. m∠1
  c. m∠2
  d. m∠D

9. NAVIGATION Lighthouses are located at points A and B on the circle of danger. If a ship is located at point X (not shown) and m∠BXA = 27°, is the ship inside or outside the circle of danger? Explain. *(APPLICATION)*

● *Practice and Apply*

In ⊙X, m$\widehat{WT}$ = 36°, m$\widehat{TU}$ = 148°, and m$\widehat{UY}$ = 70°. $\overleftrightarrow{VY}$ is tangent to ⊙X at Y. Find each of the following:

10. m∠SYV

11. m∠VSY

12. m∠SVY

In the figure, $\overline{AB}$ and $\overline{CD}$ are chords, $m\overset{\frown}{CB} = 60°$, and $m\overset{\frown}{AD} = 110°$. Find each of the following:

13. m∠1   **14.** m∠2

15. m∠3   **16.** m∠4

In the figure, $\overleftrightarrow{VA}$ is tangent to ⊙P at V.

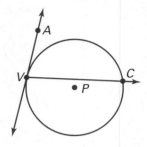

17. If $m\overset{\frown}{VC} = 150°$, find m∠AVC.

18. If m∠AVC = 80°, find $m\overset{\frown}{VC}$.

19. If $m\overset{\frown}{VC} = 2x + 4$, find m∠AVC.

20. If m∠AVC = $3x - 1$, find $m\overset{\frown}{VC}$.

Refer to ⊙O for Exercises 21–26. $\overleftrightarrow{AF}$ is tangent to ⊙O at A, $m\overset{\frown}{CD} = 105°$, $m\overset{\frown}{BC} = 47°$, and $m\overset{\frown}{AB} = m\overset{\frown}{DC}$. Find the following:

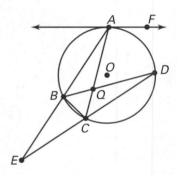

21. $m\overset{\frown}{AB}$ **22.** $m\overset{\frown}{AD}$

23. m∠AED   **24.** m∠CAF

25. m∠CQD   **26.** m∠BQC

In Exercises 27–37, you will explore cases 3a and 3c from page 588.

Case 3c: A secant and a tangent intersect outside the circle.

Refer to ⊙P for Exercises 27–30. $\overleftrightarrow{VC}$ is tangent to ⊙P at C.

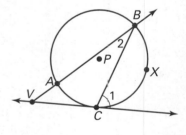

27. What is the relationship between the measure of ∠1 and the measures of ∠2 and ∠AVC?

28. Copy and complete the table. For each entry in the last row, give a reason.

$m\overset{\frown}{BXC}$	$m\overset{\frown}{AC}$	m∠1	m∠2	m∠AVC
250°	60°	**a.** ?	**b.** ?	**c.** ?
200°	40°	**d.** ?	**e.** ?	**f.** ?
130°	40°	**g.** ?	**h.** ?	**i.** ?
70°	30°	**j.** ?	**k.** ?	**l.** ?
$x_1°$	$x_2°$	**m.** ?	**n.** ?	**o.** ?

29. Write an equation that describes m∠AVC in terms of $m\overset{\frown}{BC}$ and $m\overset{\frown}{AC}$.

30. Complete the following theorem:

Theorem

The measure of a secant-tangent angle with its vertex outside the circle is _____?_____.

9.4.4

Case 3a: Two tangents intersect outside the circle.

Refer to ⊙M for Exercises 31–37. $\overleftrightarrow{VA}$ and $\overleftrightarrow{VC}$ are tangent to ⊙M at A and C, respectively.

31. What is the relationship between the measure of ∠1 and the measures of ∠2 and ∠AVC?

32. The measure of ∠1 is half the measure of its intercepted arc. Name the arc.

33. The measure of ∠2 is half the measure of its intercepted arc. Name the arc.

34. Given m$\overarc{AXC}$ = 260°, find m$\overarc{AC}$.

TABLE PROOF

35. Copy and complete the table. For each entry in the last row, give a reason.

m$\overarc{AXC}$	m$\overarc{AC}$	m∠1	m∠2	m∠AVC
300°	a. ?	b. ?	c. ?	d. ?
250°	e. ?	f. ?	g. ?	h. ?
220°	i. ?	j. ?	k. ?	l. ?
200°	m. ?	n. ?	o. ?	p. ?
x°	q. ?	r. ?	s. ?	t. ?

As you justify each entry of the table for the general case, you are also proving a theorem.

36. Write an equation that describes m∠AVC in terms of m$\overarc{AXC}$ and m$\overarc{AC}$.

37. Complete the following theorem:

Theorem

The measure of a tangent-tangent angle with its vertex outside the circle is _____?_____.

9.4.5

Algebra

38. Justify the following statement algebraically by using the diagram of ⊙M above:

$$\frac{m\overarc{AXC} - m\overarc{AC}}{2} = m\overarc{AXC} - 180°$$

(Hint: m$\overarc{AXC}$ + m$\overarc{AC}$ = ___?___)

CHALLENGE

39. Use the statement in Exercise 38 to write an alternative version of Theorem 9.4.5.

In Exercises 27–37, you completed the investigations of angle-arc relationships in circles. In Exercises 40–55, you will summarize the angles studied. Complete the table below.

SUMMARY OF ANGLE-ARC RELATIONSHIPS			
Diagram	Location of vertex	Sides of angle	Formula for m∠AVC
V is the center of the circle.	center of circle	central angle: sides formed by 2 radii	m∠AVC = m⌢AC
	40. ___?___	**41.** ___?___	**42.** ___?___
	43. ___?___	inscribed angle: sides formed by 2 secants or chords	**44.** ___?___
VC tangent at *V*	**45.** ___?___	**46.** ___?___	**47.** ___?___
48. ___?___	exterior of circle	sides formed by 2 tangents	**49.** ___?___
	50. ___?___	**51.** ___?___	**52.** ___?___
VC tangent at *C*	**53.** ___?___	**54.** ___?___	**55.** ___?___

APPLICATION

56. COMMUNICATIONS The maximum distance that a radio signal can reach directly is the length of the segment tangent to the curve of Earth's surface. If the angles formed by the tangent radio signals and the tower are 89.5°, what is the measure of the intercepted arc on Earth? Given that the radius of Earth is approximately 4000 mi, estimate how far the signal can reach.

Look Back

Find the indicated angle measures. *(LESSON 3.6)*

57.

58.

59.

APPLICATIONS

60. ARCHAEOLOGY The Great Pyramid of Giza in Egypt, built by Khufu (*Cheops* in Greek), is considered by some to be the greatest structure ever built. Its roughly square base measures an average of $775\frac{3}{4}$ ft on a side, and its original height, before erosion, was $481\frac{2}{5}$ ft. Find the volume of the original structure. [*Source: Encyclopedia Britannica*] *(LESSON 7.3)*

DESIGN A photocopier was used to reduce △ABC by a factor of 0.75.

61. Are △*ABC* and △*A′B′C′* similar? Explain your reasoning. *(LESSON 8.2)*

62. Find the following ratios: $\frac{A'B'}{AB}$, $\frac{A'C'}{AC}$, and $\frac{C'B'}{CB}$. *(LESSON 8.6)*

63. Find *A′C′*. *(LESSON 8.6)*

Look Beyond

Algebra

A *conic section* is the intersection of a right double cone and a cutting plane. In the diagram below, the intersection of the plane and the cone forms a circle. Other geometric figures are formed when the angle at which the plane cuts the double cone is adjusted.

Describe how the cutting plane should be adjusted in order to produce each conic section listed below. You may use a sketch or a verbal description.

64. ellipse ⬭

65. parabola ∪

66. one point •

67. two intersecting lines ✕

68. hyperbola ⌣⌢

69. one line ⟷

Point of Disaster

Severe Earthquake Hits Los Angeles
Collapsed Freeways Cripple City

Imagine that you are a geologist in the Los Angeles area. You have records from three different stations (A, B, and C) of waves produced by the earthquake. From these seismograms you must pinpoint the epicenter, the place on the earth's surface directly above the origin of the earthquake. To find the epicenter, you will use two types of seismic waves, S waves and P waves. Because S waves and P waves travel at different speeds, you can use the difference in their travel times to determine how far they have gone. You are now ready to begin.

Cooperative Learning

WAVE SPEED VALUES

Speed of P wave
$v_p = 6$ km/s

Speed of S wave
$v_s = 3.5$ km/s

STATION DATA
D_t (seconds)

Station A   4.4
Station B   5.5
Station C   6.3

This seismogram shows S waves and P waves.

1. First you will examine how the time difference between the arrival of the S and P waves depends on the distance from the station to the epicenter of the earthquake. Let d represent the distance from the station to the epicenter.

  a. S waves travel at 3.5 kilometers per second (km/s). Write an equation in terms of d for t_S, the time (in seconds) for the S waves to travel to the station.

$$t_S = \underline{\quad ? \quad}$$

  b. P waves travel at 6 km/s. Write an equation in terms of d for t_P, the time (in seconds) for the P waves to travel to the station.

$$t_P = \underline{\quad ? \quad}$$

  c. Use the equations from parts **a** and **b** above to write an equation in terms of d for D, the difference in time between the arrival of the S waves and P waves.

$$D = t_S - t_P = \underline{\quad ? \quad}$$

  d. Solve your equation from part **c** for d to find an equation in terms of D for the distance from the station to the epicenter.

$$d = \underline{\quad ? \quad}$$

2. Use your equation from part **d** above to find the distances in kilometers from each staion to the epicenter.

Station	D	d
A	4.4	?
B	5.5	?
C	6.3	?

3. Make a scale drawing of the locations of the three stations, with 1 km = 1 cm. Draw three circles centered at the three stations, using the distances you calculated above as the radii of the circles. The epicenter is in the region where the three circles overlap.

For more precision, draw chords connecting the intersections of each pair of circles. The epicenter is the intersection of these chords. Label the epicenter on your drawing.

B
53°
56.0 km
63.4 km
71°
56°
A
54.4 km
C

4. All of the seismograph readings used to locate the epicenter of the 1994 Los Angeles earthquake came from stations that were southeast of the epicenter. How would that make finding the location of the epicenter more difficult?

Segments of Tangents, Secants, and Chords

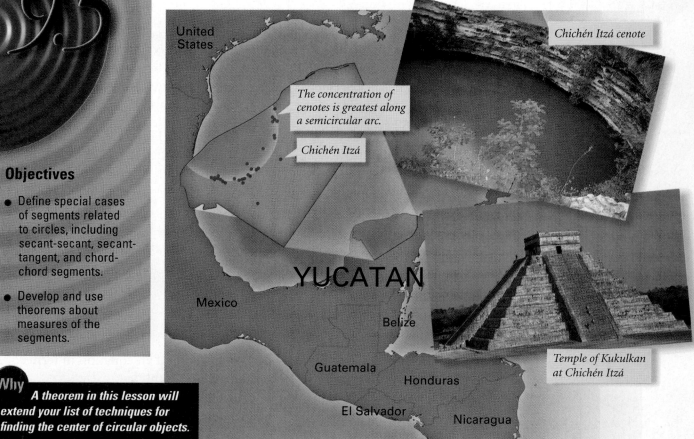

Chichén Itzá cenote

The concentration of cenotes is greatest along a semicircular arc.

Chichén Itzá

YUCATAN

United States

Mexico

Belize

Guatemala

Honduras

El Salvador

Nicaragua

Temple of Kukulkan at Chichén Itzá

Objectives

● Define special cases of segments related to circles, including secant-secant, secant-tangent, and chord-chord segments.

● Develop and use theorems about measures of the segments.

Why *A theorem in this lesson will extend your list of techniques for finding the center of circular objects.*

APPLICATION
GEOLOGY

A semicircular arc of natural wells (cenotes) like the one at Chichén Itzá stretches across the Yucatán peninsula. Some scientists believe that the cenote arc is evidence of an impact by an asteroid or comet, perhaps explaining the extinction of the dinosaurs (see page 353). The center of the arc would be the point where the asteroid or comet actually struck Earth.

Exploring Segment Relationships in Circles

In the previous lesson, you investigated special angles and arcs formed by secants and tangents of circles. As you will see, segments formed by secants and tangents of circles also have interesting relationships. The terms in the list below are used to classify such segments.

In the illustration, $\overleftrightarrow{XA}$ is a tangent line and $\overleftrightarrow{XB}$ is a secant line.

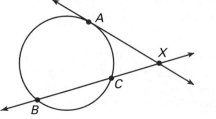

$\overline{XA}$ is a **tangent segment**.

$\overline{XB}$ is a **secant segment**.

$\overline{XC}$ is an **external secant segment**.

$\overline{BC}$ is a **chord**.

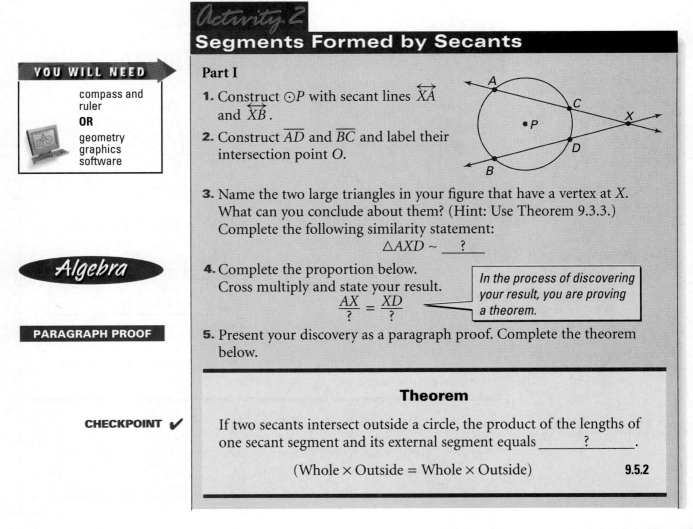

Activity 1
Segments Formed by Tangents

YOU WILL NEED

compass and ruler

OR

geometry graphics software

1. Construct ⊙P with tangent lines $\overleftrightarrow{XA}$ and $\overleftrightarrow{XB}$.

2. Measure the lengths of $\overline{XA}$ and $\overline{XB}$.

3. Make a conjecture about the lengths of two segments that are tangent to a circle from the same external point.

4. Add segments $\overline{AP}$, $\overline{BP}$, and $\overline{XP}$ to your figure. Write a paragraph proof of your conjecture.

5. Complete the theorem below.

PARAGRAPH PROOF

CHECKPOINT ✔

Theorem

If two segments are tangent to a circle from the same external point, then the segments _____?_____.

9.5.1

Activity 2
Segments Formed by Secants

YOU WILL NEED

compass and ruler

OR

geometry graphics software

Part I

1. Construct ⊙P with secant lines $\overleftrightarrow{XA}$ and $\overrightarrow{XB}$.

2. Construct $\overline{AD}$ and $\overline{BC}$ and label their intersection point O.

3. Name the two large triangles in your figure that have a vertex at X. What can you conclude about them? (Hint: Use Theorem 9.3.3.) Complete the following similarity statement:

$$\triangle AXD \sim \underline{\quad ? \quad}$$

Algebra

4. Complete the proportion below. Cross multiply and state your result.

$$\frac{AX}{?} = \frac{XD}{?}$$

> *In the process of discovering your result, you are proving a theorem.*

PARAGRAPH PROOF

5. Present your discovery as a paragraph proof. Complete the theorem below.

CHECKPOINT ✔

Theorem

If two secants intersect outside a circle, the product of the lengths of one secant segment and its external segment equals _____?_____.

(Whole × Outside = Whole × Outside)

9.5.2

Part II

1. Imagine moving point B along the circle so that $\overleftrightarrow{XB}$ becomes a tangent line. Points B and D will coincide at the point of tangency. What is the relationship between XB and XD?

2. Substitute XB for XD in your result from Step 5 of Part I.

3. Based on your result, complete the theorem below, which you will be asked to prove in Exercise 38.

B and D (relocated)

Theorem

CHECKPOINT ✔

If a secant and a tangent intersect outside a circle, then the product of the lengths of the secant segment and its external segment equals ____?____.

(Whole × Outside = Tangent Squared)

9.5.3

Activity 3

Segments Formed by Interesting Chords

YOU WILL NEED

compass and ruler

OR

geometry graphics software

Algebra

1. Construct a circle with chords $\overline{AC}$ and $\overline{DB}$ intersecting at point X.

2. Draw $\overline{AD}$ and $\overline{BC}$. Name the two triangles formed.

3. Name two angles of the triangles that intercept the same arc of the circle. What can you conclude about these angles?

4. What other angles of the triangles can you show to be congruent? What can you conclude about the triangles?

5. Complete the proportion below by relating two sides of one triangle to two sides of the other triangle. Cross multiply and state your result.

> In the process of discovering your result, you are proving a theorem.

$$\frac{DX}{XA} = \frac{?}{?}$$

PARAGRAPH PROOF

6. Present your discovery as a paragraph proof. Complete the theorem below.

Theorem

CHECKPOINT ✔

If two chords intersect inside a circle, then the product of the lengths of the segments of one chord equals ____?____.

9.5.4

Applying the Segment Theorems

EXAMPLE **1**

APPLICATION
NAVIGATION

Global positioning satellites are used in navigation. If the range of the satellite, *AX*, is 16,000 miles, what is *BX*?

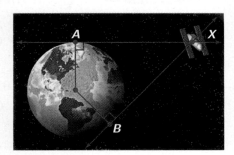

SOLUTION

$\overleftrightarrow{AX}$ and $\overleftrightarrow{BX}$ are tangents to a circle from the same external point. By Theorem 9.5.1, they are equal.

$$AX = BX = 16,000 \text{ miles}$$

EXAMPLE **2**

In the figure, *EX* = 1.31, *GX* = 0.45, and *FX* = 1.46. Find *HX*. Round your answer to the nearest hundredth.

SOLUTION

$\overleftrightarrow{EX}$ and $\overleftrightarrow{FX}$ are secants that intersect outside the circle.

By Theorem 9.5.2, Whole × Outside = Whole × Outside.

$$EX \cdot GX = FX \cdot HX$$
$$1.31 \cdot 0.45 = 1.46 \cdot HX$$
$$1.46 \cdot HX = 0.5895$$
$$HX \approx 0.40$$

EXAMPLE **3**

APPLICATION
WILDLIFE MANAGEMENT

An eagle is released from captivity at a ranger station near an approximately circular lake, and it builds a nest in a tree on the opposite shore. The distance from the station to the lake is 300 yd along a road tangent to the lake, and 50 yd along a line straight to the nest. How far is the nest from the ranger station?

SOLUTION

The road is a tangent and the line to the nest is a secant. They intersect outside the circle. Let *d* be the distance from the ranger station to the nest.

By Theorem 9.5.3, Whole × Outside = Tangent Squared.

$$d \times 50 = 300^2$$
$$d \times 50 = 90,000$$
$$d = 1800 \text{ yd} \approx 1 \text{ mi} \quad (1 \text{ mi} = 1760 \text{ yd})$$

EXAMPLE **4** In the figure, $AX = 0.26$, $XC = 0.91$, and $DX = 0.27$. Find XB.

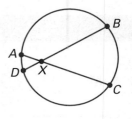

SOLUTION

$\overline{AX}$, $\overline{XC}$, $\overline{DX}$, and $\overline{XB}$ are chords that intersect inside a circle. By Theorem 9.5.4:

$$AX \cdot XC = DX \cdot XB$$
$$0.26 \cdot 0.91 = 0.27 \cdot XB$$
$$0.27 \cdot XB = 0.2366$$
$$XB \approx 0.88$$

EXAMPLE **5** The map shows the Yucatán peninsula, where A and B are points on the cenote ring. Assuming that the cenote ring defines the outer edge of an impact crater of an asteroid or comet, where did the object hit the Earth? Use a ruler and the map with the given scale to find the point of impact.

APPLICATION
GEOLOGY

SOLUTION

Construct ℓ, the perpendicular bisector of $\overline{AB}$. Label C, the midpoint of $\overline{AB}$, and D, the intersection of ℓ with the cenote ring. Somewhere on line ℓ there exists another point, E, such that $\overline{DE}$ is a diameter of the circular crater. (Why?)

The diagram below shows the position of points A, B, and D on the circle. By measuring the map, $AC = BC = 1.7$ cm, and $CD = 0.9$ cm.

By Theorem 9.5.4:
$$CD \cdot CE = AC \cdot BC$$
$$0.9 \cdot CE = 1.7 \cdot 1.7$$
$$CE \approx 3.2 \text{ cm}$$

$\overline{DE}$ is a diameter of the circle, and $\overline{DE} = 0.9 + 3.2 = 4.1$ cm.

For the given map scale, the actual diameter is about 180 km. Using a radius of 2.05 cm for the cenote ring, you can locate the center point on ℓ. This point is 0.35 cm southeast of Progreso on the map. Using the given map scale, the center of the possible impact crater is a little over 15 km southeast of Progreso. This would be where the comet or asteroid actually hit the Earth.

Communicate

For Exercises 1 and 2, explain the meaning of the statement and state which theorem is related to it.

1. Whole × Outside = Whole × Outside

2. Whole × Outside = Tangent Squared

3. Suppose that two secants or chords intersect on a circle. Can you determine anything about the lengths of the segments formed? Why or why not?

4. Suppose that a tangent and a secant intersect on a circle. Can you determine anything about the length of the segment formed? Why or why not?

5. List as many methods as you can for finding the center of a circle. How much of the circle must be known for each method?

Guided Skills Practice

6. $\overleftrightarrow{AX}$ and $\overleftrightarrow{BX}$ are tangent to ⊙P and AX = 7. Find BX.
(THEOREM 9.5.1 AND EXAMPLE 1)

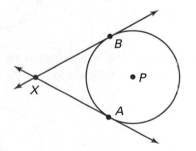

7. In ⊙Q, AX = 16, CX = 9, and BX = 18. Find DX.
(THEOREM 9.5.2 AND EXAMPLE 2)

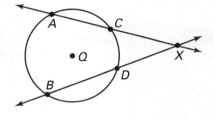

8. $\overleftrightarrow{BX}$ is tangent to ⊙R at B, AX = 8, and CX = 2. Find BX.
(THEOREM 9.5.3 AND EXAMPLE 3)

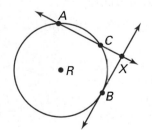

9. In the diagram, AX = 6, BX = 3, and DX = 8. Find CX.
(THEOREM 9.5.4 AND EXAMPLE 4)

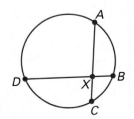

10. In the diagram at right, $\overleftrightarrow{FG}$ is the perpendicular bisector of $\overline{DE}$, $FG = 12$, and $DE = 48$. Find the diameter of the circle that contains points D, E, and G. **(EXAMPLE 5)**

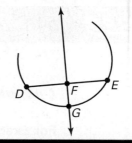

● *Practice and Apply*

$\overleftrightarrow{AB}$ **is tangent to ⊙R at C.**
Identify each of the following:

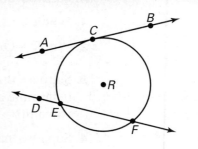

11. a tangent segment

12. a secant segment

13. an external secant segment

Identify each of the following in the figure at right:

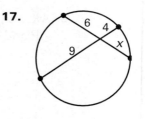

14. two pairs of congruent angles

15. a pair of similar triangles

Find x in each circle below.

16.

17.

In the figure below, $\overleftrightarrow{VA}$ and $\overleftrightarrow{VB}$ are tangent to ⊙P, the radius of ⊙P is 3 cm, and VA = 6 cm. Find the following:

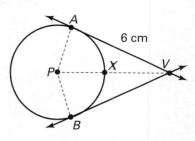

18. *VB*   **19.** *AP*

20. *PV*   **21.** *XV*

22. Name an angle congruent to ∠*AVP*.

23. Name an angle congruent to ∠*APV*.

24. Name an arc congruent to $\overparen{AX}$.

Use the figure below for Exercises 25–27. (Hint: You may need to use the quadratic formula.)

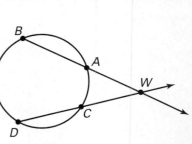

25. Given *WA* = 4, *WB* = 10, and *WC* = 5, find *CD*.

26. Given *WB* = *x*, *WA* = 6, *WD* = *x* + 3, and *WC* = 5, find *x*.

27. Given *WB* = *x*, *WA* = *x* − 16, *WD* = 8, and *WC* = 5, find *x*.

Use the figure at right for
Exercises 28 and 29.

28. Given $AE = x$, $BE = x - 2$, $CE = 3$,
and $DE = 8$, find x.

29. Given $AB = 10$, $CE = 2$, and $CD = 12$,
find AE.

In Exercises 30–37, you will summarize the properties of segments
formed by secants, tangents, and chords.

SUMMARY: SECANTS, TANGENTS, AND CHORDS			
Diagram	Location of vertex	Types of segments	Theorem
$\overrightarrow{VA}$ tangent at A, $\overrightarrow{VC}$ tangent at C	exterior of circle	2 tangent segments	$AV = CV$
	30. ___?___	**31.** ___?___	**32.** ___?___
33. ___?___	exterior of circle	1 secant segment and 1 tangent segment	**34.** ___?___
	35. ___?___	**36.** ___?___	**37.** ___?___

38. Prove Theorem 9.5.3 by using the
diagram at right. (Hint: Draw
segments $\overline{AB}$ and $\overline{BC}$. Then set
up a proportion with the
quantities in the theorem.)

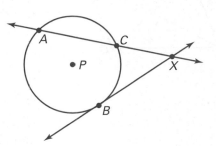

39. LUNAR EXPLORATION The diameter of a lunar crater, from measurements made on Earth, is known to be 32 km. How far is it from the lunar lander, located at L, to point X on the far rim of the crater, where a rock sample is to be taken?

40. ENGINEERING Jeff is restoring a clock and needs a new gear drive to replace the broken one shown at right. To make a new gear, he must determine the diameter of the original drive. In the picture, F is the midpoint of $\overline{BD}$. Use the product of the chord segments to find the diameter of the drive.

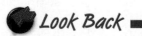

$BD = 10.1$ cm $EF = 2.5$ cm

$BD = 10.1$ cm
$EF = 2.5$ cm

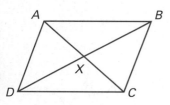 **Look Back**

PACKAGE DESIGN A manufacturer is considering the three potential package designs shown below. Draw a net for each one. Then calculate the surface area and volume for each.
(LESSONS 6.1, 7.1, 7.4, AND 7.5)

41. 5 cm / 16 cm

42. 16 cm / 10 cm / 10 cm

43. 16 cm / 5 cm

A flowchart proof of Theorem 4.5.2 is shown below. In Exercises 44–49, state the reason for each step of the proof. **(LESSON 4.5)**

Given: parallelogram $ABCD$ with diagonals intersecting at point X

Prove: $AX = CX$ and $DX = BX$

Proof:

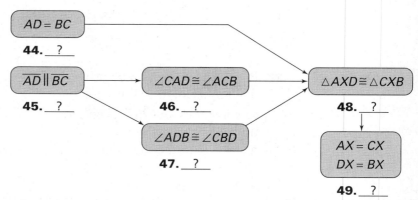

$AD = BC$

44. ___?___

$\overline{AD} \parallel \overline{BC}$ → $\angle CAD \cong \angle ACB$ → $\triangle AXD \cong \triangle CXB$

45. ___?___   **46.** ___?___   **48.** ___?___

$\angle ADB \cong \angle CBD$

47. ___?___

$AX = CX$
$DX = BX$

49. ___?___

Algebra

50. The drawing below suggests a visual proof of the Pythagorean Theorem. Use chord-segment products to explain how the proof works.

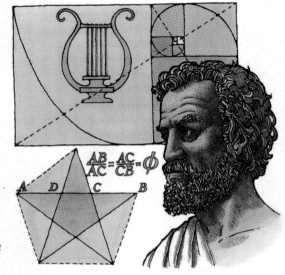

Artist's conception of Pythagoras, with related images. See Lesson 11.1.

PORTFOLIO ACTIVITY

REVERSE CURVES In the Portfolio Activity for Lesson 9.2, you constructed smooth curves by using arcs of different circles with the same tangent. If the center of one circle is outside the other circle, the direction of the curve is reversed, creating an **S** shape.

1. Using a compass and straightedge or geometry graphics software, draw ⊙P with radius $\overline{PR}$. Construct line ℓ tangent to P at R (that is, a line perpendicular to $\overline{PR}$ at R).

2. Extend $\overline{PR}$ past R, choose a point on $\overline{PR}$ in the exterior of ⊙P, and label it Q. Construct a circle centered at Q with radius $\overline{QR}$. ⊙P and ⊙Q have the same tangent at R.

3. Try constructing a "snake" of reverse curves like the one shown at right. The lines of construction have been left in the picture to help you.

WORKING ON THE CHAPTER PROJECT

You should now be able to complete Activity 3 of the Chapter Project.

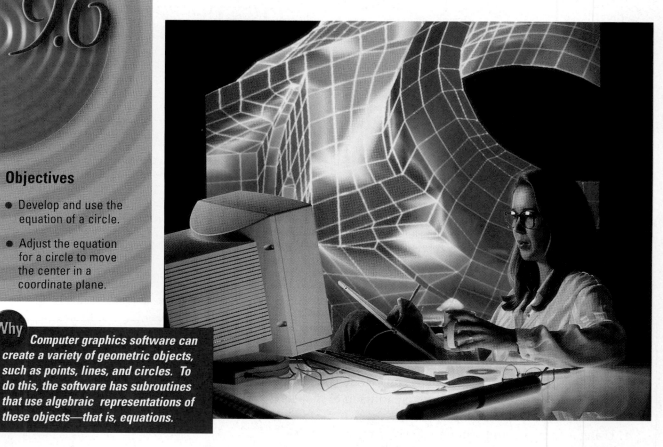

Circles in the Coordinate Plane

9.6

Objectives

- Develop and use the equation of a circle.

- Adjust the equation for a circle to move the center in a coordinate plane.

Why *Computer graphics software can create a variety of geometric objects, such as points, lines, and circles. To do this, the software has subroutines that use algebraic representations of these objects—that is, equations.*

Graphing a Circle From an Equation

In your work in algebra, you may have investigated graphs of equations such as $y = 2x - 3$ (a line), $y = x^2 - 3$ (a parabola), and $y = 3 \cdot 2^x$ (an exponential curve). In this lesson, you will investigate equations in which both x and y are squared.

EXAMPLE 1 Given: $x^2 + y^2 = 25$

Sketch and describe the graph by finding ordered pairs that satisfy the equation. Use a graphics calculator to verify your sketch.

Algebra

● **SOLUTION**

When sketching the graph of a new type of equation, it is often helpful to locate the intercepts. To find the x-intercept(s), find the value(s) of x when $y = 0$. (When a graph crosses the x-axis, $y = 0$.)

$$x^2 + 0^2 = 25$$
$$x^2 = 25$$
$$x = \pm 5$$

Thus, the graph has two x-intercepts, $(5, 0)$ and $(-5, 0)$.

To find the y-intercept(s), find the value(s) of y when $x = 0$.

$$0^2 + y^2 = 25$$
$$y^2 = 25$$
$$y = \pm 5$$

Thus, the graph has two y-intercepts, $(0, 5)$ and $(0, -5)$.

Now set x to some other value, such as 3.

$$3^2 + y^2 = 25$$
$$y^2 = 16$$
$$y = \pm 4$$

Thus, there are are two points with an x-value of 3: $(3, 4)$ and $(3, -4)$.

Similarly, by choosing other convenient values for x, a table like the one below is obtained.

x	y	Points on graph
3	±4	(3, 4), (3, −4)
−3	±4	(−3, 4), (−3, −4)
4	±3	(4, 3), (4, −3)
−4	±3	(−4, 3), (−4, −3)

Add these new points to the graph.
The graph of a circle with a radius of 5 and its center at the origin, $(0, 0)$, begins to appear. Sketch the curve.

CRITICAL THINKING How does the graph change if 25 in the equation is changed to 49? to 81? to 51?

Using Graphing Technology

TECHNOLOGY
GRAPHICS CALCULATOR

You can also graph the curve on a graphics calculator or a computer with graphing software. You can use the trace function of the calculator or software to find the coordinates of individual points on the graph.

Graphing technology normally requires an equation to be in the form $y = $ _____. You will need to solve the circle equation for y.
For example:

Algebra

$$x^2 + y^2 = 25$$
$$y^2 = 25 - x^2$$
$$y = \pm\sqrt{25 - x^2}$$

Thus, you will need to graph two separate curves:

$$y = \sqrt{25 - x^2} \qquad y = -\sqrt{25 - x^2}$$

Deriving the Equation of a Circle

CONNECTION
COORDINATE GEOMETRY

In a circle, all of the points are a certain distance, r, from a fixed point. In the simplest case, that fixed point is the origin, as shown.

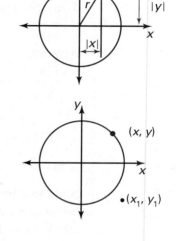

For any point (x, y) on the circle that is not on the x- or y-axis, you can draw a right triangle whose legs have lengths of $|x|$ and $|y|$. The length of the hypotenuse is the distance, r, from the point to the origin. For any such point:

$$x^2 + y^2 = r^2 \qquad \textit{Equation 1}$$

By substituting 0 for x or y in the equation, you can see that it is also true for any point (x, y) on the circle that is on the x- or y-axis.

If a point (x_1, y_1) is *not* on the circle, then its distance from the orgin is some value not equal to r, and the equation will not be true for it. That is, for any point *not* on the circle:

$$x_1{}^2 + y_1{}^2 \neq r^2$$

CHECKPOINT ✔ Explain why this is true.

Notice that equation 1 satisfies the following two conditions:

• It is true of all points (x, y) that are on the circle.
• It is not true of any point (x_1, y_1) that is not on the circle.

Thus, equation 1 is the equation of a circle.

Moving the Center of the Circle

To find the *standard form of the equation* of a circle centered at a point (h, k) that is not at the origin, study the diagram at right. For such a circle:

$$(x - h)^2 + (y - k)^2 = r^2 \qquad \textit{Equation 2}$$

It should be clear that the equation is *not* true for a point that does not lie on the circle, so equation 2 is the general equation of the circle.

CRITICAL THINKING How can you show that the relationships shown in the diagram are the same for points (h, k) in Quadrants II, III, and IV?

2 Given: $(x - 7)^2 + (y + 3)^2 = 36$
Find the center of the circle and its radius.

● **SOLUTION**

Comparing the given equation with the standard form of the equation of the circle, you find the following correspondences:

The standard form of the equation	The given equation
$(x - h)^2$	$(x - 7)^2$
$(y - k)^2$	$(y + 3)^2$ or $[y - (-3)]^2$
r^2	36

From this you can conclude that

$$h = 7, k = -3, \text{ and } r = 6.$$

That is, the center of the circle is $(7, -3)$, and the radius is 6 units.

TRY THIS For each equation below, find the radius and the center of the circle represented. Graph the equation and compare the graph with your values for the radius and the center of the circle.

a. $(x + 3)^2 + (y - 3)^2 = 49$ **b.** $(x - 3)^2 + (y + 3)^2 = 49$

c. $(x - 4)^2 + (y - 5)^2 = 30$ **d.** $(x + 2)^2 + (y - 5)^2 = 50$

Exercises

● *Communicate*

1. Explain how to find the x- and y-intercepts of $x^2 + y^2 = 4$.

2. Explain how to find the x- and y-intercepts of $(x - 2)^2 + (y + 2)^2 = 4$.

3. Is it possible for a circle to have no x- or y-intercepts? What would the graph look like?

4. **TECHNOLOGY** Most graphics calculators will graph only equations of the form $y = \_\_\_\_$. How would you solve the standard equation of a circle for y?

5. Use the distance formula (Lesson 5.6) to write an expression for the distance from point (h, k) to point (x, y). How does this expression relate to the standard equation of a circle?

Guided Skills Practice

For Exercises 6–8, refer to the equation $x^2 + y^2 = 100$. (EXAMPLE 1)

6. Find the x- and y-intercepts.

7. Complete the table below.

x	y	Points on graph
0	?	?
?	0	?
6	?	?
−6	?	?
8	?	?
−8	?	?

8. Plot the points from the table in a coordinate plane and sketch the circle.

For Exercises 9–11, refer to the equation $(x - 4)^2 + (y - 3)^2 = 25$. (EXAMPLE 2)

9. Find the x- and y-intercepts.

10. Complete the table below.

x	y	Points on graph
0	?	?
?	0	?
1	?	?
−1	?	?
4	?	?
7	?	?
8	?	?
9	?	?

11. Plot the points from the table in a coordinate plane and sketch the circle.

Practice and Apply

Algebra

Find the x- and y-intercepts for the graph of each circle.

12. $x^2 + y^2 = 64$

13. $x^2 + y^2 = 50$

14. $x^2 + (y - 4)^2 = 25$

15. $(x - 2)^2 + y^2 = 9$

16. $(x - 6)^2 + (y - 8)^2 = 100$

Write an equation for the circle with the given center and radius.

17. center: $(0, 0)$; radius $= 6$

18. center: $(0, 0)$; radius $= 2.5$

19. center: $(0, 0)$; radius $= \sqrt{13}$

20. center: $(2, 3)$; radius $= 4$

21. center: $(0, 6)$; radius $= 5$

22. center: $(4, -5)$; radius $= 7$

23. center: $(1, -7)$; radius $= 10$

24. center: $(4, -3)$; radius $= \sqrt{7}$

Find the center and radius of each circle.

25. $x^2 + y^2 = 100$

26. $x^2 + y^2 = 36$

27. $x^2 + y^2 = 101$

28. $(x - 6)^2 + y^2 = 9$

29. $x^2 + (y - 3)^2 = 4$

30. $(x + 5)^2 + (y - 2)^2 = 16$

31. $y^2 + (x + 3)^2 = 49$

32. $(x + 1)^2 + (y + 3)^2 = 19$

Write an equation for each circle.

33.

34.

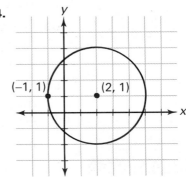

35.

Draw a circle on graph paper with the given intercepts and find its equation. Some exercises may have more than one possible answer.

	x-intercept(s)	y-intercept(s)
36.	3, -3	3, -3
37.	2, 6	none
38.	0	0, 8
39.	none	5
40.	none	none

Write an equation for the circle with the given characteristics. It may be helpful to sketch a graph.

41. center: $(2, 3)$; tangent to the x-axis

42. center: $(2, 3)$; tangent to the y-axis

43. center: $(0, 1)$; contains the point $(4, 4)$

44. center: $(2, 3)$; contains the point $(8, 3)$

45. center: $(2, 3)$; contains the point $(8, 11)$

46. has $(1, 3)$ and $(5, 3)$ as endpoints of a diameter

TECHNOLOGY Use geometry graphics software or graph paper for Exercises 47–53.

47. Sketch the graph of $(x - 3)^2 + (y - 5)^2 = 4$. Reflect the graph across the x-axis, and sketch the image. Write an equation for the image.

48. Sketch the graph of $(x - 4)^2 + (y - 2)^2 = 1$. Reflect the graph across the y-axis, and sketch the image. Write an equation for the image.

49. Sketch the graph of $(x - 2)^2 + y^2 = 9$. Translate the graph 6 units to the right, and sketch the image. Write an equation for the image.

50. Sketch the graph of $(x - 6)^2 + (y - 4)^2 = 9$. Translate the graph 2 units to the right and 1 unit down, and sketch the image. Write an equation for the image.

51. Sketch the graph of $(x - 5)^2 + (y - 4)^2 = 9$. Rotate the graph 180° about the origin, and sketch the image. Write an equation for the image.

52. Find the equation of the line tangent to the circle $x^2 + y^2 = 100$ at the point $(-6, 8)$.

CHALLENGE

53. Draw a triangle with vertices at $(0, 0)$, $(0, 6)$, and $(8, 0)$. Find the equation of the circle that circumscribes the triangle.

APPLICATION

54. COMPUTER GRAPHICS A computer screen is a coordinate plane in which each pixel is one unit. A typical computer screen measures 640 pixels horizontally and 480 pixels vertically. The origin of the plane is the upper left corner of the screen, so all y-coordinates are negative.

A computer programmer wants to create a circle that is as large as possible for a typical screen. What is the equation of the circle? (More than one answer is possible, depending on your choice of the center.)

Look Back

55. Suppose that a dart is tossed at random onto the graph of $x^2 + y^2 = 100$. What is the probability that it will land within the graph of the circle $x^2 + y^2 = 25$? **(LESSON 5.7)**

APPLICATION

56. ASTRONOMY You can view an eclipse by using a pinhole camera. If the image of the Sun measures x mm in diameter when the distance from the image to the pinhole is 50 cm, how large will the image be when the distance is 100 cm? 25 cm? 45 cm? **(LESSON 8.1)**

APPLICATION

STRUCTURAL DESIGN A manufacturer is designing a wheelchair to use in wheelchair basketball. The wheel diameter is 24.5 in. (LESSONS 5.3 AND 9.3)

57. If the wheels rotate through 45°, how far will the wheelchair travel?

58. A basketball court is 94 ft long. How many rotations of the wheels will it take to get from one end of the court to the other?

Wheelchair basketball is a highly competitive organized sport. The players' chairs are often custom-made.

59. Find *DE*. *(LESSON 9.5)*

60. Find *DE*. *(LESSON 9.5)*

Look Beyond

Algebra

In Exercises 61–63, you will prove that an angle inscribed in a semicircle measures 90°. Refer to the diagram below.

61. Use the equation for a circle with radius r centered at $(0, 0)$ to explain why the coordinates of a point on the circle are $(p, \sqrt{r^2 - p^2})$.

62. Find the slopes of the segments that form the inscribed angle in the diagram.

63. Prove that the inscribed angle measures 90°. (Hint: If two lines are perpendicular, the product of their slopes is __?__.)

Tangent Curves

The theorems you studied in this chapter will help you construct the figures in this project. You can use a compass and straightedge or geometry graphics software.

Activity 1

Make your own circle-flower designs. Begin by drawing a simple circle flower, as in the Portfolio Activity on page 572. Then extend your flower as shown at left by adding another level of circles. Write a paragraph to describe how you created the new level.

Make your figure more elaborate by adding chords and perhaps additional circles and arcs.

Activity 2

An egg shape can be constructed by connecting an arc with a relatively small radius (at one end of the egg) and an arc with a relatively large radius (at the other end) with a smooth curve or curves.

Geometric egg shapes can be classified according to the number of different radius measures involved. The egg shown at left is a three-radius egg. That is, it uses three different radii—one for each arc at the ends of the egg and one for the connecting curves. See if you can construct it. (Hint: Identify the pie-shaped sectors that fit together as in the illustration on page 579.)

The egg below is a four-radius egg. See if you can construct it. It may help to work from an enlarged photocopy of the illustration.

The egg below is a five-radius egg. See if you can construct it. The principle of the circle flower was used to construct the interior of the egg.

Activity 3

There are a number of reverse curves in the drawings below (see page 609). How many can you identify? Try constructing figures of your own that use reverse curves.

Chapter Review and Assessment

POSTULATES AND THEOREMS

Lesson	Number	Postulate or Theorem
9.1	9.1.5 Chords and Arcs Theorem	In a circle, or in congruent circles, the arcs of congruent chords are congruent.
	9.1.6 Converse of the Chords and Arcs Theorem	In a circle, or in congruent circles, the chords of congruent arcs are congruent.
9.2	9.2.2 Tangent Theorem	If a line is tangent to a circle, then the line is perpendicular to a radius of the circle drawn to the point of tangency.
	9.2.3 Radius and Chord Theorem	A radius that is perpendicular to a chord of a circle bisects the chord.
	9.2.4 Converse of the Tangent Theorem	If a line is perpendicular to a radius of a circle at its endpoint on the circle, then the line is tangent to the circle.
	9.2.5 Theorem	The perpendicular bisector of a chord passes through the center of the circle.
9.3	9.3.1 Inscribed Angle Theorem	The measure of an angle inscribed in a circle is equal to one-half the measure of the intercepted arc.
	9.3.2 Right Angle Corollary	If an inscribed angle intercepts a semicircle, then the angle is a right angle.
	9.3.3 Arc-Intercept Corollary	If two inscribed angles intercept the same arc, then they have the same measure.
9.4	9.4.1 Theorem	If a tangent and a secant (or a chord) intersect on a circle at the point of tangency, then the measure of the angle formed is one-half the measure of its intercepted arc.
	9.4.2 Theorem	The measure of an angle formed by two secants or chords that intersect in the interior of a circle is one-half the sum of the measures of the arcs intercepted by the angle and its vertical angle.
	9.4.3 Theorem	The measure of an angle formed by two secants that intersect in the exterior of a circle is one-half the difference of the measures of the intercepted arcs.

Lesson	Number	Postulate or Theorem
	9.4.4 Theorem	The measure of a secant-tangent angle with its vertex outside the circle is one-half the difference of the measures of the intercepted arcs.
	9.4.5 Theorem	The measure of a tangent-tangent angle with its vertex outside the circle is one-half the difference of the measures of the intercepted arcs, or the measure of the major arc minus 180°.
9.5	9.5.1 Theorem	If two segments are tangent to a circle from the same external point, then the segments are of equal length.
	9.5.2 Theorem	If two secants intersect outside a circle, then the product of the lengths of one secant segment and its external segment equals the product of the lengths of the other secant segment and its external segment. (Whole × Outside = Whole × Outside)
	9.5.3 Theorem	If a secant and a tangent intersect outside a circle, then the product of the lengths of the secant segment and its external segment equals the length of the tangent segment squared. (Whole × Outside = Tangent Squared)
	9.5.4 Theorem	If two chords intersect inside a circle, then the product of the lengths of the segments of one chord equals the product of the lengths of the segments of the other chord.

Key Skills & Exercises

LESSON 9.1
Key Skills

Identify parts of a circle.

In $\odot M$, $\overline{AB}$ and $\overline{CD}$ are chords. $\overline{CD}$ is also a diameter. $\overline{MC}$, $\overline{ME}$, and $\overline{MD}$ are radii. Central angle $\angle CME$ intercepts minor arc $\overparen{CE}$.

Find central angle measures.

In $\odot N$, find $m\angle ONP$.

$90° + 60° + 90° + m\angle ONP = 360°$
$m\angle ONP = 120°$

Find arc measures and lengths.

In $\odot N$, find $m\overparen{OP}$ and the length of $\overparen{OP}$.

$m\overparen{OP} = m\angle ONP = 120°$

length of $\overparen{OP} = \frac{120°}{360°} \times 2\pi(21 \text{ cm}) \approx 44 \text{ cm}$

Exercises

Refer to $\odot P$ below.

1. Name a chord, a radius, a central angle, and a major arc.

2. Find $m\angle APB$.

3. Find $m\overparen{AC}$.

4. Find the length of $\overparen{AC}$. Round your answer to the nearest centimeter.

Key Skills

Exercises

Use properties of secants and tangents to solve problems.

In the figure below, $\overleftrightarrow{EF}$ is tangent to $\odot G$ at F, $BF = 10$, $CG = 3$, and $EF = 12$. Find AD and EG.

By the Pythagorean Theorem:

$(CG)^2 + (AC)^2 = (AG)^2$

$3^2 + (AC)^2 = 5^2$

$AC = 4$

By the Chords and Arcs Theorem, $\overline{BG}$ bisects $\overline{AD}$, so $AD = 2 \cdot AC = 8$.

Because $\overleftrightarrow{EF}$ is tangent to $\odot G$, $\overline{GF} \perp \overline{EF}$.

$(GF)^2 + (EF)^2 = (EG)^2$

$12^2 + 5^2 = (EG)^2$

$EG = 13$

In $\odot M$ below, $\overleftrightarrow{NL}$ is tangent to $\odot M$ at N, $NP = 24$, $MO = 5$, and $NL = 31.2$.

5. Find the radius of $\odot M$.

6. Find ML.

In $\odot P$, $PS = 7$, $ST = 10$, $PT = 12$, $QX = 6$, and $RX = 6$. Classify each statement as true or false and explain your reasoning.

7. $\overline{PS} \perp \overline{QR}$

8. $\overline{ST}$ is tangent to $\odot P$ at S.

Key Skills

Exercises

Find the measure of an inscribed angle and its intercepted arc.

Find $m\overset{\frown}{AD}$ and $m\angle C$.

$m\overset{\frown}{AD} = 2 \times m\angle B = 100°$

$\angle B$ and $\angle C$ both intercept $\overset{\frown}{AD}$, so $m\angle C = m\angle B = 50°$.

In $\odot D$ below, $m\overset{\frown}{HE} = 40°$ and $m\angle FHG = 25°$. Find the following:

9. $m\angle F$

10. $m\overset{\frown}{FG}$

11. $m\angle E$

12. $m\angle I$

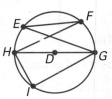

Key Skills

Exercises

Use the angles formed by secants and tangents to solve problems.

In $\odot X$, $m\angle YUZ = 66°$, $m\angle VZU = 35°$, and $m\overset{\frown}{VY} = 80°$. Find $m\overset{\frown}{YZ}$, $m\overset{\frown}{UV}$, $m\angle YTZ$, and $m\angle VYW$.

In $\odot G$ below, $\overleftrightarrow{HJ}$ is tangent to $\odot G$ at H, $\overleftrightarrow{HJ} \parallel \overleftrightarrow{NP}$, $m\overset{\frown}{HP} = 136°$, and $m\overset{\frown}{KN} = 42°$. Find the following:

13. $m\angle KMN$

14. $m\angle KHJ$

15. $m\overset{\frown}{KH}$

16. $m\overset{\frown}{NP}$

Key Skills, continued

$$m\widehat{YZ} = 2 \times m\angle YUZ = 2 \times 66° = 132°$$
$$m\widehat{UV} = 2 \times m\angle VZU = 2 \times 35° = 70°$$
$$m\angle YTZ = \frac{m\widehat{YZ} - m\widehat{UV}}{2} = \frac{132° - 70°}{2} = 31°$$
$$m\angle VYW = \frac{1}{2} \times m\widehat{VY} = \frac{1}{2} \times 80° = 40°$$

LESSON 9.5
Key Skills

Use segments formed by tangents, secants, and chords to solve problems.

In $\odot A$, $BF = 24$, $CF = 32$, and $DF = 48$. Find EF.

By Theorem 9.5.4:

$$BF \times DF = CF \times EF$$
$$24 \times 48 = 32 \times EF$$
$$EF = 36$$

Exercises

$\odot P$ has a diameter of 75, $\overleftrightarrow{SQ}$ and $\overleftrightarrow{ST}$ are tangent to $\odot P$ at Q and T, and $RS = 60$. Find the following:

17. SQ

18. ST

In $\odot X$, $SY = 12$, $SZ = 18$, $SV = 4$, $VW = 8$, $WZ = 6$, and $WT = 5$. Find the following:

19. SU

20. WY

LESSON 9.6
Key Skills

Sketch a circle from its equation.

Sketch the circle $(x + 3)^2 + (y - 4)^2 = 16$.

center: $(-3, 4)$
radius $= 4$

Write the equation of a given circle.

Write the equation of a circle with center $(5, 0)$ and radius 6.

$$(x - h)^2 + (y - k)^2 = r^2 \quad h = 5, k = 0, r = 6$$
$$(x - 5)^2 + y^2 = 36$$

Exercises

Sketch the following circles:

21. $x^2 + y^2 = 49$

22. $(x - 1)^2 + (y + 2)^2 = 25$

Write the equation of the circle with the given center and radius.

23. center: $(0, 0)$; radius $= 1$

24. center: $(6, -2)$; radius $= 8$

25. SURVEYING A simple way to get a rough estimate of property boundaries is to use a measuring wheel. Attach an 8-in. diameter wheel to a long handle, put a marker on the edge of the wheel, and count the number of rotations the wheel makes as you walk along the property line, rolling the wheel. If the wheel makes 42.5 rotations, how long is the property line?

26. COMMUNICATIONS A radio station wishes to locate its broadcasting tower an equal distance from three small towns. Trace the figure shown and locate the center of the circle that passes through the three points in order to find the location of the tower.

27. NAVIGATION A kayaker is rowing toward a lighthouse. If the light is 85 ft above sea level, how far away is the kayaker when he first sees the light? Assume that the water is calm and visibility is good. (1 mi = 5280 ft, radius of Earth = 4000 mi)

28. AGRICULTURE A field is irrigated by a pipe that extends from the center of the field to its outer edge and sweeps around in an arc. The corner of the field is blocked by the farmer's house and yard. If the area of the square field is 1 mi², what is the area of the irrigated sector? To find the area of the sector, use the following formula:

$$\text{area of sector} = \frac{C}{360°} \times \text{area of circle},$$

where C is the degree measure of the arc that bounds the sector

What percent of the entire field is the irrigated sector?

Alternative Assessment

Performance Assessment

In Lesson 1.5, you investigated inscribed and circumscribed triangles of circles. Use a compass and straightedge or geometry graphics software for the following investigations of other inscribed and circumscribed polygons of circles:

1. The figure at right is a regular 8-pointed *star polygon* inscribed in a circle. Find the measure of as many of the angles and arcs in the figure as you can by using theorems from this chapter. Construct the figure yourself and measure the angles and arcs. Do your answers agree?

2. Construct a regular hexagon inscribed in a circle. If the radius of the circle is 1 unit, what is the perimeter of the polygon? Use your answer to explain why π is greater than 3.

3. Using your figure of a hexagon inscribed in a circle, draw a segment from the center to each vertex, and construct a tangent to the circle at each vertex. The tangents will form a regular hexagon circumscribed around the circle. What is the perimeter of this hexagon? Use your answer to explain why π is less than 3.5.

Portfolio Projects

Use the clues below to "crack the code." Then decode the quote by G. H. Hardy.

A **LIRZQV** is a segment from the center of a circle to a point on the circle.

A **POGLR** is a segment whose endpoints lie on the circle.

A **HINYBNH** is a line that intersects a circle at exactly one point.

An **ZNVPLZJBR INYMB** is an angle whose vertex lies on a circle and whose sides are chords of the circle.

Quote:
I fihobfihzpzin, mzeb i uiznhbl gl i ugbh, zv i fiebl gs uihhblnv. Zs ozv uihhblnv ilb fglb ublfinbnh hoin hobzlv, zh zv jbpiqvb hobd ilb firb czho zrbiv.

G. H. Hardy (1877–1947) was a brilliant mathematician who hated to have his picture taken. Only five photographs of him are known to exist.

📶 internet**connect**

(HRW) The HRW Web site contains many resources to reinforce and expand your knowledge of circles. The Web site also provides Internet links to other sites where you can find information and real-world data for use in research projects, reports, and activities that involve circles. Visit the HRW Web site at **go.hrw.com**, and enter the keyword **MG1 CH9**.

College Entrance Exam Practice

QUANTITATIVE COMPARISON For Items 1–4, write

A if the quantity in Column A is greater than the quantity in Column B;
B if the quantity in Column B is greater than the quantity in Column A;
C if the two quantities are equal; or
D if the relationship cannot be determined from the information given.

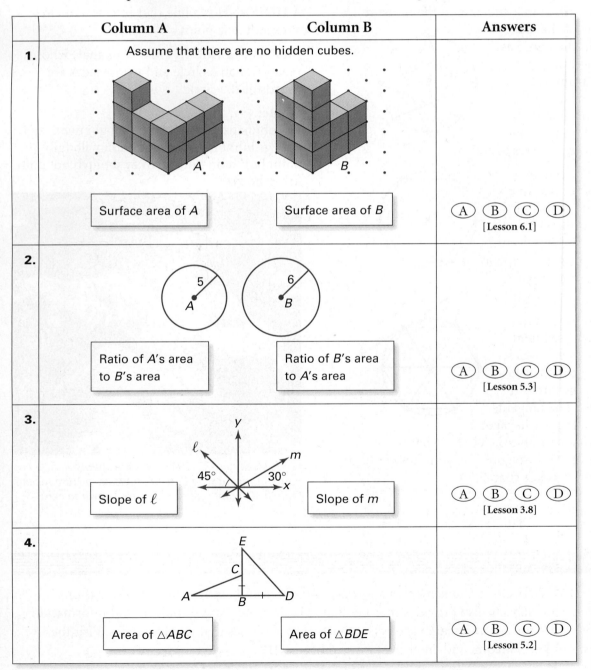

	Column A	Column B	Answers
1.	Assume that there are no hidden cubes. Surface area of A	Surface area of B	Ⓐ Ⓑ Ⓒ Ⓓ [Lesson 6.1]
2.	Ratio of A's area to B's area	Ratio of B's area to A's area	Ⓐ Ⓑ Ⓒ Ⓓ [Lesson 5.3]
3.	Slope of ℓ	Slope of m	Ⓐ Ⓑ Ⓒ Ⓓ [Lesson 3.8]
4.	Area of $\triangle ABC$	Area of $\triangle BDE$	Ⓐ Ⓑ Ⓒ Ⓓ [Lesson 5.2]

5. Choose the most complete and accurate description of the two polygons below. **(LESSON 4.6)**

   a. quadrilaterals
   b. trapezoids
   c. similar trapezoids
   d. congruent trapezoids

6. In ⊙O, which angle or arc measures 60°?
   (LESSON 9.3)
   a. ∠ABD
   b. ∠BDC
   c. $\overset{\frown}{AD}$
   d. $\overset{\frown}{BC}$

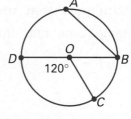

7. The oblique pyramid has a rectangular base. Find its volume. **(LESSON 7.3)**

8. The ratio of the volumes of two spheres is 27:1. If the smaller sphere has a radius of 15 in., what is the radius of the larger sphere? **(LESSONS 7.1 AND 7.6)**

For Exercises 9–10, refer to the figure below.

9. Construct a rotation of the segment. Rotate the segment 180° counterclockwise about the endpoint (0, 0). **(LESSON 4.9)**

10. Give the coordinates of the endpoints of the rotated segment from Item 9. **(LESSON 1.7)**

11. Write a paragraph proof that △PYW and △PYX in ⊙P below are congruent.
   (LESSONS 4.2 AND 4.3)

12. An equilateral triangle is inscribed in a circle with a radius of 3 units. If a point is picked at random anywhere inside the circle, what is the probability that the point will *not* be inside the triangle? **(LESSON 5.7)**

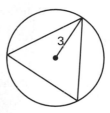

FREE-RESPONSE GRID
Items 13–15 may be answered by using a free-response grid such as that commonly used by standardized-test services.

13. Find the volume of a cylinder with a radius of 2 and a height of 7. Round your answer to the nearest tenth. **(LESSON 7.4)**

14. Find the slope of a line that passes through the points (2, 6) and (9, 12). **(LESSON 3.8)**

15. Find the area of the parallelogram below. Round your answer to the nearest hundredth.
   (LESSON 5.5)

Trigonometry

HAVE YOU EVER WONDERED HOW HIGHWAY engineers are able to make sure that a section of a freeway or overpass will correctly match up with a section that is under construction a considerable distance away? Accurate measurements and calculations are necessary to ensure success. In this kind of work, trigonometry is an indispensable tool.

Trigonometry, like much of geometry, depends on triangles. The simple study of the ratios of the sides of right triangles quickly leads to more sophisticated calculation techniques that are widely used in surveying, navigation, and the sciences.

Lessons

About the Chapter Project

Plimpton 322, the Babylonian clay tablet that you studied in Lesson 5.4, contains some very sophisticated trigonometry.

In the Chapter Project, *Plimpton 322 Revisited*, you will study the arrangement of the numbers in the table and the meaning of the values in the remaining column.

After completing the Chapter Project, you will be able to do the following:

- Read and write numbers in *cuneiform*, a system used in ancient Babylon.

- Use the Babylonian number system to analyze a cuneiform tablet.

About the Portfolio Activities

Throughout the chapter, you will be given opportunities to complete Portfolio Activities that are designed to support your work on the Chapter Project.

The theme of each Portfolio Activity and of the Chapter Project is the Plimpton 322 tablet.

- In the Portfolio Activity on page 638, you will examine the reason for the order of the values in the table.

- In the Portfolio Activity on page 646, you will discover a trigonometric identity that you will use to in examine the tablet.

Tangent Ratios

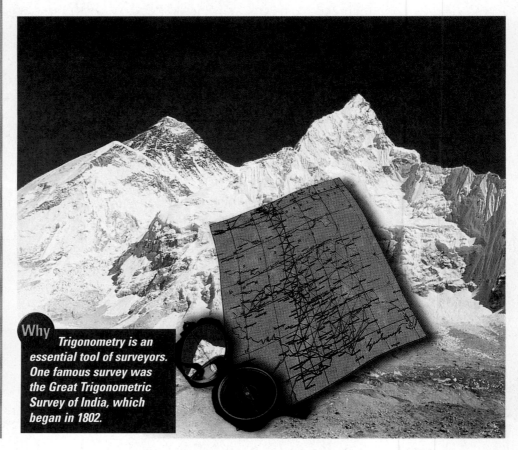

Objectives

- Develop the tangent ratio by using right triangles.

- Use a chart or graph to find the tangent of an angle or the angle for a given tangent.

- Solve problems by using tangent ratios.

Why *Trigonometry is an essential tool of surveyors. One famous survey was the Great Trigonometric Survey of India, which began in 1802.*

Mount Everest is named for Sir George Everest, who was the superintendent of the Great Trigonometric Survey of India from 1823 to 1843. The mountain's height, about 5.5 miles, was computed by using trigonometry.

Tangent Ratios

In the Activities that follow, you will examine one of the three important ratios of trigonometry, the *tangent* ratio.

Activity 1

A Familiar Ratio

1. Draw an angle between 30° and 50° such that one side of the angle is horizontal. Label the vertex *A*. Draw a vertical segment to create a right triangle.

2. Measure the leg of the triangle opposite *A*.

3. Measure the leg of the triangle adjacent to *A*.

4. Divide the length of the opposite leg by the length of the adjacent leg.

CHECKPOINT ✔   5. Repeat Steps 1–4, using the same angle but different side lengths. What do you notice?

Recall that the slope of a line is its rise divided by its run. How is the ratio you calculated in Activity 1 related to the concept of slope?

Examine the triangles at right. $\angle A$ is congruent to $\angle M$, and $\angle C$ and $\angle O$ are right angles. Thus, by AA Similarity, $\triangle ABC \sim \triangle MNO$.

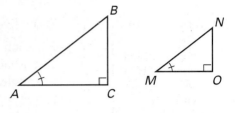

By the Polygon Similarity Postulate, $\frac{BC}{AC} = \frac{NO}{MO}$. This leads to the following definition:

> It is common in trigonometry to use the Greek letter θ (theta) to represent the measure of an angle.

Tangent Ratio

For a given acute angle $\angle A$ with a measure of $\theta°$, the **tangent** of $\angle A$, or tan θ, is the ratio of the length of the leg opposite $\angle A$ to the length of the leg adjacent to $\angle A$ in any right triangle having A as one vertex, or

$$\tan \theta = \frac{\text{opposite}}{\text{adjacent}}.$$

10.1.1

Note: In trigonometry, the letter of the vertex of an angle is often used to indicate the measure of the angle. Thus, the tangent of $\angle A$ can also be written as tan A.

E X A M P L E ❶ Measure the legs of the triangle below to find tan θ.

SOLUTION

The leg opposite the angle is 2.9 centimeters, and the leg adjacent to the angle is 3.8 centimeters.

$$\tan \theta = \frac{2.9}{3.8} \approx 0.76$$

TRY THIS Use the given measurements to find the tangent of the other acute angle in the triangle from Example 1.

CRITICAL THINKING What is the measure of an angle with a tangent of 1? Explain your reasoning.

Activity 2
Graphing the Tangent

YOU WILL NEED

ruler,
protractor, and
graph paper
OR
geometry
graphics
software

1. Draw five triangles as shown, with angles of 15°, 30°, 45°, 60°, and 75°.

2. For each triangle, measure the opposite and adjacent legs to find the tangent of the angle. Copy and complete the table below.

θ	tan θ
15°	?
30°	?
45°	?
60°	?
75°	?

3. Plot the ordered pairs $(\theta, \tan\theta)$ from the table above. Connect the points with a smooth curve.

CHECKPOINT ✔ **4.** Does your graph increase or decrease? Describe its behavior.

Activity 3
Using the Tangent

YOU WILL NEED

ruler, protractor, and
your tangent graph
from Activity 2

PROBLEM SOLVING

In this Activity you will model the calculations of a surveying crew measuring the distance across a canyon. Make a scale drawing, and let 1 centimeter = 10 meters.

1. Make a diagram showing the sides of a canyon, and draw a line across the canyon representing a line of sight from point X to point Y.

2. Draw a line through Y perpendicular to $\overleftrightarrow{XY}$. Choose a point on this line and label it Z. Draw $\overline{XZ}$.

3. Measure $\angle Z$ and $\overline{YZ}$. (Remember, $\overline{XY}$ and $\overline{XZ}$ span the canyon, so they cannot be measured directly.) Use your graph from Activity 2 to estimate the tangent of $\angle Z$.

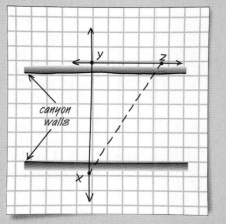

CHECKPOINT ✔ **4.** Substitute the values for tan Z and YZ into the equation below.

$$\tan Z = \frac{XY}{YZ}$$

Solve to find XY, the distance across the canyon.

5. Compare the value you calculated for XY with the actual distance on your drawing. How accurate is your answer? What could be some possible sources of error in your estimate?

Calculating Tangent Ratios

So far, you have measured the sides of triangles to find tangent ratios. However, since measurements can often be inaccurate, you will usually use a scientific or graphics calculator, or a table such as the one in the infobank in the back of this book.

Use the [TAN] function key on your scientific or graphics calculator to find the tangent of an angle. (Be sure your calculator is in degree mode.)

E X A M P L E ② Use a calculator to find tan 45°. Verify your answer by using a right triangle.

● **SOLUTION**

Using the [TAN] function key on a calculator, tan 45° = 1.

In the triangle at right, m∠A = 45°. By the Triangle Sum Theorem and the Converse of the Isosceles Triangle Theorem, the triangle is isosceles, so the opposite and adjacent legs are equal. Thus, the ratio of the opposite leg to the adjacent leg must equal 1.

Sometimes it is necessary to find the angle measure for a given tangent ratio. To find the angle with a tangent of $\frac{a}{b}$, you can draw a right triangle with legs of length a and b, and measure the angle.

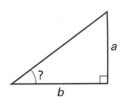

To find the angle using a calculator, use the [TAN⁻¹] key. For any positive number r, the [TAN⁻¹] key gives the measure of an angle between 0° and 90° whose tangent is r. The tan⁻¹ function is called the **inverse tangent** function. Notice that the input and output for the tangent and inverse tangent functions are reversed.

Function and key	Given (input)	Want to find (output)
tangent [TAN]	angle measure	tangent ratio
inverse tangent [TAN⁻¹]	tangent ratio	angle measure

E X A M P L E ③ Use your calculator to find an angle that has a tangent of $\frac{2}{3}$. Round your answer to the nearest degree. Verify your answer using a right triangle.

● **SOLUTION**

Using a calculator, $\tan^{-1} \frac{2}{3} = 33.69 \approx 34°$.

The triangle at right has a leg of 2 cm and a leg of 3 cm. The measure of the angle is approximately 34°.

Exercises

Communicate

1. Where do you think the word *trigonometry* comes from? (Hint: What could be another name for a tri-gon?)

2. Does the tangent ratio increase or decrease as an angle gets larger? Explain your answer.

3. What happens to the tangent ratio as an angle approaches 0°? Use your calculator the find the tangent of 0°. Does your answer make sense? Why or why not?

4. What happens to the tangent ratio as an angle approaches 90°? Try to find the tangent of 90° by using your calculator. What happens? Explain why in terms of right triangles.

5. When measuring the sides of a triangle to find the tangent ratio, does it matter what units you use? What happens to the units in your answer?

Guided Skills Practice

Measure the sides of the triangles below to find tan *A*. (EXAMPLE 1)

6.

7.

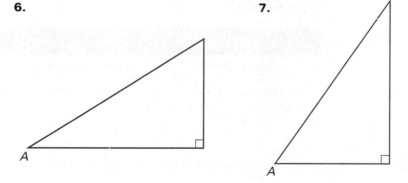

Use the graph you created in Activity 2 to estimate the tangent of each angle. Then find the tangent by using your calculator, and compare your answers. (ACTIVITY 2 AND EXAMPLE 2)

8. 20° **9.** 40° **10.** 70°

11. Use your calculator to find an angle that has a tangent of $\frac{3}{4}$. Round to the nearest degree. Verify your answer by using a right triangle. (*EXAMPLE 3*)

● **Practice and Apply**

Find tan A for each triangle below.

12.

13.

14.

15.

16.

17.

Use a scientific or graphics calculator to find the tangent of each angle below. Round to the nearest hundredth.

18. $25°$

19. $67°$

20. $19°$

21. $53°$

22. $75°$

23. $89°$

Use a scientific or graphics calculator to find the inverse tangent of each ratio below. Round to the nearest degree.

24. $\frac{3}{8}$

25. $\frac{7}{5}$

26. 3

27. 9.5

28. 1

29. 0

For Exercises 30–34, use the definition of tangent ratio to write an equation involving *x*. Find the tangent of the given angle by using a calculator, and solve the equation to find the unknown side of the triangle. Round your answer to the nearest hundredth.

30.

31.

32.

33.

CHALLENGE

34. Use the tangent ratio and the Pythagorean Theorem to find *x* and *y* in the triangle at right. Round to the nearest tenth.

35. SURVEYING Use the diagram at right to determine $\overline{AB}$, the distance across the lake.

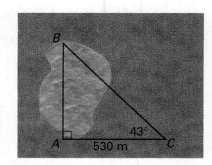

36. ENGINEERING The steepness, or grade, of a highway or railroad is expressed as a percent. In the photo of the cog railway at Pike's Peak, in Colorado, the grade is 18 percent. Thus, for every 100 ft of horizontal run, the train rises 18 ft. Find the angle of inclination of the railway.

37. ENGINEERING The maximum grade of the railway at Pike's Peak is 25 percent. Find the angle of inclination of the railway at this point.

38. INDIRECT MEASUREMENT Use the figure at left to estimate the height of the flagpole. Round to the nearest meter.

39. SURVEYING In the photo below of Glen Canyon, Utah, point P is on the north rim, point Q is on the south rim, $PR = 300$ ft, $\overline{PQ} \perp \overline{PR}$, and m$\angle R = 75°$. Find PQ, the width of the canyon.

Find the volume and surface area of each solid.

40. right prism
(LESSON 7.2)

41. cylinder
(LESSON 7.4)

42. cone
(LESSON 7.5)

43. sphere
(LESSON 7.6)

Use ⊙P, with $\overline{MN} \perp \overline{PR}$ for Exercises 44–46. (LESSONS 9.4 AND 9.5)

44. $MQ =$ _____?_____

45. $PR = 8, PQ = 3$
$MQ =$ _____?_____ $QN =$ _____?_____

46. $PR = 12, PQ = 4$
$MQ =$ _____?_____ $QN =$ _____?_____

Look Beyond

CULTURAL CONNECTION: AFRICA Trigonometry has been used by many cultures for over 4000 years. The following problem is taken from an ancient Egyptian papyrus:

If a pyramid is 250 cubits high and the side of its base is 360 cubits long, what is its seked? —Problem 56 of the Rhind papyrus

A *cubit* is a measure of length equal to about 21 inches, and the *seked* is a measure of steepness.

47. In the diagram at right, what is the ratio of the run to the rise of the pyramid? (Notice this is not the same as the slope.)

48. How is the ratio of run to rise related to the concept of the tangent ratio?

49. The seked is the ratio you computed in Exercise 47, converted into palms per cubit. Since 1 cubit = 7 palms, this is found by multiplying the ratio by 7. Find the seked of the pyramid.

TRIGONOMETRY IN BABYLONIAN MATHEMATICS

Recall from the Plimpton 322 tablet (see
Lesson 5.4) that two columns represented a
leg and the hypotenuse of Pythagorean triples.
The arrangement of the numbers may seem
random, but the reason for the arrangement
becomes clear when you consider the angles
formed in the right triangles.

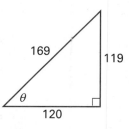

1. Complete the table below. What do you notice about the arrangement
   of the values of the angles? Why do you think this table might have
   been useful?

Column II	Column III			
Opposite leg	Hypotenuse	Adjacent leg	tan θ	θ
119	169	120	?	?
3367	4825	3456	?	?
4601	6649	4800	?	?
12,709	18,541	13,500	?	?
65	97	72	?	?
319	481	360	?	?
2291	3541	2700	?	?
799	1249	960	?	?
481	769	600	?	?
4961	8161	6480	?	?
45	75	60	?	?
1679	2929	2400	?	?
161	289	240	?	?
1771	3229	2700	?	?
56	106	90	?	?

2. The first angle is almost equal to 45°. Is it possible to find a Pythagorean
   triple that corresponds to a 45° angle? Why or why not?

WORKING ON THE CHAPTER PROJECT

You should now be able to complete Activity 1 of the Chapter Project.

Sines and Cosines

Objectives

- Explore the relationship between the measure of an angle and its sine and cosine.

- Solve problems by using sine and cosine ratios.

- Develop two trigonometric identities.

Why *Sine and cosine ratios can be used to solve many types of problems. For example, if you know the length and the approximate angle of the rope, you can estimate the height of a parasailer above the water.*

Trigonometric Ratios

In Lesson 10.1, you learned that the tangent of an angle is the ratio of two sides of a right triangle containing the given angle. However, there are other trigonometric ratios that can be formed by using different sides of the same triangle. The three most important ratios are the tangent, the sine, and the cosine.

Sine and Cosine Ratios

For a given angle $\angle A$ with a measure of $\theta°$, the **sine** of $\angle A$, or sin θ, is the ratio of the length of the leg opposite A to the length of the hypotenuse in a right triangle with A as one vertex, or

$$\sin \theta = \frac{\text{opposite}}{\text{hypotenuse}}.$$ **10.2.1**

The **cosine** of $\angle A$, or cos θ, is the ratio of the length of the leg adjacent to A to the length of the hypotenuse, or

$$\cos \theta = \frac{\text{adjacent}}{\text{hypotenuse}}.$$ **10.2.2**

scientific or graphics calculator and graph paper

Activity 1

Sines and Cosines

1. Examine the triangles in the diagram below. As θ increases, what happens to the value of sin θ? Does it increase or decrease? What happens as θ gets close to 0°? to 90°? Write a conjecture about the sine of 0° and of 90°.

In this sequence of triangles, the adjacent legs stay the same. What happens to the other sides?

2. Repeat Step 1 for cos θ.

CHECKPOINT ✔ 3. Compare the behavior of the sine and cosine ratios as the measure of an angle increases from 0° to 90°.

4. Copy and complete the table below. (Use the [SIN] and [COS] keys on your calculator.) Round your answers to the nearest hundredth.

Be sure your calculator is in degree mode.

θ	0°	10°	20°	30°	40°	50°	60°	70°	80°	90°
sin θ	?	?	?	?	?	?	?	?	?	?
cos θ	?	?	?	?	?	?	?	?	?	?

5. Plot the pairs $(\theta, \sin \theta)$ for the angles in the table. Draw a smooth curve through the points. Repeat for the pairs $(\theta, \cos \theta)$. Do your graphs verify the conclusions you made in Step 3?

CRITICAL THINKING   Could the sine or cosine of an angle ever be bigger than 1? Explain your answer in terms of right triangles.

EXAMPLE ❶   A paraglider is towed behind a boat by 400-ft ropes attached to the boat at a point 15 ft above the water. The spotter in the boat estimates the angle of the ropes to be 35° above the horizontal. Estimate the paraglider's height above the water.

APPLICATION

RECREATION

SOLUTION

Label the hypotenuse and the opposite and adjacent legs for the given angle. The hypotenuse is 400 ft. Since the height is the opposite side, use the sine ratio.

$\sin 35° = \dfrac{\text{opposite}}{\text{hypotenuse}} = \dfrac{x}{400}$ *Using a scientific calculator, sin 35° = 0.5736.*

$0.5736 = \dfrac{x}{400}$

$x = 400(0.5736) \approx 229 \text{ ft}$

Thus, the height of the paraglider is approximately 229 + 15 = 244 ft.

EXAMPLE **2**

Charles, an amateur astronomer, has been keeping records of the position of the sunrise each day as viewed from a fixed point, A. He marks the position of the sunrise on the autumnal equinox and the winter solstice. (On the autumnal equinox, there are equal hours of daylight and darkness, and on the winter solstice, there are the fewest hours of daylight.) What is the measure of $\angle A$, the angle between the positions of sunrise on these two days?

SOLUTION

Charles uses ropes to mark the lines of sight from point A in the direction of the sunrise on the two days. He uses another rope to make a line perpendicular to the equinox line as shown, and measures the distances from point A to this rope.

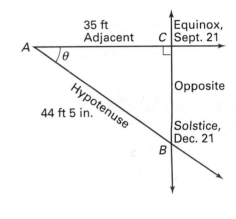

$$\cos \theta = \frac{AC}{AB} = \frac{35 \text{ ft}}{44.4 \text{ ft}} \approx 0.788$$

To find the angle whose cosine is 0.788, use the inverse cosine, or $\boxed{\cos^{-1}}$ key on your scientific or graphics calculator.

$$\cos^{-1} 0.788 \approx 38°$$

An **identity** is an equation that is true for all values of the variables in the equation for which all terms in the equation are defined. For example, the equation $(a + b)(a - b) = a^2 - b^2$ is an identity, because it is true for all values of a and b. In Activity 2, you will discover two trigonometric identities.

Activity 2

Two Trigonometric Identities

Part I

1. Copy and complete the table below by using a scientific or graphics calculator. Leave at least three digits after the decimal point for each value.

θ	$\sin \theta$	$\cos \theta$	$\dfrac{\sin \theta}{\cos \theta}$	$\tan \theta$
20°	?	?	?	?
40°	?	?	?	?
60°	?	?	?	?

CHECKPOINT ✔

2. What do you notice about the values in the tangent column? Write a trigonometric identity involving the sine, cosine, and tangent ratios.

3. What happens when you simplify the right side of the equation below?

$$\frac{\sin \theta}{\cos \theta} = \frac{\dfrac{\text{opposite}}{\text{hypotenuse}}}{\dfrac{\text{adjacent}}{\text{hypotenuse}}}$$

Does the equation prove your identity from Step 2? Explain.

This table will help you discover a Pythagorean identity (see page 646).

Part II

1. Copy and complete the table below by using a scientific or graphics calculator. Leave at least three digits after the decimal point for each value.

θ	$\sin\theta$	$\cos\theta$	$(\sin\theta)^2 + (\cos\theta)^2$
20°	?	?	?
40°	?	?	?
60°	?	?	?

CHECKPOINT ✔

2. What do you notice about the entries in the last row of the table? Write a trigonometric identity involving sine and cosine ratios.

PROOF

3. The following is a partial proof of the identity from Step 2. For each step in the proof, give a reason.

$$(\sin\theta)^2 + (\cos\theta)^2 = \left(\frac{a}{c}\right)^2 + \left(\frac{b}{c}\right)^2$$
$$= \frac{a^2}{c^2} + \frac{b^2}{c^2}$$
$$= \frac{a^2 + b^2}{c^2}$$

Compare the numerator and denominator of the fraction in the last step. How could you complete the proof of the identity?

Summary

As you saw in Activity 1, the sine of an angle increases from 0 to 1 as the measure of the angle increases from 0° to 90°, while the cosine of an angle decreases from 1 to 0 as the angle increases from 0° to 90°. The graphs are shown at right.

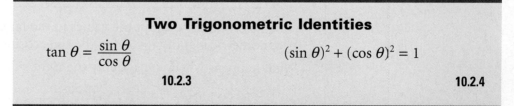

In Activity 2, you discovered and proved the following two important trigonometric identities:

Two Trigonometric Identities

$$\tan\theta = \frac{\sin\theta}{\cos\theta}$$

10.2.3

$$(\sin\theta)^2 + (\cos\theta)^2 = 1$$

10.2.4

Mathematical ideas often have intriguing connections like these, which is part of the fun of mathematics. In your later studies, you will see these identities often—they seem to pop up everywhere!

Exercises

Communicate

1. Think of a mnemonic (memory aid) to help you remember the parts of the tangent, sine, and cosine ratios.

$$\tan \theta = \frac{\text{opposite}}{\text{adjacent}} \qquad \sin \theta = \frac{\text{opposite}}{\text{hypotenuse}} \qquad \cos \theta = \frac{\text{adjacent}}{\text{hypotenuse}}$$

2. Describe three ways you could find θ in the triangle at right. Do your answers agree for the three ways?

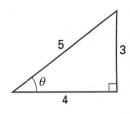

3. In Activity 2, you found that $\tan \theta = \frac{\sin \theta}{\cos \theta}$. Use the behavior of $\sin \theta$ and $\cos \theta$ to explain the behavior of $\tan \theta$ for values of θ close to $0°$ and close to $90°$.

4. Which of the following equations are identities? Explain your reasoning in each case.
   a. $2x + 5 = 7$
   b. $n + 2n = 3n$
   c. $a^2 + b^2 = c^2$
   d. $\sin \theta = \tan \theta \cdot \cos \theta$

Guided Skills Practice

Determine the height of each triangle. Round to the nearest foot.
(EXAMPLE 1)

5. 25 ft, h, 34°

6. 8 ft, h, 60°

7. 32 ft, h, 24°

Find θ in each triangle. Round to the nearest degree. *(EXAMPLE 2)*

8.

9.

87 ft, θ, 72 ft

82 ft, θ, 42 ft

Practice and Apply

For Exercises 10–17, refer to △CDE.
Find each of the following:

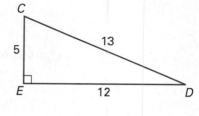

10. sin C   **11.** sin D

12. cos C   **13.** cos D

14. tan C   **15.** tan D

16. m∠C   **17.** m∠D

For Exercises 18–23, refer to △XYZ.
Find each of the following:

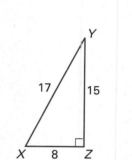

18. $\sin \underline{\ ?\ } = \frac{15}{17}$ **19.** $\cos \underline{\ ?\ } = \frac{15}{17}$

20. $\sin \underline{\ ?\ } = \frac{8}{17}$ **21.** $\cos \underline{\ ?\ } = \frac{8}{17}$

22. $\tan \underline{\ ?\ } = \frac{15}{8}$ **23.** $\tan \underline{\ ?\ } = \frac{8}{15}$

TECHNOLOGY Use a scientific or graphics calculator to find the following.
Round your answers to the nearest hundredth.

24. sin 35°   **25.** cos 72°   **26.** sin 57°

27. cos 52°   **28.** sin 45°   **29.** cos 45°

Round your answers to the nearest degree.

30. $\sin^{-1} 0.3$ **31.** $\sin^{-1} 0.875$ **32.** $\cos^{-1} 0.56$

33. $\cos^{-1} 0.125$ **34.** $\sin^{-1} 0.5$ **35.** $\cos^{-1} 0.95$

Use trigonometric ratios to find the area of each figure.

36. triangle ABC   **37.** parallelogram KLMN

TECHNOLOGY Use a scientific or graphics calculator to answer Exercises 38–40.

38. Choose an angle between 0° and 90°. Find the sine of your angle, then find the inverse sine of your answer. Repeat for four more angles. Based on your observation, complete the following identity:

$$\sin^{-1}(\sin \theta) = \underline{\quad ? \quad}$$

39. Repeat Exercise 38 for the cosine and inverse cosine. Complete the following identity:

$$\cos^{-1}(\cos \theta) = \underline{\quad ? \quad}$$

40. Repeat Exercise 38 for the tangent and inverse tangent. Complete the following identity:

$$\tan^{-1}(\tan \theta) = \underline{\quad ? \quad}$$

The prefix *co-* in cosine indicates a certain relationship of the cosine to the sine of an angle. The exercises below develop this relationship.

41. In $\triangle ABC$ below, find sin A and cos B. Using the right triangle, explain why these ratios are the same.

42. Complete the following statements. You may wish to draw right triangles to help determine your answers.

$\sin 30° = \cos$ ___?___ $\sin 65° = \cos$ ___?___

$\sin$ ___?___ $= \cos 40°$ $\sin$ ___?___ $= \cos 45°$

43. What is the relationship between the pairs of angles in Exercises 41 and 42? Use your answer to complete the following identities:

$\sin \theta = \cos$ ___?___ $\cos \theta = \sin$ ___?___

Use the identities $\tan \theta = \dfrac{\sin \theta}{\cos \theta}$ and $(\sin \theta)^2 + (\cos \theta)^2 = 1$ to simplify the following expressions:

44. $\tan \theta \cdot \cos \theta$

45. $\dfrac{\sin \theta}{\tan \theta}$

46. $1 - (\sin \theta)^2$

47. $\dfrac{1}{(\cos \theta)^2} - 1$

CHALLENGE

48. Use the identity $(\sin \theta)^2 + (\cos \theta)^2 = 1$, together with factoring and substitution, to prove the following identity:

$$(\sin \theta)^4 - (\cos \theta)^4 = 2(\sin \theta)^2 - 1$$

APPLICATIONS

49. FORESTRY A spruce tree is approximately the shape of a cone with a slant height of 20 ft. The angle formed by the tree with the ground measures 72°. Estimate the height of the tree. Round to the nearest foot.

50. RECREATION A water slide is a straight ramp 25 m long that starts from the top of a tower 21 m high. Find the angle the slide forms with the tower.

51. CONSTRUCTION According to the Americans With Disabilities Act, a ramp can rise no more than 1 ft for every 12 ft of horizontal distance. What is the maximum angle that the ramp can form with the ground?

For each length given, find the remaining two lengths.
(LESSON 5.5)

52. $x = 7$ $y = \underline{\ ?\ }$ $z = \underline{\ ?\ }$

53. $x = \underline{\ ?\ }$ $y = 14$ $z = \underline{\ ?\ }$

54. $x = \underline{\ ?\ }$ $y = \underline{\ ?\ }$ $z = 13$

55. $p = 1$ $q = \underline{\ ?\ }$ $r = \underline{\ ?\ }$

56. $p = \underline{\ ?\ }$ $q = 3$ $r = \underline{\ ?\ }$

57. $p = \underline{\ ?\ }$ $q = \underline{\ ?\ }$ $r = 16$

For each arc length, ℓ, and radius, r, given below, find the measure of the central angle. *(LESSON 9.1)*

58. $\ell = 12\pi, r = 20$ **59.** $\ell = 10\pi, r = 100$

60. $\ell = 3\pi, r = 25$

The three ratios in a right triangle that we have explored are tangent, sine, and cosine. There are three other ratios to be considered.

cotangent	secant	cosecant
$\cot \theta = \dfrac{\text{adjacent}}{\text{opposite}}$	$\sec \theta = \dfrac{\text{hypotenuse}}{\text{adjacent}}$	$\csc \theta = \dfrac{\text{hypotenuse}}{\text{opposite}}$

61. How are the ratios above related to the tangent, sine, and cosine ratios?

62. In Exercises 41–43, you discovered a relationship between the sine and cosine in the form of two identities. Is there a similar relationship between tangent and cotangent? between secant and cosecant? Explore, using right triangles or your calculator, and then complete the following identities:

a. $\tan \theta = \cot \underline{\ ?\ }$ **b.** $\cot \theta = \tan \underline{\ ?\ }$
c. $\sec \theta = \csc \underline{\ ?\ }$ **d.** $\csc \theta = \sec \underline{\ ?\ }$

PYTHAGOREAN IDENTITIES

The identity $(\sin \theta)^2 + (\cos \theta)^2 = 1$ is often called a Pythagorean identity because it is derived from the Pythagorean Theorem. There are two other Pythagorean identities that can easily be derived from this identity.

1. Divide each term in the identity $(\sin \theta)^2 + (\cos \theta)^2 = 1$ by $(\sin \theta)^2$, and simplify as much as possible. Express your identity using one of the ratios in Exercises 61 and 62.

2. Divide each term in the identity $(\sin \theta)^2 + (\cos \theta)^2 = 1$ by $(\cos \theta)^2$, and simplify as much as possible. Express your identity using one of the ratios in Exercises 61 and 62.

WORKING ON THE CHAPTER PROJECT

You should now be able to complete Activity 2 of the Chapter Project.

Extending the Trigonometric Ratios

Objectives

- Use a rotating ray in a coordinate plane to define angles measuring greater than 90° and less than 0°.

- Define *sine, cosine,* and *tangent* for angles of any size.

Why So far, trigonometric ratios have been defined in terms of acute angles of right triangles. But it is important to define the trigonometric ratios for other angles. In this lesson, you will learn how this can be done.

The curve shown on the oscilloscope is a graph of the sound wave generated by the synthesizer. Such a curve is known as a sine curve.

Extending Angle Measure

Imagine a ray with its endpoint at the origin of a coordinate plane and extending along the positive *x*-axis. Then imagine the ray rotating a certain number of degrees, say θ, counterclockwise about the origin. As the illustration shows, θ can be any number of degrees, including numbers greater than 360°. A figure formed by a rotating ray and a stationary reference ray, such as the positive *x*-axis, is called an **angle of rotation**.

Angles of Rotation

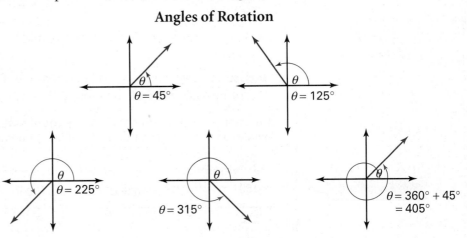

CRITICAL THINKING   How do you think an angle of rotation could have a negative measure?

The Unit Circle

To define the trigonometric ratios for all possible rotation angles you can use the *unit circle*. The **unit circle** is a circle with its center at the origin and a radius of 1. In the language of transformations, it consists of all the rotation images of the point $P(1, 0)$ about the origin.

EXAMPLE **1** Find the coordinates of P', the image of point $P(1, 0)$ rotated 150° about the origin.

SOLUTION

PROBLEM SOLVING

Use a graph. Draw a perpendicular segment from P' to the x-axis as shown. Label the intersection Q. Since $\angle P'OQ$ is supplementary to $\angle P'OP$, its measure is 30°. Thus, $\triangle P'OQ$ is a 30-60-90 triangle with its hypotenuse equal to 1.

$QO = \frac{\sqrt{3}}{2}$ and $P'Q = \frac{1}{2}$. The x-coordinate of P' is $\frac{-\sqrt{3}}{2}$ or ≈ -0.866.

The y-coordinate of P' is $\frac{1}{2} = 0.5$. Thus, the coordinates of P' are $(-0.866, 0.5)$.

Activity 1
Redefining the Trigonometric Ratios

YOU WILL NEED

scientific or graphics calculator

1. Let P' be the 30° rotation image of $P(1, 0)$ about the origin. Use the rules from the 30-60-90 Right Triangle Theorem to find the coordinates of P'.

2. Let P'' be the 210° rotation image of $P(1, 0)$ about the origin. Find the coordinates of P''.

3. Let P''' is the 330° rotation image of $P(1, 0)$ about the origin. Find the coordinates of P'''.

CHECKPOINT ✔

4. **a.** Use the results of Steps 1–3 to complete columns 2 and 3 of the table below.

   b. Use the SIN and COS keys of a scientific calculator to complete columns 4 and 5 of the table.

See Example 1.

Rotation angle, θ	x-coordinate of image point	y-coordinate of image point	$\cos \theta$	$\sin \theta$
30°	?	?	?	?
150°	$\frac{-\sqrt{3}}{2} \approx -0.866$	$\frac{1}{2} = 0.5$	?	?
210°	?	?	?	?
330°	?	?	?	?

5. What relationships do you see in the table?

6. Complete the following definition for the sine and cosine of an angle.

CHECKPOINT ✔

> ### Unit Circle Definition of Sine and Cosine
>
> Let θ be a rotation angle. Then **sin θ** is the ___?___ -coordinate of the image of point $P(1, 0)$ rotated $\theta°$ about the origin, and **cos θ** is the ___?___ -coordinate.

CRITICAL THINKING

Negative rotations are represented by *clockwise* rotations of the point $P(1, 0)$. Test the unit circle definition of sine and cosine as follows:

  a. Sketch the image point P' of $P(1, 0)$ as a result of a $-30°$ rotation (clockwise) about the origin.

  b. Find the coordinates of the image point P' and use this information to give the sine and cosine of $-30°$.

Check your answers by using a scientific calculator.

**E X A M P L E ② **

APPLICATION
ROTARY MOTION

A wheel with a 1-ft radius is turning slowly at a constant velocity of $1°$ per second and has a light mounted on its rim. A distant observer watching the wheel from the edge sees the light moving up and down in a vertical line. Write an equation for the vertical position, h, of the light starting from the horizontal position at time $t = 0$. What will be the vertical position of the light after 1 min? after 5 min? after 24 hr?

● **SOLUTION**

Imagine a coordinate system with the origin at the center of the wheel as shown. After t seconds have elapsed, the value of θ is $t°$. The coordinates of the light are $(\cos t, \sin t)$. Notice that $\sin t$, the second coordinate, is the vertical position of the light. Thus, at time t the vertical position, h, of the light is given by the equation

$$h = \sin t.$$

At $t = 1$ min, or 60 sec, $h = \sin 60° \approx 0.867$ units.

At $t = 5$ min, or 300 sec, $h = \sin 300° \approx -0.867$ units.

At $t = 24$ hr, or 86,400 sec, $h = \sin 86,400° = 0$ units.

Activity 2
Graphing the Trigonometric Ratios

1. Extend the graphs of each of the trigonometric ratios below for values of θ from 0° to 360° by plotting points at intervals of 30°.

2. Use your graphs to determine the intervals in which the sine and cosine are positive and those in which they are negative. Complete the table below.

CHECKPOINT ✔

Quadrant of image point	Sign of sin θ	Sign of cos θ
I (0° to 90°)	+	?
II (90° to 180°)	+	?
III (180° to 270°)	−	?
IV (270° to 360°)	?	?

Using a Calculator to Find an Angle

In Activity 1, you saw that sin 30° and sin 150° both equal 0.5. There are, in fact, infinitely many angles for which sin θ has a given value from 0 to 1, but in working with triangles, you will want to find angles between 0° and 180°. The example which follows illustrates how to use your calculator along with a graph to find the desired angles.

EXAMPLE ❸ Find two values of θ between 0° and 180° such that sin $\theta \approx 0.9397$.

● **SOLUTION**

Use the $\boxed{\text{SIN}^{-1}}$ (inverse sine) key to find the measure of an angle whose sin is 0.9397. The calculator will give you approximately 70°. Note: For a given value from 0 to 1, the $\sin^{-1}$ function of a calculator will always return an angle measure from 0° to 90°.

PROBLEM SOLVING

Draw a graph of the sine function. You can see that there is another angle that has the same sine value. To find that angle, subtract the first angle from 180°.

$$180° - 70° = 110°$$

Thus, the values are 70° and 110°.

70° and 110° have the same sine value.

TRY THIS Find two values of θ between 0° and 180° such that sin $\theta \approx 0.5736$.

Exercises

● **Communicate**

1. Explain how a rotating ray is used to extend the definition of an angle.

2. How is the unit circle used to extend the trigonometric ratios beyond 90°?

Explain how to use the unit circle definition of sine and cosine to find each of the following:

3. sin 90° and cos 90°
4. sin 180° and cos 180°

Explain how to use the unit circle definition of sine and cosine to find the sign of each of the following:

5. cosine in Quadrant III

6. sine in Quadrant II

7. **ENGINEERING** The drive mechanism of the oil pump converts the circular motion of point X to vertical motion in a straight line at point Y. Explain how a trigonometric function can represent the vertical component of the motion of point X.

● **Guided Skills Practice**

8. Find the coordinates of P', the image of point P(1, 0) rotated 210° about the origin.
   (EXAMPLE 1)

9. Use the unit circle to find the sine and cosine of θ from the coordinates of the given point.
   (ACTIVITY 1)

10. A wheel with a radius of 1 unit turns at a rate of 1° per second. Write an equation for the vertical position of point X starting from the horizontal position shown at t = 0. What is the vertical position of point X after 10 min? 20 min? 1 hr? 12 hr?
    (EXAMPLE 2)

    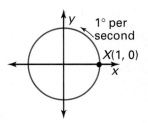

11. Use the graph of the cosine ratio to determine the sign of the cosine in Quadrants I through IV of the coordinate plane. **(ACTIVITY 2)**

12. Find two values of θ between 0° and 180° such that sin θ ≈ 0.7071.
    (EXAMPLE 3)

Practice and Apply

TECHNOLOGY In Exercises 13–20, use a calculator to find each of the following, rounded to four decimal places:

13. sin 65°   **14.** sin 115°   **15.** sin 245°   **16.** sin 295°

17. cos 65°   **18.** cos 115°   **19.** cos 245°   **20.** cos 295°

In Exercises 21–28, use graph paper and a protractor to sketch a unit circle and a ray with the given angle θ with the positive x-axis. Find the coordinates of the point on the ray at a distance of 1 from the origin, rounded to four decimal places. Use these values and the unit circle definition of sine and cosine to give the sine and cosine of each angle, rounded to four decimal places.

21. 45°   **22.** 135°   **23.** 225°   **24.** 315°

25. 30°   **26.** 150°   **27.** 210°   **28.** 330°

TECHNOLOGY In Exercises 29–36, use a calculator to find the sine and cosine of each angle. Use these values to give the x- and y-coordinates of a point at the given angle on the unit circle. Round your answers to four decimal places.

29. 30°   **30.** 60°   **31.** 90°   **32.** 120°

33. 180°   **34.** 210°   **35.** 300°   **36.** 360°

In Exercises 37–44, give two values of θ between 0° and 180° for the given value of sin θ. Express your answers to the nearest degree.

37. 0.7071   **38.** 0.8660   **39.** 0.5000   **40.** 0.9659

41. 0.3217   **42.** 0.9900   **43.** 0.9990   **44.** 0.9999

45. If sin θ = 0.4756, what are all the possible values of θ between 0° and 360°, rounded to the nearest degree?

46. If cos θ = −0.7500, what are all the possible values of θ between 0° and 360°, rounded to the nearest degree?

47. ASTRONOMY An astronomer observes a satellite that is moving around a planet at the rate of 1° per hour. Assuming that the radius of the satellite's orbit is 1 unit, what is the position, viewed from the edge, of the satellite after 2 days? after 5 days?

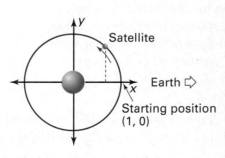

Satellite

Earth ⇨

Starting position
(1, 0)

View of planet from above

Use △*ABC* for Exercises 48–55. *(LESSONS 10.1 AND 10.2)*

48. $\sin A = \frac{?}{c}$

49. $\cos B = \frac{?}{c}$

50. $\tan A = \frac{?}{b}$

51. $\cos A = \frac{?}{c}$

52. $(\sin A)^2 + \underline{\,?\,} = 1$

53. $\frac{\sin A}{\cos A} = \underline{\,?\,}$

54. $\sin A = \cos \underline{\,?\,}$

55. $\cos A = \sin \underline{\,?\,}$

Look Beyond

ASTRONOMY The graph shows the positions of the four "Galilean" moons of Jupiter as seen from Earth at midnight on April 1–16, 1993. The parallel lines in the center of the graph represent the visible width of Jupiter.

Configuration of satellites I-IV for April 1–16, 1993 at midnight Greenwich Mean Time (GMT)

56. Galileo was able to observe Jupiter's four largest moons with his small telescope. When viewed from Earth, they appear to move back and forth in an approximately straight line through the center of the planet. The names of the four moons are listed below.

    I.   Io
    II. Europa
    III. Ganymede
    IV. Callisto

Use the graph to sketch the positions of the planet Jupiter and its four Galilean moons as they would appear to a person with a small telescope or a pair of binoculars on (a) April 4, (b) April 8, and (c) April 12. Use dots for the moons and a circle for Jupiter.

57. Use the graph to estimate the orbital periods of each of the four Galilean moons of Jupiter.

58. What kind of curve do the lines for the moons' orbits appear to be? Explain why they have this shape. (Note: The orbits of the four moons are nearly circular.)

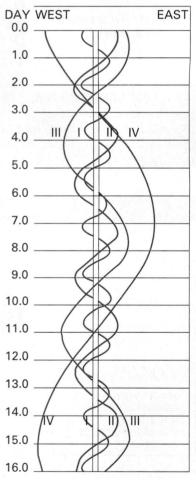

Source: *The Astronomical Almanac for the Year 1993*

The Law of Sines

Satellite photo of San Francisco Bay area

Objectives

- Develop the law of sines.

- Use the law of sines to solve triangles.

Why
The triangle formed by the three points on the satellite photo is not a right triangle, however, the distances between the points can be found by using trigonometry.

The Law of Sines

The trigonometric ratios you have studied all relate to right triangles. In the following Activity, you will explore the law of sines, a theorem involving sine ratios that applies to all triangles.

Activity 1

Law of Sines

YOU WILL NEED

geometry graphics software or ruler and protractor, and scientific or graphics calculator

1. Draw an acute triangle, a right triangle, and an obtuse triangle. Label the sides and edges as shown. Measure the sides and angles of each triangle.

2. Copy and complete the table below, using your measurements from Step 1.

	m∠A	m∠B	m∠C	a	b	c	$\frac{\sin A}{a}$	$\frac{\sin B}{b}$	$\frac{\sin C}{c}$
acute	?	?	?	?	?	?	?	?	?
right	?	?	?	?	?	?	?	?	?
obtuse	?	?	?	?	?	?	?	?	?

CHECKPOINT ✔

3. Write a conjecture using the data in the last three columns of the table.

Your conjecture from Activity 1 can be stated as the following theorem:

The Law of Sines

For any triangle $\triangle ABC$ with sides a, b, and c:

$$\frac{\sin A}{a} = \frac{\sin B}{b} = \frac{\sin C}{c}$$

10.4.1

The proof of the law of sines is divided into three cases: acute, right, and obtuse triangles. The following is a proof of the acute case; you will prove the remaining cases in Exercises 33–49.

PROOF

Given: acute $\triangle ABC$ with sides a, b, and c

Prove: $\dfrac{\sin A}{a} = \dfrac{\sin B}{b} = \dfrac{\sin C}{c}$

Proof:

Draw the altitudes from B and C. Label the intersections of the altitudes with $\overline{AB}$ and $\overline{AC}$ as D_1 and D_2, with lengths h_1 and h_2, respectively. Then $\sin A = \dfrac{h_1}{b}$ and $\sin B = \dfrac{h_1}{a}$. Solving each equation for h_1, $h_1 = b \sin A$ and $h_1 = a \sin B$. By substitution, $b \sin A = a \sin B$. Dividing both sides of the equation by ab gives:

$$\frac{b \sin A}{ab} = \frac{a \sin B}{ab}$$

$$\frac{\sin A}{a} = \frac{\sin B}{b}$$

Similarly, $\sin A = \dfrac{h_2}{c}$ and $\sin C = \dfrac{h_2}{a}$, so $c \sin A = a \sin C$, which gives:

$$\frac{\sin A}{a} = \frac{\sin C}{c}$$

Therefore, $\dfrac{\sin A}{a} = \dfrac{\sin B}{b} = \dfrac{\sin C}{c}$

Solving Triangles

The law of sines can be used to find the measures of sides and angles in a triangle, if the measures of some sides and angles are given. You can use the law of sines to solve a triangle if you are given the measures of

1. two angles and one side

or

2. two sides and an angle that is opposite one of the sides.

EXAMPLE 1 Find b and c in the triangle at right.

SOLUTION

Set up a proportion using two ratios from the law of sines. Start with the known angle whose opposite side is also known, in this case, $\angle A$ and a.

Then by the law of sines:

$$\frac{\sin A}{a} = \frac{\sin C}{c}$$

$$\frac{\sin 72°}{20} = \frac{\sin 41°}{c}$$

$$c \sin 72° = 20 \sin 41°$$

$$c = \frac{20 \sin 41°}{\sin 72°} \approx 13.8$$

$$\frac{\sin A}{a} = \frac{\sin B}{b}$$

$$\frac{\sin 72°}{20} = \frac{\sin 67°}{b}$$

$\left. \right. \qquad m\angle B = 180° - (72° + 41°)$
$\qquad\qquad\qquad = 67°$

$$b \sin 72° = 20 \sin 67°$$

$$b = \frac{20 \sin 67°}{\sin 72°} \approx 19.4$$

TRY THIS Given $\triangle ABC$ from Example 1 with $a = 15$, $m\angle A = 70°$, and $m\angle C = 50°$, find b and c.

EXAMPLE 2 Find $m\angle Q$ in the triangle at right.

SOLUTION

$$\frac{\sin R}{r} = \frac{\sin Q}{q}$$

$$\frac{\sin 82°}{34} = \frac{\sin Q}{27}$$

$$34 \sin Q = 27 \sin 82°$$

$$\sin Q = \frac{27 \sin 82°}{34} \approx 0.786$$

$$Q \approx \sin^{-1} 0.786 \approx 52°$$

TRY THIS Given $\triangle QRS$ from Example 2 with $q = 18$, $r = 26$, and $m\angle R = 75°$, find $m\angle Q$.

Note: When solving triangles by using the law of sines where two sides and a nonincluded angle are given, it may be possible to have two different answers. This is known as the *ambiguous case* of the law of sines.

In the example, $\sin^{-1}$ was used to find the angle in question. However, the inverse sine function will only give values of angles less than 90°. To find the possible obtuse angle, subtract the value from 180°.

Ambiguous case of the law of sines

CRITICAL THINKING Is the situation in Example 2 ambiguous? Why or why not?

EXAMPLE ③ A boater travels regularly from San Mateo to Alameda across San Francisco Bay, a distance of 12.8 mi. Suppose that one day he needs to stop at the baseball park. If the angle formed at the baseball park is 95° and the angle formed at Alameda is 53°, how far does the boater have to travel to get to Alameda via the baseball park?

SOLUTION

Use the law of sines

$$\frac{\sin 95°}{12.8} = \frac{\sin 53°}{b}$$

$$b \sin 95° = 12.8 \sin 53°$$

$$c = \frac{12.8 \sin 53°}{\sin 95°} \approx 10.3$$

$$\frac{\sin 95°}{12.8} = \frac{\sin 32°}{b}$$

$$b \sin 95° = 12.8 \sin 32°$$

$$c = \frac{12.8 \sin 32°}{\sin 95°} \approx 6.8$$

The total distance to Alameda is approximately $10.3 + 6.8 = 17.1$ mi.

Exercises

Communicate

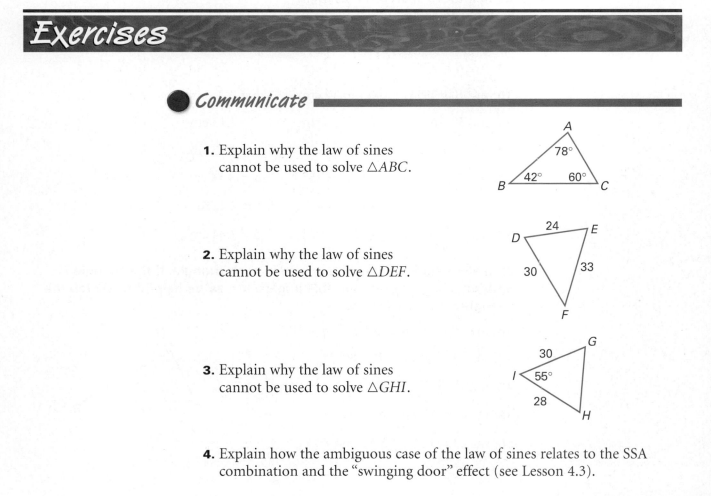

1. Explain why the law of sines cannot be used to solve $\triangle ABC$.

2. Explain why the law of sines cannot be used to solve $\triangle DEF$.

3. Explain why the law of sines cannot be used to solve $\triangle GHI$.

4. Explain how the ambiguous case of the law of sines relates to the SSA combination and the "swinging door" effect (see Lesson 4.3).

5. Find b and c in $\triangle ABC$.
(EXAMPLE 1)

6. Find $m\angle Q$ in $\triangle QRS$.
(EXAMPLE 2)

7. How much greater is the distance from point X to point Y to point Z than the distance from point X to point Z? *(EXAMPLE 3)*

Practice and Apply

In Exercises 8–15, find the indicated measures. Assume that all angles are acute. It may be helpful to sketch the triangle roughly to scale.

8. $m\angle A = 56°$ $m\angle B = 24°$ $b = 1.22$ cm $c =$ ___?___

9. $m\angle B = 29°$ $a = 3.12$ cm $b = 3.28$ cm $m\angle A =$ ___?___

10. $m\angle B = 29°$ $a = 3.12$ cm $b = 3.28$ cm $c =$ ___?___

11. $m\angle B = 67°$ $b = 7.36$ cm $c = 2.13$ cm $m\angle C =$ ___?___

12. $m\angle B = 67°$ $b = 7.36$ cm $c = 2.13$ cm $a =$ ___?___

13. $m\angle B = 73°$ $m\angle C = 85°$ $a = 3.14$ cm $b =$ ___?___

14. $m\angle A = 35°$ $m\angle B = 44°$ $c = 2.4$ cm $a =$ ___?___

15. $m\angle A = 53°$ $m\angle C = 72°$ $c = 2.34$ cm $b =$ ___?___

Find all unknown sides and angles in each triangle. If the triangle is ambiguous, give both possible angles. It may be helpful to sketch the triangle roughly to scale.

16. $m\angle P = 30°$, $m\angle Q = 40°$, $q = 10$

17. $m\angle Q = 60°$, $m\angle R = 80°$, $r = 7$

18. $m\angle P = 72°$, $m\angle R = 36°$, $q = 12$

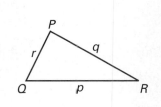

19. $m\angle P = 60°$, $p = 9$, $q = 7$

20. $m\angle Q = 120°$, $q = 12$, $r = 8$

21. $m\angle R = 45°$, $p = 11$, $r = 8$

In Exercises 22–28, you will investigate the ambiguous case of the law of sines. Suppose that two sides of a triangle, *a* and *b*, and an angle opposite one side, ∠A, are given.

22. Case I: m∠A < 90°

Draw a line and choose point *A* on the line. Choose measurements for ∠A and *b*, and use a ruler and protractor to draw side *b*.

Use a calculator to find *b* sin *A*. Explain why there is exactly one possible triangle if $a = b \sin A$.

23. If $a < b \sin A$, how many triangles are possible? Explain your reasoning.

24. If $a \geq b$, how many triangles are possible? Explain your reasoning.

25. If $b \sin A < a < b$, how many triangles are possible? Explain your reasoning.

26. Case II: m∠A ≥ 90°

Draw a line and choose point *A* on the line. Choose measurements for ∠A and *b*, and use a ruler and protractor to draw side *b*.

Explain why there is exactly one possible triangle if $a > b$.

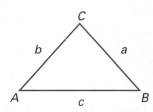

27. If $a \leq b$, how many triangles are possible? Explain your reasoning.

28. Summarize your results from Exercises 22–27 above by writing a rule in your own words for deciding whether a given set of SSA measurements determine one triangle, two possible triangles, or no triangle.

In Exercises 29–32, two sides of a triangle, *a* and *b*, and an angle opposite one side, ∠A, are given. Use your results from Exercises 22–28 to explain whether the given measurements determine one triangle, two possible triangles, or no triangle.

29. m∠A = 70°, a = 4, b = 3

30. m∠A = 110°, a = 7, b = 5

31. m∠A = 30°, a = 4, b = 6

32. m∠A = 145°, a = 3, b = 6

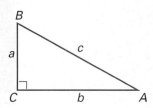

For Exercises 33–35, use the diagram to prove the law of sines for right triangles.

33. Write the sine of angles *A* and *B* in terms of *a*, *b*, and *c*.

34. Write the following ratios in terms of *a*, *b*, and *c*, and simplify:

$$\frac{\sin A}{a} \qquad \frac{\sin B}{b} \qquad \frac{\sin C}{c}$$

35. Complete the proof of the law of sines for right angles.

For Exercises 36–49, use the diagram at right to prove the law of sines for obtuse triangles.

Given: obtuse $\triangle ABC$ with sides a, b, and c

Prove: $\dfrac{\sin A}{a} = \dfrac{\sin B}{b} = \dfrac{\sin C}{c}$

Proof:

Draw altitudes from B and C. Label the intersection of the altitudes with $\overleftrightarrow{AB}$ and $\overline{AC}$ as D_1 and D_2, with lengths h_1 and h_2, respectively.

Statements		Reasons
$\sin A = \dfrac{h_1}{b}$ $\sin B = \dfrac{h_1}{a}$		Definition of sine
$h_1 = $ **36.** ? $h_1 = $ **37.** ?		**38.** ?
39. ?		Substitution Property
$\dfrac{\sin A}{a} = \dfrac{\sin B}{b}$		**40.** ?
$\sin A = \dfrac{h_2}{c}$ $\sin C = $ **41.** ?		**42.** ? **18.** ?
$h_2 = $ **43.** ? $h_2 = $ **44.** ?		**45.** ?
$c \sin A = a \sin C$		**46.** ?
47. ?		Division Property
48. ?		**49.** ?

50. Use the law of sines to prove that the identity below is true for any triangle $\triangle ABC$.

$$\frac{a-b}{\sin A - \sin B} = \frac{a+b}{\sin A + \sin B}$$

51. WILDLIFE MANAGEMENT Scientists are tracking polar bears that have been fitted with radio collars. The scientists have two stations that are 9 km apart along a straight road. At station 1, the signal from one of the collars comes from a direction 49° from the road. At station 2, the signal from the same collar comes from a direction 65° from the road. How far is the polar bear from station 2? What is the shortest distance from the road to the bear?

52. ARCHITECTURE A real-estate developer wants to build an office building on a triangular lot between Oak Street and 3rd Avenue. The dimensions of the lot are shown at right. Use the diagram to find the measures of the angles formed by the sides of the lot.

53. FORESTRY A plume of smoke is spotted from two different fire towers that are 5 mi apart. From tower A, the angle between the smoke plume and tower B is 80°. From tower B, the angle between the smoke plume and tower B is 70°. Which tower is closer to the smoke plume? How far is the smoke from the tower?

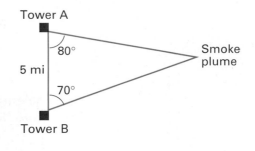

SURVEYING Suppose that you are driving toward a mountain. You stop and measure the angle of elevation from your eye to the top of the mountain, which is 11.5°. You then drive 5 mi on level ground. You stop again and measure the new angle of elevation, which is 27.5°.

54. Determine the measure of $\angle ABC$.

55. Find BC.

56. Use your result from Exercise 55 to determine the height of the mountain.

57. Find the surface area and volume of a sphere with a radius of 15. *(LESSON 7.6.)*

58. What happens to the volume and surface area of a sphere if the radius is tripled? *(LESSONS 7.6 AND 8.6)*

Find x in each diagram below. *(LESSONS 8.3 AND 8.4)*

59.

60.

Use a scientific or graphics calculator to find the following. Draw a triangle to illustrate each angle. *(LESSONS 10.1 AND 10.2)*

61. $\tan^{-1}\sqrt{5}$ **62.** $\sin^{-1}\left(\frac{\sqrt{3}}{2}\right)$ **63.** $\cos^{-1}\left(\frac{\sqrt{2}}{2}\right)$

CONNECTION

POLAR COORDINATES In a rectangular coordinate system, a point is represented by *x* and *y*, the horizontal and vertical distance from the origin. In a *polar coordinate* system, a point is represented by a distance from the origin, *r*, and an angle from the positive *x*-axis, *θ*.

64. Point *A* has coordinates (2, 30°) in a polar coordinate system. Plot point *A* on a graph and find its rectangular coordinates.

65. Point *B* has coordinates (3, 3) in a rectangular coordinate system. Plot point *B* on a graph and find its polar coordinates.

66. Use a graphics calculator with a polar graphing mode to graph some of the following functions. Draw or describe the graphs.
$r = \sin\theta$ $r = \sin(2\theta)$ $r = \sin(3\theta)$
$r = \cos\theta$ $r = \cos(2\theta)$ $r = \cos(3\theta)$

A good window to use:
$0 \le \theta \le 360$
with θ *step = 1*
$-1.5 \le x \le 1.5$
$-1 \le y \le 1$

In polar coordinates, the simplest graph is a circle. Many graphs in polar coordinates have the shapes of flowers, stars, or spirals.

The Law of Cosines

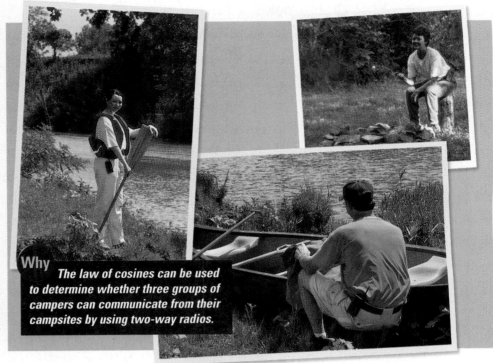

Objectives

- Use the law of cosines, together with the law of sines, to solve triangles.

- Prove the acute case of the law of cosines.

Why The law of cosines can be used to determine whether three groups of campers can communicate from their campsites by using two-way radios.

At a campground there are three groups of campers at three different campsites. The campers carry two-way radios with a range of about 1 mile. What information is needed to determine whether the three groups can all communicate directly with each other?

The Law of Cosines

In the previous lesson, you used the law of sines to solve triangles given two angles and a side or two sides and an angle opposite a given side.

Suppose that you have a triangle with two known sides in which the included angle is known. By SAS congruence, the triangle is uniquely determined. However, the triangle cannot be solved using the law of sines because every possible proportion in the law of sines will have two unknowns.

You can use the law of cosines to solve a triangle if you are given the measures of
  1. two sides and the included angle *or*
  2. three sides.

The Law of Cosines

For any triangle $\triangle ABC$ with sides a, b, and c:

$$a^2 = b^2 + c^2 - 2bc \cos A$$
$$b^2 = a^2 + c^2 - 2ac \cos B$$
$$c^2 = a^2 + b^2 - 2ab \cos C$$

10.5.1

EXAMPLE **1** Find *b* in the triangle at right.

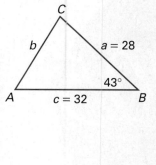

● **SOLUTION**

Using the second equation in the law of cosines,

$$b^2 = a^2 + c^2 - 2ac \cos B$$
$$b^2 = 28^2 + 32^2 - 2(28)(32) \cos 43°$$
$$b^2 \approx 497.4$$
$$b \approx 22.3$$

TRY THIS   Given $\triangle ABC$ from Example 1 with $a = 17$, $c = 25$, and m$\angle B = 55°$, find b.

CRITICAL THINKING   Explain why you cannot use the law of sines to solve the triangle in Example 1.

EXAMPLE **2** Find the measures of $\angle D$, $\angle E$, and $\angle F$ in the triangle at right.

● **SOLUTION**

Step 1

Use the law of cosines to find the first angle. It is convenient to start with the largest angle, which is opposite the longest side, *e*.

$$e^2 = d^2 + f^2 - 2df \cos E$$
$$15^2 = 7^2 + 14^2 - 2(7)(14) \cos E$$
$$225 = 245 - 196 \cos E$$
$$-20 = -196 \cos E$$
$$E = \cos^{-1}\left(\frac{-20}{-196}\right) \approx 84°$$

Step 2

Use the law of sines to find the second angle. Since the remaining angles are acute (why?), you do not have to worry about the ambiguous case.

$$\frac{\sin F}{14} = \frac{\sin 84°}{15}$$
$$\sin F = \frac{14 \sin 84°}{15} \approx 0.93$$
$$F = 68°$$

Step 3

Use the Triangle Sum Theorem to find the third angle.

$$m\angle D \approx 180° - (84° + 68°) \approx 28°$$

TRY THIS   Given $\triangle DEF$ from Example 2 with $d = 10$, $e = 18$, and $f = 11$, find the measures of $\angle D$, $\angle E$, and $\angle F$.

CRITICAL THINKING   Could the law of cosines have been used in Step 2 of Example 2? Why do you think the law of sines was used instead?

EXAMPLE 3 Suppose three campsites are positioned as shown. The campers communicate by using two-way radios with a range of about 1 mi, or 5280 ft. Will the campers at sites 1 and 3 be able to communicate directly with each other?

SOLUTION

Use the law of cosines.

$$b^2 = a^2 + c^2 - 2ac \cos B$$
$$b^2 = 3900 + 3400 - 2(3900)(3400)\cos 86°$$
$$b^2 \approx 22,670,058$$
$$b \approx 4761 \text{ ft}$$

The campers at sites 1 and 3 will be able to communicate directly by using the two-way radios.

TRY THIS   If the distance from site 1 to site 2 is 3600 ft, the distance from site 2 to site 3 is 4000 ft, and m$\angle B = 84°$, will the campers at sites 1 and 3 be able to communicate directly with each other? Why or why not?

Proof of the Law of Cosines: Acute Case

The following is a proof of the acute case of the law of cosines. You will be asked to prove the obtuse case in Exercises 26–31.

PROOF

Given: acute $\triangle ABC$ with sides a, b, and c

Prove: $a^2 = b^2 + c^2 - 2bc \cos A$

Proof:

Draw the altitude from C, with height h, which divides $\overline{AB}$ into two segments, c_1 and c_2. Then, by definition of cosine,

$$\cos A = \frac{c_1}{b}$$

$$(2bc) \cos A = \frac{c_1}{b}(2bc) \qquad \textit{Multiply both sides by 2bc.}$$

$$2bc \cos A = 2cc_1 \qquad \textit{Simplify.}$$

From the Pythagorean Theorem:

$$a^2 = h^2 + c_2{}^2$$
$$a^2 = h^2 + (c - c_1)^2 \qquad \textit{c}_2 = c - c_1$$
$$a^2 = h^2 + c^2 - 2cc_1 + c_1{}^2 \qquad \textit{Distribute.}$$
$$a^2 = (h^2 + c_1{}^2) + c^2 - 2cc_1 \qquad \textit{Rearrange terms.}$$
$$a^2 = b^2 + c^2 - 2cc_1 \qquad \textit{b}^2 = h^2 + c_1{}^2$$
$$a^2 = b^2 + c^2 - 2bc \cos A \qquad \textit{Substitution}$$

Exercises

Communicate

1. Suppose that $\triangle ABC$ is a right triangle. Use the law of cosines to find c^2, and simplify as much as possible. What do you notice?

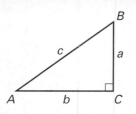

Which rule should you use, the law of sines or the law of cosines, to find each indicated measurement below? Explain your reasoning.

2.

3.

4. Suppose that three angles of a triangle are given. Could you use the law of sines or the law of cosines to solve the triangle? Explain your reasoning.

Guided Skills Practice

Use the law of cosines and the law of sines to solve each triangle.

5. *(EXAMPLE 1)*

6. *(EXAMPLE 2)*

7. Suppose that three campers have two-way radios with a range of 7920 ft. The distance between sites 1 and 2 is 5750 ft, the distance between sites 1 and 3 is 6690 ft, and the angle formed with site 1 at the vertex is 82°. Can the campers at sites 2 and 3 commumicate directly? Why or why not? *(EXAMPLE 3)*

Practice and Apply

Algebra

For Exercises 8–13, find the indicated measures. It may help to sketch the triangle roughly to scale.

8. $a = 12$ $b = \underline{\ ?\ }$ $c = 17$ $m\angle B = 33°$

9. $a = 2.2$ $b = 4.3$ $c = \underline{\ ?\ }$ $m\angle C = 52°$

10. $a = \underline{\ ?\ }$ $b = 68.2$ $c = 23.6$ $m\angle A = 87°$

11. $a = 10$ $b = 7$ $c = 8$ $m\angle A = \underline{\ ?\ }$

12. $a = 3.6$ $b = 3.6$ $c = 2.5$ $m\angle C = \underline{\ ?\ }$

13. $a = 27$ $b = 41$ $c = 15$ $m\angle B = \underline{\ ?\ }$

Solve each triangle.

14.

15.

16.

17.

18.

19.

Exercises 20–25 summarize the methods of solving triangles by using the law of sines and the law of cosines.

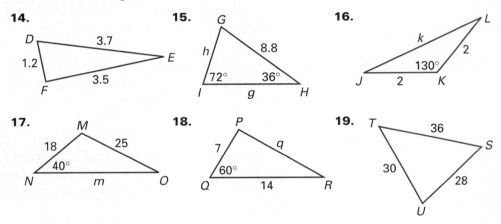

SUMMARY OF TRIANGLE-SOLVING PROCEDURES		
Given	Example	Theorem(s)
three angles		cannot be solved
two angles and one side		**20.** __?__
two sides and included angle		**21.** __?__
22. __?__	**23.** __?__	law of sines (ambiguous case)
three sides	**24.** __?__	**25.** __?__

In Exercises 26–31, you will prove the obtuse case of the law of cosines. Refer to the diagram at right.

26. By definition of cosine, $\cos \theta = \underline{\ ?\ }$.

27. Multiply both sides of the equation from Exercise 26 by b, and simplify.

28. Use the identity $\cos(180 - A) = -\cos A$ to rewrite your equation from Exercise 27 in terms of $\cos A$.

29. By the Pythagorean Theorem, $a^2 = \underline{\ ?\ } + \underline{\ ?\ }$.

30. By the Pythagorean Theorem, $b^2 = \underline{\ ?\ } + \underline{\ ?\ }$.

31. Distribute the terms in the equation from Exercise 29. Substitute the results of Exercises 28 and 30 into this equation and simplify to complete the proof of the law of cosines.

CHALLENGE

32. Using the law of cosines, prove the Triangle Inequality Theorem: For $\triangle ABC$ with sides a, b, and c, $a + b > c$. Also prove that $c > a - b$ for $a > b$.

APPLICATIONS

33. **SPORTS** On a standard baseball field, the bases form a square with 90-ft sides. The pitcher's mound is 60.5 ft from home plate on a diagonal from home plate to second base. Find the distance from the pitcher's mound to first base.

34. **NAVIGATION** Mark and Stephen walk into the woods along lines that form a 72° angle. If Mark walks at 2.8 mph and Stephen walks at 4.2 mph, how far apart will they be after 3 hr?

35. **NAVIGATION** The distance from Greenfield to Brownsville is 37 km, and the distance from Greenfield to Red River is 25 km. The angle formed by the roads at Greenfield is 42°. The state highway department plans to build a straight road from Brownsville to Red River. How long will the road be, and what angle will the new road form with the road from Greenfield to Brownsville?

36. **INDIRECT MEASUREMENT** From point A, the distance to one end of a small lake is 300 yd, the distance to the other end is 500 yd, and the angle formed by the lines of sight is 105°. Estimate the width of the lake.

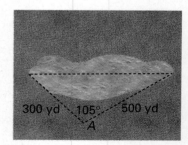

37. **LANDSCAPING** Nina wants to use three long boards as a border for a triangular garden. The boards have lengths of 10 ft, 12 ft, and 15 ft. Find the angles at the corners of the garden.

CONNECTION

COORDINATE GEOMETRY For Exercises 38–41, draw the image of △*ABC* under the given transformation. *(LESSONS 1.7 AND 8.1)*

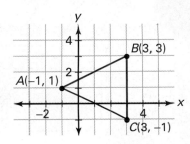

38. $T(x, y) = (3x, 3y)$

39. $T(x, y) = (x - 2, y - 2)$

40. $T(x, y) = \left(\frac{3}{4}x, \frac{3}{4}y\right)$

41. $T(x, y) = (2x + 1, 2y - 3)$

42. The perimeter of a rectangle is 80 cm. The ratio of its width to length is $\frac{1}{5}$. Find the length and width of the rectangle. *(LESSON 8.2)*

Determine whether the triangles in each pair can be proven similar. Explain your reasoning. *(LESSON 8.3)*

43. **44.**

45. In the diagram at right, $\overline{QR}$ is tangent to ⊙*P*. Find radius *PQ*. *(LESSONS 9.2 AND 10.1)*

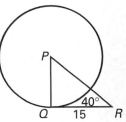

CONNECTION

TRIGONOMETRY You can use right triangles to simplify expressions involving inverse trigonometric functions. For example, consider the expression $\cos(\sin^{-1} x)$.

46. $\sin^{-1} x$ is an angle that has a sine ratio of $\frac{x}{1}$. This angle can be represented by represented by drawing a right triangle with a hypotenuse of 1 and a leg of length *x*. Use the Pythagorean Theorem to find an expression for the length of the adjacent leg.

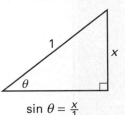

$\sin \theta = \frac{x}{1}$

$\theta = \sin^{-1} x$

47. Use the definition of cosine and your result from Exercise 47 to write an expression for $\cos \theta$, which is equal to $\cos(\sin^{-1} x)$.

Simplify the following expressions:

48. $\tan(\sin^{-1} x)$ **49.** $\sin(\cos^{-1} x)$ **50.** $\sin(\tan^{-1} x)$

Back   Forward   Home   Search   Print   Stop

ASTRONOMERS DISPUTE NASA GAUGE OF UNIVERSE'S AGE

June 2, 1999
Web posted at: 9:28 P.M. EDT (0128 GMT)

CHICAGO (AP) -- A new radio-telescope technique has established a "golden ruler" for measuring cosmic distances and raises doubts about the claim that NASA astronomers had determined the age and expansion rate of the universe.

New measuring techniques

Jim Herrnstein of the National Radio Astronomy Observatory said his method produces "the most precise distance ever measured to a remote galaxy" and suggests a 15 percent to 20 percent margin of error in the technique used by the astronomers sponsored by the National Aeronautics and Space Administration.

"Ours is a direct measurement, using geometry, and is independent of all other methods of determining cosmic distances," Herrnstein said Tuesday at a national meeting of the American Astronomical Society.

Earlier measurements (by NASA)

The NASA team concluded that the universe is 12 billion to 13 billion years old and is expanding at an accelerating rate of 70 kilometers per second for every 3.3 million light-years in distance from the Earth.

'Golden ruler' for measuring cosmic distances

Herrnstein said this calculation was off by as much as 20 percent. He based his conclusion on the difference in calculated distance to a specific galaxy when comparing his technique with the method used by the NASA group.

Herrnstein used a galaxy called NGC 4258 to establish what he called a "golden ruler" for measuring cosmic distances.

Masers: Radio hot spots

NGC 4258 is surrounded by a rotating cloud of gas. Within this gas cloud is water vapor, which tends to amplify radio signals. This creates radio "hot spots" called masers.

The orbital speed of masers between NGC 4258 and Earth was measured in 1994, and again every few months over the following three years. After determining the speed at which the masers were moving, the astronomers created a triangle with a maser position at one angle, its latest position at a second angle, and Earth at the third angle. Measuring the angles gives the distance. "It is very simple and direct," said Herrnstein.

The world's largest telescope

Herrnstein conducted his study using the Very Long Baseline Array, which is a series of radio-telescope antennae that are scattered across the United States, from the Virgin Islands to Hawaii.
The antennae operate as a single unit and are able to measure natural radio signals to an accuracy 500 times greater than the Hubble telescope can measure visible light.

Copyright 1999 The Associated Press. All rights reserved.

Cooperative Learning

Refer to the diagram at right. The angle, θ, between two positions of a given maser in NGC 4258 was measured by the Very Long Baseline Array, or VLBA. The two positions of the maser form two vertices of an isosceles triangle, and the position of Earth forms the third vertex.

Orbit of maser in NGC 4258

Position of maser in 1997

Position of maser in 1994

d d

$\theta = 0.000000025°$

Earth
Diagram not to scale

1. The maser in galaxy NGC 4258 rotates at about 1100 km per second. Use this to determine the distance that the maser traveled in 3 years between the two indicated positions in the triangle. Use 1 yr = 31536000 sec.

2. For extremely small vertex angles, such as θ, the altitude of an isosceles triangle is very nearly equal to the lengths of its legs. Thus the altitude can be used to approximate the lengths, d, of the legs of the triangle in the diagram. Find the altitude as an estimate of d.

3. According to Herrnstein's measurements, NGC 4258 is about 23.5 million light-years away. A light-year is approximately 9.46×10^{12} km. How close is your answer to Herrnstein's?

Jim Herrnstein, National Radio Astronomy Observatory

Vectors in Geometry

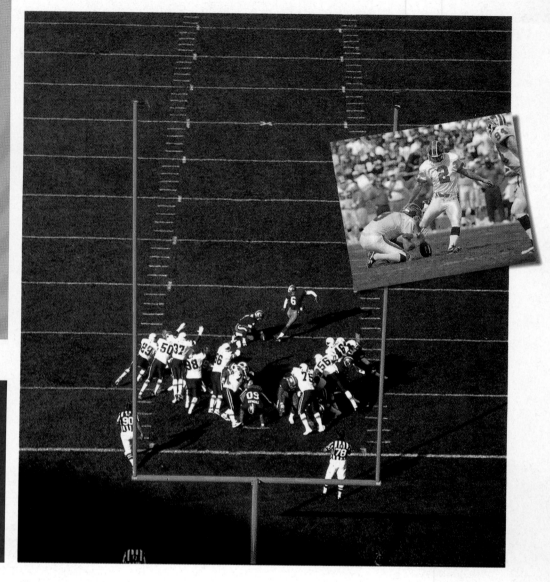

10.6

Objectives

- Define *vector*.
- Add two vectors.
- Use vectors and vector addition to solve problems.

Why

The path of a ball in a sports event is affected by the wind. For this reason, stadiums for certain sports have flags that give a visual indication of the wind's direction and its magnitude. Because wind has these two properties it can be represented by a vector.

What Is a Vector?

A **vector** is a mathematical object that has both a **magnitude** (a numerical measure) and a **direction**. Arrows are used to represent vectors, because arrows have both magnitude and direction. The length of a vector arrow represents the magnitude of the vector.

A lowercase letter with a vector arrow over it may be used to indicate a vector. The symbol $|\vec{v}|$ denotes the magnitude of $\vec{v}$.

$\vec{a}$, $\vec{b}$, and $\vec{c}$ are vectors.

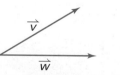

Magnitude of $\vec{v} = |\vec{v}|$

Magnitude of $\vec{w} = |\vec{w}|$

Anything that has both magnitude and direction can be modeled by a vector. In the examples below, vectors are used to represent a velocity, a force, and a displacement (a relocation).

CHECKPOINT ✔ Describe the magnitude and direction of the vector in each.

Velocity The wind sock indicates the strength and direction of the wind.

Force The tugboat applies a force on the barge in the direction the barge is pointed.

Displacement The car speeds down the quarter-mile track.

To describe the direction of a vector, you need a direction to use as a reference. On a coordinate grid this is usually the positive *x*-axis direction.

CHECKPOINT ✔ What are the reference directions in the three pictorial examples?

Vectors with the same direction, such as vectors $\vec{a}$ and $\vec{b}$ in the illustration, are said to be *parallel*. Vectors at right angles are *perpendicular*.

Vector Sums

In many situations it is appropriate to combine vectors to get a new vector called a *resultant*. The process of combining vectors represented in the example below is called *vector addition*.

A hiker travels 3 miles northeast, then 1 mile east. The resultant vector (red) is the total distance traveled in a specific direction from the starting point.

3 mi NE   1 mi E   135°

Two tractors pull on a tree stump with the forces and directions shown. The combined force exerted by the tractors in the direction shown is the resultant vector.

2500 lb
2500 lb

The Head-to-Tail Method of Vector Addition

In the hiker example, each vector represents a displacement—that is, a relocation of the hiker from a given point. In such cases, the *head-to-tail method* is a natural way to do vector addition.

EXAMPLE **1** Find the sum of vectors $\vec{a}$ and $\vec{b}$.

● **SOLUTION**

To use the head-to-tail method of vector addition, place the tail of one vector at the head of the other. The vector sum is a new displacement vector from the tail of the first vector to the head of the second vector.

TRY THIS Use the head-to-tail method to find the sum of two parallel vectors.

The Parallelogram Method of Vector Addition

In many cases, the *parallelogram method* is a natural way to think of combining vectors. This is especially true when two forces act on the same point, such as in the tractor example on the previous page.

EXAMPLE **2** Find the sum of vectors $\vec{m}$ and $\vec{v}$.

● **SOLUTION**

To find the sum of the vectors, complete a parallelogram by adding two segments to the figure. The vector sum is a vector along the diagonal of the parallelogram, starting at the common point.

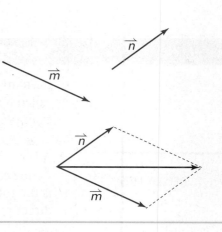

TRY THIS What is the vector sum if two equal forces are applied to a point in opposite directions? What happens to the sum if two such forces are unequal?

CRITICAL THINKING The parallelogram method and the head-to-tail method are equivalent since they produce the same resultant. How does the diagram at right show that the methods are equivalent?

EXAMPLE ③ A swimmer heads perpendicular to a 3-mph current in a river. Her speed in still water is 2.5 mph. Find the actual speed and direction of the swimmer.

APPLICATION
SPORTS

● **SOLUTION**

Because the vectors are at right angles, the parallelogram method for addition gives a rectangle. Thus, you can solve for x by using the Pythagorean Theorem.

$$x^2 = 3^2 + 2.5^2$$
$$= 15.25$$
$$x = \sqrt{15.25} \approx 3.9 \text{ mph (the actual speed of the swimmer)}$$

There is more than one way to find θ. For example:

$$\tan \theta = \frac{2.5}{3.0} \approx 0.833$$
$$\theta \approx \tan^{-1}(0.833) \approx 40°$$

The swimmer swims at an angle of 40° with the current of the river.

EXAMPLE ④ The swimmer in Example 3 changes the direction of her effort so that she is heading at an angle of 40° with the direction of the current. What is her actual speed and direction?

● **SOLUTION**

PROBLEM SOLVING

Draw a diagram. One way to solve such problems is to draw an accurate vector diagram and measure the resultant vector and its angle θ.

If you wish to solve the problem using trigonometry, begin by labeling the sides of a triangle as shown below. Then use known properties of parallelograms to find m∠C, which is 140°. Then use the law of cosines.

$$c^2 = a^2 + b^2 - 2ab \cos C$$
$$= 2.5^2 + 3^2 - 2(2.5)(3.0) \cos 140°$$
$$\approx 6.25 + 9 - 15(-0.766) = 26.74$$
$$c \approx \sqrt{26.74} \approx 5.17$$

If you are using graph paper, you can make a ruler from a strip of the same paper.

Use the law of sines to find m∠BAC, which is the same as m∠θ.

$$\frac{\sin A}{a} = \frac{\sin C}{c}$$
$$\frac{\sin A}{2.5} \approx \frac{.643}{5.17}$$
$$\sin A \approx 0.3109$$
$$\sin^{-1}(0.3109) \approx 18°$$

The swimmer's actual speed is 5.17 mph at angle of 18° with the current.

Exercises

● Communicate

1. What is the magnitude of a vector? Explain.

2. What is the direction of a vector? Explain.

Describe the magnitude and direction of the vector or vectors in each of the following:

3. an airplane flying northwest at 175 knots

4. a boat going 15 knots upstream against a current of 3 knots

5. two equal tug-of-war teams pulling in opposite directions on a rope

● Guided Skills Practice

6. Find the sum of vectors $\vec{a}$ and $\vec{b}$ by using the head-to-tail method. **(EXAMPLE 1)**

7. Find the sum of vectors $\vec{a}$ and $\vec{b}$ by using the parallelogram method. **(EXAMPLE 2)**

APPLICATIONS

8. SPORTS A swimmer heads perpendicular to the direction of a current whose speed is 1.5 mph. The speed of the swimmer in still water is 2 mph. Find the swimmer's actual speed and her direction angle, θ, with respect to the direction of the current. **(EXAMPLE 3)**

9. SPORTS A swimmer heads in a downstream direction at an angle of 15° with the direction of a 2-mph current. The speed of the swimmer in still water is 2.5 mph. Find the swimmer's actual speed and her direction angle, θ, with respect to the direction of the current. **(EXAMPLE 4)**

Practice and Apply

Copy each pair of vectors and draw the sum $\vec{a} + \vec{b}$ by using the head-to-tail method. You may need to translate one of the vectors.

10. **11.** **12.** **13.**

Copy each pair of vectors and draw the sum $\vec{a} + \vec{b}$ by using the parallelogram method. You may need to translate one of the vectors.

14. **15.**

16. **17.**

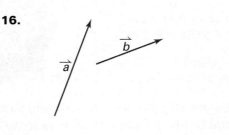

For each given pair of vectors $\vec{a}$ and $\vec{b}$ in Exercises 18–23, use the parallelogram method to find the vector sum $\vec{c}$. Use the law of cosines to find $|\vec{c}|$, the magnitude of $\vec{c}$. Use the law of sines to find the angle that $\vec{c}$ makes with $\vec{b}$.

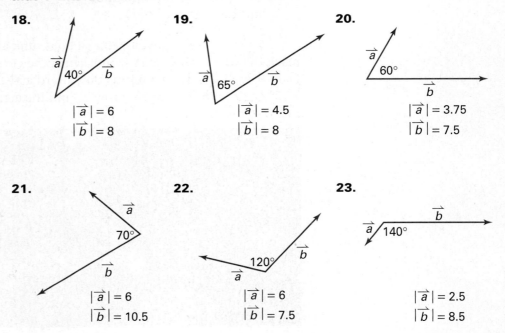

18.
$40°$
$|\vec{a}| = 6$
$|\vec{b}| = 8$

19.
$65°$
$|\vec{a}| = 4.5$
$|\vec{b}| = 8$

20.
$60°$
$|\vec{a}| = 3.75$
$|\vec{b}| = 7.5$

21.
$70°$
$|\vec{a}| = 6$
$|\vec{b}| = 10.5$

22.
$120°$
$|\vec{a}| = 6$
$|\vec{b}| = 7.5$

23.
$140°$
$|\vec{a}| = 2.5$
$|\vec{b}| = 8.5$

24. AVIATION Mina, a student pilot, is making her first solo flight from Austin to Llano, Texas. Using a plotter and an aeronautical chart, she determines that her true course direction should be 303. According to the weather report, the wind at her cruising altitude of 4500 ft has a speed of 10 knots and a heading of 220 (the direction *from* which it is blowing). The true air speed of her plane, at cruising speed, is 110 knots.

Using a mechanical computer, Mina determines that she must hold a heading of 298 to achieve her desired true course. According to the computer, her ground speed will be a little over 108 knots.

True course is set under the index.

Wind correction angle (−5°) is read under the *x*.

Ground speed is read under the center of the wheel.

a. Use trigonometry to check the computer's values for Mina's ground speed and heading. Round your answers to two decimal places.

b. Mina has established a visible landmark at the bend of a river, 14 nautical miles from her point of departure. After how many minutes into her flight should she be over this landmark? (1 knot = 1 nautical mile per hour)

> In "dead reckoning," pilots use visible landmarks to check their course.

25. SCUBA DIVING Dan is investigating a boat sunk at the mouth of a river. To reach the wreck, Dan must swim against the current of 2.7 mph. Suppose Dan dives and starts swimming downward at 4.1 mph (still water speed) at a 15° angle with the water's surface. Find the actual speed Dan will swim as a result of the current.

26. NAVIGATION A boat leaves port and sails in still water at 5.6 mph for 3.5 hr. The boat then turns at a 57° angle and sails at 4.9 mph for 4.25 hr.
 a. If the boat turns and heads directly for port at 6 mph, how long will it take it to reach port?
 b. At what angle will the boat have to turn to head directly back to port in part a?

Look Back

Use the figures at right for Exercises 27–33. *(LESSON 9.5)*

27. Given: *VA* = 5; *VB* = 8; *VC* = 6. Find *CD*.

28. Given: *VB* = *x*; *VA* = 7; *VD* = *x* − 1; *VC* = 9. Find *x*.

29. Given: *VB* = *x*; *VA* = *x* − 6; *VD* = 9; *VC* = 3. Find *x*.

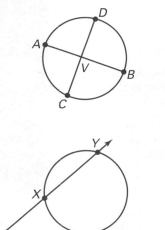

30. Given: *VY* = 9; *VX* = 4. Find *VW*.

31. Given: *VW* = 8; *VX* = 4. Find *VY*.

32. Given: *AY* = *x*; *YB* = *x* − 2; *YC* = 4, *YD* = 6. Find *x*.

33. Given: *AB* = 9; *CD* = 4; *DY* = 12. Find *AY*.

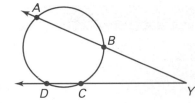

Look Beyond

COORDINATE GEOMETRY Vectors $\vec{m}$ and $\vec{n}$ have their tails at the origin, (0, 0), on a coordinate plane. The head of vector $\vec{m}$ is at point *M*(4, 5), and the head of vector $\vec{n}$ is at point *N*(−3, 8).

34. Find the magnitude of the vector sum $\vec{p}$.

35. Find the angle *θ* that $\vec{p}$ makes with the positive *x*-axis.

36. Find the coordinates of the head of $\vec{p}$.

37. What do you notice about the coordinates of the head of $\vec{p}$? Given the coordinates of the heads of two vectors that have their tails at the origin, write a rule for finding the coordinates of the head of the vector sum of the two given vectors.

38. Write a paragraph proof for your rule from Exercise 37. Use diagrams to illustrate. (Hint: Use the head-to-tail method of vector addition.)

39. Given the coordinates of the head of a vector that has its tail at the origin, write a rule for finding the angle *θ* that it makes with the positive *x*-axis.

Rotations in the Coordinate Plane

Objectives

- Use transformation equations to rotate points.

- Use a rotation matrix to rotate points or polygons.

Why *Computers use geometric transformations to show moving objects. In this lesson, you will explore rotations in a coordinate plane.*

Astronauts prepared for the Hubble Space Telescope repair mission by studying computer simulations of the planned event.

Transformation Equations

In earlier lessons, you studied translations, reflections, and dilations in a coordinate plane. You also studied the special case of a rotation by 180° about the origin. Using trigonometry, it is possible to rotate a figure in a coordinate plane by any angle measure.

The transformation $R(x, y) = (x', y')$, where

$$x' = x \cos \theta - y \sin \theta \quad \text{and}$$
$$y' = x \sin \theta + y \cos \theta,$$

is a rotation by $\theta°$ about the origin. The above equations are known as **transformation equations** above.

CRITICAL THINKING What happens to the point $(0, 0)$ under a rotation? Explain, using the transformation equations above.

Activity

Rotating a Point

YOU WILL NEED

scientific or graphics calculator, graph paper, ruler, and protractor

Part I

1. Draw coordinate axes on graph paper and choose a point P on the graph. Label the x- and y-coordinates of point P.

2. Choose a value of θ between 0° and 180°.

3. Use your values for x, y, and θ in the transformation equations to find P'. Plot P' on your graph. Draw segments connecting P and P' to the origin and measure the angle of rotation. How does the angle of rotation relate to the value of θ?

Part II

1. Copy and complete the table below.

θ	0°	90°	180°	270°	360°
$\sin \theta$	?	?	?	?	?
$\cos \theta$	?	?	?	?	?

CHECKPOINT ✔ **2.** Substitute the values of $\sin \theta$ and $\cos \theta$ into the transformation equations for each value of θ, and simplify. Use the resulting equations to complete the following transformations:

$$0° \text{ rotation: } R(x, y) = (\underline{\ ?\ }, \underline{\ ?\ })$$
$$90° \text{ rotation: } R(x, y) = (\underline{\ ?\ }, \underline{\ ?\ })$$
$$180° \text{ rotation: } R(x, y) = (\underline{\ ?\ }, \underline{\ ?\ })$$
$$270° \text{ rotation: } R(x, y) = (\underline{\ ?\ }, \underline{\ ?\ })$$
$$360° \text{ rotation: } R(x, y) = (\underline{\ ?\ }, \underline{\ ?\ })$$

3. Experiment with negative values for θ, and values greater than 360°. Describe your results.

EXAMPLE ❶ According to the treasure map below, some jewels are buried at a point 60 paces east and 35 paces north of a well. Suppose that the direction of north on the map indicates true north and that you are are using a magnetic compass, which points in a direction 5.5° west of true north. How would you find the location of the jewels?

Detail of map

● **SOLUTION**

Assume the map is on a coordinate plane, with the well at the origin. If true north is represented by the y-axis, then point X is at coordinates $(60, 35)$. Because true north is 5.5° east of the value of north on your compass, the point must be rotated about the origin clockwise by 5.5°. (Recall that clockwise rotations are negative.) Use the rotation equations.

$$x' = x \cos \theta - y \sin \theta = 60 \cos(-5.5°) - 35 \sin(-5.5°) \approx 63 \text{ paces}$$
$$y' = x \sin \theta + y \cos \theta = 60 \sin(-5.5°) + 35 \cos(-5.5°) \approx 29 \text{ paces}$$

Using your compass, walk 63 paces east and 29 paces north to find the jewels.

Rotation Matrices

Recall from algebra that a **matrix** is a set of numbers arranged in rows and columns and enclosed in brackets. Matrices are a convenient way to manipulate and display data.

A rotation in a coordinate plane can be represented in terms of the 2×2 matrix $\begin{bmatrix} \cos\theta & -\sin\theta \\ \sin\theta & \cos\theta \end{bmatrix}$, called a **rotation matrix**. When using matrices to transform points in a coordinate plane, each point is written as a column matrix $\begin{bmatrix} x \\ y \end{bmatrix}$. The image is the product of these two matrices.

To multiply two matrices, each row in the first matrix is multiplied by each column in the second matrix. The product of a row and a column is found by multiplying the terms in the row by the terms in the column and then adding.

For example, $\begin{bmatrix} 4 & 1 \end{bmatrix} \times \begin{bmatrix} 2 \\ 3 \end{bmatrix} = 4(2) + 1(3) = 11.$

For the image of a point transformed by a rotation matrix, the first row multiplied by the column is the entry in the first row of the product matrix, and the second row multiplied by the column is the entry in the second row of the product matrix.

$$\begin{bmatrix} \cos\theta & -\sin\theta \\ \sin\theta & \cos\theta \end{bmatrix} \times \begin{bmatrix} x \\ y \end{bmatrix} = \begin{bmatrix} x\cos\theta - y\sin\theta \\ x\sin\theta + y\cos\theta \end{bmatrix}$$

> The expressions in the product matrix agree with the transformation equations at the beginning of the lesson.

E X A M P L E 2 Rotate $\triangle ABC$ with vertices $A(0, 0)$, $B(5, 0)$, and $C(3, 4)$ by 60° (counterclockwise) about the origin.

SOLUTION

When rotating more than one point, it is convenient to represent the coordinates in a single matrix with several columns. The matrix representing the vertices of $\triangle ABC$ can be written as $\begin{bmatrix} 0 & 5 & 3 \\ 0 & 0 & 4 \end{bmatrix}$.

The rotation matrix is $\begin{bmatrix} \cos 60° & -\sin 60° \\ \sin 60° & \cos 60° \end{bmatrix} = \begin{bmatrix} 0.5 & -0.866 \\ 0.866 & 0.5 \end{bmatrix}$.

Multiply this matrix by the matrix representing $\triangle ABC$.

$$\begin{bmatrix} 0.5 & -0.866 \\ 0.866 & 0.5 \end{bmatrix} \times \begin{bmatrix} 0 & 5 & 3 \\ 0 & 0 & 4 \end{bmatrix} = \begin{bmatrix} 0 & 2.5 & -1.964 \\ 0 & 4.33 & 4.598 \end{bmatrix}$$
$$ A \quad B \quad C \qquad\quad A' \quad B' \quad\ C'$$

Plot points A', B', and C', and connect them to form a triangle.

TRY THIS Rotate $\triangle DEF$ with vertices $D(0, 1)$, $E(2, 0)$, and $F(1, 3)$ by 30° (counterclockwise) about the origin.

Exercises

● Communicate

1. Give examples of two rotations, one counterclockwise and one clockwise, that produce the same result. Use either transformation equations or rotation matrices to show that the results are the same.

2. Choose two supplementary angles and find their rotation matrices. How are they alike? How are they different?

3. Choose two complementary angles and find their rotation matrices. How are they alike? How are they different?

4. Find the rotation matrices for $\theta = 0°, 90°, 180°, 270°$, and $360°$ by finding the sine and cosine values. How do the matrices compare with the rotations you found in the Activity?

● Guided Skills Practice

Use transformation equations to rotate the point (5, 4) by the given angle. (EXAMPLE 1)

5. $10°$
6. $90°$
7. $-30°$

Find the rotation matrix for the given angle, and use it to rotate △ABC, with vertices A(1, 1), B(3, 0), and C(2, 2). (EXAMPLE 2)

8. $45°$
9. $72°$
10. $-60°$

● Practice and Apply

For Exercises 11–16, a point, P, and an angle of rotation, θ, are given. Determine the coordinates of the image P′.

11. $P(0, 5), \theta = 45°$
12. $P(-1, 1), \theta = 80°$

13. $P(-3, -2), \theta = 30°$
14. $P(2, 0), \theta = 140°$

15. $P(5, 5), \theta = 20°$
16. $P(2, 6), \theta = 400°$

For Exercises 17–22, the coordinates of a point P and an image P′ are given. Determine of the angle of rotation, θ.

17. $P(4, 2), P'(-2, 4)$
18. $P(3, 7), P'(-3, -7)$

19. $P(-10, -9), P'(-9, 10)$
20. $P(3, -2), P'(3, -2)$

21. $P(\sqrt{2}, \sqrt{2}), P'(0, 2)$
22. $P\left(\frac{1}{2}, \frac{\sqrt{3}}{2}\right), P'\left(\frac{\sqrt{3}}{2}, \frac{1}{2}\right)$

Find the rotation matrix for each angle below by finding the sine and cosine values. Round your answers to the nearest hundredth.

23. $45°$
24. $30°$

25. $120°$
26. $320°$

Find the image of the polygon with vertices represented by the given matrix, rotated by the angle θ. Draw the polygon and its image on the same coordinate axes. You may wish to use a graphics calculator for the matrix multiplication.

27. $\begin{bmatrix} 0 & 3 & 2 \\ 1 & 2 & 5 \end{bmatrix}$, $\theta = 90°$

28. $\begin{bmatrix} -2 & 1 & 3 \\ 1 & 2 & -4 \end{bmatrix}$, $\theta = 40°$

29. $\begin{bmatrix} 2 & 1 & 2 & 4 \\ 5 & 3 & 1 & 2 \end{bmatrix}$, $\theta = 225°$

30. $\begin{bmatrix} 0 & 2 & 4 & 6 \\ 0 & 2 & 1 & 0 \end{bmatrix}$, $\theta = 70°$

31. Find the rotation matrix for a 30° angle and the rotation matrix for a 40° angle. Show that their product is the same as the rotation matrix for a 70° angle. Explain why in terms of the rotation of a point.

32. One vertex of a regular pentagon centered at the origin is $(4, 0)$. Describe a procedure for finding the coordinates of the other four vertices. Carry out your procedure and plot all five vertices on a coordinate plane.

33. A point that is rotated by 0° stays in the same place. Find $[R_0]$, the rotation matrix for a 0° angle. This is called the *identity matrix*. Choose any rotation matrix and multiply it by the identity matrix. What do you notice?

34. Let $[R_{60}]$ denote a 60° rotation matrix. Use a graphics calculator to multiply $[R_{60}] \times [R_{60}] \times [R_{60}] \times [R_{60}] \times [R_{60}] \times [R_{60}]$. Show that the product is the same as $[R_0]$ (see Exercise 33). Explain why in terms of the rotation of a point.

35. A point that is rotated by $\theta°$, then rotated by $-\theta°$ will end up at its original position. Find the rotation matrix for a 35° angle and the rotation matrix for a −35° angle. These matrices are called *inverse matrices* of each other. Multiply the two matrices together. What do you notice?

CHALLENGE

36. Use the transformation equations on page 680 to prove that for any point A and any rotation about the origin, the distance OA is equal to OA'. (Hint: Label the coordinates of $A(x, y)$. Then $(OA)^2 = \underline{\quad ? \quad}$ and $(OA')^2 = \underline{\quad ? \quad}$. Simplify the expression for $(OA')^2$ to show that the distances are equal.)

APPLICATION

37. **ARCHITECTURE** A restaurant atop a tall building revolves 360° each hour to give patrons a panoramic view. If a certain table has a location of $(30', -42')$ relative to the center, find its coordinates:
   a. after 20 minutes
   b. after t minutes

COORDINATE GEOMETRY Graph each line and use the inverse tangent to determine the angle formed by the line and the *x*-axis. *(LESSON 10.1)*

38. $y = 2x$ **39.** $y = 0.5x$ **40.** $y = 3x + 1$

Find the tangent, sine, and cosine of each angle below. Round to the nearest hundredth. *(LESSONS 10.1 AND 10.2)*

41. $37°$ **42.** $90°$ **43.** $250°$ **44.** $-10°$

Find the angle θ between 0° and 90° for each trigonometric ratio given. Round to the nearest degree. *(LESSONS 10.1 AND 10.2)*

45. $\tan \theta = 2.75$ **46.** $\cos \theta = 0.36$ **47.** $\cos \theta = 0.81$ **48.** $\sin \theta = 0.77$

49. PUBLIC SAFETY A slide in a playground must have a maximum average slope of 30°. The dimensions of a slide are shown in the diagram at right. Does the slide meet the safety requirements? *(LESSON 10.1)*

6 ft

10 ft

TRIGONOMETRY So far you have used degrees to measure angles. Another common unit of angle measure is called a *radian*. A radian is the central angle measure of an arc in a circle with a length equal to the radius of the circle.

$m\angle\theta = 1$ radian

50. How many radians are in 360°? (Hint: Find the circumference of the circle.)

Find the radian measure for each angle below.

51. $180°$ **52.** $90°$ **53.** $60°$

PROJECT TEN

PLIMPTON 322 Revisited

In this project, you will re-examine the Babylonian tablet known as Plimpton 322.

Activity 1

THE BABYLONIAN NUMBER SYSTEM The Babylonians used a base 60 number system. Some remnants of this system that still survive are the 360° circle and our units of time, 60 minutes per hour and 60 seconds per minute. The following are the symbols for some of the Babylonian numerals written in the Babylonian alphabet, called *cuneiform*.

1	6	11	16	30
2	7	12	17	40
3	8	13	18	50
4	9	14	19	60
5	10	15	20	

Numbers greater than 60 were expressed by using these same numerals in different positions. All digits but the last position are multiplied by a power of 60. For example, the Babylonian number or 1, 16, 41, is expressed in our number system as $1 \times 60^2 + 16 \times 60 + 41 \times 1 = 4601$.

In order to make the numbers easier to read, commas are used to separate the digits and a semicolon to represent the equivalent of the decimal point. However, the Babylonians did not use gaps or punctuation marks, but relied on context to determine the positions of the digits.

A number with a fractional part, such as 1; 38, 33, 36, 36, is expressed in our number system as

$$1 + \frac{38}{60} + \frac{33}{60^2} + \frac{36}{60^3} + \frac{36}{60^4} = 1.64266944444 \ldots$$

The table below contains the transcription of the numbers on Plimpton 322. Some of the damaged numbers have been supplied by researchers. Compare the values in the transcription with columns II and III on page 638. Four entries in columns II and III below are errors. Can you find them?

Column I	Column II	Column III	Column IV
1; 59, 0, 15	1, 59	2, 49	1
1; 56, 56, 58, 14, 50, 6, 15	56, 7	3, 12, 1	2
1; 55, 7, 41, 15, 33, 45	1, 16, 41	1, 50, 49	3
1; 53, 10, 29, 32, 52, 16	3, 31, 49	5, 9, 1	4
1; 48, 54, 1, 40	1, 5	1, 37	5
1; 47, 6, 41, 40	5, 19	8, 1	6
1;43, 11, 56, 28, 26, 40	38, 11	59, 1	7
1; 41, 33, 59, 3, 45	13, 19	20, 49	8
1; 38, 33, 36, 36	9, 1	12, 49	9
1; 35, 10, 2, 28, 27, 24, 26, 40	1, 22, 41	2, 16, 1	10
1; 33, 45	45	1, 15	11
1; 29, 21, 54, 2, 15	27, 59	48, 49	12
1; 27, 0, 3, 45	7, 12, 1	4, 49	13
1; 25, 48, 51, 35, 6, 40	29, 31	53, 49	14
1; 23, 13, 46, 40	56	53	15

Activity 2

THE FIRST COLUMN Use the table above to find the values of the entries in the first column of Plimpton 322. Use as many digits after the decimal point as you can find.

Square the entries in the tan θ column on page 638. Compare them to the values in column I of Plimpton 322. What do you notice? Use the second Pythagorean identity that you found on page 646 to identify the ratio described in column I.

Chapter Review and Assessment

VOCABULARY

POSTULATES AND THEOREMS

Lesson	Number	Postulate or Theorem
10.2	10.2.3 Identity	$\tan \theta = \dfrac{\sin \theta}{\cos \theta}$
	10.2.4 Identity	$(\sin \theta)^2 + (\cos \theta)^2 = 1$
10.4	10.4.1 The Law of Sines	For any triangle $\triangle ABC$ with sides a, b, and c: $\dfrac{\sin A}{a} = \dfrac{\sin B}{b} = \dfrac{\sin C}{c}$
10.5	10.5.1 The Law of Cosines	For any triangle $\triangle ABC$ with sides a, b, and c: $a^2 = b^2 + c^2 - 2bc \cos A \qquad b^2 = a^2 + c^2 - 2ac \cos B$ $c^2 = a^2 + b^2 - 2ab \cos C$

Key Skills & Exercises

LESSON 10.1

Key Skills

Use right triangles to find tangent ratios.

Find the tangent of 45°.

The triangle at right is a 45-45-90 right triangle, so the opposite leg and adjacent leg are equal. Thus,
$\tan 45° = \dfrac{\text{opposite}}{\text{adjacent}} = 1.$

Find an angle that has a given tangent ratio.

Find an angle with a tangent of $\frac{1}{2}$.

Using a calculator, $\tan^{-1} \frac{1}{2} = 26.565 \approx 27°.$

Exercises

Use the given right triangle to find the tangent of each angle.

1. $\tan A =$ _____?_____

2. $\tan B =$ _____?_____

Use a scientific or graphics calculator to find the measure of each angle.

3. $\text{m}\angle A =$ _____?_____

4. $\text{m}\angle B =$ _____?_____

Key Skills

Exercises

Use right triangles to find sine and cosine ratios.

Find the sine and cosine of 30°.

The triangle at right is a 30-60-90 right triangle with a hypotenuse of length 2, so the opposite leg has length 1 and adjacent leg has length $\sqrt{3}$.

$$\sin 30° = \frac{\text{opposite}}{\text{hypotenuse}} = \frac{1}{2}$$

$$\cos 30° = \frac{\text{adjacent}}{\text{hypotenuse}} = \frac{\sqrt{3}}{2}$$

Find an angle that has a given sine or cosine ratio.

Give an angle with a cosine of $\frac{1}{2}$.

Using a calculator, $\cos^{-1} \frac{1}{2} = 60°$.

Use the given right triangle to find the sine and cosine of each angle.

5. $\sin A =$ ___?___ $\cos A =$ ___?___

6. $\sin B =$ ___?___ $\cos B =$ ___?___

Use a scientific or graphics calculator to find the measure of each angle.

7. $m\angle A =$ ___?___

8. $m\angle B =$ ___?___

Key Skills

Exercises

Find the coordinates of a point on the unit circle corresponding to a given angle.

Find the coordinates (x, y) of point P on the unit circle at right.

Point P has coordinates $(\cos 40°, \sin 40°) \approx (0.77, 0.64)$.

Find the measures of angles with a given sine or cosine.

Find all angles between 0° and 360° with a sine of 0.7.

Using a calculator, $\sin^{-1} 0.7 = 44.43° \approx 44°$. From the graph of $\sin \theta$, you can see that there is another angle with the same sine value, namely $180° - 44° = 136°$.

For Exercises 9–12, use a sketch to illustrate your answer.

9. Find the coordinates of a point on the unit circle corresponding to a 70° angle of rotation.

10. Find the coordinates of a point on the unit circle corresponding to a 130° angle of rotation.

11. Find all angles between 0° and 360° with a sine of 0.2.

12. Find all angles between 0° and 360° with a cosine of 0.8.

Key Skills

Solve triangles by using the law of sines.

Use the law of sines to solve the triangle below.

$\frac{\sin 36°}{37} = \frac{\sin L}{30}$, so $\sin L = 30\left(\frac{\sin 36°}{37}\right) \approx 0.477$.
Thus, $m\angle L = \sin^{-1} 0.477 \approx 28°$, so by the
Triangle Sum Theorem,
$m\angle M = 180° - (36° + 28°) = 116°$.

(Note: Because $\angle M$ is obtuse, $\angle L$ must be acute,
so the triangle is not ambiguous.)

Then $\frac{\sin 36°}{37} = \frac{\sin 116°}{m}$, so
$m = \frac{37 \sin 116°}{\sin 36°} \approx 56.6$.

Exercises

Refer to the diagram below.

13. Find the unknown angles in $\triangle PQR$.

14. Find the unknown side of $\triangle PQR$.

Refer to the diagram below.

15. Find the unknown angle in $\triangle STU$.

16. Find the unknown sides in $\triangle STU$.

Key Skills

Solve triangles by using the law of cosines.

Use the law of cosines to solve the triangle below.

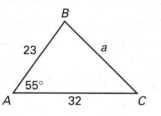

$a^2 = b^2 + c^2 - 2bc \cos A \approx 709$, so $a \approx 27$.
Use the law of sines to find $m\angle B$.
$\frac{\sin 55°}{27} = \frac{\sin B}{32}$, so $m\angle B \approx 76°$
By the Triangle Sum Theorem,
$m\angle C = 180° - (55° + 76°) = 49°$.

Exercises

Refer to the diagram below.

17. Find the unknown side in $\triangle DEF$.

18. Find the unknown angles in $\triangle DEF$.

Refer to the diagram below.

19. Find the largest unknown angle in $\triangle GHI$.

20. Find the remaining unknown angles in $\triangle GHI$.

Key Skills

Find the sum of two vectors by using the head-to-tail method and the parallelogram method.

Find the sum of vectors $\vec{a}$ and $\vec{b}$ below.

Head-to-tail method:

Parallelogram method:

Copy vectors $\vec{p}$, $\vec{q}$, and $\vec{r}$ below.

21. Find the sum of $\vec{p}$ and $\vec{q}$ by using the head-to-tail method.

22. Find the sum of $\vec{p}$ and $\vec{r}$ by using the head-to-tail method.

23. Find the sum of $\vec{p}$ and $\vec{q}$ by using the parallelogram method.

24. Find the sum of $\vec{p}$ and $\vec{r}$ by using the parallelogram method.

Key Skills

Rotate a point in a coordinate plane by a given angle.

What is the image of the point $(3, 2)$ under a $90°$ rotation?

The transformation equations are
$x' = x \cos 90° - y \sin 90° = x(0) - y(1) = -y$
$y' = x \sin 90° + y \cos 90° = x(1) + y(0) = x$
so the image is the point $(-2, 3)$.

Rotate a polygon in a coordinate plane by a given angle.

What is the image of $\triangle ABC$ with vertices $A(0, 0)$, $B(1, 2)$, and $C(1, 0)$ under a $60°$ rotation?

Multiply the rotation matrix by the triangle matrix.

$$\begin{bmatrix} 0.5 & -0.866 \\ 0.866 & 0.5 \end{bmatrix} \begin{bmatrix} 0 & 1 & 1 \\ 0 & 2 & 0 \end{bmatrix} = \begin{bmatrix} 0 & -1.232 & 0.5 \\ 0 & 1.866 & 0.866 \end{bmatrix}$$

The image $\triangle A'B'C'$ has vertices $A'(0, 0)$, $B'(-1.232, 1.866)$, and $C'(0.5, 0.866)$.

25. What is the image of the point $(1, 5)$ under a $270°$ rotation?

26. What is the image of the point $(4, -3)$ under a $45°$ rotation?

27. Find the image of $\triangle DEF$ with vertices $D(0, 0)$, $E(3, 1)$, and $F(3, 2)$ under a $30°$ rotation.

28. Find the image of $\triangle GHI$ with vertices $G(1, 1)$, $H(-1, 1)$, and $I(1, -1)$ under a $135°$ rotation.

29. INDIRECT MEASUREMENT Estimate the height of the building in the diagram at right.

30° 50°

35 m

30. NAVIGATION Two planes set off from an airport, one with a heading of 045 at 100 mph, the other with a heading of 150 at 115 mph. How far apart will the planes be after 40 min?

31. ENGINEERING A truck weighing 6000 lb is parked on a ramp that forms a 10° angle with the ground. To find the force required to move the truck uphill, the vector representing the truck's weight (pointing straight down) is broken down into two perpendicular vectors, one parallel to the ramp, labeled $\vec{a}$, and one perpendicular to the ramp, labeled $\vec{b}$. The sum $\vec{a} + \vec{b}$ equals the vector representing the weight. Find the magnitude of $\vec{a}$, the force required to push the truck uphill.

a

10°

6000 lb   b

32. ASTRONOMY The diagram at right (not to scale) shows Alpha Centauri, the closest star to the Sun, and two positions of Earth, A and B, at six months apart in its orbit around the Sun. Use the law of cosines and the information in the diagram to estimate d, the distance from Earth at position A to Alpha Centauri.

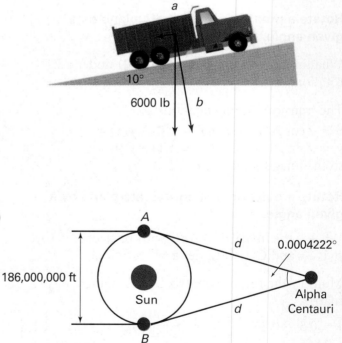

A

186,000,000 ft

d   0.0004222°

Sun

Alpha
Centauri

d

B

Alternative Assessment

Performance Assessment

Use a graphics calculator or a computer with geometry graphics software for the following exercises:

$$y = \sin \theta$$

$$y = \cos \theta$$

1. Graph the following on the same coordinate axes. Be sure to label each function.

$$y = \sin \theta$$
$$y = 2 \sin \theta$$
$$y = \frac{1}{2} \sin \theta$$
$$y = -\sin \theta$$

How are the graphs alike? How are they different? In general, describe the graph of

$$y = a \sin \theta$$
and
$$y = -a \sin \theta$$

for any positive constant a.

2. Graph the following on the same coordinate axes. Be sure to label each function.

$$y = \cos \theta$$
$$y = \cos (2\theta)$$
$$y = \cos \left(\frac{1}{2}\theta\right)$$

How are the graphs alike? How are they different? In general, describe the graph of $y = \cos (a\theta)$ for any positive constant a.

Portfolio Projects

The number of hours of daylight for each day of the year can be estimated by the sine function:

$$y = A \sin\left[\frac{360}{365}(d - 80)\right] + 12$$

where y is the number of daylight hours, d is the day of the year, and A is a constant between 0 and 12.

1. The number of daylight hours in each location varies with the latitude of the location. Find the number of daylight hours on the summer solstice, the longest day of the year, where you live. Subtract 12 from this value. This will be the value of the constant A for your latitude.

2. Using the value you found for A, graph the function on a graphics calculator or computer for values of d from 0 to 365.

3. Use your graph to estimate the number of hours of daylight on the shortest day of the year. Estimate the days when there are exactly 12 daylight hours, the vernal and autumnal equinox.

---

internet connect

The HRW Web site contains many resources to reinforce and expand your knowledge of trigonometry. The Web site also provides Internet links to other sites where you can find information and real-world data for use in research projects, reports, and activities that involve trigonometry. Visit the HRW Web site at **go.hrw.com**, and enter the keyword **MG1 CH10**.

1-10 College Entrance Exam Practice

QUANTITATIVE COMPARISON For Items 1–4, write

A if the quantity in Column A is greater than the quantity in Column B;
B if the quantity in Column B is greater than the quantity in Column A;
C if the two quantities are equal; or
D if the relationship cannot be determined from the information given.

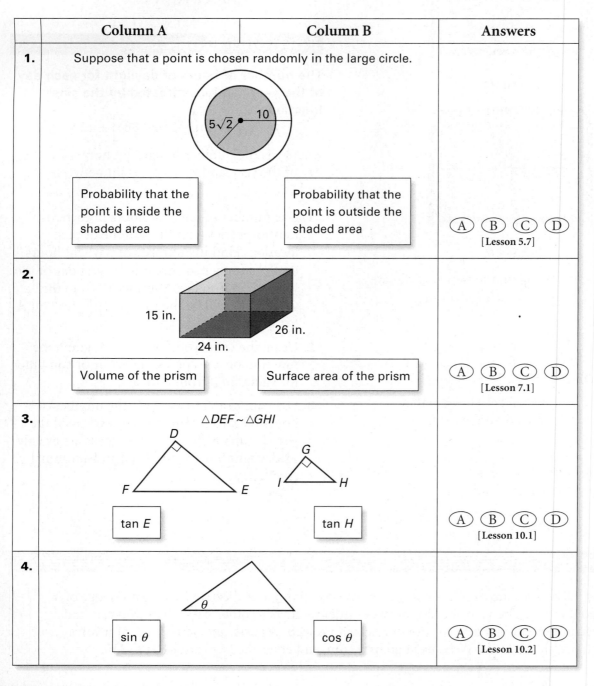

	Column A	Column B	Answers
1.	Suppose that a point is chosen randomly in the large circle.		
	Probability that the point is inside the shaded area	Probability that the point is outside the shaded area	Ⓐ Ⓑ Ⓒ Ⓓ [Lesson 5.7]
2.	Volume of the prism	Surface area of the prism	Ⓐ Ⓑ Ⓒ Ⓓ [Lesson 7.1]
3.	△DEF ~ △GHI		
	tan E	tan H	Ⓐ Ⓑ Ⓒ Ⓓ [Lesson 10.1]
4.	sin θ	cos θ	Ⓐ Ⓑ Ⓒ Ⓓ [Lesson 10.2]

5. The transformation $T(x, y) = (-x, y)$ is a
_____?_____. *(LESSONS 1.7, 8.1, AND 10.7)*
   a. translation
   b. reflection
   c. rotation
   d. dilation

6. The transformation $T(x, y) = (-x, -y)$ is a
_____?_____. *(LESSONS 1.7, 8.1, AND 10.7)*
   a. translation
   b. reflection
   c. rotation
   d. dilation

7. Which theorem can be used to find m$\angle A$ in
$\triangle ABC$? *(LESSONS 10.5 AND 10.6)*

   a. law of sines
   b. law of cosines
   c. Pythagorean Theorem
   d. none of the above

8. Suppose that $\angle A \cong \angle B$ and $\angle B \cong \angle C$. Then
$\angle A \cong \angle C$ by_____?_____. *(LESSON 2.4)*
   a. Reflexive Property
   b. Symmetric Property
   c. Transitive Property
   d. Addition Property

9. m$\angle 4 = $ _____?_____ *(LESSON 3.5)*

   a. m$\angle 1 + $ m$\angle 2$
   b. m$\angle 1 + $ m$\angle 3$
   c. m$\angle 2 + $ m$\angle 3$
   d. m$\angle 1 + $ m$\angle 2 + $ m$\angle 3$

10. The lines $y = 2x + 3$ and $y = -\frac{1}{2}x - 1$ are
_____?_____. *(LESSON 3.8)*
   a. parallel
   b. skew
   c. vertical
   d. perpendicular

11. A point with coordinates $(3, 5)$ is rotated by
$45°$ about the origin. Give the coordinates of
the image point. *(LESSON 10.7)*

12. The magnitude of $\overrightarrow{a}$ is 6 and the magnitude
of $\overrightarrow{b}$ is 10. If $\overrightarrow{a}$ and $\overrightarrow{b}$ are parallel, what is the
magnitude of $\overrightarrow{a} + \overrightarrow{b}$? *(LESSON 10.6)*

13. Give the number of vertices, edges, and faces of
an oblique pentagonal prism. *(LESSON 6.3)*

14. The rectangles at right
are similar. Find x.
(LESSON 8.2)

15. Find k in $\triangle KLM$ at right.
(LESSON 10.4)

16. Find the radius of a circle that has the same
area as a square with a side length of 5.
(LESSON 5.3)

17. Find the radius of a sphere that has the same
volume as a cube with an edge length of 5.
(LESSON 7.6)

18. Find an angle $0° < \theta < 180°$ such that
$\sin \theta = -\cos \theta$. *(LESSON 10.3)*

Taxicabs, Fractals, and More

CAN YOU IMAGINE A SYSTEM OF MATHEMATICS IN which a coffee cup is the same as a doughnut? Or one in which there are no such things as parallel lines? You will explore many strange ideas like these in this chapter.

By questioning commonly held assumptions or taking imaginative leaps, mathematicians create entirely new areas of mathematics that often turn out to be rewarding fields of study.

Before branching out to more recent discoveries in mathematics, you will first study an idea that goes back to classical times—the *golden ratio*. This ratio appears again and again in mathematics and nature. As you will see, the sunflower on the facing page has an interesting connection to the golden ratio.

About the Chapter Project

You may be familiar with the puzzle known as the tower of Hanoi, invented by Edouard Lucas in 1883. In the puzzle, a stack of disks in order from smallest to largest rests on one of three pegs. The object of the puzzle is to move the disks, one at a time, to another peg without ever placing any disk onto a smaller disk.

In the Chapter Project, you will study both the number of moves required for a given number of disks and the puzzle's relationship to the fractal known as the *Sierpinski gasket*.

After completing the Chapter Project, you will be able to do the following:

- Determine the fewest number of moves required to solve the tower of Hanoi puzzle with a given number of disks.

- Use a graph to represent the possible states of the tower of Hanoi puzzle and to find a solution.

About the Portfolio Activities

Throughout the chapter, you will be given opportunities to complete Portfolio Activities that are designed to support your work on the Chapter Project.

The theme of each Portfolio Activity and of the Chapter Project is the tower of Hanoi.

- In the Portfolio Activity on page 720, you will study how puzzles can be represented graphically and how solutions can be found by using graphs.

- In the Portfolio Activity on page 746, you will generate a fractal shape by using random processes, such as rolling a number cube or generating random points on a graphics calculator.

Golden Connections

Objectives

- Discover the relationship known as the golden ratio.

- Solve problems by using the golden ratio.

Why *The golden ratio can be used to find the dimensions of the golden rectangle, which has been considered for centuries by artists and architects to be the "ideal" rectangle.*

The Parthenon is an ancient Greek temple dedicated to the goddess Athena. Viewed from the front, the proportions of the temple suggest the golden ratio, a mathematical idea which captured the imagination of classical Greek thinkers.

Golden Rectangles

A **golden rectangle** is a rectangle with the following property: If a square is cut from one end of the rectangle, the remaining piece is similar to the original rectangle.

If the long side of the large rectangle has a length of ℓ and the short side has a length of s, then the sides of the smaller rectangle are s and $\ell - s$.

The ratio of the long side to the short side of a golden rectangle is called the **golden ratio**. Since the large and small rectangles are similar, the relationship between their sides can be expressed by the proportion below.

The smaller rectangle is similar to the entire rectangle.

$$\frac{\ell}{s} = \frac{s}{\ell - s}$$

Activity 1
The Dimensions of a Golden Rectangle

YOU WILL NEED

graph paper and calculator

PROBLEM SOLVING

1. Orient your graph paper horizontally and draw a vertical segment 10 units long at the left edge. This will be the short side of your golden rectangle.

2. **Make a table.** Your rectangle must satisfy the proportion $\frac{\ell}{s} = \frac{s}{\ell - s}$. Copy the table below and experiment with different possible values for ℓ until the values in the third and fourth columns match up to two decimal places.

ℓ	s	$\frac{\ell}{s}$	$\frac{s}{\ell - s}$
20	10	2	1
?	10	?	?
?	10	?	?
?	10	?	?

$\ell = ?$

10

3. Complete your drawing of the golden rectangle with the correct value for ℓ.

CHECKPOINT ✔ 4. Complete the following statement: The golden ratio, $\frac{\ell}{s}$, is ___?___.

Activity 2
Seashells

YOU WILL NEED

ruler, compass, and your golden rectangle from Activity 1

1. Label your golden rectangle *ABCD* and draw square *EBCF*.

```
A       E           B
|-------|-----------|
|       |           |
|       |           |
|-------|-----------|
D       F           C
```

2. Draw square *AEGH*. Use your compass and the squares on the graph paper to make your squares as accurate as possible.

```
A           E               B
|-----------|---------------|
|           |               |
H-----------G               |
|           |               |
D           F               C
```

CHECKPOINT ✔ 3. Repeat this process three more times. Do the rectangles that are formed seem to be golden rectangles? Find the ratio $\frac{\ell}{s}$ for each rectangle by measuring the sides.

```
A       E           B
|-------|-----------|
|       |           |
H---I---G           |
|   |   |           |
D   J F             C
```

4. Use a compass to make a quarter-circle in each square, as shown below. The resulting curve approximates a *logarithmic spiral*, which models the growth pattern of seashells such as the chambered nautilus.

Computing the Golden Ratio

Algebra

The proportion for the golden ratio can be solved algebraically to find the exact value for $\frac{\ell}{s}$. You must first notice that in a golden rectangle, the proportion must be true for any value of s. For convenience, let the value of s be 1, and find a value of ℓ that satisfies the proportion.

$$\frac{\ell}{s} = \frac{s}{\ell - s}$$

$$\frac{\ell}{1} = \frac{1}{\ell - 1}$$

$$\ell(\ell - 1) = 1 \qquad \textit{Cross multiply.}$$

$$\ell^2 - \ell = 1 \qquad \textit{Distribute.}$$

$$\ell^2 - \ell - 1 = 0$$

Notice that the resulting equation is a quadratic equation of the form $ax^2 + bx + c = 0$, where $a = 1$, $b = -1$, and $c = -1$. To solve, substitute these values into the quadratic formula, $x = \dfrac{-b \pm \sqrt{b^2 - 4ac}}{2a}$. Because ℓ is a length, consider only the positive value.

$$\ell = \frac{-(-1) + \sqrt{(-1)^2 - 4(1)(-1)}}{2(1)} = \frac{1 + \sqrt{5}}{2} \approx 1.618$$

Next, substitute the values of ℓ and s into the ratio.

$$\frac{\ell}{s} = \frac{\ell}{1} \approx \frac{1.618}{1}$$

This is the value of the golden ratio. It is often represented by the Greek letter ϕ (phi):

$$\phi = \frac{1 + \sqrt{5}}{2} \approx 1.618033989\ldots$$

CRITICAL THINKING

Do you think you would get the same value for $\frac{\ell}{s}$ if some value other than 1 is used for s? Try several other values for s. What do you discover?

The Parthenon, the Pentagon Building, and the chambered nautilus all involve the golden ratio.

Constructing a Golden Rectangle

CONSTRUCTION

COMPASS and
STRAIGHTEDGE

1. Construct a square and label the vertices *ABCD*. Let the length of $\overline{AB}$ be 2 units. Extend side $\overline{AB}$ as shown.

2. Construct the midpoint of $\overline{AB}$ and label it *M*. Draw $\overline{MC}$. Since *ABCD* is a square, *BC* = __?__ units. Since *M* is the midpoint of $\overline{AB}$, *MB* = __?__ units. By the Pythagorean Theorem, *MC* = __?__ units.

3. Set the point of your compass at *M*, and draw an arc through *C* that intersects $\overrightarrow{AB}$ at *E*.

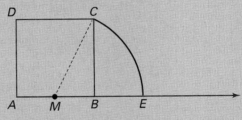

CHECKPOINT ✔

4. Extend $\overline{DC}$, and construct a perpendicular to $\overrightarrow{AB}$ at *E*. Label the intersection *F*. How do you know that rectangle *AEFD* is a golden rectangle? Complete the following statements:

$AD = $ __?__ $AE = $ __?__ $\dfrac{AE}{AD} = $ __?__

CRITICAL THINKING In the diagram for Step 4 of Activity 3, *AEFD* is a golden rectangle. Name another golden rectangle in this diagram. How do you know it is golden?

Exercises

Communicate

1. The golden rectangle is said to be *self-replicating*. Use your results from Activity 2 to explain what this means.

2. Explain the construction of the golden rectangle. How is the Pythagorean Theorem used in this construction?

3. Measure the sides of several rectangles around you, such as a sheet of paper or a poster. Find the ratios of the long side to the short side. Name some that are close to the golden ratio.

4. Find the ratio of your total height to your height up to your navel. Is your answer close to the golden ratio?

Guided Skills Practice

In Exercises 5 and 6, determine the unknown side length of the golden rectangle with the given side length. (ACTIVITY 1)

5. $s = 10$ $\ell = \underline{\quad?\quad}$

6. $s = \underline{\quad?\quad}$ $\ell = 9$

7. Use the rectangle and spiral that you created in Activity 2, or construct a new one. Draw $\overline{AC}$ and $\overline{DE}$. What seems to be true about the intersection of these segments? **(ACTIVITY 2)**

8. Use the method from Activity 3 to construct a golden rectangle whose short side has a length of 1 in. **(ACTIVITY 3)**

Practice and Apply

Determine the indicated side length of each golden rectangle. Round your answers to the nearest hundredth.

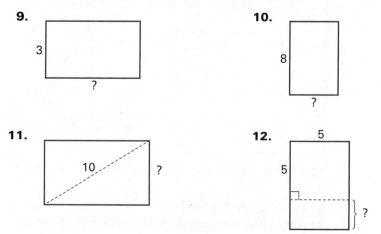

9.

3

?

10.

8

?

11.

10

?

12.

5

5

?

The value of the golden ratio, ϕ, has some interesting properties. Refer to the diagram below for Exercises 13–15.

13. Set up a proportion involving the side lengths of the two golden rectangles in the figure.

14. Enter the value of ϕ in your calculator, using as many decimal places as your calculator will hold, and find ϕ^2. Compare this value with ϕ. What do you notice? Using your proportion from Exercise 13, explain why this is true.

15. Enter the value of ϕ in your calculator, using as many decimal places as your calculator will hold, and find ϕ^{-1}. Compare this value with ϕ. What do you notice? Using your proportion from Exercise 13, explain why this is true.

CONSTRUCTION

Regular Pentagon

a. Draw $\odot A$ and diameter $\overline{BC}$. Construct M, the midpoint of $\overline{AB}$.

b. Construct line $\ell \perp \overline{BC}$ through A. Choose either point where ℓ intersects $\odot A$ and label it D.

c. Place your compass point at M and draw an arc through D that intersects $\overline{BC}$. Label this point P.

d. Set your compass to the length DP. Mark off length DP around $\odot A$. Draw segments to connect the arcs, forming a regular pentagon.

16. Use the directions above to construct regular pentagon $DEFGH$ in a circle with a radius of at least 3 cm.

17. Draw $\overline{EH}$. Measure $\overline{DE}$ and $\overline{EH}$ to the nearest millimeter.

18. Find $\dfrac{EH}{DE}$. How close to ϕ is your answer?

For Exercises 19–21, refer to the diagram of the regular pentagon at right. You will use trigonometry to verify that the ratio $\frac{EH}{DE}$ is equal to ϕ.

19. Use the fact that *DEFGH* is a regular pentagon to find m∠*D*.

20. △*DEH* is isosceles, so m∠*DEH* = m∠*DHE* = ___?___.

21. Use the law of sines and properties of proportions to write an expression for the ratio $\frac{EH}{DE}$. Evaluate this expression on your calculator. What do you get?

If you extend the sides of a regular pentagon, you get a five-pointed star known as the star of Pythagoras. This star was the symbol of the followers of Pythagoras.

22. Construct a regular pentagon and extend the sides to form a star of Pythagoras.

23. Compute the measures of the angles in △*DEI*.

24. Use the law of sines to find $\frac{EI}{DE}$. What do you notice?

25. Measure $\overline{IL}$ and $\overline{IF}$ to the nearest millimeter. Find $\frac{IL}{IF}$. What do you notice?

26. Measure $\overline{IF}$ and $\overline{IE}$ to the nearest millimeter. Find $\frac{IF}{IE}$. What do you notice?

Recall from Lesson 2.2 that the Fibonacci sequence is a sequence of numbers in which each term is the sum of the two terms preceding it. The following is the beginning of the Fibonacci sequence: 1, 1, 2, 3, 5, 8, 13, 21,…

27. Find the next five terms of the Fibonacci sequence above.

28. Copy and complete the table below by dividing each term in the sequence by the preceding term.

Term	1	1	2	3	5	8	13	21	?	?	?	?	?
Ratio	—	$\frac{1}{1}$	$\frac{2}{1}$	?	?	?	?	?	?	?	?	?	?
Value	—	1	2	?	?	?	?	?	?	?	?	?	?

29. What number do the ratios approach? Write a conjecture about the ratios of consecutive terms of the Fibonacci sequence.

30. The 28th term of the Fibonacci sequence is 317,811, and the 29th term is 514,229. Find their ratio. Does this agree with your conjecture?

31. Let $\phi' = \frac{1 - \sqrt{5}}{2}$, the negative solution to the quadratic equation on page 700. Then the *n*th term of the Fibonacci sequence is given by $\frac{\phi^n - (\phi')^n}{\sqrt{5}}$. Use this expression to find the 20th term of the Fibonacci sequence.

CHALLENGE

32. Use the expression for the *n*th term of the Fibonacci sequence from Exercise 31 to explain why the ratios in the table for Exercise 28 approach ϕ. (Hint: What happens to $(\phi')^n$ as *n* increases?)

33. PHOTOGRAPHY A photography studio offers prints in the following sizes: 3×5, 5×7, 8×10, 11×14, and 16×20. Find the ratio of the long side to the short side for each of these rectangles. Which is closest to the golden ratio?

34. MARKETING A manufacturer wishes to make a cereal box in the shape of a golden rectangle, based on the theory that this shape is the most pleasing to the average customer. If the front of the box has an area of 104 in.^{2}, what should the dimensions be? Round to the nearest inch.

35. FINE ARTS An artist wishes to make a canvas in the shape of a golden rectangle. He only has enough framing material for a perimeter of 36 in. What should the dimensions of the canvas be?

Look Back

CONNECTION

COORDINATE GEOMETRY Find the distance between each pair of points below. Round your answers to the nearest tenth. *(LESSON 5.6)*

36. $(0, 0)$ and $(4, 7)$

37. $(1, 3)$ and $(4, 2)$

38. $(-6, 0)$ and $(5, -2)$

39. $(-1, -3)$ and $(3, -5)$

40. Write an equation for the set of points in a coordinate plane that are a distance of 3 from the point $(2, 1)$. *(LESSON 9.6)*

41. Find the coordinates of a point on the unit circle that is the image of the point $(1, 0)$ under a 50° rotation about the origin. *(LESSON 10.3)*

42. The point $(-0.643, 0.766)$ is the image of the point $(1, 0)$ under a rotation about the origin. What is the angle of rotation? *(LESSON 10.3)*

43. Vector $\vec{a}$ in a coordinate plane points from the origin to $(2, 5)$ and vector $\vec{b}$ points from the origin to $(5, 2)$. What is the sum of these two vectors? Sketch a diagram to illustrate your answer. *(LESSON 10.6)*

Look Beyond

Fibonacci numbers, which are numbers in the Fibonacci sequence, occur frequently in nature. For example, the numbers of spirals in pinecones, artichokes, sunflowers, and other objects are often Fibonacci numbers.

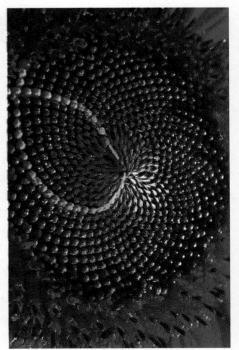

44. Examine the photo of the sunflower at right. Notice that the seeds form two sets of spirals in opposite directions. Try to count the number of spirals to the left and to the right. Both numbers should be Fibonacci numbers. What are they?

45. The numbers of petals of many types of flowers are Fibonacci numbers. Find some pictures of flowers and count their petals. Which ones are Fibonacci numbers?

Taxicab Geometry

11.2

Objectives

- Develop a non-Euclidean geometry, known as *taxicab geometry*, based on a taxicab's movements on a street grid.

- Solve problems within a taxicab geometry system.

Why *A geometry student would probably use the Pythagorean Theorem to find the distance between two points on a city map. However, a taxicab driver might have a very different idea. In this lesson, you will study a geometry based on a taxicab driver's definition of the distance between two points.*

Taxidistance

In **taxicab geometry**, points are located on a special kind of map or coordinate grid. The horizontal and vertical lines of the grid represent streets. Unlike points in a traditional coordinate plane, points in a taxicab grid can be only at intersections of two "streets." Thus, the coordinates are always integers.

In taxicab geometry, the distance between two points, known as the **taxidistance**, is the smallest number of grid units, called **blocks**, that a taxi must travel to get from one point to the other. On the map shown, the taxidistance between the two points is 5.

In this taxicab grid, the taxidistance from point A to point B is 5.

Exploring Taxidistances

YOU WILL NEED

graph paper

Part I: The taxidistance from a central terminal

Assume that all taxis leave for their destinations from a central terminal at point $O(0, 0)$.

1. Draw the six destination points in a taxicab grid, as shown in the diagram at right. Label the points A through F and give their coordinates.

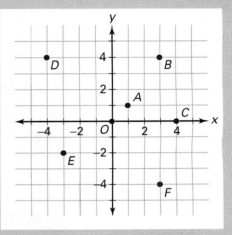

PROBLEM SOLVING

2. Make a table. Find the taxidistance from O to each of the six destination points. (Make sure that you have found the smallest number of blocks in each case.) Copy and complete the table below.

Point	Coordinates	Taxidistance from O
A	(?, ?)	?
B	(?, ?)	?
C	(?, ?)	?
D	(?, ?)	?
E	(?, ?)	?
F	(?, ?)	?

3. Based on your results, complete the following conjecture:

CHECKPOINT ✔   The taxidistance between the points $(0, 0)$ and (x, y) is ____?____.

Part II: The taxidistance between any two points

PROBLEM SOLVING

1. Make a table. Use the diagram from Part I to find the taxidistance between the pairs of points given in the table below. Copy and complete the table below.

(x_1, y_1)	(x_2, y_2)	x_1	x_2	y_1	y_2	Taxidistance
A	B	1	3	1	4	5
A	D	?	?	?	?	?
A	E	?	?	?	?	?
C	D	?	?	?	?	?
D	E	?	?	?	?	?
E	F	?	?	?	?	?
B	F	?	?	?	?	?

2. Based on your results, complete the following conjecture:

CHECKPOINT ✔   The taxidistance between the points (x_1, y_1) and (x_2, y_2) is ____?____.

Two Points Determine...?

For any two given points in Euclidean geometry, there is just one minimum-distance pathway between them. Is this true in taxicab geometry? Consider points *A* and *B* below. In both arrangements, *A* and *B* are 3 taxicab units apart. In the second arrangement, there is more than one minimum-distance pathway between them.

First arrangement:
one pathway

Second arrangement:
three pathways

CRITICAL THINKING

Under what conditions is there just one minimum-distance path between two points in taxicab geometry? Under what conditions is there more than one?

Taxicab Circles

YOU WILL NEED

graph paper

In Euclidean geometry, a circle is the set of points that are a fixed distance from a given point. What happens when this definition is applied to taxicab geometry?

1. Plot a point *P* on graph paper. Then plot all the points that are located 1 block from point *P*. The very uncircular-looking result is a **taxicab circle** with a radius of 1.

2. Draw additional taxicab circles with radii of 2, 3, 4, and 5. Count the number of points on each circle. Find the circumference of each circle by adding the taxidistances between the points on the circle.

A taxicab circle with a radius of 1

PROBLEM SOLVING

3. Make a table. Copy and complete the table below to find expressions for the number of points on a taxicab circle and the circumference in terms of its radius.

Radius	Number of points on circle	Circumference
1	4	8
2	?	?
3	?	?
4	?	?
5	?	?
r	?	?

CHECKPOINT ✔

4. Use the information in the chart to determine a taxicab equivalent for π.

$\left(\text{Hint: } \pi = \dfrac{\text{circumference}}{\text{diameter}} \right)$

Exercises

Communicate

1. Why is the geometry studied in this lesson called taxicab geometry?

2. Explain how to determine the distance between two points in taxicab geometry.

3. Why is π not used in taxicab geometry? What is the taxicab equivalent for π?

4. What are some practical applications of taxicab geometry?

5. What are some factors other than distance that a taxicab driver might take into consideration when planning a route?

Guided Skills Practice

For points A(0, 1), B(2, 3), and C(−2, 1), find the taxidistances below. (ACTIVITY 1)

6. AB **7.** BC **8.** AC

Plot the taxicab circle with the given center and radius. (ACTIVITY 2)

9. center: (3, 2); radius: 2   **10.** center: (−1, 4); radius: 4

Practice and Apply

Find the taxidistance between each pair of points.

11. (0, 0) and (7, 3)   **12.** (5, −3) and (−2, 4)

13. (1, 5) and (−2, −3)   **14.** (−9, −3) and (−3, −1)

15. (−11, 4) and (−3, 9)   **16.** (−129, 43) and (152, 236)

Find the number of points on the taxicab circle with the given radius.

17. $r = 2$ **18.** $r = 4$ **19.** $r = 12$

Find the circumference of the taxicab circle with the given radius.

20. $r = 1$ **21.** $r = 5$ **22.** $r = 10$

Plot the taxicab circle with center P and radius r.

23. $P(0, 0); r = 3$ **24.** $P(5, −2); r = 5$ **25.** $P(−1, 3); r = 2$

26. Identify two pairs of points in a taxicab grid that have a taxidistance of 4. One of these pairs should have only one minimum-distance pathway between them, and the other should have several minimum-distance pathways.

27. Sketch all of the different minimum-distance pathways between the points (1, 2) and (4, 1) in a taxicab grid. How many pathways are there?

28. Sketch all of the different minimum-distance pathways between the points (3, 0) and (5, 2) in a taxicab grid. How many pathways are there?

Recall from Chapter 4 that a point lies on a segment's perpendicular bisector if and only if it is equidistant from the segment's endpoints. Use this theorem in Exercises 29–33 to discover the taxicab equivalent of a perpendicular bisector.

29. **a.** On a taxicab grid, plot points $A(0, 0)$ and $B(4, 2)$. Locate all points that are a taxidistance of 3 from both A and B.

 b. On the same diagram, locate all points that are a taxidistance of 4 from both A and B.

 c. Continue locating points that are the same taxidistance from both A and B until you understand the form of a taxicab perpendicular bisector.

30. How is the perpendicular bisector you constructed in Exercise 29 similar to a perpendicular bisector in Euclidean geometry? How is it different?

31. Use the method from Exercise 29 to construct a taxicab perpendicular bisector for the points $(3, 3)$ and $(4, 0)$. How is it like the taxicab perpendicular bisector you constructed in Exercise 29? How is it different?

32. Use the method from Exercise 29 to construct a taxicab perpendicular bisector for the points $(1, 2)$ and $(5, 2)$. What do you notice?

33. Use the method from Exercise 29 to construct a taxicab perpendicular bisector for the points $(2, 0)$ and $(3, 2)$. What do you notice?

Algebra

CHALLENGE

34. Use your formula from Activity 1 for the taxidistance between two points to derive an equation for a taxicab circle with a radius of r.

35. For the points $A(0, 0)$ and $B(2, 2)$, find the set of all points P such that the sum of the taxidistances AP and BP is equal to 6. Describe your results.

APPLICATIONS

36. **PUBLIC SAFETY** The mayor of a city wants to install police call boxes at intersections around the city so that from every point in the city there will be a maximum distance of two blocks to a call box. Describe a way this can be done. Try to find a way that uses the minimum number of boxes. (Hint: Draw a pattern of taxicab circles with a radius of 2.)

37. **COMMUTING** Jenny works at an office during the day and then works in a retail shop in the evenings. Because she plans to ride her bicycle to work, she would like to live within 10 blocks of her day job and within 8 blocks of her night job so that she does not have to ride as far in the dark. Use the taxicab grid at left to determine all of the possible locations where Jenny should look for an apartment.

Find *y* for Exercises 38 and 39. *(LESSON 9.5)*

38.

39.

40. Let $BC = 5$, $AC = 22$, and $CD = 6$. Find DE.
(LESSON 9.5)

41. Let $MP = 18$ and $MN = 7$. Find MR.
(LESSON 9.5)

Refer to the triangle below. Find the value of each expression. *(LESSONS 10.1 AND 10.2)*

42. $\tan \theta$ **43.** $\cos \theta$ **44.** θ

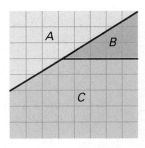

The following is a geometric "proof" that 0 = 1. Trace the steps and try to find the error in the proof.

45. **a.** On graph paper, draw an 8×8 square. Divide it into three regions, *A*, *B*, and *C*, as shown.

b. Cut out the pieces and fit them together to form a 5×13 rectangle, as shown.

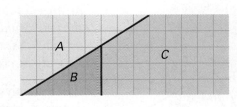

c. The area of the square is 8×8, or 64. The area of the rectangle is 5×13, or 65. However, since both areas are the sum of the areas of A, B, and C, the areas must be equal, so $64 = 65$. Subtracting 64 from both sides of the equation gives $0 = 1$. What is wrong with this proof?

Graph Theory

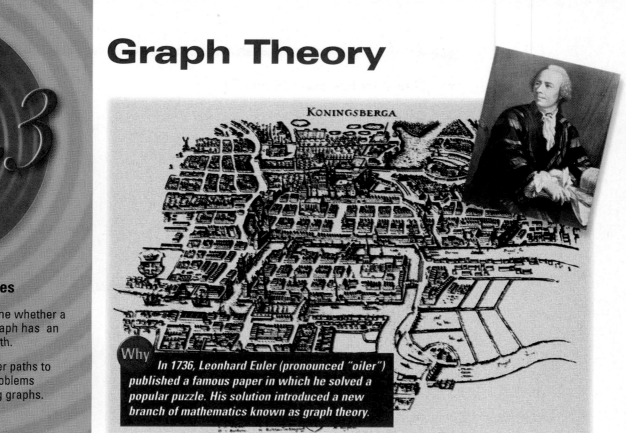

KONINGSBERGA

Objectives

- Determine whether a given graph has an Euler path.
- Use Euler paths to solve problems involving graphs.

Why In 1736, Leonhard Euler (pronounced "oiler") published a famous paper in which he solved a popular puzzle. His solution introduced a new branch of mathematics known as graph theory.

A local pastime in the city of Königsberg, Prussia (now Kaliningrad, Russia) was to try to walk across each of the seven bridges over the river without crossing any bridge twice.

The Bridge Problem

The following is an excerpt from Euler's paper, "The Solution to a Problem Relating to the Geometry of Position":

The problem, which I am told is widely known, is as follows: In Königsberg in Prussia, there is an island A, called the Kneiphoff; the river which surrounds it is divided into two branches, and these branches are crossed by seven bridges, a, b, c, d, e, f, and g. Concerning these bridges, it was asked whether anyone could arrange a route in such a way

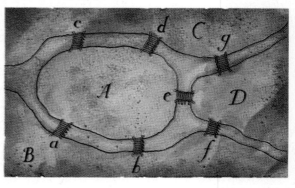

that he would cross each bridge once and only once. I was told that some people asserted that this was impossible, while others were in doubt; but nobody would actually assert that it could be done.

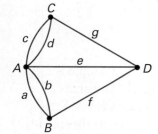

Euler's analysis of the problem can be visualized by using a diagram called a **graph**. A graph is composed of points, called **vertices**, and segments or curves linking the vertices, called **edges**. In the graph at left, each land area (refer to the map above) is represented by a vertex, and each bridge is represented by an edge. The bridge problem now becomes that of finding an **Euler path**—a continuous path that travels along each edge exactly once.

Odd and Even Vertices

You may have tried to find an Euler path in the graph of the seven bridges and decided that it was impossible. The people of Königsberg thought so, but no one had proved it mathematically until Euler presented his proof.

Euler's proof was based on the number of times each land area would have to be visited in a solution of the problem, which is related to the number of bridges to the land area. A graph makes his reasoning easier to understand.

In a graph, the number of edges at each vertex is called the **degree** of the vertex. In the graph of the bridge problem, the degree represents the number of bridges connected to each land area. **Even vertices** have an even degree and **odd vertices** have an odd degree.

In the graph at right, Q has degree 4, so it is an even vertex. R and S both have degree 3, so they are odd vertices. T has degree 2, so it is an even vertex. Does the graph contain an Euler path?

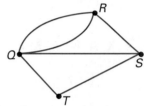

In the following Activity, you will explore the relationship between the degree of the vertices of a graph and whether the graph contains an Euler path.

Activity
Euler Paths

PROBLEM SOLVING

Which of the graphs below contain an Euler path? **Make a table** like the one below and see if you can discover the relationship.

a.   b. c. d.

e. f. g.   h.

> *In some graphs, edges may intersect at a point that is not a vertex.*

	Number of vertices	Number of odd vertices	Number of even vertices	Is there an Euler path?
a.	?	?	?	?
b.	?	?	?	?
c.	?	?	?	?
	⋮	⋮	⋮	⋮

CHECKPOINT ✔

Make a conjecture based on your completed table. Use your conjecture to complete the following theorem:

Theorem

A graph contains an Euler path if and only if there are at most ___?___ odd vertices.

11.3.1

EXAMPLE **1** Does the graph of the bridges of Königsberg contain an Euler path?

● **SOLUTION**

All four vertices in the graph are odd, so there are more than two odd vertices. Therefore, by Theorem 11.3.1, the graph does not contain an Euler path. That is, there is no route that crosses each bridge (each edge of the graph) exactly once.

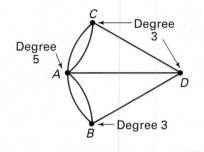

EXAMPLE **2** Refer to the floor plan of a house given at right. Is it possible to walk through all the doors in the house by going through each door exactly once?

● **SOLUTION**

PROBLEM SOLVING

Make a model. A graph can be used to model the floor plan, with a vertex for each room and for the area outside the house and an edge for each doorway.

The graph has only two odd vertices (*Q* and *R*), so it contains a Euler path. Thus, it is possible to walk through all the doors in the house by going through each door exactly once.

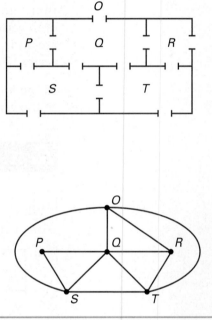

CRITICAL THINKING How could you find an Euler path in the graph above?

Starting and Stopping

The following problem may provide some insight into Theorem 11.3.1:

While hiking on a snowy day, you startle a rabbit. It runs from bush to bush, hiding under each one. It covers all paths between the bushes (see the diagram) without following the same path more than once. Then it stops. Which bush is the rabbit hiding under?

After trying a few routes, you should see that the rabbit always ends up under bush *E*, an odd vertex. The only other odd vertex is the woods, where the rabbit began its journey.

Suppose that there is an even vertex in a graph and you do not start from it. You will never get stuck at this even vertex for the following reason: *If there is an unused edge leading into the even vertex, there must be an unused edge leading out of it.*

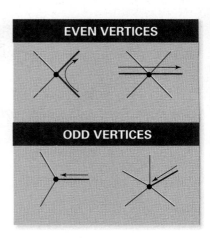

On the other hand, suppose that there is an odd vertex in the graph and that you do not start from it. If you are able to get to this vertex and keep coming back to it, then you will eventually get stuck at it. *There will be no unused paths leading away from it.*

Notice that if you start at an odd vertex, it becomes, in effect, an even vertex once you have left it; therefore, you will never get stuck at it.

If there are exactly two odd vertices in a graph, you can start at one and try to end at the other. But if there are more than two odd vertices in a graph, there is no chance of finding an Euler path through the graph.

Thus, the "only if" part of Theorem 11.3.1, which states that there is an Euler path through a graph *only if* there are at most two odd vertices, has been proven informally. And as you can see, the two odd vertices must be the starting and ending points of the graph. In the rabbit puzzle, one odd vertex (the woods), is the starting point, and the other odd vertex (bush *E*), is the ending point.

The "if" part of the theorem, which states that there actually *is* such a path when there are at most two odd vertices, has not been proven. However, in Exercises 49–51 of this lesson, you will learn a method, known as *Fleury's algorithm*, for finding such a path when the conditions are right.

CRITICAL THINKING   Do you think it is possible for a graph to have just one odd vertex?

Euler Circuits

A graph with only even vertices contains an Euler path that can start at any vertex. There is another important feature of such a graph: The vertex that is the starting point of the Euler path is also the ending point of the path. An Euler path that starts and ends at the same vertex is called an **Euler circuit**.

E X A M P L E ③   A street sweeper is cleaning the area shown at right. Is it possible to sweep both sides of every street exactly once and end at the starting point?

SOLUTION

PROBLEM SOLVING   **Make a model.** Model the problem with a graph, using an edge to represent each side of each street. The vertices represent intersections.

Because every vertex is even, the graph must contain an Euler path. The street sweeper will be able to plan the desired route.

● Communicate

1. Explain the Königsberg bridge problem in your own words. Why is there no solution?

2. Suppose that the residents of Königsberg decided to build more bridges until there was a way to cross each bridge exactly once. Discuss several ways this could be done.

3. Explain the difference between an Euler circuit and an Euler path.

4. Why do you think Euler called the Königsberg bridge problem "a problem relating to the geometry of position"?

● Guided Skills Practice

Determine whether each graph below contains an Euler path.
(EXAMPLE 1)

5.

6.

7.

8. Draw a graph to represent the floor plan at right, with a vertex for each room and the area outside the house and an edge for each doorway. Does the graph of the floor plan contain an Euler path?
(EXAMPLE 2)

9. A truck is collecting materials for recycling. The driver's route must cover both sides of every street exactly once and end at the starting point. In the map area shown at right, is such a route possible? Use a graph to illustrate your answer. **(EXAMPLE 3)**

● Practice and Apply

In Exercises 10–15, determine whether each graph contains an Euler path, an Euler circuit, or neither.

10.

11.

12.

13. **14.** **15.**

For each floor plan below, draw a graph with a vertex for each room and the area outside the house and with an edge for each doorway. Determine whether the graph contains an Euler circuit.

16. **17.**

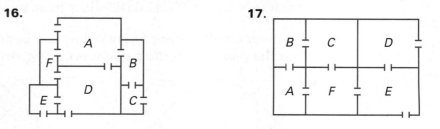

For Exercises 18–20, refer to the diagram below, taken from Euler's paper, "The Solution of a Problem Relating to the Geometry of Position."

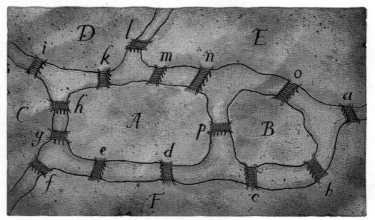

18. Draw a graph representing the land areas and bridges in the diagram.

19. Does the graph contain an Euler path? an Euler circuit?

20. If you wanted to walk across each bridge in the diagram exactly once, where would you start? Where would you finish?

In Exercises 21–29, you will explore some simple concepts of graph theory. Copy and complete the table below.

Graph	Number of edges	Sum of degrees of vertices
	21. ?	**22.** ?
	23. ?	**24.** ?
	25. ?	**26.** ?

27. Make a conjecture based on your results from Exercises 21–26. Use your conjecture to complete the following theorem:

The sum of the degrees of the vertices of a graph is __?__ the number of edges.

28. Use the fact that each edge of a graph has two endpoints to explain why the theorem in Exercise 27 is true.

29. An immediate corollary of the theorem in Exercise 27 is that the sum of the degrees of the vertices of a graph must be even. Use this fact to prove that the number of odd vertices in a graph must be even.

Determine whether each figure below can be traced completely without lifting the pencil from the page or retracing any part of the figure.

30.

31.

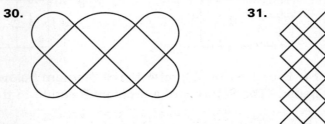

32. CULTURAL CONNECTION: AFRICA A popular game of children in the Shongo tribe in Zaire is to trace *networks*, like those shown at right, in the sand with a finger. The networks must be traced in a single stroke, without retracing. Can the network at right be traced in this way? If so, how could it be done?

CHALLENGE

33. In the diagram at right, is it possible to draw a continuous curve that intersects every segment exactly once? (The attempt shown intersects all but two segments.) If it is possible, draw the curve. If it is not possible, explain why.

APPLICATION

LAW ENFORCEMENT Refer to the diagram at right. A police officer is checking parking meters (represented by the red dots) along the streets. The officer wishes to go down each row of meters exactly once and return to the starting point.

34. Draw a graph that represents this situation.

35. Determine whether the officer can complete the desired route.

36. WILDLIFE MANAGEMENT A naturalist observing a pack of wolves comes across a set of trails on the ground, shown at right. Could the trails be traveled by a wolf in one trip, starting at the den and not retracing any of the trails? Why or why not?

37. FUND-RAISING Jo and Tamara are trying to raise money for a school trip by selling magazines door to door. Jo decides to cover each side of every street separately, while Tamara wants to alternate from one side of the street to the other, covering it in one trip. Refer to the map at right. Draw a graph for Jo's route. Is there an Euler path for Jo's route? Draw a graph for Tamara's route. Is there an Euler path for Tamara's route?

Look Back

Copy each figure and reflect it with respect to the given line. *(LESSON 1.6)*

38.

39.

In Exercises 40–44, write each statement as a conditional. Then write its converse. *(LESSON 2.2)*

40. All people who live in California live in the United States.

41. Every square is a rectangle.

42. A square is a parallelogram with four congruent sides and four congruent angles.

43. A dodecagon is a polygon with 12 sides.

CONNECTION

44. TRIGONOMETRY For any angle $\theta < 45°$, $\sin \theta < \cos \theta$.

Each solid below has a volume of 100 cubic units. Find *h* or *r* for each figure. Round your answers to the nearest tenth.

45.　　**46.**　　**47.**　　**48.**

Look Beyond

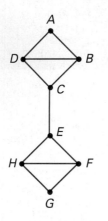

In Exercises 49 and 50, you will learn about *bridges*, which are another feature of graphs. In Exercise 51, you will apply *Fleury's algorithm*, a step-by-step procedure that uses bridges to find an Euler circuit in a graph.

49. An edge in a graph is called a *bridge* if removing it would split the graph into two separate pieces. Identify a bridge in the graph at left.

50. Draw a graph that contains an Euler circuit. Is any edge a bridge? Can a graph that contains an Euler circuit have a bridge? Why or why not?

51. *Fleury's algorithm*, given below, is a procedure that uses bridges to find an Euler circuit in a graph that has no odd vertices.

Fleury's Algorithm
Step a: Verify that the graph contains an Euler circuit. Choose a starting vertex.
Step b: Travel any edge that is not a bridge in the untraveled part of the graph.
Step c: Repeat Step **b** until you have completed an Euler circuit.

Note: In Step **b**, consider only the untraveled part of the graph. If a bridge is removed from this part of the graph, there will be edges that you are unable to get back to.

Use Fleury's algorithm to find an Euler ciruit in the graph at right. List the vertices in the order they are traveled.

USING GRAPHS TO SOLVE PUZZLES

Suppose that you have three bottles. One holds 8 quarts, one holds 5 quarts, and one holds 3 quarts. The 8-quart bottle is full of water. How can you get exactly 4 quarts of water in both the 5-quart and 8-quart bottles?

Problems like this can be solved with the aid of a graph. The amount of water in the bottles at any given time can be written as an ordered triple. For example, the initial state in the puzzle can be written as (8, 0, 0), meaning 8 quarts in the 8-quart bottle, 0 quarts in the 5-quart bottle, and 0 quarts in the 3-quart bottle. Each state is represented by a vertex in the graph, and two vertices are connected by an edge if it is possible to move from one vertex to the other in one step. The "legal moves" consist of the following: pouring all of the water from one bottle into another bottle, or pouring as much water as possible from one bottle to another bottle.

1. The first level of the graph for this problem is shown at right. Explain how to get from the initial state to each of the states represented by vertices in the graph.

2. What moves are possible from the position (3, 5, 0)? What moves are possible from the position (5, 0, 3)? Complete the next level of the graph.

3. Extend the graph until you find the position (4, 4, 0). Describe how to pour the water from the bottles to get exactly 4 quarts in both the 8-quart and 5-quart bottles.

WORKING ON THE CHAPTER PROJECT
You should now be able to complete Activity 2 of the Chapter Project.

Topology: Twisted Geometry

Objectives

● Explore and develop concepts of topology, including knots, Möbius strips, and tori.

● Use theorems of topology to solve problems.

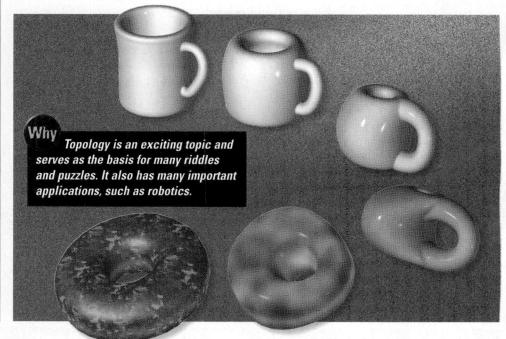

Why *Topology is an exciting topic and serves as the basis for many riddles and puzzles. It also has many important applications, such as robotics.*

Transforming a coffee cup into a doughnut

Knots, Pretzels, Molecules, and DNA

Topology is a branch of mathematics that studies the most basic properties of figures. In topology, it is important to know whether a point is inside, outside, or on the boundary of a figure, but the distance between two points or whether they are on a straight line is not significant.

In topology, figures are deformed by stretching and squashing, as in the coffee cup example above, but cutting or tearing is not allowed. Do you think it is possible to unlink the rings of the "pretzel" at right by deforming it without cutting or tearing it?

The shape at right is known as a **Möbius strip**, a classic topological figure with intriguing properties. (You will explore Möbius strips in Exercises 25–30.) The mathematical analysis of such objects has proven useful in other fields, such as chemistry and biology.

David Walba and his co-workers at the University of Colorado have synthesized a molecule in the shape of a Möbius strip. Such structures can help researchers understand why certain drugs with the same molecular structure can have vastly different effects.

Möbius molecule

Researchers have also used a branch of topology called *knot theory* to study the structures of complex, tangled structures such as DNA molecules.

Topological Equivalence

Two figures are **topologically equivalent** if one of them can be stretched, shrunk, or otherwise distorted into the other without cutting, tearing, or intersecting itself or compressing a segment or curve to a point. In a plane, all of the shapes below are topologically equivalent.

A shape that is topologically equivalent to a circle is called a **simple closed curve**. A simple closed curve does not intersect itself. The curves below are not topologically equivalent to a circle; thus, they are not simple closed curves.

CRITICAL THINKING Are any of the shapes above topologically equivalent to each other? Explain.

Jordan Curve Theorem

A fundamental theorem in topology was first stated by French mathematician Camille Jordan (1838–1922) in the nineteenth century.

Jordan Curve Theorem

Every simple closed curve in a plane divides the plane into two distinct regions, the inside and the outside. Every curve that connects a point on the inside to a point on the outside must intersect the curve.

11.4.1

The theorem, which is very difficult to prove mathematically, seems obvious for the curve above. But for some curves it is not so simple. Are points P and Q on the inside or the outside of the simple closed curve at left? For curves like the snowflake at right, which are called *fractals* (see Lesson 11.7), it can be very difficult to determine the region that contains points near the curve.

The Jordan Curve Theorem is also true for simple closed curves on a sphere, but not for those on a doughnut-shaped surface called a **torus** (plural, *tori*). Thus, a torus is not topologically equivalent to a sphere.

Sphere   Torus

Invariants

Properties that stay the same no matter how a figure is deformed without breaking are called **invariants**. For example, if the pentagon at right is distorted into some other curve, as shown, the order of the points stays the same. Thus, the order of the points is an invariant.

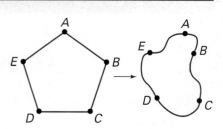

One important invariant in topology comes from Euler's formula for polyhedra, which you may recall from Lessons 6.3 and 7.3.

Euler's Formula

For any polyhedron with V vertices, E edges, and F faces,

$$V - E + F = 2.$$

11.4.2

The number $V - E + F$ is an invariant of a surface and is called the **Euler characteristic** (or Euler number). Thus, if a polyhedron is deformed into a surface that is topologically equivalent to the polyhedron, the Euler characteristic remains the same.

For example, imagine that a tetrahedron is deformed into a sphere, as shown at right. The number of vertices, edges, and faces are the same, although the edges are no longer straight and the faces are no longer flat. Thus, the Euler characteristic of a sphere is the same as that of the tetrahedron, 2.

However, consider the figure at left, which is topologically equivalent to a torus. It has 16 vertices, 32 edges, and 16 faces, so its Euler characteristic is $V - E + F = 16 - 32 + 16 = 0$. Thus, the Euler characteristic of a torus is also 0. A simpler figure for finding the Euler characteristic of a torus is shown below, with 1 vertex, 2 edges, and 1 face. Again, $V - E + F = 1 - 2 + 1 = 0$. The edges and vertices can be drawn in any way, as long as each face is topologically equivalent to a disk; that is, it can be flattened out and deformed to a filled-in circle without holes in it.

Finally, considering everything you have learned about topology, it may not surprise you to learn that the "pretzel" shown at the beginning of this lesson can indeed be unlinked.

● *Communicate*

1. Explain what it means for two shapes to be topologically equivalent.

Are the shapes below topologically equivalent? Explain your reasoning.

2.

3.

4. Explain two methods of proving that a sphere and a torus are not topologically equivalent.

5. Why do you think it might be easier to prove that figures are not topologically equivalent than to prove that figures are topologically equivalent?

● *Guided Skills Practice*

6. Which of the figures below are simple closed curves? *(TOPOLOGICAL EQUIVALENCE)*

a.   b.   c.   d.

7. Which of the figures below are topologically equivalent to ∞? *(TOPOLOGICAL EQUIVALENCE)*

a.   b.   c.   d.

Verify Euler's formula for the polyhedra below. *(EULER'S FORMULA)*

8. cube   9. octahedron

For Exercises 10–15, refer to the simple closed curve at right.

10. Is point P on the inside or the outside of the curve?

11. Is point Q on the inside or the outside of the curve?

12. Draw a ray from point P in any direction. How many times does the ray intersect the curve? (Points where the ray is tangent to the curve do not count as intersections.)

13. Repeat Exercise 12 for several rays in different directions. What can you say about the number of times each ray intersects the curve?

14. Draw several rays from point Q in several different directions. What can you say about the number of times each ray intersects the curve?

15. Complete the following conjecture: A ray from a point on the inside of a curve intersects the curve an __?__ number of times, and a ray from a point on the outside of a curve intersects the curve an __?__ number of times.

16. Use your conjecture from Exercise 15 to determine whether the point is on the inside or the outside of the simple closed curve at right.

The following is a famous graph theory problem: Can three houses be connected to three utilities (gas, water, and electricity) without any intersecting lines?

17. Draw vertices A, B, and C to represent the three houses and vertices G, W, and E to represent the three utilities.

18. Connect both A and B to both G and W with edges that do not intersect. (Remember that edges do not have to be straight.) The edges should form a simple closed curve.

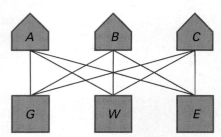

Can the houses be connected to the utilities without any intersecting lines?

19. Use your graph from Exercise 18 to explain why C and E must be both inside or both outside the simple closed curve.

20. If necessary, move vertices C and E outside the simple closed curve. Connect C to G and W with edges that do not intersect any existing edges. Use the Jordan Curve Theorem to show that either A or B cannot be connected to E without any intersecting edges.

21. Repeat Exercise 20 with C and E inside the simple closed curve. Explain why all the houses cannot be connected to all the utilities without any intersecting lines.

Suppose that a polyhedron is deformed into a sphere and then all the edges and vertices are moved to the top half of the sphere. The vertices and edges form a graph, which can be drawn in a plane. Each region in the graph is a face, and *the region outside the graph is also a face*.

22. For the graph above, count the number of vertices, edges, and faces. Verify that $V - E + F = 2$.

23. Verify that $V - E + F = 2$ for the graph at right. Remember to count the region outside the graph as a face. Based on the number of faces, what polyhedron do you think the graph represents?

24. Draw a graph of your own and use it to verify Euler's formula. Make your graph so that no two edges intersect except at a vertex.

In Exercises 25–30, you will explore the properties of the Möbius strip. You will need paper, tape, scissors, and a pencil. Cut a strip of paper about 1 in. wide and 11 in. long. Bring the ends of the strip together to form a loop, but flip one end over to form a half-twist before joining the ends. Tape both sides of the seam securely.

25. Starting at the tape, use your pencil to draw the path of an imaginary bug crawling along the center of your strip. Continue until you reach the starting point. What do you notice?

26. Create another Möbius strip, and draw a path about one-third of the way in from the edge of the strip, continuing until you reach the starting point. Describe your results.

27. Use your scissors to cut along the path that you drew on the Möbius strip in Exercise 25. Describe your results.

28. Use your scissors to cut along the path that you drew on the Möbius strip in Exercise 26. Describe your results.

Construct a double Möbius strip as follows: Hold two strips of paper together and put a half-twist in one end. Tape the ends on both sides to form two nested Möbius strips.

29. Run your finger along the inside of the two strips, keeping them nested. What do you notice? Do the strips seem to be separate?

30. Pull the strips apart. What do you notice? As a challenge, try to nest the strips back together again.

CHALLENGE

A Möbius strip is a surface with only one side. Another one-sided surface, which is theoretically a four-dimensional object, is called a *Klein Bottle*. A three-dimensional representation of a Klein Bottle is shown at right.

Construct a paper model of a Klein bottle as follows: Cut a 4-in. × 11-in. rectangle of paper and fold it lengthwise down the center. Tape the long edges together to form a flattened tube.

Hold the tube vertically, and cut a horizontal slit through the side of the tube nearest to you, about one-fourth of the way from the top.

Insert the bottom end of the tube into the slit, and align the two ends of the tube. Tape the ends together to produce one "hole" from above.

A Klein bottle in three-dimensional space must intersect itself.

31. Describe your paper model. Can any two points on the surface be connected by a path along the surface?

32. Lay your model flat and cut it in half lengthwise. Without cutting or tearing your paper, unfold the two halves of your bottle as much as possible. Describe the resulting shapes.

33. It is possible to cut the paper model into one Möbius strip. Can you discover how it is done?

APPLICATION

34. MANUFACTURING Sometimes long conveyor belts are shaped like Möbius strips. What do you think is the advantage of this shape?

Look Back

Algebra

35. Find *x* in the diagram at right.
(LESSON 2.6)

$(27x - 21)°$ $(10x + 30)°$

For Exercises 36–38, refer to the diagram at right. Assume that line ℓ ∥ line *m*. (LESSONS 3.3 AND 3.5)

36. Name two pairs of congruent angles.

37. Name three pairs of supplementary angles.

PROOF

38. Prove that m∠2 + m∠5 + m∠6 = 180° without using the Triangle Sum Theorem.

39. FOOD Suppose that you have a recipe for fruit salad that calls for one large grapefruit and that you wish to substitute an equal amount of oranges. If a large grapefruit is about 6 in. in diameter and an orange is about 4 in. in diameter, about how many oranges will you need?

Use the law of cosines and △ABC for Exercises 40 and 41. *(LESSON 10.5)*

40. Given $b = 25$, $c = 20$, and m∠$A = 55°$, find a.

41. Given $a = 104$, $c = 47$, and m∠$B = 92°$, find b.

Look Beyond

KNOT THEORY In the branch of mathematics called *knot theory*, a *knot* is a continuous loop; that is, the ends are joined so that it cannot be untied. A knot can be represented by a knot diagram, which indicates crossings with a solid curve and a broken curve. Two knots are *equivalent* if one can be deformed into the other in three-dimensional space without cutting.

The solid arc represents the strand that is on top.

Knot diagram

One way of determining whether two knots are equivalent is by using the *Reidemeister moves*. Kurt Reidemeister (1893–1971) proved that if two knots are equivalent, the diagram of one can be transformed into the diagram of the other by a combination of the three moves shown below.

42. The following is a sequence of knot diagrams. Each diagram results from performing a Reidemeister move on the preceding diagram, showing that the knots are equivalent. For each diagram, identify which of the Reidemeister moves was used.

43. Draw a knot diagram and transform it by a sequence of at least two different Reidemeister moves.

44. Use a strip of paper about 1 in. wide and 17 in. long to form a loop. Before joining the ends, put three half-twists in the strip. Then tape the ends securely. If you cut the strip down the middle, you should get a knotted loop. Flatten your shape down as much as possible, and arrange it as simply as you can. Draw a knot diagram for your shape, ignoring twists in the paper. This knot is called a *trefoil knot*.

Euclid Unparalleled

Objectives

- Explore and develop general notions for spherical and hyperbolic geometries.

- Develop informal proofs and solve problems by using concepts of non-Euclidean geometries.

Why *Many revolutionary concepts in math begin by questioning traditional assumptions. For example, a number of non-Euclidean geometries reject Euclid's assumption about parallel lines, with interesting and useful results.*

Imagine that you live on this two-dimensional surface. As you move outward from the center, everything gets smaller—including yourself. Would you ever reach the edge of your "universe"? This woodcut by M. C. Escher, known as Circle Limit 3, *is the artist's conception of a non-Euclidean geometry.*

Non-Euclidean Geometries

Euclid's geometry is based on five assumptions or postulates:

1. A line may be drawn between any two points.

2. Any segment may be extended indefinitely.

3. A circle may be drawn with any given point as the center and with any given radius.

4. All right angles are equal.

5. If two lines are met by another line and if the sum of the internal angles on one side is less than the sum of two right angles, then the two lines will meet.

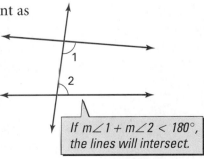

If m∠1 + m∠2 < 180°, the lines will intersect.

Because the fifth postulate in the list seemed less obvious than the others, many mathematicians wished to prove it in terms of the other four. No one ever succeeded in doing so, but along the way, a number of discoveries were made. In particular, several statements were found to be logically equivalent to the fifth postulate. Two statements are said to be **logically equivalent** if each can be derived from the other. Some examples of statements that are logically equivalent to the fifth postulate are as follows:

- If a line intersects one of two parallel lines, it will intersect the other.

- Lines that are parallel to the same line are parallel to each other.

- Two lines that intersect one another cannot be parallel to the same line.

- In a plane, if line ℓ and point P not on ℓ are given, then there exists one and only one line through P that is parallel to ℓ.

The last statement is perhaps the most useful of the four and is the version that many mathematicians refer to as the Parallel Postulate. (See Lesson 3.5, page 170.)

For years, mathematicians tried to prove the Parallel Postulate, until it was finally shown that it was impossible. Many even wondered whether the Parallel Postulate is in fact true in the real world.

If the Parallel Postulate is rejected, then any theorems that depend on it must be questioned. One such theorem is the Triangle Sum Theorem. (Recall that in the proof of the theorem, you must construct a line through one vertex that is parallel to the opposite base.)

The great mathematician Karl Friedrich Gauss (1777–1855) went so far as to measure the angles of a triangle formed by the points of three different mountain tops about 50 miles apart to see if the measures added up to 180°. (Within the limits of the accuracy of his measurements, they did.)

Does $m\angle 1 + m\angle 2 + m\angle 3 = 180°$?

Some mathematicians adopted a different attitude. They found that they could develop entirely new systems of geometry without using the Parallel Postulate. Systems in which the Parallel Postulate does not hold are examples of **non-Euclidean geometries**.

Euclidean geometry is based on figures in a plane. The figures in the non-Euclidean geometries that you will study here are in curved surfaces. Thus, the concept of a straight line no longer applies. In these geometries, a line will be defined as the shortest path between two points along the surface.

Spherical Geometry

Geometry on a sphere such as Earth's surface, or **spherical geometry**, is an example of a type of non-Euclidean geometry known as Riemannian geometry, after its discoverer, G.F.B. Riemann (1826–1866).

In spherical geometry, a line is defined as a **great circle** of the sphere; that is, a circle that divides the sphere into two equal halves. The shortest distance between two points in a sphere is always a path along a great circle. In this geometry, as in any Riemannian geometry, there are no parallel lines at all because all great circles intersect.

All great circles intersect.

Imagine two superhuman runners who start running at the equator. Their paths form right angles with the equator. What happens to their paths as they approach the North Pole? How does this differ from the result you would expect in a plane?

Theorems that depend on the Parallel Postulate for their proof may actually be false in spherical geometry. On a sphere, for example, the sum of the measures of the angles of a triangle is always greater than 180°, as in the spherical triangle at right.

CRITICAL THINKING Do Euclid's first four postulates seem to be true on the surface of a sphere? If so, should all the theorems that follow from them be true on a sphere?

In this triangle, the sum of the measures of the angles is 270°.

Hyperbolic Geometry on a "Saddle"

Lines ℓ and m are parallel to line n but not to each other.

Imagine that you are standing on the surface of a sphere. In every direction, the surface curves away from you. On the surface of a saddle, the surface curves away from you in some directions and toward you in other directions. On such a surface, there is more than one line through a point that is parallel to a given line. In fact, there are infinitely many.

The geometry of a saddle is an example of **hyperbolic geometry**, or Lobachevskian geometry. Nikolai Lobachevsky (1773–1856) was one of two mathematicians who discovered this type of geometry independently, the other being János Bolyai (1802–1860).

In hyperbolic geometry, just as in the other geometries you have studied, a line is defined as the shortest path between two points. In the illustration, lines ℓ and m pass through point P, and both are parallel to n (because they will never intersect line n).

Once again, the Triangle Sum Theorem does not hold. On a saddle, the sum of the angles in a triangle is always less than 180°.

Poincaré's Model of Hyperbolic Geometry

The "universe" in the Escher woodcut on page 731 is a representation of a three-dimensional universe imagined by the French mathematician Henri Poincaré (1854–1912). Poincaré, whose many interests included physics and thermodynamics, imagined physical reasons for the geometric properties of his imagined universe.

In the model of Poincaré's universe below, the surface is not curved, but the measurement of size and distance is defined in a different way from Euclidean geometry so that it represents a curved surface.

Temperature and size are greatest at the center.

Poincaré's universe can be represented by a circle. The temperature is greatest at the center of the circle and drops to absolute zero at the edges. According to Poincaré's rules for his universe, no one would be aware of temperature changes.

In Poincaré's universe, the size of an object is proportional to its temperature. An object would grow smaller as it moved away from the center.

Because everything, including rulers, would shrink in size, there would be no way to detect the change. In fact, distance measures would keep shrinking, so you would never get any closer to the edge of the universe, no matter how long you traveled.

Everything shrinks as you move toward the edge of Poincaré's universe.

The arcs in the circle represent the paths that rays of light would travel in Poincaré's universe. These are defined as lines in this system. Light rays that stay close to the edge of the universe appear more curved than those that pass close to the center. Lines through the center are diameters of the circle. Another feature of these lines is that they are **orthogonal** to the circle; that is, they form right angles with the circle at the intersections.

Orthogonal arcs

In Poincaré's system, a line segment connecting two points is a part of the arc that is orthogonal to the circle. This is the shortest distance between the two points. If you wanted to walk from A to B, the curved line would be shorter than the straight line connecting them because your steps would get larger as you approached the center of the circle. You would cover more distance with each step as you moved toward the center than you would by walking directly to the point.

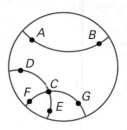

Notice also on the previous page that more than one line can be drawn through a point that is parallel to a given line. For example, $\overleftrightarrow{DE}$ and $\overleftrightarrow{FG}$ pass through the point C, and both are parallel to $\overleftrightarrow{AB}$. Thus, the geometry of Poincaré's universe is an example of a hyperbolic, or Lobachevskian, geometry.

Finally, notice that the sum of the measures of the angles in a triangle in Poincaré's universe is always less than 180°, just as it is on the surface of a saddle.

$m\angle 1 + m\angle 2 + m\angle 3 < 180°$

Years Later—Applications

In Einstein's general theory of relativity, space is non-Euclidean. In fact, due to the influence of gravity or—equivalently—acceleration, space is curved. The non-Euclidean geometries discovered years earlier proved to be both an inspiration and a useful tool to Einstein in formulating his fundamental laws of the universe.

Albert Einstein, 1879–1955

Exercises

● Communicate

1. Which postulate of Euclid was questioned by mathematicians? State it in your own words.

2. What are the alternatives to the Parallel Postulate that have yielded two non-Euclidean geometries? Complete the following statements and explain your reasoning:

   a. In a sphere, if line ℓ and point P not on ℓ are given, then there exist(s) _____?_____ line(s) through P parallel to ℓ.

   b. In a hyperbolic surface, if line ℓ and point P not on ℓ are given, then there exist(s) _____?_____ line(s) through P parallel to ℓ.

3. In spherical geometry, how does a line extend indefinitely?

4. In Poincaré's universe, how does a line extend indefinitely?

5. What can you say about the sum of the measures of the angles of a triangle in spherical geometry? in hyperbolic geometry? Express your answers as inequalities. **(SPHERICAL GEOMETRY AND HYPERBOLIC GEOMETRY)**

Refer to the diagram of a sphere at right. Determine whether each figure below is a line in spherical geometry. (SPHERICAL GEOMETRY)

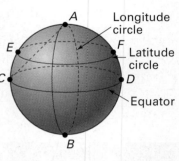

6. $\overleftrightarrow{AB}$ **7.** $\overleftrightarrow{AC}$

8. $\overleftrightarrow{EF}$ **9.** $\overleftrightarrow{CD}$

Refer to the diagram of Poincaré's model at right. Determine whether each figure below is a line in hyperbolic geometry. (HYPERBOLIC GEOMETRY)

10. $\overleftrightarrow{FG}$ **11.** $\overleftrightarrow{FJ}$

12. $\overleftrightarrow{GH}$ **13.** $\overleftrightarrow{HJ}$

Practice and Apply

14. Suppose that you cut an orange in half around its "equator" and then cut each half twice at right angles through the poles. You would have divided the orange peel into 8 congruent triangles. If you were to add the measures of the angles of one of these triangles, what would be the result?

15. Draw a 2-gon, or a polygon with two edges, on a sphere. (Note: The edges of a polygon must be segments, as they are in a plane.) What can you say about the vertices of the 2-gon?

16. In spherical geometry, what can you say about the sum of the measures of the angles of a 2-gon? Express your answer as an inequality.

17. Draw a quadrilateral on a sphere. Include one diagonal in your drawing.

18. What can you say about the sum of the measures of the angles of the quadrilateral from Exercise 17? How do you know? Express your answer as an inequality.

In Exercises 19–27, you will explore constructions in Poincaré's system.

Part I: Constructing a line through a given point

19. Construct ⊙P and choose a point, A, on the circle. To construct a line through A in Poincaré's system, first construct line ℓ tangent to ⊙P at A. Choose a point, C, on line ℓ. Place your compass point at C and draw a circle through A. The arc of the circle that lies inside ⊙P is a line through A in Poincaré's system.

20. Repeat Exercise 19, placing C at several different locations on line ℓ. What happens as C gets farther and farther away from A? What would happen if C were infinitely far from A?

21. Explain why the center of an arc that is orthogonal to a circle at a point must lie on the line tangent to the circle at that point.

22. Construct ⊙Q and choose a point, B, inside the circle. To construct a line through B in Poincaré's system, first choose a point, E, on ⊙Q. Construct line m tangent to ⊙Q at E, and construct line n, the perpendicular bisector of $\overline{BE}$. Place your compass point at the intersection of lines m and n and draw a circle through B. The arc of the circle that lies inside ⊙Q is a line through B in Poincaré's system.

23. Repeat Exercise 22, placing E at several different locations on ⊙Q. Explain how the line varies as the location of E changes.

Part II: Constructing a line through two given points

24. Construct ⊙P and choose two points, A and B, on the circle that are not endpoints of a diameter. To construct a line through A and B in Poincaré's system, construct line ℓ tangent to ⊙P at A and line m tangent to ⊙P at B. Place your compass point at the intersection of lines ℓ and m, and draw a circle through A and B. The arc of the circle that lies inside ⊙P is a line through A and B in Poincaré's system.

25. Why would the construction in Exercise 24 fail if A and B were the endpoints of a diameter? What would be the line through A and B in this case?

26. Construct ⊙Q and choose two points, C and D, inside the circle. To construct a line through C and D in Poincaré's system, first draw $\overleftrightarrow{CQ}$. Construct line n perpendicular to $\overleftrightarrow{CQ}$ at C. Line n intersects the circle at two points. Construct lines p and q tangent to ⊙Q at these points, and label as C' the intersection of p and q. Construct a circle that passes through points C, D, and C'. The arc of the circle that lies inside ⊙Q is a line through C and D in Poincaré's system.

Henri Poincaré (1854–1912)

27. Draw a line ℓ and a point A not on line ℓ in Poincaré's system. How many lines can you draw through point A that do not intersect line ℓ? How is this illustration related to the Parallel Postulate?

In Exercises 28–33, you will explore polygons in Poincaré's system.

28. Draw three lines that intersect to form a triangle in Poincaré's system. Measure the angles of the triangle.

29. What is the sum of the angles of the triangle from Exercise 28? How does this compare with the sum of the angles of a triangle in Euclidean geometry and in spherical geometry?

30. Draw a right triangle in Poincaré's system. (Hint: The diameter of the circle is a line in Poincaré's system.) What can you say about the acute angles of a right triangle in this geometry? Express your answer as an inequality.

31. Draw four lines that intersect to form a quadrilateral in Poincaré's system. Measure the angles of the quadrilateral.

32. What is the sum of the angles of the quadrilateral from Exercise 31?

33. Write a conjecture about the sum of the angles in a polygon with *n* sides in Poincaré's system. (Your conjecture should involve an inequality.)

You can use equilateral triangles to build models of surfaces in spherical, Euclidean, and hyperbolic geometries. Construct an equilateral triangle with a side length of about 3 cm. Copy the triangle carefully onto heavy paper or cardboard and cut it out. Make at least 20 copies of the triangle.

34. Tape 5 triangles together at a single vertex, as shown. This is a model of part of a surface in spherical geometry. Describe your model.

35. Tape 6 triangles together at a single vertex, as shown. This is a model of part of a surface in Euclidean geometry. Describe your model.

36. Tape 7 triangles together at a single vertex, as shown. This is a model of part of a surface in hyperbolic geometry. Describe your model.

37. How are the models you made in Exercises 34–36 alike? How are they different? Explain how the surfaces model the different geometries.

38. What happens if you extend the model in Exercise 34 to create a surface with 5 equilateral triangles at every vertex? (If you are not sure, make more triangles and experiment with them.) Could the model for spherical geometry extend to infinitely many triangles? Why or why not?

39. What happens if you extend the model in Exercise 35 to create a surface with 6 equilateral triangles at every vertex? Could the model for Euclidean geometry extend to infinitely many triangles? Why or why not?

40. What happens if you extend the model in Exercise 36 to create a surface with 7 equilateral triangles at every vertex? Model a surface in hyberbolic geometry with at least 20 triangles. Could this model extend to infinitely many triangles? Why or why not?

In Exercises 41–43, you will construct a regular octagon. *(LESSONS 4.7 AND 5.4)*

41. Construct square *ABCD*. Draw diagonals $\overline{AC}$ and $\overline{BD}$ and label their intersection *E*. Place your compass point at *A* and draw an arc through *E* intersecting $\overline{AB}$ and $\overline{AD}$.

42. Place your compass point at *B* and draw an arc through *E* in a similar fashion. Repeat at *C* and *D*. Connect the eight points where the arcs intersect the square to form an octagon.

43. Prove that the octagon you constructed in Exercises 41 and 42 is regular.

44. Draw a floor plan for your house or apartment. Draw a graph that represents the circulation of traffic in it. Remember that the rooms are the vertices and the doors are the edges of the graph. *(LESSON 11.3)*

45. Does your graph from Exercise 44 contain an Euler path? an Euler circuit? Why or why not? *(LESSON 11.3)*

In Exercises 46 and 47, you will explore topological properties of letters of the alphabet.

A B C D E F G H I J K L M N O P Q R S T U V W X Y Z

46. Identify all letters of the alphabet (drawn as shown above) that are topologically equivalent to the letter Z. *(LESSON 11.4)*

47. Are the words SIDE and CLOT topologically equivalent? Are LAST and COZY? Explain your reasoning. *(LESSON 11.4)*

The area of triangle *ABC* on a sphere with a radius of *r* is given by the formula

$$A = \pi r^2 \left(\frac{m\angle A + m\angle B + m\angle C}{180°} - 1 \right).$$

48. Verify the formula for the triangle on the sphere shown at right. (Hint: What fraction of the surface area of the sphere is covered by the triangle?)

49. In spherical geometry, the sum of the angles of a triangle is always greater than 180°. What can you say about the area of a triangle in which the sum of the angles is very close to 180°?

50. The area of a triangle on a sphere can not be greater than the surface area of the sphere. Use this fact to complete the following inequality involving the sum of the angles of a triangle on a sphere:

$$180° < m\angle A + m\angle B + m\angle C < \underline{\quad ? \quad}$$

Fractal Geometry

Objectives

- Discover the basic properties of fractals, including self-similarity and iteration.

- Build fractal designs by using iterative steps.

Why *The self-similarity of fractals enables a computer programmer to write a relatively short program for drawing such structures.*

A fractal, such as the computer-generated fern above, is a self-similar structure. Notice that each subdivision of the leaves of the fern has basically the same structure as the leaves themselves—all the way down to the curving tips.

Self-Similarity in Fractals

A **fractal**, like the **Menger sponge** at left, is a geometric object that exhibits some type of **self-similarity**. This means that the structure of the object always looks the same, whether seen in a highly magnified view, from a moderate distance, or from far away. If you were to cut off a small cube-shaped portion of the Menger sponge and examine it, you would find it to be a miniature copy of the entire sponge.

In mathematically created fractals like the Menger sponge, this process can theoretically be continued forever, and the self-similarity will always be evident.

Fractals can be created by repeating a simple procedure over and over. The Menger sponge, for example, is created by starting with a certain shape (a cube) and changing it according to a certain rule (removing a part of the cube). This same rule is applied to the newly changed shape. The process is then continued. This repetitive application of the same rule is called **iteration**.

How Long Is a Coastline?

In a theoretical fractal coastline, where the depth of self-similarity is endless, the length is considered to be infinite.

A coastline is a good example of a self-similar structure in nature. Jagged irregularities such as bays, capes, and inlets can be seen from an orbiting space shuttle. The same basic structures are evident over a smaller section of the coastline viewed from an airplane and over even smaller sections viewed during a drive or a walk.

How long is a coastline? Unlike ordinary geometric segments or curves, which can often be readily measured, the "length" of a fractal coastline depends on how closely you move in to measure it.

During the American Revolution, the British Royal Navy attempted to blockade the American coastline. Although fairly successful, the blockade was not able to prevent shipping in and out of many harbors. What does a close examination of the coastline reveal about its "length" that explains the difficulty faced by the British to completely blockade it?

Many objects found in nature have some of the properties of fractals. Notice, for example, the self-similarity of the broccoli at right.

Activity 1

Creating the "Cantor Dust"

YOU WILL NEED

a ruler and pencil with an eraser

One of the simplest fractals was discovered by Georg Cantor (1845–1918) years before fractals were defined and studied. As you will see, Cantor's fractal is a one-dimensional version of the Menger sponge.

1. Draw a line 27 cm long.

2. Erase the middle third of the segment. You should now have two segments that are 9 cm long, with a 9-cm gap between them.

3. Erase the middle third of each of the two 9-cm segments. You will now have four segments, each 3 cm long.

4. Continue erasing the middle third of each of these segments until you are left with a scattering of point-like segments, known as the **Cantor dust**.

27 cm

1 iteration   9 cm

2 iterations   3 cm   3 cm

3 iterations

4 iterations

PROBLEM SOLVING

5. Calculate the number of segments and their combined length after each iteration. **Make a table** like the one below.

Iteration	0	1	2	3	4	5	n
Number of segments	1	?	?	?	?	?	?
Combined length (cm)	27	?	?	?	?	?	?

CHECKPOINT ✔

6. As the number of iterations increases, what happens to the number of segments? What happens to the combined lengths of the segments?

CRITICAL THINKING

Describe the result if the process were repeated infinitely many times. How many segments would there be? What would be their combined length?

APPLICATION

TELECOMMUNICATIONS

When an electric current transmits data over a wire, a certain amount of "noise" occurs, which can cause errors in transmission. The noise seems to occur in random bursts, with "clean" spaces in between the bursts. Benoit Mandelbrot, the discoverer of fractal geometry, showed that the noise patterns closely matched the pattern of the Cantor dust. This geometric representation allowed the development of new strategies for reducing the transmission noise to a minimum.

The Sierpinski Gasket

The **Sierpinski gasket**, like the Menger sponge and the Cantor dust, is created by repeatedly applying a single rule to an initial shape. The rules are as follows:

1. Start with a solid triangle. Find the midpoint of each side.

2. Connect the midpoints of the sides to form four congruent isosceles triangles in the interior of the original triangle. Remove the center triangle. This is one iteration.

3. Each new iteration is performed on the remaining triangles.

Mathematical ideas often turn out to have surprising connections to seemingly unrelated fields. In Activity 2, you will learn about a connection between the Sierpinski gasket and Pascal's triangle.

Activity 2

Pascal and Sierpinski

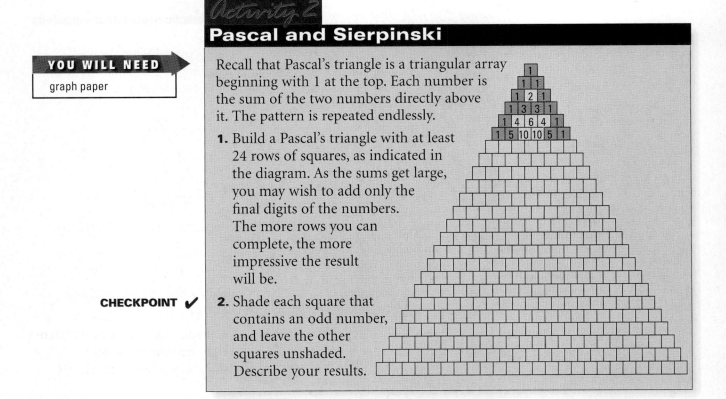

YOU WILL NEED

graph paper

Recall that Pascal's triangle is a triangular array beginning with 1 at the top. Each number is the sum of the two numbers directly above it. The pattern is repeated endlessly.

1. Build a Pascal's triangle with at least 24 rows of squares, as indicated in the diagram. As the sums get large, you may wish to add only the final digits of the numbers. The more rows you can complete, the more impressive the result will be.

CHECKPOINT ✔ **2.** Shade each square that contains an odd number, and leave the other squares unshaded. Describe your results.

Communicate

1. What is meant by self-similarity? You may wish to use an object to aid in your explanation.

2. Name an object from nature that has some type of self-similarity and describe how it is self-similar.

3. Mimi, Arnold, and Roberto measure a section of coastline by counting paces. Arnold's stride is 3 ft long, and he counts 24 paces. Roberto's stride is 2.5 ft long, and he counts 36 paces. Mimi's stride is 2 ft long, and she counts 47 paces. How long is the coastline for each person's set of measurements?

4. Explain how a fractal coastline could be considered to have an infinite length.

Three different measurements of a coastline

Guided Skills Practice

5. Draw an equilateral triangle and divide the sides into thirds. Draw segments connecting the sides to the base, forming two smaller equilateral triangles. Repeat as shown. How is the resulting fractal related to the Cantor dust? *(ACTIVITY 1)*

1 iteration   2 iterations

6. Use the following rules to explain the shading pattern in Pascal's triangle from Activity 2:

    even + even = even   odd + odd = even   even + odd = odd

    Based on these rules, devise a method for shading the squares without calculating the values in Pascal's triangle. *(ACTIVITY 2)*

7. Build a Pascal's triangle with at least 12 rows of squares. Shade each square that contains a number divisible by 3, and leave the other squares unshaded. How is the resulting pattern like the pattern from Activity 2? How is it different? *(ACTIVITY 2)*

In Exercises 8–18, you will explore the area and perimeter of the Sierpinski gasket. Copy and complete the table below.

Iterations	0	1	2	3
Shaded area	**8.** ?	**9.** ?	**10.** ?	**11.** ?
Perimeter	**12.** ?	**13.** ?	**14.** ?	**15.** ?

16. Use the terms in the table to complete the following statement: At each iteration, the area is multiplied by __?__.

17. As the number of iterations increases, what happens to the area? Does it seem to have a limit as the number of iterations approaches infinity?

18. As the number of iterations increases, what happens to the perimeter? Does it seem to have a limit as the number of iterations approaches infinity?

In Exercises 19–23, you will construct a fractal called the Koch snowflake and explore its properties.

19. Construct the first iteration of the Koch snowflake by following the directions below.

1st iteration

a. Construct an equilateral triangle with a side length of 18 cm.

b. Divide each side in thirds, and construct an equilateral triangle on each side of the triangle with the middle third as the base.

c. Erase the base of each new triangle. You should now have a six-pointed star. This is the first iteration.

20. Continue the construction of the Koch snowflake. Repeat the steps above to complete at least three iterations.

21. Find the perimeter of the Koch snowflake after the first two iterations. Remember that the snowflake starts with a perimeter of 3×18 cm, or 54 cm.

22. Does the perimeter increase or decrease? Does it increase or decrease by a greater amount or a smaller amount from each iteration to the next? What does this tell you about the perimeter?

23. Look at your snowflake and consider how it would change if you completed many more iterations. Would the area of the snowflake increase or decrease as the number of iterations increase? Would the area ever become infinite? (Hint: Can the snowflake always be enclosed by a circle?)

In Exercises 24–27 you will explore a fractal known as the dragon curve.

24. Take a long strip of paper and fold it in half and then in half again (in the same direction), repeating as many times as possible. Open each fold to 90°. The result, when viewed from the edge of the paper, is a dragon curve. Draw your dragon curve on paper.

25. The following is an algorithm for drawing a dragon curve: Start with two segments that intersect at a 90° angle. For each iteration, replace every $\vee$ with $\sqcup\!\!\ulcorner$, as shown below. Draw a dragon curve with at least five iterations.

1 iteration   2 iterations   3 iterations

26. Suppose that the dragon curve in Exercise 25 starts with a length of 2 units. What is the length after the first iteration? after two iterations? after three iterations? Divide your results from succesive iterations to complete the following statement:

   The length of a dragon curve after $n + 1$ iterations is ___?___ times the length after n iterations.

27. How does the area covered by the dragon curve change after each iteration? Do you think the area covered by the curve approaches infinity as the number of iterations increases? Explain your reasoning.

CHALLENGE

28. Refer to the drawing of the Menger sponge at the beginning of this lesson. Describe the construction of the sponge. If the volume before the first iteration is 1 cubic unit, what is the volume after one iteration? after two iterations?

APPLICATION

29. **HOBBIES** A kite may be made in the shape of a tetrahedron, with paper or fabric covering two of the faces. Four of these kites may be joined at the corners to form a larger kite, also in the shape of a tetrahedron (the first iteration). Four of these units may be joined at the corners to form a larger tetrahedral kite (the second iteration). As the units are joined, the kite will begin to resemble a three-dimensional Sierpinksi gasket. How many of the original kites are required for the second iteration? How many are required for the third iteration?

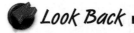
Look Back

Find the distance between each pair of points. *(LESSON 5.6)*

30. $(4, -2), (2, -1)$ 31. $(5, -10), (-2, 3)$ 32. $(15, 2), (-6, 5)$

Find the distance between each pair of points in a three-dimensional coordinate system. *(LESSON 6.4)*

33. $(4, 3, 2), (2, -3, 5)$ 34. $(18, 1, 0), (0, -1, 5)$ 35. $(5, 1, -5), (2, -12, 0)$

36. Find the measure of the diagonal of a cube in the top-front-right octant, with a vertex at $(0, 0, 0)$, and with a side length of 8. *(LESSON 6.4)*

37. CULTURAL CONNECTION: ASIA

For centuries, women of India have used fascinating curve patterns known as Kolams for ritual and decorative purposes. A fractal known as the Hilbert curve may remind you of the Kolams of India. Steps for creating a Hilbert curve are given below.

a. Start with a three-sided "square" with no bottom edge.

b. Draw a new (smaller) three-sided figure at each vertex. Be sure your new figures are oriented as shown.

c. Connect the new figures as shown to form a single continuous line. Erase the original figure. This is the first iteration.

d. For each new three-sided figure that you just created in part **b**, create a new (smaller) figure at each vertex. Connect the new figures and erase the original figures, as in part **c**. This is a new iteration.

38. The resulting curve is known as a space-filling curve or a "monster curve." If you were to continue this process indefinitely, what would the result look like?

THE CHAOS GAME

Can there be order in a random process? The following "chaos game" may lead you to ask some deep questions.

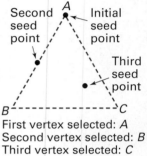

1. Draw equilateral triangle $\triangle ABC$ with a side length of 10 centimeters. Roll a number cube or use some other method to randomly select vertex A, B, or C. You can let 1 and 2 represent A, let 3 and 4 represent B, and let 5 and 6 represent C. Mark the selected vertex, which is called the *seed point*.

Second seed point — Initial seed point — A

Third seed point

First vertex selected: A
Second vertex selected: B
Third vertex selected: C

2. Roll the number cube again to select another vertex. This time, mark the point halfway between the seed point and the selected vertex. This point becomes the new seed point. Repeat the process several times, using the new seed point each time.

3. If you were to repeat this process several hundred times, a pattern would start to emerge. Combine your results with your classmates' results and see if you can guess the pattern. The program below, which will work with a graphics calculator, simulates the chaos game for a triangle with vertices at $(0, 0)$, $(0.5, 1)$, and $(1, 0)$ and an initial seed point at $(0, 0)$ for 1500 repetitions. Study the program to see how the simulation works. Then run the program and describe the result.

```
PROGRAM: CHAOS
:FnOff  :ClrDraw          } These commands clear the screen of any
:PlotsOff  :AxesOff          previous graphs.
:0→Xmin  :1→Xmax        } These commands set the viewing window.
:0→Ymin  :1→Ymax
:0→X  :0→Y                 This locates the seed point at (0, 0).
For (K,1,1500)  :rand→N    This generates 1500 random numbers.
:If N≤1/3  :Then
:.5X→X  :.5Y→Y
:End
: If 1/3<N and N≤2/3  :Then   These commands assign one of
:.5(.5+X)→X  :.5(1+Y)→Y         the vertices to each random
:End                            number and then locate the
:If 2/3<N  :Then                seed points.
:.5(1+X)→X  :.5Y→Y
:End
:Pt-On(X, Y)               This shades the seed point on the graph.
:End
```

Other Transformations and Projective Geometry

11.7

Objectives

- Understand the concepts of affine transformations and geometric projection.

- Solve problems and make conjectures by using the Theorem of Pappus and the Theorem of Desargues.

Why *When you stretch or otherwise deform a figure, will you still be able to recognize it? In what ways will it be the same? These are questions that are considered in the study of projective geometry.*

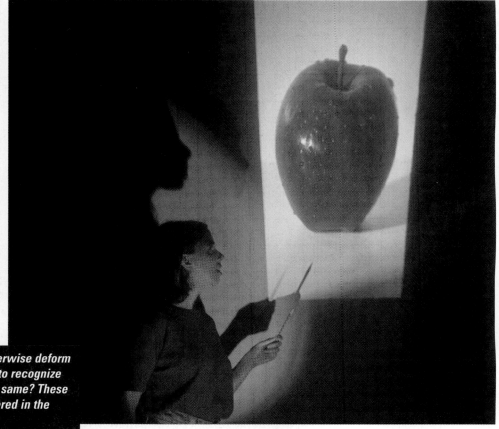

Nonrigid Transformations

The three rigid transformations—reflections, rotations, and translations—preserve the shapes and sizes of objects. Dilations, on the other hand, preserve shape but not size. There are types of transformations that shrink, expand, or stretch an object in different directions so that neither shape nor size are necessarily preserved. One of these types of nonrigid transformations is called an *affine transformation* and is defined as follows:

Definition: Affine Transformation

An **affine transformation** transforms each preimage point P in a plane to an image point P' in such a way that

   1. collinear points are transformed into collinear points,

   2. straight lines are transformed into straight lines,

   3. intersecting lines are transformed into intersecting lines, and

   4. parallel lines are transformed into parallel lines. **11.7.1**

An Affine Transformation

Part I

1. Draw a 4 × 4 square and a 4 × 4 parallelogram on graph paper as shown below. (Use more than one grid square for each unit square in order to make the figures large enough to work with.)

2. Inscribe a circle in the square. Draw a square and its diagonals inside the circle.

3. Mark the points of intersection of the circle with the grid lines. Mark the vertices of the inner square.

4. Mark the same points in the parallelogram grid. The first few are marked in the diagram below.

5. To draw the image of the figure under an affine transformation, connect the image points of the square with segments (including the diagonals) and draw a smooth curve through the image points of the circle.

CHECKPOINT ✔ 6. How does the resulting figure illustrate each of the four conditions in the definition of an affine transformation?

Part II

Affine transformations can be represented by using coordinates. For example, multiplying the x- and y-coordinates of the points of a figure by two different scale factors is an example of an affine transformation.

Preimage point	**Multiply x by 2 and y by 3.**	**Image point**
(x, y)	$\longrightarrow$	$(2x, 3y)$

1. Draw a square and its diagonals in the first quadrant of a coordinate plane.

2. Multiply the x-coordinates of the vertices of the square by 2 and the y-coordinates by 3. Plot the resulting points and connect them with segments (including the diagonals). Describe the resulting figure.

3. Multiply the x-coordinates of the vertices of the square by 3 and the y-coordinates by 2. Plot the resulting points and connect them with segments (including the diagonals). Describe the resulting figure.

CHECKPOINT ✔ 4. How do the resulting figures illustrate each of the four conditions in the definition of an affine transformation?

Projections and Projective Geometry

In the photograph at the beginning of this lesson, the image of the apple is distorted by a transformation. Is it an affine transformation? Study the two images below.

As you can see, the sides of the grid are parallel in the original image but not in the transformed image. Therefore, parallel lines do not transform to parallel lines, as required by the definition of an affine transformation. Thus, it is not an affine transformation.

The transformation of the image of the apple is an example of a class of transformations known as **central projections**. In such transformations, there is a central point known as the **center of projection**, and the **projected points** lie on rays containing the center of projection and the original points.

The diagram at right shows a projection between two lines in the same plane. The points on line ℓ are projected onto line m from the center of projection, point O. The rays drawn from the center of the projection are called the **projective rays**. These rays intersect line ℓ at points A, B, and C. The intersections of the rays with line m determine the projected points, A', B', and C'.

To make the map a Mercator projection, its vertical dimensions must be adjusted. (See page 176.)

The surface of a sphere can be projected onto a plane by means of a *cylindrical projection*. Notice that it is impossible to project the points at the North and South Poles onto the map, and that it is impractical to project points near the poles.

Geography teachers have long pointed out that Greenland is really much smaller in relation to other countries than it appears to be on cylindrical projection maps. Deformities like these arise because it is impossible to flatten Earth's spherical surface into a rectangle without some distortion of size, shape, distance, and direction.

For years, artists used the concept of projection to given their works realism and depth (see Lesson 6.6). On the other hand, mathematicians used projections to develop an entire system of geometry. The chart below summarizes its features.

Main Features of Projective Geometry

1. Projective geometry is the study of the properties of figures that do not change under a projection.

2. There is no concept of size, measurement, or congruence.

3. Its theorems state facts about such things as the positions of points and the intersections of lines.

4. An unmarked straightedge is the only tool allowed for drawing figures.   11.7.2

YOU WILL NEED

straightedge
OR
geometry
graphics
software

PROBLEM SOLVING

Activity 2

Two Projective Geometry Theorems

Part I: A Theorem of Pappus

CULTURAL CONNECTION: AFRICA Pappus was a mathematician who lived in Alexandria, Egypt, in the fourth century C.E. His work was very important in the development of projective geometry many centuries later. In this Activity, you will discover one of his theorems.

1. **Make a diagram.** Orient your paper horizontally. Mark a point, O, toward the left edge of the paper. Draw two rays from point O.

2. Mark A_1, B_1, and C_1 on one ray. In the same order, mark A_2, B_2, and C_2 on the other ray.

3. Draw $\overline{A_1B_2}$ and $\overline{A_2B_1}$. Label their intersection X.
   Draw $\overline{A_1C_2}$ and $\overline{A_2C_1}$. Label their intersection Y.
   Draw $\overline{B_1C_2}$ and $\overline{B_2C_1}$. Label their intersection Z.

CHECKPOINT ✔   4. What appears to be true about X, Y, and Z? Make a conjecture, and compare it with those of your classmates. (If you are using geometry graphics software, drag the rays in various ways to see whether your conjecture still holds.) Based on your conjecture, complete the following theorem:

The Theorem of Pappus

If A_1, B_1, and C_1 are three distinct points on one line and A_2, B_2, and C_2 are three distinct points on a second line, then the intersections of $\overline{A_1B_2}$ and $\overline{A_2B_1}$, of $\overline{A_1C_2}$ and $\overline{A_2C_1}$, and of $\overline{B_1C_2}$ and $\overline{B_2C_1}$ are ___?___.

11.7.3

Part II: A Theorem of Desargues

Girard Desargues (1593–1662) was a French mathematician whose ideas are among the most basic in projective geometry. Follow the steps below to discover one of his most important theorems. (Use as large a piece of paper as is practical.)

PROBLEM SOLVING

1. **Make a diagram.** Draw $\triangle ABC$ near the center of the paper. This can be any type of triangle, but it should be small so that the resulting construction will fit on your paper.

2. Mark a point outside $\triangle ABC$ and label it O. This will be the center of projection. Draw $\overrightarrow{OA}$, $\overrightarrow{OB}$, and $\overrightarrow{OC}$.

3. Mark a random point, A', as the projection of A on $\overrightarrow{OA}$. (For clarity of construction, it should be farther from point O than from point A.) Repeat for B', a random point on $\overrightarrow{OB}$, and C', a random point on $\overrightarrow{OC}$. Draw $\triangle A'B'C'$.

4. Extend $\overline{AB}$ and $\overline{A'B'}$ until they intersect, and label the point of intersection X. (If the extended segments are parallel or intersect at a point past the edge of your paper, reposition the image points.)

5. Extend $\overline{AC}$ and $\overline{A'C'}$ until they intersect, and label the point of intersection Y.

6. Extend $\overline{BC}$ and $\overline{B'C'}$ until they intersect, and label the point of intersection Z.

CHECKPOINT ✔

7. What appears to be true about points X, Y, and Z? Make a conjecture, and compare it with those of your classmates. (If you are using geometry graphics software, drag the rays and points in various ways to see whether your conjecture still holds.) Based on your conjecture, complete the following theorem:

The Theorem of Desargues

If one triangle is a projection of another triangle, then the intersections of the lines containing the corresponding sides of the two triangles are _____?_____.

11.7.4

Exercises

● **Communicate**

1. Describe what an affine transformation is. Which of the following are special cases of affine transformations? Explain your reasoning.
   a. dilations
   b. translations
   c. reflections
   d. rotations

2. Describe what a central projection is. Which of the following are special cases of central projections? Explain your reasoning.
   a. dilations
   b. translations
   c. reflections
   d. rotations

3. Explain how projective geometry is different from the other types of geometry you have studied.

4. In a Mercator projection, why do countries near the equator seem less distorted than countries near the North and South Poles?

● **Guided Skills Practice**

5. Copy the grids at right. Draw a right triangle on the square grid, and use the parallelogram grid to transform the triangle. **(ACTIVITY 1)**

6. Draw a square in a coordinate plane with vertices at $(0, 0)$, $(0, 1)$, $(1, 1)$, and $(1, 0)$. Use the affine transformation $T(x, y) = (4x, 3y)$ to transform the square. What is the resulting figure? **(ACTIVITY 1)**

7. Draw two parallel lines ℓ and m. On line ℓ, mark points A_1, B_1, and C_1. On line m, mark points A_2, B_2, and C_2. Find the intersections of $\overline{A_1B_2}$ and $\overline{A_2B_1}$, of $\overline{A_1C_2}$ and $\overline{A_2C_1}$, and of $\overline{B_1C_2}$ and $\overline{B_2C_1}$. Are the three points of intersection collinear? **(ACTIVITY 2)**

8. Copy $\triangle ABC$ and point O below and draw the image, $\triangle A'B'C'$, transformed by a projection centered at O. Extend the corresponding sides of the triangles until they intersect, and draw the line through the three points of intersection. **(ACTIVITY 2)**

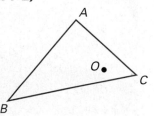

Sketch the preimage and image for each affine transformation below.

9. preimage: square $(0, 0)$, $(4, 0)$, $(4, 4)$, $(0, 4)$
transformation: $T(x, y) = (3x, -2y)$

10. preimage: rectangle $(0, 0)$, $(5, 0)$, $(5, 8)$, $(0, 8)$
transformation: $S(x, y) = (2x, 0.5y)$

11. preimage: triangle $(4, 7)$, $(-1, -1)$, $(0, 8)$
transformation: $R(x, y) = \left(\frac{5}{2}x, 5y\right)$

On a coordinate plane, draw a circle with a radius of 5 centered at the origin. Use the parallelogram grids below to transform the circle.

12.   **13.**

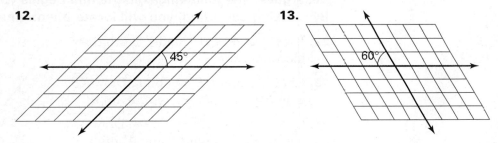

For Exercises 14–16, refer to the figures below. For each projection, identify the following:
  a. the center of projection
  b. the projective rays

14. projection of points on ℓ_1 onto ℓ_2 **15.** projection of points on m_2 onto m_1

16. projection of points on m_2 onto m_3

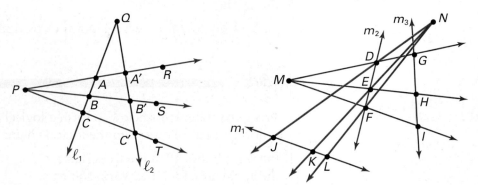

Copy the diagram below and draw each projection.

17. Using O as the center of projection, project points A, B, C, and D onto line ℓ_2. Label the projected points A', B', C', and D'.

18. Using P as the center of projection, project points A', B', C', and D' onto line ℓ_3. Label the projected points A'', B'', C'', and D''.

For Exercises 19–21, use the figure at right.

19. If point A is the center of projection, then the projection of B onto $\overline{FK}$ is ___?___, the projection of I onto $\overline{FK}$ is ___?___, and the projection of J onto $\overline{FK}$ is ___?___.

20. If point L is the center of projection, then the projection of H onto $\overline{MA}$ is ___?___, and the projection of J onto $\overline{MA}$ is ___?___.

21. If point F is the center of projection, then the projection of ___?___ onto $\overline{AG}$ is I, and the projection of ___?___ onto $\overline{AG}$ is G.

In Exercises 22–29, you will explore the converse of the Theorem of Desargues. The following construction begins where Activity 2 ended. By working backward, you will locate the original center of projection.

22. Draw ℓ and label X and Y at arbitrary locations on it.

23. Choose a point, B, not on ℓ. Draw $\overrightarrow{XB}$ and mark a random point, A, on it.

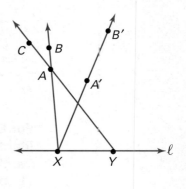

24. Choose a point, B', not on ℓ. Draw $\overrightarrow{XB'}$ and mark a random point, A', on it.

25. Draw $\overrightarrow{YA}$ and mark a random point, C, on it.

26. Draw $\overleftrightarrow{CB}$. Label the intersection of $\overleftrightarrow{CB}$ and ℓ Z.

27. Draw $\overrightarrow{YA'}$ and $\overrightarrow{ZB'}$. Label their point of intersection C'.

28. Assuming the converse of Desargue's Theorem, what can you conclude about $\triangle ABC$ and $\triangle A'B'C'$.

29. Locate the center of projection, point O, by drawing $\overleftrightarrow{AA'}$, $\overleftrightarrow{BB'}$, and $\overleftrightarrow{CC'}$.

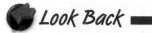

Look Back

30. Arrange the statements below to form a logical chain, and write the conditional that follows from the logical chain. *(LESSON 2.2)*

    If the wind blows, then the trees shake.
    If the apple falls, then the worm squirms.
    If the trees shake, then the apple falls.

For Exercises 31–33, tell whether each argument is an example of inductive or deductive reasoning. *(LESSON 2.5)*

31. All squares are rectangles. *ABCD* is a square. Therefore, *ABCD* is a rectangle.

32. There has never been a freeze recorded in Florida in August. Today is August 12th. Therefore, it will not freeze in Florida today.

33. Debra likes every type of fruit. Mangoes are a type of fruit. Therefore, Debra likes mangoes.

34. Describe what is meant by the golden ratio. *(LESSON 11.1)*

Find the taxidistance between each pair of points. *(LESSON 11.2)*

35. $(4, 3), (2, 1)$ **36.** $(-3, 2), (1, 1)$ **37.** $(1, 3), (5, 5)$

38. Does the figure at right contain an Euler path? an Euler circuit? Why or why not? *(LESSON 11.3)*

39. In spherical geometry, how many lines are parallel to a given line through a point not on the line? *(LESSON 11.5)*

Look Beyond

40. The Nine-Coin Puzzle In the picture of the nine coins below, you can identify eight rows with three coins each. Can you rearrange the coins to form ten rows with three coins each? (Hint: You can use the Theorem of Pappus to solve this puzzle.)

The Tower of Hanoi

In this project, you will examine a puzzle, called the tower of Hanoi, that was invented in 1883 by French mathematician Edouard Lucas.

Activity ①

THE PUZZLE The puzzle begins with three pegs, one of which has a stack of disks that increase in size from the top of the peg to the bottom. The object is to move all of the disks to another peg, following these rules:
- Only one disk may be moved at a time
- A disk must be placed on a larger disk or on an empty peg.

1. What is the least possible number of moves needed to solve the puzzle for 1 disk? for 2 disks? Complete the table at right.

Number of disks	Fewest number of moves
1	?
2	?
3	?
4	?

2. Can you find a formula for the number of moves in terms of the number of disks? Extend the table if necessary.

3. According to a popular story, 64 gold disks on 3 diamond pegs are attended by priests who move the disks according to the rules above. When they have completed the puzzle, the world will end. How many moves are required? If the priests can move one disk per second, how long will it take them to complete the puzzle?

STRANGE CONNECTIONS A graph may be used to represent the states of the tower of Hanoi. When the vertices of the graph are arranged in a certain way, the graph forms the same pattern as the Sierpinski gasket.

Start with the simplest case, a tower of Hanoi with only one disk. The pegs in the puzzle are labeled *A*, *B*, and *C*. Thus, the initial state can be represented by a vertex labeled *A*. It is possible to move the disk to either *B* or *C* in one move, so the graph for the puzzle can be drawn as shown. Notice that it is also possible to move the disk from *B* to *C* (or *C* to *B*), so *B* and *C* are connected by an edge.

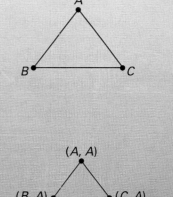

For the case with two disks, the states can be represented by ordered pairs. For example, if the smaller disk is on peg *B* and the larger disk is on peg *A*, the position can be written as (*B, A*). The graph for this puzzle is shown at right.

1. Which states in the graph for two disks represent solutions to the puzzle? Describe a possible solution to the puzzle.

2. For the case with three disks, the states may be represented as ordered triples that describe the positions of the disks in order from smallest to largest. Fill in the states in the graph at right. (Remember that if two vertices are connected by an edge, you must be able to move a disk from one to the other in a single move.) Then describe a solution to the puzzle.

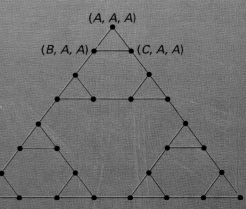

Chapter Review and Assessment

affine transformation 747	golden rectangle 698	projective rays 749
blocks 706	graph 712	self-similarity 738
Cantor dust 740	great circle 731	Sierpinski gasket 741
center of projection 749	hyperbolic geometry 731	simple closed curve 722
central projection 749	invariant 723	spherical geometry 731
degree of a vertex 713	iteration 738	taxicab circle 708
edges 712	logically equivalent 730	taxidistance 706
Euler characteristic 723	Menger sponge 738	taxicab geometry 706
Euler circuit 715	Möbius strip 721	taxicab radius 708
Euler path 712	non-Euclidean geometry ... 730	topologically equivalent ... 722
even vertex 713	odd vertex 713	topology 721
fractal 738	orthogonal 732	torus 722
golden ratio 698	projected points 749	vertices of a graph 712

POSTULATES AND THEOREMS

Lesson	Section	Postulate or Theorem
11.3	11.3.1 Theorem	A graph contains an Euler path if and only if there are at most two odd vertices.
11.4	11.4.1 Jordan Curve Theorem	Every simple closed curve divides the plane into two distinct regions, the inside and the outside. Every curve that connects a point on the inside to a point on the outside must intersect the curve.
	11.4.2 Euler's Formula	For any polyhedron with V vertices, E edges, and F faces, $V - E + F = 2$.
11.7	11.7.3 The Theorem of Pappus	If A_1, B_1, and C_1 are three distinct points on one line and A_2, B_2, and C_2 are three distinct points on a second line, then the intersections of $\overline{A_1B_2}$ and $\overline{A_2B_1}$, of $\overline{A_1C_2}$ and $\overline{A_2C_1}$, and of $\overline{B_1C_2}$ and $\overline{B_2C_1}$ are collinear.
	11.7.4 The Theorem of Desargues	If one triangle is a projection of another triangle, then the intersections of the lines containing the corresponding sides of the two triangles are collinear.

Key Skills & Exercises

LESSON 11.1
Key Skills

Determine side lengths of golden rectangles.

A golden rectangle has a short side length of 5 units. Find the length of the long side.

The ratio of the long side to the short side of a golden rectangle is the golden ratio, $\phi \approx 1.618$.

$$\frac{\ell}{5} = \phi$$
$$\ell = \phi(5) \approx 8.09$$

Exercises

1. A golden rectangle has a short side length of 3 units. Find the length of the long side.

2. A golden rectangle has a long side length of 9 units. Find the length of the short side.

Construct a golden rectangle.

Given square $ABCD$, construct a golden rectangle whose short side is $\overline{AB}$.

Extend side $\overline{BC}$. Locate M, the midpoint of $\overline{BC}$, and draw an arc centered at M that intersects $\overrightarrow{BC}$ at E. Construct a line perpendicular to $\overrightarrow{BC}$ at E and extend $\overline{AD}$ to intersect the perpendicular at F. $ABEF$ is a golden rectangle.

3. Draw square $WXYZ$ with a side length of 3 centimeters, and construct a golden rectangle whose short side is $\overline{XY}$.

4. Identify another golden rectangle in the diagram you constructed for Exercise 3.

Key Skills

Find the taxidistance between two points.

Find the taxidistance between $(3, 1)$ and $(5, 2)$.

The taxidistance between (x_1, y_1) and (x_2, y_2) is

$$|x_2 - x_1| + |y_2 - y_1| =$$

$$|5 - 3| + |2 - 1| = 3.$$

Draw a taxicab circle with a given center and radius.

Draw the taxicab circle centered at $(1, -1)$ with a radius of 3.

Exercises

Find taxidistance between the following pairs of points:

5. $(6, 1)$ and $(4, 5)$

6. $(7, 2)$ and $(-1, 0)$

Draw the taxicab circle with the given center and radius.

7. center: $(0, 0)$; radius: 2

8. center: $(2, 3)$; radius: 4

Key Skills

Determine whether a graph contains an Euler path, an Euler circuit, or neither.

Does the graph below contain an Euler path, an Euler circuit, or neither?

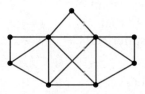

The graph contains 7 even vertices and 2 odd vertices. Since there are odd vertices, the graph does not contain an Euler circuit. However, there are not more than 2 odd vertices, so the graph contains an Euler path.

Exercises

For each graph below, determine whether the graph contains an Euler path, an Euler circuit, or neither.

9.

10.

11.

12.

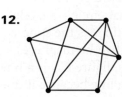

Key Skills

Determine whether two figures are topologically equivalent.

Are the figures below topologically equivalent?

The figure on the left is a simple closed curve. It encloses a region of the plane. The figure on the right does not enclose any area. There is no way to deform one into the other without cutting the figure on the left. Thus, the figures are not topologically equivalent.

Find the Euler characteristic of a figure.

Find the Euler characteristic of the figure below.

The figure is topologically equivalent to the figure below, which has 28 vertices, 56 edges, and 26 faces.

The Euler characteristic is
$V - E + F = 28 - 56 + 26 = -2$.

Exercises

Determine whether the figures in each pair below are topologically equivalent. Explain your reasoning.

13.

14.

15. Find the Euler characteristic of the figure below.

16. Use the Euler characteristic to prove that the figure above is not topologically equivalent to a sphere.

Key Skills

Identify lines in spherical geometry.

Which of the following is a line in spherical geometry?

By definition, a line in spherical geometry is a great circle—a circle that divides the sphere into two hemispheres.
Thus, figure CD is a line, but figure AB is not a line because it does not divide the circle into two hemispheres.

Exercises

Refer to the basketball below.

17. Describe a curve on the basketball that is a line in spherical geometry.

18. Describe a curve on the basketball that is not a line in spherical geometry.

Identify lines in hyperbolic geometry.

Which figures in the diagram below are lines in Poincaré's model of hyperbolic geometry?

According to Poincaré's model, lines are defined as diameters of the outer circle or arcs that are orthogonal to the outer circle. Thus, figures *XY*, *YZ*, and *YU* are lines, but figure *VW* is not a line because it is not orthogonal to the outer circle.

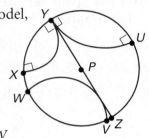

Refer to the diagram of Poincaré's model of hyperbolic geometry below.

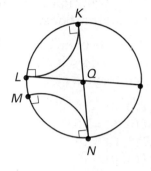

19. Name three figures in the diagram that are lines in hyperbolic geometry.

20. Draw a curve in the diagram that is not a line in hyperbolic geometry, and explain why it is not a line.

LESSON 11.6
Key Skills

Exercises

Create fractals by using iterations.

To form the Sierpinski carpet, start with a solid square. Divide the square into a grid of 9 squares, and remove the center square. Repeat on the remaining squares. Draw the first three iterations of the Sierpinski carpet.

1 iteration

2 iterations

3 iterations

The original figure and first iteration of a fractal are shown below.

21. Draw the second iteration of the fractal.

22. Draw the third iteration of the fractal.

23. Does the area of the fractal above increase or decrease with successive iterations? If it decreases, does it approach 0? If it increases, does it approach infinity?

24. Does the perimeter of the fractal above increase or decrease with successive iterations? If it decreases, does it approach 0? If it increases, does it approach infinity?

Key Skills

Transform figures by using affine transformations.

Draw the preimage and image of the rectangle with vertices at (0, 0), (0, 6), (3, 6), and (3, 0) transformed by $T(x, y) = (3x, 4y)$.

$T(0, 0) = (0, 0)$
$T(0, 6) = (0, 24)$
$T(3, 6) = (9, 24)$
$T(3, 0) = (9, 0)$

Transform figures by using central projections.

Draw a central projection of $\triangle ABC$ centered at point O.

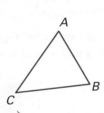

Draw projective rays $\overrightarrow{OA}$, $\overrightarrow{OB}$, and $\overrightarrow{OC}$, and choose points A', B', and C' on the rays. Draw $\triangle A'B'C'$.

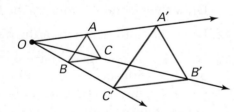

Exercises

Draw the preimage and image of the figure with the given vertices under the given transformation.

25. $(1, 3), (2, 1), (4, 0); R(x, y) = (2x, 5y)$

26. $(2, -1), (3, 0), (2, 2), (1, 1); S(x, y) = (x, 1.2y)$

27. Draw a central projection of $\triangle KLM$ centered at P.

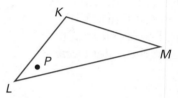

28. Draw a central projection of points D, E, and F onto line m centered at Q.

Applications

29. ARCHITECTURE An architect is designing an office building whose front face is to be in the shape of a golden rectangle. If the building will be 45 ft tall, how wide should its front face be?

30. GEOGRAPHY A pilot is flying from Tokyo to London. Use spherical geometry to describe the path the pilot should take in order to fly the shortest distance.

31. GEOGRAPHY Tani and Alex are trying to measure the coast of Oregon. Tani measures the coast from a satellite photo and finds that it is 490 km long. Alex uses a set of aerial photos and finds that it is 560 km long. Explain the difference in their results.

Alternative Assessment

Performance Assessment

1. Photocopy several reproductions of works of art. Look for examples of golden rectangles in the compositional elements of the art. If a piece of art has a prominent figure or space, try drawing the smallest rectangle that will completely enclose it. How close are its proportions to the golden ratio?

Seurat, Georges. *Sunday Afternoon on the Island of La Grande Jatte.* 1884-1885. The Art Institute of Chicago.

Use a graphics calculator for the following exercises:

2. Set your calculator to a square viewing window with a *y*-value range of at least $-3 \leq y \leq 3$. Graph the circle given by the equations below and describe the result.

$$Y1 = \sqrt{(9 - X^2)} \qquad Y2 = -\sqrt{(9 - X^2)}$$

3. Under ZOOM **MEMORY**, set the zoom factor for *x* to 1 and the zoom factor for *y* to 3, and zoom out. Describe the results. What type of transformation does this represent? What happens if you zoom out again? Does zooming in again reverse the transformation?

4. Describe what happens to the *x*- and *y*-values in the viewing window at each step as you zoom out twice and then zoom in twice.

Portfolio Projects

Copy a part of a city map that shows an area of at least 25 square blocks.

1. Choose two points on the map at least 5 blocks apart that are not on the same street, and draw several paths between them. Find the shortest distance between the two points. Are there streets on your map that do not lie on a square grid? How does this affect the distances between points as traveled by a taxicab?

2. Make a graph of the streets on the map, using vertices to represent intersections and edges to represent streets. (Ignore the parts of streets that go off the edge of the map.) Does the graph contain an Euler path? an Euler circuit?

Detail from a map of San Francisco.

QUANTITATIVE COMPARISON For Items 1–4, write

A if the quantity in Column A is greater than the quantity in Column B;
B if the quantity in Column B is greater than the quantity in Column A;
C if the two quantities are equal; or
D if the relationship cannot be determined from the information given.

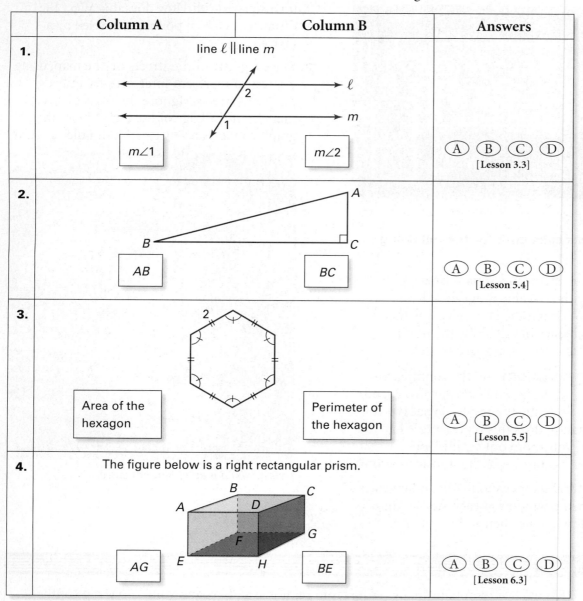

	Column A	Column B	Answers
1.	line ℓ ∥ line m $m\angle 1$	$m\angle 2$	Ⓐ Ⓑ Ⓒ Ⓓ [Lesson 3.3]
2.	AB	BC	Ⓐ Ⓑ Ⓒ Ⓓ [Lesson 5.4]
3.	Area of the hexagon	Perimeter of the hexagon	Ⓐ Ⓑ Ⓒ Ⓓ [Lesson 5.5]
4.	The figure below is a right rectangular prism. AG	BE	Ⓐ Ⓑ Ⓒ Ⓓ [Lesson 6.3]

5. Find the *x*-intercept of the line given by the parametric equations below. *(LESSON 6.5)*

$$x = 4t - 8$$
$$y = 2t - 3$$
$$z = t + 4$$

a. (0, 2, 6)
b. (0, 1, 6)
c. (0, 0, 0)
d. none of the above

6. As the side length of a cube is increased, the ratio of surface area to volume ___?___.
(LESSON 7.1)
a. increases
b. remains constant
c. is equal to 1
d. decreases

7. If the radius of a sphere is doubled, the volume is increased by a factor of ___?___.
(LESSON 8.6)
a. 2
b. 4
c. 6
d. 8

8. The value of sin θ is given by the ratio ___?___.
(LESSON 10.2)

a. $\dfrac{\text{opposite}}{\text{adjacent}}$

b. $\dfrac{\text{adjacent}}{\text{hypotenuse}}$

c. $\dfrac{\text{opposite}}{\text{hypotenuse}}$

d. $\dfrac{\text{hypotenuse}}{\text{opposite}}$

9. For the two similar quadrilaterals below, write a proportionality statement by using the ratios between the sides. *(LESSON 8.2)*

10. Determine whether the graph at right contains an Euler path, an Euler circuit, or neither. *(LESSON 11.3)*

11. How many axes of symmetry does the figure below have? *(LESSON 3.1)*

12. Find the slope of the segment with endpoints at (2, 4) and (3, −1). *(LESSON 3.8)*

13. Find *x* in the triangle below. *(LESSON 4.4)*

14. Find *s* in the golden rectangle below.
(LESSON 11.1)

15. Find the taxidistance between the points (1, 4) and (7, 10). *(LESSON 11.2)*

12 A Closer Look at Proof and Logic

e4 c6 2. d4 d5 3. Nc3 dxe4

CAN A MACHINE THINK? AT TIMES IT CERTAINLY seems as if they can—as when the computer Deep Blue beat the world chess champion Gary Kasparov in a chess match. Whether they can actually think or not, machines can clearly follow rules and make "decisions" based on given conditions. The decisions made by chess-playing computers and other "smart" machines are made possible by formal logic.

Formal logic reduces logical procedures to their essential elements, which can be implemented by computers or other machines. For example, the following rule in chess can be programmed into a chess computer.

If my king is in check, I must
- move out of check,
- capture the attacking piece,
- interpose a piece between the attacker and my king, or
- resign.

About the Chapter Project

Number theory is a branch of mathematics that is at least as old as Euclid. In fact, three books of Euclid's *Elements* are devoted to number theory. In the *Elements*, Euclid uses segments of different lengths to represent positive whole numbers and proves theorems about numbers geometrically.

In the Chapter Project, you will study two famous theorems from number theory. The first is that there are infinitely many prime numbers, and the second is that the square root of 2 is irrational.

After completing the Chapter Project you will be able to do the following:

- Have a deeper understanding of the techniques of proof.
- Understand the need for a class of numbers that are not rational.

About the Portfolio Activities

Throughout the chapter, you will be given opportunities to complete Portfolio Activities that are designed to support your work on the Chapter Project.

The theme of each Portfolio Activity and of the Chapter Project is number theory.

- In the Portfolio Activity on page 775, you will explore the Euclidean algorithm for finding the greatest common divisor (gcd) of two numbers
- In the Portfolio Activity on page 797, you will learn how to prove the Euclidean algorithm.

Truth and Validity in Logical Arguments

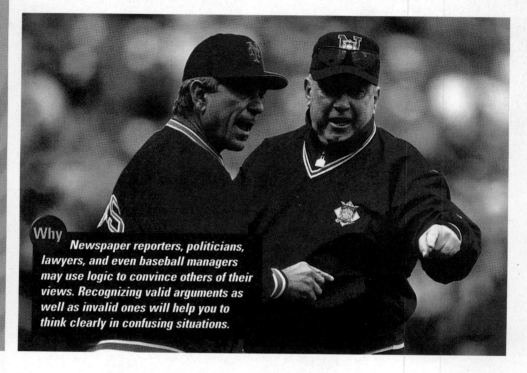

Objectives

- Define and use the valid argument forms *modus ponens* and *modus tollens*.

- Define and illustrate the invalid argument forms of *affirming the consequent and denying the antecedent.*

Why Newspaper reporters, politicians, lawyers, and even baseball managers may use logic to convince others of their views. Recognizing valid arguments as well as invalid ones will help you to think clearly in confusing situations.

Valid Arguments

In logic, an **argument** consists of a sequence of statements. The final statement of the argument is called the **conclusion**, and the statements that come before it are known as **premises**. The following is an example of a logical argument:

If an animal is an amphibian, then it is a vertebrate.
Frogs are amphibians.   *Premises*

Therefore, frogs are vertebrates.   *Conclusion*

In this argument, the conclusion is said to follow logically from the premises. The premises force the conclusion. An argument of this kind is known as a **valid argument**, and the conclusion of such an argument is said to be a **valid conclusion**.

A valid argument makes the following "guarantee": *If the premises are all true, then the conclusion is true.* In the valid argument above, both premises are true. Therefore, the conclusion must be true.

Invalid Arguments

Invalid argument (Don't use one like this!)

Now consider a different argument:

Some vertebrates are warm-blooded.
Frogs are vertebrates.   *Premises*

Therefore, frogs are warm-blooded.   *Conclusion(?)*

This new argument is invalid. Both of the premises are true, but the conclusion is false. *This is never the case in a valid argument.*

CRITICAL THINKING   Consider the second argument. Suppose you did not know that a frog is not a warm-blooded animal. Would you have questioned the conclusion anyway? Is there something basically wrong with the argument? If you think there is, try to describe what it is.

TRY THIS   Write your own examples of valid and invalid arguments.

A Valid Argument Form: *Modus Ponens*

Logicians can tell whether an argument is valid or invalid without knowing anything about the truth of its premises or conclusion. They are able to do this by analyzing the form of the argument. The valid argument on the previous page, for example, has the following form:

Note: p and q represent **statements**: that is, sentences that can be true or false—not questions, commands, etc.

Argument Form: *Modus Ponens*

If p then q
p *Premises*

Therefore, q *Conclusion*   **12.1.1**

This argument form is sometimes referred to by its Latin name, ***modus ponens***, or the "proposing mode." Any argument that has this form is valid, regardless of the statements that are substituted for p and q. The following nonsense argument, for example, is valid because it has a valid form:

If flivvers twiddle, then bokes malk.
Flivvers twiddle.   *Premises*

Therefore, bokes malk.   *Conclusion*

This argument's form guarantees that if the first two statements should somehow turn out to be true, then the third statement (the conclusion) would be true as well.

Another Valid Argument Form: *Modus Tollens*

Consider the following argument:

> If a shirt is a De Morgan, then it has a blackbird logo.
> This shirt does not have a blackbird logo.   *Premises*
>
> Therefore, this shirt is not a De Morgan.   *Conclusion*

Does this argument seem valid to you? If you knew that the premises were true, would you be certain that the conclusion was true? The argument above has the following form:

The symbol ~ means "not."

Argument Form: *Modus Tollens*

If p then q
~q *Premises*

Therefore, ~p *Conclusion*   **12.1.2**

This is a valid argument form. It is sometimes referred to by its Latin name, ***modus tollens***, or the "removing mode." In more recent times, it has come to be known as the law of indirect reasoning.

False Premises

If the premises of an argument are false, then there is no guarantee that the conclusion is true, even though the argument might be valid. The following *modus ponens* argument is valid, but its conclusion is false.

> If an animal is an amphibian, then it can fly.
> A frog is an amphibian.   *Premises*
>
> Therefore, frogs can fly.   *Conclusion*

The conclusion, though false, is a valid conclusion of the argument because the form of the argument is valid. Remember, a valid argument guarantees that its conclusion is true *only if its premises are true*. There is no guarantee if one or more of the premises are false.

CRITICAL THINKING   The valid argument above has a false premise and a false conclusion. Do you think it is possible for a valid argument to have a false premise and a true conclusion? If so, give an example. If not, explain why not.

Invalid Forms

You should be careful not to confuse the *modus ponens* form with the following *invalid* form:

Invalid Form: Affirming the Consequent

If p then q	
q	*Premises*
Therefore, p	*Not a valid conclusion* **12.1.3**

Form 12.1.3 is the form of a common logical mistake, or fallacy, known as "affirming the consequent." The conclusion does not follow logically from the premises even if it is a true statement. Be sure that you understand the difference between this form and the *modus ponens* form, which it closely resembles. An example of affirming the consequent is as follows:

> If Sancho is having plum pudding for dessert,
>   then he is happy.   *Premises*
> Sancho is happy.
>
> Therefore, Sancho is having plum pudding for dessert.   *Conclusion (?)*

Clearly, the argument doesn't work, because Sancho may be happy for some other reason.

Picasso, Pablo. *Don Quixote. 1955.* Musee d'Art et d'Histoire, St. Denis, France. © 2001 Estate of Pablo Picasso/Artists Rights Society (ARS), New York. Scala/Art Resource.

Don Quixote's squire, Sancho Panza, loved good food.

Another important invalid form that you will need to recognize closely resembles the *modus tollens* form.

Invalid Form: Denying the Antecedent

If p then q	
$\sim p$	*Premises*
Therefore, $\sim q$	*Not a valid conclusion* **12.1.4**

An example of denying the antecedent is as follows:

> If Susan overslept, then she is running late.
> Susan did not oversleep.   *Premises*
>
> Therefore, Susan is not running late.   *Conclusion (?)*

Again, the argument isn't valid because Susan might be running late for any number of reasons.

Communicate

In Exercises 1–4, determine whether the argument is valid. Explain why or why not.

1. If today is Wednesday, then the cafeteria is serving beef stew.
Today is Wednesday. *Premises*

Therefore, the cafeteria is serving beef stew. *Conclusion (?)*

2. If pigs fly, then today is February 30.
Today is February 30. *Premises*

Therefore, pigs fly. *Conclusion (?)*

3. If Jan is a man, then Jan is mortal.
Jan is not a man. *Premises*

Therefore, Jan is not mortal. *Conclusion (?)*

4. If $y = x$, then $a = b$.
$a \neq b$ *Premises*

Therefore, $y \neq x$. *Conclusion (?)*

5. Is it possible for a valid argument to have a false conclusion? Explain your reasoning.

Guided Skills Practice

In Exercises 6–9, analyze the form of each argument and give its traditional name. Then state whether it is valid or invalid.

6. If the weather takes a turn for the worse, then the local farmers will suffer a loss of income.
The weather takes a turn for the worse.

Therefore, the local farmers will suffer a loss of income.
(ARGUMENT FORM 12.1.1)

7. If Stokes was in top form, then he won the competition.
Stokes did not win the competition.

Therefore, Stokes was not in top form.
(ARGUMENT FORM 12.1.2)

Dusenberg

8. If the car is a Dusenberg, then it is a classic.
The car is a classic.

Therefore, the car is a Dusenberg.
(ARGUMENT FORM 12.1.3)

9. If Sean stuffed himself at lunch, then he is feeling sleepy now.
Sean did not stuff himself at lunch.

Therefore, Sean is not feeling sleepy now.
(ARGUMENT FORM 12.1.4)

In Exercises 10–11, write a valid conclusion from the given premises, and identify the form of the argument.

10. If the team won on Saturday, then the team is in the playoffs.
The team is not in the playoffs.

11. If Sabrina finished her assignment on time, then she did a stupendous amount of work at the last minute.
Sabrina finished her work on time.

In Exercises 12–15, arrange the sentences to form an argument. Identify the argument form, and state whether the argument is valid or invalid.

12. Therefore, Samantha is ill.
Samantha is absent.
If Samantha is ill, then she is absent.

13. Sims is not a man of good moral character.
If Sims is a man of good moral character, then he is innocent.
Therefore, Sims is not innocent.

14. Hedgehogs are tone deaf.
Therefore, hedgehogs will seldom be seen at symphony concerts.
If hedgehogs are tone deaf, they will seldom be seen at symphony concerts.

15. There were unpleasant surprises.
If the plan was foolproof, then there were no unpleasant surprises.
Therefore, the plan was not foolproof.

(See Exercise 14.)

In Exercises 16–19, use the following conditional: If the building was of sound construction, then it survived the storm.

16. Write a *modus ponens* argument, using the given conditional as one of its premises.

17. Write a *modus tollens* argument, using the given conditional as one of its premises.

18. Write an invalid argument in the form of affirming the consequent, using the given conditional as one of its premises. Label the argument as invalid.

19. Write an invalid argument in the form of denying the antcedent, using the given conditional as one of its premises. Label the argument as invalid.

You are given the following premises:

If a student studies, the student will succeed.
Eleanor studies.
Tamara does not study.
Jose will succeed.
Mary will not succeed.

Which of the following conclusions are valid?

20. Eleanor will succeed.

21. Eleanor will not succeed.

22. Tamara will succeed.

23. Tamara will not succeed.

24. Jose studies.

25. Jose does not study.

26. Mary studies.

27. Mary does not study.

In Exercises 28–31, consider the following argument:
> **If a quadrilateral is a parallelogram, then its diagonals are congruent.**
> **Quadrilateral *PQRS* is a parallelogram.**
> **Therefore, the diagonals of quadrilateral *PQRS* are congruent.**

28. Is the argument valid? Explain your reasoning.

29. Is the first premise true or false? Explain your reasoning.

30. If quadrilateral *PQRS* is a rectangle or a square, is the conclusion of the argument true or false? Explain your reasoning.

31. If quadrilateral *PQRS* is not a rectangle or a square, is the conclusion of the argument true or false? Explain your reasoning.

A valid argument has premises *a*, *b*, *c*, and *d* and conclusion *r*.

32. Does the validity of the argument guarantee that the premises and conclusion must be true? Explain your reasoning.

33. If the four premises are true, does the validity of the argument guarantee that *r* is true? Explain your reasoning.

34. Under what circumstances might *r* be false?

In Exercises 35 and 36, use two of the given premises to write a valid conclusion. Identify the form of the argument that you used.

35. SPORTS In football, if a team does not move the ball 10 yd in 4 downs, then they lose possession of the ball.
The Mammoths did not lose possession of the ball.
The Voyageurs moved the ball 10 yd in 4 downs.
The Cheetahs did not move the ball 10 yd in 4 downs.

36. LANDSCAPING If tulips are not planted in the fall, then they will not flower in the spring.
Adrianne's tulips did not flower in the spring.
Nuna's tulips flowered in the spring.
Geraldo planted tulips in the fall.

Look Back

Can each pair of triangles below be proven congruent? Why or why not?
(LESSONS 4.2 AND 4.3)

37.   **38.**   **39.**

For Exercises 40–42, refer to the figure below, in which
m∠1 = 20°, m∠2 = 35°, and m$\overset{\frown}{SR}$ = 80°.
(LESSONS 9.3 AND 9.4)

40. m$\overset{\frown}{QR}$ = _____?_____

41. m$\overset{\frown}{PS}$ = _____?_____

42. m$\overset{\frown}{PQ}$ = _____?_____

θ is an angle in the first quadrant with a tangent of $\sqrt{3}$. *(LESSON 10.3)*

43. Find θ, sin θ, and cos θ.

44. Can an angle in the second quadrant have the same tangent as θ? Why or why not?

45. Give the measure of an angle in the third quadrant with the same tangent as θ.

Look Beyond

Recall the If-Then Transitive Property from Lesson 2.2:

> **If p then q**
> **If q then r**
> **Therefore, if p then r**

List all conclusions that can be drawn from the premises given below.

46. If x then y
If y then k
$\sim k$

47. If n then m
If q then r
If m then q
n

THE EUCLIDEAN ALGORITHM

An algorithm for finding the greatest common divisor (gcd) of two numbers is found in Book VII of Euclid's *Elements*. Trace the steps in the following example of finding the gcd of 1320 and 546:

$$1320 = 546(2) + 228 \quad \longleftarrow \text{\textit{To find the numbers in this row, note}}$$
$$\textit{that } 1320 \div 546 = 2 \text{ remainder } 228.$$
$$546 = 228(2) + 90$$
$$228 = 90(2) + 48$$
$$90 = 48(1) + 42$$
$$48 = 42(1) + \textcircled{6} \quad \longleftarrow \text{\textit{The last nonzero entry in this column}}$$
$$\textit{is the gcd.}$$
$$42 = 6(7) + 0$$

1. Follow the steps used in the example to find the gcd of 630 and 165, of 280 and 150, and of 462 and 120.

2. Two numbers are *relatively prime* if their gcd is 1. Use the Euclidean algorithm to show that 330 and 91 are relatively prime and that 560 and 429 are relatively prime.

And, Or, and Not in Logic

Objectives

- Define *conjunction*, *disjunction*, and *negation*.

- Solve logic problems by using conjunction, disjunction, and negation.

Why *The words and, or, and not are used frequently in everyday situations. These words have precise mathematical meanings in logical arguments.*

In logic, a statement is a sentence that is either true or false. The sentence "Belinda ordered pepperoni on her pizza" is a statement because it must be either true or false. A **compound statement** is formed when two or more statements are connected. The sentence "John had a soda, and Belinda had tea" is a compound statement. A compound statement, like a simple statement, is either true or false.

Conjunctions

A compound statement that uses the word *and* to connect simple statements is called a **conjunction**.

Sentence *p*:	Today is Tuesday.
Sentence *q*:	Tonight is the first varsity track meet.
Conjunction *p* AND *q*:	Today is Tuesday, and tonight is the first varsity track meet.

A conjunction is true *if and only if* both of its statements are true. If one or both of its statements are false, the conjunction is false. The four possibilities for a conjunction can be illustrated in a **truth table.**

All possible combinations of truth values for the two statements that form the conjunction are placed in the first two columns. The last column indicates the truth values for the conjunction. In the first combination, for example, both of the statements that make up the conjunction are true. In this case, the conjunction is true.

p	*q*	*p* AND *q*
T	T	T
T	F	F
F	T	F
F	F	F

Determine whether the following conjunctions are true:

    a. George Washington was the first president of the United States, and John Adams was the second.

    b. The sum of the measures of the angles of a triangle is 200°, and blue is a color.

● **SOLUTION**

    a. The conjunction is true because both of its statements are true.

    b. The conjunction is false because one of its statements is false.

Disjunctions

Two statements may also be combined into a single compound statement by the word *or*. This type of compound statement is known as a **disjunction**.

When used in everyday language, the word *or* often means "one or the other, but not both." For example, if a waitress says, "You may have soup or salad with your dinner," she means that you may choose just one of the two. This use of *or* is known as the **exclusive *or***.

In mathematics and logic, *or* means "one or the other, or both." This use of *or* is known as the **inclusive *or***. If someone asks how John will spend his Saturday afternoons, the answer might be, "He will go swimming or bowling." This sentence can be written in logical form as *p* OR *q*, where *p* and *q* are identified as shown below.

$$\underbrace{\text{He will go swimming}}_{p} \quad \text{or} \quad \underbrace{[\text{he will go}] \text{ bowling.}}_{q}$$

               p   OR   *q*

The statement is false only if John does neither one. There is nothing about the statement which implies that John won't do both. Notice that in the truth table, only the fourth combination gives a value of false for the disjunction. If John goes swimming or bowling or both, the disjunction is true, as the values for the first three combinations show.

p	*q*	*p* OR *q*
T	T	T
T	F	T
F	T	T
F	F	F

Determine whether each disjunction below is true.

    a. A square is a rectangle, or a pentagon has five sides.

    b. Dogs can play golf, or $5 - 3 = 2$.

● **SOLUTION**

    a. The disjunction is true because both of the statements are true.

    b. The disjunction is true because one of the statements is true.

Negations

Scene from the film
Singing in the Rain

Consider each of the following statements:

It is raining outside.   It is not raining outside.

The second statement is the *negation* of the first one. If p is a statement, then NOT p is its **negation**. The negation of p can also be written as ~p.

It is raining outside.   It is not raining outside.
p ~p

Examine the truth table for the negation. Notice the following: When a statement, p, is true, its negation, ~p, is false. When a statement, p, is false, its negation, ~p, is true.

p	~p
T	F
F	T

In the following Activity, you will explore one of De Morgan's laws, which are named after British mathematician Augustus De Morgan (1806–1871).

Activity
The Negation of a Conjunction

1. Copy and complete the truth table for the negation of a conjunction. Notice that the fourth column represents the negation of the third column.

p	q	p AND q	~(p AND q)
T	T	?	?
T	F	?	?
F	T	?	?
F	F	?	?

2. Copy and complete the truth table for a disjunction of two negations. Notice that the values for ~p and ~q are used to determine the truth values for ~p OR ~q.

p	q	~p	~q	~p OR ~q
T	T	?	?	?
T	F	?	?	?
F	T	?	?	?
F	F	?	?	?

CHECKPOINT ✔

3. Compare the last column in the first table with the last column in the second table. Explain what you observe.

4. When two logic statements have the same truth values, they are said to be **truth functionally equivalent**. Complete the statement below, which is one of De Morgan's laws.

~(p AND q) is truth functionally equivalent to ____?____.

CRITICAL THINKING

In the two truth tables in the Activity, why should the combinations of T and F be listed in exactly the same order in the first two columns of each table? If you list T and F values for the three statements p, q, and r, how many different combinations of T and F will there be? What is a good order for listing these combinations?

Exercises

Communicate

1. Explain the conditions necessary for a conjunction to be true.

2. Explain the conditions necessary for a disjunction to be true.

3. Explain the difference between an inclusive *or* and an exclusive *or*.

4. Give three different ways for the statement "Kimba likes pizza and Lin likes spaghetti" to be false.

Guided Skills Practice

Indicate whether each compound statement is true or false. Explain your reasoning. *(EXAMPLES 1 AND 2)*

5. $4 + 5 = 9$ and $4 \cdot 5 = 9$

6. All triangles have three angles, or all triangles have four angles.

7. Three noncollinear points determine a unique plane, and a segment has two endpoints.

8. All squares are hexagons, or all triangles are squares.

Write a statement that is truth functionally equivalent to each statement below. *(ACTIVITY)*

9. $\sim r$ OR $\sim s$

10. $\sim(t$ AND $u)$

Practice and Apply

Write a conjunction for each pair of statements. State whether the conjunction is true or false.

11. A carrot is a vegetable. Florida is a state.

12. A ray has only one endpoint. Kangaroos can sing.

13. The sum of the measures of the angles of a triangle is 180°. Two points determine a line.

Write a disjunction for each pair of statements. State whether the disjunction is true or false.

14. Triangles are circles.
    Squares are parallelograms.

15. Points in a plane equidistant from a given point form a circle. The sides of an equilateral triangle are congruent.

16. An orange is a fruit.
    Cows have kittens.

Write the negation of each statement.

17. The figure is a rectangle.

18. My client is not guilty.

19. Rain makes the road slippery.

20. Triangles have six sides.

21. a. Copy and complete the truth table for $\sim(\sim p)$.

p	$\sim p$	$\sim(\sim p)$
T	?	?
F	?	?

b. What statement is equivalent to the statement $\sim(\sim p)$? Explain your reasoning.

For Exercises 22–29, write the statement expressed by the symbols, where p, q, r, and s represent the statements below.

p: $\triangle ABC$ is isosceles.
q: $\triangle ABC$ has two equal angles.
r: $\angle 1$ and $\angle 2$ are adjacent.
s: $\angle 1$ and $\angle 2$ are acute angles.

22. $\sim p$

23. q OR p

24. p AND q

25. $\sim q$

26. $\sim s$

27. r OR s

28. s AND $\sim r$

29. q OR $\sim s$

For Exercises 30–32, construct a truth table for the given compound statement. When is the compound statement false?

30. $(p$ AND $q)$ AND r

31. $(p$ OR $q)$ OR r

CHALLENGE

32. $(p$ AND $q)$ OR $(r$ AND $s)$

p	q	r	p AND q	$(p$ AND $q)$ AND r
T	T	T	?	?
T	T	F	?	?
T	F	T	?	?
T	F	F	?	?
F	T	T	?	?
F	T	F	?	?
F	F	T	?	?
F	F	F	?	?

Use this table for Exercise 30.

33. Explain all of the logical possibilities that would make this sentence true: Flora will cook or wash the dishes, and Vernon will vacuum or wash the windows.

APPLICATION

34. ADVERTISING An advertisement for a set of holiday lights contains the following statement: "Not all bulbs go out when one bulb burns out." Is this the same as "When one bulb burns out, all of the remaining bulbs continue burning"?

35. COMPUTER DATABASES Computer database software can perform AND and OR operations on stored data. Study the sample database printout below, which contains 10 records with 5 fields each.

Last name	First name	State	Year of birth	Annual income
Craighead	Alicia	TX	1955	$ 25,000
Nicar	Bill	MN	1942	$ 45,000
Tuggle	Lawrence	LA	1972	$ 20,000
Mallo	Elizabeth	TX	1956	$ 50,000
Torres	Ernest	AK	1940	$ 38,000
Tong	Jun	TX	1952	$ 18,000
Jurek	Chandra	AZ	1944	$ 31,000
Brookshier	Mary	OH	1960	$ 62,000
Lamb	Charles	TX	1951	$ 41,000
Raemsch	Martin	OK	1965	$ 32,000

a. List all the records in which the individual was born after 1950 AND the annual income is greater then $30,000.

b. List all the records in which the individual lives in Texas OR the annual income is less than $30,000.

Look Back

Can each pair of triangles below be proven similar? Explain why or why not. *(LESSONS 8.3 AND 8.4)*

36. **37.** **38.**

Find the indicated ratios for the triangle below. *(LESSONS 10.1 AND 10.2)*

39. $\sin A$ **40.** $\cos B$

41. $\tan B$ **42.** $\cos A$

Look Beyond

43. Can you determine whether the sentence below is true or false? Explain your reasoning.

This sentence is false.

44. A woman is on an island and is trying to determine whether she should go east or west in order to get back to the mainland. Two different groups of people live on the island. One group always tells the truth. The other group always lies. The groups dress differently, but the woman does not know which is which. She approaches two islanders who are dressed differently to ask the direction to the mainland. What one question can she ask one of the islanders to determine the correct direction?

CHALLENGE

MESSAGE OF THE MAYA
In Modern Translation

by Greg Stec
Special to the *Christian Science Monitor*

Discovering the titles of Mayan royalty, the names of their gods and their food, the dates of important events, and all the other things great and small that make up an advanced society has taken more than a hundred years of digging and probing. At last, though, most of the Maya's intricate writing is yielding to modern techniques of linguistics and computer technology.

So many dates are involved in Mayan inscriptions that at one time archaeologists theorized that the Maya worshipped time and numbers. The current thinking holds that the Maya were slaves to ritual and ceremony. By dating an elite's birth, accomplishments, and death with unimpeachable accuracy, the person's position and rank would be permanently established.

Computer programs wade through a sea of Mayan dates, saving investigators effort that can be used to examine other translations.

Imagine having two calendars on your wall, with a different number of days in each year. Every day would have two dates, usually different. Your birthday, for example, might be August 4 and April 29.

The Mayan system was similar to this in that it had two different calendars, one of them 260 days long and the other 365 days long. To get a better idea of how it worked, you can use the diagram of three wheels at left.

Cooperative Learning

The Mayan ceremonial calendar, or *tzolkin*, was 260 days long. It was composed of two independent cycles, a 13-day cycle and a 20-day cycle. The days in the 13-day cycle were numbered from 1 to 13, and the days in the 20-day cycle were named, like our days of the week. The two smaller wheels in the diagram can be used to explain the order of the days. As the wheels turn, each cycle advances by one day, much like our calendar going from Monday the 1st to Tuesday the 2nd, and so on.

For convenience, let the days in the 20-day cycle be represented by letters instead of the Mayan names. Then each day has a date composed of a number from 1 to 13 followed by a letter from A to T. For example, the first day of the year would be 1A.

1. How many possible dates are there? Explain your reasoning.

2. The date shown in the diagram below is 7G. What would be the date one day later? 13 days later? 20 days later?

3. To help you understand the arrangements of the dates, consider some "calendars" with fewer days.

   a. Suppose that a calendar has cycles of 8 days and 5 days. The days in the 5-day cycle will be represented by letters from A to E. How many dates are possible? To find the order of the dates, repeat the numbers from 1 to 8 in one row and the letters from A to E in another, as follows:

   1 2 3 4 5 6 7 8 1 2 3 4 5 6 7 8 1 2...
   A B C D E A B C D E A B C D E A B C...

   Continue until the date 1A repeats. Does every possible date occur?

   b. Suppose that a calendar has cycles of 6 days and 4 days. The days in the 4-day cycle will be represented by letters from A to D. How many dates are possible? To find the order of the dates, repeat the numbers from 1 to 6 in one row and the letters from A to D in another, as follows:

   1 2 3 4 5 6 1 2 3 4 5...
   A B C D A B C D A B C...

   Continue until the date 1A repeats. Does every possible date occur?

   c. For every possible date to occur, what must be true of the number of days in the cycles? (Hint: Look at the factors of the number of days in each cycle.)

The Mayan solar calendar, or *haab*, was 365 days long. It was composed of 18 months of 20 days each (numbered from 0 to 19) and one month of 5 days; the days of the 5-day month were considered unlucky. The order of dates in the solar calendar was similar to our calendar. Using lowercase letters to represent the names of the months, the dates would start at 0a, then 1a, 2a, and so on. The solar calendar is represented by the large wheel in the diagram.

4. In the diagram, the date on the solar calendar is 13a. The combined date is 7G 13a. What will be the combined date 1 day later? 20 days later? 260 days later? 365 days later?

5. How many possible combined dates are there? Based on your results from part **c** of Step 3, do you think every possible date will occur? How many years will it be before the date 7G 13a occurs again?

A Closer Look at If-Then Statements

12.3

Objectives

- Create truth tables for conditionals and for converses, inverses, and contrapositives of conditionals.

- Use if-then statements and forms of valid argument for problems involving logical reasoning.

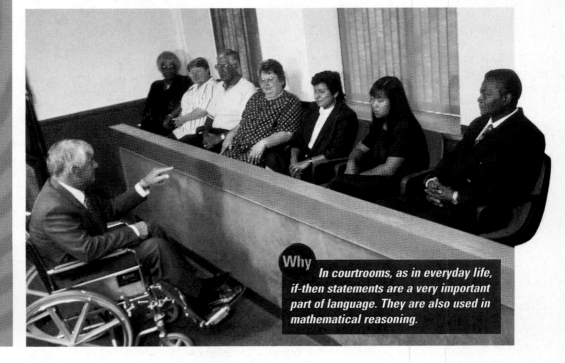

Why *In courtrooms, as in everyday life, if-then statements are a very important part of language. They are also used in mathematical reasoning.*

Lawyers use logic when presenting their cases to juries. For example, a lawyer might argue, "If the defendant committed the crime, then he could not have been at the north shopping mall between 10 A.M. and 10:30 A.M." How do you think the argument would continue?

The Truth Table for a Conditional

Various forms of if-then statements have been used throughout this book. You may recall from Lesson 2.2 that if-then statements are called conditionals. You can think of a conditional as a promise. In logic, if the "promise" is broken, the conditional is said to false. Otherwise, it is said to be true.

Suppose that your neighbor makes the following promise:

$$\underbrace{\text{If you mow my lawn,}}_{p} \text{ then } \underbrace{\text{I will give you \$10.}}_{q}$$

Four possible situations can occur.

1. *You mow the lawn and your neighbor gives you $10.*
   The promise is kept. Therefore, the conditional is said to be true.

2. *You mow the lawn and your neighbor does not give you $10.*
   The promise is broken. Therefore, the conditional is said to be false.

3. *You do not mow the lawn. Your neighbor gives you $10.*
   The promise is not broken. Therefore, the conditional is said to be true.

4. *You do not mow the lawn. Your neighbor does not give you $10.*
   The promise is not broken. Therefore, the conditional is said to be true.

The truth table below summarizes the truth values for the conditional $p \Rightarrow q$. Recall that the logical notation for "if p then q" is $p \Rightarrow q$ (read as "p implies q").

p	q	$p \Rightarrow q$
T	T	T
T	F	F
F	T	T
F	F	T

The first two columns of the truth table list all possible combinations of T and F for the two statements p and q. The third column list the truth values of the conditional $p \Rightarrow q$. Notice that the only time $p \Rightarrow q$ is false is when p is true and q is false—that is, when the promise is broken.

The Converse of a Conditional

Recall from Lesson 2.2 that the converse of a conditional results from interchanging the statements following *if* (the hypothesis) and *then* (the conclusion). Consider the following conditional and converse:

Conditional
If Tamika lives in Montana, then she lives in the United States.

CHECKPOINT ✔ Is this statement true?

Converse of the conditional
If Tamika lives in the United States, then she lives in Montana.

CHECKPOINT ✔ Is this statement true?

The truth table below summarizes the truth values for the converse of the conditional $p \Rightarrow q$.

p	q	$q \Rightarrow p$
T	T	T
T	F	T
F	T	F
F	F	T

CRITICAL THINKING   Recall that two statements are truth functionally equivalent if and only if they have the same values in their truth tables. Compare the truth values for the conditional and its converse. Are they truth functionally equivalent? Explain your reasoning.

The Inverse of a Conditional

The **inverse of a conditional** is formed by negating both the hypothesis and the conclusion. Below is the inverse of the conditional in the previous example.

Inverse of the conditional
If Tamika does not live in Montana, then she does not live in the United States.

CHECKPOINT ✔ Is this statement true?

The truth table below represents the inverse of the conditional $p \Rightarrow q$. Notice that extra columns are required for the negations of p and q.

p	q	$\sim p$	$\sim q$	$\sim p \Rightarrow \sim q$
T	T	F	F	T
T	F	F	T	F
F	T	T	F	T
F	F	T	T	T

CRITICAL THINKING Are a conditional and its inverse truth functionally equivalent? Is the converse of a conditional truth functionally equivalent to its inverse? Explain your reasoning.

The Contrapositive of a Conditional

The **contrapositive of a conditional** is formed by interchanging the hypothesis and the conclusion of the conditional and then negating each part. Below is the converse of the original conditional.

Contrapositive of the conditional
If Tamika does not live in the United States, then she does not live in Montana.

CHECKPOINT ✔ Is the statement true?

The truth table below summarizes the truth values for the contrapositive of the conditional $p \Rightarrow q$.

p	q	$\sim q$	$\sim p$	$\sim q \Rightarrow \sim p$
T	T	F	F	T
T	F	T	F	F
F	T	F	T	T
F	F	T	T	T

Notice that the final columns for the truth tables of the original conditional and those of its contrapositive are the same. Thus, the two statements are truth functionally equivalent. If a conditional is true, its contrapositive must also be true. Moreover, if a conditional is false, its contrapositive must also be false.

Thus, every theorem or postulate that can be written in if-then form can be rewritten in contrapositive form—which will also be true.

Summary of Conditionals

THREE RELATED FORMS OF AN IF-THEN STATEMENT OR CONDITIONAL		
Conditional	If p then q	$p \Rightarrow q$
Converse	If q then p	$q \Rightarrow p$
Inverse	If $\sim p$ then $\sim q$	$\sim p \Rightarrow \sim q$
Contrapositive	If $\sim q$ then $\sim p$	$\sim q \Rightarrow \sim p$

TRY THIS   Write the converse, inverse, and contrapositive of each conditional below. Determine whether the original statement and each new statement are true or false.

    a. If a triangle is equilateral, then the triangle is isosceles.

    b. If a quadrilateral is a rhombus, then the quadrilateral is a square.

Exercises

● Communicate

In Exercises 1–4, write the converse, inverse, and contrapositive of each conditional.

    1. If today is February 30, then the moon is made of green cheese.

    2. If all three sides of a triangle are congruent, then the triangle is equilateral.

    3. If I do not go to the market, then I will not buy cereal.

    4. If the car starts, then I will not be late for school.

    5. Describe the circumstances that would make the statement "If a then b" false.

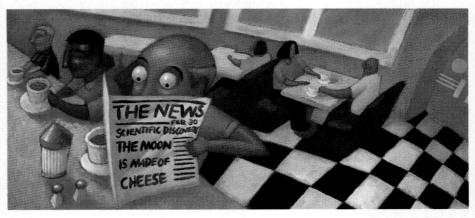

(See Exercise 1.)

6. Complete the truth table for the conditional $p \Rightarrow q$.
(TRUTH TABLE OF A CONDITIONAL)

p	q	$p \Rightarrow q$
T	T	?
T	F	?
F	T	?
F	F	?

7. Complete the truth table for the converse of the condtional $p \Rightarrow q$.
(CONVERSE OF A CONDITIONAL)

p	q	$q \Rightarrow p$
T	T	?
T	F	?
F	T	?
F	F	?

8. Complete the truth table for the inverse of the conditional $p \Rightarrow q$.
(INVERSE OF A CONDITIONAL)

p	q	~p	~q	$\sim p \Rightarrow \sim q$
T	T	?	?	?
T	F	?	?	?
F	T	?	?	?
F	F	?	?	?

9. Complete the truth table for the contrapositive of a conditional.
(CONTRAPOSITIVE OF A CONDITIONAL)

p	q	~p	~q	$\sim q \Rightarrow \sim p$
T	T	?	?	?
T	F	?	?	?
F	T	?	?	?
F	F	?	?	?

Practice and Apply

For each conditional in Exercises 10–15, write the converse, inverse, and contrapositive. Decide whether each is true or false and explain your reasoning.

10. If a figure is a square, then it is a rectangle.

11. If $a = b$, then $a^2 = b^2$.

12. If $a < b$, then $a^2 < b^2$.

13. If three angles of a triangle are congruent to three angles of another triangle, then the triangles are congruent.

14. If p and q are even numbers, then $p + q$ is an even number.

APPLICATION

15. PHYSICS If water is frozen at normal atmospheric pressure, then its temperature is less than or equal to 32°F.

16. Given: If p then q. Write the contrapositive of the statement. Then write the contrapositive of the contrapositive. What can you conclude about the contrapositive of the contrapositive of an if-then statement?

17. Suppose that the following statement is true: If the snow exceeds 6 in., then school will be canceled. Which of the following statements must also be true?
   a. If the snow does not exceed 6 in., then school will not be canceled.
   b. If school is not canceled, then the snow does not exceed 6 in.
   c. If school is canceled, then the snow exceeds 6 in.

18. State the Pythagorean Theorem and its converse, inverse, and contrapositive. Determine whether each is true or false, and explain your reasoning.

19. Choose a postulate or theorem from Chapter 3 that is written in if-then form. Write its converse, inverse, and contrapositive, decide whether each is true or false, and explain your reasoning.

20. Choose a postulate or theorem from Chapter 9 that is written in if-then form. Write its converse, inverse, and contrapositive, decide whether each is true or false, and explain your reasoning.

Some statements that are not written in if-then form can be rewritten in if-then form. For example, "Every rectangle is a parallelogram" can be rewritten as "If a figure is a rectangle, then it is a parallelogram." In Exercises 21–24, rewrite each statement in if-then form.

21. All seniors must report to the auditorium.

22. A point on the perpendicular bisector of a segment is equidistant from the endpoints of the segment.

23. She will call me if she is going to be late.

24. Doing mathematics homework every night will improve your grade in mathematics.

25. The statement "p if and only if q," written as $p \Leftrightarrow q$, is equivalent to the following two statements:

$$\text{If } p \text{ then } q$$
$$\text{and} \qquad \text{If } q \text{ then } p$$

Suppose that the statement $r \Leftrightarrow s$ is true. Which of the following must also be true? Explain your reasoning.
   a. If r then s
   b. If s then r
   c. If $\sim r$ then $\sim s$
   d. If $\sim s$ then $\sim r$

26. Statements using "if and only if" are known as **biconditionals**. Based on your results from Exercise 25, what can you conclude about certain theorems in this book? Explain your reasoning and give two examples.

27. ACADEMICS Consider the following statement: You will make the honor roll only if you get at least a B in mathematics.

Which of the following statements appear to convey the same meaning as the original statement? Explain your reasoning.

a. If you make the honor roll, then you must have gotten at least a B in mathematics.

b. If you get at least a B in mathematics, then you will make the honor roll.

c. If you do not make the honor roll, then you did not get at least a B in mathematics.

d. If you do not get at least a B in mathematics then you will not make the honor roll.

Look Back

Use the triangle below and the law of sines to find each missing measure. *(LESSON 10.5)*

28. Given m∠B = 37°, m∠A = 50°, and b = 100, find c.

29. Given m∠C = 65°, m∠A = 47°, and c = 3.45, find b.

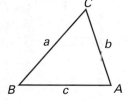

Copy the vectors below and draw the resultant vector by using the head-to-tail method. You may need to translate one of the vectors. *(LESSON 10.7)*

30.   **31.**   **32.**

Look Beyond

33. Suppose you know that statement *p* is true and you want to prove that statement *q* is true. You begin by proving that the conditional $p \Rightarrow q$ is true, which leads immediately to the conclusion.

Write an outline of the argument in terms of what is given, what is to be proved, and the logical principles that allow you to draw your conclusion.

34. Suppose that, as in Exercise 33, you know that statement *p* is true and you want to prove that statement *q* is true. You are unable to prove directly that the conditional $p \Rightarrow q$ is true, so you prove instead that its contrapositive, $\sim q \Rightarrow \sim p$, is true. If the contrapositive is true, can $\sim q$ possibly be true? (Remember that you know *p* is true.)

Write an outline of the argument in terms of what is given, what is to be proved, and the logical principles that allow you to draw your conclusion. You will study this type of argument in the next lesson.

Indirect Proof

Objectives

- Develop the concept of indirect proof (*reductio ad absurdum*, or proof by contradiction).

- Use indirect proof with problems involving logical reasoning.

Why *Lewis Carroll, the author of* Alice in Wonderland *and* Through the Looking-Glass, *was a logician who was fond of absurdity as a form of entertainment. But does absurdity have any real place in logic or mathematics? In this lesson, you will see that it can, in fact, be quite useful.*

'The time has come,' the Walrus said,
'To talk of many things:
Of shoes—and ships—and sealing-wax—
Of cabbages—and kings—
And why the sea is boiling hot—
And whether pigs have wings.'
—Lewis Carroll, *Through the Looking-Glass*

Modus Tollens Revisited

You may have heard expressions such as "If he's twenty-one, then pigs have wings!"

The speaker, perhaps without realizing it, is inviting you to use the law of indirect reasoning *(modus tollens)*, which you studied in Lesson 12.1. It is certainly not true that pigs have wings, so the statement in question (the hypothesis of the conditional) must be false.

If he's twenty-one, then pigs have wings.	*(If p then q)*
Pigs do not have wings.	*(~q)*
Therefore, he's not twenty-one.	*(Therefore, ~p)*

Indirect Proofs

A form of argument closely related to *modus tollens* is known by its Latin name **reductio ad absurdum**—literally, "reduction to absurdity." In this form, an assumption is shown to lead to an absurd or impossible conclusion, which means that the assumption must be rejected.

In formal logic and mathematics, certain proofs use the strategy of *reductio ad absurdum*, but with an important twist. In such proofs, you assume the opposite or, in logical terms, the negation of the statement that you want to prove. If this assumption leads to an impossible result, then it must be concluded that the assumption was false. Hence, the original statement was true. Such proofs are known as **indirect proofs** or **proofs by contradiction**.

Using a Contradiction to Prove Your Point

What is meant by an "absurd" or "impossible" result? In logic, a *contradiction* is such a result. A **contradiction** has the following form:

$$p \text{ AND } \sim p$$

That is, a contradiction asserts that a statement and its negation are both true. The following compound statement is a contradiction:

> A horse is a vegetarian, and a horse is not a vegetarian.

In formal logic and in mathematics, an assumption that leads to a contradiction must be rejected. Thus, contradictions turn out to be very useful.

Proof by Contradiction

To prove a statement is true, assume that it is false and show that this leads to a contradiction. **12.4.1**

An Indirect Proof

The following proof uses a contradiction to prove the converse of the Corresponding Angles Postulate. Recall that the converse of this postulate (which is itself a theorem) states the following:

> If two lines are cut by a transversal in such a way that corresponding angles are congruent, then the two lines are parallel. (Theorem 3.4.1)

To prove Theorem 3.4.1, use the "if" part of the theorem as the given.

Given: Line ℓ is a transversal that intersects lines m and n, and $\angle 1 \cong \angle 2$.

Prove: $m \parallel n$

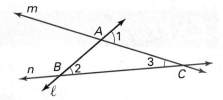

Proof: Assume that m is not parallel to n.

By assumption, m is not parallel to n, so the two lines will meet at some point, C, as shown in the redrawn figure above.

Because $\angle 1$ is an exterior angle of $\triangle ABC$, m$\angle 1$ = m$\angle 2$ + m$\angle 3$. But this means that m$\angle 1$ > m$\angle 2$ (because m$\angle 3$ > 0°). Therefore, $\angle 1$ is not congruent to $\angle 2$. Thus, the assumption that m is not parallel to n has led to the following contradiction:

$$(\angle 1 \cong \angle 2) \text{ AND } (\angle 1 \ncong \angle 2)$$

Therefore, the assumption must be false. The conclusion is $m \parallel n$.

Alibis and *Modus Tollens*

Arguments using the law of indirect reasoning (*modus tollens*) are more common than you might think. In a court of law, for example, a lawyer might want to show that a statement made by the prosecutor contradicts the accepted evidence, or the given. Arguments such as the following are quite common:

"If the defendant set the fire, then she must have been at the restaurant between 7:30 P.M. and 11:00 P.M. But three witnesses have testified that the defendant was not at the restaurant during those hours—she was in fact at a party on the other side of town. Therefore, the defendant did not set the fire."

Courtroom artists make sketches of trial scenes

The form of this argument can be represented as follows:

1. If the defendant set the fire, then she was at the restaurant between 7:30 P.M. and 11:30 P.M. (If p then q)

2. The defendant was not at the restaurant between 7:30 P.M. and 11:30 P.M. ($\sim q$)

3. Therefore, the defendant did not set the fire. (Therefore, $\sim p$)

TRY THIS   Show how the argument above could be made into a proof by contradiction.

Exercises

Communicate

1. What is a contradiction? Give the logical form and explain what it means.

Are the following statements contradictions? Explain your reasoning.

2. A tiger is a cat, and a tiger is a mammal.

3. A crocodile is a reptile, and a crocodile is not a reptile.

4. Some insects are butterflies, and some insects are not butterflies.

5. Some dogs are not pets, and some pets are not dogs.

6. Summarize the steps for writing an indirect proof.

Guided Skills Practice

Form a contradiction by using each statement and its negation. *(USING A CONTRADICTION)*

7. Lines ℓ and m are parallel.　　**8.** $\triangle ABC$ is isosceles.

9. All squares are rectangles.　　**10.** $ABCD$ is a square.

Suppose that two lines are cut by a transversal in such a way that the same-side interior angles are supplementary. Complete the following proof that the lines are parallel. *(INDIRECT PROOFS)*

Suppose that **11.** ? . Then lines ℓ and m must meet at some point, C, forming $\triangle ABC$. The sum of the measures of the angles of $\triangle ABC$ must be 180°. However, this is a contradiction because **12.** ? . Thus, lines ℓ and m must be parallel.

Practice and Apply

For Exercises 13–17, determine whether the given argument is an example of indirect reasoning. Explain why or why not.

13. If it were snowing, there would be snowflakes in the air. There are no snowflakes in the air. Therefore, it is not snowing.

14. If you were not ill, then you would eat a large dinner. You did not eat a large dinner. Therefore, you must be ill.

15. If I see my shadow, the sun must be shining. I see my shadow. Therefore, the sun is shining.

16. If I am not in the United States, then I am not in New York. I am in the United States. Therefore, I am in New York.

17. If the dog knocked over the trash, then he would look guilty. The dog does not look guilty. Therefore, the dog did not knock over the trash.

The following indirect proof, based on Euclid's proof of Proposition 6 in Book I of the *Elements*, has the classical *reductio ad absurdum* form. From a certain assumption an absurd result is produced—namely, that a larger triangle is congruent to a smaller one contained in it. No contradiction is stated formally.

Proposition 6 If in a triangle two sides equal one another, then the sides opposite the equal angles also equal one another.

Given: In $\triangle ABC$, $\angle B \cong \angle C$.

Prove: $\overline{AB} \cong \overline{AC}$

Proof: Suppose that **18.** ? . Then one of the two sides $\overline{AB}$ or $\overline{AC}$ must be longer than the other. Let $\overline{AB}$ be the longer side. Then there exists a point, D, on $\overline{AB}$ such that **19.** ? . Since $\overline{BC} \cong$ **20.** ? , $\overline{DB} \cong \overline{AC}$, and $\angle B \cong$ **21.** ? , $\triangle ABC \cong$ **22.** ? by **23.** ? , which is absurd. Therefore, **24.** ? .

Complete the indirect proof below.

Given: $\overline{JM}$ bisects $\angle KJL$, and $\overline{JM}$ is not a median of $\triangle JKL$.

Prove: $\overline{JK} \not\cong \overline{JL}$

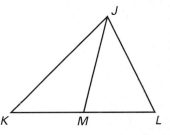

Proof: Suppose that **25.** ? . Then $\triangle JKL$ is **26.** ? . By the Isosceles Triangle Theorem, **27.** ? . Since $\overline{JM}$ bisects $\angle KJL$, **28.** ? , and so **29.** ? by **30.** ? . Thus, **31.** ? because CPCTC. This gives a contradiction because **32.** ? . Therefore, **33.** ? .

NUMBER THEORY Write an indirect proof of each of the following:

34. There is no largest integer. (Hint: If n is an integer, then $n + 1$ is an integer.)

35. There is no smallest positive real number. (Hint: $\frac{1}{2}x < x$ for all positive values of x)

36. For two integers m and n, if m^2 does not divide n^2 with no remainder, then m does not divide n with no remainder. (Hint: If m divides n with no remainder, then n can be factored into $k \cdot m$ for some integer k.)

37. If a fraction has a terminating decimal expansion, then the denominator divides some power of 10 (10, 100, 1000, etc.) with no remainder. (Hint: If a decimal terminates after n places, then multiplying it by 10^n gives an integer.)

38. LAW A defense attorney begins an argument with the following words: "Suppose that the defendent is guilty of the crime. Then he was capable of carrying a heavy load up a steep hill for nearly a mile…" How might the attorney continue the argument? Explain how this would be an example of an indirect proof. State what the contradiction would be.

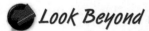

Look Back

39. Solve for x in the figure below. *(LESSON 3.8)*

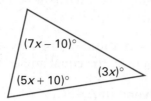

$(7x - 10)°$

$(5x + 10)°$ $(3x)°$

In Exercises 40–43, state whether the given side lengths determine an acute triangle, a right triangle, an obtuse triangle, or no triangle. *(LESSON 5.4)*

40. $a = 10, b = 12, c = 14$ **41.** $a = 7, b = 9, c = 15$

42. $a = 4, b = 7, c = 12$ **43.** $a = 8, b = 12, c = 17$

Find the sine, cosine, and tangent of each angle below. *(LESSONS 10.1 AND 10.2)*

44. $72°$ **45.** $45°$ **46.** $140°$ **47.** $5°$

APPLICATION

48. TELECOMMUNICATIONS Suppose that you wish to measure the height of a radio tower. From point A, you measure the angle of elevation to the top of the tower to be $55°$. Then you walk 90 feet toward the tower to point B and measure the new angle of elevation, which is $68°$. Estimate the height of the tower. *(LESSON 10.4)*

$55°$ $68°$
A 90 B

Look Beyond

Mathematical Induction: The principle of mathematical induction is useful in many proofs. The principle can be stated as follows: Suppose that S_n is a statement in terms of n, which can be any positive integer. Suppose that the statement is true for $n = 1$ and that if the statement is true for n, then it is always true for $n + 1$. Then the statement is true for all values of n.

49. Show that the statement "$5^n - 1$ is divisible by 4" is true for $n = 1$.

50. Assume that "$5^n - 1$ is divisible by 4" is true for some value of n. Show that it must be true for $n + 1$; that is, show that "$5^{n+1} - 1$ is divisible by 4." (Hint: If a number is divisible by 4, then it can be written as $4k$ for some positive integer k. Since $5^n - 1$ is divisible by 4, write 5^n in terms of some positive integer k.)

51. According to the principle of mathematical induction, what have you proved?

MORE ON THE EUCLIDEAN ALGORITHM

The Euclidean algorithm you studied in the Portfolio Activity in Lesson 12.1 can be proved by using an indirect proof. There are two parts to the proof—the first part is to prove that the result is a divisor of both numbers, and the second is to prove that it is the largest such number.

Examine the steps of the Euclidean algorithm on page 775. The process for finding the gcd of any two numbers a and b, where $a > b$, can be written as follows:

$$a = b(x_1) + r_1 \qquad a \div b = x_1 \text{ remainder } r_1$$
$$b = r_1(x_2) + r_2$$
$$r_1 = r_2(x_3) + r_3$$
$$\vdots \qquad \vdots \qquad \vdots$$
$$r_{n-2} = r_{n-1}(x_n) + r_n \qquad r_n \text{ is the gcd of } a \text{ and } b.$$
$$r_{n-1} = r_n(x_{n+1}) + 0$$

Note: Since the remainders are always nonnegative integers, and the remainder gets smaller at each step, the sequence of remainders will always end in 0.

1. Use the last equation above to explain why r_n must be a divisor of r_{n-1}. Use the next-to-last equation to explain why r_n must be a divisor of r_{n-2}. (Hint: Is the right side of the equation divisible by r_n?) Use the entire sequence of equations to explain why r_n must be a divisor of both a and b.

2. Suppose that there is a number $r > r_n$ which is a divisor of both a and b. Use the first equation above to explain why r must be a divisor of r_1. Use the entire sequence of equations to explain why r must be a divisor of r_n. How does this lead to a contradiction?

Computer Logic

OUTPUT   INPUT

INPUT

C49A

POWER

Objectives

- Explore on-off tables, logic gates, and computer logic networks.

- Solve problems by using computer logic.

Why Logic provides the foundation for the arithmetic and decision-making processes of many "smart" electronic devices. The fundamental units of logical circuits are logic gates, which function like logical operators such as AND, OR, and NOT.

APPLICATION
COMPUTER ARCHITECTURE

A single computer chip in this "motherboard" contains hundreds of thousands of logic gates like the one shown greatly magnified here.

The Binary Number System

Computers use the **binary number system**. *Binary* means "having two parts," and the binary number system is based on the two numbers 1 and 0. You can think of a computer as a system of electrical switches that can be in two possible conditions, *on* and *off*. A switch that is on is represented by a 1, and a switch that is off is represented by a 0.

Activity

On-Off Tables

The tables in this Activity simulate the working of three different devices. The columns of the tables represent the different states of the devices and their components. Let 1 = ON and 0 = OFF. Work in pairs to complete each table.

Part I

Pushing the power button on a TV remote control will turn the TV on if it is off and turn it off if it is on. Determine whether the TV will be on or off after pressing the power button, and fill in the blanks in the table below.

CHECKPOINT ✔

TV before pressing POWER button	TV after pressing POWER button
1	?
0	?

Part II

To record on a particular video recorder, the user must press both PLAY and RECORD. Determine whether the video recorder will record or not for each combination of buttons listed in the table below, and fill in the blanks.

CHECKPOINT ✔

PLAY button	RECORD button	Video recorder
1	1	?
1	0	?
0	1	?
0	0	?

Part III

The student driver car at Dover High School has been equipped with two brake pedals so that either the student driver or the instructor can stop the car. Determine whether the car brakes will be on or off for each combination of pedals listed in the table below, and fill in the blanks.

CHECKPOINT ✔

Student pedal	Instructor pedal	Brakes
1	1	?
1	0	?
0	1	?
0	0	?

CRITICAL THINKING   How do the tables in the Activity compare with truth tables?

Logic Gates

Each table in the Activity above corresponds to a particular type of electronic circuitry called a **logic gate**. Logic gates are the building blocks of "smart" electronic devices. Each logic gate is named for a special logical function, such as NOT, AND, or OR.

In the diagram below, p represents the electrical input. The input-output table shows the resulting output for the given input. If the input, p, of a NOT gate has a value of 1, the output, NOT p, will have a value of 0—and vice versa. Notice that the input-output table corresponds to the table in Part I of the Activity.

NOT Logic Gate

p —|NOT ▷∘— NOT p

Input   Gate   Output

Input-Output Table	
Input	**Output**
p	NOT p
1	0
0	1

An AND logic gate needs two inputs, which are represented by p and q in the diagram below. Notice that in order to get an output value of 1, both inputs must have a value of 1. The input-output table corresponds to the table in Part II of the Activity.

AND Logic Gate

Input   Gate   Output

Input-Output Table		
Input		Output
p	q	p AND q
1	1	1
1	0	0
0	1	0
0	0	0

An OR logic gate is represented below, along with its input-output table. Notice that the input-output table corresponds to the table in Part III of the Activity.

OR Logic Gate

$\begin{matrix} p \\ q \end{matrix}$ — OR — p OR q

Input   Gate   Output

Input-Output Table		
Input		Output
p	q	p OR q
1	1	1
1	0	1
0	1	1
0	0	0

Network of Logic Gates

Logic gates can be combined to form networks. You can use input-output tables to determine how a network operates.

EXAMPLE **1** Construct an input-output table for the network at right.

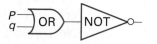

SOLUTION

Read from left to right. The first gate is p OR q. Determine the output from this gate. Then perform the NOT operation: NOT (p OR q). To construct the input-output table, consider all possible input combinations. Fill in the values as you would for a truth table, where 1 is T and 0 is F.

Input-Output Table			
Input		Output	
p	q	p OR q	NOT (p OR q)
1	1	1	0
1	0	1	0
0	1	1	0
0	0	0	1

EXAMPLE **2** Create a logical expression that corresponds to the network at right.

SOLUTION

Read from left to right, one branch at a time. Combine the results of the branches when they flow together.

1. NOT appears first: NOT *p*.

2. The AND gate gives (NOT *p*) AND *q*.

3. The bottom branch gives NOT *r*.

4. The OR gate combines the output from Steps 2 and 3:
((NOT *p*) AND *q*) OR (NOT *r*)

TRY THIS How many rows and how many columns would be in an input-output table for the network in Example 2? Explain your reasoning.

Exercises

Communicate

1. What is a logic gate? What do logic gates represent?

2. Does the OR logic gate represent an inclusive or exclusive *or*? How can you tell?

3. How are input-output tables for networks of logic gates related to truth tables?

4. In a network of logic gates with two inputs, there are four possible combinations of input values. How many combinations are possible with three inputs? with four inputs? Explain your reasoning.

Guided Skills Practice

Complete the input-output table for each network. *(EXAMPLE 1)*

5.

Input-Output Table			
Input		**Output**	
p	*q*	NOT *p*	NOT *p* OR *q*
1	1	?	?
1	0	?	?
0	1	?	?
0	0	?	?

6.

Input-Output Table			
Input		**Output**	
p	*q*	*p* AND *q*	NOT (*p* AND *q*)
1	1	?	?
1	0	?	?
0	1	?	?
0	0	?	?

Create a logical expression that corresponds to each network.
(EXAMPLE 2)

7.

8.

Practice and Apply

Draw the symbol for the gate that corresponds to each of the following:

9. NOT   **10.** AND   **11.** OR

Use the logic gates below to answer each question.

12. If $p = 1$, what is the output?   **13.** If $p = 1$, what is the output?

p—NOT— p—NOT—NOT—

14. If $p = 1$ and $q = 0$, what is the output?   **15.** If $p = 1$ and $q = 0$, what is the output?

For Exercises 16–19, complete the input-output table for the network of logic gates at right.

	p	q	NOT p	NOT q	(NOT p) OR (NOT q)
16.	1	1	?	?	?
17.	1	0	?	?	?
18.	0	1	?	?	?
19.	0	0	?	?	?

For Exercises 20–27, complete the input-output table for the network of logic gates at right.

	p	q	r	NOT p	(NOT p) AND q	NOT r	((NOT p) AND q) OR (NOT r)
20.	1	1	1	?	?	?	?
21.	1	1	0	?	?	?	?
22.	1	0	1	?	?	?	?
23.	1	0	0	?	?	?	?
24.	0	1	1	?	?	?	?
25.	0	1	0	?	?	?	?
26.	0	0	1	?	?	?	?
27.	0	0	0	?	?	?	?

Create a logical expression that corresponds to each network below.

28.

29.

30.

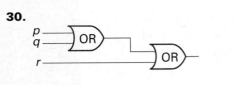

31.

Construct a network of logic gates for the following expressions:

32. p AND (NOT q)

33. NOT (p OR q)

34. (p AND (NOT q)) OR r

35. (p OR q) OR (NOT r)

Two logical expressions are functionally equivalent if, given the same input, they produce the same output. For Exercises 36 and 37, identify the two expressions in each list that are functionally equivalent.

36. a. NOT (p AND q)
  b. (NOT p) AND (NOT q)
  c. (NOT p) OR (NOT q)

37. a. p OR (NOT q)
  b. NOT (p OR q)
  c. (NOT p) AND (NOT q)

CHALLENGE

38. Create a network of logic gates that corresponds to the input-output table below.

Input			Output
p	q	r	?
1	1	1	1
1	1	0	1
1	0	1	1
1	0	0	1
0	1	1	1
0	1	0	0
0	0	1	1
0	0	0	1

APPLICATION

39. ELECTRICITY In electrical circuit diagrams, a switch is represented as shown below. An open switch does not allow current to flow through the circuit.

open switch   closed switch

In the circuits at right, the bulb will light up if there is a closed path through the circuit. Tell what logical function is represented by each arrangement of switches. Explain how switches A and B work together to form a logic gate in each case.

a.

b.

battery: ⊣||⊢ bulb: ─⦿─

APPLICATION

40. ENVIRONMENTAL SCIENCE
The aerial photograph at right shows workers cleaning up an oil spill. Use objects in the photo to estimate the size of the squares in the yellow grid. Then estimate the area of the spill. *(LESSON 5.1)*

41. A 50-ft pipe carries water from a well to a house. Suppose that the pipe springs a leak. If the leak is equally likely to occur at any point along the pipe, what is the probability that the leak will be within 5 ft of the house? *(LESSON 5.7)*

42. A point is chosen at random in the diagram at right.Whatis the probability that the point lies inside the shaded triangle? *(LESSON 5.7)*

In Exercises 43 and 44, determine whether the conditional is true for all values of the variables for which the proportions are defined. If the conditional is false, give a numerical counterexample. *(LESSON 8.2)*

43. If $\frac{x}{y} = \frac{r}{s}$, then $\frac{x+c}{y} = \frac{r+c}{s}$.   **44.** If $\frac{x}{y} = \frac{r}{s}$, then $\frac{x}{y} = \frac{x+r+m}{y+s+m}$.

45. If the perimeters of two similar triangles are 18 cm and 25 cm, what is the ratio of their corresponding sides? *(LESSON 8.6)*

🔶 *Look Beyond*

In Exercises 46–48, you will discover a network of logic gates that is functionally equivalent to the statement "If *p* then *q*."

46. Create a truth table for the statement "If *p* then *q*." Use 1 for true statements and 0 for false statements.

47. Create an input-output table for each network below.

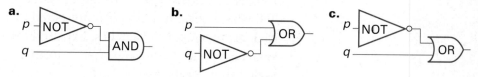

48. Which of the networks in Exercise 47 is functionally equivalent to the statement "If *p* then *q*"? Create a logical expression for this network.

Two Famous Theorems

Two of the most famous theorems in mathematics come from classical times. Both of them involve indirect proof, which you studied in Lesson 12.4. Both of them also involve the prime factorization theorem, which is known as the Fundamental Theorem of Number Theory. It is not difficult to prove, but here it will be assumed true.

Every number has a unique prime factorization.

The Infinity of the Primes

Euclid's proof that there are infinitely many primes has captured the imagination of people over the ages. To help you understand the proof, which is actually very brief, a numerical example is given first.

The proof will assume that there is only a finite number of primes. Suppose, for example, that there are only 5 primes, and that 11 is the largest prime number. Then the list of all the prime numbers would read:

2, 3, 5, 7, 11

Now form a new number, m, by multiplying the prime numbers together and adding 1.

$m = (2)(3)(5)(7)(11) + 1 = 2311$

The new number m must be composite, according to the assumption, which states that 11 is the largest prime number. Therefore, m must be factorable into a combination of prime numbers, which (according to the assumption) range from 2 to 11. However, none of the numbers in the list of primes will divide the number exactly because you will always get a remainder of 1.

For example: $\dfrac{462 \text{ R } 1}{5\overline{)2311}}$

If 2311 is not divisible by one of the primes in the list, then it must be either prime, or divisible by some other prime greater than 11, which contradicts the assumption that 11 is the largest prime.

Once you understand this example, you should be able to follow the generalization of it, which is given below.

Prove: There are infinitely many primes.

Proof: Assume that there are a finite number of primes, say, n of them. Then the list of the primes would be $p_1, p_2, p_3, \dots, p_n$. Form a new number, $m = (p_1)(p_2)(p_3) \cdots (p_n) + 1$.

By assumption, m must be composite. However, m is not divisible by any of the numbers in the list of primes, so there must be a prime larger than p, which contradicts the assumption. Thus, there are infinitely many primes.

The "Incommensurability" of the Square Root of 2

In the early history of mathematics, it seems to have been widely believed that any number could be represented as a fraction—that is, as a ratio of two integers. This was certainly believed by the early Pythagoreans. Thus, the proof that the square root of 2 *cannot* be represented by such a ratio (which means that it is irrational or *incommensurable*) came as a profound shock to them and shook the foundations of their beliefs.

Before studying the incommensurability theorem, you will need to know three simple theorems from number theory. The three theorems, given below, use the prime factorization theorem.

Prove: The square of an even number is even.

Proof: If a number is even, then 2 is one of its prime factors. When the number is squared, the number 2 will appear at least twice in the prime factorization of the square. Therefore, the square is divisible by 2, which means that it is an even number.

Prove: The square of an even number is divisible by 4.

Proof: The proof is left as an exercise for you in Activity 2.

Prove: The square of an odd number is odd.

Proof: The proof is left as an exercise for you in Activity 2.

You are now ready to tackle the incommensurability theorem.

Prove: The square root of 2 cannot be written as a ratio of two integers.

Proof: Assume that the square root of 2 can be represented as the ratio of two numbers—that is, as a fraction. If this fraction is not in lowest terms, then there is a fraction in lowest terms to which it can be reduced. Thus, the assumption can be written in the following way, which is equivalent to the original statement: Assume that the square root of 2 can be represented as a fraction in lowest terms.

Let p be the numerator and q be the denominator of the fraction that is assumed to exist.

$$\frac{p}{q} = \sqrt{2}$$ p and q have no common factors.

$$\left(\frac{p}{q}\right)^2 = (\sqrt{2})^2 = 2$$ Square both sides of the equation.

$$p^2 = 2q^2$$ Multiply to clear the fraction.

The last equation implies that p^2 is an even number and so p is an even number. (Why?) Therefore, p^2 is the square of an even number. This implies that p^2 is divisible by 4. But if p^2 is divisible by 4, then q^2 must be an even number. (Why?) Therefore, q is even, as well.

If p and q are both even, then they must have a common factor of 2, which contradicts the original assumption. Therefore, the assumption must be false, and the theorem is proven.

Activity ❶ **The Infinity of the Primes**

1. Repeat the numerical example in the proof for different numbers of primes. Explain why you always get a remainder of 1 when you divide m by one (or more) of the primes in your list.

2. If you subtract 1 instead of adding 1 to obtain the number m in the proof, how would the proof be affected?

3. Do some research on number theory and learn about some of the unproven conjectures such as the *Goldbach Conjecture* and the *Twin Primes Conjecture*. Explain them in your own words and give numerical illustrations for each one.

Activity ❷ **The "Incommensurability" of the Square Root of 2**

1. Prove that the square of an even number is divisible by 4.

2. Prove that the square of an odd number is odd.

3. In the incommensurability proof, there are two points at which the reader is asked to explain why a certain result occurs. Write these explanations in your own words.

Chapter Review and Assessment

ARGUMENT FORMS

Lesson	Section	Argument Form
12.1	12.1.1 Argument Form: *Modus Ponens*	If p then q p Therefore, q
	12.2.2 Argument Form: *Modus Tollens*	If p then q $\sim q$ Therefore, $\sim p$
	12.1.3 Invalid Form: Affirming the Consequent	If p then q q Therefore, p
	12.1.4 Invalid Form: Denying the Antecedent	If p then q $\sim p$ Therefore, $\sim q$
12.4	12.4.1 Proof by Contradiction	To prove statement s, assume $\sim s$. Then the following argument form is valid: If $\sim s$ then (t AND $\sim t$) Therefore, s

Key Skills & Exercises

Key Skills

Determine whether an argument is valid or invalid.

Is the following argument valid or invalid?

If the Memorial Day parade passes the house, then the house is on Main Street.
The Memorial Day parade does not pass the house.

Therefore, the house is not on Main Street.

This argument is invalid. It contains the invalid argument form known as denying the antecedent.

Use the law of indirect reasoning (*modus tollens*).

Give a valid conclusion based on the following premises:

If the oven is on, then the oven door feels warm.
The oven door does not feel warm.

Using indirect reasoning (*modus tollens*), the statement "The oven is not on" is a valid conclusion.

Exercises

Determine whether each argument is valid or invalid.

1. If a polygon has 5 sides, then it is a pentagon.
   Polygon *ABCDE* has 5 sides.
   Therefore, polygon *ABCDE* is a pentagon.

2. If the car is out of gas, then the car will not start. The car will not start.
   Therefore, the car is out of gas.

Use the law of indirect reasoning to give valid conclusions based on the following premises:

3. If an animal is a rodent, then it has incisors that grow continuously.
   A cat does not have incisors that grow continuously.

4. If a number is divisible by 6, then it is divisible by 3.
   25 is not divisible by 3.

Key Skills

Write the conjuction and disjunction of two statements.

Write the conjunction and disjunction of the following statements:

The sky is green.
Your pants are on fire.

Conjunction: The sky is green, and your pants are on fire.

Disjunction: The sky is green, or your pants are on fire.

Exercises

For Exercises 5–8, refer to the statements below.
John is my brother.
17 is prime.

5. Write the conjunction of the statements above.

6. Write the disjunction of the statements above.

Create a truth table for a compound statement.

Create a truth table for the conjuction of the statements above.

Let p represent the statement "The sky is green" and let q represent "Your pants are on fire." Then the truth table for the conjunction p AND q is shown below.

p	q	p AND q
T	T	T
T	F	F
F	T	F
F	F	F

7. Create a truth table for the compound statement from Exercise 5.

8. Create a truth table for the compound statement from Exercise 6.

Key Skills

Create a truth table for a conditional.

Create a truth table for the following conditional: If it is 1:00 P.M., then Jill is in geometry class.

The conditional can be separated into its hypothesis, p, "It is 1:00 P.M.," and its conclusion, q, "Jill is in geometry class." The truth table is below.

p	q	$p \Rightarrow q$
T	T	T
T	F	F
F	T	T
F	F	T

Write a converse, inverse, and contrapositive of a conditional.

Write the converse, inverse, and contrapositive of the conditional above.

Converse: If Jill is in geometry class, then it is 1:00 P.M.

Inverse: If it is not 1:00 P.M., then Jill is not in geometry class.

Contrapositive: If Jill is not in geometry class, then it is not 1:00 P.M.

Exercises

For Exercises 9–12, refer to the following conditional: If it is raining, then the ground is wet.

9. Write the converse, inverse, and contrapositive of the given conditional.

10. Create a truth table for the converse of the given conditional.

11. Create a truth table for the inverse of the given conditional.

12. Create a truth table for the contrapositive of the given conditional.

Key Skills

Use indirect reasoning in a proof.

In △*PQR*, point *S* is on $\overline{PR}$ and point *T* is on $\overline{QR}$. $\overline{QS}$ and $\overline{PT}$ intersect at point *X* inside △*PQR*. Prove that $\overline{QS}$ and $\overline{PT}$ do not bisect each other.

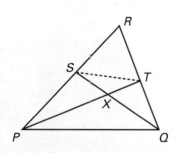

Assume that $\overline{QS}$ and $\overline{PT}$ bisect each other. Since $\overline{QS}$ and $\overline{PT}$ are diagonals of quadrilateral *PQTS*, if they bisect each other, then *PSTQ* is a parallelogram. If *PSTQ* is a parallelogram, then $\overline{PS}$ and $\overline{QT}$ are parallel. But $\overline{PS}$ and $\overline{QT}$ are not parallel because they lie on intersecting lines. This contradiction proves indirectly that $\overline{QS}$ and $\overline{PT}$ do not bisect each other.

Exercises

Given: △*ABC* is a scalene triangle, and $\overrightarrow{AD}$, the bisector of ∠*A*, intersects $\overline{BC}$ at point *D*.

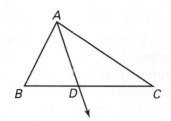

Prove: $\overrightarrow{AD}$ is not perpendicular to $\overline{BC}$.

Proof:

Assume **13.** ? . Then ∠*ADB* and ∠*ADC* both measure 90°. Since $\overrightarrow{AD}$ is the bisector of ∠*A*, ∠*BAD* ≅ ∠*CAD* and $\overline{AD}$ ≅ $\overline{AD}$, so **14.** ? by ASA. Thus, **15.** ? because CPCTC. However, this contradicts **16.** ? , thus proving indirectly that $\overrightarrow{AD}$ is not perpendicular to $\overline{BC}$.

Key Skills

Write a logical expression for a network.

Write a logical expression for the network below.

(NOT *p*) AND *q*

Create input-output tables.

Create an input-output table for the network above.

p	*q*	(NOT *p*) AND *q*
1	1	0
1	0	0
0	1	1
0	0	0

Exercises

For Exercises 17 and 18, refer to the diagram below.

17. Write a logical expression for the given network.

18. Create an input-output table for the given network.

19. Write a logical expression for the network below.

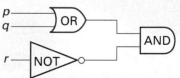

20. Create an input-output table for the network from Exercise 19.

21. LAW Determine whether the argument below is valid or invalid. Explain your reasoning.

If Mr. Smith committed the crime, then he would have been at the bank on Tuesday morning.
Mr. Smith was not at the bank on Tuesday morning.

Therefore, Mr. Smith did not commit the crime.

22. ELECTRICITY Imagine that you are planning your dream house and you want certain appliances to work in an interconnected way. Suppose that you want the living room TV to be off and the kitchen radio to be on if and only if the coffee maker and toaster are on or the kitchen light is on. Draw a network of logic gates to illustrate this situation.

Alternative Assessment

Performance Assessment

1. Write each of the following folk sayings, or *aphorisms*, as a conditional statement. Then write the converse, inverse, and contrapositive of each.

   a. The squeaky wheel gets the grease.
   b. Beggars can't be choosers.
   c. He who lives by the sword shall die by the sword.
   d. People who live in glass houses shouldn't throw stones.
   e. Out of sight, out of mind.
   f. Absence makes the heart grow fonder.

Write two true statements and label them *p* and *q*. Write two false statements and label them *r* and *s*.

2. Write the negation of each statement. Which negations are true?

3. Write the conjunction of *p* and *q*, of *p* and *r*, and of *r* and *s*. Which conjunctions are true?

4. Write the disjunction of *p* and *q*, of *p* and *r*, and of *r* and *s*. Which disjunctions are true?

Portfolio Projects

Create a network of logic gates like the one below. Your network should have at least three inputs, *p, q,* and *r*, and at least two logic gates.

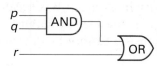

1. Flip a coin to determine values of *p, q,* and *r*. Let heads be 1 and tails be 0. Use these values to determine the output value of the network.

2. Create a network with three inputs and at least two logic gates that gives an output value of 0 if all three inputs are 0 and an output value of 1 otherwise.

3. Play this game with a friend. Each of you must create a network with at least three, but no more than five, inputs and as many logic gates as you want. Then flip a coin to determine the value of each input. Use these input values to determine the output value of each network. You win if your network has an output value of 1 and your opponent's network has an output value of 0.

📶 internet**connect**

🌐 **HRW** The HRW Web site contains many resources to reinforce and expand your knowledge of formal logic. The Web site also provides Internet links to other sites where you can find information and real-world data for use in research projects, reports, and activities that involve formal logic. Visit the HRW Web site at **go.hrw.com**, and enter the keyword **MG1 CH12**.

1-12 College Entrance Exam Practice

QUANTITATIVE COMPARISON For Items 1–3, write

A if the quantity in Column A is greater than the quantity in Column B;
B if the quantity in Column B is greater than the quantity in Column A;
C if the two quantities are equal; or
D if the relationship cannot be determined from the information given.

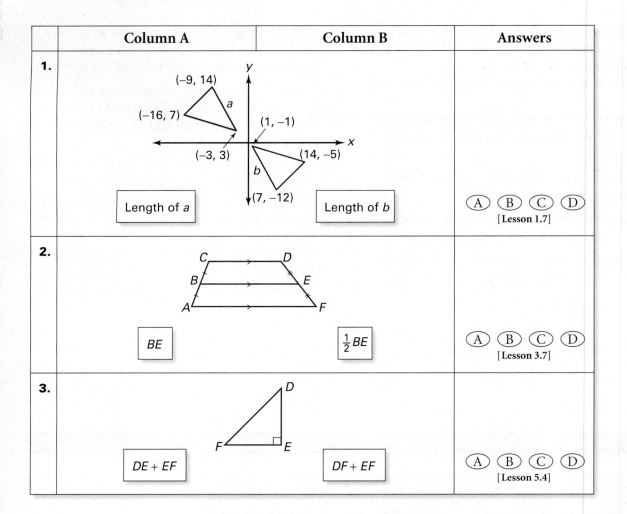

	Column A	Column B	Answers
1.	Length of *a*	Length of *b*	Ⓐ Ⓑ Ⓒ Ⓓ [Lesson 1.7]
2.	BE	$\frac{1}{2}$ BE	Ⓐ Ⓑ Ⓒ Ⓓ [Lesson 3.7]
3.	DE + EF	DF + EF	Ⓐ Ⓑ Ⓒ Ⓓ [Lesson 5.4]

4. Which of the following statements is a definition? *(LESSON 2.3)*
  a. A monarch is an orange butterfly.
  b. A line segment is the shortest path between two points.
  c. A pile of loose rubble is a hazard.
  d. A rhombus is a parallelogram.

5. $\overline{AF}$ has endpoints at (13, 15) and (9, 20). The endpoints of a segment parallel to $\overline{AF}$ are at
  ___?___. *(LESSON 3.8)*
  a. (0, 0) and (25, 16)
  b. (15, 13) and (20, 9)
  c. (4, −6) and (8, −1)
  d. (4, 3) and (12, −13)

6. Identify the correct expression for *EC*. *(LESSON 9.5)*

a. $ED \times \dfrac{EB}{EA}$

b. $EB \times \dfrac{EA}{ED}$

c. $EB \times \dfrac{CD}{AB}$

d. $EB \times \dfrac{AB}{CD}$

7. Which statement is true in hyperbolic geometry? *(LESSON 11.5)*
a. No lines are parallel.
b. Two-sided polygons exist.
c. The sum of the angles in a triangle is less than 180°.
d. The shortest path between two points is along a great circle.

For Exercises 8–10, refer to the statements below. *(LESSONS 12.1 AND 12.3)*
p: If the fog has lifted, the boat can leave the harbor.
q: The boat cannot leave the harbor.

8. What conclusion can be drawn from statements *p* and *q*?
a. The fog has lifted.
b. The fog has not lifted.
c. If the fog has not lifted, the boat cannot leave the harbor.
d. No valid conclusion can be drawn from statements *p* and *q*.

9. Write the negation of statement *q*.

10. Write the contrapositive of statement *p*.

11. In the diagram at right, △*ABC* is equilateral. What are the coordinates of point *C*? *(LESSON 5.7)*
a. $(0, a)$
b. (a, a)
c. $\left(\dfrac{a}{2}, \dfrac{a\sqrt{3}}{2}\right)$
d. $(a, a\sqrt{3})$

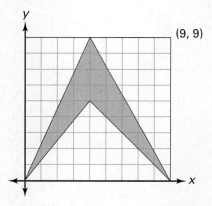

12. What is the central angle of a regular 13-gon? *(LESSON 3.1)*

13. A point is chosen at random inside the square in the coordinate plane below. What is the probability that the point lies within the shaded area? *(LESSON 5.7)*

(9, 9)

14. Find *MN*. *(LESSON 10.1)*

15. Find *AB*. *(LESSONS 10.5 AND 10.6)*

Info Bank

Index

Glossary

Postulates, Theorems

Tables

Extra Practice

CHAPTER 1

LESSON 1.1

In Exercises 1–3, refer to the rectangle at right.

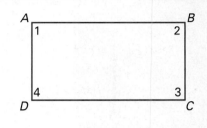

1. Name all of the segments in the rectangle.

2. Name the plane that contains the rectangle.

3. Name the rays that form each side of the angles in the rectangle.

In Exercises 4–6, classify each statement as true or false, and explain your reasoning.

4. Two planes intersect in a line.

5. Two lines are contained in exactly one plane.

6. Any three points are contained in exactly one line.

Refer to the figure at right for Exercises 7 and 8.

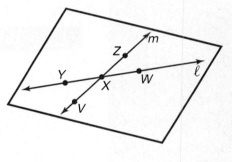

7. Name the intersection of lines ℓ and m.

8. Name an angle at the intersection of lines ℓ and m. Name the vertex of this angle and the two rays that form the sides of the angle.

LESSON 1.2

In Exercises 1–3, find the length of $\overline{AB}$.

1.

2.

3.

In Exercises 4 and 5, point C is between points A and B on $\overline{AB}$. Sketch the figure and find the missing lengths.

4. $AC = 12$; $CB = $ _____?_____; $AB = 34$

5. $AC = 11$; $CB = 18$; $AB = $ _____?_____

Find the indicated value.

6. $WY = 20$; $v = $ _____?_____

In Exercises 7 and 8, explain why each statement does or does not make sense.

7. $\overline{AB} + BC = 15$ cm

8. $mVW = 8$ in.

LESSON 1.3

Find the measure of each angle. Refer to the diagram at right.

1. m∠AVB

2. m∠AVC

3. m∠BVC

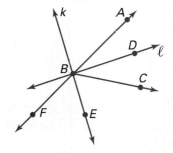

In the figure at right, m∠BVC = 55° and ∠AVB and ∠BVD form a linear pair. Find the following:

4. m∠CVD

5. m∠AVB

6. m∠AVC

In the diagram at right, m∠WXY = (7x + 14)°. Find the value of x, and then find each indicated angle measure.

7. m∠WXZ

8. m∠ZXY

9. m∠WXY

LESSON 1.4

Use folding paper to construct the figures described in Exercises 1–4. Do not use a ruler or protractor. Trace over each figure and label all relevant parts.

1. a 135° angle   **2.** a 112.5° angle

3. a parallelogram that is not a rectangle   **4.** a rhombus

For Exercises 5–7, suppose that ℓ is the angle bisector of ∠ABC, line k is perpendicular to ℓ, and m∠ABD = 35°.

5. Find m∠ABC and m∠DBC.

6. Find m∠EBC.

7. If ∠FBC and ∠CBA are supplementary angles, find m∠FBE.

LESSON 1.5

Trace each of the triangles at right onto a separate piece of folding paper or draw them by using geometry graphics software. Triangles are named with the △ symbol and the names of the vertices, such as △*ABC*.

Find the following:

1. the perpendicular bisectors of △*ABC*
2. the angle bisectors of △*ABC*
3. the circumscribed circle of △*ABC*
4. the inscribed circle of △*ABC*
5. the midpoints of each side of ∠*DEF*
6. the medians of ∠*DEF*

LESSON 1.6

For Exercises 1–4, trace the figures onto folding paper.

Reflect each figure across the given line.

1.

2.

Rotate each figure about the given point.

3.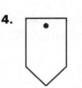

4.

5. Make a list of capital letters that stay the same when reflected across a vertical line. What is the longest word you can write with the letters in your list?

LESSON 1.7

In Exercises 1–4, use graph paper to draw the transformations of the figure as indicated.

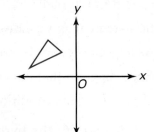

1. Rotate the figure 180° about the origin.

2. Reflect the figure across the y-axis.

3. Reflect the figure across the x-axis.

4. Reflect the figure across the x-axis and then reflect the image across the y-axis. How does this figure relate to the transformed figure in Exercise 1?

In Exercises 5–8, describe the result of applying each rule to a figure in the coordinate plane.

5. $H(x, y) = (x, y - 3)$

6. $Z(x, y) = (x + 2, y)$

7. $G(x, y) = (-x, y - 5)$

8. $F(x, y) = (y, x)$

CHAPTER 2

LESSON 2.1

Use the table below to answer Exercises 1–4. The numbers in the table are perfect squares: $1^2 = 1$, $2^2 = 4$, $3^2 = 9$, $4^2 = 16$, . . .

1. Suppose you want to know which column contains the number 10,201. Ten thousand two hundred one (10,201) is the square of what number?

A	B	C	D	E
1	4	9	16	25
36	49	64	81	100
⋮	⋮	⋮	⋮	⋮

2. Look at the numbers that occur in each column of the table. What is true of every number in column E?

3. In what column does the number 100^2 occur?

4. Prove that 10,201 occurs in column A.

Suppose that you know the square of a positive integer n and you want to find the square of $n + 2$.

5. Draw a square of dots to represent the square of the number n.

6. Increase the side lengths of your square by 2 to represent the square of the number $n + 2$.

7. Use your diagram to prove that the square of $n + 2$ is found by adding $2n$ and $2n + 4$ to the square of n. Your proof should work for all values of n.

LESSON 2.2

For Exercises 1–4, refer to the following statement:

All animals that live in the Everglades live in Florida.

1. Rewrite the statement as a conditional.

2. Identify the hypothesis and the conclusion of the conditional.

3. Draw an Euler diagram that illustrates the conditional.

4. Write the converse of the conditional.

For Exercises 5–8, identify the hypothesis and conclusion of each conditional. Write the converse of each conditional. If the converse is false, give a counterexample.

5. If two angles are supplementary, then the sum of their measures is 180°.

6. If your transportation is a motorcycle, then you ride on two wheels.

7. If you live in Baltimore, then you live in Maryland.

8. If you have a rose, then you have a flower.

LESSON 2.3

In Exercises 1–5, use the following steps to determine whether the given sentence is a definition:
   a. Write the sentence as a conditional statement.
   b. Write the converse of the conditional.
   c. Write a biconditional statement.
   d. Decide whether the sentence is a definition, and explain your reasoning.

1. Mercury is the planet closest to our Sun.

2. School is the place where kids learn mathematics.

3. A multiple of ten has a ones digit of zero.

4. The set of points in a plane equidistant from a given point form a circle.

5. An equilateral triangle is a closed figure with three congruent angles.

6. The following are pentagons: The following are not pentagons:

Which of the figures at right are not pentagons?

a.   b.   c.   d.

LESSON 2.4

In Exercises 1 and 2, identify the Algebraic Properties of Equality that justify the steps.

1.
$$x - 7 = 15 \qquad \text{Given}$$
$$x - 7 + 7 = 15 + 7 \qquad \underline{\quad ? \quad}$$
$$x = 22$$

2.
$$4x + 6 = 38 \qquad \text{Given}$$
$$4x + 6 - 6 = 38 - 6 \qquad \underline{\quad ? \quad}$$
$$4x \div 4 = 32 \div 4 \qquad \underline{\quad ? \quad}$$
$$x = 8$$

Refer to the diagram at right, in which m∠ABC = m∠EBD. Use the Overlapping Angles Theorem to complete the following:

3. m∠ABC = 75°   m∠ABD = 110°   m∠CBD = ___?___

4. m∠ABC = (3x + 4)°   m∠CBD = (x − 5)°   m∠CBE = 67°

    x = ___?___   m∠ABD = ___?___   m∠EBD = ___?___

LESSON 2.5

For Exercises 1–4, refer to the diagram at right, which consists of three intersecting lines.

1. Which angle is congruent to ∠FBC?

2. Which angle is congruent to ∠GBC?

3. Which angle is congruent to ∠EBD?

4. Find the measure of ∠ABC.

5. Find the measures of all the angles in the diagram at right.

Tell whether each of the following is an example of inductive or deductive reasoning. Is the argument a proof? Why or why not?

6. Every Saturday for the last six weeks, it has rained.
Tomorrow is Saturday.
Therefore, it will rain tomorrow.

7. Multiples of 5 are not prime.
The number 35 is a multiple of 5.
Therefore, 35 is not prime.

CHAPTER 3

LESSON 3.1

For Exercises 1–3, copy each figure, and draw all axes of symmetry.

1.

2.

3.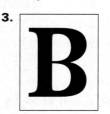

Each figure below shows part of a shape with reflectional symmetry. Copy and complete each figure.

4.

5.

6.

7. Which of the completed figures from Exercises 4–6 also has rotational symmetry?

8. Graph $y = 2(x - 3)^2 + 1$ on a graphing calculator or graph paper. Then write an equation for the axis of symmetry of the graph.

LESSON 3.2

For Exercises 1–8, use your conjectures from Activity 2 in Lesson 3.2 and the diagram below to find the indicated measurements. In rhombus *PQRS*, *PQ* = 5, *PR* = 6, and m∠*PQR* = 74°.

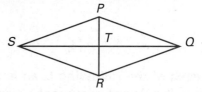

1. *QR* **2.** *RS*

3. *PT* **4.** m∠*PSR*

5. m∠*QPS* **6.** m∠*PTQ*

7. In rectangle *KLMN*, diagonal *KM* = *x* + 4 and diagonal *LN* = 3*x*. Find the value of *x* and the length of the diagonals of *KLMN*.

Use the definitions of quadrilaterals and your conjectures from Activities 1–4 in Lesson 3.2 to decide whether each statement is true or false. If the statement is false, give a counterexample.

8. If a figure is not a rectangle, then it is not a square.

9. If a figure is a rhombus, then it is a square.

10. If a figure is not a parallelogram, then it cannot be a rhombus.

LESSON 3.3

In the figure at right, lines _k_ and _ℓ_ are parallel.

1. List all angles that are congruent to ∠3.

2. List all angles that are congruent to ∠4.

3. Are there any angles in the figure that are not congruent to ∠3 or ∠4? Explain.

4. If m∠3 = 140°, find the measure of each angle in the figure.

5. If m∠2 = 2x° and m∠6 = (3x − 50)°, find the measure of each angle in the figure.

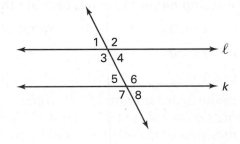

In △ FGH, $\overline{IJ} \| \overline{HG}$ and ∠FIJ ≅ ∠FJI. Find the indicated angle measures.

6. m∠HIG

7. m∠FIJ

8. m∠FJI

9. m∠JGH

10. m∠GIJ

11. m∠FGI

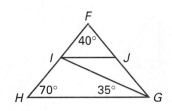

LESSON 3.4

For Exercises 1–4, refer to the diagram below, and fill in the name of the appropriate theorem or postulate.

1. If m∠EDB = m∠DBC, then $\overline{ED} \| \overline{BC}$ by the converse of the _____.

2. If m∠AED = m∠ABC, then $\overline{ED} \| \overline{BC}$ by the converse of the _____.

3. If m∠DEB + m∠EBC = 180°, then $\overline{ED} \| \overline{BC}$ by the converse of the _____.

4. If m∠ADE = m∠ACB, then $\overline{ED} \| \overline{BC}$ by the converse of the _____.

5. If m∠DEC = m∠ECB, then $\overline{ED} \| \overline{BC}$ by the converse of the _____.

6. If m∠EDC + m∠DCB = 180°, then $\overline{ED} \| \overline{BC}$ by the converse of the _____.

7. Given m∠CBE = 43° and m∠BED = 137°, write a two-column proof that $\overline{ED} \| \overline{BC}$.

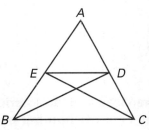

LESSON 3.5

For Exercises 1–3, two angle measurements of a triangle are given. Find the missing angle measure, or state that the triangle does not exist.

1. m∠A = 20°   m∠B = 70°   m∠C = ___?___

2. m∠A = 30°   m∠B = ___?___   m∠C = 160°

3. m∠A = ___?___   m∠B = 66°   m∠C = 66°

4. In the figure at right, $\overline{LJ}\|\overline{HI}$, $\overline{IJ}\|\overline{HL}$, m∠GKJ = 20°, and m∠GIH = 55°. Find the measures of the other 8 angles in the figure.

In Exercises 5–7, find the missing angle measures.

5.

35°

6.
93° 30°

7.
28°
76°

LESSON 3.6

In Exercises 1–3, find the unknown angle measures.

1.
?
120°
110°
80°

2.
160° ?
150°

3.
95°
105°
? 100°

In Exercises 4–6, an interior angle measure of a regular polygon is given. Find the number of sides of the polygon.

4. 108°   6. 140°

5. 156°

In Exercises 7–9, an exterior angle measure of a regular polygon is given. Find the number of sides of the polygon.

7. 72°   8. 45°   9. 30°

10. In quadrilateral FGHI, m∠F = (2x)°, m∠G = x°, m∠H = (x + 16)°, and m∠I = (2x + 20)°. Find the value of x and the measure of each angle in the quadrilateral.

LESSON 3.7

Use the conjectures you made in Activities 1–3 in Lesson 3.7 to find the indicated measures.

1. $DE =$ ___?___

2. $GF =$ ___?___

3. $NO =$ ___?___

A young girl is using a 15-ft ladder to rescue her kitten from the branches of a tree. The ladder's base is 6 ft from the base of the tree.

4. Determine the distance from the ladder to the tree at a point halfway up the ladder.

5. Determine the distance from the ladder to the tree at a point three-quarters of the way up the ladder.

6. The little girl can reach the tree from 1 ft away or less. Estimate the percent of the ladder's height that the little girl must climb to reach the tree.

LESSON 3.8

In Exercises 1–4, the endpoints of a segment are given. Determine the slope and midpoint of the segment.

1. $(1, 1)$ and $(3, 2)$

2. $(-2, 1)$ and $(2, 3)$

3. $(1, -2)$ and $(-2, 5)$

4. $(0, -4)$ and $(3, -1)$

In Exercises 5–8, the endpoints of two segments are given. Determine whether the segments are parallel, perpendicular, or neither.

5. $(4, 1)$ and $(7, 6)$; $(2, 5)$ and $(12, -1)$

6. $(-1, 2)$ and $(2, 4)$; $(-3, 1)$ and $(-1, -2)$

7. $(2, 3)$ and $(6, 2)$; $(0, 5)$ and $(-1, 9)$

8. $(-2, -2)$ and $(5, 2)$; $(4, 3)$ and $(11, 7)$

For Exercises 9–12, the vertices of a triangle are given. Use slopes to determine whether each is a right triangle.

9. $(1, 2), (3, 1), (2, 4)$

10. $(0, 1), (4, 2), (-3, 1)$

11. $(-3, -1), (-1, 1), (-2, -4)$

12. $(-1, 3), (2, 1), (1, 6)$

CHAPTER 4

LESSON 4.1

In Exercises 1–6, determine whether the figures in each pair are congruent. Explain your reasoning.

1.

2.

3.

4.

5.

6.

Suppose that heptagon _LKJMNOP_ ≅ heptagon _AEFXWCB_.

7. Name the segment that is congruent to each segment below.

   a. $\overline{FX}$ **b.** $\overline{NO}$ **c.** $\overline{PL}$ **d.** $\overline{CB}$

8. List all pairs of congruent angles.

LESSON 4.2

In Exercises 1–3, determine whether each pair of triangles can be proven congruent by using the SSS, SAS, or ASA Congruence Postulate. If so, identify which postulate is used.

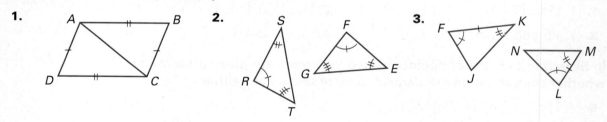

1.

2.

3.

In Exercises 4–9, some measurements of a triangle are given. Is there a unique triangle that can be constructed with the given measurements? If so, identify the postulate that applies.

4. $\triangle PQR$: $PQ = 3$, $PR = 1$, $m\angle P = 30°$ **5.** $\triangle EFG$: $FG = 1$, $EG = 7$, $m\angle E = 12°$

6. $\triangle ABC$: $BC = 4$, $m\angle B = 15°$, $m\angle C = 100°$ **7.** $\triangle RST$: $m\angle R = 20°$, $m\angle S = 100°$, $m\angle T = 60°$

8. $\triangle MNO$: $MN = 2$, $NO = 2$, $MO = 3$ **9.** $\triangle JKL$: $JK = 6$, $m\angle J = 15°$, $m\angle K = 30°$

LESSON 4.3

For each pair of triangles given in Exercises 1–10, is it possible to prove that the triangles are congruent? If so, write a congruence statement and name the postulate or theorem used.

1.   **2.** **3.**

4. △ABC and △PQR with ∠A ≅ ∠P, ∠B ≅ ∠Q, and $\overline{AB} \cong \overline{PQ}$

5. △RST and △JKL with ∠R ≅ ∠L, $\overline{RS} \cong \overline{JL}$, and $\overline{RT} \cong \overline{LK}$

6. △WXY and △ERG with ∠X ≅ ∠R, ∠Y ≅ ∠G, and $\overline{WY} \cong \overline{EG}$

7. △ALS and △BMK with ∠A ≅ ∠B, ∠L ≅ ∠K, and ∠S ≅ ∠M

8. △KIP and △EMU with ∠K ≅ ∠M, $\overline{IP} \cong \overline{EU}$, and $\overline{KP} \cong \overline{UM}$

9. △MNO and △GHI with $\overline{MN} \cong \overline{GH}$, $\overline{MO} \cong \overline{GI}$, and $\overline{NO} \cong \overline{HI}$

10. △RVU and △OLP with ∠R ≅ ∠L, ∠V ≅ ∠P, and $\overline{VU} \cong \overline{OP}$

LESSON 4.4

Given the following information about a triangle, find the missing measure.

1. △MRS: $\overline{MR} \cong \overline{RS}$, m∠S = 40°, m∠M = ___?___

2. △UVW: $\overline{VW} \cong \overline{UV}$, m∠U = 85°, m∠W = ___?___

3. △ABC: ∠A ≅ ∠B, AC = 12, CB = ___?___

4. △DLJ: $\overline{DJ} \cong \overline{JL}$, m∠D = 15°, m∠J = ___?___

5. △KXQ: $\overline{KX} \cong \overline{QK}$, m∠X = 50°, m∠Q = ___?___

6. △RST: ∠R ≅ ∠T, RS = 5, ST = ___?___

7. △DEF: $\overline{DE} \cong \overline{DF}$, m∠E = 45°, m∠D = ___?___

8. △LMN: ∠M ≅ ∠N, LM = 13, LN = ___?___

9. △KUD: $\overline{KU} \cong \overline{UD}$, m∠K = (3x)°, m∠D = 90°, x = ___?___

10. △COL: ∠C ≅ ∠L, CO = 23, OL = ___?___

LESSON 4.5

For Exercises 1–6, use the parallelogram shown at right. Find the indicated measures.

1. Given m∠A = (3x)° and m∠D = (x + 100)°, find m∠D.
2. Given AB = t and CD = (9 − 2t), find CD.
3. Given m∠CDB = 40°, find m∠A.
4. Given m∠B = 50°, m∠C = (2x)°, and BD = (x − 4), find BD.
5. Given m∠ABC = 15° and m∠ACD = 45°, find m∠CBD.
6. Given m∠C = 80° and m∠D = x°, find m∠A.

Determine whether each pair of triangles could fit together to form a parallelogram, and justify your answer.

7.   8.   9.

LESSON 4.6

Exercises 1–4 refer to quadrilateral *WXYZ* with diagonals $\overline{WY}$ and $\overline{XZ}$ intersecting at point *P*. For each set of conditions given, determine whether the quadrilateral is a parallelogram. If so, give the theorem that justifies your answer.

1. $\overline{WP} \cong \overline{PY}, \overline{XP} \cong \overline{PZ}$
2. $\overline{WX} \parallel \overline{ZY}, \overline{WX} \cong \overline{ZY}$
3. $\overline{WX} \cong \overline{ZY}, \overline{WZ} \cong \overline{XY}$
4. $\overline{XY} \parallel \overline{WZ}, \overline{WP} \cong \overline{PX}$

Exercises 5–10 refer to parallelogram *ABCD* with diagonals $\overline{AC}$ and $\overline{BD}$ intersecting at point *P*. For each set of conditions given, determine whether the parallelogram is a rhombus, a rectangle, or neither. Give the theorem that justifies your answer.

5. m∠ABP = m∠CBP, m∠BCP = m∠DCP
6. $\overline{AC} \cong \overline{BD}$
7. $\overline{AD} \cong \overline{AB}$
8. m∠ABC = 90°
9. $\overline{AP} \cong \overline{AD}$
10. m∠DPA = 90°

LESSON 4.7

Construct a figure congruent to each figure below.

1.

2.

3.

Trace each triangle below, and construct the angle bisector of each angle. Using the intersection of the angle bisectors, construct the inscribed circle of each triangle.

4.

5.

6.

Trace each triangle below, and construct the perpendicular bisector of each side. Using the intersection of the perpendicular bisectors, construct the circumscribed circle of each triangle.

7.

8.

9.

LESSON 4.8

Trace each figure below and translate it by the direction and distance of the given translation vector.

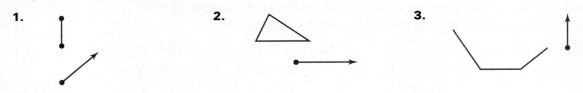

1.

2.

3.

Which of the following triangles are possible? Explain your reasoning.

4. $WX = 1, XY = 2, YW = 3$

5. $MN = 14, NO = 17, OM = 15$

6. $PQ = 3, QR = 4, RP = 5$

7. $AB = 10, BC = 5, CA = 4$

8. $GH = 14, HI = 20, GI = 24$

9. $DE = 12, EF = 5, DF = 28$

CHAPTER 5

LESSON 5.1

In Exercises 1–5, find the area of the rectangle with vertices at the given points. You may find it helpful to sketch a graph.

1. $(3, 0), (0, 0), (3, 4), (0, 4)$

2. $(4, 0), (4, -5), (0, 0), (0, -5)$

3. $(2, 3), (6, 4), (2, 4), (6, 3)$

4. $(-1, 2), (-1, -4), (2, 2), (2, -4)$

5. $(-7, -3), (-2, 0), (-7, 0), (-2, -3)$

6. The perimeter of a rectangle is 42 in. The length is twice the width. What are the dimensions? What is the area?

7. The area of a rectangle is 36 cm². The width is 4 times the length. What are the dimensions? What is the perimeter?

8. The area of a rectangle is $64x^2$. The length is 4 times the width. In terms of x, what are the dimensions? What is the perimeter?

LESSON 5.2

In Exercises 1–3, find the area of each triangle.

1.

2.

3.

In Exercises 4–7, find the area of a parallelogram with vertices at the given points. You may find it helpful to sketch a graph.

4. $(3, 4), (8, 4), (1, 1), (6, 1)$

5. $(-5, -1), (-3, -1), (5, 1), (3, 1)$

6. $(-1, -3), (7, -7), (7, -3), (-1, -7)$

7. $(2, -1), (0, -5), (5, -1), (3, -5)$

In Exercises 8–11, find the area of a trapezoid with vertices at the given points. You may find it helpful to sketch a graph.

8. $(-3, 2), (2, 5), (6, 2), (1, 5)$

9. $(2, 3), (6, -1), (6, 6), (2, 5)$

10. $(4, 3), (2, -1), (8, -1), (6, 3)$

11. $(4, 0), (4, 4), (7, -1), (7, 5)$

LESSON 5.3

In Exercises 1–8, find the circumference and area of each circle whose radius or diameter is given. Use 3.14 for π. Round your answers to the nearest tenth.

1. $r = 3$ **2.** $r = 8$

3. $d = 14$ **4.** $d = 36$

Use $\frac{22}{7}$ for π. Leave your answers in fractional form.

5. $r = 20$ **6.** $d = 32$

7. $r = 12$ **8.** $d = 15$

Find the radius of the circle with the given area or circumference. Give your answers both in terms of π and rounded to the nearest tenth.

9. $A = 20\pi$ **10.** $A = 628$ **11.** $C = 24$

12. $C = 2\pi$ **13.** $A = 117$ **14.** $A = 17.4$

LESSON 5.4

For Exercises 1–5, two lengths of sides of a right triangle are given. Find the missing length. Leave your answers in radical form.

1. $a = 5, b = 12, c = \underline{\ ?\ }$ **2.** $a = 7, b = 9, c = \underline{\ ?\ }$

3. $a = 10, b = \underline{\ ?\ }, c = 25$ **4.** $a = \underline{\ ?\ }, b = 6, c = 10$

Each of the following triples represent the sides of a triangle. Determine whether the triangle is right, acute, or obtuse.

5. 15, 17, 3   **6.** 6, 8, 10   **7.** 11, 10, 13

8. 5, 12, 13   **9.** 18, 28, 40   **10.** 7, 9, 11

Find the area and perimeter of each triangle. Round your answers to the nearest tenth.

11.   **12.**   **13.**

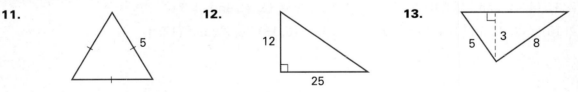

LESSON 5.5

For the given length, find the two remaining lengths. Give your answers in simplest radical form.

1. $a = 7$

2. $b = 3$

3. $c = 4$

4. $b = 6\sqrt{3}$

In Exercises 5–10, find the perimeter and area of each figure. Give your answers in simplest radical form.

5. an equilateral triangle with side lengths of 22

6. a square with a diagonal of 6

7. a 30-60-90 triangle with a hypotenuse of 16

8. a regular hexagon with side lengths of 10

9. a 45-45-90 triangle with leg lengths of 3

10. a regular octagon with side lengths of 6

LESSON 5.6

Find the distance between each pair of points. Round your answers to the nearest hundredth.

1. $(1, 3)$ and $(2, -1)$

2. $(-5, 6)$ and $(3, -2)$

3. $(3, 4)$ and $(0, 0)$

4. $(2, 2)$ and $(5, 11)$

5. $(6, 3)$ and $(2, 1)$

6. $(-5, -7)$ and $(-1, 7)$

7. $(0, 1)$ and $(4, 5)$

8. $(1, -2)$ and $(12, 3)$

In Exercises 9–14, use the converse of the Pythagorean Theorem to determine whether the triangle with the given vertices is a right triangle. You may wish to plot the points and draw each triangle.

9. $(1, 5)$, $(2, -1)$, and $(7, 6)$

10. $(-1, 3)$, $(4, -2)$, and $(-4, 0)$

11. $(2, 3)$, $(4, -2)$, and $(-4, 0)$

12. $(2, 3)$, $(8, 4)$, and $(1, 7)$

13. $(2, 10)$, $(12, 0)$, and $(-4, 4)$

14. $(-16, 18)$, $(4, 2)$, and $(12, 8)$

LESSON 5.7

Determine the coordinates of the unknown vertex or vertices of each figure below. Use variables to represent any coordinates that are not completely determined by the given information.

1. isosceles triangle ABC with $\overline{AC} \cong \overline{BC}$
$A(0, 0)$, $B(0, 2a)$, $C(?, ?)$

2. parallelogram $DEFG$
$D(0, 0)$, $E(?, ?)$, $F(a, b)$, $G(?, ?)$

For Exercises 3–8, refer to right triangle ABC shown at right with vertices at $A(0, 0)$, $B(b, 0)$, and $C(0, c)$. Using the coordinates of A, B, and C, find the following lengths:

3. AB **4.** AC **5.** BC

6. $(AB)^2 + (AC)^2$ **7.** $(BC)^2$

8. Using the results from Exercises 3–7, draw a conclusion about right triangle ABC.

LESSON 5.8

Find the theoretical probability that a dart tossed at random onto each figure will land in the shaded area.

1.   **2.**   **3.**

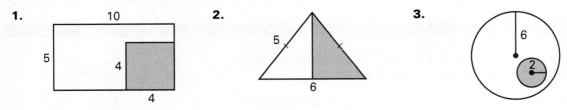

Convert the following probabilities to percents. Round your answers to the nearest tenth of a percent.

4. 0.61 **5.** $\frac{4}{5}$ **6.** $\frac{3}{7}$ **7.** 1

Convert the following percents to decimal probabilities:

8. 40% **9.** 29% **10.** 73% **11.** 9%

12. 12% **13.** 0.1% **14.** 5.8% **15.** 45.3%

Convert the following percents to fractional probabilities. Give your answers in lowest terms.

16. 10% **17.** 25% **18.** 50% **19.** 75%

20. 2% **21.** 91% **22.** 44% **23.** 28%

CHAPTER 6

LESSON 6.1

For Exercises 1–5, refer to the isometric drawing at right.

1. Give the volume in cubic units.

2. Give the surface area in square units.

3. Draw six orthographic projections of the solid.

4. Draw the solid on isometric dot paper from a different viewpoint.

5. Draw the solid on isometric dot paper with a cube added at each shaded face.

The three solids at right each have a volume of 5 cubic units.

6. Find the surface area of each solid in square units. Which has the least surface area?

7. Draw six orthographic projections of each solid.

a.

b.

c.

LESSON 6.2

In the figure at right, $\overline{XY}$, $\overline{ZY}$, and $\overline{AB}$ are perpendicular to ℓ, and $\overline{BC}$ is not perpendicular to ℓ.

1. Is m∠ABC the measure of the dihedral angle formed by planes M and N?

2. Is m∠XYZ the measure of the dihedral angle formed by planes M and N?

In the figure at right, line m is perpendicular to plane $\mathcal{P}$.

3. Is line m perpendicular to line n? Explain.

4. What is the relationship between line m and line r?

5. Indicate whether the statement below is true or false for a figure in space. Explain your answer by using sketches.

If two lines never intersect, then they are parallel.

LESSON 6.3

Which of the figures below appear to be prisms? Give the name for each prism. If the figure is not a prism, explain why.

1.

2.

3.

Use the oblique triangular prism at right for Exercises 4–6.

4. Which face is congruent to triangle *ABC*? Explain.

5. Name all segments congruent to $\overline{AE}$.

6. What type of quadrilateral is *ACDE*?

For Exercises 7–10, find the length of a diagonal of a right rectangular prism with the given dimensions.

7. $\ell = 10, w = 10, h = 5$

8. $\ell = 3, w = 4, h = 5$

9. $\ell = 5, w = 6, h = 8$

10. $\ell = 8, w = 24, h = 6$

LESSON 6.4

Name the octant, coordinate plane, or axis where each point is located.

1. $(2, 1, 3)$

2. $(-14, 0, 0)$

3. $(-1, 1, 8)$

4. $(53, 1.2, 0)$

5. $(-8, -1, -3)$

6. $(8, 0, -4)$

Find the distance between each pair of points.

7. $(1, 0, 2)$ and $(4, 1, -2)$

8. $(2, 4, 0)$ and $(4, 3, 2)$

9. $(-1, -2, -3)$ and $(1, 4, -1)$

10. $(4, 1, 6)$ and $(10, -4, 14)$

LESSON 6.5

Use intercepts to sketch a graph of the plane represented by each equation.

1. $3x - 6y - 6z = -12$

2. $2x - 2y - 2z = 8$

3. $-4x - 2y + 8z = 16$

4. $x - 3y = 3$

In a coordinate plane, plot the lines represented by each pair of parametric equations.

5. $x = t + 1$
   $y = t$

6. $x = -2$
   $y = t + 3$

Recall that a *trace* of a plane is its intersection with the *xy*-plane. Find the equation of the trace for each equation of a plane below.

7. $7x + 2y - z = 1$

8. $2x + 4y - 18z = 2$

LESSON 6.6

In Exercises 1–6, locate the vanishing point and horizon line for each figure.

1.

2.

3.

4.

5.

6.

7. Make a one-point perspective drawing of a triangular prism.

CHAPTER 7

LESSON 7.1

Determine the surface-area-to-volume ratio for a rectangular prism with the following dimensions:

1. $7 \times 1 \times 1$

2. $3 \times 2 \times 5$

3. $6 \times 6 \times 6$

4. $12 \times 12 \times 12$

5. $57 \times 2 \times 18$

6. $31 \times 94 \times 95$

Find the surface-area-to-volume ratio for each of the following:

7. a cube with a volume of 125 cubic units

8. a cube with a surface area of 96 square units

9. a rectangular prism with dimensions $1 \times n \times 2n$

For each situation, determine whether you would want to maximize the volume or minimize the surface area.

10. designing a bowl out of a limited amount of clay

11. building a carton with a fixed amount of cardboard

LESSON 7.2

Find the volume of a prism with the given dimensions.

1. $B = 6 \text{ in.}^2, h = 4 \text{ in.}$

2. $B = 14 \text{ cm}^2, h = 3 \text{ cm}$

3. $B = 10 \text{ m}^2, h = 13 \text{ m}$

4. $B = 42 \text{ ft}^2, h = 10 \text{ ft}$

Find the surface area and volume of a right rectangular prism with the given dimensions.

5. $\ell = 4, w = 12, h = 5$

6. $\ell = 13, w = 16, h = 3.1$

7. $\ell = 42, w = 30, h = 1$

8. $\ell = 0.05, w = 1.1, h = 2$

9. $\ell = 21, w = 21, h = 21$

10. $\ell = 4.3, w = 3.7, h = 6.8$

11. Find the height of a rectangular prism with a surface area of 2880 in.^{2} and a base measuring 5 in. by 8 in.

12. Find the width of a rectangular prism with a volume of 144 cm^3 and a base measuring 3 cm by 6 cm.

LESSON 7.3

Find the surface area of each regular pyramid with side length *s* and slant height ℓ given below. The number of sides of the base is given by *n*.

1. $n = 6, s = 4, \ell = 4$

2. $n = 4, s = 2, \ell = 3$

3. $n = 3, s = 5, \ell = 10$

4. $n = 6, s = 14, \ell = 7$

Find the volume of a rectangular pyramid with height *h* and base dimensions ℓ and *w*.

5. $h = 4, \ell = 7.2, w = 6.8$

6. $h = 12, \ell = 18, w = 10$

7. $h = 10, \ell = 24, w = 20$

8. $h = 17, \ell = 13, w = 5$

9. $h = 42, \ell = 15, w = 20$

10. $h = 23, \ell = 19, w = 12$

Find the height of each pyramid described below.

11. a regular triangular pyramid with a base area of 16 square units and a volume of 48 cubic units

12. a regular square pyramid with a base length of 24 units and a volume of 14,400 cubic units

LESSON 7.4

Find the unknown value for a right cylinder with radius *r*, height *h*, and surface area *S* given below. Round your answers to the nearest tenth.

1. $r = 6, h = 3, S =$ ___?___

2. $r = 14, h = 7, S =$ ___?___

3. $r = 0.5, h = 1.2, S =$ ___?___

4. $r =$ ___?___$, h = 3, S = 18$

5. $r = 2, h =$ ___?___$, S = 72$

6. $r =$ ___?___$, h = 3, S = 320$

Find the unknown value for a right cylinder with radius *r*, height *h*, and volume *V*. Round your answers to the nearest tenth.

7. $r = 4.5, h = 3.2, V =$ ___?___

8. $r =$ ___?___$, h = 12, V = 150$

9. $r = 15, h = 25, V =$ ___?___

10. $r = 4, h =$ ___?___$, V = 24\pi$

11. $r =$ ___?___$, h = 20, V = 180\pi$

12. $r = 10, h = 8, V =$ ___?___

13. A cylinder with a diameter of 4 in. and a height of 8 in. is replaced by a cylinder that has the same volume. The new cylinder has a diameter of 6 in. What is its height?

LESSON 7.5

Find the surface area and volume of each right cone. Express your answers in terms of π.

1.

5.6
5.1
2.4

2.

11.7
10
6

3.

20.2
17
11

4.

23
19
13

5.

17.9
16
8

6.

56.6
38
42

7.

7.3
5.9
4.3

8.

6.8
5.8
3.5

9.

8.5
6.4
5.6

10. A right cone has a radius of 17 in. and a height of 12 in. What is the slant height of the cone?

LESSON 7.6

Find the surface area and volume of each sphere, with the given radius or diameter. Round your answers to the nearest hundredth.

1. $r = 5$

2. $d = 1.4$

3. $d = 66$

4. $r = 2.8$

5. $r = 2.9$

6. $d = 24$

7. $d = 4.02$

8. $r = 62$

9. $d = 16$

In Exercises 10–15, find the surface area and volume of each sphere with the given radius or diameter. Express your answers exactly in terms of π and a variable.

10. $r = 3x$

11. $d = 7y$

12. $d = x$

13. $r = 1.1y$

14. $r = 2y$

15. $d = \frac{x}{2}$

What are the coordinates of the reflection image when each point below is reflected across the _xy_-plane?

1. (1, 1, 2)　　　　　　**2.** (3.1, −2.5, −7)　　　　　**3.** (2, 1, −3)

4. (4, −4, 1.2)　　　　　**5.** (−2, 15, 0)　　　　　　**6.** (−12, −3, −8)

What are the coordinates of the reflection image when each point below is reflected across the _xz_-plane?

7. (32, 32, 32)　　　　　**8.** (0, −3, −6)　　　　　　**9.** (14, −4, 6.2)

10. (1.3, 5.2, −4)　　　　**11.** (−12.1, −3, 6)　　　　**12.** (−53, −64, −0.2)

What are the coordinates of the reflection image when each point below is reflected across the _yz_-plane?

13. (15, 15, 15)　　　　　**14.** (−2, 3, 3)　　　　　　**15.** (−7.8, −7.8, −10)

16. (3, −6, 8.9)　　　　　**17.** (2, −2, −2)　　　　　　**18.** $\left(\frac{1}{2}, -4, 3\right)$

CHAPTER 8

Find the image of the point transformed by the given dilation. Plot the point and its image on a coordinate plane.

1. (3, −1); $D(x, y) = (2x, 2y)$　　　　**2.** (4, 1); $D(x, y) = (−4x, −4y)$

3. (−2, −6); $D(x, y) = (0.5x, 0.5y)$　　**4.** (−5, 3); $D(x, y) = (−x, −y)$

For Exercises 5 and 6, the dashed figures represent preimages of dilations and the solid figures represent images. Find the scale factor of each dilation.

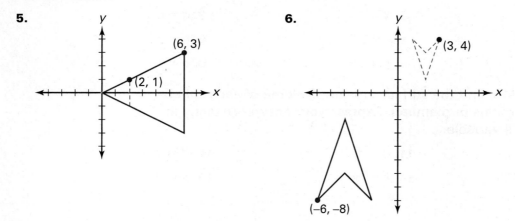

5.　(6, 3)　(2, 1)

6.　(3, 4)　(−6, −8)

For Exercises 7–10, a point and a scale factor _n_ are given. Find the line that passes through the preimage and image, and show that the origin is on this line.

7. (−1, −3); $n = \frac{3}{4}$　　**8.** (4, 1); $n = -\frac{1}{2}$　　**9.** (1, −5); $n = 2$　　**10.** (2, 3); $n = −1$

LESSON 8.2

In Exercises 1 and 2, determine whether the polygons are similar. Explain your reasoning. If the polygons are similar, write a similarity statement.

1.

2.

In Exercises 3–6, solve for *x*.

3. $\dfrac{2}{x} = \dfrac{6}{9}$

4. $\dfrac{2}{x+2} = \dfrac{1}{x-1}$

5. $\dfrac{\frac{1}{3}}{x} = \dfrac{\frac{1}{4}}{9}$

6. $\dfrac{3.5}{x} = \dfrac{5.2}{15.6}$

LESSON 8.3

Determine whether each pair of triangles can be proven similar by using the AA, SSS, or SAS Similarity Postulates and Theorems. If so, write a similarity statement and identify the postulate or theorem used.

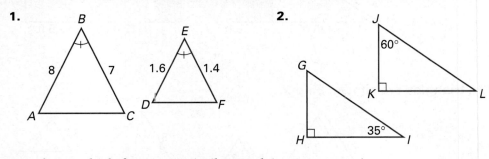

1.

2.

3. Indicate which figures are similar. Explain your reasoning.

LESSON 8.4

Use the Side-Splitting Theorem to find *x*. Some exercises may have more than one possible answer for *x*.

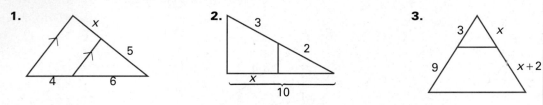

1.   **2.**   **3.**

Name all of the similar triangles in each figure. State the postulates or theorems that justify each similarity.

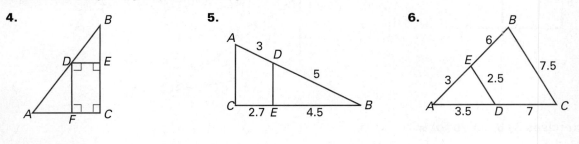

4.   **5.**   **6.**

LESSON 8.5

In Exercises 1–3, use the diagrams to find the height of each rectangle.

1.   **2.**   **3.**

In Exercises 4–6, the triangles are similar. Find *x*.

4.   **5.**   **6.**

LESSON 8.6

In the figure at right, $\overline{DE} \parallel \overline{AB}$.

1. Find the ratio of the perimeters of $\triangle ABC$ and $\triangle DEC$.

2. Find the ratio of the areas of $\triangle ABC$ and $\triangle DEC$.

The ratios of the sides of two similar hexagons is $\frac{2}{7}$. Find the ratio of

3. their perimeters.

4. their areas.

The ratios of the sides of two similar triangular prisms is $\frac{3}{4}$. Find the ratio of

5. the perimeters of their triangular faces.

6. the areas of their triangular faces.

7. their volumes.

CHAPTER 9

LESSON 9.1

Determine the length of an arc with the given central angle measure, $m\angle P$, in a circle with radius r. Round your answers to the nearest hundredth.

1. $m\angle P = 40°$; $r = 6$

2. $m\angle P = 20°$; $r = 8$

3. $m\angle P = 75°$; $r = 20$

4. $m\angle P = 100°$; $r = 16$

5. $m\angle P = 55°$; $r = 13$

6. $m\angle P = 118°$; $r = 30$

7. $m\angle P = 66°$; $r = 40$

8. $m\angle P = 130°$; $r = 61$

9. $m\angle P = 80°$; $r = 39$

10. $m\angle P = 82°$; $r = 5$

Determine the degree measure of an arc with the given length, L, in a circle with radius r.

11. $L = 52$; $r = 14$

12. $L = 27$; $r = 5$

13. $L = 35$; $r = 11$

14. $L = 56$; $r = 30$

15. $L = 8$; $r = 2$

16. $L = 2.3$; $r = 85$

17. $L = 25$; $r = 25$

18. $L = 100$; $r = 79$

LESSON 9.2

For Exercises 1–6, refer to circle _J_, in which $\overline{JP} \perp \overline{KL}$ at _S_.

1. $\overline{SL} \cong$ _____?_____

2. If $JL = 4$ and $JS = 1$, what is KS? What is SL?

3. If $JL = 10$ and $JS = 3$, what is KS? What is SL?

4. If $JK = 26$ and $JS = 11$, what is KS? What is SL?

5. If $JK = 60$ and $JS = 12$, what is KS? What is SL?

6. If $JP = 42$ and $JS = 30$, what is KS? What is SL?

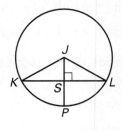

LESSON 9.3

In circle _P_, m∠_LPJ_ = 30° and m∠_KMJ_ = 45°. Find the following:

1. m∠_LMP_

2. m∠_JPK_

3. m∠_MJK_

4. m∠_LPM_

5. m$\widehat{MLJ}$

6. m∠_MPK_

7. m∠_JKP_

8. m$\widehat{LJ}$

9. m$\widehat{KM}$

10. m∠_PLM_

11. m$\widehat{JK}$

12. m∠_KJP_

13. m$\widehat{ML}$

14. m∠_PKM_

LESSON 9.4

In circle *E*, m$\widehat{BD}$ = 20°, m$\widehat{DF}$ = 180°, and m$\widehat{AF}$ = 45°. $\overleftrightarrow{GA}$ is tangent to circle *E* at *A*. Find each of the following:

1. m∠*CAG*

2. m∠*GCA*

3. m∠*CGA*

In the figure, $\overleftrightarrow{XY}$ is tangent to circle *Z* at *X*.

4. If m$\widehat{XW}$ = 95°, find m∠*YXW*.

5. If m∠*YXW* = 100°, find m$\widehat{XW}$.

6. If m$\widehat{XW}$ = *x* + 15, find m∠*YXW*.

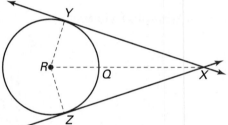

LESSON 9.5

$\overleftrightarrow{XY}$ and $\overleftrightarrow{XZ}$ are tangent to circle *R*, *XY* = 4, and the radius of circle *R* is 2. Find the following:

1. *XZ*

2. *YR*

3. *QX*

4. *RX*

5. Name an angle congruent to ∠*YXR*.

6. Name an angle congruent to ∠*YRX*.

7. Name an arc congruent to $\widehat{YQ}$.

Find the value of *x* in each figure.

8.

9.

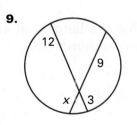

LESSON 9.6

Find the *x*- and *y*-intercepts for the graph of each equation below.

1. $x^2 + y^2 = 49$

2. $x^2 + y^2 = 12$

3. $(x-1)^2 + y^2 = 16$

4. $x^2 + (y+2)^2 = 36$

5. $(x-3)^2 + (y+4)^2 = 81$

6. $(x-7)^2 + (y+1)^2 = 121$

Find the center and radius of each circle.

7. $x^2 + y^2 = 4$

8. $x^2 + y^2 = 64$

9. $(x+1)^2 + y^2 = 14$

10. $x^2 + (y-1)^2 = 27$

11. $(x+3)^2 + (y-5)^2 = 31$

12. $(x-7)^2 + (y+1)^2 = 12$

Write an equation for the circle with the given center and radius.

13. center: $(1, 2)$; radius $= 4$

14. center: $(-1, -6)$; radius $= 12$

CHAPTER 10

LESSON 10.1

Find tan *A* for each triangle below.

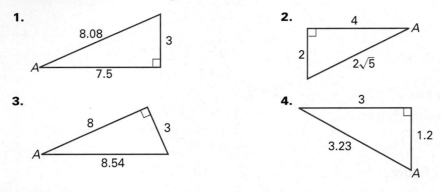

1.

8.08   3   7.5

2.

4   A   2 $2\sqrt{5}$

3.

8   3   8.54

4.

3   1.2   3.23   A

Use a scientific calculator to find the tangent of each angle below. Round your answers to the nearest hundredth.

5. $15°$

6. $23°$

7. $47°$

8. $69°$

9. $42°$

10. $54°$

LESSON 10.2

For Exercises 1–6, refer to $\triangle XYZ$. Find each of the following:

1. $\sin \underline{\quad ? \quad} = \dfrac{2}{\sqrt{13}}$

2. $\cos \underline{\quad ? \quad} = \dfrac{2}{\sqrt{13}}$

3. $\cos \underline{\quad ? \quad} = \dfrac{3}{\sqrt{13}}$

4. $\tan \underline{\quad ? \quad} = \dfrac{3}{2}$

5. $\tan \underline{\quad ? \quad} = \dfrac{2}{3}$

6. $\sin \underline{\quad ? \quad} = \dfrac{3}{\sqrt{13}}$

Use a scientific calculator to find the following. Round your answers to the nearest hundredth.

7. $\sin 15°$

8. $\cos 23°$

9. $\tan 17°$

10. $\cos 69°$

11. $\sin 42°$

12. $\tan 47°$

LESSON 10.3

Use a scientific calculator to find the following. Round your answers to four decimal places.

1. $\sin 55°$

2. $\sin 169°$

3. $\sin 340°$

4. $\cos 55°$

5. $\cos 169°$

6. $\cos 340°$

In Exercises 7–9, use a scientific calculator to find the sine and cosine of each angle. Use these values to give the x- and y-coordinates of a point at the given angle on the unit circle. Round your answers to four decimal places.

7. $20°$

8. $160°$

9. $200°$

In Exercises 10–15, give two values for angle θ between 0° and 180° for the given value of sin θ. Round your answers to the nearest degree.

10. 0.2250

11. 0.6157

12. 0.8746

13. 0.3907

14. 0.3090

15. 0.5150

LESSON 10.4

For Exercises 1–4, find the indicated measures. Assume that all angles are acute. It may be helpful to sketch the triangles roughly to scale.

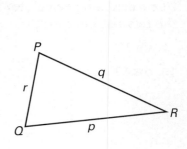

1. $m\angle A = 52°$ $m\angle B = 68°$ $b = 4.2$ cm $a =$ ___?___
2. $m\angle A = 72°$ $m\angle C = 32°$ $a = 1.4$ cm $c =$ ___?___
3. $m\angle B = 64°$ $a = 2.34$ cm $b = 3.5$ cm $m\angle A =$ ___?___
4. $m\angle A = 25°$ $m\angle C = 65°$ $c = 5$ cm $a =$ ___?___

Find all unknown sides and angles in each triangle. If the triangle is ambiguous, give both possible angles. It may be helpful to sketch the triangles roughly to scale.

5. $m\angle P = 36°$ $m\angle Q = 68°$ $q = 7$
6. $m\angle R = 75°$ $p = 10$ $r = 14$
7. $m\angle Q = 40°$ $m\angle P = 25°$ $q = 12$

LESSON 10.5

For Exercises 1–4, find the indicated measures. It may be helpful to sketch the triangles roughly to scale. Round to scale. Round your answers to the nearest tenth.

1. $a = 1$ $b = 13$ $c =$ ___?___ $m\angle C = 20°$
2. $a = 12$ $b = 5$ $c =$ ___?___ $m\angle C = 68°$
3. $a = 4$ $b = 7$ $c = 5$ $m\angle C =$ ___?___
4. $a = 9$ $b = 3$ $c = 8$ $m\angle C =$ ___?___

Solve each triangle.

5.

6.

LESSON 10.6

Copy each pair of vectors below and draw the vector sum, $\vec{a} + \vec{b}$, by using the head-to-tail method. You may need to translate one of the vectors.

1.

2.

Copy each pair of vectors below and draw the vector sum, $\vec{a} + \vec{b}$, by using the parallelogram method. You may need to translate one of the vectors.

3.

4.
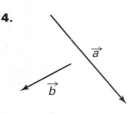

LESSON 10.7

For Exercises 1–8, a point, *P,* and an angle of rotation, θ, are given. Determine the coordinates of the image, *P′*. Round your answers to the nearest tenth.

1. $P(0, 5); \theta = 48°$

2. $P(4, 2); \theta = 270°$

3. $P(1, 4); \theta = 17°$

4. $P(-2, -5); \theta = 65°$

5. $P(2, -2); \theta = 30°$

6. $P(-1, 2); \theta = 180°$

7. $P(3, 0); \theta = 90°$

8. $P(-3, -2); \theta = 200°$

Find the rotation matrix for each angle below by filling in the sine and the cosine values. Round your answers to the nearest hundredth.

9. $135°$

10. $215°$

11. $35°$

12. $345°$

13. $40°$

14. $180°$

15. $300°$

16. $120°$

LESSON 11.1

Determine the indicated side length of each golden rectangle. Round your answers to the nearest hundredth.

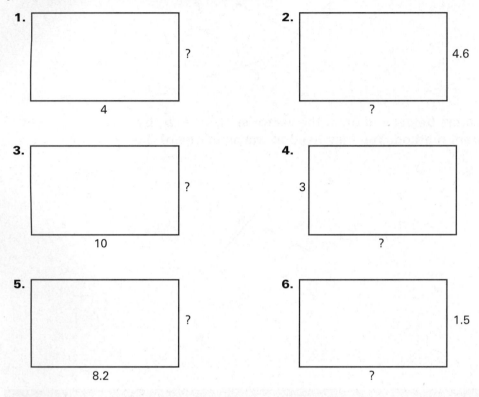

1. ? 4

2. 4.6 ?

3. ? 10

4. 3 ?

5. ? 8.2

6. 1.5 ?

LESSON 11.2

Find the taxidistance between each pair of points.

1. (2, 4) and (1, 6)

2. (–5, 8) and (0, –2)

3. (–9, –5) and (–4, –7)

4. (–10, 10) and (10, –10)

Find the number of points on the taxicab circle with the given radius.

5. $r = 3$

6. $r = 7$

7. $r = 10$

8. $r = 8$

9. $r = 5$

10. $r = 20$

Find the circumference of the taxicab circle with the given radius.

11. $r = 3$

12. $r = 6$

13. $r = 8$

14. $r = 9$

15. $r = 11$

16. $r = 20$

LESSON 11.3

Determine whether the graphs below contain an Euler path, an Euler circuit, or neither.

1.

2.

3.

4.

LESSON 11.4

1. Which, if any, of the following figures are topologically equivalent?

a.
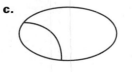

b.

c.

Verify Euler's formula for each polyhedron below.

2. tetrahedron

3. dodecahedron

LESSON 11.5

Refer to the diagram of a sphere at right. Determine whether the following figures are lines in spherical geometry:

1. $\overleftrightarrow{XY}$

2. $\overleftrightarrow{NL}$

3. $\overleftrightarrow{MY}$

4. $\overleftrightarrow{LM}$

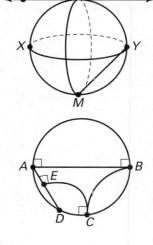

Refer to the diagram at right. Determine whether the following figures are lines in hyperbolic geometry:

5. $\overleftrightarrow{AE}$

6. $\overleftrightarrow{AB}$

7. $\overleftrightarrow{EC}$

8. $\overleftrightarrow{BC}$

LESSON 11.6

Using the technique shown in the construction of the Sierpinski gasket, construct a Sierpinski gasket with the following triangles:

1.

2.

3. Build Pascal's triangle with at least 12 rows. Shade around each number that is divisible by 2, and leave the other numbers unshaded. How is the pattern different from the pattern you found in Activity 2 of Lesson 11.6? How is it the same?

LESSON 11.7

Draw a circle with a radius of 3 centered at the origin on a coordinate plane. Use the grids of parallelograms below to transform the circle.

1.   2.

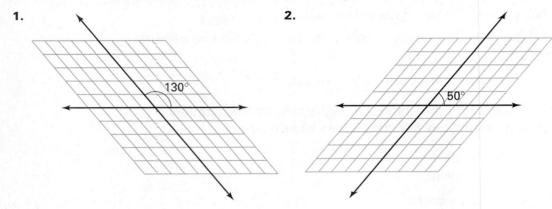

Copy the diagram at right and draw each projection.

3. Using O as the center of projection, project points X, Y, and Z onto line ℓ_2. Label the projected points X', Y', and Z'.

4. Using P as the center of projection, project points X', Y', and Z' onto line ℓ_3. Label the projected points X'', Y'', and Z''.

CHAPTER 12

LESSON 12.1

In Exercises 1–4, write a valid conclusion from the given premises. Identify the form of the argument.

1. If the baby is hungry, then he cries. The baby is hungry.

2. If the baby is hungry, then he cries. The baby is not crying.

3. If the baby is not hungry, then he throws his cereal on the floor. The baby did not throw his cereal on the floor.

4. If the baby is not hungry, then he will play in his playpen. The baby is not hungry.

For Exercises 5–7, refer to the following premises:

> **If Harry is a purple hippopotamus, then Larry is a blue dinosaur.**
> **If Larry is not a blue dinosaur, then Barry is a pink alligator.**
> **Larry is not a blue dinosaur.**

State whether each conclusion below is valid. Explain.

5. Harry is not a purple hippopotamus.

6. Harry is a purple hippopotamus.

7. Barry is a pink alligator.

LESSON 12.2

1. Write a conjunction for the pair of statements below. Determine whether the conjunction is true or false.

   Some flowers are red. Cucumbers are vegetables.

2. Write a disjunction for the pair of statements below. Determine whether the disjunction is true or false.

   Parallel lines intersect. All rectangles are squares.

For Exercises 3–8, write the statement expressed by the symbols, where *p*, *q*, and *r* represent the statements shown below.

p: Harry is older than Larry.

q: Larry and Barry are twins.

r: Harry is younger than Barry.

3. $\sim p$

4. $\sim q$ OR $\sim r$

5. $\sim q$

6. p AND $\sim q$

7. $\sim (p$ AND $r)$

8. $\sim (p$ OR $\sim r)$

LESSON 12.3

For each conditional in Exercises 1–3, write the converse, inverse, and contrapositive. Decide whether each is true or false and explain your reasoning.

1. If Harry and Larry are twins, then they are the same age.

2. If two triangles are congruent, then they are similar.

3. If $m \times n = 0$, then $m = 0$.

For Exercises 4–6, write each statement in if-then form.

4. All blue pigs can play the banjo.

5. No odd numbers are divisible by 2.

6. I will buy you a camera if they go on sale.

Extra Practice

LESSON 12.4

In Exercises 1–4, determine whether the given argument is an example of an indirect reasoning. Explain why or why not.

1. If Bobby ate some cookies from the cookie jar, then the cookie jar is not full. The cookie jar is not full. Therefore, Bobby ate some cookies from the cookie jar.

2. If Juan eats strawberries, then he breaks out in hives. Juan did not break out in hives. Therefore, he did not eat strawberries.

3. I was in an accident. If I was in an accident, then I must have been driving my car. Therefore, I was driving my car.

4. If the alarm rings, then the dogs bark. The dogs did not bark. Therefore, the alarm did not ring.

Complete the indirect proof below.

Given: $\triangle ABC$, $m\angle A < 30°$, and $m\angle B < 45°$

Prove: $\angle C$ is obtuse.

Proof: Suppose that $\angle C$ is **5.** ___?___ . Then $m\angle C \leq$ **6.** ___?___ .
Thus $m\angle A + m\angle B + m\angle C \leq$ **7.** ___?___ + **8.** ___?___ + **9.** ___?___ = **10.** ___?___ .
This contradicts the property of triangles that **11.** ___?___ . Therefore, **12.** ___?___ .

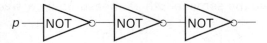

LESSON 12.5

Use the logic gates below to answer each question.

1. If $p = 0$, what is the output?

2. If $p = 1$ and $q = 1$, what is the output?

For Exercises 3–10, complete the input-output table for the network of logic gates at right.

	p	q	r	~p	~p AND q	(~p AND q) OR r
3.	1	1	1			
4.	1	1	0			
5.	1	0	1			
6.	1	0	0			
7.	0	1	1			
8.	0	1	0			
9.	0	0	1			
10.	0	0	0			

Postulates, Theorems, and Definitions

Def 1.1.1 **Segment** A **segment** is a part of a line that begins at one point and ends at another. The points are called the **endpoints** of the segment.

Def 1.1.2 **Ray** A **ray** is a part of a line that starts at a point and extends infinitely in one direction. The point is called the **endpoint** of the ray.

Def 1.1.3 **Angle** An **angle** is a figure formed by two rays with a common endpoint. The common endpoint is called the **vertex of the angle**, and the rays are the **sides of the angle**. An angle divides a plane into two regions: the **interior** and the **exterior** of the angle. If two points, one from each side of an angle, are connected by a segment, the segment passes through the interior of the angle.

Post 1.1.4 **Postulate** The intersection of two lines is a point.

Post 1.1.5 **Postulate** The intersection of two planes is a line.

Post 1.1.6 **Postulate** Through any two points there is exactly one line.

Post 1.1.7 **Postulate** Through any three noncollinear points there is exactly one plane.

Post 1.1.8 **Postulate** If two points are in a plane, then the line containing them is in the plane.

Def 1.2.1 **Length of $\overline{AB}$** Let A and B be points on a number line, with coordinates a and b. Then the measure of $\overline{AB}$, which is called its **length**, is $|a - b|$ or $|b - a|$.

$\text{m}\overline{AB}$, or $AB =$ $|a - b|$ or $|b - a|$

Post 1.2.2 **Segment Congruence Postulate** If two segments have the same length as measured by a given fair ruler, then the segments are congruent. Also, if two segments are congruent, then they have the same length as measured by a given ruler.

Post 1.2.3 **Segment Addition Postulate** If point R is between points P and Q on a line, then $PR + RQ = PQ$.

Def 1.3.1 **Measure of an Angle** Suppose that the vertex, V, of $\angle AVB$ is placed on the center point of a half-circle with coordinates from $0°$ to $180°$ so that $\overrightarrow{VA}$ and $\overrightarrow{VB}$ intersect the half-circle. Let a and b be the coordinates of the intersections. Then the **measure of the angle**, written as $\text{m}\angle AVB$, is $|a - b|$ or $|b - a|$.

Post 1.3.2 **Angle Addition Postulate** If point S is in the interior of $\angle PQR$, then $\text{m}\angle PQS + \text{m}\angle SQR = \text{m}\angle PQR$.

Post 1.3.3 **Angle Congruence Postulate** If two angles have the measure, then they are congruent. If two angles are congruent, then they have the same measure.

Def 1.3.4 **Special Angle Pairs**
Complementary angles are two angles whose measures have a sum of $90°$. Each angle is called the **complement** of the other.
Supplementary angles are two angles whose measures have a sum of $180°$. Each angle is called the **supplement** of the other.

Thm 1.3.5 **Linear Pair Property** If two angles form a linear pair, then they are supplementary.

Def 1.3.6 **Three Types of Angles**
A **right angle** is an angle whose measure is 90°.
An **acute angle** is an angle whose measure is less than 90°.
An **obtuse angle** is an angle whose measure is greater than 90°.

Def 1.4.1 **Perpendicular and Parallel Lines** **Perpendicular lines** are two lines that intersect to form a right angle. **Parallel lines** are two coplanar lines that do not intersect.

Def 1.4.2 **Bisectors and Midpoint** A **segment bisector** is a line that divides a segment into two congruent parts. The point where a bisector intersects a segment is the **midpoint** of the segment. A bisector that is perpendicular to a segment is called a **perpendicular bisector**. An **angle bisector** is a line or ray that divides an angle into two congruent angles.

Def 1.6.1 **Translation** A **translation** is a transformation in which every point of the preimage is moved the same distance in the same direction.

Def 1.6.2 **Rotation** A **rotation** is a transformation in which every point of the preimage is moved by the same angle through a circle centered at a given fixed point known as the *center of rotation*.

Def 1.6.3 **Reflection** A **reflection** is a transformation in which every point of the preimage is moved across a line known as the *mirror line* so that the mirror is the perpendicular bisector of the segment connecting the point and its image.

Def 1.7.1 **Horizontal and Vertical Coordinate Translations**
Horizontal translation of h units: $H(x, y) = (x + h, y)$
Vertical translation of v units: $V(x, y) = (x, y + v)$

Def 1.7.2 **Reflection Across the x- or y-axis**
Reflection across the x-axis: $M(x, y) = (x, -y)$
Reflection across the y-axis: $N(x, y) = (-x, y)$

Def 1.7.3 **180° Rotation About the Origin** $R(x, y) = (-x, -y)$

Post 2.2.1 **If-Then Transitive Property** Given: "If A, then B, and if B, then C." You can conclude: "If A, then C."

Def 2.3.1 **Adjacent Angles** Adjacent angles are angles in a plane that have their vertex and one side in common but have no interior points in common.

Post 2.4.1 **Addition Property** If $a = b$, then $a + c = b + c$.

Post 2.4.2 **Subtraction Property** If $a = b$, then $a - c = b - c$.

Post 2.4.3 **Multiplication Property** If $a = b$, then $ac = bc$.

Post 2.4.4 **Division Property** If $a = b$ and $c \neq 0$, then $\frac{a}{c} = \frac{b}{c}$.

Post 2.4.5 **Substitution Property** If $a = b$, you may replace a with b in any true equation containing a and the resulting equation will still be true.

Thm 2.4.6 **Overlapping Segments Theorem** Given a segment with points A, B, C, and D (in order), the following statements are true:
1. If $AB = CD$, then $AC = BD$.
2. If $AC = BD$, then $AB = CD$.

Post 2.4.7 **Reflexive Property of Equality** For any real number a, $a = a$.

Post 2.4.8 **Symmetric Property of Equality** For all real numbers a and b, if $a = b$, then $b = a$.

Post 2.4.9 **Transitive Property of Equality** For all real numbers a, b, and c, if $a = b$ and $b = c$, then $a = c$.

Post 2.4.10 **Reflexive Property of Congruence** figure $A \cong$ figure A

Post 2.4.11 **Symmetric Property of Congruence** If figure $A \cong$ figure B, then figure $B \cong$ figure A.

Post 2.4.12 **Transitive Property of Congruence** If figure $A \cong$ figure B and figure $B \cong$ figure C, then figure $A \cong$ figure C.

Thm 2.4.13 **Overlapping Angles Theorem** Given $\angle AOD$ with points B and C in its interior, the following statements are true:
1. If m$\angle AOB =$ m$\angle COD$, then m$\angle AOC =$ m$\angle BOD$.
2. If m$\angle AOC =$ m$\angle BOD$, then m$\angle AOB =$ m$\angle COD$.

Thm 2.5.1 **Vertical Angles Theorem** If two angles form a pair of vertical angles, then they are congruent.

Thm 2.5.2 **Theorem** Reflection across two parallel lines is equivalent to a translation of twice the distance between the lines and in a direction perpendicular to the lines.

Thm 2.5.3 **Theorem** Reflection across two intersecting lines is equivalent to a rotation about the point of intersection through twice the measure of the angle between the lines.

Def 3.1.1 **Polygon** A polygon is a plane figure formed from three or more segments such that each segment intersects exactly two other segments, one at each endpoint, and no two segments with a common endpoint are collinear. The segments are called the **sides of the polygon**, and the common endpoints are called the **vertices of the polygon**.

Def 3.1.2 **Reflectional Symmetry** A figure has **reflectional symmetry** if and only if its reflected image across a line coincides exactly with the preimage. The line is called an **axis of symmetry**.

Def 3.1.3 **Triangles Classified by Number of Congruent Sides**
Three congruent sides: equilateral
At least two congruent sides: isosceles
No congruent sides: scalene

Def 3.1.4 **Rotational Symmetry** A figure has **rotational symmetry** if and only if it has at least one rotation image, not counting rotation images of 0° or multiples of 360°, that coincides with the original image.

Def 3.3.1 **Transversal** A transversal is a line, ray, or segment that intersects two or more coplanar lines, rays, or segments, each at a different point.

Post 3.3.2 **Corresponding Angles Postulate** If two lines cut by a transversal are parallel, then corresponding angles are congruent.

Thm 3.3.3 **Alternate Interior Angles Theorem** If two lines cut by a transversal are parallel, then alternate interior angles are congruent.

Thm 3.3.4 **Alternate Exterior Angles Theorem** If two lines cut by a transversal are parallel, then alternate exterior angles are congruent.

Thm 3.3.5 **Same-Side Interior Angles Theorem** If two lines cut by a transversal are parallel, then same-side interior angles are supplementary.

Thm 3.4.1 **Theorem: Converse of the Corresponding Angles Postulate** If two lines are cut by a transversal in such a way that corresponding angles are congruent, then the two lines are parallel.

Thm 3.4.2 **Converse of the Same-Side Interior Angles Theorem** If two lines are cut by a transversal in such a way that same-side interior angles are supplementary, then the two lines are parallel.

Thm 3.4.3 **Converse of the Alternate Interior Angles Theorem** If two lines are cut by a transversal in such a way that alternate interior angles are congruent, then the two lines are parallel.

Thm 3.4.4 **Converse of the Alternate Exterior Angles Theorem** If two lines are cut by a transversal in such a way that alternate exterior angles are congruent, then the two lines are parallel.

Thm 3.4.5 **Theorem** If two coplanar lines are perpendicular to the same line, then the two lines are parallel.

Thm 3.4.6 **Theorem** If two lines are parallel to the same line, then the two lines are parallel.

Post 3.5.1 **The Parallel Postulate** Given a line and a point not on the line, there is one and only one line that contains the given point and is parallel to the given line.

Thm 3.5.2 **Triangle Sum Theorem** The sum of the measures of the angles of a triangle is 180°.

Thm 3.5.3 **Exterior Angle Theorem** The measure of an exterior angle of a triangle is equal to the sum of the measures of the remote interior angles.

Thm 3.6.1 **Sum of the Interior Angles of a Polygon** The sum, s, of the measures of the interior angles of a polygon with n sides is given by $s = (n - 2)180°$.

Thm 3.6.2 **The Measure of an Interior Angle of a Regular Polygon** The measure, m, of an interior angle of a regular polygon with n sides is given by $m = 180° - \frac{360°}{n}$.

Thm 3.6.3 **Sum of the Exterior Angles of a Polygon** The sum of the measures of the exterior angles of a polygon is 360°.

Def 3.7.1 **Midsegment of a Triangle** A midsegment of a triangle is a segment whose endpoints are the midpoints of two sides.

Def 3.7.2 **Midsegment of a Trapezoid** A midsegment of a trapezoid is a segment whose endpoints are the midpoints of the nonparallel sides.

Def 3.8.1 **Slope** The slope of a nonvertical line that contains the points (x_1, y_1) and (x_2, y_2) is equal to the ratio $\frac{y_2 - y_1}{x_2 - x_1}$.

Thm 3.8.2 **Parallel Lines Theorem** Two nonvertical lines are parallel if and only if they have the same slope.

Thm 3.8.3 **Perpendicular Lines Theorem** Two nonvertical lines are perpendicular if and only if the product of their slopes is −1.

Post 4.1.1 **Polygon Congruence Postulate** Two polygons are congruent if and only if there is a way of setting up a correspondence between their sides and angles, in order, such that (1) each pair of corresponding angles is congruent and (2) each pair of corresponding sides is congruent.

Post 4.2.1 **SSS (Side-Side-Side) Postulate** If the sides of one triangle are congruent to the sides of another triangle, then the two triangles are congruent.

Post 4.2.2 **SAS (Side-Angle-Side) Postulate** If two sides and their included angle in one triangle are congruent to two sides and their included angle in another triangle, then the two triangles are congruent.

Post 4.2.3   **ASA (Angle-Side-Angle) Postulate** If two angles and their included side in one triangle are congruent to two angles and their included side in another triangle, then the two triangles are congruent.

Thm 4.3.1   **AAS (Angle-Angle-Side) Congruence Theorem** If two angles and a nonincluded side of one triangle are congruent to the corresponding angles and nonincluded side of another triangle, then the triangles are congruent.

Thm 4.3.2   **HL (Hypotenuse-Leg) Congruence Theorem** If the hypotenuse and a leg of a right triangle are congruent to the hypotenuse and corresponding leg of another right triangle, then the two triangles are congruent.

Thm 4.4.1   **Isosceles Triangle Theorem** If two sides of a triangle are congruent, then the angles opposite those sides are congruent.

Thm 4.4.2   **Converse of the Isosceles Triangle Theorem** If two angles of a triangle are congruent, then the sides opposite those angles are congruent.

Cor 4.4.3   **Corollary** The measure of each angle of an equilateral triangle is 60°.

Cor 4.4.4   **Corollary** The bisector of the vertex angle of an isosceles triangle is the perpendicular bisector of the base.

Thm 4.5.1   **Theorem** A diagonal of a parallelogram divides the parallelogram into two congruent triangles.

Thm 4.5.2   **Theorem** The opposite sides of a parallelogram are congruent.

Thm 4.5.3   **Theorem** The opposite angles of a parallelogram are congruent.

Thm 4.5.4   **Theorem** The diagonals of a parallelogram bisect each other.

Thm 4.5.5   **Theorem** A rhombus is a parallelogram.

Thm 4.5.6   **Theorem** A rectangle is a parallelogram.

Thm 4.5.7   **Theorem** The diagonals and sides of a rhombus form four congruent triangles.

Thm 4.5.8   **Theorem** The diagonals of a rhombus are perpendicular.

Thm 4.5.9   **Theorem** The diagonals of a rectangle are congruent.

Thm 4.5.10 **Theorem** The diagonals of a kite are perpendicular.

Thm 4.5.11 **Theorem** A square is a rectangle.

Thm 4.5.12 **Theorem** A square is a rhombus.

Thm 4.5.13 **Theorem** The diagonals of a square are congruent and are the perpendicular bisectors of each other.

Thm 4.6.1   **Theorem** If two pairs of opposite sides of a quadrilateral are congruent, then the quadrilateral is a parallelogram.

Thm 4.6.2   **Theorem** If one pair of opposite sides of a quadrilateral are parallel and congruent, then the quadrilateral is a parallelogram.

Thm 4.6.3   **Theorem** If the diagonals of a quadrilateral bisect each other, then the quadrilateral is a parallelogram.

Thm 4.6.4   **Theorem** If one angle of a parallelogram is a right angle, then the parallelogram is a rectangle.

Thm 4.6.5   **Housebuilder Theorem** If the diagonals of a parallelogram are congruent, then the parallelogram is a rectangle.

Thm 4.6.6   **Theorem** If one pair of adjacent sides of a parallelogram are congruent, then the quadrilateral is a rhombus.

Thm 4.6.7   **Theorem** If the diagonals of a parallelogram bisect the angles of the parallelogram, then the parallelogram is a rhombus.

Thm 4.6.8   **Theorem** If the diagonals of a parallelogram are perpendicular, then the parallelogram is a rhombus.

Thm 4.6.9   **Triangle Midsegment Theorem** A midsegment of a triangle is parallel to a side of the triangle, and its length is equal to half the length of that side.

Post 4.8.1   **Betweenness Postulate** Given three points P, Q, and R, if $PQ + QR = PR$, then Q is between P and R on a line.

Thm 4.8.2   **Triangle Inequality Theorem** The sum of the lengths of any two sides of a triangle is greater than the length of the third side.

Def 5.1.1   **Perimeter** The **perimeter** of a closed plane figure is the distance around the figure.

Def 5.1.2   **Area** The **area** of a closed plane figure is the number of non-overlapping squares of a given size that will exactly cover the interior of the figure.

Post 5.1.3   **Postulate: Sum of Areas** If a figure is composed of non-overlapping regions A and B, then the area of the figure is the sum of the areas of regions A and B.

Thm 5.1.4   **Perimeter of a Rectangle** The perimeter of a rectangle with base b and height h is $P = 2b + 2h$.

Post 5.1.5   **Postulate: Area of a Rectangle** The area of a rectangle with base b and height h is $A = bh$.

Def 5.2.1   **Parts of a Triangle** Any side of a triangle can be called the **base of the triangle**. The **altitude of the triangle** is a perpendicular segment from a vertex to a line containing the base of the triangle. The **height of the triangle** is the length of the altitude.

Thm 5.2.2   **Area of a Triangle** The area of a triangle with base b and height h is $A = \frac{1}{2}bh$.

Def 5.2.3   **Parts of a Parallelogram** Any side of a parallelogram can be called the **base of the parallelogram**. An **altitude of a parallelogram** is a perpendicular segment from a line containing the base to a line containing the side opposite the base. The **height of the parallelogram** is the length of the altitude.

Thm 5.2.4   **Area of a Parallelogram** The area of a parallelogram with base b and height h is $A = bh$.

Def 5.2.5   **Parts of a Trapezoid** The two parallel sides of a trapezoid are known as the **bases of the trapezoid**. The two nonparallel sides are called the **legs of the trapezoid**. An **altitude of a trapezoid** is a perpendicular segment from a line containing one base to a line containing the other base. The **height of a trapezoid** is the length of an altitude.

Thm 5.2.6   **Area of a Trapezoid**   The area of a trapezoid with bases b_1 and b_2 and height h is
$A = \frac{1}{2}(b_1 + b_2)h$.

Def 5.3.1   **Circle**   A **circle** is the set of all points in a plane that are the same distance, r, from a given point in the plane known as the **center** of the circle. The distance r is known as the **radius of the circle**. The distance $d = 2r$ is known as the **diameter of the circle**.

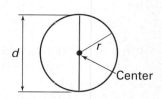

Thm 5.3.2   **Circumference of a Circle**   The circumference of a circle with diameter d and radius r is $C = \pi d$ or $C = 2\pi r$.

Thm 5.3.3   **Area of a Circle**   The area of a circle with radius r is $A = \pi r^2$.

Thm 5.4.1   **Pythagorean Theorem**   For any right triangle, the square of the length of the hypotenuse is equal to the sum of the squares of the lengths of the legs; that is, $c^2 = a^2 + b^2$.

Thm 5.4.2   **Converse of the Pythagorean Theorem**   If the square of the length of one side of a triangle equals the sum of the squares of the lengths of the other two sides, then the triangle is a right triangle.

Thm 5.4.3   **Pythagorean Inequalities**   For any triangle ABC, with c as the length of the longest side:
$\quad\quad$ If $c^2 = a^2 + b^2$, then $\triangle ABC$ is a right triangle.
$\quad\quad$ If $c^2 > a^2 + b^2$, then $\triangle ABC$ is an obtuse triangle.
$\quad\quad$ If $c^2 < a^2 + b^2$, then $\triangle ABC$ is an acute triangle.

Thm 5.5.1   **45-45-90 Triangle Theorem**   In any 45-45-90 triangle, the length of the hypotenuse is $\sqrt{2}$ times the length of a leg.

Thm 5.5.2   **30-60-90 Triangle Theorem**   In any 30-60-90 triangle, the length of the hypotenuse is 2 times the length of the shorter leg, and the length of the longer leg is $\sqrt{3}$ times the length of the shorter leg.

Thm 5.5.3   **Area of a Regular Polygon**   The area of a regular polygon with apothem a and perimeter p is $A = \frac{1}{2}ap$.

Thm 5.6.1   **Distance Formula**   On a coordinate plane, the distance between two points (x_1, y_1) and (x_2, y_2) is $d = \sqrt{(x_2 - x_1)^2 + (y_2 - y_1)^2}$.

Def 6.2.1   **Polyhedron**   A **polyhedron** is a closed spatial figure composed of polygons, called the **faces** of the polyhedron. The intersections of the faces are the **edges** of the polyhedron. The vertices of the faces are the **vertices** of the polyhedron.

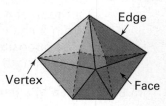

Def 6.2.2   **Parallel Planes**   Two planes are parallel if and only if they do not intersect.

Def 6.2.3   **A Line Perpendicular to a Plane**   A line is perpendicular to a plane at a point P if and only if it is perpendicular to every line in the plane that passes through P.

Def 6.2.4   **A Line Parallel to a Plane**   A line that is not contained in a given plane is parallel to the plane if and only if it is parallel to a line contained in the plane.

Def 6.2.5 **Dihedral Angle** A **dihedral angle** is the figure formed by two half-planes with a common edge. Each half-plane is called a **face** of the angle, and the common edge of the half-planes is called the **edge** of the angle.

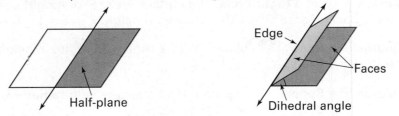

Def 6.2.6 **Measure of a Dihedral Angle** The measure of a dihedral angle is the measure of an angle formed by two rays that are on the faces and that are perpendicular to the edge.

Thm 6.3.1 **Diagonal of a Right Rectangular Prism** The length of the diagonal, d, of a right rectangular prism is given by $d = \sqrt{\ell^2 + w^2 + h^2}$.

Thm 6.4.1 **Distance Formula in Three Dimensions** The distance, d, between the points (x_1, y_1, z_1) and (x_2, y_2, z_2) in space is given by $d = \sqrt{(x_2 - x_1)^2 + (y_2 - y_1)^2 + (z_2 - z_1)^2}$.

Thm 6.4.2 **Midpoint Formula in Three Dimensions** The midpoint of a segment with endpoints at (x_1, y_1, z_1) and (x_2, y_2, z_2) in space is given by $\left(\frac{x_1 + x_2}{2}, \frac{y_1 + y_2}{2}, \frac{z_1 + z_2}{2} \right)$.

Thm 6.6.1 **Theorem: Sets of Parallel Lines** In a perspective drawing, all lines that are parallel to each other, but not parallel to the picture plane, will seem to meet at a single point known as the vanishing point.

Thm 6.6.2 **Theorem: Lines Parallel to the Ground** In a perspective drawing, a line that is on the plane of the ground and is not parallel to the picture plane will meet the horizon of the drawing. Any line parallel to this line will meet the horizon at the same point.

Thm 7.1.1 **Surface Area and Volume of a Right Rectangular Prism** The surface area, S, and volume, V, of a right rectangular prism with length ℓ, width w, and height h are $S = 2\ell w + 2wh + 2\ell h$ and $V = \ell wh$.

Thm 7.1.2 **Surface Area and Volume of a Cube** The surface area, S, and volume, V, of a cube with side length s are $S = 6s^2$ and $V = s^3$.

Thm 7.2.1 **Surface Area of a Right Prism** The surface area, S, of a right prism with lateral area L, base area B, perimeter p, and height h is $S = L + 2B$ or $S = hp + 2B$.

Thm 7.2.2 **Cavalieri's Principle** If two solids have equal heights and the cross sections formed by every plane parallel to the bases of both have equal areas, then the two solids have equal volumes.

Thm 7.2.3 **Volume of a Prism** The volume, V, of a prism with height h and base area B is $V = Bh$.

Thm 7.3.1 **Surface Area of a Regular Pyramid** The surface area, S, of a regular pyramid with lateral area L, base area B, perimeter of the base p, and slant height ℓ is $S = L + B$ or $S = \frac{1}{2}\ell p + B$.

Thm 7.3.2 **Volume of a Pyramid** The volume, V, of a pyramid with height h and base area B is $V = \frac{1}{3}Bh$.

Thm 7.4.1 **Surface Area of a Right Cylinder** The surface area, S, of a right cylinder with lateral area L, base area B, radius r, and height h is $S = L + 2B$ or $S = 2\pi rh + 2\pi r^2$.

Thm 7.4.2   **Volume of a Cylinder**   The volume, V, of a cylinder with radius r, height h, and base area B is $V = Bh$ or $V = \pi r^2 h$.

Thm 7.5.1   **Surface Area of a Right Cone**   The surface area, S, of a right cone with lateral area L, base of area B, radius r, and slant height ℓ is $S = L + B$ or $S = \pi r \ell + \pi r^2$.

Thm 7.5.2   **Volume of a Cone**   The volume, V, of a cone with radius r, height h, and base area B is $V = \frac{1}{3}Bh$ or $V = \frac{1}{3}\pi r^2 h$.

Thm 7.6.1   **Volume of a Sphere**   The volume, V, of a sphere with radius r is $V = \frac{4}{3}\pi r^3$.

Thm 7.6.2   **Surface Area of a Sphere**   The surface area, S, of a sphere with radius r is $S = 4\pi r^2$.

Def 8.2.1   **Similar Figures**   Two figures are similar if and only if one is congruent to the image of the other by a dilation.

Post 8.2.2   **Polygon Similarity Postulate**   Two polygons are similar if and only if there is a way of setting up a correspondence between their sides and angles such that the following conditions are met:
1. Each pair of corresponding angles is congruent.
2. Each pair of corresponding sides is proportional.

Thm 8.2.3   **Cross-Multiplication Property of Proportions**   For real numbers a, b, c, and d such that b and $d \neq 0$, if $\frac{a}{b} = \frac{c}{d}$, then $ad = bc$.

Thm 8.2.4   **Reciprocal Property of Proportions**   For real numbers a, b, c, and d such that a, b, c, and $d \neq 0$, if $\frac{a}{b} = \frac{c}{d}$, then $\frac{b}{a} = \frac{d}{c}$.

Thm 8.2.5   **Exchange Property of Proportions**   For real numbers a, b, c, and d such that a, b, c, and $d \neq 0$, if $\frac{a}{b} = \frac{c}{d}$, then $\frac{a}{c} = \frac{b}{d}$.

Thm 8.2.6   **"Add-One" Property of Proportions**   For real numbers a, b, c, and d such that b and $d \neq 0$, if $\frac{a}{b} = \frac{c}{d}$, then $\frac{a+b}{b} = \frac{c+d}{d}$.

Post 8.3.1   **AA (Angle-Angle) Similarity Postulate**   If two angles of one triangle are congruent to two angles of another triangle, then the triangles are similar.

Thm 8.3.2   **SSS (Side-Side-Side) Similarity Theorem**   If the three sides of one triangle are proprtional to the three sides of another triangle, then the triangles are similar.

Thm 8.3.3   **SAS (Side-Angle-Side) Similarity Theorem**   If two sides of one triangle are proprtional to two sides of another triangle and their included angles are congruent, then the triangles are similar.

Thm 8.4.1   **Side-Splitting Theorem**   A line parallel to one side of the triangle divides the other two sides proportionally.

Cor 8.4.2   **Two-Transversal Proportionality Corollary**   Three or more parallel lines divide two intersecting transversals proportionally.

Thm 8.5.1   **Proportional Altitudes Theorem**   If two triangles are similar, then their corresponding altitudes have the same ratio as their corresponding sides.

Thm 8.5.2   **Proportional Medians Theorem**   If two triangles are similar, then their corresponding medians have the same ratio as their corresponding sides.

Thm 8.5.3   **Proportional Angle Bisectors Theorem**   If two triangles are similar, then their corresponding angle bisectors have the same ratio as the corresponding sides.

Thm 8.5.4   **Proportional Segments Theorem**   An angle bisector of a triangle divides the opposite side into two segments that have the same ratio as the other two sides.

Def 9.1.1 **Circle** A circle is the set of all points in a plane that are equidistant from a given point in the plane known as the center of the circle. A **radius** (plural, *radii*) is a segment from the center of the circle to a point on the circle. A **chord** is a segment whose endpoints line on a circle. A **diameter** is a chord that contains the center of a circle.

Def 9.1.2 **Central Angle and Intercepted Arc** A **central angle** of a circle is an angle in the plane of a circle whose vertex is the center of the circle. An arc whose endpoints lie on the sides of the angle and whose other points lie in the interior of the angle is the **intercepted arc** of the central angle.

Def 9.1.3 **Degree Measure of Arcs** The degree measure of a minor arc is the measure of its central angle. The degree measure of a major arc is 360° minus the degree measure of its minor arc. The degree measure of a semicircle is 180°.

Thm 9.1.4 **Arc Length** If r is the radius of a circle and M is the degree measure of an arc of the circle, then the length, L, of the arc is given by the following: $L = \frac{M}{360°}(2\pi r)$.

Thm 9.1.5 **Chords and Arcs Theorem** In a circle, or in congruent circles, the arcs of congruent chords are congruent.

Thm 9.1.6 **Converse of the Chords and Arcs Theorem** In a circle, or in congruent circles, the chords of congruent arcs are congruent.

Def 9.2.1 **Secants and Tangents** A **secant** to a circle is a line that intersects the circle at two points. A **tangent** is a line in the plane of the circle that intersects the circle at exactly one point, which is known as the **point of tangency**.

Thm 9.2.2 **Tangent Theorem** If a line is tangent to a circle, then the line is perpendicular to a radius of the circle drawn to the point of tangency.

Thm 9.2.3 **Radius and Chord Theorem** A radius that is perpendicular to a chord of a circle bisects the chord.

Thm 9.2.4 **Converse of the Tangent Theorem** If a line is perpendicular to a radius of a circle at its endpoint on the circle, then the line is tangent to the circle.

Thm 9.2.5 **Theorem** The perpendicular bisector of a chord passes through the center of the circle.

Thm 9.3.1 **Inscribed Angle Theorem** The measure of an angle inscribed in a circle is equal to one-half the measure of the intercepted arc.

Cor 9.3.2 **Right Angle Corollary** If an inscribed angle intercepts a semicircle, then the angle is a right angle.

Cor 9.3.3 **Arc-Intercept Corollary** If two inscribed angles intercept the same arc, then they have the same measure.

Thm 9.4.1 **Theorem** If a tangent and a secant (or a chord) intersect on a circle at the point of tangency, then the measure of the angle formed is one-half the measure of its intercepted arc.

Thm 9.4.2 **Theorem** The measure of an angle formed by two secants or chords that intersect in the interior of a circle is one-half the sum of the measures of the arcs intercepted by the angle and its vertical angle.

Thm 9.4.3 **Theorem** The measure of an angle formed by two secants that intersect in the exterior of a circle is one-half the difference of the measures of the intercepted arcs.

Thm 9.4.4   **Theorem** The measure of a secant-tangent angle with its vertex outside the circle is one-half the difference of the measures of the intercepted arcs.

Thm 9.4.5   **Theorem** The measure of a tangent-tangent angle with its vertex outside the circle is one-half the difference of the measures of the intercepted arcs, or the measure of the major arc minus 180°.

Thm 9.5.1   **Theorem** If two segments are tangent to a circle from the same external point, then the segments are of equal length.

Thm 9.5.2   **Theorem** If two secants intersect outside a circle, then the product of the lengths of one secant segment and its external segment equals the product of the lengths of the other secant segment and its external segment. (Whole × Outside = Whole × Outside)

Thm 9.5.3   **Theorem** If a secant and a tangent intersect outside a circle, then the product of the lengths of the secant segment and its external segment equals the length of the tangent segment squared. (Whole × Outside = Tangent Squared)

Thm 9.5.4   **Theorem** If two chords intersect inside a circle, then the product of the lengths of the segments of one chord equals the product of the lengths of the segments of the other chord.

Def 10.1.1   **Tangent Ratio** For a given acute angle $\angle A$ with a measure of $\theta°$, the **tangent** of $\angle A$, or tan θ, is the ratio of the length of the leg opposite $\angle A$ to the length of the leg adjacent to $\angle A$ in any right triangle with A as one vertex, or $\tan \theta = \frac{\text{opposite}}{\text{adjacent}}$.

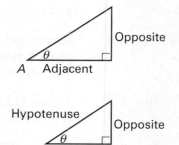

Def 10.2.1   **Sine Ratio** For a given angle $\angle A$ with a measure of $\theta°$, the **sine** of $\angle A$, or sin θ, is the ratio of the length of the leg opposite A to the length of the hypotenuse in a right triangle with A as one vertex, or $\sin \theta = \frac{\text{opposite}}{\text{hypotenuse}}$.

Def 10.2.2   **Cosine Ratio** The cosine of $\angle A$, or cos θ, is the ratio of the length of the leg adjacent to A to the length of the hypotenuse, or $\cos \theta = \frac{\text{adjacent}}{\text{hypotenuse}}$.

Thm 10.2.3   **Identity** $\tan \theta = \frac{\sin \theta}{\cos \theta}$

Thm 10.2.4   **Identity** $(\sin \theta)^2 + (\cos \theta)^2 = 1$

Def 10.3.1   **Unit Circle Definition of Sine and Cosine** Let θ be a rotation angle. Then **sin θ** is the y-coordinate of the image of point $P(1, 0)$ rotated $\theta°$ about the origin, and **cos θ** is the x-coordinate.

Thm 10.4.1   **The Law of Sines** For any triangle $\triangle ABC$ with sides a, b, and c:
$$\frac{\sin A}{a} = \frac{\sin B}{b} = \frac{\sin C}{c}$$

Thm 10.5.1   **The Law of Cosines** For any triangle $\triangle ABC$ with sides a, b, and c:
$$a^2 = b^2 + c^2 - 2bc \cos A \qquad\qquad b^2 = a^2 + c^2 - 2ac \cos B$$
$$c^2 = a^2 + b^2 - 2ab \cos C$$

Thm 11.3.1   **Theorem** A graph contains an Euler path if and only if there are at most two odd vertices.

Thm 11.4.1   **Jordan Curve Theorem** Every simple closed curve in a plane divides the plane into two distinct regions, the inside and the outside. Every curve that connects a point on the inside to a point on the outside must intersect the curve.

Thm 11.4.2   **Euler's Formula** For any polyhedron with V vertices, E edges, and F faces, $V - E + F = 2$.

Def 11.7.1 **Affine Transformation** An **affine transformation** transforms each preimage point P in a plane to an image point P' in such a way that
   1. collinear points are transformed into collinear points,
   2. straight lines are transformed into straight lines,
   3. intersecting lines are transformed into intersecting lines, and
   4. parallel lines are transformed into parallel lines.

Def 11.7.2 **Main Features of Projective Geometry**
   1. Projective geometry is the study of the properties of figures that do not change under a projection.
   2. There is no concept of size, measurement, or congruence.
   3. Its theorems state facts about things such as the positions of points and the intersections of lines.
   4. An unmarked straightedge is the only tool allowed for drawing figures.

Thm 11.7.3 **The Theorem of Pappus** If A_1, B_1, and C_1 are three distinct points on one line and A_2, B_2, and C_2 are three distinct points on a second line, then the intersections of $\overline{A_1B_2}$ and $\overline{A_2B_1}$, of $\overline{A_1C_2}$ and $\overline{A_2C_1}$, and of $\overline{B_1C_2}$ and $\overline{B_2C_1}$ are collinear.

Thm 11.7.4 **The Theorem of Desargues** If one triangle is a projection of another triangle, then the intersections of the lines containing the corresponding sides of the two triangles are collinear.

Thm 12.1.1 **Argument Form: *Modus Ponens***
   If p then q
   p
   Therefore, q

Thm 12.1.2 **Argument Form: *Modus Tollens***
   If p then q
   $\sim q$
   Therefore, $\sim p$

Thm 12.1.3 **Invalid Form: Affirming the Consequent**
   If p then q
   q
   Therefore, p

Thm 12.1.4 **Invalid Form: Denying the Antecedent**
   If p then q
   $\sim p$
   Therefore, $\sim q$

Thm 12.4.1 **Proof by Contradiction** To prove statement s, assume $\sim s$. Then the following argument form is valid:
   If $\sim s$ then (t AND $\sim t$)
   Therefore, s

Table of Trigonometric Ratios

Angle	sin	cos	tan	Angle	sin	cos	tan
0°	0.0000	1.0000	0.0000	45°	0.7071	0.7071	1.0000
1°	0.0175	0.9998	0.0175	46°	0.7193	0.6947	1.0355
2°	0.0349	0.9994	0.0349	47°	0.7314	0.6820	1.0724
3°	0.0523	0.9986	0.0524	48°	0.7431	0.6691	1.1106
4°	0.0698	0.9976	0.0699	49°	0.7547	0.6561	1.1504
5°	0.0872	0.9962	0.0875	50°	0.7660	0.6428	1.1918
6°	0.1045	0.9945	0.1051	51°	0.7771	0.6293	1.2349
7°	0.1219	0.9925	0.1228	52°	0.7880	0.6157	1.2799
8°	0.1392	0.9903	0.1405	53°	0.7986	0.6018	1.3270
9°	0.1564	0.9877	0.1584	54°	0.8090	0.5878	1.3764
10°	0.1736	0.9848	0.1763	55°	0.8192	0.5736	1.4281
11°	0.1908	0.9816	0.1944	56°	0.8290	0.5592	1.4826
12°	0.2079	0.9781	0.2126	57°	0.8387	0.5446	1.5399
13°	0.2250	0.9744	0.2309	58°	0.8480	0.5299	1.6003
14°	0.2419	0.9703	0.2493	59°	0.8572	0.5150	1.6643
15°	0.2588	0.9659	0.2679	60°	0.8660	0.5000	1.7321
16°	0.2756	0.9613	0.2867	61°	0.8746	0.4848	1.8040
17°	0.2924	0.9563	0.3057	62°	0.8829	0.4695	1.8807
18°	0.3090	0.9511	0.3249	63°	0.8910	0.4540	1.9626
19°	0.3256	0.9455	0.3443	64°	0.8988	0.4384	2.0503
20°	0.3420	0.9397	0.3640	65°	0.9063	0.4226	2.1445
21°	0.3584	0.9336	0.3839	66°	0.9135	0.4067	2.2460
22°	0.3746	0.9272	0.4040	67°	0.9205	0.3907	2.3559
23°	0.3907	0.9205	0.4245	68°	0.9272	0.3746	2.4751
24°	0.4067	0.9135	0.4452	69°	0.9336	0.3584	2.6051
25°	0.4226	0.9063	0.4663	70°	0.9397	0.3420	2.7475
26°	0.4384	0.8988	0.4877	71°	0.9455	0.3256	2.9042
27°	0.4540	0.8910	0.5095	72°	0.9511	0.3090	3.0777
28°	0.4695	0.8829	0.5317	73°	0.9563	0.2924	3.2709
29°	0.4848	0.8746	0.5543	74°	0.9613	0.2756	3.4874
30°	0.5000	0.8660	0.5774	75°	0.9659	0.2588	3.7321
31°	0.5150	0.8572	0.6009	76°	0.9703	0.2419	4.0108
32°	0.5299	0.8480	0.6249	77°	0.9744	0.2250	4.3315
33°	0.5446	0.8387	0.6494	78°	0.9781	0.2079	4.7046
34°	0.5592	0.8290	0.6745	79°	0.9816	0.1908	5.1446
35°	0.5736	0.8192	0.7002	80°	0.9848	0.1736	5.6713
36°	0.5878	0.8090	0.7265	81°	0.9877	0.1564	6.3138
37°	0.6018	0.7986	0.7536	82°	0.9903	0.1392	7.1154
38°	0.6157	0.7880	0.7813	83°	0.9925	0.1219	8.1443
39°	0.6293	0.7771	0.8098	84°	0.9945	0.1045	9.5144
40°	0.6428	0.7660	0.8391	85°	0.9962	0.0872	11.4301
41°	0.6561	0.7547	0.8693	86°	0.9976	0.0698	14.3007
42°	0.6691	0.7431	0.9004	87°	0.9986	0.0523	19.0811
43°	0.6820	0.7314	0.9325	88°	0.9994	0.0349	28.6363
44°	0.6947	0.7193	0.9657	89°	0.9998	0.0175	57.2900
45°	0.7071	0.7071	1.0000	90°	1.0000	0.0000	∞

Glossary

A

acute angle An angle whose measure is less than 90°. (29)

acute triangle A triangle with three acute angles. (654)

adjacent angles Two angles in a plane that share a common vertex and a common side but have no interior points in common. (101)

affine transformation A transformation in which all preimage points are mapped to image points in such a way that collinear points, straight lines, intersecting lines, and parallel lines are transformed as such. (747)

alternate exterior angles Two nonadjacent exterior angles that lie on opposite sides of a transversal. (156)

alternate interior angles Two nonadjacent interior angles that lie on opposite sides of a transversal. (156)

altitude of a cone A segment from the vertex perpendicular to the plane of the base. (460)

altitude of a cylinder A segment that has endpoints in the planes containing the bases and is perpendicular to both planes. (453)

altitude of a parallelogram A segment from a line containing the base to a line containing the side opposite the base that is perpendicular to both lines. (305)

altitude of a prism A segment that has endpoints in the planes containing the bases and is perpendicular to both planes. (437)

altitude of a pyramid A segment from the vertex perpendicular to the plane of the base. (445)

altitude of a trapezoid A segment from a line containing one base to the line containing the other base, which is perpendicular to both lines. (306)

altitude of a triangle A segment from a vertex perpendicular to the line containing the opposite side. (303)

angle A figure formed by two rays with a common endpoint. (11)

angle bisector A ray that divides an angle into two congruent angles. (38)

angle of rotation A figure formed by a rotating ray and a stationary reference ray. (647)

annulus The region between two circles in a plane that have the same center but different radii. (469)

apothem A segment from the center of a regular polygon perpendicular to a side of the polygon. (334)

arc An unbroken part of a circle. (566)

arc length The measure of the arc of a circle in terms of linear units, such as centimeters. (567)

arc measure The measure of an arc in a circle in terms of degrees. (567)

area The number of nonoverlapping unit squares of a given size that will exactly cover the interior of a figure. (295)

argument A sequence of statements that lead to a logical conclusion. (768)

axis of cylinder The segment joining the centers of the two bases. (453)

axis of symmetry A line that divides a planar figure into two congruent reflected halves. (139)

B

base angles of an isosceles triangle The angles whose vertices are the endpoints of the base of an isosceles triangle. (237)

base edge of a pyramid An edge that is part of the base of a pyramid; each lateral face has one edge in common with the base. (445)

base of a cone The circular face of the cone. (460)

base of a parallelogram Any side of a parallelogram. (305)

base of a prism The faces formed by the polygonal region and its image. (388)

base of a pyramid The polygonal face that is opposite the vertex. (445)

bases of a trapezoid The two parallel sides of a trapezoid. (306)

base of a triangle Any side of a triangle. (303)

base of isosceles triangle The side opposite the vertex angle. (237)

bases of a cylinder The faces formed by the circular region and its translated image. (453)

betweenness Given three points, A, B, and C, if $AB + BC = AC$, then A, B, and C are collinear and B is between A and C. (273)

biconditional A statement using "if and only if." (99)

binary number system A number system based on the digits 0 and 1. (798)

blocks The grid units in taxicab geometry. (706)

Cantor's dust A one-dimensional fractal created from a segment by removing the center one-third of segments at all levels of the structure. (740)

center of a circle The point inside a circle that is equidistant from all points on the circle. (314)

center of a regular polygon The point that is equidistant from all vertices of a polygon. (139)

center of dilation The point in a dilation through which every line connecting a preimage point to an image point passes. (500)

center of projection The central point in a class of transformations known as central projections. (749)

central angle of a circle An angle formed by two rays originating from the center of a circle. (566)

central angle of a regular polygon An angle whose vertex is the center of the polygon and whose sides pass through adjacent vertices. (139)

central projections A class of transformations in which every projected point of an image lies on a ray containing the center of projection and a point in the image. (749)

centroid The point where the three medians of a triangle intersect. (47)

chord A segment whose endpoints lie on a circle. (564)

circle The set of points in a plane that are equidistant from a given point known as the center of the circle. (314)

circumcenter The center of a circumscribed circle; the point where the three perpendicular bisectors of the sides of a triangle intersect; it is equidistant from the three vertices of the triangle. (45)

circumference The distance around a circle. (315)

circumscribed circle A circle that is drawn around the outside a triangle and contains all three vertices; a circle is circumscribed about a polygon if each vertex of the polygon lies on the circle. (44)

collinear Lying on the same line. (10)

complementary angles Two angles whose measures have a sum of 90°. (28)

compound statement A statement formed when two statements are connected by AND or OR. (776)

concave polygon A polygon that is not convex. (177)

conclusion The phrase following the word *then* in a conditional statement; the final statement of an argument. (90)

concurrent Literally, "running together"; of three or more lines, intersecting at a single point. (45)

conditional statement A statement that can be written in the form "If *p*, then *q*," where *p* is the hypothesis and *q* is the conclusion. (90)

cone A three-dimensional figure that consists of a circular base and a curved lateral surface that connects the base to a single point not in the plane of the base, called the vertex. (460)

congruence The relationship between figures having the same shape and same size; congruent segments are segments that match exactly. (19)

congruent polygons Two polygons are congruent if and only if there is a correspondence between their sides and angles such that each pair of corresponding angles is congruent and each pair of corresponding sides is congruent. (212)

conjecture A statement that is believed to be true. (36)

conjunction A compound statement that uses the word *AND*. (776)

contraction A dilation in which the preimage is reduced in size. (500)

contradiction A contradiction asserts that a statement and its negation are both true. (792)

converse of a conditional The statement formed by interchanging the hypothesis and conclusion of a conditional statement. (92)

convex polygon A polygon in which any line segment connecting two points of the polygon has no part outside the polygon. (177)

coordinate of a point The real number represented by a point on a number line. (17)

coordinate plane A grid formed by two or more coordinatized lines, known as *axes*, that intersect at right angles at a point known as the *origin*. (397)

contrapositive of a conditional The statement formed by interchanging the hypothesis and conclusion of a conditional statement and negating each part. (786)

coplanar Lying in the same plane. (10)

corresponding angles Two nonadjacent angles, one interior and one exterior, that lie on the same side of a transversal. (156)

corresponding angles of a polygon Angles of a polygon that are matched up with angles of another polygon with the same number of angles. (211)

corresponding sides of a polygon Sides of a polygon that are matched up with sides of another polygon with the same number of sides. (211)

corollary of a theorem A theorem that follows directly from another theorem and that can easily be proved from that theorem. (237)

cosine In a right triangle, the ratio of the length of the side adjacent to an acute angle to the length of the hypotenuse. (639)

counterexample An example that proves that a statement, often a conjecture, is false. (92)

CPCTC Abbreviation for "corresponding parts of congruent triangles are congruent." (235)

cylinder A solid that consists of a circular region and its translated image in a parallel plane with a lateral surface connecting the circles. (453)

deductive reasoning The process of drawing conclusions by using logical reasoning in an argument. (91)

degree In a graph, the number of edges at a vertex. (713)

degree measure of arcs The measure of a minor arc is the measure of its central angle. The degree measure of a major arc is 360° minus the degree measure of its central angle. (567)

diagonal of a polygon A segment that joins two nonadjacent vertices of a polygon. (390)

diagonal of a polyhedron A segment whose endpoints are vertices of two different faces of the polyhedron. (390)

diameter A chord that passes through the center of a circle; twice the length of the radius of the circle. (314)

dihedral angle An angle formed by two half-planes with a common edge. (382)

dilation A transformation in which every point P has an image point P' such that a line connecting the two points passes through a point O, known as the *center of dilation*, and $OP' = k \cdot OP$, where k is the *scale factor* of the dilation. (498)

direction of vector The orientation of a vector, generally indicated by an arrowhead. (672)

disjunction A compound statement that uses the word *OR*. (777)

displacement vector A vector that represents the change in position of an object. (674)

distance formula In a coordinate plane, the distance, d, between two points, (x_1, y_1) and (x_2, y_2), is given by the formula $d = \sqrt{(x_2 - x_1)^2 + (y_2 - y_1)^2}$. (340)

edge Segments or curves that connect vertices of a graph. (712)

edge of a dihedral angle The common edge of the half-planes of the angle. (382)

edge of a polyhedron The segment formed by the intersection of two faces of a polyhedron. (379)

endpoint A point at an end of a segment or the starting point of a ray. (10)

equiangular polygon A polygon in which all angles are congruent. (139)

equilateral polygon A polygon in which all sides are congruent. (139)

equilateral triangle A triangle in which all three sides are congruent. (196)

Equivalence Properties of Equality or Congruence The Reflexive, Symmetric, and Transitive Properties. (110)

equivalence relation Any relation that satisfies the Reflexive, Symmetric, and Transitivie Properties. (110)

Euler characteristic The Euler number $V - E + F$; an invariant of a surface. (723)

Euler circuit An Euler path that starts and ends at the same vertex and passes through all vertices of the graph. (715)

Euler path A continuous path that travels along each edge of a graph exactly once. (712)

even vertices The vertices of a graph that have an even number of edges leading to them. (713)

exclusive OR Indicating either one or the other, but not both. (777)

expansion A dilation in which the preimage is enlarged in size. (500)

exterior angle An angle formed between one side of polygon and the extension of an adjacent side. (174)

external secant segment The portion of a secant segment that lies outside the circle. (600)

face of a dihedral angle One of the half-planes of a dihedral angle. (382)

face of a polyhedron One of the polygons that form a polyhedron. (379)

face of a prism Each flat surface of a prism. (388)

fractal A structure that is self-similar; each subdivision has the same structure as the whole and the structure of the object looks the same from any view. (738)

glide reflection A combination of a translation and a reflection. (56)

golden ratio The ratio of the long side to the short side of a golden rectangle. (698)

golden rectangle A rectangle in which the length ℓ and the width w satisfy the proportion $\frac{w}{\ell} = \frac{\ell - w}{}$. (698)

graph A diagram of vertices and edges. (712)

great circle A circle on a sphere that divides the sphere into two equal parts. (172)

half-plane The portion of a plane that lies on one side of a line in the plane and includes the line. (382)

head-to-tail method In vector addition, finding the sum of two vectors by placing the tail of one vector at the head of the other; the vector drawn from the tail of the first vector to the head of the second is the vector sum. (674)

height The length of an altitude of a polygon. (303)

hyperbolic geometry The geometry of a surface that curves like a saddle. (731)

hypotenuse The side opposite the right angle in a right triangle. (190)

hypothesis The phrase following the word *if* in a conditional statement. (90)

identity An equation that is true for all values of the variables in the equation for which the terms in the equation are defined. (641)

image A shape that results from a transformation of a figure known as the *preimage*. (50)

indirect proof A proof in which the statement that you want to prove is assumed to be false and a contradiction or other "absurdity" is shown to follow from the assumption. (792)

incenter The center of an inscribed circle; the point where the three angle bisectors of a triangle intersect; it is equidistant from the three sides of the triangle. (45)

inclusive OR Indicating either one or the other or both. (777)

inductive reasoning Forming conjectures on the basis of observations. (118)

input-output table A table that gives the outputs of a logic gate for different input combinations. (799)

inscribed angle An angle whose vertex lies on the circle and whose sides are chords of the circle. (580)

inscribed circle An inscribed circle in a triangle is inside the triangle and touches each side at one point; a circle is inscribed in a polygon if each side of the polygon is tangent to the circle. (44)

intercepted arc An arc whose endpoints lie on the sides of an inscribed angle. (566)

intercepts The points where a line in a coordinate plane passes through the x- and y-axes or where a plane in a three-dimensional space crosses the x-, y-, and z-axes. (403)

intersect To one or more points in common. (11)

intersection point A point that two or more geometric figures have in common. (11)

invariant Properties of a figure that stay the same regardless of how the figure is deformed. (723)

inverse of a conditional The statement formed by negating both the hypothesis and conclusion of a conditional statement. (786)

inverse tangent The function used to find the angle with a given tangent ratio; abbreviated $\tan^{-1}$. (633)

isometric drawing A type of three-dimensional drawing. (372)

isosceles triangle A triangle with at least two congruent sides. (237)

iteration The repetitive application of the same rule. (738)

kite A quadrilateral with exactly two pairs of adjacent congruent sides. (310)

lateral area The sum of the areas of the lateral faces of a polyhedron. (437)

lateral edge The intersection of two lateral faces of a polyhedron. (445)

lateral edges of a prism The edges of the lateral faces of a polyhedron that are not edges of either base. (388)

lateral faces The faces of a prism or pyramid that are not bases. (388)

lateral surface The curved surface of a cylinder or cone. (453)

legs of a right triangle The sides adjacent to the right angle. (229)

legs of a trapezoid The two nonparallel sides of a trapezoid. (306)

legs of an isosceles triangle The two congruent sides of an isosceles triangle. (237)

length The length of a segment is the measure of the distance from one endpoint to the other. (17)

line An undefined term in geometry, a line is understood to be perfectly straight, contain an infinite number of points, extend infinitely in two directions, and have no thickness. (9)

linear pair of angles The two angles formed by the endpoint of a ray when the endpoint falls on a line; two adjacent angles whose noncommon sides are opposite rays. (28)

logic gate An electronic circuit that represents *not*, *and*, or *or*. (799)

logical chain A series of logically linked conditional statements. (92)

logically equivalent statements Two statements, each of which can be logically derived from the other. (730)

magnitude The numerical measure of a vector. (672)

major arc An arc of a circle that is longer than a semicircle of that circle. (566)

matrix Data arranged in rows and columns and enclosed in brackets. (682)

median A segment from a vertex to the midpoint of the opposite side in a triangle. (47)

Menger sponge A three-dimensional fractal created from a cube by removing cubes from the centers of cubic structures at all levels. (738)

midpoint of a segment The point that divides a segment into two congruent segments. (349)

midsegment of a trapezoid A line connecting the midpoints of the two nonparallel segments of a trapezoid. (309)

midsegment of a triangle A segment whose endpoints are the midpoints of two sides. (183)

minor arc An arc of a circle that is shorter than a semicircle of that circle. (566)

Möbius strip A one-sided surface formed by taking a long rectangular strip and joining the ends together after giving it half a twist. (721)

modus ponens In logic, a valid argument of the following form:
    If *p* then *q*
    p
    Therefore, *q* (769)

modus tollens In logic, a valid argument of the following form:
    If *p* then *q*
    Not *q*
    Therefore Not *p* (770)

negation If *p* is a statement, then not *p* is its negation. (778)

non-Euclidean geometry A system of geometry in which the Parallel Postulate does not hold. (729)

nonoverlapping Having no points in common (except for common boundary points). (295)

number line A line whose points correspond with the set of real numbers. (17)

oblique cone A cone that is not a right cone. (460)

oblique cylinder A cylinder that is not a right cylinder. (453)

oblique prism A prism that has at least one nonrectangular lateral face. (389)

obtuse angle An angle whose measure is greater than 90°. (29)

obtuse triangle A triangle that has one obtuse angle. (654)

octant One of the eight spaces into which a three-dimensional coordinate system is divided by the *xy*-, *yz*-, and *xz*-planes. (397)

odd vertices The vertices of a graph that have an odd number of edges leading to them. (713)

orthogonal Pertaining to right angles. (732)

orthographic projection A view of an object in which the line of sight is perpendicular to the plane of the picture. (374)

paragraph proof A form of proof in which one's reasoning is explained in paragraph form, as opposed to flow chart or two-column proofs. (109)

parallel Of coplanar lines or any two planes in space, that they do not meet, no matter how far they might be extended. (35)

parallelogram A quadrilateral with two pairs of parallel sides. (148)

parallelogram method A method of adding vectors. Two vectors are represented as acting on a common point, and a parallelogram is formed by adding two sides to the figure. The vector sum is a vector along the diagonal of the parallelogram, starting from the common point. (674)

parametric equations Rules for the coordinates of points in a coordinate system, given in terms of a certain *parameter*, such as the time, *t*. (404)

perimeter The distance around a closed plane figure. (294)

perpendicular bisector A line that is perpendicular to a segment at its midpoint. (38)

perpendicular lines Lines that intersect to form right angles. (35)

plane An undefined term in geometry; a plane is understood to be a flat surface that extends infinitely in all directions. (10)

point An undefined term in geometry; a point can be thought of as a dot that represents a location on a plane or in space. Geometric points have no size. (9)

point of tangency The point of intersection of a circle or sphere with a tangent line or plane. (573)

polygon A closed plane figure formed from three or more segments such that each segment intersects exactly two other segments, one at each endpoint, and no two segments with a common endpoint are collinear. (138)

polyhedron A geometric solid with polygons as faces. (379)

postulate A statement that is accepted as true without proof. (11)

preimage A shape that undergoes a motion or transformation. (50)

premise A statement which is given or accepted as true in a logical argument and is used to establish a conclusion. (768)

prism A polyhedron that consists of a polygonal region and its translated image in a parallel plane, with quadrilateral faces connecting the corresponding edges. (388)

probability A number from 0 to 1 (or from 0 to 100%) that indicates how likely an event is to occur. (353)

projected points In projective geometry, the points that lie on rays containing the center of projection and the original points. (749)

projective geometry The study of the properties of figures that do not change under projection. (749)

projective rays Rays drawn from the center of a projection. (749)

proof A convincing argument that uses logic to show that a statement is true. (82)

proof by contradiction An indirect proof in which the statement that you want to prove is assumed to be false; the assumption is shown to lead to a contradiction, which indicates that the original statement must be true rather then false. (792)

proportion A statement of the equality of two ratios. (508)

proportional sides The sides of two polygons are proportional if all of the ratios of the corresponding sides are equal. (508)

pyramid A polyhedron in which all but one of the polygonal faces intersect at a single point known as the *vertex* of the pyramid. (445)

Pythagorean triple A set of positive integers a, b, and c such that $a^2 + b^2 = c^2$. (322)

quadrilateral A polygon with four sides. (148)

radius A segment that connects the center of a circle with a point on the circle; one-half the diameter of a circle. (314)

ray A part of a line that starts at a point and extends infinitely in one direction. (10)

rectangle A quadrilateral with four right angles. (148)

reductio ad absurdum A form of argument in which an assumption is shown to lead to an absurd or impossible conclusion so that the assumption must be rejected; literally "reduction to the absurd." (792)

reflection A transformation such that every point of the preimage may be connected to its image point by a segment that (a) is perpendicular to the line or plane that is the "mirror" of the reflection and (b) has its midpoint on the mirror of the reflection. (53)

reflectional symmetry A plane figure has reflectional symmetry if its reflection image across a line coincides with the preimage, the original figure. (139)

reflex angles Angles with measures greater than 180°. (29)

Reflexive Property of Equality For any real number a, $a = a$. (110)

regular polygon A polygon that is both equilateral and equiangular. (139)

regular polyhedron A polyhedron in which all faces are congruent and the same number of polygons meet at each vertex. (379)

regular pyramid A pyramid whose base is a regular polygon and whose lateral faces are congruent isosceles triangles. (445)

remote interior angle An interior angle of a triangle that is not adjacent to a given exterior angle. (174)

resultant vector The vector that represents the sum of two given vectors. (673)

rhombus A quadrilateral with four congruent sides. (148)

right angle An angle with a measure of 90°. (29)

right cone A cone in which the altitude intersects the base at its center point. (460)

right cylinder A cylinder whose axis is perpendicular to the bases. (453)

right-handed system A three-dimensional system of coordinates, named from a mnemonic device involving the fingers of the right hand. (396)

right prism A prism in which all of the lateral faces are rectangles. (389)

rigid transformation A transformation that does not change the size or shape of a figure. (50)

rise The vertical distance between two points in a coordinate plane. (190)

rotation A transformation in which every point of the preimage is rotated by a given angle about a point (in two dimensions) or a line (in three dimensions). (51)

rotation matrix A matrix used to rotate a figure about the origin through a given angle. (682)

rotational symmetry A figure has rotational symmetry if and only if it has at least one rotation image, not counting rotation images of 0° or multiples of 360°, that coincides with the original figure. (141)

run The horizontal distance between two points in a coordinate plane. (190)

same-side interior angles Interior angles that lie on the same side of a transversal. (156)

scale factor In a transformation, the number by which the distance of the preimage from the center of dilation is multiplied to determine the distance of the image point from the center. (498)

secant A line that intersects a circle at two points. (573)

secant segment A segment that contains a chord of a circle and has one endpoint exterior to the circle and the other endpoint on the circle. (600)

sector of a circle A region of a circle bounded by two radii and their intercepted arc. (316)

segment A part of a line that begins at one point and ends at another; a segment of a line has two endpoints. (10)

segment bisector A line that divides a segment into two congruent segments. (38)

self-similarity The property, possessed by fractals, that every subdivision of the fractal has a structure similar to the structure of the whole. (738)

semicircle The arc of a circle whose endpoints are the endpoints of a diameter. (566)

sides of an angle The two rays that form an angle. (11)

Sierpinski gasket A two-dimensional fractal created from a triangle by removing triangles from the centers of triangular structures at all levels. (741)

similar figures Two figures that have the same shape, but not necessarily the same size; two figures are similar if and only if one is congruent to the image of the other by a dilation. (507)

simple closed curve A shape that is topologically equivalent to a circle; a shape that does not intersect itself. (722)

sine In a right triangle, the ratio of the length of the side opposite an acute angle to the length of the hypotenuse. (639)

skew lines Lines that are not coplanar and do not intersect. (380)

slant height In a regular pyramid, the length of an altitude of a lateral face. (445)

slope The ratio of rise to run for a segment; the slope of a nonvertical line that contains the points (x_1, y_1) and (x_2, y_2) is the ratio $\frac{y_2 - y_1}{x_2 - x_1}$. (191)

solid Closed spatial figures. (379)

solid of revolution An object formed by rotating a plane figure about an axis in space. (481)

sphere The set of points in space that are equidistant from a given point known as the center of the sphere. (469)

spherical geometry Geometry on a sphere; a geometry in which line is defined as a great circle and there are no parallel lines. (731)

square A quadrilateral with four congruent sides and four right angles. (148)

straight angle An angle with a measure of 180°. (29)

supplementary angles Two angles whose measures have a sum of 180°. (28)

surface area of a prism The sum of the areas of all faces of a prism. (438)

Symmetric Property of Equality For all real numbers a and b, if $a = b$, then $b = a$. (110)

tangent In a right triangle, the ratio of the length of the side opposite an acute angle to the length of the side adjacent to it. (631)

tangent segment A segment that is contained by a line tangent to a circle and has one of its endpoints on the circle. (600)

tangent to a circle A line in the plane of a circle that intersects a circle at a exactly one point. (573)

taxicab geometry A geometry in which points are located on a special kind of map or coordinate grid whose horizontal and vertical lines represent streets and whose coordinates are always integers. (706)

taxicab radius The distance between the center of a taxicab circle and any point on the taxicab circle. (706)

taxidistance In taxicab geometry, the smallest number of grid units that must be traveled to move from one point to another point. (706)

theorem A statement that has been proven to be true deductively. (109)

topology A branch of mathematics that studies the most basic properties of figures. (721)

topologically equivalent Able to be stretched, shrunk, or otherwise distorted into another figure without cutting, tearing, or intersecting itself or compressing a segment or curve to a point. (722)

torus A three-dimensional donut-shaped surface. (723)

trace of a plane The intersection of a plane with the xy-plane in a three-dimensional coordinate system. (406)

transformation The movement of a figure in a plane from its original position, the preimage, to a new position, the image. (50)

transformation equations The equations $x' = x \cos \theta - y \sin \theta$ and $y' = x \sin \theta + y \sin \theta$ are known as transformation equations. (680)

Transitive Property of Equality For all real numbers a, b, and c, if $a = b$ and $b = c$, then $a = c$. (110)

translation A transformation in which every point of the preimage moves in the same direction by the same amount to form the image. (51)

transversal A line, ray, or segment that intersects two or more coplanar lines, rays, or segments, each at a different point. (155)

trapezoid A quadrilateral with one and only one pair of parallel sides. (148)

45-45-90 triangle A right triangle whose base angles have measures of 45°. (332)

30-60-90 triangle A right triangle whose acute angles have measures of 30° and 60°. (332)

triangle rigidity A property of triangles which states that if the sides of a triangle are fixed, the triangle can have only one shape. (217)

truth functionally equivalent When two logic statements have the same truth tables. (778)

truth table A table that lists all possible combinations of truth values for a given statement or combinations of statements. (776)

two-column proof A proof in which the statements are written in the left-hand column and the reasons are given in the right-hand column. (109)

unit circle A circle with a radius of 1 centered at the origin of the coordinate plane. (648)

valid argument An argument in which the premises force a conclusion; if the premises are true, then the conclusion is true. (768)

valid conclusion The conclusion in a valid argument that follows logically from the premises. (768)

vanishing point In a perspective drawing, the point at which parallel segments of a depicted object will meet if they are extended. (410)

vector A mathematical quantity that has both magnitude (a numerical measure) and direction. (672)

vector addition The process of combining two vectors to create a resultant vector. (673)

vertex angle of an isosceles triangle The angle opposite the base of the triangle. (237)

vector sum The resultant vector created by vector addition. (674)

vertex A point where the edges of a figure intersect; plural, *vertices*. (379)

vertex of a cone The point opposite the base of the cone. (460)

vertex of an angle The point in common of the two rays that form an angle. (11)

vertical angles The opposite angles formed by two intersecting lines. (117)

Selected Answers

Chapter 1

LESSON 1.1

Exercises

6. Sample answer: point A; $\overleftrightarrow{AB}$, $\overline{AB}$, $\overrightarrow{AB}$ **7.** $\angle 1$, $\angle Q$, $\angle PQR$, or $\angle RQP$ **8.** Sample answer: plane MNO, plane NOM, or plane $\mathcal{R}$. **9.** $\overline{AB}$, $\overline{BC}$, $\overline{AC}$ **11.** $\angle A$: $\overrightarrow{AB}$, $\overrightarrow{AC}$; $\angle B$: $\overrightarrow{BA}$, $\overrightarrow{BC}$; $\angle C$: $\overrightarrow{CA}$, $\overrightarrow{CB}$ **13.** line
15. plane **17.** point **19.** False. Planes are infinite, extending without bound and having no edges.
21. True. For example, two opposite sides of a box intersect the bottom plane of a box, but they don't intersect each other. **23.** False. There are an infinite number of planes through any two points. **25.** False. Three noncollinear points are contained in exactly one plane, but the fourth point might be in a different plane. **27.** Sample answer: point, E, A, or C
29. Sample answer: $\angle DAC$; point A; $\overrightarrow{AD}$, $\overrightarrow{AC}$
31. line ℓ **33.** point M **35.** point M **37.** Sample answer: $\overline{KP}$ and $\overline{MU}$ **39.** line m **41.** 1; $\overline{AB}$ **43.** 6, $\overline{AB}$, $\overline{AC}$, $\overline{AD}$, $\overline{BC}$, $\overline{BD}$, $\overline{CD}$ **45.** 1; $\angle AVB$ **47.** 6; $\angle AVD$, $\angle AVC$, $\angle BVD$, $\angle AVB$, $\angle BVC$, and $\angle CVD$
49. 16 **51.** 15 **53.** 49 **55.** -164 **57.** 74

LESSON 1.2

TRY THIS (p. 18)

7

Exercises

6. 5 **7.** 4 **8.** 9 **9. a.** $AB = CD$ **b.** $\overline{AB} \cong \overline{CD}$
10. $FD = 5$ miles and $BF = 35$ miles **11.** 4 **13.** 3
15. 5 **17.** $AB = 2$; $BC = 4$; $AC = 6$; The order of the coordinates does not matter in subtraction, because of the absolute value signs. For example, $|-3 - (-1)| = |-1 - (-3)| = 2$. **19.** $\overline{AF} \cong \overline{ED}$; $\overline{FG} \cong \overline{EG}$; $\overline{BG} \cong \overline{CG}$; $\overline{FC} \cong \overline{BE}$

21.

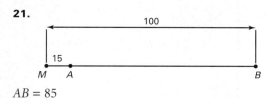

$AB = 85$

23. 5 **25.** 37 **27.** $XC = 35$ miles
29. Does not make sense; $\overline{PQ}$ is not a number.
31. Makes sense; XY, XZ, and 32 are all numbers.
33. Does not make sense; $\overline{XY}$ and $\overline{XZ}$ are not numbers. **35.** Sample answer: The Egyptian royal cubit is longer. **39.** positive **41.** negative; The result is a positive number. **43.** negative **45.** negative; The result is a positive number.

LESSON 1.3

Exercises

8. 60° **9.** 25° **10.** 20° **11.** 45°
12. a. $\angle UVW \cong \angle XYZ$ **b.** $m\angle UVW = m\angle XYZ$
13. Complementary angle pair: $\angle CAD$ and $\angle DAE$. Supplementary angle pair: $\angle BAD$ and $\angle DAE$ or $\angle BAC$ and $\angle CAE$ **15.** 85° **17.** 40° **19.** 35° **21.** 120°
23. 30° **25.** 32° **27.** 25° **29.** 155° **31.** 30° **33.** 40°
35. $m\angle SLA = 90°$; $\angle SLE$ and $\angle ALE$ are complementary angles. **37.** 22° **39.** 15 **41.** 22°
43. 30° **45.** 90° **47.** A gradian is smaller than 1 degree since one gradian is $\frac{9}{10}$ of one degree. **49.** 000
51. 180 **53.** 045 **55.** 022.5 **57.** Less than 355. The pilot would have to fly more to the west. **59.** $\angle CAR$
61. Lines ℓ, m, and n **63.** 6 **65.** $AC = 21$; $BC = 6$

LESSON 1.4

Exercises

5. parallel **6.** PB **7.** equal **8.** equal

9.

11.

13. $\angle 1$ and $\angle 2$ are complementary.

15. Lines ℓ and m are parallel.

17.

19.

21.

23. Triangles that have a vertex located on the perpendicular bisector of one of the sides have at least two sides of equal length. **25.** $AC = BC = 5$ **27.** bisector **29.** The bisectors of the angles formed by intersecting lines are perpendicular. **31.** The shape appears to be a square. **33.** line **35.** line **37.** $\overleftrightarrow{XY}$, $\overleftrightarrow{YX}$, or line ℓ **39.** $\angle MAP$; $\angle PAM$; $\angle A$ **41.** Plane $\mathcal{F}$, Plane PQR, Plane QRP, or Plane RQP

LESSON 1.5

Exercises

4. The perpendicular bisectors meet in a single point. **5.** The angle bisectors meet in a single point.

	Intersecting lines	Type of circle formed	Name of center
6.	perpendicular bisectors	circumscribed	circumcenter
7.	angle bisectors	inscribed	incenter

8.

9.

11.

13.

15.

17.

19. Inside the triangle **21.** On the triangle **23.** The center of the circumscribed circle of a right triangle is the midpoint of the longest side of the triangle. **25.** Sample answer: The distance from a vertex to the centroid is twice the distance from the centroid to the opposite side. **27.** Y is the same distance from $\overline{DE}$ as it is from $\overline{EF}$; Any point on the angle bisector

of $\angle D$ is the same distance from $\overline{DF}$ as it is from $\overline{DE}$; Any point on the perpendicular bisector of $\angle F$ is the same distance from $\overline{EF}$ as it is from $\overline{DF}$. The intersection of angle bisectors is the same distance from all 3 sides of the triangle. **29.** collinear **31.** No. Through any two points there is exactly one line. **33.** Yes. If two non-parallel lines do not intersect, they are not in the same plane. **35.** 25 **37.** 13°

LESSON 1.6, PAGES 50–58

TRY THIS (p. 53)

The figures are congruent. The line of reflection is the perpendicular bisector of the segment connecting each point with its image, not including points on the line of reflection.

Exercises

8.

9.

10.

11.

13.

15.

17.

19.

21.

23. The letters that stay the same in a reflection across a vertical line are A, H, I, M, O, T, U, V, W, X, and Y. The letters that stay the same in a reflection across a horizontal line are B, C, D, E, H, I, K, O, and X. **25.** The angles all measure 70°. **27. a.** Footprints made by the right and left foot, the pair on the left **b.** Footprints made by the same foot, the pair on the right **29.** Translate square *ABCD* from its preimage position one unit up, one unit down, and one unit to the left. Then translate the square twice to the right by one unit. **31.** Reflect square *ABCD* from its preimage position through $\overline{CB}$, $\overline{CD}$, $\overline{DA}$, and $\overline{BA}$. Then reflect the square at *BEFA* through $\overline{EF}$.

33. Transform the second image, then the third, and so on.

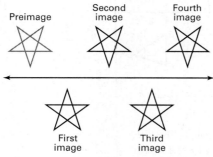

35. a. $\overline{DE}$, $\overline{DF}$, $\overline{EF}$, $\angle D$, $\angle E$, $\angle F$ **b.** *DE* = 2.2 cm, *DF* = 3 cm, *EF* = 1.6 cm **c.** m$\angle D$ = 33°, m$\angle E$ = 103°, m$\angle F$ = 44°

37.

LESSON 1.7

TRY THIS (p. 61)

If rules do not work, check your answers to the Activities.

Exercises

5. $H(x, y) = ((x + 5), y)$;

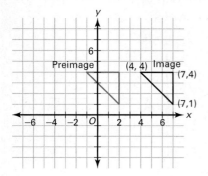

6. $N(x, y) = (-x, y)$;

7. $R(x, y) = (-x, -y)$;

9. horizontal translation;

11. vertical translation;

13. reflection across the *x*-axis;

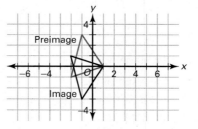

15. reflection across the *y*-axis

17. 180° rotation;

19.

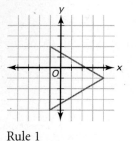

$N(x, y) = (-x, y)$

21. The result is a 180° rotation about the origin as in Ex. 20. **23.** translation 7 units to the right
25. translation 6 units to the left and 7 units up
27. translation up 7 units **29.** translation 7 units to the left

31.

Rule 1   Rule 2

Rule 3   Rule 4
The triangle ends where it started.

33. $\frac{3}{4}, \frac{3}{4}, \frac{3}{4}$ **35.** If K' is the result of the transformation of K under $T(x, y) = (x + h, y + k)$ then $\overleftrightarrow{KK'}$ has slope $\frac{k}{h}$. **37.** 45°

39.

41. $R(x, y) = (y, x)$ **43.** above the line; below the line: below the line; above the line: on the line; on the line

45. It flips back and forth across the y-axis and moves up to form a pattern.

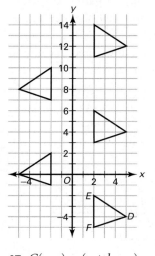

47. $G(x, y) = (x + h, -y)$
49. newpath
    72 72 moveto
    216 324 lineto
stroke
showpage

51. 315 **53.** 202.5 **55.** Find the point where the angle bisectors meet. **57.** In a rotation, a figure turns around a given point called the center of rotation.
59. In a glide transformation, a figure is reflected across a line, while being translated in a direction parallel to that line.

CHAPTER REVIEW AND ASSESSMENT

1. $\overline{KJ}, \overline{JL}, \overline{LK}$ **3.** $\overleftrightarrow{SQ}, \overleftrightarrow{RC}, \overleftrightarrow{EN}$ **5.** $CH = 4$; $CJ = 8$; $CK = 16$; $HJ = 4$; $HK = 12$; $JK = 8$

7.

$RP = 38$

9. $m\angle PVT = 125°$; $m\angle QVT = 95°$; $m\angle RVT = 65°$; $m\angle SVT = 30°$; $m\angle PVS = 95°$; $m\angle QVS = 65°$; $m\angle RVS = 35°$; $m\angle PVR = 60°$; $m\angle QVR = 30°$; $m\angle PVQ = 30°$ **11.** $m\angle LKN = 38°$

13.   **15.**

17. Sample answer: **19.**

21. a. reflection **b.** translation **c.** rotation

23.

Image   Preimage

25. **27.**

29. It is 49 miles from Smithville to LaGrange and 71 miles from Bastrop to LaGrange.

Chapter 2

LESSON 2.1

Exercises

6. Two squares of each color need to be removed.
7. 100; 10,000 **8.** Area of $B = 36$; area of $C = 16$; area of $D = 16$; area of $E = 36$; area of $F = 16$; area of $G = 4$; area of $H = 4$; overall area = 196 **9.** No The squares must be removed in 2×2 groups of four (the same shape and size as the tiles). **11.** Sample answer: If the board is completely covered by the 16 tiles, the tiles will form a 4×4 grid. The 4 removed chessboard squares must all come from just one of the squares in the 4×4 grid. **13.** 625; 5625; 11,025, yes **15.** Sample answer: The shortcut says that $(10n + 5)^2 = n(n + 1) \cdot 100 + 25$. The diagram can be rearranged as a rectangle which is $10n$ by $10(n + 1)$ and a square which is 5 by 5. **17.** 1000^3 **19.** Every entry in column C is the cube of a multiple of 3. No, columns A and B do not contain any cubes of multiples of 3. **21.** Because 999^3 will occur in column C, the cube of the number immediately following 999 must occur in column A. Therefore, 1000^3 must occur in column A. **23.** The numbers in the sequence are getting smaller. The numerator of the fraction remains 1, while the numbers in the denominators are getting larger, meaning that 1 is

being divided into more and more parts. The value of the sequence fraction is therefore decreasing.
25. The sums seem to be getting closer and closer to 1. If the sum were continued infinitely, the sum would be 1.

27.

29. The number of dots is found by multiplying the number of rows (5) by the number of columns (5), so the number of dots represents the number 5^2.

31. A square array of dots with n rows of n dots represents the number n^2. When the square is increased to $n + 1$ rows of $n + 1$ dots, $n + n + 1$ dots will have been added to the number of dots in the original square.

33. For $n = 1$: $1 = 1$ and $\frac{1(1 + 1)}{2} = \frac{1(2)}{2} = 1$.
For $n = 2$: $1 + 2 = 3$ and $\frac{2(2 + 1)}{2} = \frac{2(3)}{2} = 3$.
For $n = 3$: $1 + 2 + 3 = 6$ and $\frac{3(3 + 1)}{2} = \frac{3(4)}{2} = 6$.
For $n = 4$: $1 + 2 + 3 + 4 = 10$ and
$\frac{4(4 + 1)}{2} = \frac{4(5)}{2} = 10$.
For $n = 5$: $1 + 2 + 3 + 4 + 5 = 15$ and
$\frac{5(5 + 1)}{2} = \frac{5(6)}{2} = 15$.

35. 21; yes **37.** The number of dots in the triangle is one-half the number of dots in the rectangle. The diagram suggests that the sum of the integers from 1 to n can be related to the pattern of the dots in the triangle. This triangle encloses half the number of dots in the rectangle with dimensions n and $n + 1$, therefore it contains $\frac{n(n + 1)}{2}$ dots. **39.** The murder was committed in the study with the candlestick by Colonel Mustard. A disjunctive syllogism can often be used when at least one of two things must be true.

If one of the two can be shown to be false, then the other must be true.

41.

a triangle

43. 23 **45.** m∠*KIY* = 45°, m∠*TIY* = 35°

LESSON 2.2

TRY THIS (p. 92)

If an animal is a snake, then the animal is a reptile. Converse: If an animal is a reptile, then it is a snake. The conditional is true; the converse is false.

TRY THIS (p. 93)

Your answer should show connections between several "If…then" statements to prove an "If…then" statement.

Exercises

6. If a worker is a United States Postal worker, then the person is a federal employee. John is a United States Postal worker. Therefore, John is a federal employee.

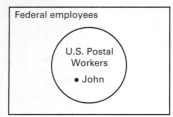

7. Conditional: If two lines are parallel, then the two lines do not intersect. Converse: If two lines do not intersect, then the two lines are parallel. The conditional is true. The converse is not true.
8. Form a logical chain: If a number is divisible by 4, then the number is divisible by 2. If a number is divisible by 2, then the number is even. If a number is even, then the last digit is 0, 2, 4, 6, or 8. Therefore, if a number is divisible by 4, then the last digit is 0, 2, 4, 6, or 8 by the If-Then Transitive Property. **9.** If a person lives in Ohio, then the person lives in the United States.

11.

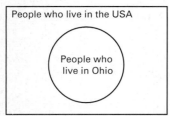

13. If a plant is a tulip, then it is a flower. **15.** If a person is a flutist, then that person is a musician.
17. Hypothesis: it is snowing in Chicago
Conclusion: it is snowing in Illinois
Converse: If it is snowing in Illinois, then it is snowing in Chicago. The converse is false. Sample counterexample: It is snowing in Springfield, Illinois, but not in Chicago, Illinois. **19.** Hypothesis: the measure of each angle in a triangle is less than 90°
Conclusion: the triangle is acute Converse: If a triangle is acute, then the measure of each angle is less than 90°. The converse is true. **21.** If ∠*AXB* and ∠*BXD* form a linear pair, then ∠*AXB* and ∠*BXD* are supplementary. **23.** If m∠*BXC* + m∠*CXD* = 90°, then m∠*AXB* = 90°

25. Conclusion: Socrates is a mortal.

27. Conclusion: Ingrid lives in Scandinavia.

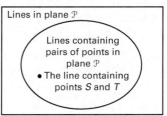

29. Conclusion: The line containing points *S* and *T* is in the plane.

31. If Tim drives a car, then Tim drives too fast. If Tim drives too fast, then the police catch Tim speeding. If the police catch Tim speeding, then Tim gets a ticket. Conditional: If Tim drives a car, then Tim gets a ticket. **33.** If you clean your room, then you will go to the movie. If you go to the movie, then you will spend all of your money. If you spend all of your money, then you cannot buy gas for your car. If you cannot buy gas for your car, then you will be stranded. Conclusion: If you clean your room, then you will be stranded. **35.** If a nail is lost, then the shoe is lost. If a shoe is lost, then the horse is lost. If a horse is lost, then the rider is lost. If a rider is lost, then the battle is lost. If a battle is lost, then the war is lost. Conclusion: If a nail is lost, then the war is lost. **37.** Using the first conditional to lay the foundations, your argument should include statements that support the claim that the object displays form, beauty, and unusual perception on the part of its creator. Using the second conditional to lay the foundations, your argument should include statements that support the claim that the object displays creativity on the part of its creator. **39.** a plane **41.** Fold over one side of the angle until it meets the edge of the opposite side of the angle. The line formed is the bisector of the angle. **43.** incenter, or center of the inscribed circle of the triangle **45.** reflection, rotation, and translation

LESSON 2.3

TRY THIS (p. 101)

Your definition should accurately describe the object in clear and concise terms.

Exercises

6. Sample answer: A glosh is a figure which has 6 sides, 2 of which are parallel, and the remaining 4 sides are equal in length. According to this definition, **a, c, d,** and **f** are gloshes. **7.** ∠1 and ∠3; ∠1 and ∠2; ∠2 and ∠4; ∠3 and ∠4 **9. a.** Conditional: If a person is a teenager, then the person is from 13 to 19 years old. **b.** Converse: If a person is from 13 to 19 years old, then the person is a teenager. **c.** Biconditional: A person is a teenager if and only if the person is from 13 to 19 years old. **d.** The statement is a definition because the conditional and the converse are both true. **11. a.** Conditional: If a number is even, then it is divisible by 2. **b.** Converse: If a number is divisible by 2, then it is even. **c.** Biconditional: A number is even if and only if it is divisible by 2. **d.** The statement is a definition because the conditional and the converse are both true. **13. a.** Conditional: If an angle is a right angle, then it has a measure of 90°. **b.** Converse: If an angle has a measure of 90°, then it is a right angle. **c.** Biconditional: An angle is a right angle if and only if it has a measure of 90°. **d.** The statement is a definition because the conditional and the converse are both true. **15. a.** Conditional: If a substance is hydrogen, then it is the lightest of all known substances. **b.** Converse: If a substance is the lightest of all known substances, then it is hydrogen. **c.** Biconditional: A substance is hydrogen if and only if it is the lightest of all known substances. **d.** The statement is a definition because the conditional and the converse are both true. **17.** ∠WVX and ∠XVY; ∠XVY and ∠YVZ; ∠WVY and ∠YVZ; ∠WVX and ∠XVZ **19.** They do not share a common side. **21.** They do not share a common vertex. **23.** shapes **b** and **d 25.** shapes **b** and **c 27.** Sample Answer: A polygon is a regular polygon if and only if all of its sides and angles are congruent. **29.** Ciconiiformes: Great Blue Heron, Roseate Spoonbill; Cuculiformes: Greater Roadrunner, Yellow-Billed Cuckoo; Apodiformes: White-Collared Swift, Ruby-Throated Hummingbird

31.

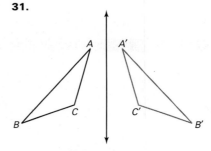

33. They are parallel.

35.

37. They are parallel and the same length.

LESSON 2.4

TRY THIS (p. 111)

Sample answer: Line segments $\overline{AB}$, $\overline{BC}$, and $\overline{CD}$ such that $AB = 10$, $\overline{AB} \cong \overline{BC}$, and $\overline{BC} \cong \overline{CD}$. $\overline{AB} \cong \overline{AB}$ by the Reflexive Property. Since $\overline{AB} \cong \overline{BC}$, $\overline{BC} \cong \overline{AB}$ by the Symmetric Property. Since $\overline{AB} \cong \overline{BC}$ and $\overline{BC} \cong \overline{CD}$, then $\overline{AB} \cong \overline{CD}$ by the Transitive Property.

Exercises

5. 19 **6.** 30 **7.** $\triangle CDE \cong C''D''E''$ **8.** Transitive Property of Congruence **9.** Subtraction **11.** Transitive **13.** m∠MLN + m∠NLP **15.** Angle Addition Postulate **17.** m∠MLP = m∠NLQ **19.** m∠MLN = m∠PLQ **21.** 15 **23.** 85° **25.** GH **27.** Transitive Property **29.** Yes; Transitive Property **31.** Transitive Property **33.** m∠BAC = m∠DEC **35.** Substitution Property **37.** m∠1 + m∠2 + m∠3 = 180° **39.** 90° **41.** m∠CDE + m∠CDB = m∠EDB **43.** Transitive (or Substitution) Property **45.** Substitution Property **47.** Transitive; the relation is not an equivalence relation. Counterexample to the Reflexive Property: $5 \not< 5$. Counterexample to the Symmetric Property: $2 < 3$ but $3 \not< 2$. **49.** Reflexive and Transitive; the relation is not an equivalence relation.

Counterexample to the Symmetric Property: 10 is divisible by 5, but 5 is not divisible by 10. **51.** Reflexive, Symmetric, and Transitive; it is an equivalence relation. (Note: Any figure is a rotation of itself, by an amount of 360°). **53.** Reflexive, Symmetric, and Transitive; it is an equivalence relation. **55.** Postulate 1.1.7: Through any three noncollinear points there is exactly one plane. **57.** Postulate 1.1.6: Through any two points there is exactly one line. **59.** Postulate 1.1.5: The intersection of two planes is a line.

LESSON 2.5

Exercises

7.

8.

9.

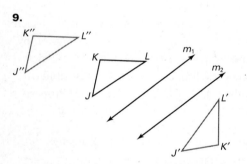

The first set of reflections translates the figure right and downward. The second set of reflections translates the figure left and upward.

11. ∠LEP **13.** 160° **15.** 90° **17.** x = 15; m∠ABC = 60° **19.** x = −5; m∠ABC = 73° **21.** Inductive reasoning; not a proof; it has not been shown for all cases. **23.** Deductive reasoning; a proof; The angles in the diagram form a linear pair. **25.** Transitive (or Substitution) Property **27.** Substitution Property (Angle Congruence Postulate) **29.** 10 cm **31.** 2x cm **33.** The parallel lines should be drawn perpendicular to the direction of the arrow.

35.

37.

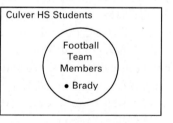

39. 31

41.

Number of Regions	1	2	4	8	16	31	57	99
1st differences →		1	2	4	8	15	26	42
2nd differences →			1	2	4	7	11	16
3rd differences →				1	2	3	4	5

43. No, an induction proof cannot be a proof for all cases. **45.** Sample answer: $\overline{AB}, \overline{BC}, \overline{CD}, \overline{AD}$ **47.** If it is Saturday, then I sleep until 8:00 A.M. If I sleep until 8:00 A.M., then I am well rested during the afternoon. If I am well rested in the afternoon, then I am in a good mood.

49.

Culver HS Students

Football Team Members

• Brady

CHAPTER REVIEW AND ASSESSMENT

1. 7; 10 **3.** The curve is divided by n horizontal lines into $1 + 3n$ sections. **5.** The evening star is a planet. **7.** c, b, a **9.** c and d **11.** No. The converse is not true because a figure with four sides is not always a square. **13.** Linear Pair Property **15.** Vertical Angles Theorem **17.** Let $a = 3$, $b = 4$, $c = 5$. Then $a^2 + b^2 = c^2$ gives: $3^2 + 4^2 = 5^2$, which is true, so 3, 4 and 5 form a Pythagorean triple. Let $a = 5$, $b = 12$, $c = 13$. Then $a^2 + b^2 = c^2$ gives: $5^2 + 12^2 = 13^2$, which is true, so 5, 12, and 13 form a Pythagorean triple. **19.** Conjecture: If a, b, and c are a Pythagorean triple such that $a^2 + b^2 = c^2$ and c, d and e are another Pythagorean triple such that $c^2 + d^2 = e^2$, then $a^2 + b^2 + d^2 = e^2$. **21.** $3^2 − 4 \times 1 \times 4 = 9 − 16 = −7$. The program would display "NO REAL SOLUTIONS."

Chapter 3

LESSON 3.1

Exercises

6. Yes, the axis of symmetry bisects both the side and the angle through which it passes.

7. square: 90°, pentagon: 72°, hexagon: 60°
8. 4 blades: 4-fold, 6 blades: 6-fold

9. **11.**

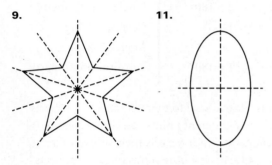

13. A circle has an infinite number of axes of symmetry. Any line through the center of the circle is an axis of symmetry; an infinite number of such lines can be drawn.

15.

17. Exercise 16

19. **21.**

23. Sample answer: an isosceles triangle that is not equilateral. **25.** Sample answer: an equilateral triangle. **27.** Sample answer: a regular octagon. **29.** point B is the image of point A, point A is the image of point B, $\overrightarrow{BA}$ is the image of $\overrightarrow{AB}$ **31.** $\overrightarrow{BC}$ is the image of $\overrightarrow{BA}$, $\overrightarrow{BA}$ is the image of $\overrightarrow{BC}$, $\angle CBA$ is the image of $\angle ABC$ **33.** either **35.** rotation **37.** reflection **39.** reflection

41.

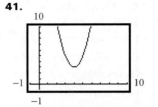

Axis of symmetry: $x = 4$

43.

Axis of symmetry: $x = -5$

45.

Axis of symmetry: $x = -3$

47. Sample answer:

49. The extended table is not regular. It is not equilateral but it is equiangular. **51.** The Egyptian bowl has 5 axes of symmetry. They are located through the middle of every row of diamond shaped figures. There are nontrivial rotational symmetries for rotations of 72°, 144°, 216°, and 288°. **53.** The Egyptian bowl has 2 axes of symmetry—vertically and horizontally through the center. There is a nontrivial rotational symmetry for a rotation of 180°. **55.** The sign is a regular octagon with 8 axes of symmetry—4 lines through the midpoints of opposite sides and 4 lines through opposite vertices. It has nontrivial rotational symmetries for rotations of 45°, 90°, 135°, 180°, 225°, 270° and 315°. **57.** The sign is a regular triangle or equilateral triangle. It has 3 axes of symmetry—each axis passes through a vertex and the midpoint of the opposite side. It has nontrivial rotational symmetries for rotations of 120° and 240°. **59.** point O **61.** $\overleftrightarrow{AB}$ **63.** 85°

LESSON 3.2

Exercises

5. a. 3 in. **b.** 50° **c.** 1 in. **d.** 130° **6. a.** 12 cm **b.** 30° **c.** 150° **d.** 90° **7.** 4 ft **8.** m$\angle PTQ$ = m$\angle QTR$ = 90°

9. 10 **11.** 6.5 **13.** 130° **15.** 50° **17.** 21 **19.** 7.5 **21.** 110° **23.** 90° **25.** 8 **27.** 5 **29.** 50 **31.** 90° **33.** $x = 60$, m∠P = m∠R = 120°, m∠Q = m∠S = 60° **35.** False; rectangle **37.** True **39.** True **41.** False; square **43.** If a figure is a rhombus, then its opposite sides are congruent, so it must be a parallelogram. The statement is true. **45.** If the creature is a whale, then it is a mammal. Converse: If the creature is a mammal, then it is a whale. This is not a definition because not all mammals are whales. **47.** If a figure is a square, then it is a rectangle. Converse: If a figure is a rectangle, then it is a square. This is not a definition because not all rectangles are squares.

49.

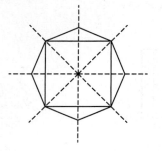

LESSON 3.3

TRY THIS (p. 157)

Sample answer: alternate exterior angles 2 and 7, alternate interior angles 3 and 6, and same-side interior angles 4 and 6

Exercises

5. ∠3 and ∠6; ∠4 and ∠5 **6.** ∠1 and ∠8; ∠2 and ∠7 **7.** ∠3 and ∠5; ∠4 and ∠6 **8.** Sample answer: ∠1 and ∠5; ∠2 and ∠6 **9.** ∠4, ∠5, ∠8 **11.** No. Every angle either is vertical to, corresponds to, or is vertical to a corresponding angle of ∠1 or ∠2. Hence every other angle is congruent to ∠1 or ∠2. **13.** ∠1 ≅ ∠6 by the Vertical Angles Theorem. ∠1 ≅ ∠9 by the Corresponding Angles Postulate. ∠9 ≅ ∠14 by the Vertical Angle Theorem, so ∠1 ≅ ∠14 by the Transitive Property of Congruence. **15.** ∠3 ≅ ∠4 by the Vertical Angles Theorem. ∠3 ≅ ∠8 by the Corresponding Angles Postulate. ∠8 ≅ ∠11 by the Vertical Angles Theorem, so ∠3 ≅ ∠11 by the Transitive Property of Congruence. **17.** By the Vertical Angles Theorem, the angle formed by ∠2 and ∠3 is congruent to the angle formed by ∠4 and 5; and by the Corresponding Angles Postulate, is congruent to ∠10. By the Vertical Angles Theorem, ∠10 ≅ ∠13, so by the Transitive Property of Congruence, ∠13 is congruent to the angle formed by ∠2 and ∠3. **19.** transversal; intersects lines r and s **21.** transversal;

intersects lines p and q **23.** 50° **25.** 25° **27.** 25° **29.** 150° **31.** 150° **33.** 150°

35. Given: Line ℓ ∥ line m. Line p is a transversal.
Prove: m∠2 + m∠1 = 180°

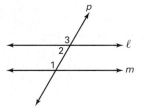

Statements	Reasons
Line ℓ ∥ line m Line p is a transversal.	Given
m∠2 + m∠3 = 180°	Linear Pair Property
m∠3 = m∠1	Corresponding Angles Postulate
m∠2 + m∠1 = 180°	Substitution Property

37. It is a transversal to them because it intersects them at two different points. **39.** It is a transversal to them because it intersects them at two different points. **41.** $\overleftrightarrow{AB}$ is a transversal since it intersects the parallel mirrors. It is the path that light travels between the periscope mirrors and it represents the line of sight of the periscope. **43.** By optics, ∠1 ≅ ∠2. ∠2 ≅ ∠3 because they are alternate interior angles. ∠1 ≅ ∠3, by the Transitive Property of Congruence. By optics, ∠3 ≅ ∠4. Thus, ∠1 ≅ ∠4, by the Transitive Property of Congruence. **45.** ray **47.** parallel **49.** polygon **51.** rhombus

LESSON 3.4

Exercises

5. If alternate interior angles are congruent, then the two lines cut by the transversal are parallel. **6.** The indicated angles are corresponding and congruent, so the Converse of the Corresponding Angles Postulate states lines m and n are parallel.

7.

9. Alternate Exterior Angles Theorem **11.** Alternate Interior Angles Theorem **13.** Corresponding Angles Postulate **15.** Same-Side Interior Angles Theorem

17. Given: $m \parallel n$; m$\angle RSU$ = m$\angle RTU$ = 70°
Prove: $RSUT$ is a parallelogram.

Statements	Reasons
$m \parallel n$; m$\angle RSU$ = m$\angle RTU$ = 70°	Given
$\angle RSU \cong \angle RTU$	Angle Congruence Postulate
$\angle RTU \cong \angle 1$ where $\angle 1$ is its alternate interior angle with transversal $\overline{RT}$.	Alternate Interior Angles Theorem
$\angle RSU \cong \angle 1$	Transitive Property of Congruence
$\overline{RT} \parallel \overline{SU}$	Converse of the Corresponding AnglesPostulate
$RSTU$ is a parallelogram.	definition of parallelogram

19. Definition of supplementary angles
21. Transitive (or Substitution) Property
23. Converse of the Corresponding Angles Postulate
25. Given: $\angle 1 \cong \angle 2$; Prove: $\ell_1 \parallel \ell_2$

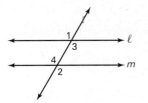

Statements	Reasons
$\angle 1 \cong \angle 2$	Given
$\angle 2 \cong \angle 4$	Vertical Angles Theorem
$\angle 1 \cong \angle 4$	Transitive Property of Congruence
$\ell_1 \parallel \ell_2$	Converse of the Corresponding Angles Postulate

27. both are right angles **29.** line p **31.** $\angle 3$ **33.** $\angle 2$
35. $\angle 2$ **37.** Converse of Corresponding Angles
Postulate **39.** Using the T square as the transversal,
each line drawn with the triangle will have
congruent corresponding angles to every other line
drawn with the triangle. Thus, by the Converse of
the Corresponding Angles Postulate, the lines will be
parallel. **41.** If a figure is a rectangle, then it is a
parallelogram.

43.

45. It is false. For example, a rhombus is a
parallelogram, but it is not necessarily a rectangle.
47. 145 and 325

LESSON 3.5

Exercises
5. $\angle 1$ and $\angle 4$; $\angle 3$ and $\angle 5$ **6.** 180° **7.** 65° **9.** 45°
11. No such triangle exists. **13.** 60° **15.** 50° **17.** 130°
19. 70° **21.** 30° **23.** 100° **25.** 5, 90°, 45°, 45° **27.** 80°
29. 110° **31.** 110° **33.** 60° **35.** 130° **37.** 130°
39. the sum of the measures of the remote interior
angles **41.** Linear Pair Property **43.** Transitive (or
Substitution) Property **45.** 115 **47.** 2 **49.** linear pair

LESSON 3.6

Exercises
4. 1080° **5.** 1980° **6.** 360° **7.** 360° **9.** 110° **11.** 110°
13. 120° **15.** 90°; 90° **17.** 150°; 30° **19.** 8 **21.** 24
23. 10 **25.** 36° **27.** 108° **29.** 70° **31.** 50°
33. 140° **35.** 71° **37.** 58° **39.** 3; Sample answer: a
quadrilateral with angle measures 60°, 70°, 70°, 160°
has 3 acute angles, but if a quadrilateral had 4 acute
angles, then the sum of their measures would be less
than 360°. Yes; for example, a rectangle has 4 right
angles.
41. 180°

43.

$x = 143$, $y = 100$

45. $\angle 1$ and$\angle 2$, $\angle 2$ and $\angle 3$, $\angle 3$ and $\angle 4$, and $\angle 4$ and
$\angle 1$; linear pair angles **47.** $\angle 1$ and $\angle 3$, $\angle 2$ and $\angle 4$;
vertical angles

LESSON 3.7

Exercises

5. 6 **6.** 14 miles **7.** 11 **8.** 114 feet **9.** Length of triangle midsegment $= \frac{\text{base}}{2}$; Length of trapezoid midsegment $= \frac{\text{base}_1 + \text{base}_2}{2}$ the formula for the length of the trapezoid midsegment can be used to find the length of a triangle midsegment by letting the short base $= 0$. **11.** 25 **13.** 80 **15.** 10 **17.** The lengths of the segments are multiples of the length of DE.
19. If parallel segments divide two sides of a triangle into three congruent segments, the length of the shortest segment is $\frac{1}{3}$ of the length of the base and the length of the other segment is $\frac{2}{3}$ of the length of the base. If parallel segments divide two sides of a triangle into eight congruent segments, the length of the shortest segment is $\frac{1}{8}$ of the length of the base and the length of each of the remaining segments is $\frac{m}{8}$ of the length of the base where $m = 2, 3, 4, 5, 6, 7$. If parallel segments divide two sides of a triangle into n congruent segments, the length of each segment is $\frac{m}{n}$ of the length of the base where m is the segment number, from 1 to $n - 1$. **21.** By the argument of Exercise 19, each successive decreases by $\frac{1}{n} = \frac{1}{3}$ if the base length KL is taken as one unit. Therefore $MN = \frac{1}{3}KL$ and $PQ = \frac{2}{3}KL$.

23.

Parallelogram; The two sides of the figure are triangle midsegments making opposite sides parallel, therefore a parallelogram is formed inside the triangle.

25.

Rectangle; Two sides of the figure are triangle midsegments making opposite sides parallel, therefore a parallelogram is formed. Consecutive interior angles are supplementary, so all angles measure 90°.

27. Midsegment lengths are $\frac{1}{2}, \frac{1}{4}, \frac{1}{16}, \cdots$ Looking at the bottom edge of the box, we can figure out the lengths of the segments along the bottom by the lengths of the midsegments. The first segment is $1 - \frac{1}{2} = \frac{1}{2}$. The 2nd segment is $\frac{1}{2} - \frac{1}{4} = \frac{1}{4}$. The 3rd segment is $\frac{1}{4} - \frac{1}{8} = \frac{1}{8}$ and so on. Thus

$\frac{1}{2} + \frac{1}{4} + \frac{1}{8} + \cdots$ must add up to the length of bottom edge of the square, or 1. **29.** 2.5 ft **31.** 40%

33.

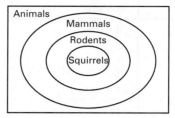

35. True; all rectangles have two sets of parallel sides, thus all rectangles are parallelograms. **37.** False; parallelograms need not have all four sides congruent. **39.** 70

LESSON 3.8

TRY THIS (p. 193)

$\overline{GD}$ midpoint: $(2, 1)$, $\overline{EF}$ midpoint: $(8, 1)$, $DE = 3$, $GF = 9$, length of midsegment $= 6$. So the length of the midsegment is $\frac{1}{2}(DE + GF) = \frac{1}{2}(3 + 9) = \frac{1}{2}(12) = 6$, confirming the trapezoid midsegment conjecture.

Exercises

6.

slope $= \frac{6}{2} = 3$

7. 1 **8.** $\frac{2}{5}$ **9.** $-\frac{7}{2}$

10.

This is a parallelogram because the slopes of opposite sides are equal, so opposite sides are parallel.

11.

Midsegment from $(0, 2)$ to $(4, 2)$ has length 4. Using the bases, the midpoint length $= \frac{1}{2}(\text{base 1} + \text{base 2}) = \frac{1}{2}(3 + 5)$. Also, since the bases and midsegment are all horizontal, they are parallel.

13. -1; $(0, 0)$ **15.** $-\frac{5}{6}$; $\left(-2, -\frac{1}{2}\right)$ **17.** perpendicular
19. neither **21.** $\frac{3}{2}$ **23.** Yes **25.** Yes **27.** No

29.

Rectangles; The slopes of $\overline{CD}$ and $\overline{AB}$ are 0 and the slopes of $\overline{DA}$ and $\overline{BC}$ are undefined. Therefore the opposite sides of the polygon are parallel. Vertical lines have undefined slopes and horizontal lines have zero slopes. Thus adjacent sides are perpendicular and $ABCD$ is a rectangle.

31.

Parallelogram; Sides $\overline{AB}$ and $\overline{CD}$ have slope 2, hence are parallel. Sides $\overline{AD}$ and $\overline{BC}$ have slope $\frac{-1}{3}$, hence are parallel. Since it has two pairs of parallel sides, $ABCD$ is a parallelogram.

33.

Rectangle; Sides $\overline{BC}$ and $\overline{AD}$ have slope -1, hence are parallel. Sides $\overline{AB}$ and $\overline{CD}$ have slope 1, hence are parallel. There are two sets of parallel sides, and

the adjacent sides have slopes whose product is -1, therefore the adjacent sides are perpendicular. $ABCD$ is a rectangle.

35.

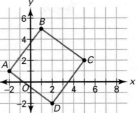

The slope of diagonal $\overline{AC} = \frac{1-2}{-2-5} = \frac{1}{7}$, the slope of diagonal $\overline{BD} = \frac{5-(-2)}{1-2} - 7$. Since the product of the two slopes is -1, the diagonals are perpendicular.

37.

The slope of the diagonal $\overline{EG} = \frac{a-0}{a-0} = \frac{a}{a} = 1$, and the slope of the diagonal $\overline{EF} = \frac{0-a}{a-0} = \frac{-a}{a} = -1$. Since the product of the slopes is -1, the diagonals are perpendicular.

39. Two possibilities are $(7, 1)$ and $(4, -5)$. **41.** Two possibilities are $(3, 4)$ and $(5, 2)$. **43.** The endpoints of the midsegment are $(2, 0)$ and $(2, 8)$. The midsegment is vertical, as are the bases, hence it is parallel to them.

Sample Answers, 45-49

45.

47.

49.

51. The roof rises $23.0 - 10.5 = 12.5$ ft over a run of $\frac{25}{2} = 12.5$ ft. The slope or pitch is then $\frac{12.5}{12.5} = 1.0$. The house violates the building codes. Sample answer: Adjust the roof so that $\frac{\text{rise}}{12.5} = .7$ rise $= .7(12.5) = 8.75$ Adjust the peak of the roof to $(12.5, 19.25)$. **53.** 20° **55.** 35° **57.** 108°

CHAPTER REVIEW AND ASSESSMENT

1. **3.**

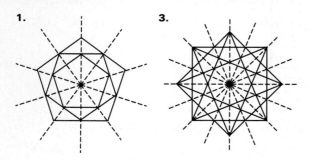

5. 48° **7.** 90° **9.** Sample answer: ∠2 and ∠4
11. ∠3, ∠6, ∠8 **13.** Yes; by the Converse of the Corresponding Angles Postulate they are parallel.
15. No; because if so, then same-side exterior angles would be supplementary, but $115° + 70° = 185° \neq 180°$. **17.** m∠TRS = 25°, m∠RST = 30°, m∠STR = 125° **19.** m∠QPR = 33°, m∠PQR = 57°, m∠QRP = 90° **21.** 35°
23. $\left(\frac{2700}{17}\right)° \approx 158.8°$ **25.** 6 **27.** 40 **29.** Yes
31. c. rectangle; Slopes of opposite sides are equal and slopes of consecutive sides have product −1.
33. The quilt has 2 axes of symmetry—2 lines through the center along the diagonals. It has a nontrivial rotational symmetry for a rotation of 180°. **35.** 28.6%, 40%

Chapter 4

LESSON 4.1

TRY THIS (p. 211)

KLMNO	ONMLK
LMNOK	KONML
MNOKL	LKONM
NOKLM	MLKON
OKLMN	NMLKO

Exercises

7. QPTSR, QRSTP, PTSRQ, RSTPQ, TSRQP, STPQR, SRQPT, TPQRS, RQPTS, PQRST
8. Sample answer: $PQRST \cong VZYXW$ **9.** It is given that $\angle A \cong \angle D$, $\angle ABC \cong \angle DBC$, $\angle ACB \cong \angle DCB$, $\overline{AB} \cong \overline{DB}$, and $\overline{AC} \cong \overline{DC}$. By the Reflexive Property of Congruence, $\overline{BC} \cong \overline{BC}$, so $\triangle ABC \cong \triangle DBC$ by the Polygon Congruence Postulate. **11.** ∠QRS, ∠TUQ, ∠RST, ∠UQR, ∠STU **13. a.** $\overline{XY}$ **b.** $\overline{RS}$ **c.** $\overline{VZ}$
15. Yes; the segments have the same length.
17. $\triangle ABC \cong \triangle DBC$; All corresponding angles and sides are congruent. **19.** Yes; the measure of the two angles is equal. **21. a.** $\overline{ED}$ **b.** $\overline{DF}$ **c.** $\overline{EF}$ **23. a.** $\overline{OP}$
b. $\overline{MP}$ **c.** $\overline{LM}$ **d.** $\overline{MP}$ **25. a.** $\overline{FG}$ **b.** $\overline{DE}$ **c.** $\overline{CD}$
d. $\overline{GH}$ **e.** $\overline{BC}$ **27.** 8 **29.** $(180 - 2x)°$; $(180 - 2x)°$
31. $\triangle XYZ \cong \triangle JKL$ **33.** $\triangle ABC$ is isosceles; $\overline{AB}$ and $\overline{AC}$ are corresponding sides, so $\overline{AB} \cong \overline{AC}$.
35. a. $\triangle RNO \cong \triangle AKC$, $FRNT \cong BAKX$, $FRAB \cong TNKX$, $FBXT \cong RAKN$, $RACO \cong NKCO$, $FRONT \cong BACKX$ **b.** Sample answer: $\angle ORN \cong \angle CAK$, $\angle ONR \cong \angle CKA$, $\angle RON \cong \angle ACK$, $\angle RFT \cong \angle ABX$, $\angle RFB \cong \angle NTX$ **c.** Sample answer: $\overline{FR} \cong \overline{TN}$, $\overline{BA} \cong \overline{XK}$, $\overline{RA} \cong \overline{NK}$, $\overline{FB} \cong \overline{TX}$, $\overline{FT} \cong \overline{BX}$

37. Sample answer:

$\triangle ABC \cong \triangle DEF$

39. definition of parallelogram **41.** Alternate Interior Angles Theorem **43.** Reflexive Property of Congruence **45.** False; parallelograms have two pairs of parallel sides but trapezoids only have 1 pair of parallel sides. **47.** True; every rectangle is a quadrilateral with two pairs of parallel sides.

LESSON 4.2

Exercises

5. SAS **6.** SSS **7.** ASA **9.** $\triangle FDR \cong \triangle WXY$; ASA
11. $\triangle FGH \cong \triangle FDH$; ASA **13.** Can't be proven congruent **15.** Yes; SAS **17.** Yes; ASA **19.** No, more than one triangle can be constructed.

21.

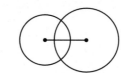

Any segment that connects this endpoint to a point on the circle is congruent to the third segment, because the compass setting used to create the circle was the length of the third segment.

23. Converse of Same-Side Interior Angles Theorem
25. Converse of Same-Side Interior Angles Theorem
27. Reflexive Property of Congruence **29.** ASA

31. Given: rhombus *ABCD*
Prove: Diagonal $\overline{DB}$ divides rhombus *ABCD* into two congruent triangles.

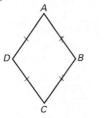

Statements	Reasons
$\overline{AB} \cong \overline{CD}$	definition of a rhombus
$\overline{AD} \cong \overline{BC}$	
$\overline{BD} \cong \overline{DB}$	Reflexive Property of Congruence
$\triangle ABD \cong \triangle CDB$	SSS

33. The diagonal board turns the rectangle into two congruent, rigid triangles. **35. a.** Yes; Because vertical angles are congruent, $\angle FOA \cong \angle COD$. It is given that $\overline{FO} \cong \overline{CO}$ and $\overline{OA} \cong \overline{OD}$. By SAS, $\triangle FOA \cong \triangle COD$. **b.** Yes; Because it is a regular hexagon, $\overline{AB} \cong \overline{DC}$. It is given that $\overline{BO} \cong \overline{CO}$ and $\overline{OA} \cong \overline{OD}$. By SSS, $\triangle BOA \cong \triangle COD$.
c. $\triangle FOE \cong \triangle COB$ by either SAS or SSS. **37.** Yes, by SSS **39.** Yes, by ASA **41.** Given **43.** Transitive Property of Congruence **45.** $\angle X$ **47.** Possible

LESSON 4.3

TRY THIS (p. 228)

a and c

Exercises

6. No, AAA is not a valid test for congruence. **7.** Yes, AAS is a valid test for congruence. **8.** No, SSA is not a valid test for congruence. **9.** 2×10^8 km and 0.5×10^8 km **11.** $\triangle DEF \cong \triangle FGD$; SSS
13. $\triangle PQR \cong \triangle TSU$; SAS **15.** Can't be proven congruent **17.** $\triangle WXY \cong \triangle WZY$; ASA **19.** Can't be proven congruent **21.** $\triangle XWY$ has hypotenuse $\overline{XY}$; $\triangle XWZ$ has hypotenuse $\overline{XZ}$ **23.** Reflexive Property of Congruence

25.

Statements	Reasons
$\angle A \cong \angle D$	Given
$\overline{AF} \cong \overline{DC}$	
$\angle BFA \cong \angle ECD$	
$\triangle AFB \cong \triangle DCE$	ASA

27. AAS **29.** Transitive Property of Congruence
31. No

33. Yes; HL   **35.**

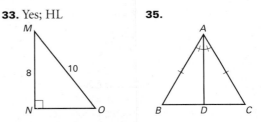

a. $\triangle ABD \cong \triangle ACD$ by SAS. **b.** $\angle ABD \cong \angle ADC$ because CPCTC. Because they form a linear pair and are congruent, $\angle ADB$ and $\angle ADC$ are right angles, so $m\angle ADB = m\angle ADC = 90°$.

37. Given: $\angle B \cong \angle E$, $\angle A \cong \angle D$, $\overline{BC} \cong \overline{EF}$
Prove: $\triangle ABC \cong \triangle DEF$ (AAS is a valid theorem)

Statements	Reasons
$\angle B \cong \angle E$ ($m\angle B = m\angle E$)	Given
$\angle A \cong \angle D$ ($m\angle A = m\angle D$)	
$\overline{BC} \cong \overline{EF}$	
$m\angle F = 180° - m\angle E - m\angle D$ $m\angle C = 180° - m\angle B - m\angle A$	Triangle Sum Theorem
$m\angle C = 180° - m\angle E - m\angle D$	Substitution Property
$m\angle C = m\angle F$ ($\angle C \cong \angle F$)	Substitution Property
$\angle C \cong \angle F$	Angle Congruence Postulate
$\triangle ABC \cong \triangle DEF$	ASA

39. a. 1 **b.** 1 **c.** 1 **d.** longer than

41.

Height ≈ 461 ft

43. a.

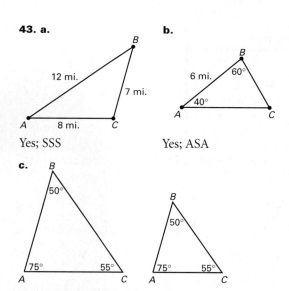

Yes; SSS

b.

Yes; ASA

c.

No. AAA is not a valid congruence postulate.

45. Sample answer:

$\ell_1 \quad \ell_2 \quad \ell$

47. $\angle 1 \cong \angle 3 \cong \angle 5 \cong \angle 7$, $\angle 2 \cong \angle 4 \cong \angle 6 \cong \angle 8$
49. 40°, 60°, 80°

LESSON 4.4

TRY THIS (p. 237)

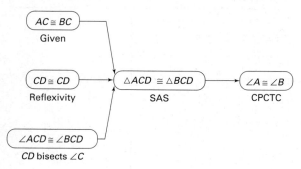

5. $\triangle ABC \cong \triangle AED$ **6.** CPCTC
7. Overlapping Segments Theorem **8.** HL
9. $\angle A$ is supplementary to an angle measuring 50°
so $m\angle A = 180° - 50° = 130°$. By the Triangle Sum
Theorem $m\angle B = 180° - m\angle A - m\angle C$.
$$= 180° - 130° - 25°$$
$$= 25°$$
By the Converse of the Isosceles Triangle Theorem
and the definition of an isosceles triangle, $\triangle ABC$ is
isosceles; 100 feet. **11.** $KL = 23$ **13.** $m\angle F = 60°$
15. $GH = 24$ **17.** $m\angle L = 40°$

19.

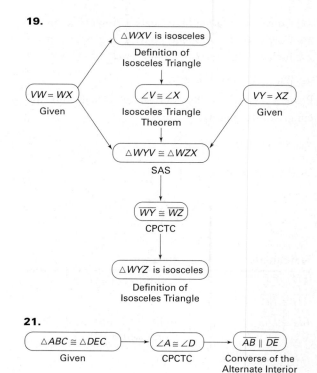

21.

$\triangle ABC \cong \triangle DEC$ (Given) → $\angle A \cong \angle D$ (CPCTC) → $\overline{AB} \parallel \overline{DE}$ (Converse of the Alternate Interior Angles Theorem)

23. Reflexive Property of Congruence **25.** AAS
27. Reflexive Property of Congruence **29.** CPCTC
31. bisects the side opposite the angle **33.** Let ABC
be an isosceles triangle with $\angle A \cong \angle C$ and $\overline{BX}$ be
the bisector of $\angle B$. By the proof in Exercises 27–29,
$\overline{BX}$ bisects $\overline{AC}$ and $\triangle ABX \cong \triangle CBX$. Thus, by
CPCTC, $\angle AXB \cong \angle CXB$. Since $\angle AXB$ and $\angle CXB$
form a linear pair, $m\angle AXB + m\angle CXB = 180°$. Since
$m\angle AXB \cong \angle CXB$, $m\angle AXB = m\angle CXB = 90°$. Thus
$\overline{BX}$ is the perpendicular bisector of $\overline{AC}$. **35.** The
surveyor constructs $\triangle ZYX \cong \triangle TBX$ by SAS. Because
CPCTC, the surveyor knows that $YZ = TB$. Thus, by
measuring $\overline{YZ}$, he can determine the distance across
the pond. **37.** Sum of interior angles = 360° Sum of
exterior angles = 360° **39.** Sum of interior angles =
1800°; Sum of exterior angles = 360°
41. perpendicular **43.** parallel

LESSON 4.5

Exercises

5. 25° **6.** 50° **7.** 2.2 **8.** The ramp, handrails, and the
two upright posts form a parallelogram. Opposite
sides of a parallelogram are congruent. **9.** 5 **11.** 50°
13. 60° **15.** 80° **17.** 10 **19.** 6 **21.** 20° **23. a.** No;
The triangles fit together to form a kite. **b.** Yes; The
opposite sides are parallel by the Alternate Interior
Angles Theorem. **c.** No; The sides of the triangles
will not match up. **25.** Alternate Interior Angles
Theorem **27.** Definition of a parallelogram
29. Alternate Interior Angles Theorem **31.** ASA

33. △ABD ≅ △CDB **35.** △CDB **37.** △DCA
39. ∠BAD ≅ ∠DCB

41.

Given: ABCD is a parallelogram.
Prove: ∠A and ∠B, ∠B and ∠C, ∠C and ∠D, and ∠D and ∠A are supplementary. Proof: By definition of a parallelogram, $\overline{AB} \parallel \overline{DC}$ and $\overline{AD} \parallel \overline{BC}$. By the Same-Side Interior Angles Theorem, ∠A and ∠B, ∠B and ∠C, ∠C and ∠D, and ∠D and ∠A are supplementary. **43.** alternate interior **45.** congruent **47.** CPCTC **49.** Definition of a rhombus **51.** SSS **53.** CPCTC **55.** Definition of a rectangle **57.** Converse of the Same-Side Interior Angles Theorem **59.** Converse of the Same-Side Interior Angles Theorem **61.** diagonals of a parallelogram bisect each other **63.** SSS **65.** 180°

67. Given: rectangle RSTU with diagonals $\overline{RT}$ and $\overline{US}$ interesting at V; Prove: $\overline{RT} \cong \overline{US}$

Statements	Reasons
RSTU is a rectangle.	Given
RSTU is a parallelogram.	Theorem 4.5.7
$\overline{ST} \cong \overline{UR}$	Theorem 4.5.2
$\overline{TU} \cong \overline{TU}$	Reflexive Property of Congruence
m∠RUT = 90° m∠STU = 90°	definition of rectangle
m∠RUT = m∠STU (∠RUT ≅ ∠STU)	Theorem 4.5.3
△RTU ≅ △SUT	SAS
$\overline{RT} \cong \overline{SU}$	CPCTC

69. Theorem: A square is a rectangle. A square is a quadrilateral with the property that all of its sides are equal and every angle is a right angle. Since every angle is a right angle, a square is a rectangle by the definition of a rectangle. **71.** Theorem: The diagonals of a square are congruent and are the perpendicular bisectors of each other. A square is a rhombus and hence a parallelogram by theorems 4.5.12 and 4.5.6, respectively. Therefore its diagonals are perpendicular bisectors of each other by Theorems 4.5.8 and 4.5.5. A square is also a rectangle by Theorem 4.5.11 and thus its diagonals are congruent by Theorem 4.5.9.

73. a.

b.

c.

d.

e.

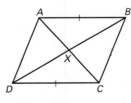

75. No. AAA does not imply that triangles are congruent. **77.** △GHI ≅ △JKL, HL **79.** x = 15; y = 20

LESSON 4.6

Exercises
5. none **6.** parallelogram **7.** parallelogram, rectangle **8.** parallelogram, rhombus **9.** The diagonals must be congruent. **11.** Yes, Theorem 4.6.3 **13.** Yes, Theorem 4.6.2 **15.** Rhombus, Theorem 4.6.6 **17.** Neither **19.** Neither **21.** No, KLMN is a rhombus by Theorem 4.6.6. But suppose ∠KLM > 90°. KLMN is not necessarily a square. **23.** KLMN is a rhombus by Theorem 4.6.6. KLMN is a rectangle by Theorem 4.6.5. Therefore KLMN is a square. **25.** KLMN is a rhombus by Theorem 4.6.6, since $\overline{NK} \cong \overline{KL}$ by CPCTC. Also by CPCTC, $\overline{LN} \cong \overline{MK}$, so KLMN is a rectangle by Theorem 4.6.5. KLMN is a square by definition. **27.** Reflexive Property **29.** CPCTC **31.** definition of parallelogram **33.** SSS **35.** Same-Side Interior Angles Theorem **37.** Opposite angles of a parallelogram are congruent.

39. Given: quadrilateral ABCD, $\overline{AB} \cong \overline{CD}$, $\overline{AB} \parallel \overline{CD}$
Prove: ABCD is a parallelogram

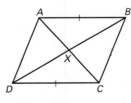

Statements	Reasons
$\overline{AB} \parallel \overline{CD}, \ \overline{AB} \cong \overline{CD}$	Given
$\angle XAB \cong \angle XCD$	Alternate Interior
$\angle XBA \cong \angle XDC$	Angles Theorem
$\triangle AXB \cong \triangle CXD$	ASA
$\overline{XB} \cong \overline{XD}$	CPCTC
$\overline{XC} \cong \overline{XA}$	
$\angle AXD \cong \angle BXC$	Vertical angles
$\triangle AXD \cong \triangle BXC$	SAS
$\angle DAX \cong \angle BCX$	CPCTC
$\overline{AD} \parallel \overline{BC}$	Converse of the Alternate Interior Angles Theorem
$ABCD$ is a paralleloram.	Definition of a parallelogram

Statements	Reasons
$ABCD$ is a parallelogram.	Given
$m\angle B = m\angle D$	Opposite angles of a parallelogram are congruent.
$\overline{BD}$ bisects $\angle B$ and $\angle D$	Given
$m\angle ABD = \frac{1}{2}m\angle B$ $m\angle ADB = \frac{1}{2}m\angle D$	Definition of angle bisector
$m\angle ADB = \frac{1}{2}m\angle B$	Substitution
$m\angle ABD = m\angle ADB$	Substitution
$\overline{AD} \cong \overline{AB}$	Converse of the Isosceles Triangle Theorem
$ABCD$ is a rhombus.	Theorem 4.6.6

45. Vertical Angles Theorem **47.** CPCTC
49. Theorem 4.6.2 **51.** Opposite sides of a parallelogram are congruent. **53.** Division Property
55. The 2-ft boards must be opposite each other, and the 3-ft boards must be opposite each other. She must also make sure the diagonals have the same measure. **57.** 5 sides **59.** 72° **61.** Since $\overline{BA} \cong \overline{BC}$, $\triangle ABC$ is isosceles, so $\angle BAC \cong \angle BCA$ by the Isosceles Triangle Theorem. Also, since $\overline{BD} \cong \overline{BE}$, by the Segment Addition Postulate $\overline{AD} \cong \overline{CE}$. By the Reflexive Property of Congruence $\overline{AC} \cong \overline{AC}$, so $\triangle ADC \cong \triangle CEA$ by SAS.

41. Given: parallelogram $ABCD$ with $m\angle A = 90°$
Prove: $ABCD$ is a rectangle

43. Given: $ABCD$ is a parallelogram and $\overline{AC}$ bisects $\angle A$ and $\angle C$ and $\overline{BD}$ bisects $\angle B$ and $\angle D$.
Prove: $ABCD$ is a rhombus.

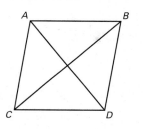

LESSON 4.7

TRY THIS (p. 263)

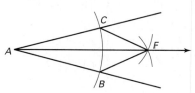

Statements	Reasons
$\overline{AC} \cong \overline{AB}$	Same compass setting used
$\overline{CF} \cong \overline{BF}$	Same compass setting used
$\overline{AF} \cong \overline{AF}$	Reflexive property of congruence
$\triangle ACF \cong \triangle ABF$	SSS
$\angle CAF \cong \angle BAF$	CPCTC

Exercises
5.
6.

7.

8.

9.

11.

13.

15.

17.

19.

21.

23.

25.

27.

29. a.

b.

c.

The circle drawn is the inscribed circle of the triangle.

31.

33.

35.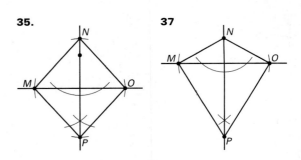

37

39. Same compass setting used **41.** CPCTC

43.

Statements	Reasons
$\overline{AB} \cong \overline{AD}$	Same compass
$\overline{CB} \cong \overline{CD}$	setting
$ABCD$ is a kite.	Definition of kite
$\overline{AC} \perp \overline{BD}$	Diagonals of a kite
$\overleftrightarrow{AC} \perp$ line ℓ	are perpendicular.

45.

47. rotation **49.** translation **51.** parallelogram and rhombus **53.** parallelogram, rectangle, square, and rhombus

LESSON 4.8

TRY THIS (p. 273)

R is between T and S, and the points are collinear by Theorem 4.9.1.

Selected Answers

TRY THIS (p. 274)

a. not possible, since $14 + 8 < 25$

b. not possible, since $16 + 7 = 23$

c. possible, since $18 + 8 > 24$

Exercises

5.

6.

7.

9.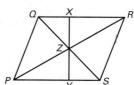

The angles have the same measure.

11.

13.

15.

17.

19.

21. $\triangle DEF$ is impossible since $DE + EF < DF$.

23. $\triangle JKL$ is possible.

25. $\begin{cases} x + 8 > 10 \Rightarrow x > 2 \\ x + 10 > 8 \Rightarrow x > -2 \\ 10 + 8 > x \Rightarrow x < 18 \end{cases} \Rightarrow 2 < x < 18$

27. Overlapping Angles Theorem **29.** CPCTC
31. definition of a rectangle **33.** Opposite sides of a rectangle are congruent. **35.** Segment Addition Postulate **37.** CPCTC **39.** Sample answer: They have to walk the same distance in the same direction.
41. Reflectional symmetry across a vertical axis
43. Reflectional symmetry across a vertical and horizontal axis and 180° rotational symmetry
45. a, b, and d

47. Prove: $XZ = YZ$

Statements	Reasons
$\angle XZQ \cong \angle YZS$	Vertical Angles Theorem
$\angle XQZ \cong \angle YSZ$	Alternate Interior Angles Theorem
$\overline{QZ} \cong \overline{ZS}$	Diagonals of a parallelogram bisect each other.
$\triangle QXZ \cong \triangle SYZ$	ASA
$\overline{XZ} \cong \overline{YZ}$ $(XZ = YZ)$	CPCTC

CHAPTER REVIEW AND ASSESSMENT

1. $\overline{QR}$ **3.** $\angle K$ **5.** Yes; SAS **7.** Yes; SSS **9.** Yes; AAS
11. Yes; AAS **13.** Definition of isosceles triangle
15. CPCTC

17. Given: $CDEF$ is a rectangle and $\overline{CE}$ and $\overline{DF}$ intersect at G. Prove: $\triangle CDG \cong \triangle EFG$

Statements	Reasons
$CDEF$ is a rectangle.	Given
$CDEF$ is a parallelogram.	Theorem 4.5.7
$\overline{CG} \cong \overline{GE}$	Theorem 4.5.5
$\overline{DG} \cong \overline{GF}$	
$\overline{CD} \cong \overline{EF}$	Theorem 4.5.2
$\triangle CDG \cong \triangle WXY$	SSS

INFO BANK

19. Given: $VWXY$ is a square.
Prove: $\triangle VWX \cong \triangle WXY$

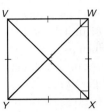

Statements	Reasons
$VWXY$ is a square.	Given
$\overline{VW} \cong \overline{WX}$	Definition of a square
$\overline{WX} \cong \overline{XY}$	
$VWXY$ is a rectangle.	Theorem 4.5.11
$m\angle W = 90°$	Definition of a rectangle
$m\angle X = 90°$	
$m\angle W = m\angle X$	Transitive or Substitution Property
$\triangle VWX \cong \triangle WXY$	SAS

21. parallelogram by definition; rhombus by Theorem 4.6.6 **23.** parallelogram by HL and Theorem 4.6.2; rectangle by Theorem 4.6.4 or 4.6.5

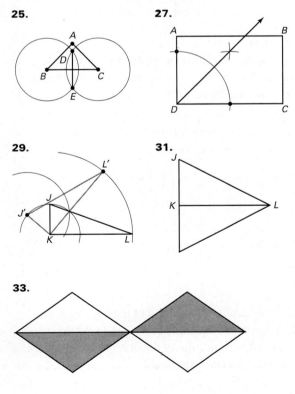

25.

27.

29.

31.

33.

Chapter 5

Exercises

5. 21 units **6.** 16.75 units^2 **7.** 144 sq. ft **8.** 120 ft
9. 52 in. **11.** 26 in. **13.** 42 in. **15.** 24 in. **17.** 16 in^2
19. Let b be the base and h be the height of a rectangle. Since opposite sides of a rectangle are the same length, two sides have length b and two sides have length h. The perimeter is
$b + b + h + h = 2b + 2h$. **21.** 36 units^2 **23.** 18 units^2
25. $h = 3$ cm, $b = 9$ cm, $P = 24$ cm **27.** $b = 3$, $h = 6$,
$A = 18$ units^2
29. $b = \frac{100}{h}$, or $h = \frac{100}{b}$

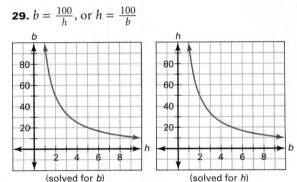

(solved for b)   (solved for h)

The relationship is a nonlinear function. Any value which causes division by zero will make the function undefined, and a negative value for b or h doesn't make sense for a side length—that is, $b \le 0$ or $h \le 0$ are sets of values that don't make sense in the equation.

31. ≈ 7.5 cm^2 **33.** 12 ft^2 **35.** 2500 ft^2 **37.** 5000 ft^2
39. 690 ft^2 **41.** \$107.00 **43.** 120° **45.** $(n - 2)180°$
47. −2 **49.** 0 **51.** (1, 5)

Exercises

6. 300 units^2 **7.** 35 units^2 **8.** 51 units^2; 102 units^2
9. a. 204 units^2 **b.** 480 units^2 **c.** 684 units^2
11. 12 units^2 **13.** 312 units^2 **15.** 48 units^2
17. 6 units^2 **19.** 98 units^2 **21.** 70 units^2
23. 140 units^2 **25.** 56 units^2 **27.** 143.5 units^2
29. 238 units^2 **31.** 6 units^2 **33.** 6 units^2 **35.** 8 units^2
37. 16 units^2 **39.** 9 units^2 **41.** 16 units^2
43. $b = 20$ cm **45.** $A = b_m h$; $b_m = $ midsegment
47. The parallelogram with the largest area is the rectangle. Conjecture: For parallelograms with a given perimeter, the parallelogram with the largest area is a rectangle. Sample answer: The shortest distance between a point and a line is along a perpendicular. This perpendicular distance is larger in the rectangle than in the other two parallelograms.

49. Given a parallelogram $ABCD$ with base b and height h, let $\overline{AE}$ be a segment perpendicular to the base, so that $\triangle AED$ is formed. Translate $\triangle AED$ by b units so that points D and C coincide, forming a rectangle with base b and height h. Its area is bh, so the area of the parallelogram is bh.

51. Prove: Area $ABCD = \frac{1}{2}(BD)(AC)$

Statements	Reasons
$ABCD$ is a kite with $\overline{AC} \perp \overline{BD}$	Given
Area kite $ABCD$ $=$ Area $\triangle ABD +$ Area $\triangle BCD$	Sum of Areas Postulate
Area kite $ABCD$ $= \frac{1}{2}(BD)(AX + XC)$	Area of a Triangle Formula
Area kite $ABCD$ $= \frac{1}{2}(AX)(BD) + \frac{1}{2}(XC)(BD)$	Distributive Property
Area kite $ABCD$ $= \frac{1}{2}(BD)(AC)$	Segment Addition Postulate

53. $A = 4,900,000 \text{ ft}^2 \approx 112.5$ acres; 48,937.5 lb of fertilizer

55.

Statements	Reasons
$\angle A \cong \angle D$, $\angle EFD \cong \angle BCA$, $\overline{AF} \cong \overline{CD}$	Given
$\overline{FC} \cong \overline{FC}$	Reflexive Property of Congruence
$\overline{AC} \cong \overline{FD}$	Overlapping Segments Theorem
$\triangle ABC \cong \triangle DEF$	ASA

57. $x^2 - x - 6$

LESSON 5.3

Exercises

4. $6\pi \approx 18.85$ units **5.** $25\pi \approx 78.54$ units

6. $25\pi \approx 78.54 \text{ units}^2$ **7.** $196\pi \approx 615.75 \text{ units}^2$

9. 62.8 units; 314 units^2 **11.** $\frac{264}{7}$ units; $\frac{792}{7} \text{ units}^2$

13. $\frac{55}{4}$ units; $\frac{1925}{128} \text{ units}^2$ **15.** $\frac{31.4}{\pi} \approx 10$ units

17. $\sqrt{\frac{314}{\pi}} \approx 10$ units **19.** 10 units

21. $65\pi \approx 204.20 \text{ units}^2$ **23.** $4\pi - 7.84 \approx 4.73 \text{ units}^2$

25. $360 + 112.5\pi \approx 713.43 \text{ units}^2$

27. $56 - 8\pi \approx 30.87 \text{ units}^2$ **29.** the area is multiplied by 4 **31.** The 18-inch pizza is the better deal. The 18-inch pizza gives $5.4\pi \text{ in}^2$ per dollar, while the 10-inch pizza gives $5\pi \text{ in}^2$ per dollar. **33.** 44 ft **35.** The inside tires do not have to go as far, so they are not moving as fast. **37.** 10 revolutions **39.** $6\sqrt{2}$ **41.** $30\sqrt{5}$

43. $x = 5$

45. Sample answer:

Statements	Reasons
$ABCD$ is a parallelogram.	Given
$\overline{AB} \cong \overline{CD}$, $\overline{BC} \cong \overline{AD}$	Opposite sides of a parallelogram are $\cong$.
$\overline{AC} \cong \overline{AC}$	Reflexive Property
$\triangle ABC \cong \triangle CDA$	SSS

47. 17.5 cm^2

LESSON 5.4

TRY THIS (p. 325)

$8^2 < 8^2 + 8^2$ so the triangle is acute.

Exercises

5. 64 **6.** No **7.** obtuse **9.** $5\sqrt{13}$ **11.** $2\sqrt{7}$ **13.** $\sqrt{2}$

15. 129.5 units **17.** 63 units **19.** acute **21.** obtuse

23. acute **25.** 72 units^2 **27.** 11.31 units **29.** 157.1 units^2

31. Triples: 3, 4, 5; 5, 12, 13; 7, 24, 25; 9, 40, 41; 11, 60, 61; Algebraic proof:

$$m^2 + \left(\frac{m^2 - 1}{2}\right)^2 = m^2 + \left(\frac{m^4 - 2m^2 + 1}{4}\right)$$
$$= \frac{4m^2 + m^4 - 2m^2 + 1}{4}$$
$$= \frac{m^4 + 2m^2 + 1}{4}$$
$$= \frac{(m^2 + 1)^2}{4}$$
$$= \left(\frac{m^2 + 1}{2}\right)^2$$

Since the sum of the squares of the lengths of two sides equals the square of the longest side, any triple generated with this method will represent the sides of a right triangle. **33.** Answers may vary. **35.** Let $AB = a$, $AR = EH = b$, and $BR = c$. Then area $ABCD = a^2$, area $EFGH = b^2$, and area $BRST = c^2$. Since triangle BAR is a right triangle, $a^2 + b^2 = c^2$, showing that $BRST$ is the square whose area is equal to the sum of the areas of the two given squares. **37.** Use the formula for the area of a trapezoid,

$A = \frac{1}{2}h(b_1 + b_2) = \frac{1}{2}(a + b)(a + b) = \frac{1}{2}(a + b)^2$.

Find the sum of the area of the three right triangles, $A = \frac{1}{2}ab + \frac{1}{2}ab + \frac{1}{2}c^2 = ab + \frac{1}{2}c^2$. Then set the areas equal to one another and simplify.

$$\frac{1}{2}(a + b)^2 = ab + \frac{1}{2}c^2$$
$$(a + b)^2 = 2ab + c^2$$
$$a^2 + b^2 + 2ab = 2ab + c^2$$
$$a^2 + b^2 = c^2,$$

which is the Pythagorean Theorem. **39. a.** The area of each inner right triangle $= \frac{1}{2}ab$ and, by segment subtraction, the side of square $IJKL$ is $(b - a)$. By the Sum of Areas postulate, area $EFGH$ = area $IJKL$ + 4 × (area of one inner right triangle):
$c^2 = (b - a)^2 + 4\left(\frac{1}{2}ab\right) = (b - a)^2 + 2ab$
b. The area of each outer right triangle $= \frac{1}{2}ab$ and the side of square $ABCD = (a + b)$. By the Sum of Areas postulate, area $ABCD$ = 4 × (area of one outer right triangle) + area $EFGH$. Thus, area $EFGH$ = area $ABCD$ − 4 × (area of one outer right triangle). Algebraically, $c^2 = (a + b)^2 - 4\left(\frac{1}{2}ab\right) = (a + b)^2 - 2ab$. Simplifying each:

Part **a.**
$c^2 = (b - a)^2 + 2ab$
$\quad = b^2 - 2ab + a^2 + 2ab$
$\quad = b^2 + a^2$
So $c^2 = a^2 + b^2$

Part **b.**
$c^2 = (a + b)^2 - 2ab$
$\quad = a^2 + 2ab + b^2 - 2ab$
$\quad = a^2 + b^2$
So $c^2 = a^2 + b^2$

41. 13.5 miles **43.** $2\sqrt{2}$ **45.** $\frac{\sqrt{35}}{5}$ **47.** True. Opposite sides of a rhombus have equal lengths. Therefore, the rhombus must be a parallelogram.

LESSON 5.5

Exercises

5. $7\sqrt{2}$ **6.** $14; 7\sqrt{3}$ **7.** 34.6 ft **8.** 374.12 inches^2
9. 276.9 units^2 **11.** $x = 2\sqrt{3}; z = 4\sqrt{3}$ **13.** $x = 4$;
$z = 8$ **15.** $p = 3\sqrt{2}; q = 3\sqrt{2}$ **17.** $p = 5\sqrt{2}; q = 5\sqrt{2}$
19. $h = 18; g = 12\sqrt{3}$ **21.** $k = 2; g = 4$
23. 25.3 units^2 **25.** 112.5 units^2 **27.** 147.8 units^2
29. 45-45-90 **31.** neither **33.** 30-60-90
35. $P = 24\sqrt{2} + 24$ units; $A = 144$ units^2
37. $P = 78$ units; $A = 253.5\sqrt{3}$ units^2
39. $P = 20\sqrt{3}$ units; $A = 50\sqrt{3}$ units^2
41. $5\sqrt{2} + 5$ units **43.** $A = 2s^2(\sqrt{2} + 1)$ **45.** 69.28 ft
47. The lines are parallel because they both form 60° angles with the same line. **49.** 6 units^2; 11.25 units^2
51. $8\sqrt{5}$ units^2; 20 units **53.** $A = \frac{1}{2}ab$;
$P = a + b + \sqrt{a^2 + b^2}$

LESSON 5.6

Exercises

5. $\sqrt{13}$ **6.** $2\sqrt{10}$ **7.** 13.98 units^2 **8.** 9.98 units^2
9. 9.43 units **11.** 5.39 units **13.** 13 units **15.** 6.08 units **17.** 3.61 units **19.** 22.37 units **21.** The triangle is isosceles. **23.** The triangle is not equilateral.
25. The triangle is a right triangle. **27.** $\sqrt{\frac{74}{2}}$ units
29. $\frac{5}{2}$ units; $\frac{5}{2} \cdot \sqrt{3}$ units **31.** The midpoint of the hypotenuse is $(3, 4)$. The distances are the same, 5 units. **33.** Conjecture: The distances from the midpoint of the hypotenuse to any vertex of a right triangle are equal. **35.** Sample answer: 16.375 units^2
37. Sample answer: $6\frac{2}{3}$ units^2 **39.** 9.85 miles
41. Let $W = (-1, 1)$, $X = (7, -5)$, $Y = (10, -1)$, $Z = (2,5)$

slope $\overline{AB} = \frac{-5 - 1}{7 - (-1)} = \frac{-6}{8} = -\frac{3}{4}$

slope $\overline{BC} = \frac{-5 - (-1)}{7 - 10} = \frac{-4}{-3} = \frac{4}{3}$

slope $\overline{CD} = \frac{-1 - 5}{10 - 2} = \frac{-6}{8} = -\frac{3}{4}$

slope $\overline{AD} = \frac{1 - 5}{-1 - 2} = \frac{-4}{-3} = \frac{4}{3}$

Since slope $\overline{AB}$ = slope $\overline{CD}$ and slope $\overline{BC}$ = slope $\overline{AD}$, $\overline{AB} \parallel \overline{CD}$, and $\overline{BC} \parallel \overline{AD}$ and $ABCD$ is a parallelogram. Since (slope $\overline{AB}$)(slope $\overline{BC}$) = −1, $\overline{AB} \perp \overline{BC}$ and $ABCD$ is a rectangle.

43. The square has the greater perimeter. **45.** 3 cm
47. $A = \frac{s^2}{4}\sqrt{3}$ units^2; $P = 3s$ units

LESSON 5.7

Exercises

5. $\sqrt{p^2 + q^2}$ **6.** $\sqrt{(r - p)^2 + (s - q)^2}$ **7.** (p, q)
8. $(p + r, q + s)$ **9.** $\frac{q}{p}$ **10.** $\frac{s - q}{r - p}$ **11.** $C(q, p)$
13. $I(r + p, q)$ **15.** $R(a, q); S(b, 0)$ **17.** $C(2r, 2q)$ and $D(2s, 0)$ **19.** $AD = 2s; BC = 2r - 2p; MN = s + r - p$
$\frac{(AD + BC)}{2} = \frac{(2s + 2r - 2p)}{2} = s + r - p = MN$

21. $M\left(\frac{q}{2}, \frac{p}{2}\right)$ **23.** M is equidistant from the three vertices. Therefore, a circle can be drawn by joining $J, K,$ and L and its radius is $JM = KM = LM$.

25.

27. Let $ABCD$ be a parallelogram formed by the following vertices: $A(0, 0)$, $B(p, q)$, $C(r, q)$, and $D(s, 0)$ where $\overline{AB}$ and $\overline{CD}$ are opposite sides and $\overline{BC}$

and $\overline{AD}$ are opposite sides. Slopes of opposite sides are equal:

$$\frac{q}{p} = \frac{q}{r-s}$$
$$p = r - s$$
$$AB = \sqrt{p^2 + q^2}$$
$$CD = \sqrt{(s-r)^2 + q^2} = \sqrt{p^2 + q^2}$$
$$AB = CD$$
$$BC = \sqrt{(r-p)^2} = \sqrt{[r-(r-s)]^2} = \sqrt{s^2}$$
$$AD = \sqrt{s^2}$$
$$BC = AD$$

29. Let $ABCD$ be a rectangle formed by $A(0, 0)$, $B(0, p)$, $C(p, q)$, and $D(q, 0)$.

diagonal $AC = \sqrt{(p-0)^2 + (q-0)^2}$
$\qquad = \sqrt{p^2 + q^2}$

diagonal $BD = \sqrt{(q-0)^2 + (0-p)^2}$
$\qquad = \sqrt{p^2 + q^2}$

Thus, the diagonals of a rectangle are congruent.

31. Let $\triangle ABC$ be a triangle formed by $A(0, 0)$, $B(2p, 2q)$, and $C(2r, 0)$. Then the midpoints of the sides of the triangle are $M(p, q)$, $N(p + r, q)$, and $P(r, 0)$.

The slope of median $\overleftrightarrow{AN}$ is $m = \frac{q-0}{p+r-0} = \frac{q}{p+r}$.

The equation of $\overleftrightarrow{AN}$ is $y = \frac{q}{p+r}x$

The slope of median $\overleftrightarrow{CM}$ is $m = \frac{q-0}{p+r-0} = \frac{q}{p+r}$.

The equation of $\overleftrightarrow{CM}$ is $y = \frac{q}{p-2r}(x - 2r)$

The slope of median $\overleftrightarrow{BP}$ is $m = \frac{2q-0}{2p-r} = \frac{2q}{2p-r}$.

The equation of $\overleftrightarrow{BP}$ is $y = \frac{2q}{2p-r}(x - r)$

Set the first two equations equal and solve for x:

$$\frac{q}{p+r}x = \frac{q}{p-2r}(x - 2r)$$
$$qx(p - 2r) = q(x - 2r)(p + r)$$
$$xp - 2rx = xp + xr - 2rp - 2r^2$$
$$2r + 2r^2 p = 3rx$$
$$x = \tfrac{2}{3}(p + r)$$

Substitute into the first equation to find y:

$$y = \frac{q}{p+r}\,\tfrac{2}{3}(p + r) = \tfrac{2}{3}q$$

So the first two equations intersect at $\left(\tfrac{2}{3}(p + r), \tfrac{2}{3}q\right)$.

Substitute these values into the third equation to verify that the three lines intersect at a single point:

$$\tfrac{2}{3}q = \frac{2q}{2p-r}\left(\tfrac{2}{3}(p + r) - r\right)$$
$$= \frac{2q}{2p-r}\,\frac{2p + 2r - 3r}{3}$$
$$= \frac{2q}{2p-r}\,\frac{2p - r}{3}$$
$$= \tfrac{2}{3}q$$

Since this results in a true statement, the three medians intersect at a single point.

33. 5 **35.** 5 **37.** $5\sqrt{2}$ **39.** $\frac{5\sqrt{2}}{2}$ **41.** $\frac{5\sqrt{2}}{2}$ **43.** 90°
45. 24.85

LESSON 5.8

6. $\frac{14}{25}$ or 0.56 **7.** area of unshaded region:
$4 - \pi$ units^2; probability of penny not touching intersection: $\frac{4-\pi}{4} = 1 - \frac{\pi}{4} \approx 0.21$ **8.** Sample answer: The width of the penny is considered 1 unit, so the shaded area is π, and the square has area 4. The probability that the penny will cover a vertex is $\frac{\pi}{4}$, theoretically. Repeated tosses should produce a number close to $\frac{\pi}{4}$. The more tosses, the more accurate the estimate should be. Multiply by 4 to give an estimate for π. **9.** $\frac{8}{9}$ **11.** $\frac{8}{9}$ **13.** $\frac{1}{3}$ **15.** 75%
17. $66\frac{2}{3}$% **19.** 0.5 **21.** $\frac{9}{20}$ **23.** $\frac{2}{3}$ **25.** $\frac{3}{5}$ or 0.6
27. $0.6 + 0.4 = 1.0$. There is a 100% chance of landing on either an odd or an even number.

29. Sample answer:

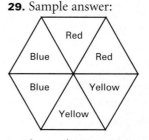

31. $\frac{1}{8}$ **33.** $\frac{4-\pi}{11} \approx 0.11$ **35.** 0.0089 **37.** 0.089
39. 0.717 **41.** right **43.** not a triangle **45.** acute
47. 15 units^2 **49.** 315 units^2

CHAPTER REVIEW AND ASSESSMENT

1. 90 units **3.** 1120 units^2 **5.** 24 units^2 **7.** 36 units^2
9. 11π units **11.** 100π units **13.** 29 **15.** acute
17. $50\sqrt{2}$ units **19.** $9\sqrt{3}$ units; $18\sqrt{3}$ units **21.** 10
23. $5\sqrt{2}$ **25.** $D(a + b, c)$

27. $BD = AC$

$$\sqrt{(a+b)^2 + c^2} = \sqrt{(b-a)^2 + c^2}$$
$$(a+b)^2 + c^2 = (b-a)^2 + c^2$$
$$(a+b)^2 = (b-a)^2$$
$$a^2 + 2ab + b^2 = b^2 - 2ab + a^2$$
$$4ab = 0$$
$$a = 0 \text{ or } b = 0$$

29. 0.25 **31.** 0.1875 **33.** ≈16.52 feet; yes; ≈4.13 feet

Chapter 6

LESSON 6.1

Exercises

5.

6.

7. Sample answer:

8.

9. 6 units³ **10.** 26 units² **11.** *a* **13.** *c* **15.** *f*

17.

19. 5 units³

21.

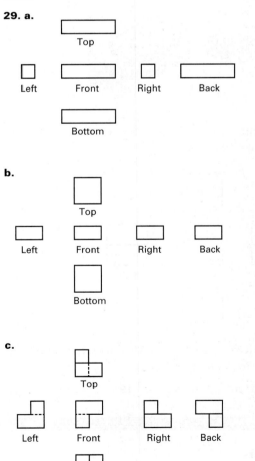

Top

Left Front Right Back

Bottom

23. **25.**

27. Yes, the top view has 2 distinct faces which are congruent because they have the same size and shape. They are rectangles which are 1 unit by 2 units.

29. a.

Top

Left Front Right Back

Bottom

b.

Top

Left Front Right Back

Bottom

c.

Top

Left Front Right Back

Bottom

31.

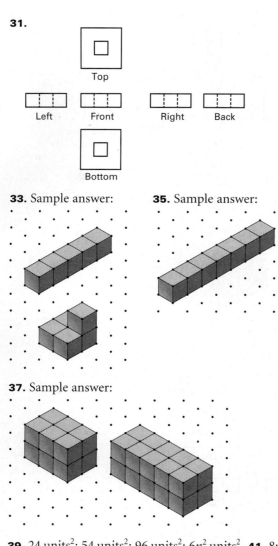

Top

Left   Front   Right   Back

Bottom

33. Sample answer: **35.** Sample answer:

37. Sample answer:

39. 24 units^2; 54 units^2; 96 units^2; $6n^2$ units^2 **41.** 8; 24; 24; 8

43.

Top

Left   Front   Right   Back

Bottom

These views might be used when writing instructions on how to assemble the tent.

45. 24.61 m^2 **47.** ≈ 78.54 m^2 **49.** ≈ 170.29 yd

LESSON 6.2

Exercises

7. Sample answer: $\overline{HI}$ and $\overline{KG}$; $\overline{LM}$ and $\overline{ON}$
8. Sample answer: $\overline{HI}$ and $\overline{MN}$; $\overline{HL}$ and $\overline{IG}$
9. Sample answer: $HIML$ and $KGNO$; $HKOL$ and $IGNM$ **10.** $\overline{HI}$ and $IGNM$; $\overline{HL}$ and $HIGK$

11. Sample answer: $\overline{HI}$ and $\overline{ON}$; $\overline{LO}$ and $\overline{IG}$
12. perpendicular; perpendicular **13.** There exists a line in S that is parallel to line m. **14.** The measure of the dihedral angle is m∠EFG. **15.** Line p and plane M are parallel. **17.** Line p is not perpendicular to plane R.

19. True

21. True

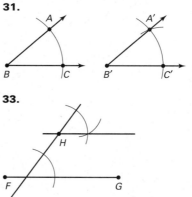

25. The angle formed by the insert in $\overline{AB}$.
27. intersect **29.** plane

31.

33.

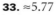

35. 196 units^2 **37.** 504 units^2 **39.** ≈ 97.4 units^2

LESSON 6.3

Exercises

5. parallelogram **6.** parallelogram **7.** rectangle
8. ≈ 186 pm **9.** hexagonal prism **11.** not a prism; prisms do not have curved edges **13.** not a prism; no translated bases **15.** $\triangle ABC \cong \triangle DEF$; in a prism, the bases are translated images of each other.
17. rectangle **19.** $ABED$ and $BCFE$ **21.** $\overline{HM}$
23. parallelogram **25.** ∠KNJ, ∠JGK, ∠LMI, ∠IHL, ∠NJI, ∠IMN, ∠KGH, ∠HLK **27.** ≈ 16.40 **29.** $a\sqrt{3}$
31. ≈ 8.72

33. ≈ 5.77

	Number of faces	Number of vertices	Number of edges
35.	6	8	12
37.	8	12	18

39. faces: $20 + 2 = 22$; vertices: $2(20) = 40$; edges: $3(20) = 60$ **41.** $d = \sqrt{4a^2 + h^2}$ **43.** $\frac{4a\sqrt{3}}{3}$

45. True because the diagonals of a parallelogram bisect each other.

47. False

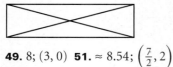

49. 8; $(3, 0)$ **51.** ≈ 8.54; $\left(\frac{7}{2}, 2\right)$

LESSON 6.4

TRY THIS (p. 398)

$\sqrt{33} \approx 5.74$

Exercises

6.

(2, 0, −1)

7.

(−1, −2, −3)

8.

(3, −1, 4)

9. $(0, 0, 0)$; $(2, 0, 0)$; $(0, 7, 0)$; $(2, 7, 0)$; $(0, 0, 6)$; $(2, 0, 6)$; $(0, 7, 6)$; $(2, 7, 6)$ **10.** $\sqrt{46} \approx 6.78$
11. $\sqrt{21} \approx 4.58$ **12.** $\sqrt{61} \approx 7.81$

13.

(4, 1, 5)

15.

(0, −5, 2)

17. top-front-right **19.** xz-plane **21.** top-back-right
23. x-axis **25.** Sample answer: $(1, 2, 3)$ **27.** Sample
answer: $(−1, −2, 3)$ **29.** $\sqrt{12} \approx 3.46$ **31.** 11
33. $(0, 10, 7)$ **35.** $(6, 10, 0)$ **37.** 6 **39.** $\sqrt{185} \approx 13.6$
41. 42 units^2 **43.** $\left(\frac{11}{2}, -\frac{9}{2}, \frac{7}{2}\right)$ **45.** $\left(\frac{5}{2}, -\frac{3}{2}, -\frac{5}{2}\right)$
47. $(0, 0, 0)$ **49.** Sample answer: $\left(\frac{\sqrt{3}}{3}, \frac{\sqrt{3}}{3}, \frac{\sqrt{3}}{3}\right)$
51. $\frac{2}{3}$ **53.** 6

55.
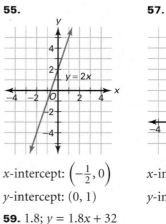

x-intercept: $\left(-\frac{1}{2}, 0\right)$
y-intercept: $(0, 1)$

57.

$y - 3 = 4(x + 1)$

x-intercept: $\left(-\frac{7}{4}, 0\right)$
y-intercept: $(0, 7)$

59. 1.8; $y = 1.8x + 32$

LESSON 6.5

Exercises

6.

(0, 0, 3)

(0, 2, 0)

(4, 0, 0)

7.
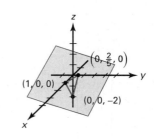

$\left(0, \frac{2}{5}, 0\right)$

(1, 0, 0)

(0, 0, −2)

8.

t	x	y
1	3	4
2	4	8
3	5	12

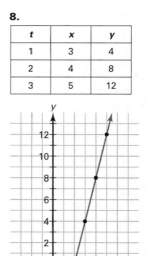

9.

t	x	y
1	0	5
2	1	7
3	2	9

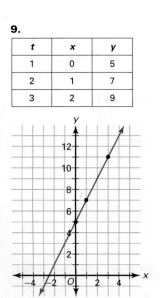

10.

t	x	y	z
1	1	2	2
2	2	4	3
3	3	6	4

11.

t	x	y	z
1	1	7	−1
2	4	8	−2
3	7	9	−3

13.

15.

17.

19.

21.

23.

25.

27. trace: $5x - 2y = 2$ **29.** trace: $-4x - 2y = 1$

31. Sample answer: $x = t$, $y = -\frac{5}{8}t - \frac{1}{4}$, $z = \frac{1}{4}t + \frac{11}{2}$

33.

Parallel. The lines are both vertical, perpendicular to the xy-plane at $(1, 2, 0)$ and $(3, 5, 0)$.

35. Yes. Set the coordinates of the two lines equal to each other and solve the system of equations. $(0, 3, -1)$

37. $7x + 4y - z = 50$

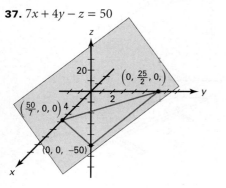

$7x + 4y = 50$; $0

39. parallel   **41.** neither

43.

Statements	Reasons
$\angle 1 \cong \angle 2$; $\overline{EH} \cong \overline{FG}$	Given
$\overline{EH} \parallel \overline{FG}$	Converse of the Corresponding Angles Postulate
$EFGH$ is a parallelogram.	Theorem 4.6.2
$\overline{EF} \cong \overline{HG}$	Theorem 4.5.2

45.

Statements	Reasons
$ABCD$ is a rectangle.	Given
$ABCD$ is a parallelogram.	Theorem 4.5.6
$\overline{AD} \cong \overline{BC}$	Theorem 4.5.2
$AE \cong CE$	Theorem 4.5.4
$BE \cong DE$	
$\triangle AED \cong \triangle CEB$	SSS

LESSON 6.6

Exercises

8. Parallel to each other but not to the picture plane. Parallel to each other but not to the picture plane.
9. The lines containing the vertical segments will not meet because they are parallel to each other and parallel to the picture plane.   **10.** The lines containing the nonvertical sides of the building meet at the horizon of the drawings.   **11.** Follow the steps to produce a one-point perspective drawing of a cube.

13.

15.

Vanishing point

17. Follow the steps to produce a two-point perspective drawing of a cube.

19.

21. The cube becomes distorted.

23.

27. Sample answer: Make a perspective drawing of a square with horizontal and vertical diagonals. Divide the horizontal diagonal equally. Pass lines from the vanishing points through the division points.

29. Sample answer:

31.   **33.**

35. ≈17.03 **37.** $\left(-\frac{1}{2}, 5, \frac{9}{2}\right)$

CHAPTER REVIEW AND ASSESSMENT

1. Sample answer:

3. 11 units³ **5.** Sample answer: *OPQRST* and

UVWXYZ, PQWV and *TSYZ* **7.** Sample answer: $\overline{PO}$ and $\overline{WX}$ **9.** Sample answer: *DEF* **11.** Sample answer: $\overline{CF}$ **13.** top-front-right **15.** 7

17.

19.

21.

23.

25.

27.

Chapter 7

LESSON 7.1

Exercises

6. 4.4 **7.** 4.2 **8.** 1.2 **9.** 0.6 **10.** 6682.72 in³;
7.45 in. × 7.45 in. **11.** box A **13.** 4 **15.** 48 **17.** 3
19. 105 **21.** 7 **23.** 1.45 **25.** 60 **27.** 1.5 **29.** $2 + \frac{4}{n}$
31. Sample answer: Maximize the volume; assuming
you use all available lumber, the surface area is fixed.
You want to create the maximum amount of storage
space. **33.** Sample answer: Minimize the surface
area; assuming the item(s) to be mailed have a fixed
volume, you would want to minimize the amount
of packaging to reduce the weight of the package.
35. $2 + \frac{4}{n}$; $\frac{6}{n}$ **37.** 6 **39.** Sample answer: The flat
shape of the worm maximizes the surface area of
its body exposed to the air which maximizes the
amount of oxygen it can take in through its skin.
41. Sample answer: Tall trees with large, broad leaves
have a large surface area exposed. The large surface
area is important if there is competition for light, but
it also increases water loss. In desert conditions there
is plenty of light but little water, so conservation of
water is important; broad-leafed trees would soon
die. **43.** Sample answer: When broken into smaller
pieces, a block of ice will have a larger surface-area-
to-volume ratio which maximizes heat exchange.
45. Sample answer: The short, squat form maximizes
the volume of the cactus and therefore the amount
of water it can hold. Also the surface area is kept to
a minimum which minimizes water loss through
transpiration and evaporation. **47.** 435 units²
49. 67.1 units²
51. No. Sample answer:

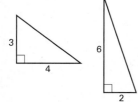

These two triangles have the same area, but are not
congruent.
53. 25π units² **55.** ≈8.49 cm

LESSON 7.2

Exercises

5. 888 units² **6.** 176 units² **7.** 840 units³
8. 27,811.74 units³

9. Sample answer:

11. Sample answer:

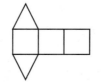

13. 35 cm³ **15.** 391 in³ **17.** 118 units²; 70 units³
19. 3 units²; $\frac{1}{3}$ units³ **21.** 5 in. **23.** 1292.55 cm²
25. 16.8 units³; 41.2 units² **27.** 882 m³
29. 197.87 in²; 103.18 in³ **31.** The surface area is
multiplied by $2^2 = 4$, and the volume is multiplied by
$2^3 = 8$. **33.** The volume is multiplied by $2^3 = 8$, and
the surface area is multiplied by $2^2 = 4$. **35.** Feline
Feast: surface area = 198 in²; volume = 162 in³: Kitty
Krunchies: surface area = 190 in²; volume = 126 in³
37. 896 ft²; 2.24 gal **39.** 23,040 ft² **41.** 6 **43.** 24 **45.**
$5\sqrt{3}$ **47.** 8.66 in.

LESSON 7.3

Exercises

6. 126.22 units² **7.** 132 units² **8.** 140 ft²
9. 13,142,640,826 pounds

11.   **13.**

15. 120 units² **17.** 30.4 m³ **19.** 468 in³
21.   **23.**

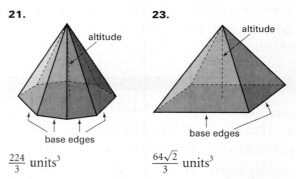

$\frac{224}{3}$ units³ $\frac{64\sqrt{2}}{3}$ units³

25. 5 units **27.** ≈14.85 units³ **29.** 114.38 units²
31. 13 units **33.** ≈3.46 units **35.** 5 **37.** 5 **39.** 10
41. $n + 1$ **43.** $n + 1$ **45.** 8400 units³; 3348 units²
47. 503.1 m² **49.** parallelogram **51.** 30 units²
53. 21.46 units² **55.** 8.29 units²

LESSON 7.4

Exercises

6. 836.58 mm² **7.** 579.82 mm² **8.** 1057.86 mm²

9. 688.98 mm³ **10.** 340.11 mm³ **11.** 808.93 mm³

13.

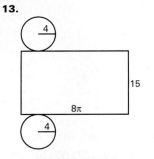

15. 477.5 units² **17.** 9 units **19.** 5 units

21. 240π units³ **23.** $\frac{24}{\pi}$ units **25.** $\frac{4\sqrt{5}}{3}$ units

27. 37.7 mm **29.** It doubles the volume of the cylinder. **31.** It multiplies the volume by 8.

33.
$$S = V$$
$$2\pi r^2 + 2\pi rh = \pi r^2 h$$
$$2r + 2h = rh$$
$$\frac{2r + 2h}{rh} = 1$$
$$\frac{2}{h} + \frac{2}{r} = 1$$

If either h or r is less than or equal to 2, then $\frac{2}{h} \geq 1$ or $\frac{2}{r} \geq 1$, and so $\frac{2}{h} + \frac{2}{r} > 1$, so both h and r must be greater than 2.

35. ≈0.75 cm **37.** 60 times more **39.** $r \approx 2.64$ in., $S \approx 131.29$ in², $h \approx 5.28$ in.

41. **43.**

The symbol has 180° rotational symmetry.   The symbol has 180° rotational symmetry.

45. 337.5 units²; 421.88 units³ **47.** 360 units²; 400 units³

LESSON 7.5

Exercises

6. 282.74 units² **7.** 628.32 units² **8.** $S = \pi r\ell + \pi r^2$

9. 1809.56 units³ **11.** 301.59 cm² **13.** 2.83 units²

15. 138.54 units² **17.** 7.81 in. **19.** 1413.72 units²

21. 3171.72 units² **23.** 40.72 units²

25. 2309.07 units³ **27.** 19,396.19 units³

29. 5598.32 units³ **31.** 6031.86 units³; 2513.27 units³

33. 6.46 in. **35.** 9.77 cm **37.** $\sqrt{91}$ **39.** $\sqrt{84}$

41. $\sqrt{75}$ **43.** 8.2 units **45.** ≈141, 372 ft³

47. 56.55 cm²; 37.70 cm² **49.** 23.70 cm³ **51.** $-\frac{3}{5}$

53. 8.06 units

LESSON 7.6

Exercises

6. 268,083 ft³ **7.** 7238.23 ft²; $10,857

9. a. 256π units²; $\frac{2048}{3}$ units³ **b.** 804.25 units²; 2144.66 units³ **11. a.** 4356π units²; 47,916π units³

b. 13,684.78 units²; 150,532.55 units³

13. a. 597.3136π units²; 2433.06π units³

b. 1876.52 units²; 7643.68 units³ **15. a.** 256π units²; $\frac{2048}{3}$π units³ **b.** 804.25 units²; 2144.66 units³

17. a. 131.7904π units²; 252.16π units³

b. 414.03 units²; 792.18 units³

19. a. 39,984.0016π units²; 1,332,533.49π units³

b. 125,613.45 units²; 4,186,277.43 units³ **21.** 16y^2π; $\frac{32}{3}y^3$π **23.** 16y^2π; $\frac{32}{3}y^3$π **25.** $\frac{y^2}{4}$π; $\frac{y^3}{48}$π **27.** 500 units²

29. 44.88 units² **31.** 2827.43 units² **33.** 600 units²; 483.60 units² **35.** 10 in.; 20 in.; 6.67 in.

37. 26.42 in²; 12.77 in³ **39.** 67,020,643 mi²

41. 13.09 in³ **43.** 2.60 in. **45.** 2827.43 cm²

47. 420 ft³ **49.** 3534.29 m³

LESSON 7.7

TRY THIS (p. 481)

sphere; cone; doughnut

Exercises

5. (2, 1, 1) **6.** (2, −1, −1) **7.** (−2, 1, −1) **8.** about the x-axis

9.

11.

13.

15.

17. $(-2, 3, 1)$ **19.** $(4, -2, -3)$ **21.** $(1, -2, 3)$
23. back-right-bottom, $(-4, 4, -1)$ **25.** in the *xy*-plane, $(1, -1, 0)$ **27.** front-left-bottom, $(2, -2, -8)$
29. The area of this circle $= \pi(10)^2$
$$= 100\pi$$
$$\approx 314.16 \text{ units}^2$$

31. $250\pi \approx 785.40 \text{ units}^3$ **33.** $\frac{64}{3}\pi \approx 67.02 \text{ units}^3$

35.   **37.**

39.

41. 445.48 units^3; 456.68 units^2 **43.** $1535\frac{1}{3} \text{ ft}^3$
45. about 5 ft^3

CHAPTER REVIEW AND ASSESSMENT

1. $\frac{3}{2}$ **3.** cube, $\frac{3}{2}$; prism, $\frac{61}{42}$; The ratio is larger for the cube. **5.** 288 units^2 **7.** 748.25 units^3 **9.** 18 units^2
11. 5.20 units^3 **13.** $72\pi \approx 226.19 \text{ units}^2$
15. $144\pi \approx 452.39 \text{ units}^3$ **17.** $580\pi \approx 1822.12 \text{ units}^2$
19. $2800\pi \approx 8796.46 \text{ units}^3$ **21.** $100\pi \approx 314.16 \text{ units}^2$

23. 3 units

25.

27.

29. $\approx 55,850 \text{ ft}$ **31.** 6.30 cm

Chapter 8

LESSON 8.1

Exercises

7. $(3, 15)$

8. $(-2, 8)$

9. $(1.5, -0.5)$

10. $(-4, -6)$

11.

12.

13. −0.5

15. $\left(-1, \frac{5}{3}\right), \left(\frac{8}{3}, 3\right), \left(\frac{2}{3}, -2\right)$

17. (1.6, 1.6), (4.8, −1.6), (−3.2, −4.8)

19. $\frac{1}{2}$ **21.** $-\frac{1}{3}$

23.

25.

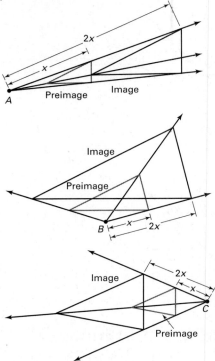

Each of the three images is twice as large as its corresponding image. The difference is that each image is placed in a different position relative to the preimage.

27. $m = \frac{3-0}{5+1} = \frac{3}{4}$; Endpoints of image are (2, 0) and (10, 6). So, $m = \frac{6-0}{10-2} = \frac{6}{8} = \frac{3}{4}$

29. $m = \frac{8-4}{4-(-2)} = \frac{2}{3}$; Endpoints of image are (1, −2) and (−2, −4). So, $m = \frac{-4-(-2)}{-2-1} = \frac{-2}{-3} = \frac{2}{3}$

31. $y = \frac{1}{5}x. \ 0 = \frac{1}{5}(0) \Rightarrow 0 = 0$. So, the origin is on this line. **33.** $y = \frac{-5}{3}x. \ 0 = -\frac{5}{3}(0) \Rightarrow 0 = 0$. So, the origin is on this line.

35.

37. 4

39.

Scale factor is 2. Center of dilation is (4, 3).
$D(x, y) = (4x - 2, 4y - 1)$.

41. The small opening in the center plate. **43.** The image is inverted because the light from point A travels in a straight line to A'; the light from point B travels in a straight line to point B'; and so on for each point on the preimage. **45.** Perimeter = 22 m; Area $\cong$ 22.25 m^2 **47.** 6019.29 in^2
49. 6366.2 km

LESSON 8.2

TRY THIS (p. 511)

Cross-Multiplication Property
$3 \neq 0, 6 \neq 0, (1)(6) = (2)(3) \Rightarrow 6 = 6$
Reciprocal Property
$1 \neq 0, 3 \neq 0, 2 \neq 0, 6 \neq 0, \frac{3}{1} = \frac{6}{2} \Rightarrow 3 = 3$
Exchange Property
$1 \neq 0, 3 \neq 0, 2 \neq 0, 6 \neq 0, \frac{1}{2} = \frac{3}{6} \Rightarrow \frac{1}{2} = \frac{1}{2}$
"Add-one" Property
$3 \neq 0, 6 \neq 0, \frac{1+3}{3} = \frac{2+6}{6} \Rightarrow \frac{4}{3} = \frac{8}{6} \Rightarrow \frac{4}{3} = \frac{4}{3}$

Exercises

6. Yes. Corresponding angles are congruent, and corresponding sides are proportional: $\frac{20}{25} = \frac{24}{30} = \frac{28}{35}$.
7. No. $\frac{27}{36} \neq \frac{32}{48}$, corresponding sides are not proportional. **8.** $EH = 19.8$ **9.** $JK = 15$ **10.** 90 ft
11. $\triangle SGT \sim \triangle MWR$ **13.** $\frac{AB}{XY} = \frac{BC}{YZ} = \frac{CA}{ZX}$
15. $\frac{JK}{PQ} = \frac{KL}{QR} = \frac{LM}{RS} = \frac{MN}{ST} = \frac{NJ}{TP}$ **17.** $\angle J \cong \angle P$, $\angle K \cong \angle M$, and $\angle L \cong \angle N$, so corresponding angles are congruent. $\frac{20}{15} = \frac{27}{20.25} = \frac{30}{22.5}$, so corresponding sides are proportional. Thus, $\triangle JKL \sim \triangle PMN$.
19. $\frac{25}{18} \neq \frac{30}{20}$, so corresponding sides are not proportional. Thus, pentagon $KLMNJ \not\sim$ pentagon $PQRST$. **21.** $\frac{2}{5} = \frac{4}{10} \Rightarrow \frac{2}{5} = \frac{2}{5}$ **23.** $\frac{x}{y} = \frac{1}{2}$ **25.** Ratio of sides: 3 Ratio of areas: 9 Observe that $(3)^2 = 9$.
27. 1.2 **29.** 5 **31.** $x = 12$ **33.** $x = 14$ **35.** $x = 10$
37. If the ratio of the sides is n, then the ratio of the areas is n^2. **39.** False. It is true that $\frac{3}{4} = \frac{6}{8}$, but $\frac{3+1}{4+1} = \frac{4}{5}$ and $\frac{6+1}{8+1} = \frac{7}{9}$. So, $\frac{3+1}{4+1} \neq \frac{6+1}{8+1}$.
41. True. If $\frac{a}{b} = \frac{c}{d}$ then

$ad = bc \Rightarrow ac + ad = ac + bc \Rightarrow a(c + d) = c(a + b)$
$\Rightarrow \frac{a}{a+b} = \frac{c}{c+d}$.
43. 3750 fish **45.** 25.5 ft **47.** $5\frac{5}{8}$ in. tall and $3\frac{3}{4}$ in. wide **49.** 25°, 112°, and 43° **51.** valid **53.** valid
55. $\approx$8.6 cm

LESSON 8.3

Exercises

5. $\triangle ABC \sim \triangle LKJ$; SSS Similarity Theorem
6. $\triangle EFD \sim \triangle QPR$; AA Similarity Postulate
7. $\triangle GHI \sim \triangle MOA$; SAS Similarity Theorem
9. $\triangle GHF \sim \triangle ECD$; AA Similarity Postulate
11. $\triangle LJK \sim \triangle CBA$; SSS Similarity Theorem
13. $\triangle RTS \sim \triangle NLM$; SSS or SAS Similarity Theorem
15. Yes, by the SAS Similarity Theorem
17. $\triangle XYZ \sim \triangle RST$; by the Angle-Sum Theorem, m$\angle S = 32°$. Also, $\angle Z > \angle T$ since both are right angles. Then, apply the AA Similarity Postulate.
19. Yes. AB and AC are proportional to AD and AE, respectively. Also, the included angle, $\angle A$, is shared by both $\triangle ABC$ and $\triangle ADE$, so by the SAS Similarity Theorem, $\triangle ABC \sim \triangle ADE$. **21.** not enough; need $\angle D \cong \angle B$ to ensure that $\triangle ABC \sim \triangle ADE$ **23.** No, because $\triangle QRS$ is not a right triangle. Therefore, $\triangle QRS$ is not similiar to $\triangle KLM$. **25.** Corresponding angles are congruent; since $\overline{GH} \parallel \overline{BC}$, it follows that $\angle AGH \cong \angle ABC$. Also, $\angle A$ is shared by both triangles, so by the AA Similarity Postulate, $\triangle AGH \sim \triangle ABC$. **27.** Since $\triangle AGH \cong \triangle DEF$, then $\triangle AGH \sim \triangle DEF$. But $\triangle AGH \sim \triangle ABC$, so that $\triangle ABC \sim \triangle DEF$. **29.** $\triangle UST \sim \triangle UVW$; this means that $\overline{US}$ and $\overline{UT}$ are proportional to $\overline{UV}$ and $\overline{UW}$ respectively. That is, $\frac{US}{UV} = \frac{UT}{UW}$. But it is given that $\frac{UV}{XY} = \frac{UW}{XZ}$. So, $\frac{US}{UT} = \frac{XY}{XZ}$. But $US = XY$ which means that $UT = XZ$. Also, $\angle U \cong \angle X$, so $\triangle UST \cong \triangle XYZ$ by SAS. **31.** $\angle A \cong \angle A$ and $\angle B \cong \angle B$ by the Symmetric Property of Congruence, so by the AA Similarity Postulate, $\triangle ABC \sim \triangle ABC$. **33.** If $\triangle ABC \sim \triangle DEF$, then $\angle A \cong \angle D$ and $\angle B \cong \angle E$ by the Polygon Similarity Postulate, and if $\triangle DEF \sim \triangle GHI$, then $\angle D \cong \angle G$ and $\angle E \cong \angle H$. So, it follows that $\angle A \cong \angle G$ and $\angle B \cong \angle H$ by the Transitive Property of Congruence. Thus, by the AA Similarity Postulate, $\triangle ABC \sim \triangle GHI$. **35.** Yes. Since Tony copied two angles from the original triangle, then by the AA Similarity Postulate, the two triangles are similar. **37.** Yes. Since Miki copied an angle that is adjacent to two sides of her triangle, of which the lengths are 5 times the lengths of two sides of the original triangle, then by the SAS Similarity

Theorem, the two triangles are similar. **39.** 135
41. 60 **43.** $\frac{5}{3}$ **45.** $\frac{5}{3}$ **47.** Yes; both have a right angle and share a non-right angle. Thus, by the AA Similarity Postulate, both are similar.

LESSON 8.4

Exercises

6. 15 **7.** 6 **8.** 4.5 **9.** $\frac{4}{5}$ **10.** $\frac{8}{5}$ **11.** $\frac{18}{5}$ **12.** $\frac{24}{5}$
13. $\frac{48}{5} = 9.6$ **15.** $\frac{21}{2} = 10.5$ **17.** 6 **19.** 1 or 2
21. By the Side Splitting Theorem and the SAS Similarity Theorem, $\triangle BED \sim \triangle BAC$. Similarly, $\triangle ACB \sim \triangle FCD$. Then, by the Transitive Property, $\triangle EBD \sim \triangle FDC$. **23.** $x = 16$; $y = 15$ **25.** $x = \frac{2}{5}$
and $y = 2$

27. Sample answer:

29. By the Two-Transversal Proportionality Corollary, the arrowed lines divide and the ray proportionally. But the ray was divided into congruent parts, so $\overline{AB}$ is divided into congruent parts. **31.** Each line has a slope of 1. **33.** (0, 0) and (−1, 1); $\sqrt{2} \approx 1.41$

35. Given: $\triangle ABC$ with $\overline{DE}$ such that $\frac{DB}{AD} = \frac{EC}{AE}$
Prove: $\overline{DE} \parallel \overline{BC}$
Proof: Since $\frac{DB}{AD} = \frac{EC}{AE}$, by the Add-one Property
$\frac{DB + AD}{AD} = \frac{EC + AE}{AE}$, so by the Segment Addition
Postulate $\frac{AB}{AD} = \frac{AC}{AE}$. Also, $\angle A \cong \angle A$, so
$\triangle ABC \sim \triangle ADE$ by SAS Similarity. By the Polygon Similarity Postulate, $\angle ADE \cong \angle ABC$, so $\overline{DE} \parallel \overline{BC}$ by the Converse of the Corresponding Angles Postulate.

37. The string for the note C should be placed $13\frac{1}{3}$ cm from the string for the lower F. **39.** 1716 cm

41.

43.

$D\left(\frac{9}{2}, 3\right)$; $E(2, -3)$; $F\left(\frac{1}{2}, -1\right)$

45.

$\sqrt{65} \approx 8.06$

47.

$\frac{\sqrt{569}}{2} \approx 11.93$

LESSON 8.5

Exercises

5. $\frac{x}{10} = \frac{5.4}{3}$; $x = 18$ km **6.** $\frac{x}{x + 10} = \frac{9}{14}$; $x = 18$ km
7. ≈ 460.65 cm **8.** 4.8 **9.** 2.2 **11.** $16\frac{2}{3}$ ft **13.** 21 ft
15. 10 **17.** 25 **19.** 14.4 **21.** Yes. $\frac{MK}{JK} = \frac{KJ}{KL} = \frac{4}{3}$ and $\angle MKJ \approx \angle JKL$, since both are right angles; so the triangles are similar by the SAS Similarity Theorem.
23. $\triangle JKL$: 54; $\triangle MKJ$: 96 **25.** Yes. $\frac{9.6}{7.2} = \frac{MK}{JK} = \frac{4}{3}$
27. Given **29.** Polygon Similarity Postulate
31. m$\angle BAD$ = m$\angle YXW$ **33.** $\frac{AD}{XW} = \frac{AB}{XY}$
35. Definition of median **37.** Substitution Property and Division Property **39.** Substitution Property
41. $\frac{EH}{SV} = \frac{EF}{ST}$ **43.** Since vertical angles are congruent, $\angle BAC \cong \angle B'AC'$. Since $\overline{BC} \perp \overline{B'C'}$ $\angle BCA$ and $\angle B'C'A$ are both right angles, $\angle BCA \cong \angle B'C'A$. Thus, by the AA Similarity Postulate, $\triangle ABC \sim \triangle AB'C'$. **45.** 2 cm

47. Arrange them so that the distance from the image to the lens is 20 times greater than the distance from the object to the lens. **49.** 4 **51.** $\frac{13}{2}$ **53.** 10

LESSON 8.6

TRY THIS (p. 545)

a. $\frac{25}{81}$ **b.** $\frac{3}{2}$ **c.** $\frac{1}{27}$

Exercises

6. $\frac{4}{1}$ **7.** $\frac{8}{1}$ **8.** The volume would be multiplied by 27, while the cross-sectional area would only be multiplied by 9. **9.** $\frac{10}{3}$ **11.** 30 m
13. $P = 28$ m; $A \approx 42.67$ m² **15.** $\frac{49}{25}$ **17.** $\frac{5}{7}$ **19.** $\frac{125}{343}$
21. $\frac{343}{729}$ **23.** $\frac{4}{5}$ **25.** $\frac{64}{125}$ **27.** $\frac{1728}{2197}$ **29.** 12 cm
31. 686 cm³ **33.** $\frac{4}{5}$ **35.** $\frac{16}{25}$ **37.** 57.6 cm² **39.** 14.14 cm
41. $\frac{201}{50}$; $\frac{8,120,601}{125,000}$ **43.** No. Doubling all the dimensions would increase the volume of the box by a factor of $2^3 = 8$. To double the volume, all the dimensions would be increased by a factor of $\sqrt[3]{2} \approx 1.26$. **45.** 1050 m² **47.** ≈1420 tons **49.** 9
51. $y = -\frac{5}{4}x + \frac{47}{4}$ **53.** $x = 2$ **55.** $x = 4$ **57.** $x = 17$

CHAPTER REVIEW AND ASSESSMENT

1. The endpoints of the image are $(-6, 3)$ and $(9, 12)$.

3.

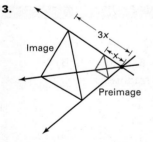

5. If the triangles were similar, the proportion $\frac{30}{18} = \frac{36}{21}$ would be true. By cross-multiplying, we obtain $30 \cdot 21 = 36 \cdot 18$ or $630 = 648$, so the triangles are not similar. **7.** 5.4 **9.** SAS Similarity Theorem
11. AA Similarity Postulate **13.** 12 **15.** 9.6 **17.** 8 m
19. 24 **21.** $\frac{25}{1}$ **23.** $\frac{1}{1000}$ **25.** 14.4 in.²
27. $\frac{1,767,172,329}{42,875} \approx \frac{41217}{1}$

Chapter 9

LESSON 9.1

TRY THIS, (p. 568)

$L = \frac{5\pi}{2} \approx 7.85$ units

Exercises

8. 70°; 160°; 130° **9.** 3.67

10.

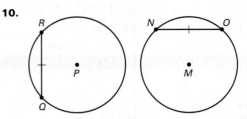

It is given that $\odot P$ and $\odot M$ are congruent, and that $\overline{QR}$ and $\overline{NO}$ are congruent. Form two triangles by adding the radii $\overline{PR}, \overline{PQ}, \overline{MN},$ and $\overline{MO}$. Then $\triangle PQR$ and $\triangle MNO$ are congruent by SSS. Thus, $\angle QPR$ and $\angle NMO$ are congruent and hence the arcs $\overset{\frown}{QR}$ and $\overset{\frown}{NO}$ are congruent. **11.** point P **13.** $\overline{ED}$ and $\overline{EC}$
15. Sample answer: $\angle BPC$ **17.** Sample answer: $\overset{\frown}{AB}$ and $\overset{\frown}{BC}$ **19.** Radius **21.** Chord **23.** 130° **25.** 126°
27. 116° **29.** 244° **31.** 15.71 **33.** 62.83
35. $\frac{36°}{\pi} \approx 11.46°$ **37.** $\frac{36°}{\pi} \approx 11.46°$ **39.** 72° **41.** 288°
43. $\frac{16\pi}{5} \approx 10.05$ **45.** $\frac{11\pi}{30} \approx 1.15$ ft **47.** Sample answer: 240 ft **49.** ≈12.77 in³ **51.** $26\pi \approx 81.68$ in.

LESSON 9.2

Exercises

6. $\sqrt{3}$ **7.** 3 **8.** 24
9. $10^2 = 8^2 + 6^2$
$(AC)^2 = (BC)^2 + (AB)^2$

Thus, $\triangle ABC$ is a right triangle and $\overline{BC} \perp \overline{AB}$. Thus $\overleftrightarrow{AB}$ is tangent to $\odot C$ by the Converse of the Tangent Theorem. **11.** $3\sqrt{5}$; $3\sqrt{5}$ **13.** 10 **15.** Let $\overline{PQ}$ be a chord in $\odot M$ and consider the line ℓ, which is the perpendicular bisector of $\overline{PQ}$ intersecting $\overline{PQ}$ at point X. Let $\overline{MN}$ be the radius of $\odot M$, which is perpendicular to $\overline{PQ}$. By the Radius and Chord Theorem, $\overline{MN}$ bisects $\overline{PQ}$, so $\overline{MN}$ also passes through point X. Thus both $\overline{MN} \perp \overline{PQ}$ and $\ell \perp \overline{PQ}$ so $\ell \parallel \overline{MN}$. Since both ℓ and $\overline{MN}$ pass through point X, they must coincide. Thus, the perpendicular bisector of $\overline{PQ}$, ℓ, passes through the center of the circle, M.

17. Sample answer:

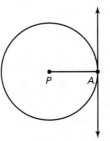

The tangent line at the point A must be perpendicular to the radius $\overline{AP}$.

19. 2 **21.** $\sqrt{3} \approx 1.73$ **23.** 4 **25.** $d \approx 1124.29$ miles
27. $P = 48$ cm; $A = 96$ cm^2 **29.** $6666\frac{2}{3}$ in^3

LESSON 9.3

Exercises

7. 35° **8.** 35° **9.** m∠$GFH = 50°$, m∠$GKH = 25°$, m∠$JFK = 90°$, m∠$KHJ = 45°$ **10.** m∠$KGH = 90°$
11. 30° **13.** 120° **15.** 180° **17.** 120° **19.** 100°
21. 29° **23.** 61° **25.** 50° **27.** 80° **29.** 34°; 34°
31. 43.5°; 43.5° **33.** 200° **35.** 80° **37.** 180°; The sum of the measures of the interior angles of a quadrilateral is 360°. **39.** 200° **41.** 180° **43.** $(2x)°$; The measure of an inscribed angle is one-half the measure of the intercepted arc. **45.** $(180° - x)°$; The measure of the inscribed angle is one-half the measure of the intercepted arc. **47.** 180°; The sum of the measures of the interior angles of a quadrilateral is 360°. **49.** The longest sides of the triangles intersect at the center of the circle. Both right angles intercept an arc that measures 180°, so the intercepted arcs are semicircles. Thus, the longest sides of the triangles are diameters of the circles, which intersect at the center of the circle. **51.** line m is the angle bisector of ∠CXD. **53.** m∠2 = m∠5
55. Angle Addition Postulate **57.** Linear pair property **59.** Division Property of Equality
61. Definition of perpendicular.

LESSON 9.4

Exercises

5. 69° **6.** 105° **7.** 22° **8. a.** 35° **b.** 100° **c.** 80° **d.** 20°
9. The ship is outside the circle of danger since m∠$BXA <$ m∠BCA. **11.** 17° **13.** 85° **15.** 85°
17. 75° **19.** $(x + 2)°$ **21.** 105° **23.** 28° **25.** 105°
27. m∠1 = m∠2 + m∠AVC
29. m∠$AVC = \frac{1}{2}\left(\text{m}\widehat{BC} - \text{m}\widehat{AC}\right)$
31. m∠1 = m∠AVC + m∠2 **33.** $\widehat{AC}$
35. a. 60° **b.** 150° **c.** 30° **d.** 120°;
m∠ABC = m∠1 − m∠2 **e.** 110° **f.** 125° **g.** 55°
h. 70°; m∠AVC = m∠1 − m∠2 **i.** 140° **j.** 110°
k. 70° **l.** 40°; m∠AVC = m∠1 − m∠2 **m.** 160°
n. 100° **o.** 80° **p.** 20°; m∠AVC = m∠1 − m∠2
q. $360° - x°$ **r.** $\frac{x°}{2}$ **s.** $180° - \frac{x°}{2}$ **t.** $x° - 180°$;
m∠AVC = m∠1 − m∠2 **37.** one-half the difference of the measures of the intercepted arcs **39.** The measure of a tangent-tangent angle with its vertex outside the circle is the measure of the major arc minus 180°. **41.** Two secants or chords **43.** On the circle **45.** On the circle **47.** m∠$AVC = \frac{1}{2}\text{m}\widehat{AV}$

49. m∠$AVC = \frac{1}{2}\left(\text{m}\widehat{AXC} - \text{m}\widehat{AC}\right) = \text{m}\widehat{AXC} - 180°$
51. Two secants **53.** Exterior of a circle
55. m∠$AVC = \frac{1}{2}\left(\text{m}\widehat{BC} - \text{m}\widehat{AC}\right)$ **57.** 70° **59.** 95°
61. Yes; everything was reduced by a common factor.
63. 9

LESSON 9.5

Exercises

6. 7 **7.** 8 **8.** 4 **9.** 4 **10.** 60 **11.** $\overline{AC}$ or $\overline{BC}$ **13.** $\overline{DE}$
15. Sample answer: △SVR and △TVU; △USW and △RTW **17.** 6 **19.** 3 cm **21.** ≈ 3.71 cm **23.** ∠BPV
25. 3 **27.** $\frac{16 + \sqrt{416}}{2} \approx 18.198$ **29.** $\frac{10 \pm \sqrt{20}}{2}$; 2.76 or 7.24

31. Two secant segments

33.

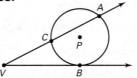

35. Interior of circle **37.** $AV \cdot VB = CV \cdot VD$
39. 45.22 km **41.** $210\pi \approx 659.73$ cm^2;
$400\pi \approx 1256.64$ cm^3
43. $5\pi\sqrt{281} + 25\pi \approx 341.85$ cm^2;
$\frac{400\pi}{3} \approx 418.88$ cm^3 **45.** Definition of a parallelogram. **47.** Alternate Interior Angles Theorem **49.** CPCTC

LESSON 9.6

TRY THIS, (p. 613)

a. Center $(-3, 3)$;
radius 7

b. Center $(3, -3)$;
radius 7

c. Center $(4, 5)$;
radius $\sqrt{30} \approx 5.48$

d. Center $(-2, 5)$;
radius $\sqrt{50} \approx 7.07$

Exercises

6. *x*-intercepts: $(10, 0)$ and $(-10, 0)$
y-intercepts: $(0, 10)$ and $(0, -10)$

7.

x	*y*	Points on graph
0	±10	$(0, 10), (0, -10)$
±10	0	$(10, 0), (-10, 0)$
6	±8	$(6, 8), (6, -8)$
−6	±8	$(-6, 8), (-6, -8)$
8	±6	$(8, 6), (8, -6)$
−8	±6	$(-8, 6), (-8, -6)$

8.

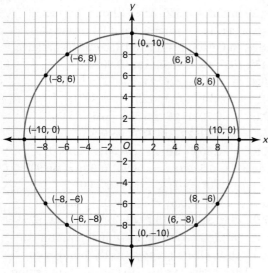

9. *x*-intercepts: $(8, 0), (0, 0)$
y-intercepts: $(0, 6), (0, 0)$

10.

x	*y*	Points on graph
0	6, 0	$(0, 6), (0, 0)$
8, 0	0	$(8, 0), (0, 0)$
1	7, −1	$(1, 7), (1, -1)$
−1	3	$(-1, 3)$
4	8, −2	$(4, 8), (4, -2)$
7	7, −1	$(7, 7), (7, -1)$
8	6, 0	$(8, 6), (8, 0)$
9	3	$(9, 3)$

11.

13. $\left(5\sqrt{2}, 0\right); \left(-5\sqrt{2}, 0\right), \left(0, 5\sqrt{2}\right); \left(0, -5\sqrt{2}\right)$

15. $(5, 0); (-1, 0), \left(0, \sqrt{5}\right); \left(0, -\sqrt{5}\right)$
17. $x^2 + y^2 = 36$ **19.** $x^2 + y^2 = 13$
21. $x^2 + (y - 6)^2 = 25$ **23.** $(x - 1)^2 + (y + 7)^2 = 100$
25. $(0, 0); r = 10$ **27.** $(0, 0); r = \sqrt{101}$ **29.** $(0, 3);$
$r = 2$ **31.** $(-3, 0); r = 7$ **33.** $x^2 + y^2 = 16$
35. $(x + 2)^2 + (y - 3)^2 = 4$

37. Sample answer: **39.** Sample answer:
$(x - 4)^2 + y^2 = 4$ $(x - 1)^2 + (y - 5)^2 = 1$

41. $(x - 2)^2 + (y - 3)^2 = 9$ **43.** $x^2 + (y - 1)^2 = 25$
45. $(x - 2)^2 + (y - 3)^2 = 100$
47. $(x - 3)^2 + (y + 5)^2 = 4$

49. $(x - 8)^2 + y^2 = 9$

51. $(x + 5)^2 + (y + 4)^2 = 9$

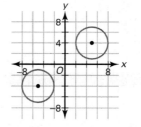

53. $(x - 4)^2 + (y - 3)^2 = 25$

55. $\frac{1}{4}$ **57.** 3.06π in. ≈ 9.6 in. **59.** 9

CHAPTER REVIEW AND ASSESSMENT

1. Sample answer: $\overline{BC}$, $\overline{AP}$ $\angle APC$, $\overset{\frown}{CDA}$ **3.** 72°
5. 13 **7.** True; $\triangle QPX \cong \triangle RPX$ **9.** 20° **11.** 25° **13.**
47° **15.** 94° **17.** 90 **19.** 6

21.

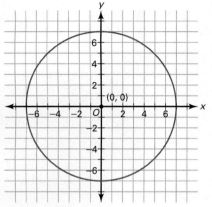

23. $x^2 + y^2 = 1$ **25.** $340\pi \approx 1068$ in. **27.** 11.35 mi

Chapter 10

LESSON 10.1

TRY THIS (p. 631)

Let β be the other angle. Then $\tan \beta = \frac{3.8}{2.9} \approx 1.3$.

Exercises

6. 0.62. Answers will vary due to inaccuracies in
measurement. **7.** 1.45. Answers will vary due to
inaccuracies in measurement. **8.** 0.36 **9.** 0.84
10. 2.75 **11.** 37° **13.** 2.5 **15.** 2.4 **17.** 1 **19.** 2.36
21. 1.33 **23.** 57.29 **25.** 54° **27.** 84° **29.** 0°
31. $\tan 45° = \frac{x}{2}$; $x = 2$ **33.** $\tan 28° = \frac{12}{x}$; $x \approx 22.57$
35. 494.2 m **37.** 14.0° **39.** 1119.6 ft
41. $V \approx 1099.56$ units³; $S \approx 596.90$ units²
43. $V \approx 65,449.85$ units³; $S \approx 7853.98$ units²
45. $MQ \approx 7.42$ units;
$QN \approx 7.42$ units

LESSON 10.2

Exercises

5. 14 feet **6.** 7 feet **7.** 13 feet **8.** 34° **9.** 59° **11.** $\frac{5}{13}$
13. $\frac{12}{13}$ **15.** $\frac{5}{12}$ **17.** 22.62° **19.** Y **21.** X **23.** Y
25. 0.31 **27.** 0.62 **29.** 0.71 **31.** 61° **33.** 83°
35. 18° **37.** 67.5 units² **39.** $\cos^{-1}(\cos \theta) = \theta$
41. $\sin 70° = 0.9397$; $\cos 20° = 0.9397$;
$\sin 70° = \cos 20°$; they are the same because the side
opposite the 70° angle is the side adjacent to the 20°
angle. **43.** $(90 - \theta)$; $(90 - \theta)$ **45.** $\cos \theta$ **47.** $(\tan \theta)^2$
49. 19 feet **51.** 4.8° **53.** $x = \frac{14}{\sqrt{3}} = \frac{14\sqrt{3}}{3}$; $z = \frac{28\sqrt{3}}{3}$

55. $p = q = 1$; $r = \sqrt{2}$ **57.** $q = \frac{16}{\sqrt{2}} = 8\sqrt{2}$;
$p = q = 8\sqrt{2}$ **59.** 18°

LESSON 10.3

TRY THIS (p. 650)

$\sin^{-1}(0.5736) \approx 35°$
The second angle is $180° - 35° \approx 145°$.

Exercises

8. $\left(-\frac{\sqrt{3}}{2}, -\frac{1}{2}\right)$ **9.** $\sin \theta = 0.6428$; $\cos \theta = 0.7660$
10. $y = \sin t$: After 10 minutes (600 sec): -0.866.
After 20 minutes: 0.866. After 1 hour: 0. After 12
hours: 0. **11.** In Quadrants I and IV cosine is
positive. In Quadrants II and III, cosine is negative.
12. 45°; 135° **13.** 0.9063 **15.** −0.9063
17. 0.4226 **19.** −0.4226 **21.** $\sin 45° \approx 0.7071$;
$\cos 45° \approx 0.7071$ **23.** $\sin 225° \approx -0.7071$;
$\cos 225° \approx -0.7071$ **25.** $\sin 30° = 0.5$;
$\cos 30° \approx 0.8660$ **27.** $\sin 210° = -0.5$;
$\cos 210° \approx -0.8660$ **29.** $x = \cos 30° \approx 0.8660$;
$y = \sin 30° = 0.5$ **31.** $x = \cos 90° = 0$; $y = \sin 90° = 1$
33. $x = \cos 180° = -1$; $y = \sin 180° = 0$
35. $x = \cos 300° = 0.5$; $y = \sin 300° \approx -0.8660$
37. 45°; 135° **39.** 30°; 150° **41.** 19°; 161° **43.** 87°;
93° **45.** 28°; 152° **47.** The horizontal position is
given by $\cos t$, where t represents the number of
hours. After 2 days (48 hrs), the horizontal position
is 0.6691. After 5 days, the horizontal position is
−0.5. **49.** a **51.** b **53.** $\tan A$ **55.** B (or $90° - A$)

LESSON 10.4

TRY THIS (p. 656, Ex. 1)

$b \approx 13.8$
$c \approx 12.2$

TRY THIS (p. 656, Ex. 2)

$Q \approx 42°$

Exercises

5. $b \approx 10.1$; $c \approx 11.9$ **6.** 111.74° **7.** ≈ 2.59 ft **9.** 27.46°
11. 15.45° **13.** 8.02 cm **15.** 2.02 cm
17. $m\angle P = 40°$; $q \approx 6.16$; $p \approx 4.57$
19. $m\angle Q \approx 42.34°$; $m\angle R \approx 77.66°$; $r \approx 10.15$
21. *Case 1:* $m\angle P \approx 76.48°$; $m\angle Q \approx 58.52°$; $q \approx 9.65$
Case 2: $m\angle P \approx 103.52°$; $m\angle Q \approx 31.48°$; $q \approx 5.91$
23. Sample answer: If $a < b \sin A$, then side a is too
short to form any triangle, so no triangle is possible.
25. Sample answer: If $b \sin A < a < b$, there are 2
triangles possible, one by "swinging" side a to the
left and the other by "swinging" side a to the right.
27. Sample answer: If $a \leq b$ then no triangle is
possible since for an obtuse or a right triangle, the

side across from the obtuse or right angle must be the longest side of the triangle. **29.** 1 triangle possible **31.** 2 triangles possible **33.** $\sin A = \frac{a}{c}$; $\sin B = \frac{b}{c}$ **35.** $\frac{\sin A}{a} = \frac{\sin B}{b} = \frac{\sin C}{c}$, since they all are equal to the same quantity. **37.** $a \sin B$
39. $b \sin A = a \sin B$ **41.** $\frac{h_2}{a}$ **43.** $c \sin A$
45. Multiplication Property of Equality
47. $\frac{\sin A}{a} = \frac{\sin C}{c}$ **49.** Transitive or Substitution Property of Equality **51.** 7.44 km; 6.74 km
53. Tower A is closer to the smoke plume. The smoke is 9.40 miles from the tower. **55.** 3.62 mi
57. $S = 2827.4 \text{ m}^2$; $V = 14{,}137.2 \text{ m}^3$ **59.** ≈ 11.3 units
61. $65.91°$ **63.** $45°$

LESSON 10.5

TRY THIS (p. 664, Ex. 1)

$b \approx 20.65$

TRY THIS (p. 664, Ex. 2)

$m\angle E \approx 117.9°$; $m\angle F \approx 32.7°$; $m\angle D \approx 29.4°$

TRY THIS (p. 665)

$b \approx 5094.07$ mi
Since 5094.07 < 5280 ft, campers at Site 1 and 3 will be able to communicate directly with each other.

Exercises

5. $b \approx 28.42$; $C \approx 53.97°$; $m\angle A \approx 76.03°$
6. $E \approx 119.66°$; $F \approx 27.61°$; $m\angle D \approx 32.73°$
7. Campers at sites 2 and 3 cannot communicate directly since their two-way radios have a range of only 7920 feet, and the distance between them is about 8192 feet. **9.** $c \approx 3.42$ units **11.** $m\angle A \approx 83.33°$
13. $m\angle B \approx 153.84°$ **15.** $m\angle G \approx 72°$; $g = 8.8$; $h \approx 5.44$ **17.** $m\angle O \approx 27.57°$; $m\angle M \approx 112.43°$; $m \approx 35.95$ **19.** $m\angle U \approx 76.65°$; $m\angle S \approx 54.17°$; $m\angle T \approx 49.18°$ **21.** law of cosines

23. Sample answer:

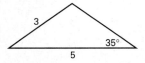

25. law of cosines **27.** $b \cos \theta = c_1$
29. $a^2 = h^2 + (c_1 + c)^2$

31. $a^2 = h^2 + (c_1 + c)^2$
$a^2 = h^2 + c_1^2 + 2c_1 c + c^2$
$a^2 = b^2 + 2(-b \cos A) \cdot c + c^2$
$a^2 = b^2 - 2bc \cos A + c^2$
$a^2 = b^2 + c^2 - 2bc \cos A$

33. 63.72 feet **35.** The road will be 24.88 km long, and the angle formed is 42.24°. **37.** $m\angle A = 41.65°$; $m\angle B \approx 52.89°$; $m\angle C \approx 85.46°$
39. **41.**

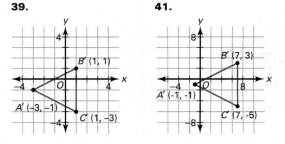

43. Yes, SAS-similarity **45.** $PQ \approx 12.59$

LESSON 10.6

TRY THIS (p. 674, Ex. 1)

Resultant Vector

TRY THIS (p. 674, Ex. 2)

If two equal forces are applied in opposite directions, the resultant vector would have magnitude 0. If the forces are unequal, then the resultant vector would have a magnitude which is the difference of the magnitude of the two vectors.

Exercises

6. **7.**

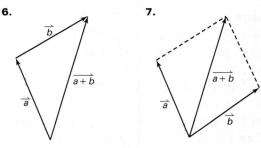

8. 2.5 mph; 53.13° **9.** 4.46 mph; 8.34°

11. **13.**

15.

17.

19. $|\vec{c}| \approx 10.71$; 22.39° **21.** $|\vec{c}| \approx 13.76$; 24.19°
23. $|\vec{c}| \approx 6.78$; 13.71°

25.

1.65 mph

27. $12\frac{2}{3}$ **29.** $x = 9$ **31.** 16 **33.** $AY \approx 15.28$

LESSON 10.7

TRY THIS (p. 682)

$$\begin{bmatrix} \cos 30° & -\sin 30° \\ \sin 30° & \cos 30° \end{bmatrix} \times \begin{bmatrix} 0 & 2 & 1 \\ 1 & 0 & 3 \end{bmatrix}$$

$$= \begin{bmatrix} -0.5 & 1.73 & -0.63 \\ 0.87 & 1 & 3.10 \end{bmatrix}$$

The new vertices are: $D' = (-0.5, 0.87)$,
$E' = (1.73, 1)$, $F' = (-0.63, 3.10)$

Exercises
5. $x' \approx 4.23$; $y' \approx 4.81$ **6.** $x' = -4$; $y' = 5$
7. $x' \approx 6.33$; $y' \approx 0.96$ **8.** $A' = (0, 1.41)$,
$B' = (2.12, 2.12)$, $C' = (0, 2.83)$
9. $A' = (-0.64, 1.26)$, $B' = (0.93, 2.85)$,
$C' = (-1.28, 2.52)$ **10.** $A' = (1.37, -0.37)$,
$B' = (1.5, -2.60)$, $C' = (2.73, -0.73)$ **11.** $(-3.54, 3.54)$ **13.** $(-1.60, -3.23)$ **15.** $(2.99, 6.41)$ **17.** 90°

19. 270° or −90° **21.** 45° **23.** $\begin{bmatrix} 0.71 & -0.71 \\ 0.71 & 0.71 \end{bmatrix}$

25. $\begin{bmatrix} -0.5 & -0.87 \\ 0.87 & -0.5 \end{bmatrix}$

27.

29.

$$\begin{bmatrix} \cos 30° & -\sin 30° \\ \sin 30° & \cos 30° \end{bmatrix} \begin{bmatrix} \cos 40° & -\sin 40° \\ \sin 40° & \cos 40° \end{bmatrix}$$

31.

$$= \begin{bmatrix} 0.34 & -0.94 \\ 0.94 & 0.34 \end{bmatrix};$$

$$\begin{bmatrix} \cos 70° & -\sin 70° \\ \sin 70° & \cos 70° \end{bmatrix} = \begin{bmatrix} 0.34 & -0.94 \\ 0.94 & 0.34 \end{bmatrix};$$

A rotation of 30° followed by a rotation of 40° is
equivalent to a rotation of 70°.

33. $\begin{bmatrix} R_0 \end{bmatrix} = \begin{bmatrix} \cos 0° & -\sin 0° \\ \sin 0° & \cos 0° \end{bmatrix} = \begin{bmatrix} 1 & 0 \\ 0 & 1 \end{bmatrix}$

The product of any rotation matrix and the identity
matrix gives the rotation matrix. **35.** The product of
two inverse matrices gives the identity matrix.

$$\begin{bmatrix} R_{35} \end{bmatrix} = \begin{bmatrix} 0.82 & -0.57 \\ 0.57 & 0.82 \end{bmatrix}$$

$$\begin{bmatrix} R_{-35} \end{bmatrix} = \begin{bmatrix} 0.82 & 0.57 \\ -0.57 & 0.82 \end{bmatrix}$$

$$\begin{bmatrix} R_{35} \end{bmatrix} \times \begin{bmatrix} R_{-35} \end{bmatrix} = \begin{bmatrix} 1 & 0 \\ 0 & 1 \end{bmatrix}$$

37. a. (21.37, 46.98) **b.** The coordinates are
$(30 \cos(6t°) + 42 \sin(6t°), 30 \sin(6t°) - 42 \cos(6t°))$
after rotating for t minutes. 63.43°

39.

26.57°

41. $\tan 37° \approx 0.75$; $\sin 37° \approx 0.60$; $\cos 37° \approx 0.80$
43. $\tan 250° \approx 2.75$; $\sin 250° \approx -0.94$;
$\cos 250° \approx -0.34$ **45.** 70° **47.** 36° **49.** The slope is too steep. $\theta \approx 30.96°$

CHAPTER REVIEW AND ASSESSMENT

1. $\frac{3}{7}$ **3.** 23.2° **5.** $\sin A = \frac{55}{73}$; $\cos A = \frac{48}{73}$ **7.** 48.9°

9. $(0.34, 0.94)$ **11.** 11.54°; 168.46°

13. $m\angle R = 21.26°$; $m\angle Q = 43.74°$ **15.** 107°
17. 18.06 **19.** $m\angle I \approx 107.24°$

21. **23.**

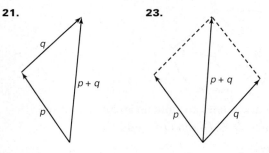

25. $(5, -1)$

27. $D'(0, 0)$, $E'(2.1, 2.4)$, and $F'(1.6, 3.2)$

29. 39.2 m **31.** A force of 1041.9 lb is required to push the truck uphill.

Chapter 11

LESSON 11.1

Exercises
5. 16.18 **6.** 5.56 **7.** Sample answer: $\overline{AC} \perp \overline{DE}$; The intersection seems to be the center of the spiral.
8. The long side of the rectangle should be 1.618 inches. **9.** 4.85 **11.** 5.26
13. $\frac{\phi}{1} = \frac{1}{\phi - 1}$

15. $\phi^{-1} \approx 0.6180339887$; From Exercise 13,
$\frac{\phi}{1} = \frac{1}{\phi - 1} \Rightarrow \frac{1}{\phi} = \phi - 1 \Rightarrow \phi^{-1} = \phi - 1$

17. Sample answer: In a circle of radius 3 cm, $DE \approx 3.5$ and $EH \approx 5.7$. **19.** 108°
21. $\frac{EH}{DE} = \frac{\sin 108°}{\sin 36°} \approx 1.618033989 \approx \phi$
23. $m\angle IDE = 72°$; $m\angle IED = 72°$; $m\angle I = 36°$
25. $\frac{IL}{IF} \approx 1.62 \approx \phi$ **27.** 34, 55, 89, 144, 233 **29.** The ratios seem to approach ϕ.
31. 6765
33. $\frac{5}{3} \approx 1.67$
$\frac{7}{5} = 1.4$
$\frac{10}{8} = 1.25$
$\frac{14}{11} \approx 1.27$
$\frac{20}{16} = 1.25$

3×5 is closest to the golden ratio. **37.** 3.2 **39.** 4.5
41. $(0.6428, 0.7660)$

43.

LESSON 11.2

Exercises
6. 4 **7.** 6 **8.** 2

9. **10.**

11. 10 **13.** 11 **15.** 13 **17.** 8 **19.** 48 **21.** 40

23. **25.**

27. 4 pathways

29. a. $(1, 2)$, $(2, 1)$, $(3, 0)$ **b.** $(3, -1)$, $(1, 3)$

c.

31.

The bisector has a bend in it. However, the bend consists of only two points, rather than the three formed in Exercise 29.

33. Because the taxicab distance between $(2, 0)$ and $(3, 2)$ is odd, a perpendicular bisector cannot be formed.

35. The points are: $(-1, 0)$, $(-1, 1)$, $(-1, 2)$, $(0, -1)$, $(0, 3)$, $(1, -1)$, $(1, 3)$, $(2, -1)$, $(2, 3)$, $(3, 0)$, $(3, 1)$, and $(3, 2)$.

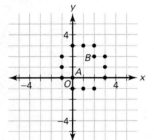

Sample answer: The points form two sets of parallel line segments of length 3; or a square with side length 4 which has its corner points deleted.

37. The set of points (x, y) will be the intersection of the circle of radius 10 centered about $(0, 0)$ and its interior and the circle of radius 8 centered about $(5, 4)$ and its interior. Jenny should look for places within the region shown, including the edge points and sides.

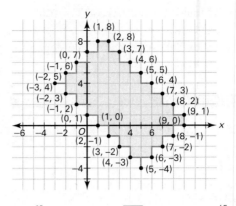

39. $\frac{60}{7} \approx 8.57$ **41.** $\sqrt{126} \approx 11.22$ **43.** $\frac{45}{53} \approx 0.85$

LESSON 11.3

Exercises

5. Yes **6.** No **7.** Yes

8.

The graph does not contain an Euler path since vertices B, C, D, and O are odd.

9. Yes

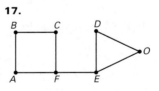

11. neither **13.** Euler path and circuit **15.** Euler path and circuit

17.

There is no Euler circuit.

19. Euler path; no Euler circuit **21.** 5 **23.** 6 **25.** 8
27. The sum of the degrees of the vertices of a graph is twice the number of edges. **29.** Let S_e be the sum of the degrees of the even vertices and let S_o be the sum of the degrees of the odd vertices with E the number of edges. Then $S_e + S_o = 2E$. Note that S_e is even since the sum of even numbers is even. Then $S_o = 2E - S_e$ is even, since the difference of two even numbers is even. Therefore, S_o is an even number, that is, the sum of degrees of odd vertices

is an even number. But the only way a sum of odd numbers is even is if there were an even number of those odd numbers. Therefore the number of odd vertices must be even. **31.** Yes

33. Sample answer: It is not possible. Consider each box and the outside to be a vertex and each segment between two boxes as an edge.

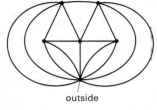

outside

The vertices representing the three larger boxes will have degree 5, so there can be no Euler circuit.

35. Yes; all vertices of the graph are even.

37.

Graph for Jo

In Jo's graph, every vertex is even, so there is an Euler path.

Graph for Tamara

In Tamara's graph, there are two odd vertices, so there is an Euler path.

39.

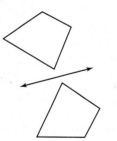

41. If a shape is a square, then the shape is a rectangle. Converse: If a shape is a rectangle, then the shape is a square. **43.** If a shape is a dodecagon, then the shape is a polygon with 12 sides. Converse: If a shape is a polygon with 12 sides, then the shape is a dodecagon. **45.** 8.3 **47.** 4.4

LESSON 11.4

Exercises

6. a, b, d **7.** b, c **8.** $8 - 12 + 6 = 2$
9. $6 - 12 + 8 = 2$ **11.** outside **13.** The number of times the ray from P crosses the curve is always odd.
15. odd; even

17.

19. If one were inside and one were outside, there would be no way to connect C to E without intersecting the simple closed curve.

21.

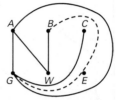

Sample answer: note that B can be connected to G. But then $BGAW$ forms a simple closed curve with C on the inside and E on the outside. Every curve connecting C to E intersects an already connected utility line. (The other cases follow for similar reasons). **23.** $20 - 30 + 12 = 2$; the polyhedron has 12 faces, so it is a dodecahedron. **25.** The path covers both sides of the Möbus strip. **27.** The result is a loop of paper with four half-twists. **29.** Yes
31. Yes **33.** Cut the bottle "diagonally." Start at any point on the surface and return to it without crossing itself. By avoiding the joined edges entirely, this produces a surface without dividing it into more than one piece. Then cut around the self-intersection. **35.** $x = 3$
37. Sample answer: $\angle 1$ and $\angle 4$, $\angle 3$ and $\angle 7$, $\angle 6$ and $\angle 7$ **39.** $3\frac{3}{8}$ oranges **41.** $b \approx 115.6$

LESSON 11.5

Exercises

5. If A, B and C are angles of a triangle, $m\angle A + m\angle B + m\angle C > 180°$ in spherical geometry and $m\angle A + m\angle B + m\angle C < 180°$ in hyperbolic geometry. **6.** Yes **7.** No **8.** No **9.** Yes **10.** Yes
11. No **12.** Yes **13.** No

15. Sample answer:

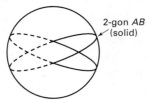

2-gon *AB* (solid)

The vertices must be opposite each other on the sphere.

17. Sample answer:

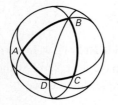

The diagonal is $\overline{BD}$.

19. Sample answer:

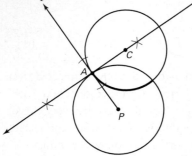

ℓ

21. If an arc is orthogonal to a circle at a point, then the tangent of the arc is perpendicular to the tangent of the circle at that point. By the Tangent Theorem, the tangent of the arc is also perpendicular to a radius of the circle drawn to the point of tangency. The tangent of the circle at the given point must contain the radius of the arc, and so it must contain the center of the arc.

23.

As *E* gets closer to the line $\overline{BQ}$, the radii of the arcs get very large.

25. If *A* and *B* are endpoints of a diameter, then $\ell \parallel m$ and they never intersect. In this case the line through *A* and *B* is the circle *P* itself. **27.** Infinitely many lines can be drawn through *A* that do not intersect ℓ. This shows that the Parallel Postulate does not hold in Poincaré's system, because there are infinitely many lines through *A* that are parallel to ℓ. **29.** The sum of the measures of the angles in the triangle is less than 180°. (Measure the angles with a protractor to get a fairly accurate reading.) In Euclidean Geometry, the sum of the angle measures of a triangle is 180°, while in spherical geometry it would be greater than 180°.

31.

Quadrilateral *ABCD*

33. In hyperbolic geometry, the sum of the angles of an $n - gon < (n - 2)180°$ **35.** The triangles form a flat surface. **37.** Sample answer: The models are all composed of a number of equilateral triangles about a given vertex. But in the first model, the angle measures of the triangles about the vertex add to less than 360°, so the model will not lay flat: it is convex. In the second model the sum of the angle measures equals 360°, so the model will lay flat. In the third model the sum of the angle measures is greater than 360°, so it will not lay flat: it "bunches up." **39.** A Euclidean plane will be formed. This is called a tessellation of the plane by equilateral triangles and will extend to infinitely many triangles.

41.

43.

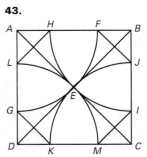

$ABCD$ is a square, so $AB = BC = CD = DA$ and $\angle A \cong \angle B \cong \angle C \cong \angle D$. Also, the same compass setting was used to draw all four arcs, so $AE = AF = AG = BH = BI = CJ = CK = DL = DM$. By the Overlapping Segments Theorem, $AH = FB = BJ = IC = CM = KD = DG = LA$. By the Segment Addition Postulate, $HF = JI = MK = GL$. By SAS, $\triangle AHL \cong \triangle BJF \cong \triangle CMI \cong \triangle DGK$, so $HL = FJ = IM = KG$ because CPCTC. Let $AB = 1$ unit. Since $\triangle ABE$ is a 45-45-90 triangle, $AE = AF = BH = \frac{\sqrt{2}}{2}$. By the Segment Addition Postulate, $AF + BH - HF = AB = 1$, so $HF = 1 - \frac{\sqrt{2}}{2} - \frac{\sqrt{2}}{2} = 1 - \sqrt{2}$. Also by the Segment Addition Postulate, $AH = 1 - \frac{\sqrt{2}}{2}$. $\triangle AHL$ is a 45-45-90 triangle, so $HL = \sqrt{2}\left(1 - \frac{\sqrt{2}}{2}\right) = 1 - \sqrt{2}$, so $AH = HL$. Thus, all sides of $HFJIMKGL$ are congruent. Also, because CPCTC, $\angle AHL \approx \angle ALH \approx \angle BFJ \approx BLF \approx CIM \approx CMI \approx DKG \approx \angle DGK$. By the Linear Pair of Property, $\angle LHF \approx \angle HFJ \approx \angle FJI \approx \angle JIM \approx \angle IMK \approx \angle MKG \approx \angle KGL \approx \angle GLH$. Since all of the sides and angles are congruent, $HFJIMKGL$ is a regular octagon. **45.** Sample answer: the graph has four odd vertices, so no Euler path or circuit exists. **47.** SIDE and CLOT are topologically equivalent. LAST and COZY are not topologically equivalent since A is not topologically equivalent to O.

LESSON 11.6

Exercises

5. The points where segments connect to the base form the Cantor dust. **6.** If the two boxes above a box are both shaded or unshaded, then the box is left unshaded. However, if one of the boxes is shaded and one is unshaded, the box beneath them is shaded. **7.** The shaded areas are inverted pyramids with increasingly longer bases arranged in a self-similar patterns, as in Activity 2.

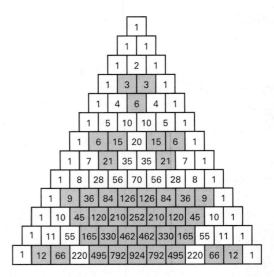

9. 18 **11.** 10.125 **13.** 36 **15.** 81 **17.** The area decreases, and its limit is 0.

21. $P = 3 \times 18 = 54$ Step 0
$P = 12 \times 6 = 72$ Step 1
$P = 48 \times 2 = 96$ Step 2

23. The area of the snowflake is always increasing. The area never becomes infinite because the snowflake can always be enclosed in a circle of radius 10.4 cm.

radius $= 6\sqrt{3} \approx 10.4$

25. Sample answers:

4 iterations   5 iterations

27. The area increases as iterations increase, but not without bound, because the amount it increases by is always less than half of the increase of the previous iterations. **29.** 16; 64 **31.** 14.8 **33.** 7 **35.** 14.2

LESSON 11.7

Exercises

5.

6. The resulting figure is a rectangle with vertices (0, 0), (0, 3), (4, 3), and (4, 0).

7. Yes

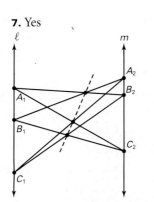

8. The points of intersection should be collinear.

9.

11.

13.

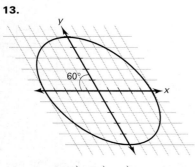

15. a. N **b.** $\overrightarrow{NJ}, \overrightarrow{NK}, \overrightarrow{NL}$

17.

19. $C; H; K$ **21.** $B; D$ or E

23–27, 29.

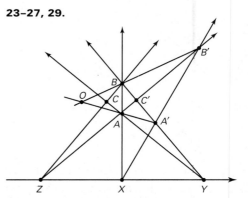

31. deductive **33.** deductive **35.** 4 **37.** 6 **39.** 0

CHAPTER REVIEW AND ASSESSMENT

1. 4.85

3.

5. 6

7.

9. Euler path; no Euler circuit **11.** neither **13.** Yes. Both are simple closed curves. **15.** −4 **17.** The two curves that intersect at the endpoints of a diameter of the ball are both lines in spherical geometry.
19. KL, KN, MN

21.

23. The area of the nth iteration is $\left(\frac{4}{9}\right)^{n} s^{2}$ more than the area of the $(n-1)$st iteration. The area of the fractal increases, but it is bounded.

25.

27.

29. 73 ft or 28 ft **31.** They were measuring from different heights. Alex's photos have more detail, so his measurements may include small inlets or peninsulas that are not visible in the satellite photo.

Chapter 12

LESSON 12.1

TRY THIS, (p. 769)

Sample answer: Valid argument: If an animal is an amphibian, then it is a vertebrate. Frogs are amphibians. Therefore, frogs are vertebrates. Invalid argument: Some reptiles are venomous. Chameleons are reptiles. Therefore, Chameleons are venomous.

Exercises

6. Argument form: *modus ponens* (valid)
7. Argument form: *modus tollens* (valid) **8.** Invalid form: affirming the consequent **9.** Invalid form: denying the antecedent **11.** By the *modus ponens* argument form: Therefore, she did a stupendous amount of work at the last minute. **13.** If Sims is a man of good moral character, then he is innocent. Sims is not a man of good moral character. Therefore, Sims is not innocent. Argument form: denying the antecedent; invalid **15.** If the plan was foolproof there were no unpleasant surprises. There were unpleasant surprises. Therefore, the plan was not foolproof. Argument form: *modus tollens;* Valid
17. If the building was of sound construction, then it survived the storm. The building did not survive the storm. Therefore, it was not of sound construction.

19. If the building was of sound construction, then it survived the storm. The building was not of sound construction. Therefore, it did not survive the storm. Invalid **21.** Invalid **23.** Invalid **25.** Invalid **27.** Valid **29.** False **31.** False **33.** Yes, since a valid argument guarantees that its conclusion is true if its premises are true. **35.** Sample answer: In football, if a team does not move the ball 10 yards in 4 downs, then they lose possession of the ball. The Mammoths did not lose possession of the ball. Therefore, the Mammoths moved the ball at least 10 yards in 4 downs. (*modus tollens*) **37.** Yes, using the SSS Postulate **39.** Yes, using the SAS Postulate **41.** 110°
43. $\theta = 60°$; $\sin\theta = \frac{\sqrt{3}}{2} \approx 0.866$; $\cos\theta = 0.5$
45. 240°

LESSON 12.2

Exercises

5. False, because only one of the statements is true.
6. True, because one of the statement is true.
7. True, because both of the statements are true.
8. False, because both statements are false.
9. $\sim(r \text{ AND } s)$ **10.** $\sim t$ OR $\sim s$ **11.** A carrot is a vegetable, and Florida is a state. The conjunction is true because both statements are true. **13.** The sum of the measures of the angles of a triangle is 180°, and two points determine a line. The conjunction is true because both statements are true. **15.** Points in a plane equidistant from a given point form a circle, or the sides of an equilateral triangle are congruent. The disjunction is true because both of the statements are true. **17.** The figure is not a rectangle.
19. Rain does not make the road slippery.

21. a.

p	~p	~(~p)
T	F	T
F	T	F

b. $\sim(\sim p)$ is logically equivalent to p because they both have the same truth values. **23.** $\triangle ABC$ has two equal angles, or $\triangle ABC$ is isosceles. **25.** $\triangle ABC$ does not have two equal angles. **27.** $\angle 1$ and $\angle 2$ are adjacent, or $\angle 1$ and $\angle 2$ are acute angles. **29.** $\triangle ABC$ has two equal angles, or $\angle 1$ and $\angle 2$ are not acute angles.

31.

p	*q*	*r*	*p* OR *q*	(*p* OR *q*) OR *r*
T	T	T	T	T
T	T	F	T	T
T	F	T	T	T
T	F	F	T	T
F	T	T	T	T
F	T	F	T	T
F	F	T	F	T
F	F	F	F	F

All three statements must be false in order for (*p* OR *q*) OR *r* to be false.

33. The sentence is true if any of the following combinations are true:

—Flora will cook; Vernon will vacuum.

—Flora will cook; Vernon will wash the windows.

—Flora will wash the dishes; Vernon will vacuum.

—Flora will wash the dishes; Vernon will wash the windows.

35.

Last name	First name	State	Year of birth	Annual income
Mallo	Elizabeth	TX	1956	50,000
Brookshier	Mary	OH	1960	62,000
Lamb	Charles	TX	1951	41,000
Raemsch	Martin	OK	1965	32,000

Last name	First name	State	Year of birth	Annual income
Craighead	Alicia	TX	1955	25,000
Tuggle	Lawrence	LA	1972	20,000
Mallo	Elizabeth	TX	1956	50,000
Tony	Jun	TX	1952	18,000
Lamb	Charles	TX	1951	41,000

37. Yes, by the AA Similarity Postulate. **39.** $\frac{4}{5}$ **41.** $\frac{3}{4}$

LESSON 12.3

TRY THIS, (p. 787)

a. Original statement: If a triangle is equilateral, then the triangle is isosceles. (true) Converse: If a triangle is isosceles, then the triangle is equilateral. (false) Inverse: If a triangle is not equilateral, then the triangle is not isosceles. (false) Contrapositive: If the triangle is not isosceles, then the triangle is not equilateral. (true)

b. Original statement: If a quadrilateral is a rhombus, then the quadrilateral is a square. (false) Converse: If a quadrilateral is a square, then the quadrilateral is a rhombus. (true) Inverse: If a quadrilateral is not a rhombus, then the quadrilateral is not a square. (true) Contrapositive: If a quadrilateral is not a square, then the quadrilateral is not a rhombus. (false)

Exercises

6.

p	*q*	*p* ⇒ *q*
T	T	T
T	F	F
F	T	T
F	F	T

7.

p	*q*	*q* ⇒ *p*
T	T	T
T	F	T
F	T	F
F	F	T

8.

p	*q*	~*p*	~*q*	~*p* ⇒ ~*q*
T	T	F	F	T
T	F	F	T	T
F	T	T	F	F
F	F	T	T	T

9.

p	*q*	~*q*	~*p*	~*q* ⇒ ~*p*
T	T	F	F	T
T	F	T	F	F
F	T	F	T	T
F	F	T	T	T

11. Conditional: True, since squaring both sides maintains equality due to the Multiplication Property of Equality. Converse: If $a^2 = b^2$ then $a = b$. False; if $a = -3$ and $b = 3$, then $(-3)^2 = (3)^2$ but a and b are not equal. Inverse: If $a \neq b$, then $a^2 \neq b^2$ False; if $a = -3$ and $b = 3$, they are not equal, but $(-3)^2 = (3)^2$ Contrapositive: If $a^2 \neq b^2$ then $a \neq b$. True; if $a \neq b$ then taking the square root of each side will not yield the same number, so $a \neq b$.

13. Conditional: False, since AAA does not guarantee triangle congruence. Converse: If two triangles are congruent, then the three angles of one triangle are congruent to the three angles of the other triangle. True; given that two triangles are congruent, their corresponding angles are congruent by definition. Inverse: If the three angles of one triangle are not congruent to the three angles of another triangle, then the triangles are not congruent. True; if all the corresponding angles are not congruent, then the triangles are not congruent. Contrapositive: If two triangles are not congruent, then the three angles of one triangle are not congruent to the three angles of the other triangle. False; if the triangles are not congruent they may still have congruent corresponding angles. They could be similar.

15. Conditional: True, by the laws of physics. Converse: If the water temperature is less than or equal to 32°F, then it will freeze at normal atmospheric pressure. True, by the laws of physics. Inverse: If water does not freeze at normal atmospheric pressure, then its temperature is greater than 32°F. True, by the laws of physics. Contrapositive: If water's temperature is greater than

32°F, then it will not freeze at normal atmospheric pressure. True, by the laws of physics. **17. b** is true because it is the contrapositive of the original conditional. Since the statement is true, the contrapositive is also true. The other two could be false because school could be canceled for other reasons besides snow. **19.** Sample answer: Theorem 3.8.2: If two nonvertical lines are parallel, then they have the same slope. Converse: If two nonvertical lines have the same slope, then they are parallel. True; since Theorem 3.8.2 is actually an if-and-only-if statement, the converse is true. Inverse: If two nonvertical lines are not parallel, then they do not have the same slope. True, because the inverse has the same truth value as the converse. Contrapositive: If two nonvertical lines do not have the same slope, then they are not parallel. True, because the contrapositive has the same truth value as the conditional. **21.** If you are a senior, then you must report to the auditorium. **23.** If she is going to be late, then she will call me. **25.** All of them are true. r if and only if s can be written as $r \Leftrightarrow s$. In other words, r implies s and s implies r. So if the statement r if and only if s is true then: **a.** If r is true then s must also be true, because r implies s. **b.** The statement is true because s implies r. **c.** The statement is true because it is the contrapositive of **a.**, and since **a.** is true, then this statement is also true. **d.** The statement is true because it is the contrapositive of **b.**, and **b.** is true. **27. a.** Same meaning; implies that getting at least a B in mathematics is a requirement to make the honor role. **b.** Different meaning; this is the converse of the original statement. **c.** Same meaning; implies that getting at least a B in mathematics is a requirement to make the honor roll. **d.** Same meaning; implies that getting at least a B in math is a requirement to make the honor roll. **29.** 3.53 units

31.

resultant

LESSON 12.4

TRY THIS (p. 793)

Assume that the defendant set the fire. Then the defendant was at the restaurant between 7:30 P.M. and 11:30 P.M. But three witnesses have testified that the defendant was not at the restaurant during those hours. Thus the assumption that the defendant set

the fire leads to the following contradiction: (The defendant was at the restaurant between 7:30 P.M. and 11:30 P.M.) AND (the defendant was not at the restaurant between 7:30 P.M. and 11:30 P.M.). Therefore, the assumption must be false. The conclusion is the defendant did not set the fire.

Exercises

7. (Lines ℓ and m are parallel) AND (lines ℓ and m are not parallel). **8.** ($\triangle ABC$ is isosceles) AND ($\triangle ABC$ is not isosceles). **9.** (All squares are rectangles) AND (all squares are not rectangles). **10.** ($ABCD$ is a square) AND ($ABCD$ is not a square). **11.** lines ℓ and m are not parallel **12.** the same-side interior angles are supplementary **13.** Yes. The argument has *modus tollens* form: if p then q; $\sim q$; therefore, $\sim p$. **15.** No. The argument has the *modus ponens* form: if p then q; p; therefore, q. **17.** Yes. The argument has the *modus tollens* form; if p then q; $\sim q$; therefore, $\sim p$. **19.** $\overline{DB} \cong \overline{AC}$ **21.** $\angle ACB$ **23.** SAS Postulate **25.** $\overline{JK} \cong \overline{JL}$ **27.** $\angle K \cong \angle L$ **29.** $\triangle KJM \cong \triangle LJM$ **31.** $\overline{KM} \cong \overline{LM}$ **33.** $\overline{JK} \not\cong \overline{JL}$ **35.** Suppose that there is a smallest positive real number x. But $\frac{1}{2}x < x$, so x is not the smallest positive real number. This is a contradiction. Therefore there is no smallest positive real number. **37.** Given: a fraction $\frac{x}{y}$ with a decimal expansion that terminates after n places. Suppose the denominator y does not divide any power of 10 with no remainder. Then y does not divide 10^n. But $10^n\left(\frac{x}{y}\right) = m$ an integer, and so $10^n x = my$. Then y divides my so y divides $10^n x$. But y does not divide x, so y divides 10^n, a contradiction. Therefore y divides 10^n for some n. **39.** x = 12 **41.** obtuse **43.** obtuse **45.** sin 45° ≈ 0.707; cos 45° ≈ 0.707; tan 45° = 1 **47.** sin 5° ≈ 0.087; cos 5° ≈ 0.996; tan 5° ≈ 0.087

LESSON 12.5

TRY THIS (p. 801)

An input-output table would require 8 rows because there are three inputs; the table would require seven columns: three for the inputs and four for the combinations.

Exercises

5.

Input-Output Table			
Input		Output	
p	*q*	Not *p*	Not *p* OR *q*
1	1	0	1
1	0	0	0
0	1	1	1
0	0	1	1

6.

Input-Output Table			
Input		Output	
p	q	p AND q	Not (p AND q)
1	1	1	0
1	0	0	1
0	1	0	1
0	0	0	1

7. p AND (NOT q) **8.** (p AND q) OR r

9.   **11.**

13. 1 **15.** 1

	p	q	NOT p	NOT q	(NOT q) OR (NOT q)
17.	1	0	0	1	1
19.	0	0	1	1	1

	p	q	r	NOT p	(NOT p) AND q	NOT r	((NOT p) AND q) OR (NOT r)
21.	1	1	0	0	0	1	1
23.	1	0	0	0	0	1	1
25.	0	1	0	1	1	1	1
27.	0	0	0	1	0	1	1

29. (NOT p) AND (NOT q)
31. NOT ((NOT p) OR (q AND r))

33.

35.

37. NOT (p OR q) and (NOT p) AND (NOT q)
39. a. The arrangement corresponds to the logic function AND. Both A and B have to be closed for the bulb to burn. **b.** The arrangement corresponds to the logic function OR. Either A or B must be closed for the bulb to burn. **41.** 10% **43.** False; example: $\frac{2}{3} = \frac{4}{6}$ but $\frac{(2+1)}{3} \neq \frac{(4+1)}{6}$ because $\frac{3}{3} \neq \frac{5}{6}$
45. $\frac{18}{25}$ or $\frac{25}{18}$ = the ratio of the perimeters $\frac{25}{18}$ or $\frac{18}{25}$ = the ratio of the sides. The ratio of the perimeters is equal to the ratio of the sides in similar triangles.

CHAPTER REVIEW AND ASSESSMENT

1. Valid **3.** Therefore, a cat is not a rodent. **5.** John is my brother and 17 is prime.

7.

p	q	p AND q
T	T	T
T	F	F
F	T	F
F	F	F

9. Converse: If the ground is wet, then it is raining. Inverse: If it is not raining, then the ground is not wet. Contrapositive: If the ground is not wet, then it is not raining.

11.

p	q	$\sim p$	$\sim q$	$\sim p \Rightarrow \sim q$
T	T	F	F	T
T	F	F	T	T
F	T	T	F	F
F	F	T	T	T

13. $\overline{AD} \perp \overline{BC}$ **15.** $\overline{AC} \cong \overline{AB}$ **17.** p OR (NOT q)
19. (p OR q) AND (NOT r) **21.** It is a valid argument. It has the *modus tollens* form.

Extra Practice

CHAPTER 1

Lesson 1.1
1. $\overline{AB}, \overline{BC}, \overline{CD}, \overline{DA}$ **3.** $\overrightarrow{AB}$ and $\overrightarrow{AD}$ form angle 1, $\overrightarrow{BA}$ and $\overrightarrow{BC}$ form angle 2, $\overrightarrow{CB}$ and $\overrightarrow{CD}$ form angle 3, and $\overrightarrow{DC}$ and $\overrightarrow{DA}$ form angle 4. **5.** False. The two lines might not be coplanar; they could be skew lines. **7.** point X

Lesson 1.2
1. 5 **3.** 8 **5.** 29 **7.** Does not make sense because $\overline{AB}$ refers to the segment, not the measure of the segment.

Lesson 1.3
1. 42° **3.** 57° **5.** 70° **7.** $x = 5.5$, $m\angle WXZ = 35°$
9. $m\angle WXY = 52.5°$

Lesson 1.4
1. **3.**

5. 70°, 35° **7.** 55°

Lesson 1.5

1.

3.

5.

Lesson 1.6

1. **3.**

5. A H I M O T U V W X Y,
Sample answer: YAHOO

Lesson 1.7

1. **3.**

5. vertical shift down 3 units **7.** reflection across the y-axis, vertical shift down 5 units

---

CHAPTER 2

Lesson 2.1

1. 101 **3.** column E

5.

7. The representation of $(n + 2)^2$ has an additional $2n$ dots directly below the original n columns and $2n + 4$ dots added to the right of the original square. The total dots in the new array is then $n^2 + 2n + (2n + 4)$.

Lesson 2.2

1. If an animal lives in the Everglades, then it lives in Florida.

3.

> Animals in Florida
>
> Animals in Everglades

5. Hypothesis: Two angles are supplementary. Conclusion: The sum of their measures is 180°. Converse: If the sum of the measures of two angles is 180°, then the angles are supplementary. Converse is true. **7.** Hypothesis: You live in Baltimore. Conclusion: You live in Maryland. Converse: If you live in Maryland, then you live in Baltimore. Converse is false; you may live in Annapolis.

Lesson 2.3

1. a. If a planet is Mercury, then it is the planet closest to our sun. **b.** If a planet is the one closest to our sun, then it is Mercury. **c.** A planet is the closest to our sun if and only if it is Mercury. **d.** Definition: The biconditional is true.

3. a. If a number is a multiple of ten, then it has a ones digit of zero. **b.** If a number has a ones digit of zero, then it is a multiple of ten. **c.** A number is a multiple of ten if and only if it has a ones digit of zero. **d.** Definition: The biconditional is true. **5. a.** If a closed figure is an equilateral triangle, then it has three congruent angles. **b.** If a closed figure has three congruent angles, then it is an equilateral triangle. **c.** A closed figure is an equilateral triangle if and only if it has 3 congruent angles. **d.** Not a definition: A pentagon may contain 3 congruent angles and two other angles.

Lesson 2.4

1. Addition Property of Equality **3.** 35°

Lesson 2.5

1. ∠GBD **3.** ∠ABC **5.** m∠ABE = m∠CBD = 137°; m∠ABC = m∠EBD = 43° **7.** Deductive reasoning; is a proof; The argument is true for all multiples of 5.

CHAPTER 3

Lesson 3.1

1.

3.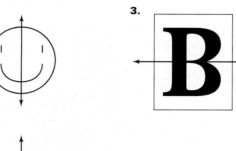

5.

7. The figure in Exercise 5 has 90°, 180°, and 270° rotational symmetry. 6 has 180° rotational symmetry.

Lesson 3.2

1. 5 **3.** 3 **5.** 106° **7.** $x = 2$; 6 **9.** False; a rhombus may have interior angles ≠ 90°.

Lesson 3.3

1. ∠2, ∠6, and ∠7 **3.** No; all the angles are listed in either Exercise 1 or Exercise 2. **5.** m∠1, m∠4, m∠5, and m∠8 = 80°; m∠2, m∠3, m∠6, and m∠7 = 100° **7.** 70° **9.** 70° **11.** 35°

Lesson 3.4

1. Alternate Interior Angles Theorem **3.** Same-Side Interior Angles Theorem **5.** Alternate Interior Angles Theorem

7.

Proof Statements	Reasons
m∠CBE = 43° m∠BED = 137°	Given
m∠CBE + m∠BED = 180°	Addition
∠CBE and ∠BED are supplementary.	Definition of Supplementary Angles
$\overline{ED} \parallel \overline{BC}$	Converse of Same-Side Interior Angles Theorem

Lesson 3.5

1. 90° **3.** 48° **5.** 55° **7.** 76°

Lesson 3.6

1. 50° **3.** 150° **5.** 15 **7.** 5 **9.** 12

Lesson 3.7

1. 30 **3.** 45 **5.** 1.5 feet

Lesson 3.8

1. $\frac{1}{2}$; (2, 1.5) **3.** $-\frac{7}{3}$; (−0.5, 3.5) **5.** Perpendicular **7.** Neither **9.** Right triangle **11.** Not a right triangle

CHAPTER 4

Lesson 4.1

1. Not congruent. They are different lengths.
3. Not congruent. They have different side lengths.
5. Congruent. All corresponding parts are congruent.
7. a. $\overline{JM}$ **b.** $\overline{WC}$ **c.** $\overline{BA}$ **d.** $\overline{OP}$

Lesson 4.2

1. SSS **3.** ASA **5.** No **7.** No **9.** ASA

Lesson 4.3

1. △CMQ ≅ △DFB by SSS **3.** △ABD ≅ △CBD by SAS **5.** △RST ≅ △LJK by SAS **7.** No **9.** △MNO ≅ △GHI by SSS

Lesson 4.4

1. 40° **3.** 12 **5.** 50° **7.** 90° **9.** 15

Lesson 4.5

1. 150° **3.** 40° **5.** 30° **7.** △JKL is congruent to △OMN by ASA, so they can be put together to form a parallelogram. **9.** △XYZ is congruent to △DEF by SSS, so they can be put together to form a parallelogram.

Lesson 4.6

1. WXYZ is a parallelogram. If the diagonals of a quadrilateral bisect each other, then the quadrilateral is a parallelogram. **3.** WXYZ is a parallelogram. If two pairs of opposite sides of a quadrilateral are congruent, then the quadrilateral is a parallelogram. **5.** ABCD is a rhombus. If the diagonals of a parallelogram bisect the angles of the parallelogram, then the parallelogram is a rhombus. **7.** ABCD is a rhombus. If one pair of adjacent sides of a parallelogram are congruent, then the parallelogram is a rhombus. **9.** Neither

Lesson 4.7

1.

3.

5.

7.

9.

Lesson 4.8

1. **3.**

5. Possible triangle. The sum of the lengths of any two sides is greater than the length of the third side. **7.** Not possible, since $BC + CA < AB$ **9.** Not possible, since $DE + EF < DF$

CHAPTER 5

Lesson 5.1
1. 12 **3.** 4 **5.** 15 **7.** $h = 3$ cm, $b = 12$ cm, $P = 30$ cm

Lesson 5.2
1. $A = 24$ **3.** $A = 73.5$ **5.** $A = 4$ **7.** $A = 12$ **9.** $A = 18$
11. $A = 15$

Lesson 5.3
1. $C = 18.8, A = 28.3$ **3.** $C = 44.0, A = 153.9$
5. $C = \frac{880}{7} A = \frac{8800}{7}$ **7.** $C = \frac{528}{7} A = \frac{3168}{7}$
9. $r = 2\sqrt{5}, 4.5$ **11.** $r = \frac{12}{\pi}, 3.8$ **13.** $r = \sqrt{\frac{117}{\pi}}, 6.1$

Lesson 5.4
1. 13 **3.** $5\sqrt{21}$ **5.** obtuse **7.** acute **9.** obtuse
11. $A = 10.8, P = 15$ **13.** $A = 17.1, P = 24.4$

Lesson 5.5
1. $b = 7\sqrt{3}, c = 14$ **3.** $a = 2, b = 2\sqrt{3}$
5. $A = 121\sqrt{3}, P = 66$ **7.** $A = 32\sqrt{3}, P = 24 + 8\sqrt{3}$
9. $A = 4.5, P = 6 + 3\sqrt{2}$

Lesson 5.6
1. 4.12 **3.** 5 **5.** 4.47 **7.** 5.66 **9.** yes **11.** no **13.** yes

Lesson 5.7
1. $C(b, a)$ **3.** b **5.** $\sqrt{b^2 + c^2}$ **7.** $b^2 + c^2$

Lesson 5.8
1. $P = \frac{8}{25}$ **3.** $P = \frac{1}{9}$ **5.** 80% **7.** 100% **9.** 0.29
11. 0.09 **13.** 0.001 **15.** 0.453 **17.** $\frac{1}{4}$ **19.** $\frac{3}{4}$ **21.** $\frac{91}{100}$
23. $\frac{7}{25}$

CHAPTER 6

Lesson 6.1
1. 4 units^3

3.

5.

7. a.

b.

c.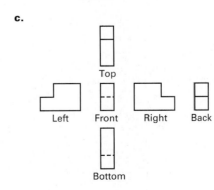

Lesson 6.2

1. No. **3.** Yes. Since line m is perpendicular to plane ℓ, line m is perpendicular to all lines in $\mathcal{P}$ that it intersects.

5. False. The two lines could be skew. For example, in this sketch, the lines ℓ and m lie in parallel planes.

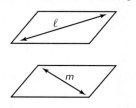

Lesson 6.3

1. pentagonal prism **3.** rectangular prism **5.** $\overline{BF}$ and $\overline{CD}$ **7.** $d = 15$ **9.** $d = 5\sqrt{5}$

Lesson 6.4

1. first octant **3.** top-back-right octant **5.** bottom-back-left octant **7.** $\sqrt{26}$ **9.** $2\sqrt{11}$

Lesson 6.5

1.

3.

5.

7. $7x + 2y = 1$

Lesson 6.6

1.

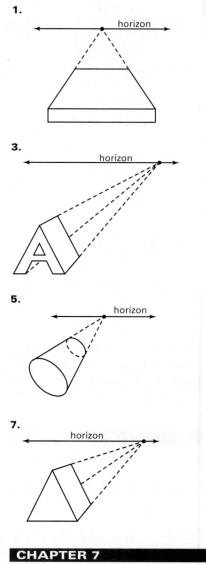

3.

5.

7.

CHAPTER 7

Lesson 7.1

1. 4.29 **3.** 1 **5.** 1.15 **7.** 1.2 **9.** $\frac{3 + 2n}{n}$ **11.** Maximize volume.

Lesson 7.2

1. 24 in^3 **3.** 130 m^3 **5.** $S = 256$; $V = 240$
7. $S = 2664$; $V = 1260$ **9.** $S = 2646$; $V = 9261$
11. 107.69 in.

Lesson 7.3

1. 89.6 **3.** 85.8 **5.** 65.3 **7.** 1600 **9.** 4200 **11.** 9

Lesson 7.4

1. 339.3 **3.** 5.3 **5.** 3.7 **7.** 203.6 **9.** 17,671.5 **11.** 3
13. $\frac{32}{9}$ in.

Lesson 7.5

1. $S = 19.2\pi$; $V = 9.792\pi$ **3.** $S = 343.2\pi$;
$V = 685.67\pi$ **5.** $S = 207.2\pi$; $V = 341.33\pi$

7. $S = 49.88\pi$; $V = 36.36\pi$ **9.** $S = 78.96\pi$; $V = 66.90\pi$

Lesson 7.6
1. $S = 314.16$; $V = 523.60$ **3.** $S = 13,684.78$; $V = 150,532.55$ **5.** $S = 105.68$; $V = 102.16$ **7.** $S = 50.77$; $V = 34.02$ **9.** $S = 804.25$; $V = 2144.66$ **11.** $S = 49\pi y^2$; $V = \frac{343}{6}\pi y^3$ **13.** $S = 4.84\pi y^2$; $V = \frac{5.324}{3}\pi y^3$ **15.** $S = \frac{1}{4}\pi x^2$; $V = \frac{1}{48}\pi x^3$

Lesson 7.7
1. $(1, 1, -2)$ **3.** $(2, 1, 3)$ **5.** $(-2, 15, 0)$ **7.** $(32, -32, 32)$ **9.** $(14, 4, 6.2)$ **11.** $(-12.1, 3, 6)$ **13.** $(-15, 15, 15)$ **15.** $(7.8, -7.8, -10)$ **17.** $(-2, -2, -2)$

CHAPTER 8

Lesson 8.1
1. $(6, -2)$

3. $(-1, -3)$

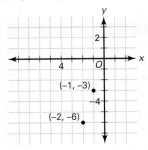

5. 3 **7.** $y = 3x$; Plugging in $O(0, 0)$ gives $0 = 0$, which is true. Thus, the origin is on this line. **9.** $y = -5x$; Plugging in $O(0, 0)$ gives $0 = 0$, which is true. Thus, the origin is on this line.

Lesson 8.2
1. Yes. $\angle A \cong E$, $\angle B \cong F$, $\angle D \cong H$ and $\angle C \cong G$. Also, $\frac{AD}{EH} = \frac{AB}{EF} = \frac{BC}{FG} = \frac{DC}{HG} = \frac{2}{3}$. quadrilateral $ABCD \sim$ quadrilateral $EFGH$ **3.** $x = 3$ **5.** $x = 12$

Lesson 8.3
1. Yes, since $m\angle B = m\angle E$ and since $\frac{AB}{DE} = \frac{BC}{EF} = 5$, $\triangle ABC \sim \triangle DEF$ by SAS Similarity Theorem. **3.** Quadrilateral $ABCD \sim$ quadrilateral $IJKL$ since corresponding angles are congruent and

$\frac{AB}{IJ} = \frac{BC}{JK} = \frac{AD}{IL} = \frac{DC}{LK} = \frac{2}{5}$. Quadrilateral $EFGH$ is not similar to quadrilaterals $ABCD$ and $IJKL$ since $\frac{AB}{EF} = \frac{1}{2}$ but $\frac{AD}{EH} = \frac{2}{3} \neq \frac{1}{2}$.

Lesson 8.4
1. $x = \frac{10}{3}$ **3.** $x = 1$ **5.** $\triangle ABC \sim \triangle DBE$ by the SAS Similarity Theorem.

Lesson 8.5
1. $h = 37.5$ ft **3.** $h = 28.8$ ft **5.** $x = 2.2$

Lesson 8.6
1. $\frac{3}{2}$ **3.** $\frac{2}{7}$ **5.** $\frac{3}{4}$ **7.** $\frac{27}{64}$

CHAPTER 9

Lesson 9.1
1. 4.19 **3.** 26.18 **5.** 12.48 **7.** 46.08 **9.** 54.45 **11.** 212.8° **13.** 182.3° **15.** 229.2° **17.** 57.3°

Lesson 9.2
1. $\overline{KS}$ **3.** 9.54, 9.54 **5.** 58.79, 58.79

Lesson 9.3
1. 15° **3.** 45° **5.** 180° **7.** 45° **9.** 90° **11.** 90° **13.** 150°

Lesson 9.4
1. 122.5° **3.** 45° **5.** 160°

Lesson 9.5
1. 4 **3.** 2.47 **5.** $\angle ZXR$ **7.** $\overgroup{ZQ}$ **9.** 4

Lesson 9.6
1. x-intercepts: $(7, 0)$, $(-7, 0)$; y-intercepts: $(0, 7)$, $(0, -7)$ **3.** x-intercepts: $(5, 0)$, $(-3, 0)$; y-intercepts: $(0, 3.87)$, $(0, -3.87)$ **5.** x-intercepts: $(11.06, 0)$, $(-5.06, 0)$; y-intercepts: $(0, 4.49)$, $(0, -12.49)$ **7.** center: $(0, 0)$, $r = 2$ **9.** center: $(-1, 0)$, $r = \sqrt{14}$ **11.** center: $(-3, 5)$, $r = \sqrt{31}$ **13.** $(x - 1)^2 + (y - 2)^2 = 16$

CHAPTER 10

Lesson 10.1
1. $\tan A = \frac{2}{5} = 0.4$ **3.** $\tan A = \frac{3}{8} = 0.375$ **5.** 0.27 **7.** 1.07 **9.** 0.90

Lesson 10.2
1. X **3.** X **5.** X **7.** 0.26 **9.** 0.31 **11.** 0.67

Lesson 10.3
1. 0.8192 **3.** −0.3420 **5.** −0.9816 **7.** $(0.9397, 0.3420)$ **9.** $(-0.9397, -0.3420)$ **11.** 38° and 142° **13.** 23° and 157° **15.** 31° and 149°

Lesson 10.4

1. 3.57 cm **3.** 36.9° **5.** m∠R = 76°; r ≈ 7.33; p ≈ 4.44 **7.** m∠R = 115°; r ≈ 16.92; p ≈ 7.89

Lesson 10.5

1. 12.1 **3.** 44.4° **5.** m∠D ≈ 84.1°; m∠E ≈ 63.7°; m∠F ≈ 32.2°

Lesson 10.6

1.

3.

Lesson 10.7

1. (−3.7, 3.3) **3.** (−0.2, 4.1) **5.** (2.7, −0.7) **7.** (0, 3)

9. $\begin{bmatrix} -.71 & -.71 \\ .71 & -.71 \end{bmatrix}$

11. $\begin{bmatrix} .82 & -.57 \\ .57 & .82 \end{bmatrix}$

13. $\begin{bmatrix} .77 & -.64 \\ .64 & .77 \end{bmatrix}$

15. $\begin{bmatrix} .5 & .87 \\ -.87 & .5 \end{bmatrix}$

CHAPTER 11

Lesson 11.1

1. 2.47 **3.** 6.18 **5.** 5.07

Lesson 11.2

1. 3 **3.** 7 **5.** 12 **7.** 40 **9.** 20 **11.** 24 **13.** 64 **5.** 88

Lesson 11.3

1. Euler circuit **3.** Euler path

Lesson 11.4

1. a. and **b.** are topologically equivalent.
3. $V − E + F = 2$; $20 − 30 + 12 = 2$

Lesson 11.5

1. $\overleftrightarrow{XY}$ is a line. **3.** $\overleftrightarrow{MY}$ is not a line. **5.** $\overleftrightarrow{AE}$ is not a line. **7.** $\overleftrightarrow{EC}$ is not a line.

Lesson 11.6

1.

3.

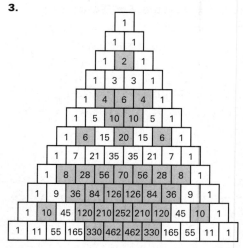

The shaded boxes in Activity 2 are unshaded in this Pascal Triangle. The unshaded boxes in Activity 2 are shaded in this Pascal Triangle. In each case, a Sierpinski gasket is suggested.

Lesson 11.7

1.

CHAPTER 12

Lesson 12.1

1. Therefore, he cries. (*modus ponens*) **3.** Therefore, he is hungry. (*modus tollens*) **5.** Valid by *modus tollens*. **7.** Valid by *modus ponens*.

Lesson 12.2

1. Some flowers are red, and cucumbers are vegetables. True, since both statements are true.
3. Harry is not older than Larry. **5.** Larry and Barry are not twins. **7.** It is not the case that Harry is both older than Larry and younger than Barry.

Lesson 12.3

1. *Conditional:* True. Twins are always the same age. *Converse:* If Harry and Larry are the same age, then they are twins. False. They can be the same age without even being related. *Inverse:* If Harry and Larry are not twins, then they are not the same age. False: They could be the same age without being twins. *Contrapositive:* If Harry and Larry are not the same age, then they are not twins. True. To be twins, they must be the same age. **3.** *Conditional:* False. It could be the case that $m ≠ 0$ and $n = 0$. *Converse:* If

$m = 0$, then $m \times n = 0$. True. Zero multiplied by any number is zero. *Inverse:* If $m \times n \neq 0$ then $m \neq 0$. True: If m were zero, $m \times n$ would be zero. *Contrapositive:* If $m \neq 0$, then $m \times n \neq 0$. False. If $n = 0$ and $m \neq 0$, $m \times n$ would still equal zero. **5.** If a number is odd, then it is not divisible by 2.

Lesson 12.4

1. No. To prove that Bobby ate some cookies by an indirect argument, it should first be assumed that Bobby did not eat some cookies from the cookie jar. Also, the argument assumes that the converse of a conditional has the same truth value as the conditional, which is not the case. **3.** No. The proof does not start by assuming the opposite of the statement to be proved. **5.** not obtuse (acute or right) **7.** 30° **9.** 90° **11.** The sum of the measures of the three angles is 180°

Lesson 12.5

1. 1

	~*p*	(~*p*) AND *q*	(~*p* AND *q*) OR *r*
3.	0	0	1
5.	0	0	1
7.	1	1	1
9.	1	0	1

INDEX

Credits

PHOTOS

Abbreviations used: (t) top, (c) center, (b) bottom, (l) left, (r) right, (bckgd) background, (bdr) border

FRONT COVER: Dale Sanders/Masterfile. **TABLE OF CONTENTS:** Page vi(tl), Michael Newman/PhotoEdit; vi(bc), Gary Vestal/Tony Stone Images(butterfly on daisy); vi(br), Tim Davis/Tony Stone Images(tail); vi(bl), Breck P. Kent/Earth Scenes; vi(bc), Alfred B. Thomas/Earth Scenes(fern); vi(br), Lester Lefkowitz/The Stock Market(Nautilus shell); vii(tl), Joe McDonald/Corbis-Bettmann; viii(tl), Gary Randall/FPG International; viii(br), John Blaustein/Liaison Agency; ix(tl), Michael P. Gadomski/Photo Researchers Inc.; ix(br), Cosmo Condina/Tony Stone Images; x(tl),(b), Christine Galida/HRW Photo; xi(tl), Don Couch/HRW Photo; xi(br), Artbase Inc.; xii(tl), M. Antman/The Image Works; xii(tr), Michael Keller/Index Stock Imagery Inc.(beach and ball); xii(bl), ChromaZone Images/Index Stock Imagery Inc.(Iris); xii(br), Artbase Inc.(starfish); xii(bc), Vee Sawyer/MertzStock; xiii(tl), Peter Van Steen/HRW Photo; xiii(br), Michael Collier/Stock Boston; xiv(tl), Chad Ehlers/Tony Stone Images; xiv(br), Dugald Bremner/Tony Stone Images; xv(tl), Don Couch/HRW Photo; xv(br), Steven C. Amstrup Photo; xvi(tl), Tomas Pantin/HRW Photo; xvi(br), Don Couch/HRW Photo; xvii(tl), Tim Davis/Photo Researchers Inc.; xvii(br), Allsport USA/Al Bello; vii(br), Dennis Fagan/HRW Photo. **CHAPTER ONE:** Page 2(br), The Granger Collection, New York; 2(bc), Tim Davis/Science Photo Library/Photo Researchers Inc.; 3(tl), Artbase Inc.; 3(c), Corbis-Bettmann/Leonard de Selva; 3(bckgd), Corbis-Bettmann/Richard T. Nowitz; 3(bl), Artbase Inc.; 4(bckgd), Georgia Tblisi/Sovfoto/Eastphoto(background); 4(c), Artbase Inc.(woman); 5(tc), Gary Vestal/Tony Stone Images(butterfly on daisy); 5(tr), Tim Davis/Tony Stone Images(tail); 5(tl), Breck P. Kent/Earth Scenes; 5(tc), Alfred B. Thomas/Earth Scenes(fern); 5(tr), Lester Lefkowitz/The Stock Market(Nautilus shell); 5(cr), Herbert Bayer, *Structure with Three Squares*. 1967. Painted aluminum. 62.9 cm h x 62.9 cm w x 5.7 cm d. Gift of Jan van der Marck in loving memory of his wife Ingeborg. Photograph ©1999 The Detroit Institute of Arts.; 5(br), Miwako Ikeda/International Stock; 5(bl), Stefano Amantini/Bruce Coleman Inc.; 5(bc), Dale Knuepfer/Bruce Coleman Inc.; 6(tr), Planet Art Collection; 6(b),(br) Peter Van Steen/HRW Photo; 7(bl), Dennis Fagan/HRW Photo; 7(br), Christine Galida/HRW Photo; 7(cr), Robert Sieck Flandes, 1939, National Institute of Anthropology, Mexico; 8(cr), Erich Lessing/Art Resource; 8(cl), The Granger Collection, New York; 8(tr), Addison Geary/Stock Boston; 8(bc),(br), Artbase Inc.; 9(tr), Celestial Image Picture Co./Science Photo Library/Photo Researchers Inc.; 9(br), Artbase Inc.; 10(tr),(cr),(br), Artbase Inc.; 10(bl), Michael J. Howell/International Stock; 11(tl),(c), Artbase Inc.; 13(cr), Alan Levenson/Tony Stone Images; 14(cr), E.R. Degginger/Bruce Coleman Inc.; 16(br), Dennis Fagan/HRW Photo; 17(tc), ©2001 PhotoDisc, Inc./HRW; 17(tr), Michael Abbey/Photo Researchers Inc.; 19(tr), Peter Van Steen/HRW Photo; 24(tr), Elliott Smith/International Stock; 25(tr), George Hall/Corbis-Bettmann; 29(bl), Eric Sanford/International Stock; 33(bl), Sam Dudgeon/HRW Photo; 34(br), The Granger Collection, New York; 34(tl), Peter Van Steen/HRW Photo; 35(tr),(cr), H.DE.Marcilliac/GLMR/Liaison Agency; 35(c), Frederic Reglain/Gamma Liaison/Liaison Agency; 36(all) Don Couch/HRW Photo; 37(tr), Gabe Palmer/The Stock Market; 37(c),(cr),(br), 38(cr),(br), Don Couch/HRW Photo; 39(tr), Peter Van Steen/HRW Photo; 40(tr), Dennis Fagan/HRW Photo; 41(cr), Michael Newman/PhotoEdit; 42(all), Christine Galida/HRW Photo; 43(tr),(tc), Martha Cooper/Peter Arnold, Inc.; 43(br), Don Couch/HRW Photo; 44(br), Rogers Fund, 1918/The Metropolitan Museum of Art; 45(tr), John Langford/HRW Photo; 47(br), Peter Van Steen/HRW Photo; 49(bc),(br), Dennis Fagan/HRW Photo; 50(tr), Jean Y. Ruszniewski/Tony Stone Images; 50(br), Peter Van Steen/HRW Photo; 52(tr), Dennis Fagan/HRW Photo; 54(tr), Craig Aurness/Corbis-Bettmann(clock); 54(tr), Bob Firth/International Stock(boy in canoe); 54(tr), Brian Vikander/Corbis-Bettmann(palace); 54(cl),(c), Dennis Fagan/HRW Photo; 56(cr), Courtesy of Quarto Publishing, England(butterfly); 58(cr), Carl Zeiss/Bruce Coleman Inc.; 58(cr), Nuridsany Et Pérennou/Science Source/Photo Researchers Inc.; 63(br), Tom Till Photography(skyline); 67(tr), Christine Galida/HRW Photo; 68(bl), Walter Hodges/Tony Stone Images; 68-69, Dennis Fagan/HRW Photo(paper folding steps); 69(br), Artbase Inc.; 75(tr), Dennis Fagan/HRW Photo. **CHAPTER TWO:** Page 78(bl), Hulton Getty/Liaison Agency; 78(br), Bruce Ayres/Tony Stone Images; 78(tr), The Granger Collection, New York; 78(cr), Corbis-Bettmann; 78(cr), Chris Thomaidis/Tony Stone Images(chess); 79(tr), (tl), Artbase Inc.; 79(c),80(tr),82(cr),(br), Dennis Fagan/HRW Photo; 86(c), Christine Galida/HRW Photo; 87(br), Dennis Fagan/HRW Photo; 88(bckgd), Artbase Inc.; 89(tr), Corbis-Bettmann/Bob Rowan/Progressive Image(Alaska); 89(cr),(br), Corbis-Bettmann/Joseph Sohm/ChromoSohm Inc.(Utah), (California), (Illinois), (Arizona); 90(tr), Bertram G. Murray Jr./Earth Scenes; 90(br), Steve Solum/Bruce Coleman Inc.; 90(c), John Shaw/Bruce Coleman Inc.; 90(cr), D. Madison/Bruce Coleman Inc.; 90(cl), Bruce MacDonald/Earth Scenes; 90(tl), Robert Lubeck/Earth Scenes; 91(t), Gerald Gusthall/FPG International; 94(tr), European Space Agency/Science Photo/Photo Researchers Inc.; 95(cl), Henry Holiday; 96(tl), Andy Sacks/Tony Stone Images; 97(c), Tate Gallery, London/Art Resource; 97(cr), Alexandre Calder, *Fond jaune*, 1972, Galerie Maeght, © Alexandre Calder/ARS (New-York)1999; 97(br), ©1988 Dan Lamont; 98(tr), Joseph Sachs/Still Life Stock; 99(tr), Dennis Fagan/HRW Photo; 103(cl), Joe McDonald/Corbis-Bettmann; 107(tr), Corbis-Bettmann; 111(tr), Ernest Manewal/FPG International; 114(all), Christine Galida/HRW Photo; 115(br), Coco McCoy/Rainbow; 116(bl),117(tr), Dennis Fagan/HRW Photo; 117(br), Artbase Inc.; 120(bc), Everett Collection(*ET*),(*Jurassic Park*),(*Close Encounters*); 120(br), Paramount Pictures/Shooting Star; 122(br), Robert Franz/Corbis-Bettmann; 132(br), Gerard Lacz/Publiphoto; 133(cr), The Granger Collection, New York.

CHAPTER THREE: Page 136(tr), Craig D. Wood; 136(cr), Jackie Foryst/Bruce Coleman Inc.; 136(bc), Greg Hursley; 137(cr), Giraudon/Art Resource, NY; 137(tr), Paul Avis/Liaison Agency; 138(tr), John Blaustein/Liaison Agency; 138(tc), Suzanne A. Vlamis/International Stock; 140(tl),(tc),(tr), Don Couch/HRW Photo; 140(cr), Dennis Fagan/HRW Photo; 141(tr), John Shaw/Bruce Coleman Inc.; 142(tr), H. Thonig/H. Armstrong Roberts/Comstock; 142(bl), Joy Spurr/Bruce Coleman Inc.; 142(br), *Symmetry Drawing E85* by M.C. Escher ©1995 Cordon Art-Baarn-Holland. All rights reserved; 146(tc),(tr), Artbase Inc.; 146(b), MapQuest.com, Inc.; 147(tl),(tc),(tr), Copyrighted material of Meredith Corporation, used with their permission. All rights reserved.; 147(cr), The Quilt Complex; 147(br), M. Eastcott/The Image Works; 148(tr), Spencer Jones/Bruce Coleman Inc.; 148(tc), David Madison/Bruce Coleman Inc.; 153(tr), Peter Van Steen/HRW Photo; 154(tr), *Symmetry Drawing E121* by M.C. Escher ©1999 Cordon Art-Baarn-Holland. All rights reserved; 155(tr), Gary Randall/FPG International; 157(cr), Milwaukee Art Museum, Gift of Frederick Layton Art League in

Memory of Miss Charlotte Partridge and Miss Miriam Frink, Photographed by Richard Beauchamp 1/92; 159(tl), Peter Pearson/Tony Stone Images; 162(tc), Courtesy of NASA; 162(tr), Space Frontiers/Masterfile; 164(tl),166(cr),167(tl), Randal Alhadeff/HRW Photo; 167(br), Daryl Benson/Masterfile; 168(tl),(bl),(cl), Annette Del Zoppo; 168(c), Robin White/Fotolex Associates; 168(tc), Artbase Inc.; 170(cr), Mark E. Gibson Photography; 170(tr), Richard Pharaoh/International Stock; 170(tc), Gene Ahrens/Bruce Coleman Inc.; 171(tr), Randal Alhadeff/HRW Photo; 172(bl), Tony Craddock/Science Photo Library/Photo Researchers Inc.; 173(tr), Randal Alhadeff/HRW Photo; 175(br), Neil Rabinowitz/Corbis-Bettmann; 176(cr), Corbis-Bettmann/The Mariner's Museum; 177(tr), Tom Sanders/The Stock Market; 182(br), Brownie Harris/The Stock Market; 182(tl), Rosemary Weller/Tony Stone Images; 182(cr), Ric Ergenbright/Corbis-Bettmann; 183(tr), Paul G. Adam/Publiphoto; 184(tr), Alex S. MacLean/Landslides; 185(t), Norman Owen Tomalin/Bruce Coleman Inc.; 188(bl), Lester Lefkowitz/Tony Stone Images; 189(tr), Jeff Greenberg/Photo Researchers Inc.; 189(cr), Maresa Pryor/Animals Animals; 190(tr), Alan Thornton/Tony Stone Images; 195(br), Peter Van Steen/HRW Photo; 197(tr), *Symmetry Drawing E25* by M.C. Escher ©1999 Cordon Art-Baarn-Holland. All rights reserved; 197(br), Tomas Pantin/HRW Photo; 204(c), Copyrighted material of Meredith Corporation, used with their permission. All rights reserved; 204(br), Danny Lehman/Corbis-Bettmann. **CHAPTER FOUR:** Page 208(cr), Corbis-Bettmann(Fuller); 208(cr), Robert Fried/Stock Boston(dome); 208(bckgd), D.P.Hershkowitz/Bruce Coleman Inc.; 208(bl), Keith Gunnar/Bruce Coleman Inc.; 209(tr), Dick Dickinson/ International Stock; 210(bckgd), SEF/Art Resource, NY; 210(tc), Rick Smolan/Stock Boston; 210(c), Jean-Claude Lejeune/Stock Boston; 215(tr), John Langford/HRW Photo; 216(tr), John Langford/HRW Photo(earrings); 216(br), Dennis Fagan/HRW Photo; 216(tr), Randal Alhadeff/HRW Photo(girl); 217(tr), Hilary Wilkes/International Stock; 217(br), Randal Alhadeff/HRW Photo; 220(br), Cosmo Condina/Tony Stone Images; 223(tr), J C Carton/Bruce Coleman Inc.; 223(br), Mark E. Gibson Photography; 224(br), Sam Dudgeon/HRW Photo; 225(b), Don Couch/HRW Photo; 225(tr), Peter Van Steen/HRW Photo; 226(tc),(tr), Sullivan/Texastock; 226(cr), Daniel Brody/Stock Boston; 229(cr), Jerry Schad/Photo Researchers Inc.; 230(cr), Michael P. Gadomski/Photo Researchers Inc.; 233(cr), Cliff Hollenbeck/International Stock; 234(bl), David Madison/Bruce Coleman Inc.; 234(br), Chuck Mason/International Stock; 235(tr), *Square Limit* by M.C. Escher ©1999 Cordon Art - Baarn - Hollland. All rights reserved.; 242(tr),(cr), Artbase Inc.; 243(tr), Eric Beggs/HRW Photo; 243(bl), Randal Alhadeff/HRW Photo; 253(tr), Peter Van Steen/HRW Photo; 254(tl), Randal Alhadeff/HRW Photo; 261(tr), Scott Van Osdol/HRW Photo; 269(bl), Courtesy Big Bend National Park; (br), U. S. Geological Survey; 270(bl), Todd Gipstein/National Geographic; 271(tr), William Swartz/Index Stock Imagery Inc.; 274(all), Randal Alhadeff/HRW Photo; 279(tr), Kevin R. Morris/Corbis-Bettmann; 280(cr),282(all),283(all),289(br), Randal Alhadeff/HRW Photo. **CHAPTER FIVE:** Page 292(b), James Montgomery/Bruce Coleman Inc.; 292(cr), Mark E. Gibson Photography; 293(tl), Barrett & Mackay Photography; 293(cl), J-C Carton/Bruce Coleman Inc.; 293(tr), Wolfgang Volz/Bilderberg; 294(tl), NASA/Corbis-Bettmann; 294(cr), O. Louis Mazzatenta/National Geographic Society; 294(tr), David Boyer/National Geographic Society; 297(tr), Peter Van Steen/HRW Photo; 298(tr), Steve Elmore/The Stock Market; 298(tr), Artbase Inc.; 300(br), James Amos/National Geographic Society; 301(br), Jacqui Hurst/The Garden Picture Library; 302(tl), Mark Douet/Tony Stone Images; 303(tr),(tl), Janice K. Wright/Down East Books, Camden, Maine, USA; 304(br), Christine Galida/HRW Photo; 306(br), Barrett & Mackay Photography; 311(tr), Gary Irving/Tony Stone Images; 312(b), James Montgomery/Bruce Coleman Inc.; 314(tr), W. Muller/Peter Arnold, Inc.; 314(cr), Peter Van Steen/HRW Photo; 317(cr),(tr), Randal Alhadeff/HRW Photo; 318(br), F. Gohier/Photo Researchers Inc.; 318(bl), Peter Stephenson/Masterfile; 318(c), Peter Van Steen/HRW Photo; 319(cl), W. Muller/Peter Arnold, Inc.; 320(cr), Robert & Linda Mitchell; 321(c), Columbia University Rare Book and Manuscript Library. Gift of George A. Plimpton; 321(tr), Erich Lessing/Art Resource; 322(bl), Vee Sawyer/Mertzstock; 325(tc), G. Gardner/The Image Works; 328(cl), National Portrait Gallery, Smithsonian Institution/Art Resource, NY; 328(tl), Dinodia Picture Agency; 329(cr), Frank Siteman/Monkmeyer Press Photo; 331(tr),332(b), Christine Galida/HRW Photo; 334(bl), Artbase Inc.; 337(br), Barrett & MacKay Photography; 337(br), Christine Galida/HRW Photo; 339(tr), L. Kolvoord/The Image Works; 341(tr),342(tr), Don Couch/HRW Photo; 345(tr), Clark James Mishler/The Stock Market; 347(tc), Artbase Inc.; 347(tr), Mike Clemmer; 349(tl), Dennis Fagan/HRW Photo; 352(tl), Steve Myers Studios/International Stock; 354(bl), Peter Van Steen/HRW Photo; 358(tl), Jump Run Productions/The Image Bank (c) 2000; 359(tr), G. Schiele/Publiphoto; 360(bckgd),(tr),361(bckgd),(br),366(br), John Langford/HRW Photo. **CHAPTER SIX:** Page 370(tr), Laguna Design/Science Photo Library/Photo Researchers Inc.; 370(bl), Prof. K. Seddon & Dr. T. Evans, Queen's University Belfast/Science Photo Library/Photo Researchers Inc.; 370(bl), Alfred Pasieka/Science Photo Library/Photo Researchers Inc.; 372(tr), Jon Feingersh/Stock Boston; 373(bl), Randal Alhadeff/HRW Photo; 377(cr), Scott Van Osdol/HRW Photo; 377(cr), Christine Galida/HRW Photo; 379(tr), Breck P. Kent/Earth Scenes; 381(tr),(cr),385(tr), Randal Alhadeff/HRW Photo; 385(br), Bair/Monkmeyer Press Photo; 386(b),(br), John Langford/HRW Photo; 387(all), Peter Van Steen/HRW Photo; 388(tr), Charles D. Winters/Photo Researchers Inc.; 389(all),391(br), Dennis Fagan/HRW Photo; 393(cl), Michael Nicholson/Corbis-Bettmann; 393(br),(bc), Peter Van Steen/HRW Photo; 394(br), David Parker/Science Photo Library/Photo Researchers Inc.; 395(bl),(br), Scott Van Osdol/HRW Photo; 396(tr), Artbase Inc.; 396(br), Dennis Fagan/HRW Photo; 401(tl), Michael P. Gadomski/Photo Researchers Inc.; 402(tc),(cl), Roman E. Maeder; 402(cr), ©Wolfram Research Inc.; 406(bl), Peter Van Steen/HRW Photo; 407(tr), Mark E. Gibson Photography; 408(bl), Don Couch/HRW Photo(crocheted ball); 408(bl), Artbase Inc.(soccer ball); 409(tr), Scala/Art Resource; 409(tc), Erich Lessing/Art Resource; 409(bc), Dennis Fagan/HRW Photo; 409(bl), Sam Dudgeon/HRW Photo; 410(tr), Fotomas Index; 410(tr), Donovan Reese/Tony Stone Images; 410(br), Dennis Fagan/HRW Photo; 410(bc), Cindy Verheyden/HRW Photo; 411(br), H. L. Romberg; 412(tl), Roger Ressmeyer/Corbis-Bettmann; 412(b), Brunelleschi, "Interno verso l'altare S. Lorenzo"/Art Resource; 413(br), Kathy Bushue/Tony Stone Images; 417(tr), Scott Van Osdol/HRW Photo; 418(bl), Sam Dudgeon/HRW Photo; 418-419(bckgd),419(bl), Don Couch/HRW Photo; 419(tl), Corbis-Bettmann(representation of orbits); 419(tl), Science Photo Library/Photo Researchers Inc. (portrait); 421(tl), Randal Alhadeff/HRW Photo; 424(cr), Gary Retherford/Photo Researchers Inc. **CHAPTER SEVEN:** Page 428(br), Chris Drury; 428(bl), Courtesy Danese, New York; 428(tr), Houston Museum of Natural

ILLUSTRATIONS